Grant's
ATLAS OF ANATOMY
Eighth Edition

HERCULES AND ANTAEUS

Grant's
ATLAS OF ANATOMY

Eighth Edition

JAMES E. ANDERSON, M.D.

Professor of Anatomy
Associate Professor of Psychiatry
Faculty of Health Sciences
McMaster University
Hamilton, Ontario, Canada

WILLIAMS & WILKINS
Baltimore • London • Los Angeles • Sydney

Copyright ©, 1983
Williams & Wilkins
428 East Preston Street
Baltimore, Maryland 21202, U.S.A.

First edition, 1943
Reprinted 1944, 1945

Second edition, 1947
Reprinted 1948, 1949

Third edition, 1951
Reprinted 1953, 1954

Fourth edition, 1956
Reprinted 1958, 1960, 1961

Fifth edition, 1962
Reprinted 1963, 1964, 1966, 1967, 1968, 1969, 1970

Sixth edition, 1972
Reprinted 1973 (twice) , 1974, 1977

Seventh edition, 1978
Reprinted 1979 (twice), 1980, 1982

Eighth edition, 1983

Foreign editions:
    Italian
    Japanese
    Portuguese (1946)
    Spanish
    Turkish

English language co-editions:
    Asian, 1972
    Reprinted 1973, 1974 (twice), 1978
    Indian, 1973
    Reprinted 1975, 1976

Library of Congress Cataloging in Publication Data

Grant, J. C. Boileau (John Charles Boileau), 1866–1973
    Grant's Atlas of anatomy.

    Includes index.
    1. Anatomy, Human—Atlases.   I. Anderson, James Edward, 1926–
II. Title.   III. Title: Atlas. of anatomy. [DNLM: 1. Anatomy Regional—Atlases.
QS 17 G767a]
QM25·G7   1983   611'.0022'2   82-10877
ISBN 0-683-00211-2

TITLE PAGE ILLUSTRATION:
Hercules and Anatalus.
Reproduced by permission from a
photograph of the statue at
Haussner's Museum,
Baltimore, Maryland.

88 89   10 9 8

For Helen, Jonathan, and Mary Kate,
and remembering Ronnie

Preface to the Eighth Edition

The first edition of Grant's *An Atlas of Anatomy* appeared in two volumes in 1943. In the Preface to that edition, the author explained:

"The collection of illustrations depicts the structures of the human body, region by region, in much the same order as the student displays them by dissection.

"In the execution of these illustrations the following preliminary steps were taken: each specimen was posed and photographed; from the negative film so obtained an enlarged positive film was made; with the aid of a viewing box the outlines of the structures on the enlarged film were traced on tracing paper; and these outlines were scrutinized against the original specimen, in order to ensure that the shapes, positions, and relative proportions of the various structures were correct. The outline tracing was then presented to the artist who transferred it to suitable paper and, having the original dissection beside her, proceeded to work up a plastic drawing in which the important features were brought out. Thus, little, if any, liberty has been taken with the anatomy; that is to say, the illustrations profess a considerable accuracy of detail.

"In order that the student may be able to turn the pages and study figure after figure without requiring to re-orient himself, all illustrations of bilaterally symmetrical structures are from the right half of the body, unless it is stated otherwise.

"Most of the specimens here depicted are in the Anatomy Museum of the University of Toronto; the others were specially prepared for this atlas.

"The observations and comments that accompany the illustrations are designed to attract attention to salient points and to points of significance that might otherwise escape notice. Their purpose is to interpret the illustrations. They are not, nor are they intended to be, exhaustive descriptions."

Subsequent editions followed: the second in 1947 (with 200 new illlustrations, the addition of color, and the useful schemes of distribution of cranial nerves), the third in 1951, the fourth in 1956, the fifth in 1962 (at which time the plates were re-engraved and small diagrams from *Grant's Method* added as secondary or supporting figures), and the sixth in 1972. A change in terminology was made in the fifth edition.

"As to terminology, the international nomenclature, *Nomina Anatomica* (1955), is employed in place of the Birmingham Revision (B.R., 1933) of the *Basle Nomina Anatomica* (B.N.A., 1895) which was originally in use. Where, however, the adopted terms differ substantially from the revised B.N.A. terms— and this is in a small minority of instances—both terms are given, the discarded terms being set within parentheses ( ). Square brackets [ ] indicate a synonym approved at the International Congress of Anatomists held periodically.

"The often recurring noun, musculus or muscle, is omitted from the names of the muscles, but by way of compensation the initial letters of the qualifying adjectives are printed in capitals; e.g., musculus trapezius and musculus rectus abdominis are printed thus: Trapezius and Rectus Abdominis."

Dr. Grant recorded his indebtedness to the many students and colleagues who assisted him in the preparation of the *Atlas* over the years. In his words:

"I have been most fortunate in the expert assistance rendered me by a number of medical artists. I owe a particular debt to Mrs. Dorothy I. Chubb, a pupil of Max Brödel, who worked with me from the beginning, and also to Miss Nancy Joy, now Director of Art as Applied to Medicine, in the University of Toronto, who soon joined her, for without their expert skill, patience in a laborious task, and the desire to achieve accuracy and effect, this book could hardly have been made. Mrs. Chubb was mainly responsible for the artwork of the first two editions and almost solely responsible for that of this, the sixth edition. Miss Joy for those in between. To them I record again my grateful appreciation of their work so cheerfully and carefully done.

"I gladly acknowledge the assistance of Miss Elizabeth Blackstock, of the Department of Art as Applied to Medicine of this University, whose talent is apparent in the cross-sections of the upper limb, the paranasal sinuses, and the arteries of the stomach, pancreas, and bile passages. Mrs.

E. Hopper Ross and Miss Marguerite Drummond have provided me with occasional and excellent illustrations, for which I am very grateful.

"To Mr. James B. Francis and Mr. Douglas Baker, for their early and valued assistance with the bones and muscles of the lower limb, I express my thanks.

"To Mr. Charles E. Storton, who has throughout given me general expert assistance, including the preliminary photographic work, which has been of very high order, and to Mr. H. Whittaker, who at times has undertaken the photographic work, my thanks are due for their willing and very skillful help."

Those who prepared dissections, provided specimens, gathered data, or assisted with the text have been given credit sometimes in the Preface and sometimes in relation to figures in the *Atlas*. Their valuable contributions are recorded here:

C. A. Armstrong	W. J. Horsey
P. G. Ashmore	G. F. Lewis
D. A. Barr	Ian B. MacDonald
J. V. Basmajian	D. L. MacIntosh
Sylvia Bensley	Ross G. MacKenzie
Douglas Bilbey	K. O. McCuaig
William Boyd	W. R. Mitchell
J. Callaghan	K. Nancekivell
H. A. Cates	A. J. A. Noronha
S. A. Crooks	W. M. Paul
Milne Dickie	C. H. Sawyer
J. W. A. Duckworth	Allene I. Scott
F. B. Fallis	J. S. Simpson
J. S. Fraser	C. G. Smith
R. K. George	J. S. Thompson
M. G. Gray	I. Maclaren Thompson
B. L. Guyatt	N. A. Watters
C. W. Hill	Roger W. Wilson

Grant's Atlas of Anatomy is:

1. A companion in the dissecting room, guiding the dissector through each step of exploring and exposing structures. Rather than providing an idealized view of human anatomy it presents real dissections that are readily related to the actual specimen being studied.

2. A portable anatomical museum useful in determining relationships when deep dissection has destroyed superficial structures, in reviewing (at home) regions by studying the illustrations guided by the brief text, and in preparing for practical examination (cover the leader labels and practice identifying structures).

3. One member of a trio of books designed by Dr. Grant particularly for beginning students, the others being *Grant's Method of Anatomy* by John V. Basmajian and *Grant's Dissector* by Eberhardt K. Sauerland.

4. A lifetime companion as a reference book in all fields of medicine.

5. A primary source of data on the occurrence of variations. Unless otherwise noted, data presented were gathered by Dr. Grant or under his direction.

The seventh edition, the first following Dr. Grant's death in 1973 and the first entrusted to my care, saw major reorganization of the *Atlas*. The sequence of regions was changed to conform with that of *Grant's Dissector*, and the book was divided into ten color-coded sections for easier reference. For the first time radiographs were added and the text was rewritten. I acknowledged my gratitude to Drs. John Basmajian, Eberhardt Sauerland, and Wazir Pallie for their advice and assistance; to Mr. John Simpkins for artistic and technical help; to Mrs. Lorna Tranter for typing the manuscript; and to Mr. Wayne Nusca for taking major responsibility in the preparation of the index. At Williams & Wilkins, Mr. Donald S. Slokan, Mr. Norman Och, and Mr. Robert Och contributed greatly to the production of the *Atlas*. In particular, I expressed my gratitude to Ms. Sara A. Finnegan, then Vice President and Editor-in-Chief.

This, the eighth edition, appears on the fortieth anniversary of the first edition in 1943. It is 32 pages longer than the previous edition and contains 91 new figures. For the first time it includes photographs of living anatomy. There is an increasing emphasis on clinical applications. Introductory material on the central nervous system has been added to provide a bridge to the later study of neuroanatomy. To facilitate cross-referencing to its companion volumes, there has been no significant change in figure numbers.

It is important that I share credit for this edition with those who have helped me so much: Mr. Joseph Bottos who created the new line drawings; Mr. Ian Matheson who patiently photographed and rephotographed; Mrs. Wendy Vriesma who typed the manuscript and fielded telephone calls; Mr. B. S. Jadon who prepared specimens for this and the last edition; and Ms. Nan Curtis Tyler who made valuable contributions during the planning stage. In the weeks leading to the manuscript deadline my resident, Dr. Jim Quan, carried a heavier than usual patient load and my friend, Dr. Stan Eaman, helped in many ways. I do not acknowledge the assistance of the loudest rock group in North America which practices at our house.

I am grateful to the subjects who posed for the living anatomy photographs: Dwayne Baldwin, Rob Clapham, Scott Dewar, Paul Nauman, Mike McNeill, Chris Newhouse, and Joe Thornley. Figure 7-66 is a contrast ocular photograph kindly provided to our Department by Dr. Scott Harris.

At Williams & Wilkins, a cast of hundreds contributes to the production of the *Atlas*. Most of them I will never meet, but I hope they know that I am grateful to them. Those who worked more closely with me were Mr. Don Slokan, Illustration Planner; Mr. George Stamathis, Associate Editor; and Ms. Diana Welch, Production Editor. In particular I express my gratitude to Ms. Toni M. Tracy, Vice President and Editor-in-Chief, who served as guide, tyrant, and good friend.

My personal gratitude is due my colleagues, students, and patients from whom I have learned. Dr. Grant was my undergraduate and graduate teacher, supervised my work on the fourth and fifth editions of the *Atlas*, and stimulated my interest in anthropology. He leaves happy memories.

J.E.A.

SECTION 1

The Thorax

CONTENTS

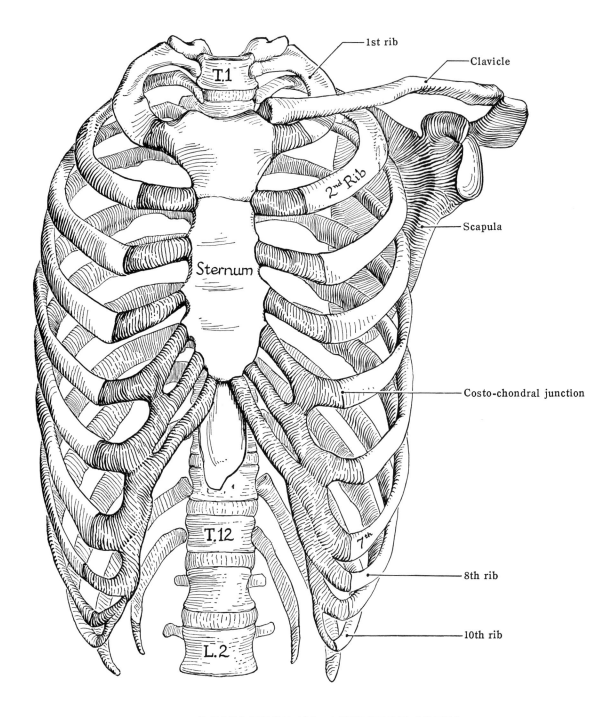

1st rib

Clavicle

T.1

2nd Rib

Scapula

Sternum

Costo-chondral junction

T.12

7th

8th rib

L.2

10th rib

1-1 BONY THORAX, ANTERIOR VIEW

Study both views of the bony Thorax and observe that:
1. The skeleton of the Thorax consists of 12 Thoracic vertebrae, 12 pairs of ribs and costal cartilages, and the sternum.
2. Each rib articulates posteriorly with the vertebral column.
3. In front, the upper seven costal cartilages articulate with the sternum. The 8th, 9th, and (10th) cartilages articulate with the cartilages next above. The 11th and 12th are "floating" ribs; their cartilages do not articulate anteriorly.
4. Posteriorly, all ribs incline downward. Anteriorly, the 3rd to 10th costal cartilages incline upward.
5. Because of the downward inclination of ribs, the sternal end of the first costal cartilage is about 4 cm below the level of the head of its rib, and the tip of the 12th costal cartilage is at the level of the second lumbar vertebra.

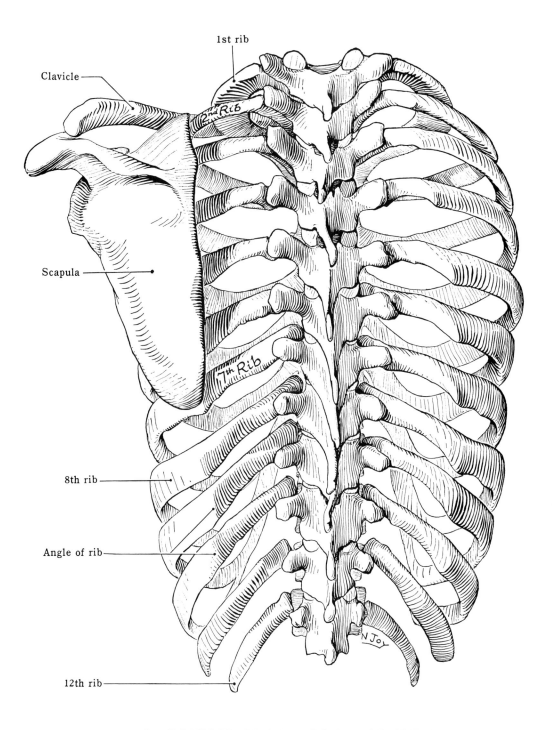

Clavicle

1st rib

2nd Rib

Scapula

7th Rib

8th rib

Angle of rib

12th rib

1-2 BONY THORAX, POSTERIOR VIEW

6. The first seven ribs and their costal cartilages increase progressively in length, so that the seventh is the longest rib.
7. The oval doorway between the Thoracic cavity and the neck region is bounded by the first Thoracic vertebra, the first ribs and their cartilages, and the manubrium of the sternum.
8. The clavicle and scapula are also shown. The clavicle lies over the first rib, making it difficult to palpate in the living subject. The second rib, however, is easy to locate because its costal cartilage articulates at the junction of manubrium and body of the sternum, the sternal angle.
9. The scapula crosses 50 per cent of the ribs: the 2nd to the 7th.

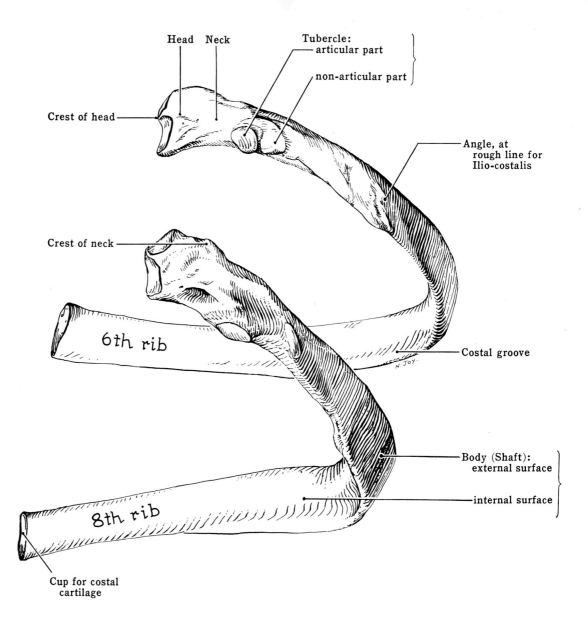

Head   Neck

Tubercle:
— articular part

non-articular part

Crest of head —

Angle, at
rough line for
Ilio-costalis

Crest of neck —

6th rib

Costal groove

Body (Shaft):
external surface

internal surface

8th rib

Cup for costal
cartilage

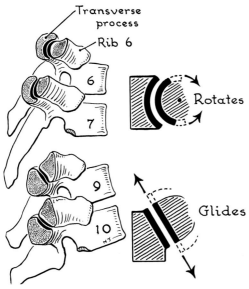

Transverse
process

Rib 6

6

7

Rotates

9

10

Glides

1-4 COSTO-TRANSVERSE JOINTS

This diagram shows that at the upper joints the
ribs *rotate;* at the 8th, 9th, and 10th they *glide,*
increasing the transverse diameter of the upper
abdomen.

1-3 TYPICAL RIBS, FROM BEHIND

Ribs 3 to 10 inclusive are considered "typical" and represented
here by ribs 6 and 8. Observe:

1. The wedge-shaped head, the posterior part of the shaft which
   is round in cross-section, the anterior part which is flattened,
   and the anterior slightly enlarged articular end.
2. Two facets on the head: a larger, lower one for articulation
   with the vertebral body of its own number, and a smaller
   upper facet for the vertebral body above. The crest of the
   head is joined to the intervertebral disc by a ligament.
3. The tubercle articulates with the transverse process of its
   own numbered vertebra.
4. Rough markings: on the posterior surface of the neck for the
   costo-transverse ligament, on the crest of the neck for the
   superior costo-transverse ligament, on the nonarticular part
   of the tubercle for the lateral costo-transverse ligament, and
   at the angle for the attachment of Ilio-costalis.
5. The costal groove which shelters the intercostal vein, artery,
   and nerve.

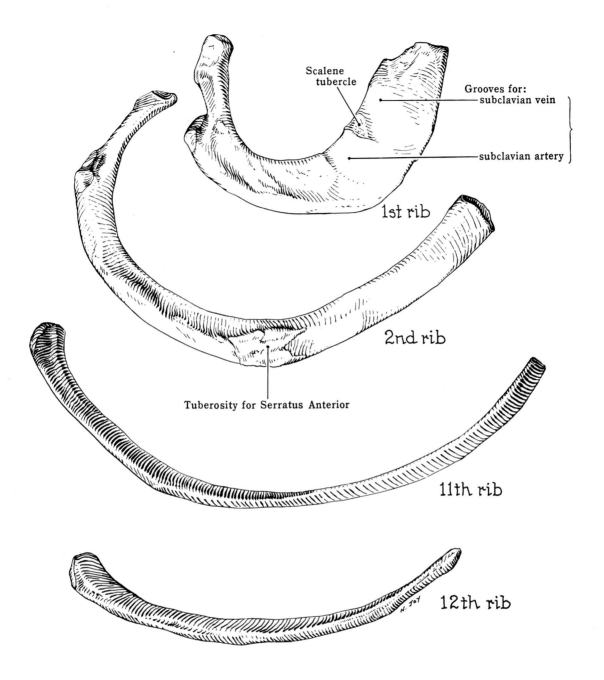

Scalene tubercle

Grooves for:
subclavian vein

subclavian artery

1st rib

2nd rib

Tuberosity for Serratus Anterior

11th rib

12th rib

1-5 ATYPICAL RIBS, FROM ABOVE

1. The first rib is the shortest, broadest, and most curved. The head has a single facet which articulates with the body of T1 and slightly with the disc above. Its prominent tubercle articulates with the transverse process of T1. The tubercle for Scalenus Anterior separates the groove for the subclavian vein in front from the groove for the subclavian artery behind.

2. The second rib has a poorly marked costal groove and a rough tuberosity for Serratus Anterior.

3. The 11th and 12th ribs are "floating" ribs, have a single facet on the head for articulation with the body of their own numbered vertebra, no tubercle and thus no costo-transverse articulation, and a tapering anterior end.

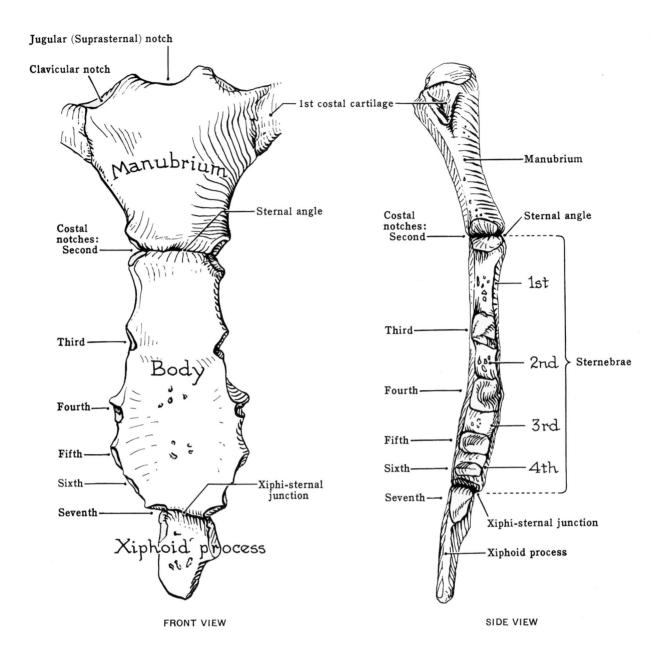

Jugular (Suprasternal) notch

Clavicular notch

Manubrium

1st costal cartilage

Sternal angle

Costal notches:
Second

Third

Body

Fourth

Fifth

Sixth

Seventh

Xiphi-sternal junction

Xiphoid process

FRONT VIEW

Manubrium

Sternal angle

Costal notches:
Second

Third

1st

2nd

Sternebrae

Fourth

3rd

Fifth

Sixth

4th

Seventh

Xiphi-sternal junction

Xiphoid process

SIDE VIEW

1-6 STERNUM

Observe:
1. The great thickness of the upper third of the manubrium.
2. The medial ends of the clavicles deepen the suprasternal notch. A finger placed in this notch palpates the trachea.
3. The sternal angle, at the junction of manubrium and body, is a palpable landmark guiding your fingers to the second costal cartilage.
4. The sharp lower edge of the body at the Xiphi-sternal junction is palpable and is the lowest point at which pressure is applied in external cardiac compression. Forceful displacement of the Xiphoid endangers the underlying liver.
5. Seven costal cartilages articulate with the sternum: the 1st to the manubrium and the 6th to the side of the 4th sternebra. All others articulate at junctions of the six elements of the sternum: manubrium, 4 sternebrae, and Xiphoid process.

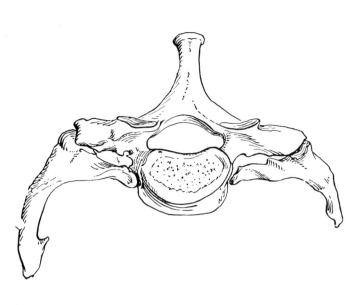

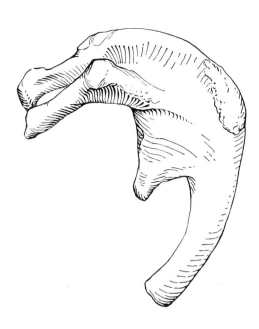

1-7 CERVICAL RIBS

A cervical rib is an enlarged costal element of the 7th cervical vertebra (see Fig. 5-3C). It may be large and palpable or detectable only radiologically; unilateral or bilateral; asymptomatic or through pressure on the lowest root of the brachial plexus may produce sensory and motor changes over the distribution of the ulnar nerve.

1-8 BICIPITAL RIB

In this specimen there has been partial fusion of the first two thoracic ribs. A similar condition results from the partial fusion of a cervical rib with the first thoracic rib.

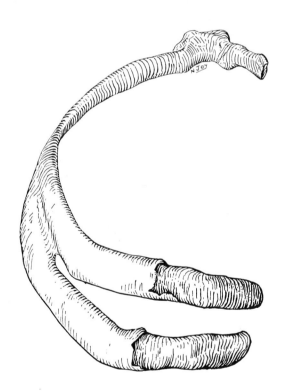

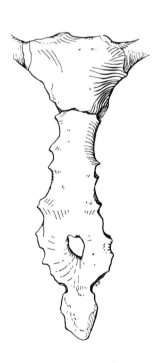

1-9 BIFID RIB

The upper component of this third rib is supernumerary and articulated with the side of the first sternebra. Bifid ribs occur in 1 to 2 per cent of most populations, but have been found in 8.4 per cent of 1000 Samoan radiographs.
See Martin, E. J. (1960) Incidence of bifidity and related rib abnormalities in Samoans. *Am. J. Phys. Anthropol.*, *18*: 179–187.

1-10 STERNAL FORAMEN

This relatively common anomaly results from a defect of ossification and is diagnosed as a bullet wound by the unwary. This specimen also shows synostosis of the Xiphisternal joint.

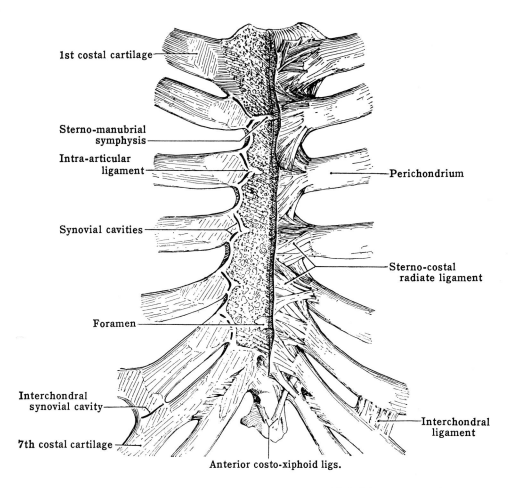

1st costal cartilage

Sterno-manubrial
symphysis

Intra-articular
ligament

Synovial cavities

Foramen

Interchondral
synovial cavity

7th costal cartilage

Perichondrium

Sterno-costal
radiate ligament

Interchondral
ligament

Anterior costo-xiphoid ligs.

1-11 STERNO-COSTAL AND INTERCHONDRAL JOINTS, ANTERIOR VIEW

Observe:
1. On the right side, the cortex of the sternum and costal cartilages has been shaved away. To obtain a specimen of bone marrow, a sternal puncture is done through the thin cortical bone into the area of spongy bone.
2. On the left side, dissection shows that the fibers of the perichondrium terminate as sterno-costal radiate ligaments.
3. Three types of joints are demonstrated:
   a) *Synchondroses:* between 1st costal cartilage and manubrium, and between 7th costal cartilage and sternum (in this case).
   b) *Symphysis:* the sterno-manubrial joint.
   c) *Synovial joints:* the other sterno-costal joints and the interchondral joints.

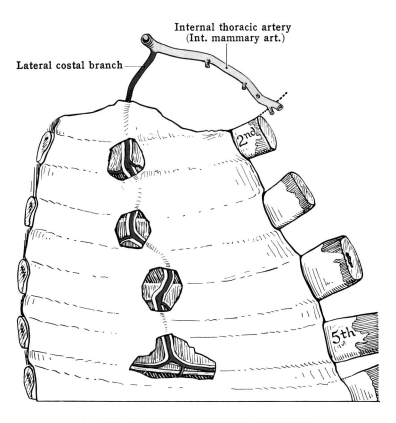

Internal thoracic artery
(Int. mammary art.)

Lateral costal branch

2nd

5th

This variant branch of the internal thoracic artery (see Figs. 1-13 and 1-15) occurs in about 25 per cent of cases.

1-12 LATERAL COSTAL ARTERY

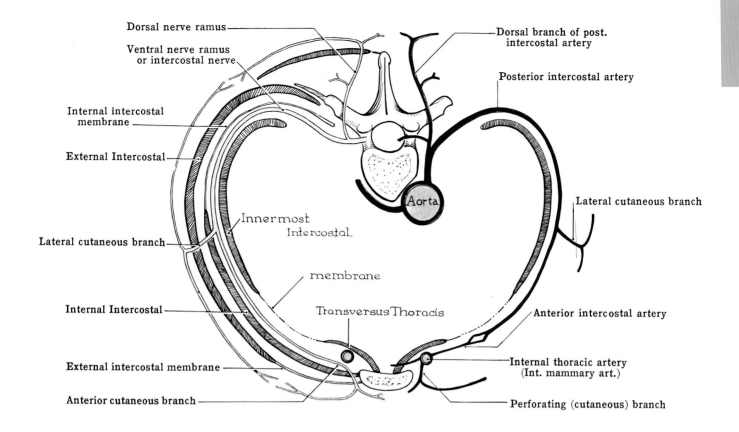

Dorsal nerve ramus

Ventral nerve ramus or intercostal nerve

Internal intercostal membrane

External Intercostal

Innermost Intercostal.

Lateral cutaneous branch

Internal Intercostal

membrane

Transversus Thoracis

External intercostal membrane

Anterior cutaneous branch

Dorsal branch of post. intercostal artery

Posterior intercostal artery

Aorta

Lateral cutaneous branch

Anterior intercostal artery

Internal thoracic artery (Int. mammary art.)

Perforating (cutaneous) branch

1-13 CONTENTS OF AN INTERCOSTAL SPACE

This diagram is simplified by showing nerves on the *right* and arteries on the *left*. Observe:

1. Three muscular layers: (a) External Intercostal muscle and membrane, (b) Internal Intercostal muscle and membrane, (c) Innermost Intercostal and Transversus Thoracis muscles and the membrane connecting them.
2. The intercostal vessels and nerves run in the plane between the middle and innermost layers of muscles. Subsequently, the lower intercostal vessels and nerves occupy a corresponding plane in the abdominal wall as shown in Figures 1–17 and 2–8.
3. Posterior intercostal arteries are branches of the aorta (the upper two spaces are supplied from the superior intercostal branch of the costo-cervical trunk); anterior intercostal arteries are branches of the internal thoracic artery.

1-14 AN INTERCOSTAL SPACE

This diagram shows relationships:
1. From above downward: vein, artery, nerve (VAN).
2. Order of entry from medial to lateral: artery, vein, nerve.

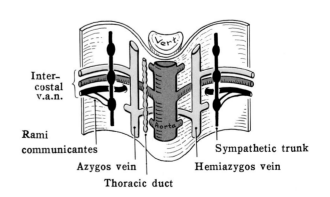

Inter-costal v.a.n.

Rami communicantes

Azygos vein

Thoracic duct

Sympathetic trunk

Hemiazygos vein

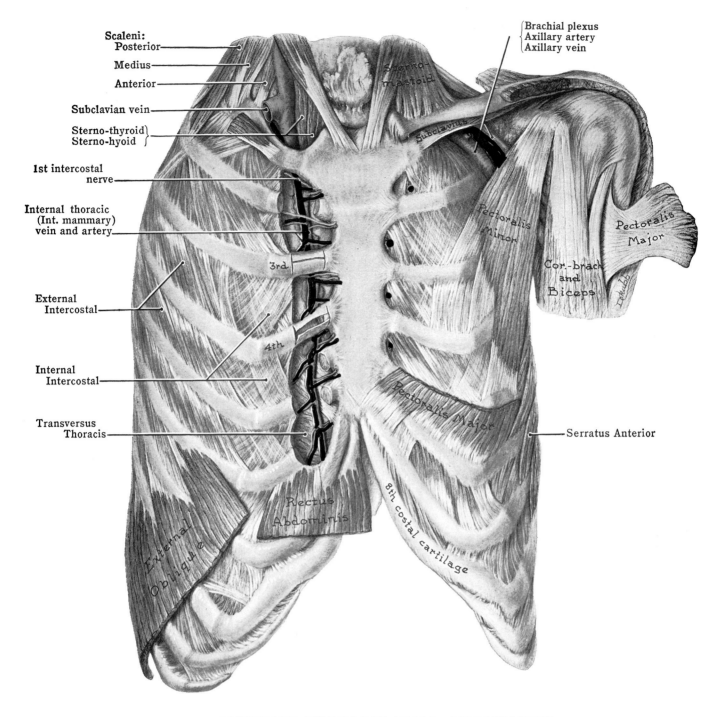

Scaleni:
Posterior
Medius
Anterior
Subclavian vein
Sterno-thyroid
Sterno-hyoid
1st intercostal nerve
Internal thoracic (Int. mammary) vein and artery
External Intercostal
Internal Intercostal
Transversus Thoracis

Brachial plexus
Axillary artery
Axillary vein

Sterno-mastoid
Subclavius
Pectoralis Minor
Pectoralis Major
Cor-brach and Biceps
3rd
4th
Pectoralis Major
Rectus Abdominis
8th costal cartilage
External Oblique

Serratus Anterior

1-15 ANTERIOR THORACIC WALL, FRONT VIEW

The muscles superficial to this dissection may be studied in Figures 9-3 and 6-14.
Observe:

1. The internal thoracic vessels running downward about 1 cm from the edge of the sternum and providing intercostal branches.
2. The parasternal lymph nodes (*green*) which receive lymphatic vessels from the intercostal spaces, from the costal pleura and diaphragm, and from the medial part of the breast. It is by this route that cancer of the breast may spread to lungs and mediastinum.
3. The subclavian vessels are sandwiched between the first rib and the clavicle (padded by subclavius).
4. Usually the seventh costal cartilage is the last to reach the sternum, although it is not uncommon, as in this specimen, for the eighth to do so.
5. The H-shaped cut through the perichondrium of the 3rd and 4th cartilages was used to shell out segments of cartilage. Similarly, in performing a thoracotomy the surgeon may shell a segment of rib out of its periosteum. Later, bone regenerates from this periosteum.

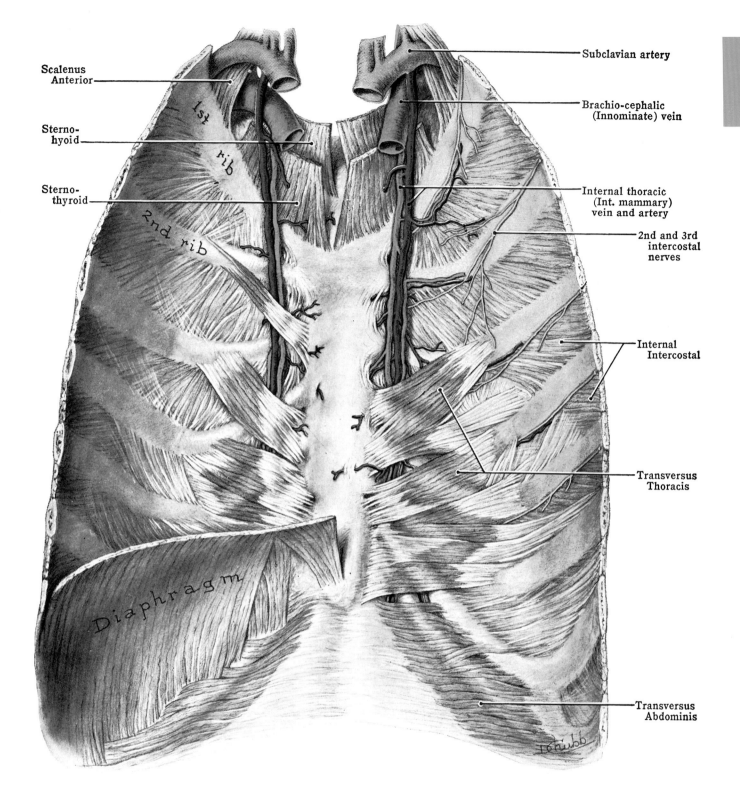

Scalenus Anterior

Sterno-hyoid

Sterno-thyroid

1st rib

2nd rib

Diaphragm

Subclavian artery

Brachio-cephalic (Innominate) vein

Internal thoracic (Int. mammary) vein and artery

2nd and 3rd intercostal nerves

Internal Intercostal

Transversus Thoracis

Transversus Abdominis

1-16 ANTERIOR THORACIC WALL, FROM BEHIND

Observe:

1. The continuity of Transversus Thoracis with Transversus Abdominis, these being the innermost layer of the three flat muscles of the thoraco-abdominal wall (Fig. 1-13).

2. The internal thoracic artery, arising from the first part of the subclavian artery, accompanied by two veins (venae comitantes) up to the 3rd or 2nd intercostal space and above this by a single vein (internal thoracic vein), which proceeds to the brachio-cephalic vein.

3. The lower portions of the internal thoracic vessels, covered posteriorly with Transversus Thoracis; the upper portions in contact with parietal pleura (removed).

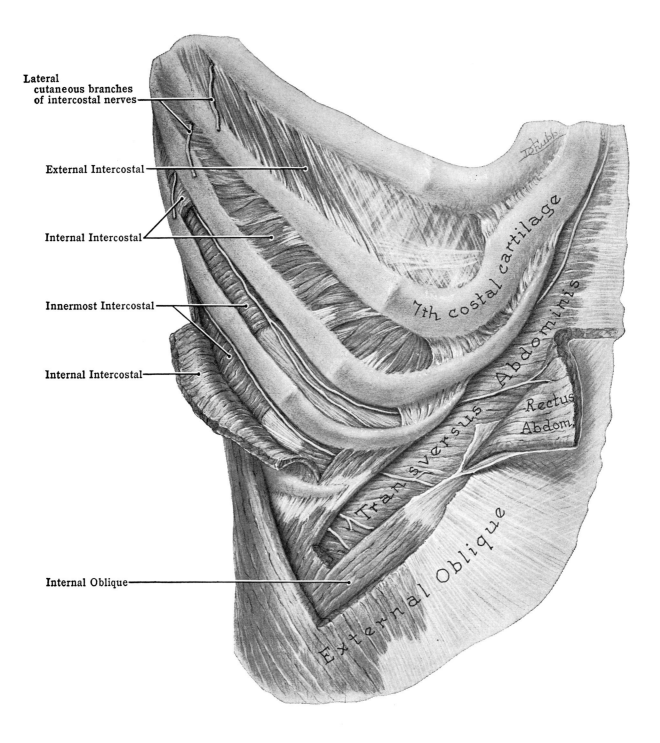

Lateral
cutaneous branches
of intercostal nerves

External Intercostal

Internal Intercostal

Innermost Intercostal

Internal Intercostal

Internal Oblique

7th costal cartilage

Transversus Abdominis

Rectus Abdom.

External Oblique

1-17 ANTERIOR ENDS OF LOWER INTERCOSTAL SPACES

Observe:

1. The fibers of External Intercostal and External Oblique run in the same direction: downward and medially.
2. The Internal Intercostal and Internal Oblique are in continuity at the ends of the 9th, 10th, and 11th intercostal spaces.
3. As explained in Figure 1-13, an intercostal nerve lies deep to Internal Intercostal but superficial to the Innermost Intercostal and to either Transversus Thoracis or Transversus Abdominis.
4. An intercostal nerve runs parallel to its rib and then to its costal cartilage. Thus, on reaching the abdominal wall, Nerves T7 and T8 continue upward, T9 continues nearly horizontally, and T10 continues downward and medially toward the umbilicus. These nerves provide cutaneous enervation in overlapping segmental bands as shown in Figure 5-42.

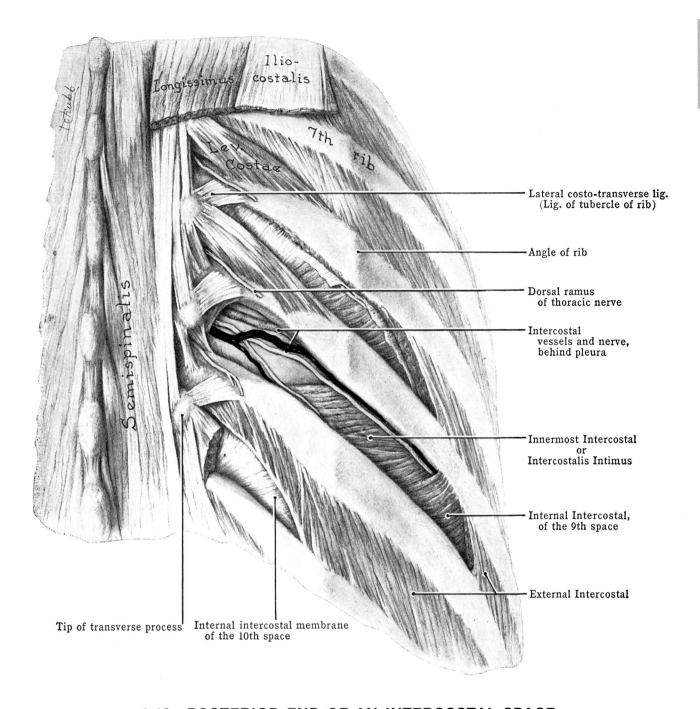

Lateral costo-transverse lig.
(Lig. of tubercle of rib)

Angle of rib

Dorsal ramus
of thoracic nerve

Intercostal
vessels and nerve,
behind pleura

Innermost Intercostal
or
Intercostalis Intimus

Internal Intercostal,
of the 9th space

External Intercostal

Tip of transverse process

Internal intercostal membrane
of the 10th space

1-18 POSTERIOR END OF AN INTERCOSTAL SPACE

In this view from behind, Ilio-costalis and Longissimus have been removed, exposing the Levatores Costarum. Of the five intercostal spaces shown:

a) The upper two (6th and 7th) are intact.

b) The 10th space reveals the internal intercostal membrane, following removal of the Levator Costae and underlying part of the External Intercostal.

c) From the 8th space more of the External Intercostal has been removed, exposing the Internal Intercostal membrane.

d) In the 9th space the Internal Intercostal membrane has been removed to show the intercostal vessels and nerve.

Note:

1. The vessels and nerve appear between the superior costo-transverse ligament and the pleura (*light blue*).

2. The intercostal nerve is lowest of the trio and least sheltered in the intercostal groove.

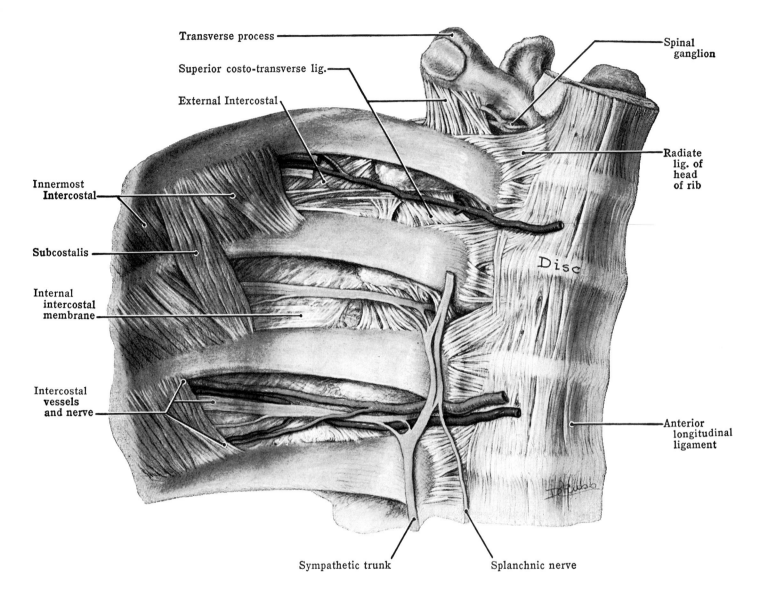

Transverse process

Superior costo-transverse lig.

External Intercostal

Innermost Intercostal

Subcostalis

Internal intercostal membrane

Intercostal vessels and nerve

Spinal ganglion

Radiate lig. of head of rib

Disc

Anterior longitudinal ligament

Sympathetic trunk

Splanchnic nerve

1-19 VERTEBRAL END OF AN INTERCOSTAL SPACE

In this anterior view observe:
1. Portions of the Innermost Intercostal muscle that bridge two intercostal spaces are called Subcostal muscles.
2. An External Intercostal muscle in the uppermost space.
3. An Internal Intercostal membrane in the middle space, continuous medially with a superior costo-transverse ligament.
4. In the lowest space, the order of the structures—intercostal vein, artery, and nerve. Note their collateral branches.
5. Near the top of the illustration, a thoracic nerve. The ventral ramus crosses in front of the superior costo-transverse ligament and the dorsal ramus behind it.
6. The attachment of intercostal nerves to the sympathetic trunk as in Figure 1-14. The splanchnic nerve is a visceral branch of the trunk.

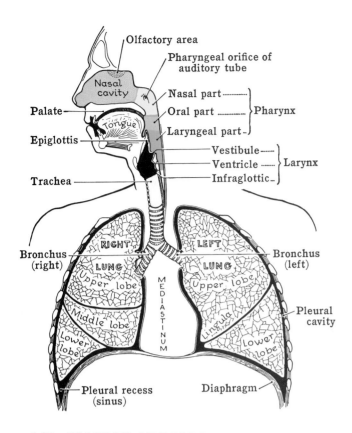

1-20 DIAGRAM OF RESPIRATORY SYSTEM

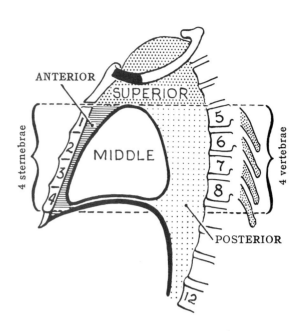

1-21 SUBDIVISIONS OF MEDIASTINUM

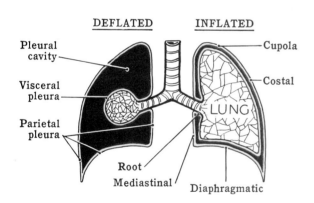

1-22 PLEURAL CAVITY AND PLEURA

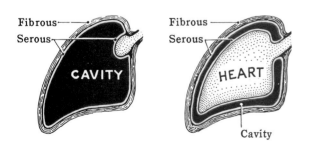

1-23 FIBROUS AND SEROUS PERICARDIA

Parietal pericardium is fibrous pericardium lined with serous. Visceral pericardium, called epicardium, is serous only.

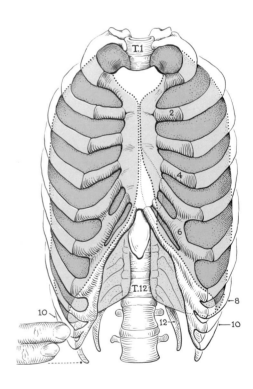

1-24 EXTENT OF PLEURA

The Pleura rises to, but not above, the neck of the first rib. Right and left Sternocostal Reflexions meet behind sternum above level of 2nd ribs, descend together to 4th ribs where left pleura deviates variably to 6th or 7th rib, is in midclavicular line at 8th rib, in midlateral line at 10th rib; thence as Vertebral Reflexion it ascends on vertebral bodies Th. 12th to 1st.

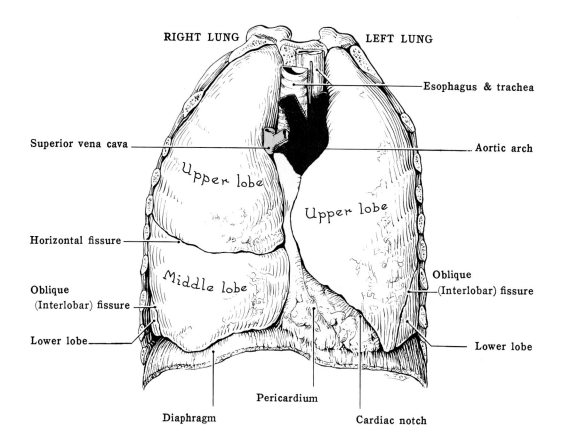

RIGHT LUNG   LEFT LUNG

Esophagus & trachea

Superior vena cava

Aortic arch

Upper lobe

Upper lobe

Horizontal fissure

Oblique
(Interlobar) fissure

Oblique
(Interlobar) fissure

Middle lobe

Lower lobe

Lower lobe

Pericardium

Diaphragm

Cardiac notch

1-25 LUNGS AND PERICARDIUM, FRONT VIEW

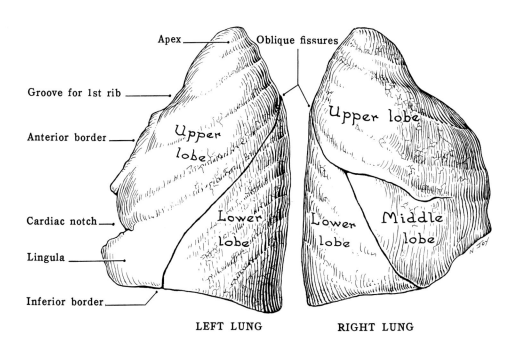

Apex   Oblique fissures

Groove for 1st rib

Upper lobe

Upper lobe

Anterior border

Cardiac notch

Lower lobe

Lower lobe

Middle lobe

Lingula

Inferior border

LEFT LUNG   **RIGHT LUNG**

1-26 LUNGS, LATERAL VIEWS

Observe:
1. The three lobes of the right lung and the two lobes of the left.
2. The middle lobe (of the right lung) lying at the front of the thorax; *i.e.*, it is entirely anterior to the midlateral line.
3. The deficiency of the upper lobe of the left lung, called the cardiac notch, allowing the pericardium to appear.
4. A horizontal fissure (of the right lung), complete in Figure 1-25 and incomplete in Figure 1-26.

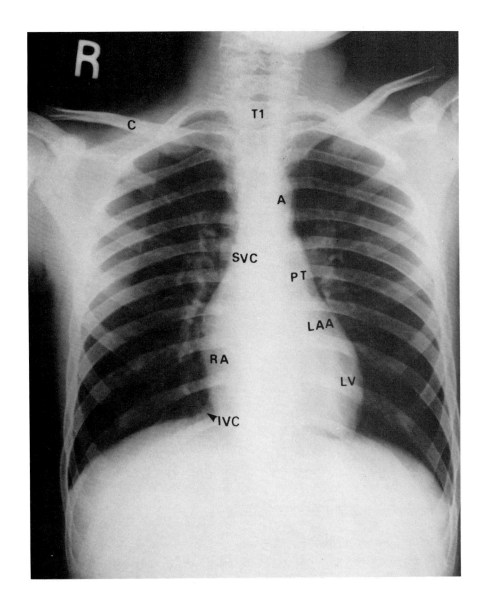

1-27 RADIOGRAPH OF CHEST

Observe in this postero-anterior projection:

1. The body of the first thoracic vertebra (*T1*). Follow it laterally to the first rib which curves outward, then medially crossing the clavicle (*C*).
2. The dome of the diaphragm is somewhat higher on the right.
3. The convexity of the right mediastinal border is formed by the right atrium (*RA*). The lesser convexity above this is produced by the superior vena cava (*SVC*). In the angle between the right atrium and upper border of the diaphragm, an *arrow* points to the inferior vena cava (*IVC*).
4. The left mediastinal border is formed by the aortic arch (*A*) or "aortic knob," the pulmonary trunk (*PT*), the left auricular appendage (*LAA*), and the left ventricle (*LV*).

See Figures 1-53 and 1-54 for details of the heart's borders.

For a review of the radiology of the heart see Shulman, H. S. (1980) *Med. Clin. North Am. 1:* 34–57.

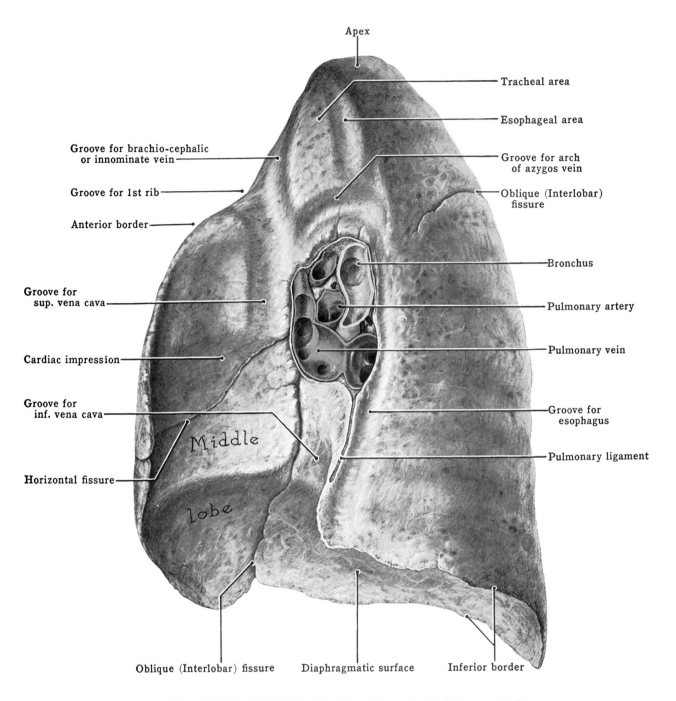

Apex

Tracheal area

Esophageal area

Groove for brachio-cephalic or innominate vein

Groove for arch of azygos vein

Groove for 1st rib

Oblique (Interlobar) fissure

Anterior border

Bronchus

Groove for sup. vena cava

Pulmonary artery

Pulmonary vein

Cardiac impression

Groove for inf. vena cava

Groove for esophagus

Pulmonary ligament

Middle

Horizontal fissure

lobe

Oblique (Interlobar) fissure   Diaphragmatic surface   Inferior border

1-28 MEDIASTINAL SURFACE OF RIGHT LUNG

Observe:
1. The lungs, resembling inflated balloons in that they take the impressions of the structures with which they come into contact. Thus, the base is fashioned by the cupola of the diaphragm; the costal surface bears the impressions of the ribs; distended vessels leave their mark; empty vessels and nerves do not.
2. The somewhat pear-shaped root of the lung near the center of the mediastinal surface, and the pulmonary ligament descending like a stalk from the root.
3. The groove for (or line of contact with) the esophagus throughout the length of the lung, except where the arch of the azygos vein intervenes. This groove passes behind the root and therefore behind the pulmonary ligament, which separates it from the groove for the inferior vena cava.
4. The oblique (interlobar) fissure, here incomplete, but complete in Figure 1-29.
5. The two pulmonary veins, here uniting unusually close to the lung.

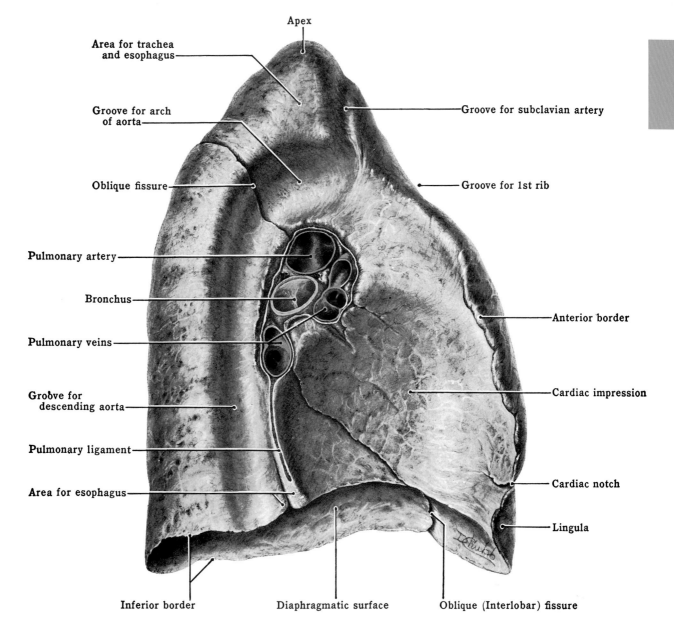

Apex

Area for trachea and esophagus

Groove for arch of aorta

Oblique fissure

Pulmonary artery

Bronchus

Pulmonary veins

Groove for descending aorta

Pulmonary ligament

Area for esophagus

Groove for subclavian artery

Groove for 1st rib

Anterior border

Cardiac impression

Cardiac notch

Lingula

Inferior border

Diaphragmatic surface

Oblique (Interlobar) fissure

1-29 MEDIASTINAL SURFACE OF LEFT LUNG

Observe:
1. Near the center, the root and the pulmonary ligament descending from it.
2. The site of contact with the esophagus, between the aorta and the lower end of the ligament.
3. The oblique (interlobar) fissure, cutting completely through the lung substance.
4. In both the right root and the left, the artery is above; the bronchus is behind; one vein is in front, and the other is below. In the right root, the bronchus (eparterial) to the upper lobe is the highest structure.

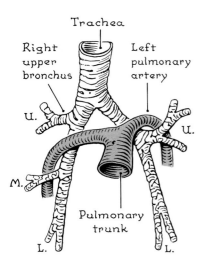

Trachea

Right upper bronchus

Left pulmonary artery

U.

U.

M.

Pulmonary trunk

L.

L.

1-30 LUNG ROOTS

ote the gross difference between right and left
ots. On the *right,* the upper and middle bron-
i are spread apart by the artery; on the *left,*
ey are pressed together.

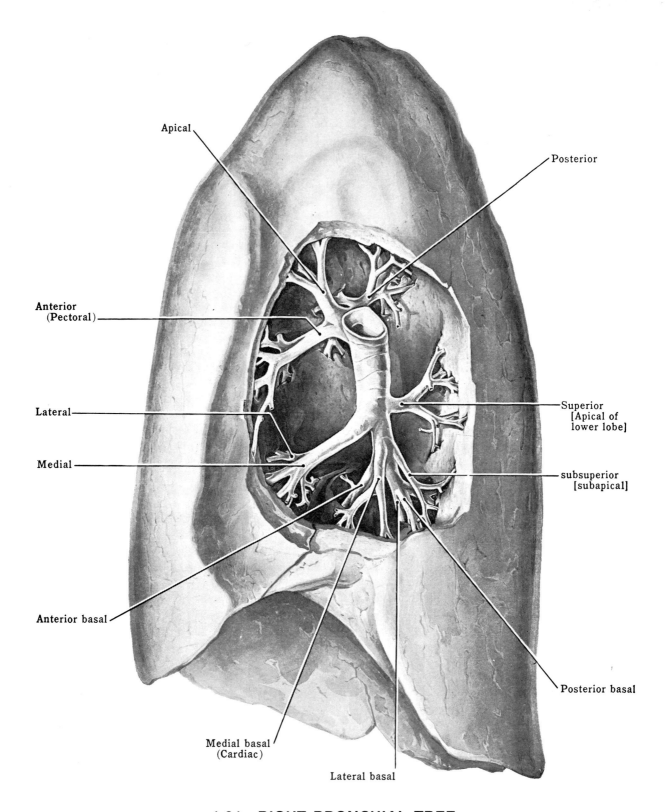

Apical

Posterior

Anterior
(Pectoral)

Lateral

Medial

Superior
[Apical of
lower lobe]

subsuperior
[subapical]

Anterior basal

Posterior basal

Medial basal
(Cardiac)

Lateral basal

1-31 RIGHT BRONCHIAL TREE

Fresh lungs were kept inflated with air under low pressure until thoroughly dry (about a week) and their natural form thereby assured. The tissues surrounding the bronchi were then moistened and cut away.

Terminology (after Jackson and Huber). There are usually 10 right and 8 left tertiary or segmental bronchi. They are approximately symmetrical in the two lungs. The reduced number in the left lung is accounted for by the fact that the left apical and posterior bronchi arise from a common stem, as do also the left anterior basal and medial basal.

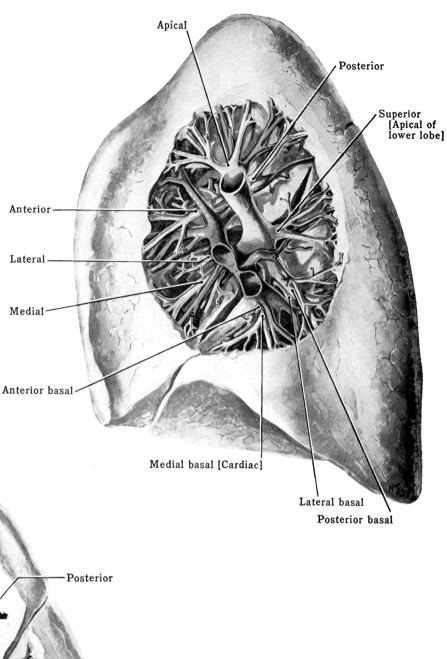

Apical

Posterior

Superior
[Apical of
lower lobe]

Anterior

Lateral

Medial

Anterior basal

Medial basal [Cardiac]

Lateral basal

Posterior basal

1-32 RIGHT BRONCHI AND PULMONARY VEINS

The veins of fresh lungs were filled with blue latex and the bronchi kept inflated and treated as for Figure 1-31.

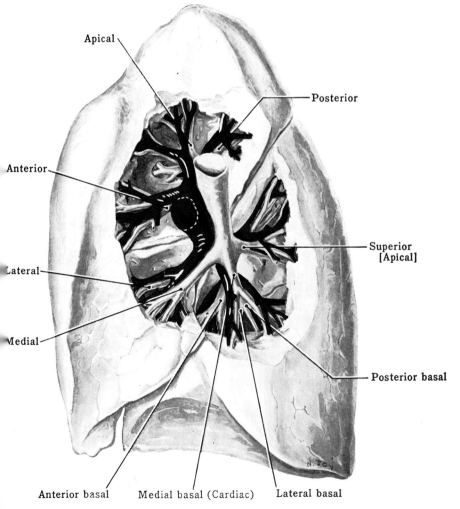

Apical

Posterior

Anterior

Superior
[Apical]

Lateral

Medial

Posterior basal

Anterior basal

Medial basal (Cardiac)

Lateral basal

1-33 RIGHT BRONCHI AND PULMONARY ARTERIES

The arteries of fresh lungs were filled with red latex and the bronchi kept inflated and treated as for Figure 1-31.

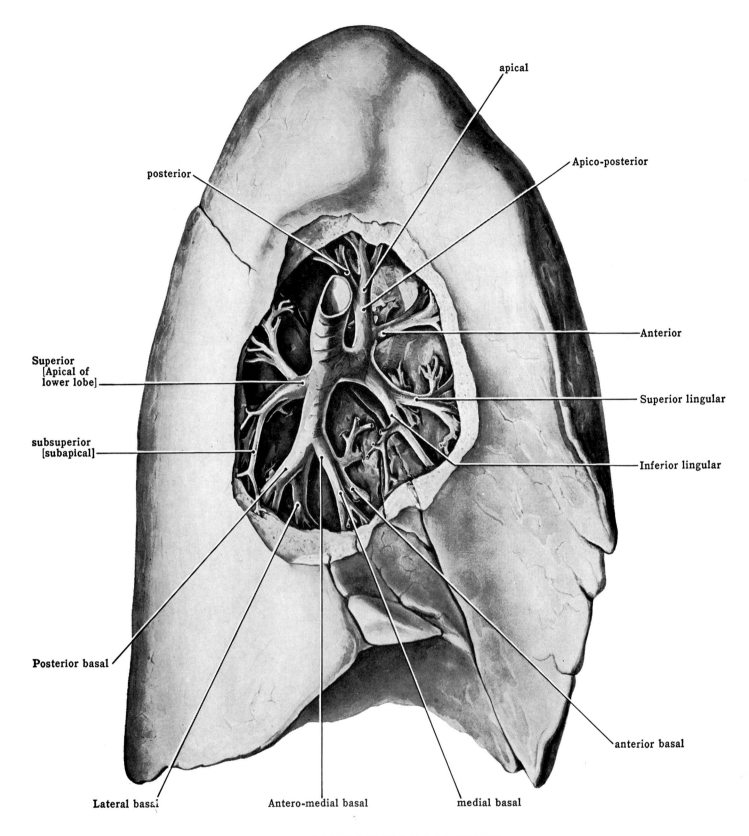

apical

Apico-posterior

posterior

Anterior

Superior
[Apical of
lower lobe]

Superior lingular

subsuperior
[subapical]

Inferior lingular

Posterior basal

anterior basal

Lateral basal

Antero-medial basal

medial basal

1-34 LEFT BRONCHIAL TREE

Fresh lungs were kept inflated with air under low pressure until thoroughly dry (about a week) and their natural form thereby assured. The tissues surrounding the bronchi were then moistened and cut away.

Terminology (after Jackson and Huber). There are usually 10 right and 8 left tertiary or segmental bronchi. They are approximately symmetrical in the two lungs. The reduced number in the left lung is accounted for by the fact that the left apical and posterior bronchi arise from a common stem, as do also the left anterior basal and the medial basal.

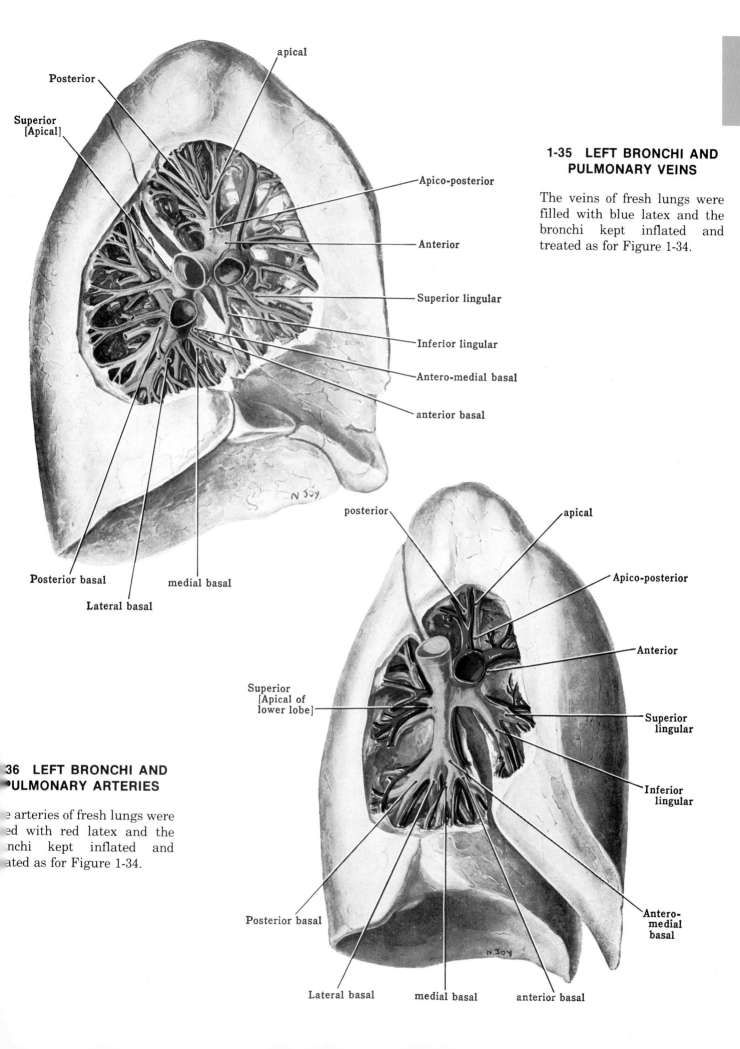

apical

Posterior

Superior
[Apical]

Apico-posterior

Anterior

Superior lingular

Inferior lingular

Antero-medial basal

anterior basal

The veins of fresh lungs were filled with blue latex and the bronchi kept inflated and treated as for Figure 1-34.

Posterior basal

Lateral basal

medial basal

posterior

apical

Apico-posterior

Anterior

Superior
[Apical of
lower lobe]

Superior
lingular

Inferior
lingular

36 LEFT BRONCHI AND
PULMONARY ARTERIES

e arteries of fresh lungs were
ed with red latex and the
nchi kept inflated and
ated as for Figure 1-34.

Antero-
medial
basal

Posterior basal

Lateral basal

medial basal

anterior basal

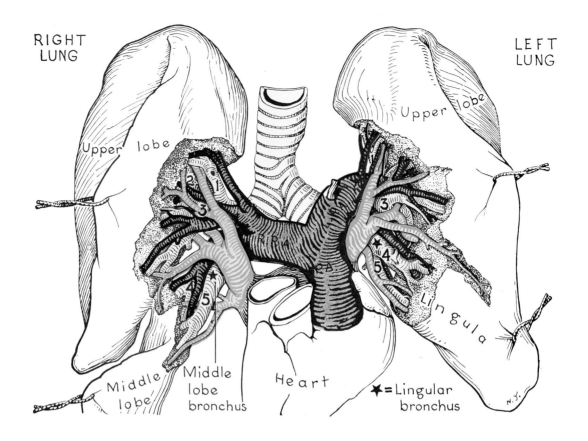

RIGHT
LUNG

LEFT
LUNG

Upper lobe

Upper lobe

Lingula

Middle lobe

Middle lobe bronchus

Heart

★ = Lingular bronchus

1-37 DISSECTION OF THE HILI OF THE LUNGS, FROM THE FRONT
(The bronchi and the pulmonary veins and arteries were injected.)

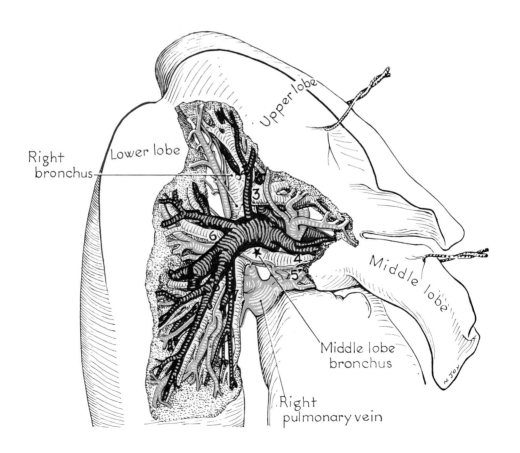

Upper lobe

Lower lobe

Right bronchus

Middle lobe

Middle lobe bronchus

Right pulmonary vein

1-38 DISSECTION OF THE HILUS OF THE RIGHT LUNG, AFTER OPENING THE OBLIQUE FISSURE

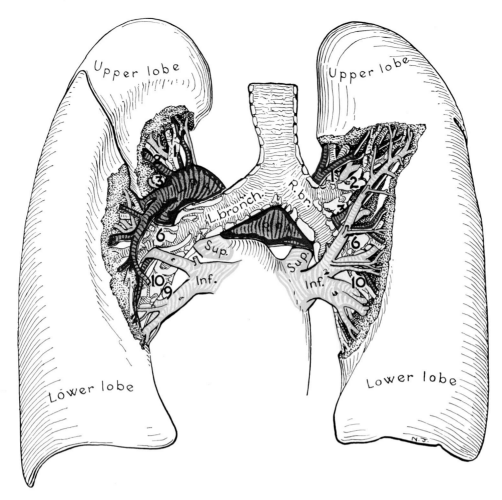

1-39 DISSECTION OF THE HILI OF THE LUNGS, FROM BEHIND
(The bronchi and the pulmonary veins and arteries were injected.)

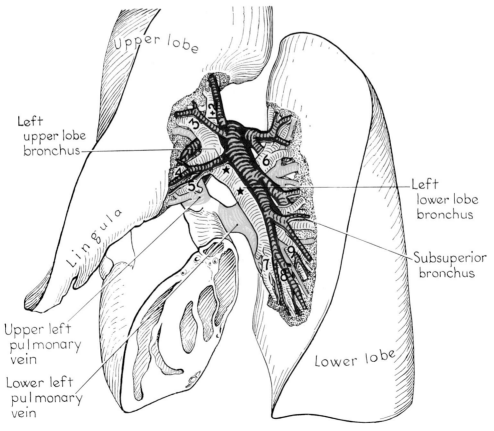

**1-40 DISSECTION OF THE HILUS OF THE LEFT LUNG, AFTER
OPENING THE OBLIQUE FISSURE**

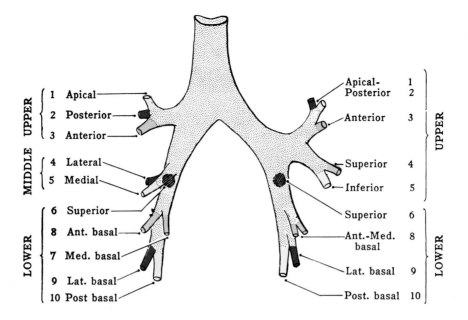

1-41 SEGMENTAL BRONCHI

The right lung has three lobes; the left has two. There are 10 tertiary or segmental bronchi on the right, 8 on the left. Note that on the *left* the apical and posterior bronchi arise from a single stem, as do the anterior basal and medial basal.

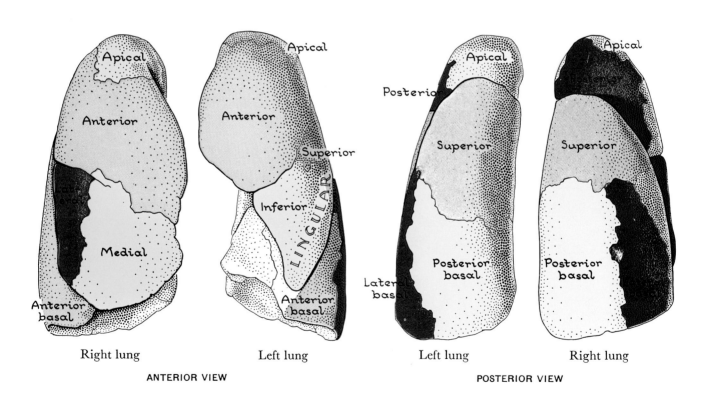

Right lung   Left lung

ANTERIOR VIEW

Left lung   Right lung

POSTERIOR VIEW

1-42 BRONCHO-PULMONARY SEGMENTS

A broncho-pulmonary segment consists of a tertiary bronchus, the portion of lung it ventilates, an artery, and a vein. These are surgically separable. To prepare Figures 1-42, 1-43, and 1-44, the tertiary bronchi of fresh lungs were isolated within the hilus and injected with latex of various colors. Minor variations in the branching of the bronchi result in variations in the surface patterns.

For detailed information consult Jackson, C. L., and Huber, J. F. (1943). Correlated applied anatomy of the bronchial tree and lungs with a system of nomenclature. *Dis. Chest, 9:* 319; Boyden, E. A. (1954) *Segmental Anatomy of the Lungs.* McGraw-Hill Book Company, New York; Boyden, E. A. (1961) The nomenclature of the broncho-pulmonary segments and their blood supply. *Dis. Chest, 39:* 1.

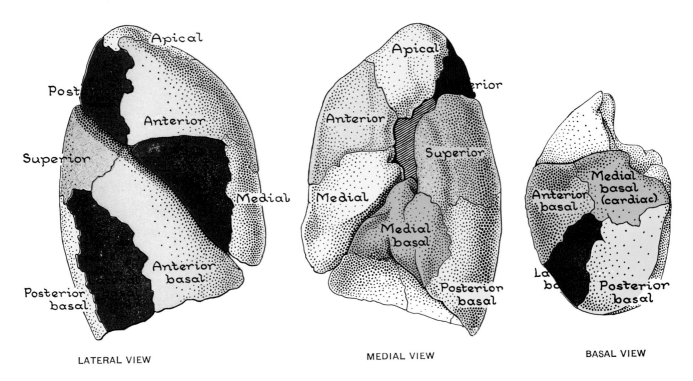

LATERAL VIEW

MEDIAL VIEW

BASAL VIEW

1-43 RIGHT BRONCHO-PULMONARY SEGMENTS

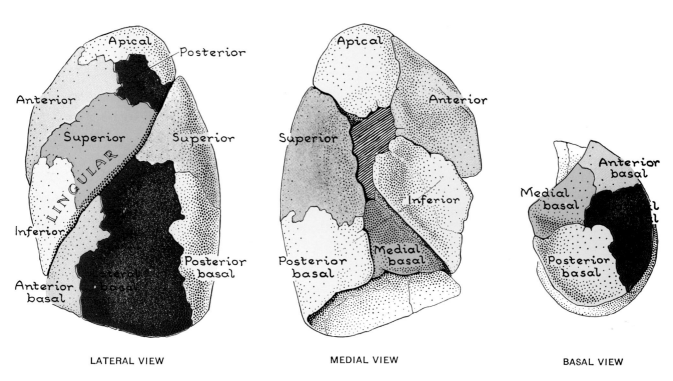

LATERAL VIEW

MEDIAL VIEW

BASAL VIEW

1-44 LEFT BRONCHO-PULMONARY SEGMENTS

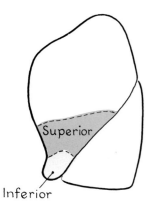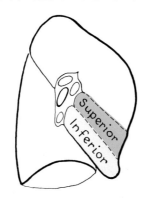

1-45 SEGMENTS OF THE LINGULA

This is the usual pattern. Compare with Figure 1–44 above.

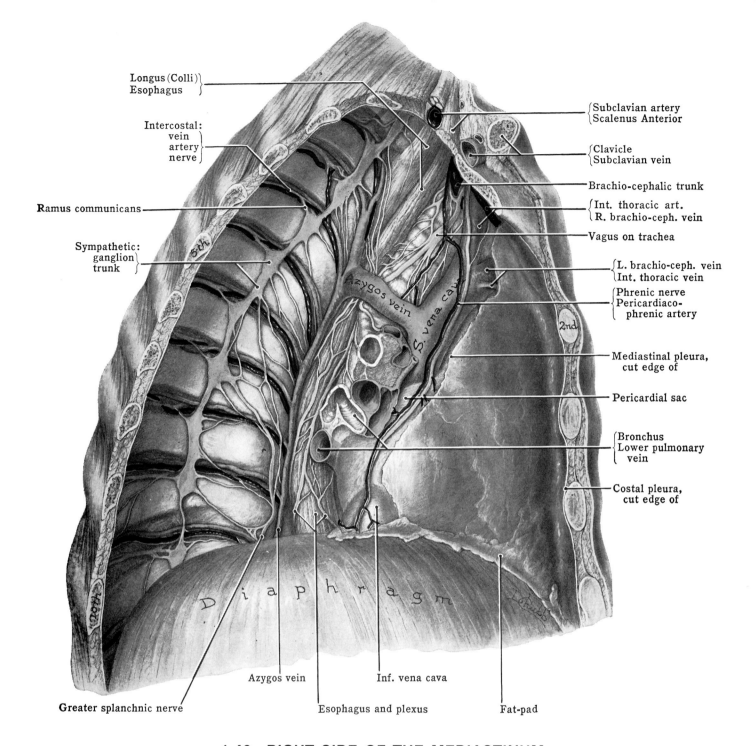

Longus (Colli)
Esophagus

Intercostal:
vein
artery
nerve

Ramus communicans

Sympathetic:
ganglion
trunk

5th

Subclavian artery
Scalenus Anterior

Clavicle
Subclavian vein

Brachio-cephalic trunk

Int. thoracic art.
R. brachio-ceph. vein

Vagus on trachea

L. brachio-ceph. vein
Int. thoracic vein

Phrenic nerve
Pericardiaco-
phrenic artery

Azygos vein

S. vena cava

2nd

Mediastinal pleura,
cut edge of

Pericardial sac

Bronchus
Lower pulmonary
vein

Costal pleura,
cut edge of

Diaphragm

Greater splanchnic nerve

Azygos vein

Inf. vena cava

Esophagus and plexus

Fat-pad

1-46   RIGHT SIDE OF THE MEDIASTINUM

The costal and mediastinal pleura has mostly been removed, exposing the underlying structures. Compare with the mediastinal surface of the right lung in Figure 1–28.

In this important dissection observe:

1. The right side of the mediastinum is the *blue* side, dominated by the arch of the azygos vein, the superior vena cava, and the right atrium.
2. When mediastinal pleura is removed, the phrenic nerve is free. Follow its medial relationships to the diaphragm.
3. Trachea and esophagus are visible.
4. The right vagus nerve enters on the trachea and falls back upon the esophagus. It is medial to the azygos arch.
5. The sympathetic trunk and its ganglia.

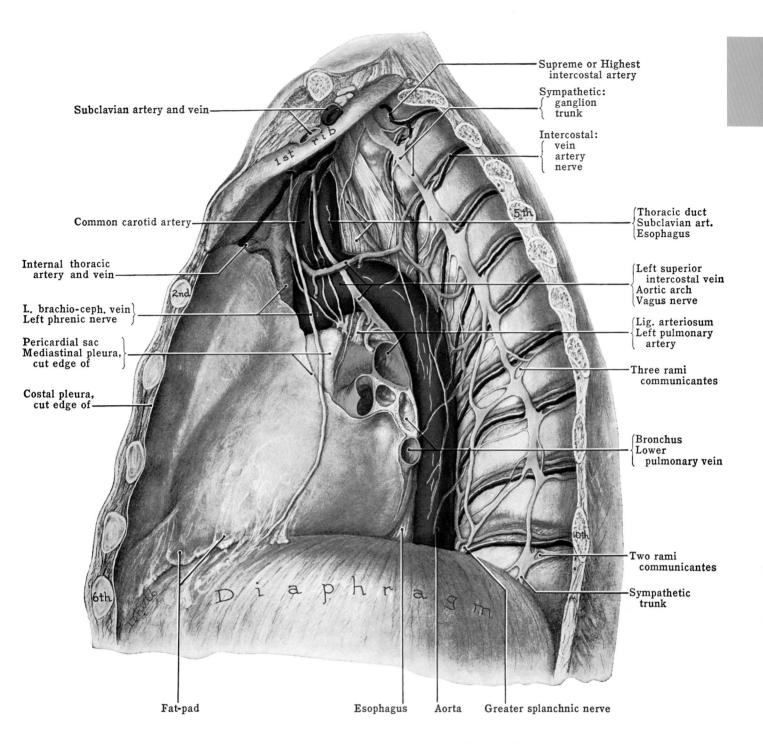

Subclavian artery and vein

Common carotid artery

Internal thoracic
artery and vein

L. brachio-ceph. vein
Left phrenic nerve

Pericardial sac
Mediastinal pleura,
cut edge of

Costal pleura,
cut edge of

1st rib

2nd

6th

Diaphragm

Fat-pad

Esophagus

Aorta

Greater splanchnic nerve

Supreme or Highest
intercostal artery

Sympathetic:
ganglion
trunk

Intercostal:
vein
artery
nerve

5th

Thoracic duct
Subclavian art.
Esophagus

Left superior
intercostal vein
Aortic arch
Vagus nerve

Lig. arteriosum
Left pulmonary
artery

Three rami
communicantes

Bronchus
Lower
pulmonary vein

10th

Two rami
communicantes

Sympathetic
trunk

1-47 LEFT SIDE OF THE MEDIASTINUM

The costal and mediastinal pleura has mostly been removed, exposing the
underlying structures. Compare with the mediastinal surface of the left lung in
Figure 1–29.

In this important dissection observe:

1. The left side of the mediastinum is the *red* side, dominated by the arch and
   descending portion of the aorta, the left common carotid and subclavian
   arteries.

2. The phrenic nerve, freed by removal of pleura, passing in front of the root of
   the lung.

3. The thoracic duct on the side of the esophagus.

4. The left vagus nerve on the side of arteries, passing behind the root of the
   lung, and sending its recurrent laryngeal branch around the ligamentum
   arteriosum.

5. The sympathetic trunk attached to intercostal nerves by rami communicantes.

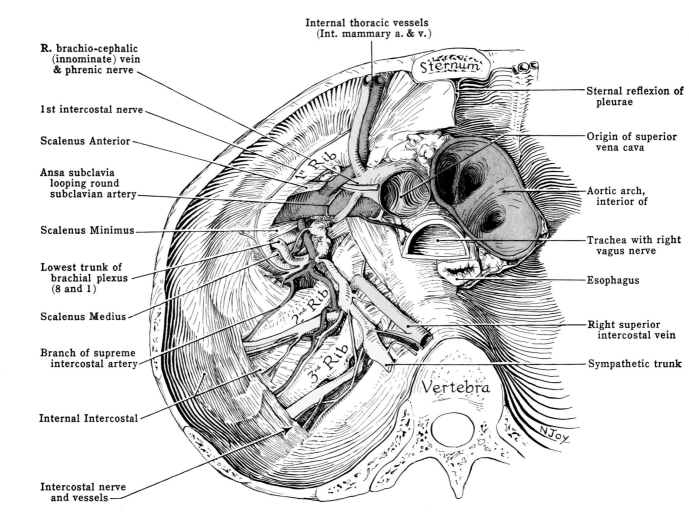

R. brachio-cephalic
(innominate) vein
& phrenic nerve

1st intercostal nerve

Scalenus Anterior

Ansa subclavia
looping round
subclavian artery

Scalenus Minimus

Lowest trunk of
brachial plexus
(8 and 1)

Scalenus Medius

Branch of supreme
intercostal artery

Internal Intercostal

Intercostal nerve
and vessels

Internal thoracic vessels
(Int. mammary a. & v.)

Sternum

Sternal reflexion of
pleurae

Origin of superior
vena cava

Aortic arch,
interior of

Trachea with right
vagus nerve

Esophagus

Right superior
intercostal vein

Sympathetic trunk

Vertebra

1-48 ROOF OF PLEURAL CAVITY

This view, from below, shows the roof of the pleural cavity (or the floor of the root of the neck) with the pleura removed. See Figure 9–38.

Observe:
1. The first part of the subclavian artery, arching over the cupola and disappearing between Scalenus Anterior and Scalenus Minimus, which is an occasional muscle.
2. The internal thoracic artery and the supreme intercostal branch of the costo-cervical trunk.
3. Two nerves—the ansa subclavia from the sympathetic trunk and the recurrent laryngeal nerve from the vagus—looping round the subclavian artery.
4. The supreme intercostal artery crossing the neck of the first rib between the sympathetic trunk, which is on its medial side, and the ventral ramus of T1 to the brachial plexus, which is on its lateral side.

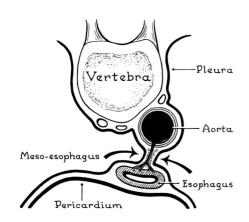

Vertebra

Pleura

Aorta

Meso-esophagus

Esophagus

Pericardium

1-49 THE MESO-ESOPHAGUS

Between the lower part of the esophagus and the aorta, right and left layers of mediastinal pleura meet and form a dorsal meso-esophagus.

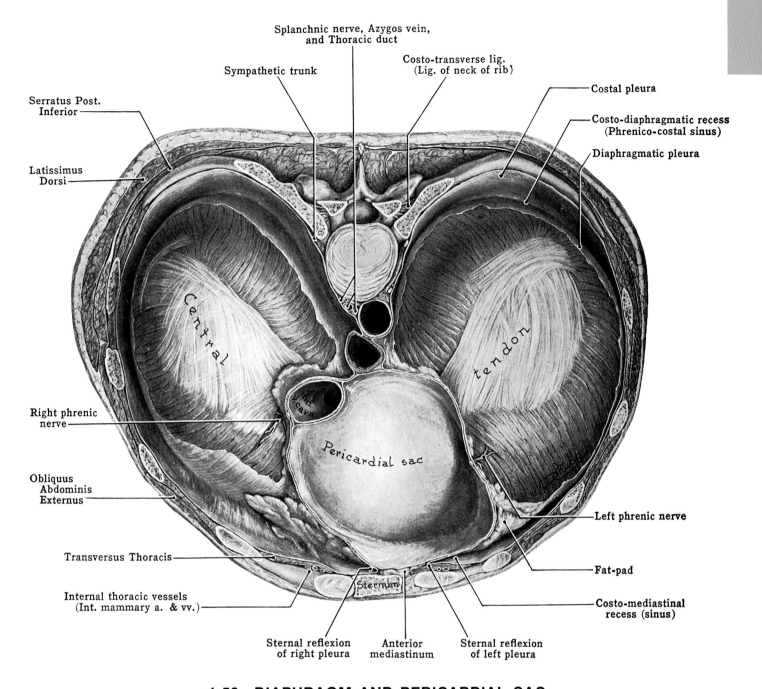

Splanchnic nerve, Azygos vein, and Thoracic duct

Sympathetic trunk

Costo-transverse lig. (Lig. of neck of rib)

Costal pleura

Costo-diaphragmatic recess (Phrenico-costal sinus)

Diaphragmatic pleura

Serratus Post. Inferior

Latissimus Dorsi

Central   tendon

Right phrenic nerve

Pericardial sac

Obliquus Abdominis Externus

Left phrenic nerve

Transversus Thoracis

Fat-pad

Internal thoracic vessels (Int. mammary a. & vv.)

Sternum

Costo-mediastinal recess (sinus)

Sternal reflexion of right pleura

Anterior mediastinum

Sternal reflexion of left pleura

1-50 DIAPHRAGM AND PERICARDIAL SAC

In this view from above, the diaphragmatic pleura is mostly removed. Observe:
1. The pericardial sac, situated on the ventral half of the diaphragm; one-third being to the right of the median plane and two-thirds being to the left; the most caudal point being ventrally and to the left, like the apex of the heart.
2. The mouths of the large hepatic veins, opening into the inferior vena cava and directed upward, toward the heart.
3. The sternal reflexion of the left pleural sac, failing to meet that of the right sac in the median plane, ventral to the pericardium.
4. The right and left pleural sacs almost meeting between esophagus and aorta to form a mesoesophagus.
5. The costo-diaphragmatic recess, deepest about the midlateral line (Fig. 2–27).
6. The costal pleura, on reaching the vertebral column imperceptibly becoming the mediastinal pleura.

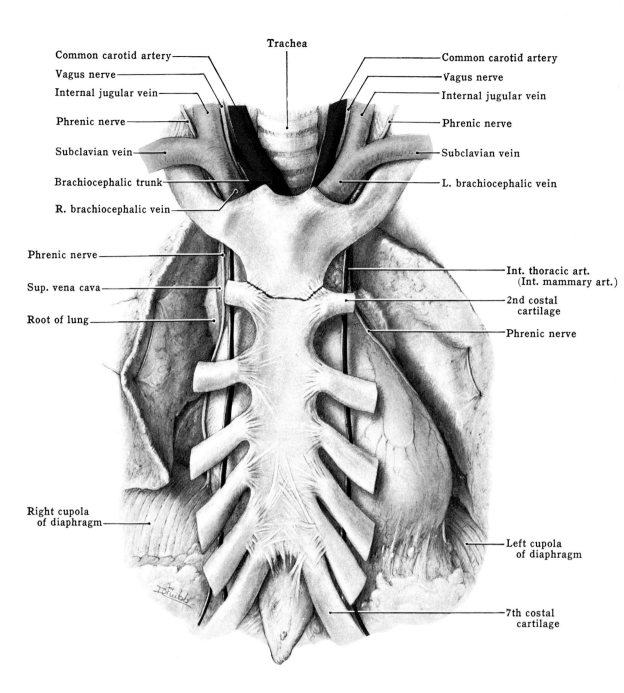

Common carotid artery

Trachea

Common carotid artery

Vagus nerve

Vagus nerve

Internal jugular vein

Internal jugular vein

Phrenic nerve

Phrenic nerve

Subclavian vein

Subclavian vein

Brachiocephalic trunk

L. brachiocephalic vein

R. brachiocephalic vein

Phrenic nerve

Int. thoracic art.
(Int. mammary art.)

Sup. vena cava

2nd costal
cartilage

Root of lung

Phrenic nerve

Right cupola
of diaphragm

Left cupola
of diaphragm

7th costal
cartilage

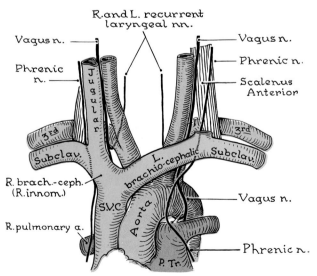

R. and L. recurrent
laryngeal nn.

Vagus n.

Phrenic
n.

Vagus n.

Phrenic n.

Scalenus
Anterior

3rd

Jugular

3rd

Subclav.

L.
brachio-cephalic

Subclav.

R. brach.-ceph.
(R. innom.)

S.V.C.

Aorta

Vagus n.

R. pulmonary a.

Phrenic n.

P. Tr.

1-52 GREAT VESSELS

This diagram shows the great vessels of the
superior mediastinum in front view.

1-51 PERICARDIAL SAC IN RELATION TO STERNUM

Observe:
1. The pericardial sac lies behind the body of the sternum
   from just above the sterno-manubrial joint to the level of
   the Xiphi-sternal joint. About one-third lies to the right of
   the median plane and two-thirds to the left.
2. The heart thus lies between the sternum in front and the
   vertebral column (plus posterior mediastinum) behind. In
   cardiac compression, the sternum is depressed 4 to 5 cm.
   forcing blood out of the heart and into the great vessels.
3. The internal thoracic arteries lateral to the borders of the
   sternum.
4. The right and left phrenic nerves applied to the pericardial
   sac.

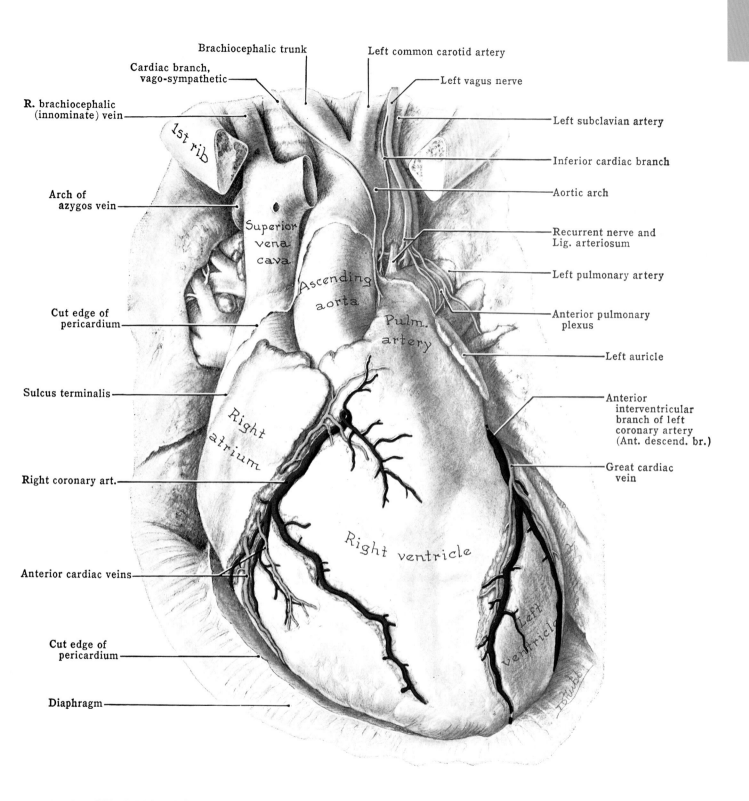

Brachiocephalic trunk

Cardiac branch,
vago-sympathetic

Left common carotid artery

Left vagus nerve

R. brachiocephalic
(innominate) vein

1st rib

Left subclavian artery

Inferior cardiac branch

Arch of
azygos vein

Superior
vena
cava

Aortic arch

Recurrent nerve and
Lig. arteriosum

Ascending
aorta

Left pulmonary artery

Cut edge of
pericardium

Pulm.
artery

Anterior pulmonary
plexus

Left auricle

Sulcus terminalis

Right
atrium

Anterior
interventricular
branch of left
coronary artery
(Ant. descend. br.)

Right coronary art.

Great cardiac
vein

Anterior cardiac veins

Right ventricle

Left
ventricle

Cut edge of
pericardium

Diaphragm

1-53 STERNOCOSTAL SURFACE OF THE HEART AND GREAT VESSELS, *IN SITU*

Observe:
1. The entire right auricle and much of the right atrium are visible from the front, but only a slight portion of the left auricle is visible. The auricles, like two closing claws, grasp the pulmonary artery and ascending aorta from behind.
2. The ligamentum arteriosum, continuing the direction of the pulmonary trunk (artery), and passing from the root of the left pulmonary artery to the aortic arch beyond the site of origin of the left subclavian artery.

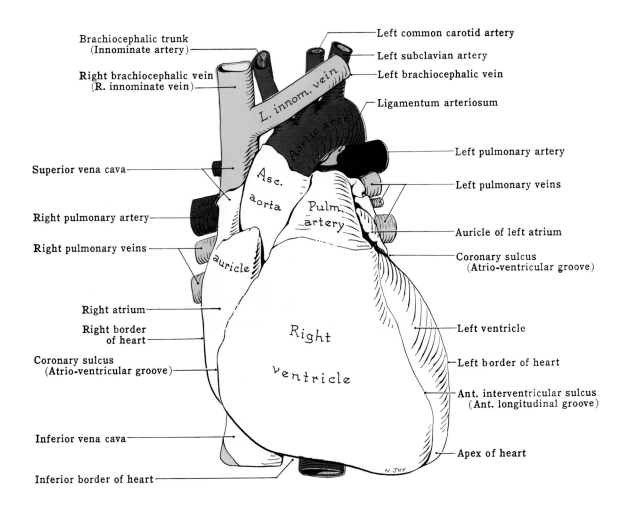

Brachiocephalic trunk (Innominate artery)
Right brachiocephalic vein (R. innominate vein)
Superior vena cava
Right pulmonary artery
Right pulmonary veins
Right atrium
Right border of heart
Coronary sulcus (Atrio-ventricular groove)
Inferior vena cava
Inferior border of heart

Left common carotid artery
Left subclavian artery
Left brachiocephalic vein
Ligamentum arteriosum
Left pulmonary artery
Left pulmonary veins
Auricle of left atrium
Coronary sulcus (Atrio-ventricular groove)
Left ventricle
Left border of heart
Ant. interventricular sulcus (Ant. longitudinal groove)
Apex of heart

L. innom. vein
Aortic arch
Asc. aorta
Pulm. artery
auricle
Right ventricle

1-54A   HEART AND GREAT VESSELS, HARDENED *IN SITU* AND REMOVED EN MASSE, STERNOCOSTAL ASPECT

Observe:
1. The *right border*, formed by the right atrium, slightly convex and almost in line with the superior and inferior caval veins. Enlargement of the right atrium shows as a bulging of the right border of the heart.
2. The *inferior border* is formed by the right ventricle and slightly by the left ventricle. Dilation of the right ventricle is directed toward the pulmonary artery and causes the heart to rotate so that more of the right ventricle forms the left border of the heart.
3. The *left border* is formed by the left ventricle and very slightly by the left auricle. When the left ventricle is dilated, the apex of the heart extends to the left.

1-54B   HEART CHAMBERS

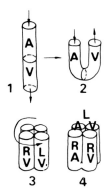

These four tiny diagrams are not meant as a literal description of the embryology of the heart, but as a visual pneumonic to help understand the relationship of the chambers to each other.
1. Front view of the heart as a simple tube, atrium above and ventricle below.
2. Side view of the heart curved into a U shape with ventricle anterior to atrium.
3. Front view of the curved tube now divided into right and left sides, atria behind, ventricles in front. *Arrow* shows direction of partial rotation to the left.
4. The finished product in front view with right atrium forming right border, and left ventricle forming left border.

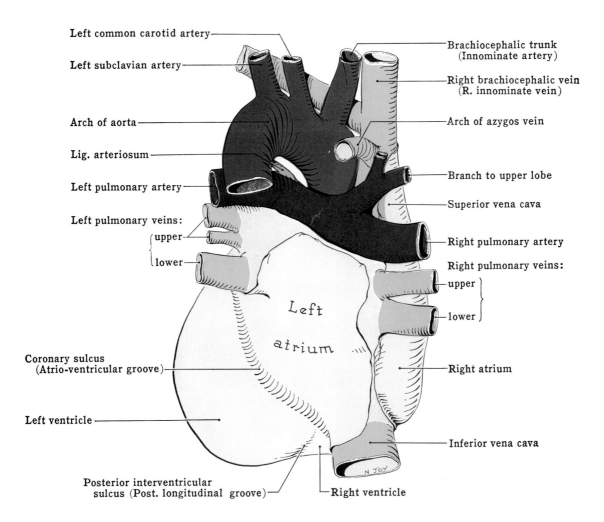

Left common carotid artery

Left subclavian artery

Arch of aorta

Lig. arteriosum

Left pulmonary artery

Left pulmonary veins:
  upper
  lower

Coronary sulcus
  (Atrio-ventricular groove)

Left ventricle

Posterior interventricular
sulcus (Post. longitudinal groove)

Brachiocephalic trunk
(Innominate artery)

Right brachiocephalic vein
(R. innominate vein)

Arch of azygos vein

Branch to upper lobe

Superior vena cava

Right pulmonary artery

Right pulmonary veins:
  upper
  lower

Right atrium

Inferior vena cava

Right ventricle

Left atrium

1-55 HEART AND GREAT VESSELS, HARDENED *IN SITU* AND REMOVED EN MASSE, POSTERIOR ASPECT

Observe:
1. Visible from behind: most of the left atrium, much of the left ventricle, a little of the right atrium, and almost none of the right ventricle.
2. The right and left pulmonary veins, converging to open into the left atrium. The part of the atrium between them lies to the left of the line of the inferior vena cava, occupies the median plane, and forms the anterior wall of the pericardial *cul-de-sac* called the oblique pericardial sinus, which admits two fingers (Fig. 1–60).
3. The right and left pulmonary arteries, just above and parallel to the pulmonary veins and inclining from the left side downward and to the right. Hence, the root of the right lung is lower than that of the left.
4. The aorta, arching over the left pulmonary vessels (and bronchus); the azygos vein arching over the right pulmonary vessels (and bronchus).
5. The aortic arch, arched in two planes: (a) upward, (b) to the left. The convexity to the *left* is molded on the esophagus and trachea.

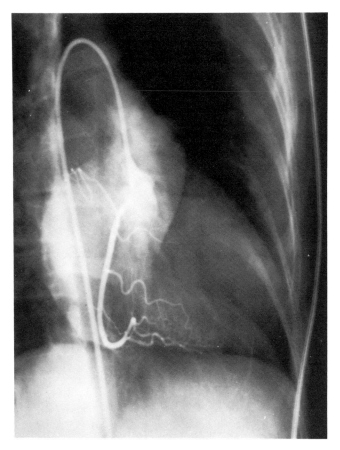

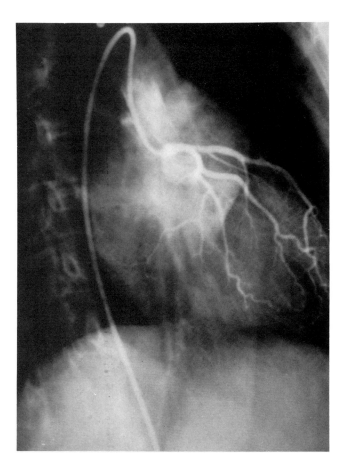

1-56A RIGHT CORONARY ARTERIOGRAM   **1-56B LEFT CORONARY ARTERIOGRAM**

BLOOD SUPPLY OF THE HEART

Consult Figures 1-56, *A* and *B*, and 1-57.

1. The coronary arteries travel in the grooves between atria and ventricles and between the ventricles.
2. The *right coronary artery* travels in the coronary sulcus to reach the posterior surface of the heart where it will anastomose with the circumflex branch of the left coronary artery. Early in its course it gives off the *sinus node artery* which supplies the right atrium and reaches the SA Node. Major branches are a *marginal branch* supplying much of the anterior wall of the right ventricle, an *AV Nodal artery* given off near the posterior border of the interventricular septum, and a *posterior descending* branch in the interventricular groove which anastomoses with the anterior descending branch of the left coronary artery.
3. The *left coronary artery* divides soon into a *circumflex* branch which passes posteriorly to anastomose with the right coronary on the back of the heart, and an *anterior descending* branch in the interventricular groove.
4. The distribution of blood to the heart becomes of clinical significance when occlusion of a vessel occurs, resulting in a myocardial infarction. It can be seen that the *right atrium* is supplied by the right coronary artery and the *left atrium* by the left coronary artery. The *right ventricle* is supplied by the right coronary artery except for the left part of the anterior wall which receives supply from the interventricular branch of the left coronary. The *left ventricle* is supplied by the left coronary artery except for the right part of the posterior wall which receives supply from the right coronary.
5. The *interventricular septum* receives supply from septal branches of the two descending branches: the anterior two-thirds from the left coronary, the posterior one-third from the right.
6. The *conducting system* is usually supplied as follows: SA Node and AV Node from the right coronary artery, the AV Bundle from branches of the right coronary, the Bundle branches from the left coronary except for the posterior limb of the left bundle branch which receives supply from both coronary arteries.

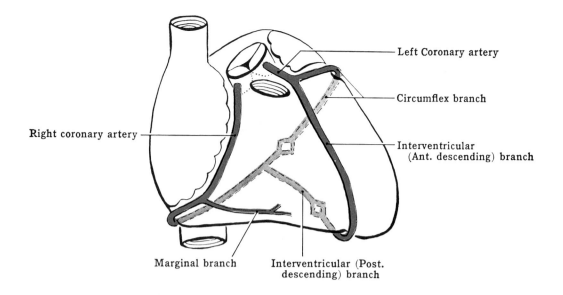

1-57 DIAGRAM OF THE CORONARY (CARDIAC) ARTERIES

The coronary circulation is extremely variable in detail. In most cases, the right and left coronary arteries share equally in the blood supply to the heart. In about 15 per cent of hearts the left coronary artery is said to be dominant in that the posterior interventricular branch comes off the circumflex as in Figure 1-59A

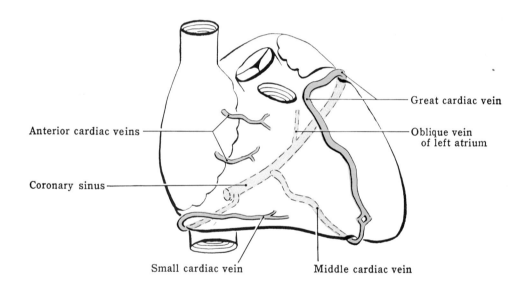

1-58 DIAGRAM OF THE CARDIAC VEINS

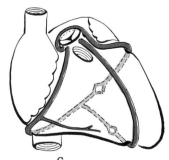

A.
Left coronary artery supplying part of right coronary artery territory.

B.
A single coronary artery.

C.
Circumflex branch springing from right aortic sinus.

1-59 SOME VARIETIES OF CORONARY ARTERIES

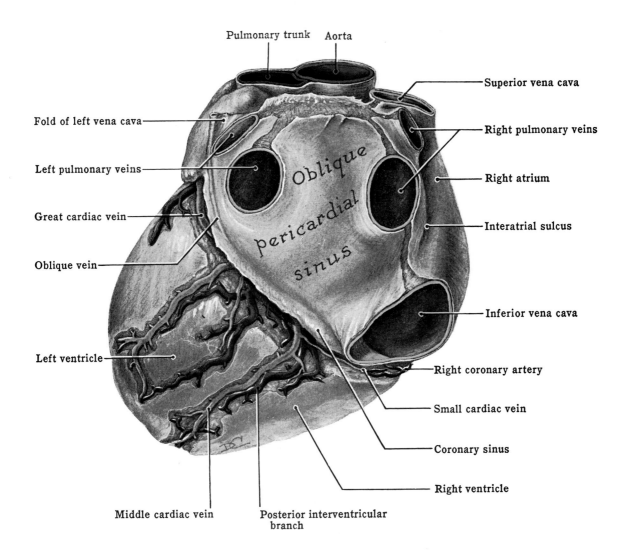

Pulmonary trunk   Aorta

Superior vena cava

Fold of left vena cava

Right pulmonary veins

Left pulmonary veins

Right atrium

Oblique pericardial sinus

Great cardiac vein

Interatrial sulcus

Oblique vein

Left ventricle

Inferior vena cava

Right coronary artery

Small cardiac vein

Coronary sinus

Right ventricle

Middle cardiac vein

Posterior interventricular branch

1-60 HEART, VIEWED FROM BEHIND

This heart was removed from the specimen in Figure 1-61.

Observe:

1. The entire base, or posterior surface, and part of the diaphragmatic surface are in view.
2. The slight appearance made by the right atrium.
3. The superior and the much larger inferior caval vein joining the upper and lower limits of the right atrium.
4. The left atrium, forming the greater part of the posterior surface.
5. The coronary arteries, here irregular in that the left one supplies the posterior interventricular branch.
6. Branches of the cardiac veins, when crossing branches of the coronary arteries, mostly do so superficially.
7. The fold of the left caval vein. For its significance consult Figure 1-89.
8. The right pulmonary artery (removed) lay on the bare strip at the upper border of the atria, and intervened between the oblique sinus and the transverse sinus (not labeled).

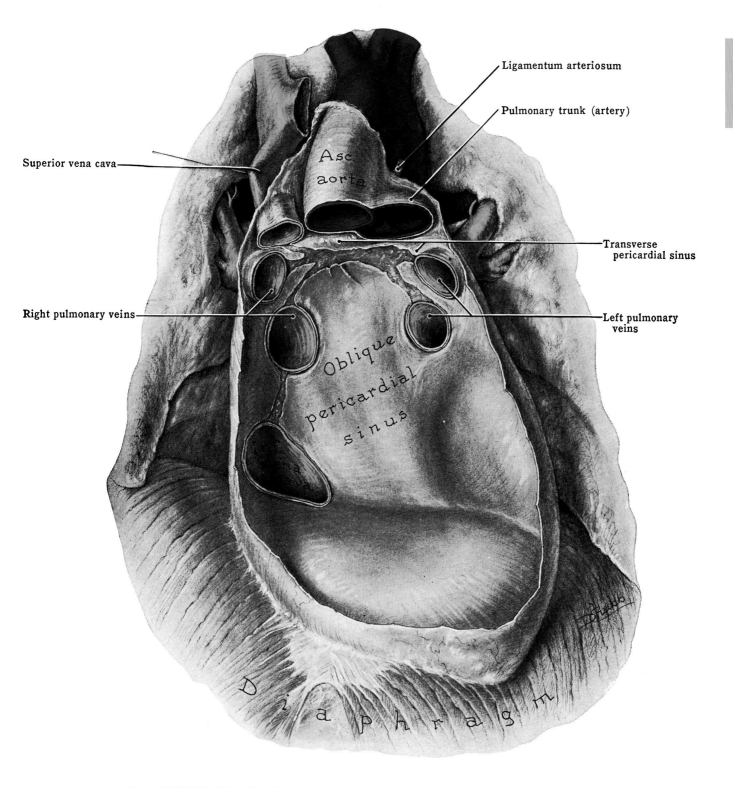

Labels on figure:
- Ligamentum arteriosum
- Pulmonary trunk (artery)
- Superior vena cava
- Asc aorta
- Transverse pericardial sinus
- Right pulmonary veins
- Left pulmonary veins
- Oblique pericardial sinus
- Diaphragm

1-61 INTERIOR OF THE PERICARDIAL SAC, ANTERIOR VIEW

Observe:
1. The 8 vessels severed on excising a heart: 2 caval veins, 4 pulmonary veins, and 2 arteries.
2. The oblique sinus, circumscribed by 5 veins, open below and to the left, and rising to the level of the right pulmonary artery which separates it from the transverse sinus.
3. The peak of the pericardial sac, near the junction of the ascending aorta and the aortic arch.
4. The superior vena cava, partly inside and partly outside the pericardium, and the ligamentum arteriosum entirely outside.

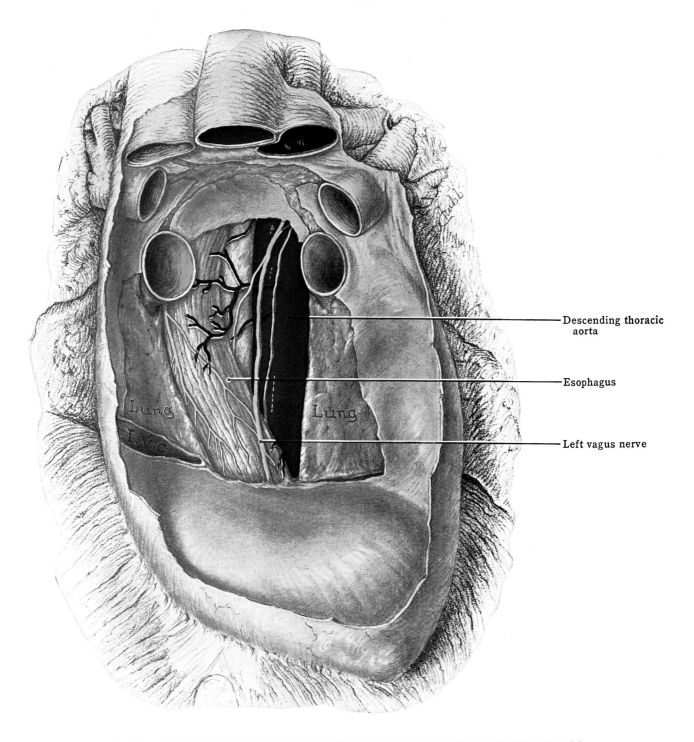

Descending thoracic aorta

Esophagus

Left vagus nerve

1-62 POSTERIOR RELATIONS OF HEART AND PERICARDIUM

The parietal pericardium (fibrous and serous layers) has been removed behind the oblique sinus and also on each side of it.

Observe:
1. The posterior relations are: part of the right lung and the esophagus grooving it; part of the left lung and the aorta grooving it; and the vagus nerves forming a plexus on the esophagus.
2. The esophagus is here unduly deflected to the right. It usually lies in contact with the aorta. (For abdominal relations of pericardium and heart, see Fig. 2-27.)

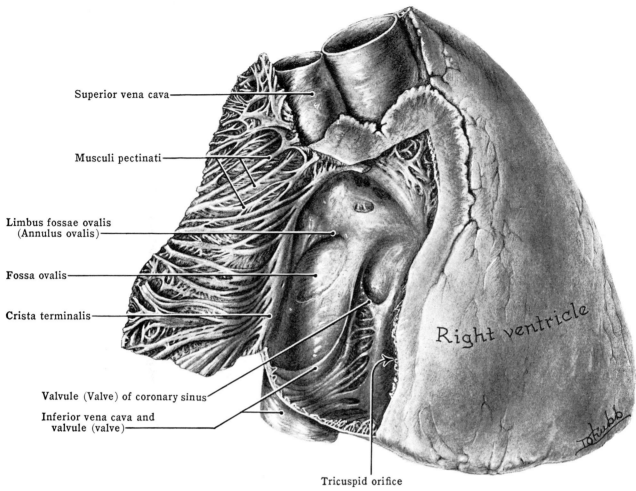

Superior vena cava

Musculi pectinati

Limbus fossae ovalis
(Annulus ovalis)

Fossa ovalis

Crista terminalis

Right ventricle

Valvule (Valve) of coronary sinus

Inferior vena cava and
valvule (valve)

Tricuspid orifice

1-63A   INTERIOR OF THE RIGHT ATRIUM, ANTERO-LATERAL VIEW

Observe:

1. The smooth part of the atrial wall (primitive sinus venarum) and the rough part (primitive atrium).
2. Crista terminalis, the valvule of the inferior vena cava, and the valvule of the coronary sinus, separating the smooth from the rough.
3. The two caval veins and the coronary sinus opening onto the smooth part.
4. The anterior cardiac veins (Fig. 1-53) and the venae minimae (thebesian veins), not visible, also open into the atrium.
5. The crista terminalis, descending from the front of the superior vena cava to the front of the inferior vena cava; and the musculi pectinati, passing forward from the crista-like teeth from the back of a comb. The crista underlies the sulcus terminalis (Fig. 1-53).
6. The right atrio-ventricular or tricuspid orifice, situated at the anterior aspect of the atrium.
7. In Figure 1-63B that the inflow from the superior vena cava is directed toward the tricuspid orifice, while blood from the inferior vena cava directed toward the fossa (foramen) ovale. Consult Figure 10-1C which shows the fetal circulation.

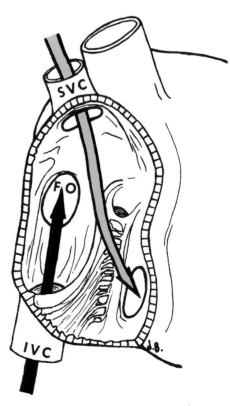

1-63B   INFLOW: RIGHT ATRIUM

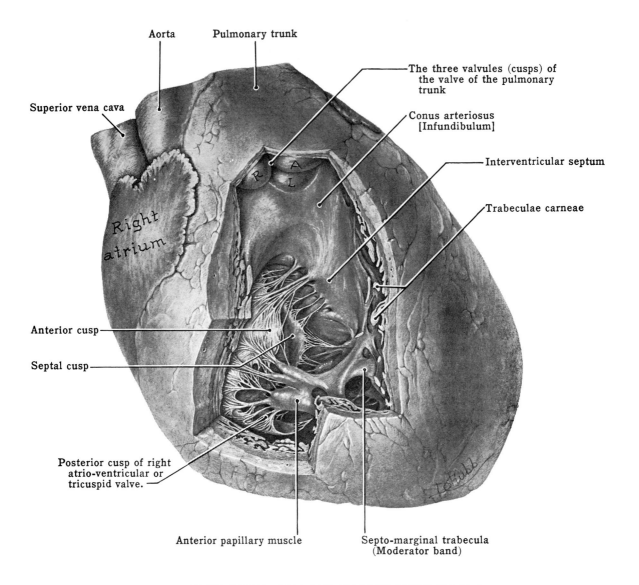

Aorta Pulmonary trunk

The three valvules (cusps) of
the valve of the pulmonary
trunk

Superior vena cava

Conus arteriosus
[Infundibulum]

Interventricular septum

Trabeculae carneae

Right
atrium

Anterior cusp

Septal cusp

Posterior cusp of right
atrio-ventricular or
tricuspid valve.

Anterior papillary muscle

Septo-marginal trabecula
(Moderator band)

1-64 INTERIOR OF THE RIGHT VENTRICLE

Observe:

1. The entrance to this chamber (right atrio-ventricular or tricuspid orifice) situated behind; the exit (orifice of the pulmonary trunk) situated above.
2. The smooth funnel-shaped wall (conus arteriosus) below the pulmonary orifice; the remainder of the ventricle, rough with fleshy trabeculae.
3. Three types of trabeculae: (a) mere ridges, (b) bridges, attached only at each end, and (c) fingerlike projections called papillary muscles. The anterior papillary muscle rising from the anterior wall; the posterior (not labeled) from the posterior wall; and a series of small septal papillae from the septal wall.
4. The septo-marginal trabecula, here very thick, extending from the septum to the base of the anterior papillary muscle. It is a bridge.
5. The chordae tendineae (not labeled), passing from the tips of the papillary muscles to the free margins and ventricular surfaces of the three cusps of the tricuspid valve.
6. Each papillary muscle controlling the adjacent sides of two cusps (Fig. 1-67).

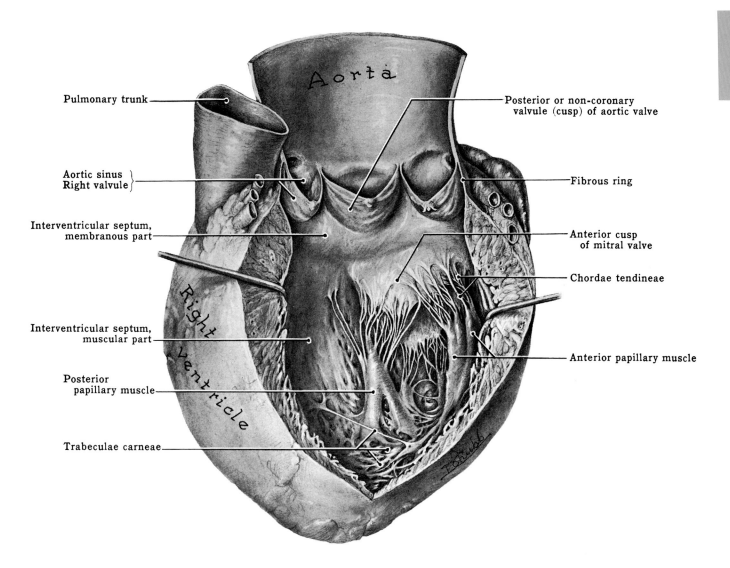

Pulmonary trunk

Aortic sinus }
Right valvule }

Interventricular septum,
membranous part

Interventricular septum,
muscular part

Posterior
papillary muscle

Trabeculae carneae

Aorta

Right Ventricle

Posterior or non-coronary
valvule (cusp) of aortic valve

Fibrous ring

Anterior cusp
of mitral valve

Chordae tendineae

Anterior papillary muscle

1-65 INTERIOR OF THE LEFT VENTRICLE

Observe:
1. The conical shape of the left chamber.
2. The entrance (left atrio-ventricular, bicuspid, or mitral orifice), situated posteriorly. The exit (aortic orifice), situated superiorly.
3. The wall, thin and muscular near the apex; thick and muscular above; thin and fibrous (nonelastic) at the aortic orifice.
4. Trabeculae carneae, as in the right ventricle, forming ridges, bridges, and papillary muscles.
5. Two large papillary muscles—the anterior from the anterior wall and the posterior from the posterior wall—each controlling, via chordae tendineae, the adjacent halves of two cusps of the mitral valve.
6. The anterior cusp of the mitral valve intervening between the inlet (mitral orifice) and the outlet (aortic orifice).

Note the structure of the mitral and of the aortic valves (Figs. 1-66 to 1-68).

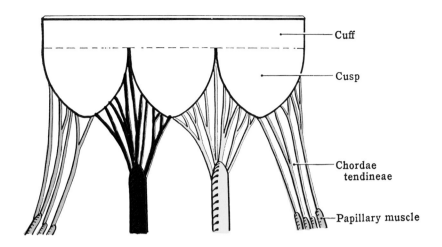

Cuff

Cusp

Chordae tendineae

Papillary muscle

1-66 DIAGRAM OF RIGHT ATRIOVENTRICULAR VALVE, SPREAD OUT

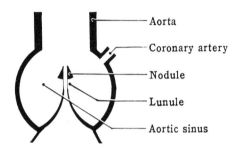

Aorta

Coronary artery

Nodule

Lunule

Aortic sinus

A. Closed, on longitudinal section.

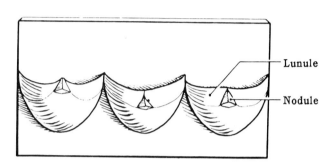

Lunule

Nodule

B. Spread out.

1-67 DIAGRAM OF THE AORTIC VALVE

This, like the valve of the pulmonary trunk, has three semilunar cusps, each with a fibrous nodule at the midpoint of its free edge. When the valve is closed, the nodules meet in the center.

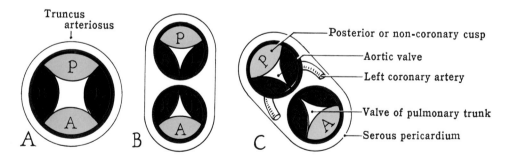

Truncus arteriosus

Posterior or non-coronary cusp

Aortic valve

Left coronary artery

Valve of pulmonary trunk

Serous pericardium

1-68 PULMONARY AND AORTIC VALVE NAMES

The names applied to these cusps are explained on a developmental basis: The truncus arteriosus with four cusps (A) splits to form two valves, each with three cusps (B). The heart undergoes partial rotation to the left on its axis resulting in the arrangement of cusps shown in (C). Inability of the valve to close completely is called insufficiency and results in regurgitation. Fusion of the cusps to each other produces stenosis.

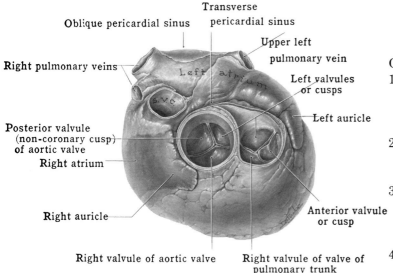

Oblique pericardial sinus

Transverse pericardial sinus

Upper left pulmonary vein

Right pulmonary veins

Left valvules or cusps

Left auricle

Posterior valvule (non-coronary cusp) of aortic valve

Right atrium

Right auricle

Anterior valvule or cusp

Right valvule of aortic valve

Right valvule of valve of pulmonary trunk

1-69A EXCISED HEART, VIEWED FROM ABOVE

Observe:

1. The ascending aorta and the pulmonary trunk, which conduct blood from the ventricles, accordingly placed anterior to the atria and to the superior vena cava and pulmonary veins, which conduct blood to the atria.
2. These two stems, enclosed within a common tube of serous pericardium and partly embraced by the auricles of the atria.
3. The transverse pericardial sinus, curving behind the enclosed stems of the aorta and pulmonary trunk, and in front of the superior vena cava and upper limits of the atria.
4. The three cusps of the aortic valve (as seen from the aorta) and of the pulmonary valve (as seen from the pulmonary trunk). To understand the developmental origin of the names of these cusps consult Figure 1–70.

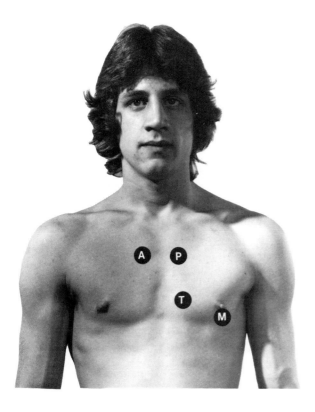

1-69B AUSCULTATION POINTS

Auscultation points are the areas where sounds from each of the heart's valves may be heard most distinctly through a stethoscope. They do not represent the location of the valves projected on the surface of the chest, although for the tricuspid and pulmonary valves location and sound are quite close. Aortic and mitral valves are deep in the chest and their sounds are heard best at the points where the direction of blood flow is closer to the chest wall. Aortic (A) and Pulmonary (P) areas are in the second interspace to the right and left of the sternal border. The Tricuspid area (T) is near the left sternal border in the 5th or 6th interspace. The Mitral valve (M) is heard best near the apex of the heart in the 5th intercostal space in the midclavicular line.

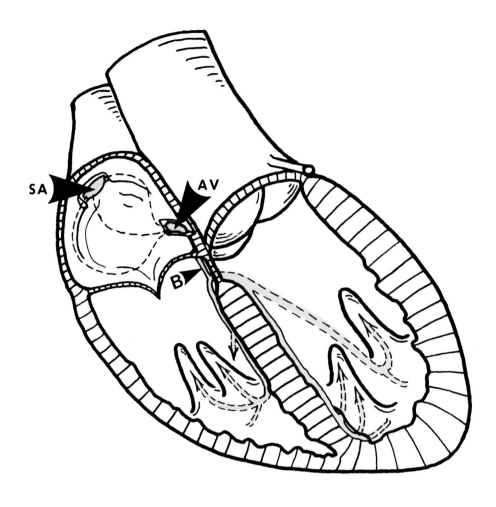

1-70 CONDUCTING SYSTEM OF HEART

Observe:

1. The *Sinu-atrial (SA) Node* in the wall of the right atrium near the upper end of the sulcus terminalis and extending over the front of the opening of the superior vena cava. The SA Node is the "pacemaker" of the heart because it initiates cardiac muscle contraction and determines the heart rate. It is supplied by the *sinus node artery*, usually a branch of the right coronary artery.

2. Contraction spreads through the atrial wall until it reaches the *Atrio-ventricular (AV) Node* in the right atrial side of the interatrial septum just above the opening of the Coronary sinus. After a brief delay contraction passes to the ventricles. The AV Node is usually supplied by the distal right coronary artery.

3. The *AV Bundle (B)* passes from the AV Node in the membranous part of the interventricular septum and divides into right and left *Bundle Branches* on either side of the muscular part of the septum.

4. The *right Bundle Branch* travels down the septum to the anterior wall of the ventricle, enters the base of the anterior papillary muscle, and excitation spreads to the right ventricular wall.

5. The *left Bundle Branch* is a thin, broad band which usually divides into two divisions which enter papillary muscles and excitation of the muscular walls of the left ventricle occurs.

6. Damage to the conducting system (often by compromised blood supply as in coronary artery disease) leads to disturbances of cardiac muscle contraction. Damage to the AV Node results in "heart block" as the atrial excitation wave does not reach the ventricles which begin to contract independently at their own rate which is slower than that of the atria. Damage to one of the branches results in "bundle branch block" in which excitation goes down the unaffected branch to cause systole of that ventricle. The impulse then spreads to the other ventricle producing later asynchronous contraction.

The diagram is a synthetic view from various specimens; no single view allows visualization of all parts of the conduction system and its dissection is extremely difficult. See Duckwork, J.W.A. (1952) *The Development of the Sinu-atrial and Atrioventricular Nodes of the Human Heart*, M.D. Thesis, Univ. of Edinburgh.

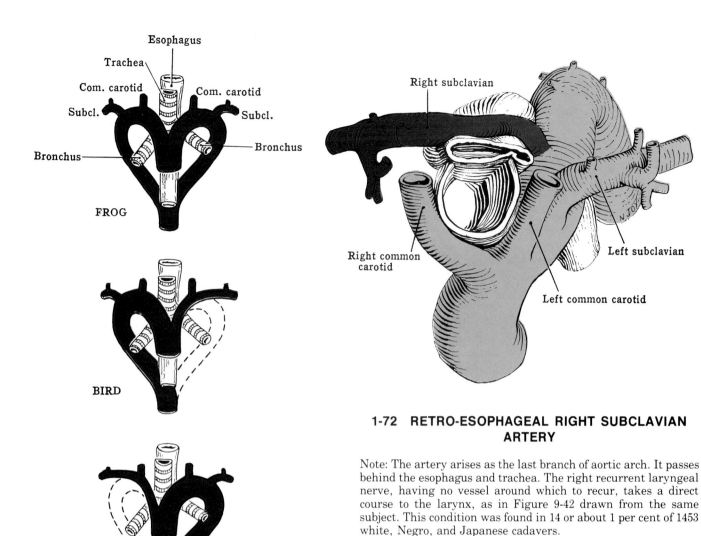

FROG

BIRD

MAN

1-71 VARIETIES OF AORTIC ARCHES

Note: The double aortic arch of the frog; the right aortic arch of
the bird; the left aortic arch of the mammal, including man.

**1-72 RETRO-ESOPHAGEAL RIGHT SUBCLAVIAN
ARTERY**

Note: The artery arises as the last branch of aortic arch. It passes
behind the esophagus and trachea. The right recurrent laryngeal
nerve, having no vessel around which to recur, takes a direct
course to the larynx, as in Figure 9-42 drawn from the same
subject. This condition was found in 14 or about 1 per cent of 1453
white, Negro, and Japanese cadavers.

See McDonald, J. J., and Anson, B. J. (1940) Variations in origin
of arteries derived from the aortic arch in American whites and
Negroes. *Am. J. Phys. Anthropol., 27:* 91.

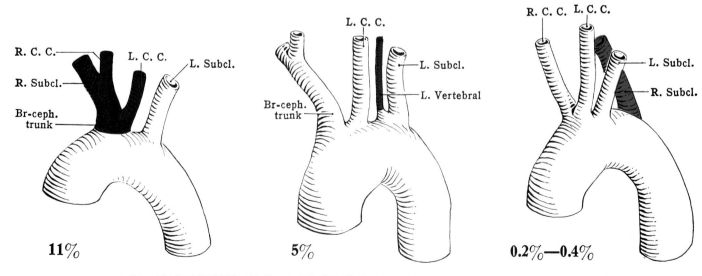

1-73 VARIATIONS IN THE ORIGINS OF THE BRANCHES OF THE AORTIC ARCH

Approximate incidence: 11 per cent of 1220 arches, 5 per cent of 1000 arches, and 0.2 to 0.4 per cent of 1500 arches.
See Adachi, B. (1928) *Das Arteriensystem der Japaner.* Kyoto.

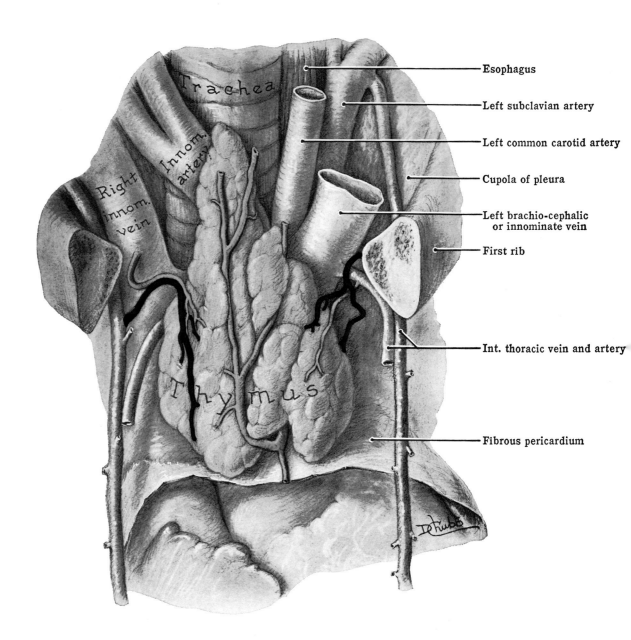

- Esophagus
- Left subclavian artery
- Left common carotid artery
- Cupola of pleura
- Left brachio-cephalic or innominate vein
- First rib
- Int. thoracic vein and artery
- Fibrous pericardium

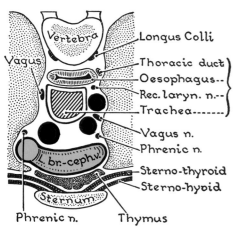

- Longus Colli
- Thoracic duct
- Oesophagus
- Rec. laryn. n.
- Trachea
- Vagus n.
- Phrenic n.
- Sterno-thyroid
- Sterno-hyoid
- Phrenic n.
- Thymus

1-75 SUPERIOR MEDIASTINUM

Diagram of a cross-section through the superior mediastinum above the level of the aortic arch. Consulting this and Figure 1-74, note the posterior relations of the thymus.

1-74 SUPERIOR MEDIASTINUM—I: THE THYMUS

The sternum and ribs have been excised and the pleurae removed. It is unusual in an adult dissecting room specimen to see so discrete a thymus, which is impressive during puberty but subsequently regresses and becomes largely replaced by fat and fibrous tissue.

Observe:

1. The thymus lying in the superior mediastinum; overlapping the upper limit of the pericardial sac below; and extending into the neck, here farther than usual, above.
2. The longitudinal fissure that divides the thymus into two asymmetrical lobes, a larger right and a smaller left. These two developmentally separate parts are easily separated from each other by blunt dissection.
3. The blood supply: arteries from the internal thoracic arteries; veins to the brachio-cephalic and internal thoracic veins and communicating above with the inferior thyroid veins.

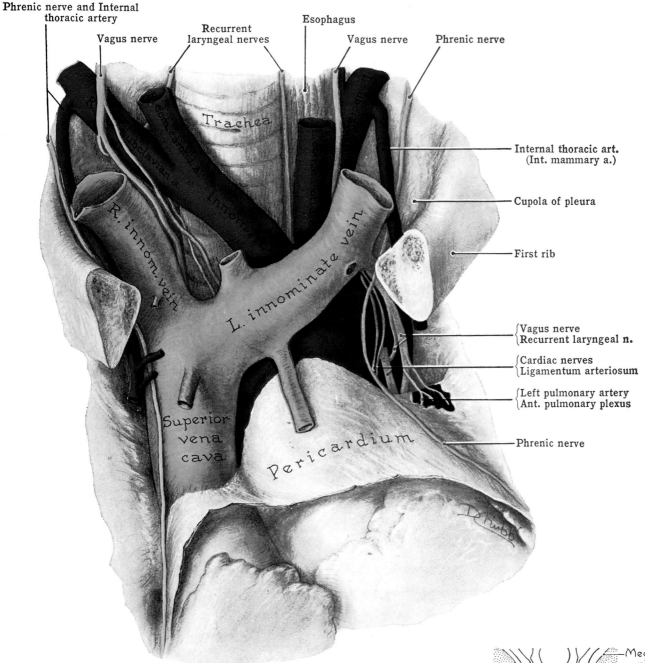

Phrenic nerve and Internal thoracic artery
Vagus nerve
Recurrent laryngeal nerves
Esophagus
Vagus nerve
Phrenic nerve

Trachea

R. innom. vein

L. innominate vein

Superior vena cava

Pericardium

Internal thoracic art. (Int. mammary a.)

Cupola of pleura

First rib

{ Vagus nerve
Recurrent laryngeal n.

{ Cardiac nerves
Ligamentum arteriosum

{ Left pulmonary artery
Ant. pulmonary plexus

Phrenic nerve

1-76 SUPERIOR MEDIASTINUM – II: ROOT OF THE NECK

This is Figure 1-74 with the thymus removed.

Observe:

1. The great veins, anterior to the great arteries.
2. The backward direction of the aortic arch and the nerves crossing its left side.
3. The ligamentum arteriosum, outside the pericardial sac and having the left recurrent nerve on its left side and the vagal and sympathetic branches to the superficial cardiac plexus on its right.
4. The right vagus, crossing the right subclavian artery (4th right primitive aortic arch), there giving off its recurrent branch, and passing medially to reach the trachea and esophagus.
5. The left vagus, crossing the aortic arch (4th left primitive aortic arch), there giving off its recurrent branch, and passing medially to reach the esophagus (Fig. 1-78).
6. The left phrenic nerve, crossing the path of the vagus, but 1 cm anterior to it.

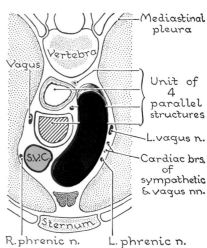

Mediastinal pleura
Vertebra
Vagus
Unit of 4 parallel structures
L. vagus n.
Cardiac brs. of sympathetic & vagus nn.
S.V.C.
Sternum
R. phrenic n.
L. phrenic n.

1-77 SUPERIOR MEDIASTINUM

Diagram of a cross-section through the superior mediastinum at the level of the aortic arch. The *4 parallel structures* are trachea, esophagus, recurrent laryngeal nerve, and thoracic duct (see Fig. 1-80).

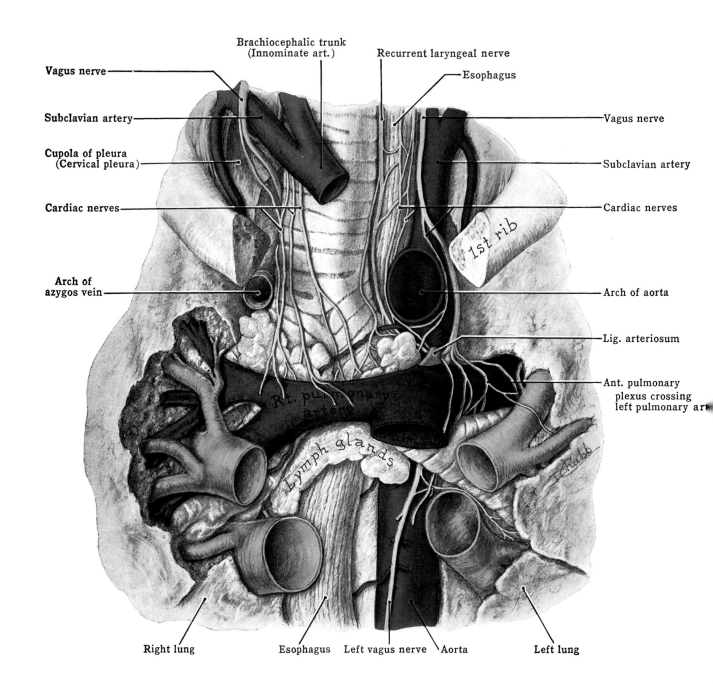

Brachiocephalic trunk
(Innominate art.)

Recurrent laryngeal nerve

Esophagus

Vagus nerve

Subclavian artery

Cupola of pleura
(Cervical pleura)

Cardiac nerves

Arch of
azygos vein

Vagus nerve

Subclavian artery

Cardiac nerves

1st rib

Arch of aorta

Lig. arteriosum

Ant. pulmonary
plexus crossing
left pulmonary art

Rt. pulmonary artery

Lymph glands

Right lung

Esophagus

Left vagus nerve

Aorta

Left lung

1-78 SUPERIOR MEDIASTINUM–
III: PULMONARY ARTERIES

Observe:

1. The pulmonary trunk, dividing into rig
and left pulmonary arteries which run
oblique course (Fig. 1-54). The right arte
crossing below the bifurcation of the trach
and separated from the esophagus by lym
nodes (glands).

2. Cardiac branches of the vagus and symp
thetic, streaming down the sides of the tr
chea and forming the cardiac plexuses.

Esophagus

A. Lymph
nodes

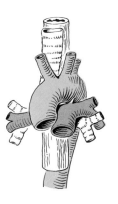

B. Pulmonary
arteries

C. Asc. aorta and
arch

1-79 RELATIONS AT BIFURCATION OF TRACHEA

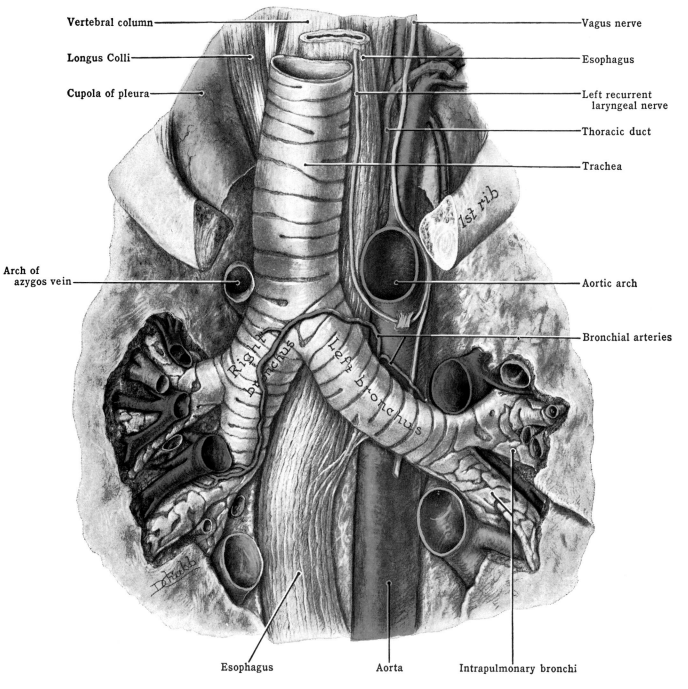

Vertebral column ⎯ Vagus nerve

Longus Colli ⎯ Esophagus

Cupola of pleura ⎯ Left recurrent laryngeal nerve

⎯ Thoracic duct

⎯ Trachea

1st rib

Arch of azygos vein ⎯ Aortic arch

⎯ Bronchial arteries

Right bronchus   Left bronchus

Esophagus   Aorta   Intrapulmonary bronchi

1-80 SUPERIOR MEDIASTINUM-IV: BRONCHI

Observe:
1. Four parallel structures–trachea, esophagus, left recurrent laryngeal nerve, and thoracic duct. The esophagus bulges to the left of the trachea, the recurrent nerve lies in the angle between the trachea and esophagus, and the duct is at the side of the esophagus.
2. The aortic arch runs backward on the left of these 4 structures, and the arch of the azygos vein passes forward on the right.
3. The trachea inclines to the right. Hence the right bronchus is more vertical than the left and its stem is shorter and wider, its first branch arises about 2.5 cm from the bifurcation, whereas the first left branch arises about 5 cm from the bifurcation.
4. The U-shaped rings of the trachea, commonly bifurcated; the ring at the bifurcation of the trachea being V-shaped; the intrapulmonary rings forming a mosaic around the bronchi.
5. The occasional broncho-esophageal muscle, attaching the esophagus to the left bronchus.

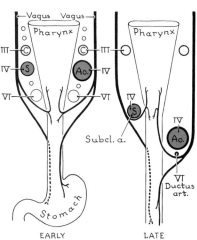

1-81 RECURRENT NERVES

Scheme to explain asymmetrical courses of right and left recurrent laryngeal nerves. (*III*, *IV*, and *VI* are embryonic aortic arches.)

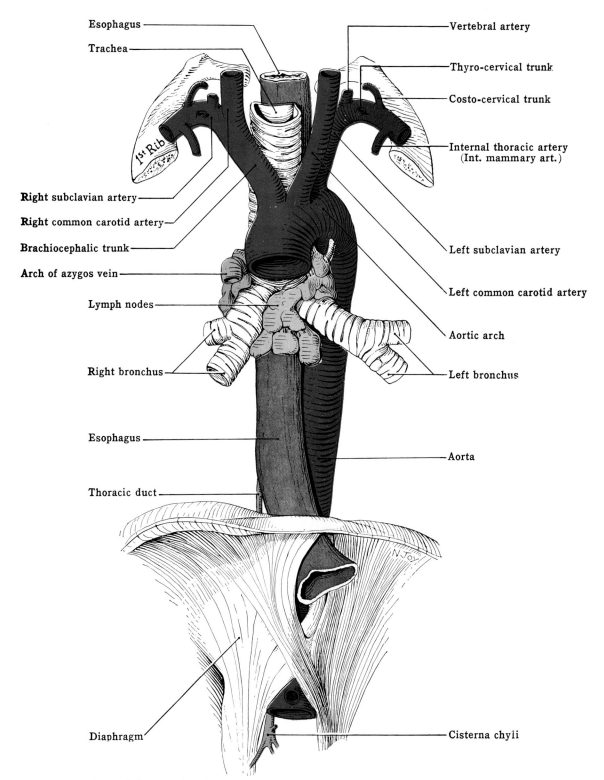

Esophagus

Trachea

Vertebral artery

Thyro-cervical trunk

Costo-cervical trunk

Internal thoracic artery
(Int. mammary art.)

1st Rib

Right subclavian artery

Right common carotid artery

Brachiocephalic trunk

Arch of azygos vein

Lymph nodes

Right bronchus

Left subclavian artery

Left common carotid artery

Aortic arch

Left bronchus

Esophagus

Aorta

Thoracic duct

Diaphragm

Cisterna chyli

1-82 ESOPHAGUS, TRACHEA AND AORTA, ANTERIOR VIEW

1. The arch of the aorta, arching backward on the left sides of the trachea and esophagus, and the arch of the azygos vein arching forward on their right sides. Each arches above the root of a lung (Figs. 1-28 and 1-29).

2. The posterior relation of the trachea is — esophagus.

3. The anterior relations of the thoracic part of the esophagus from above downward are: trachea (throughout its entire length), left recurrent nerve, right and left bronchi, the inferior tracheobronchial lymph nodes (in front of which lies the right pulmonary artery, Fig. 1-78), the pericardium (removed, in front of which lies the oblique pericardial sinus and the left atrium of the heart, Figs. 1-61 and 1-62), and, finally, the diaphragm.

4. Above the level of the aortic arch, the esophagus bulges to the left beyond the trachea.

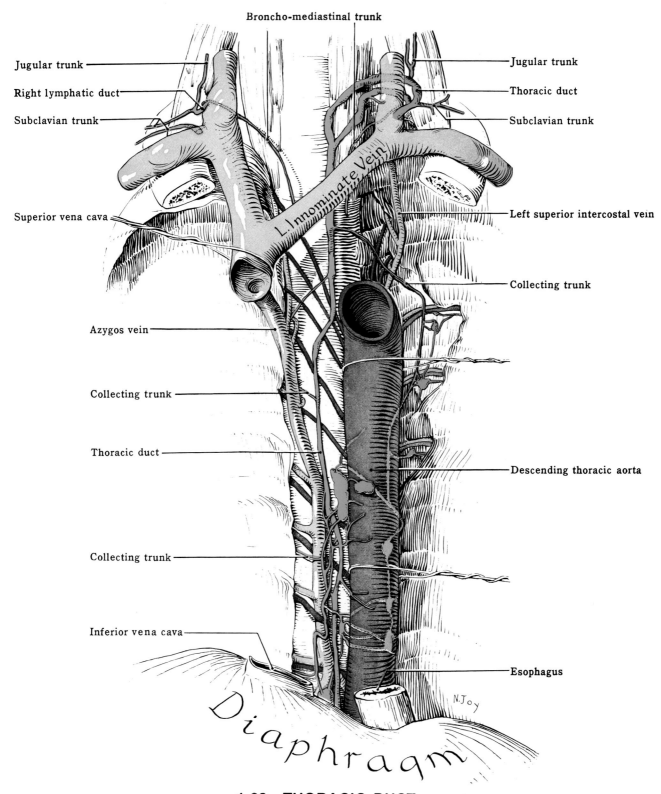

Broncho-mediastinal trunk

Jugular trunk

Right lymphatic duct

Subclavian trunk

Superior vena cava

L. Innominate Vein

Azygos vein

Collecting trunk

Thoracic duct

Collecting trunk

Inferior vena cava

Jugular trunk

Thoracic duct

Subclavian trunk

Left superior intercostal vein

Collecting trunk

Descending thoracic aorta

Esophagus

N.Joy

Diaphragm

1-83 THORACIC DUCT

Observe:

1. The thoracic duct (a) ascending on the vertebral column between the azygos vein and the descending aorta and (b) at the junction of the posterior and superior mediastina, passing to the left and continuing its ascent to the neck where (c) it arches laterally to open near, or at, the angle of union of the internal jugular and subclavian veins.

2. The duct, here, as commonly (a) plexiform in the posterior mediastinum and (b) splitting in the neck.

3. The duct receiving branches from the intercostal spaces of both sides via several collecting trunks and also branches from posterior mediastinal structures.

4. The duct finally receiving the jugular, subclavian, and broncho-mediastinal trunks.

5. The right lymph duct, very short and formed by the union of the right jugular, subclavian, and broncho-mediastinal trunks.

For origin of duct in cisterna chyli see Figure 1-82; for termination of duct in the neck see Figures 9-38 and 9-40.

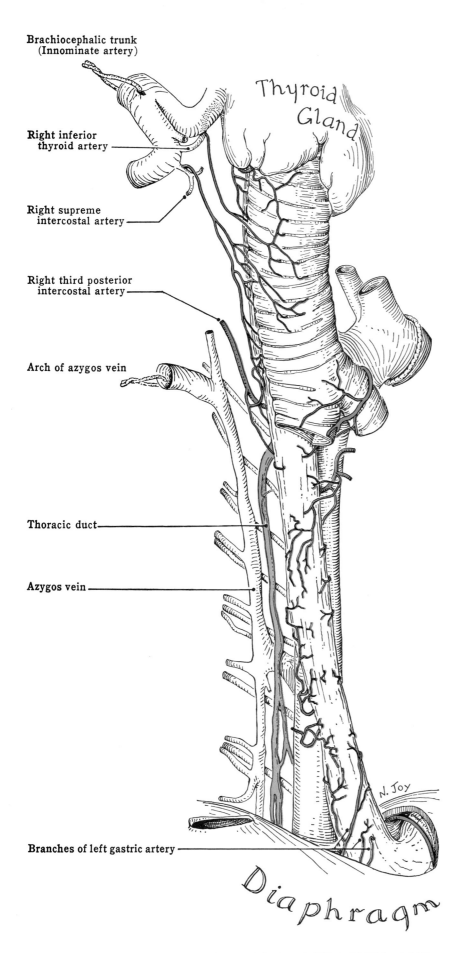

Brachiocephalic trunk
(Innominate artery)

Thyroid Gland

Right inferior thyroid artery

Right supreme intercostal artery

Right third posterior intercostal artery

Arch of azygos vein

Thoracic duct

Azygos vein

N. JOY

Branches of left gastric artery

Diaphragm

Observe:

1. The four unpaired median branches of the descending thoracic aorta (2 bronchial and 2 esophageal), which supply the trachea, bronchi, and esophagus. They are in series with the three unpaired median branches of the abdominal aorta (celiac, superior mesenteric, and inferior mesenteric arteries), which supply the gastrointestinal tract.

1-84 ARTERIAL SUPPLY TO THE TRACHEA AND ESOPHAGUS (RIGHT THREE-QUARTER VIEW)

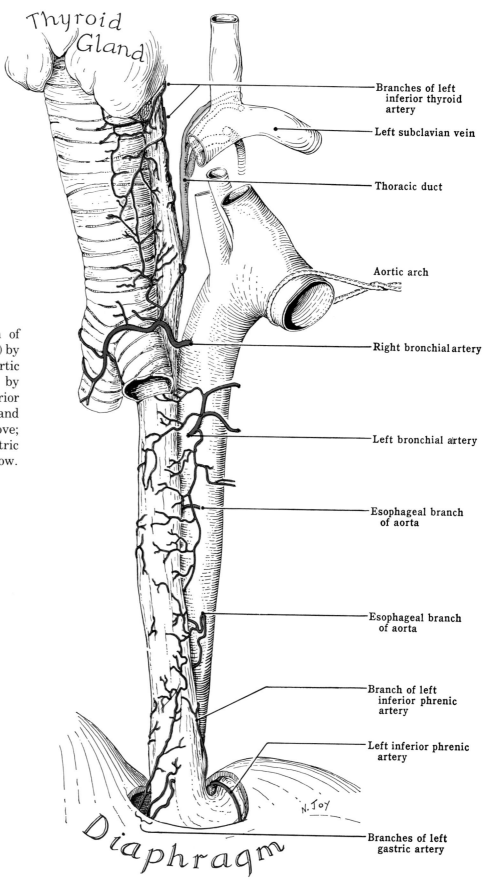

2. The continuous anastomotic chain of arteries on the esophagus formed: (a) by the four unpaired median aortic branches with each other; (b) by branches of the right and left inferior thyroid, right supreme intercostal, and right third intercostal arteries above; and (c) by branches of the left gastric and left inferior phrenic arteries below.

Thyroid Gland

Branches of left inferior thyroid artery

Left subclavian vein

Thoracic duct

Aortic arch

Right bronchial artery

Left bronchial artery

Esophageal branch of aorta

Esophageal branch of aorta

Branch of left inferior phrenic artery

Left inferior phrenic artery

Branches of left gastric artery

Diaphragm

N. Joy

1-85 ARTERIAL SUPPLY TO THE TRACHEA AND ESOPHAGUS (LEFT THREE-QUARTER VIEW)

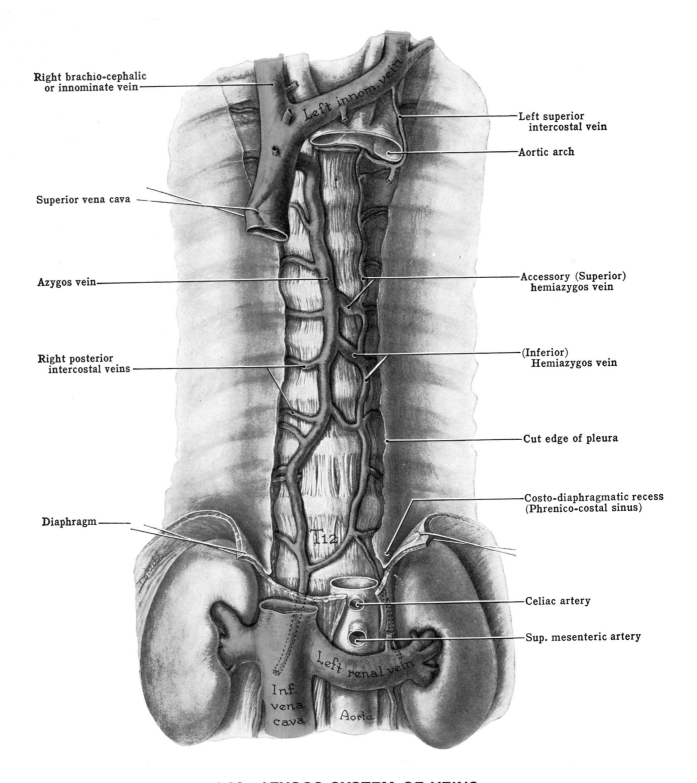

Right brachio-cephalic or innominate vein

Left innom. vein

Left superior intercostal vein

Aortic arch

Superior vena cava

Azygos vein

Accessory (Superior) hemiazygos vein

Right posterior intercostal veins

(Inferior) Hemiazygos vein

Cut edge of pleura

Costo-diaphragmatic recess (Phrenico-costal sinus)

Diaphragm

T12

Celiac artery

Sup. mesenteric artery

Left renal vein

Inf. vena cava

Aorta

1-86 AZYGOS SYSTEM OF VEINS

While consulting Figure 1-87 on the facing page observe:

1. The left renal vein ventral to the aorta; the left brachio-cephalic (innominate) vein ventral to the three branches of the aortic arch. These two cross-channels conduct blood from the left side of the body to the right side and so to the right atrium.

2. The paired and approximately symmetrical, right and left, longitudinal veins ventral to the vertebral column. The right, or azygos, vein communicating caudally with the inferior vena cava; the left vein communicating with the left renal vein; and each receiving the respective right and left posterior intercostal veins. The right vein ending cranially in the superior vena cava; the left in the left brachio-cephalic, or innominate, vein.

3. The left vein being a chain of veins—hemiazygos, accessory hemiazygos, and left superior intercostal—which here are continuous, but commonly are discontinuous. It is united to the right, or azygos, vein by several (here 4) cross-connecting channels.

See Seib, G. A. (1934) The azygos system of veins in American whites and American Negroes, including observations on inferior caval venous system. *Am. J. Phys Anthropol., 19:* 39.

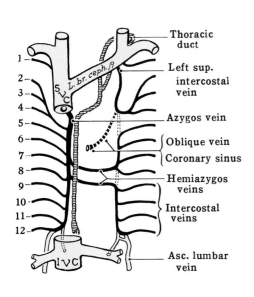

1-87 AZYGOS SYSTEM

Labels in figure 1-87:
- Thoracic duct
- Left sup. intercostal vein
- Azygos vein
- Oblique vein
- Coronary sinus
- Hemiazygos veins
- Intercostal veins
- Asc. lumbar vein
- S.V.C.
- L. br. ceph. V.
- I.V.C.

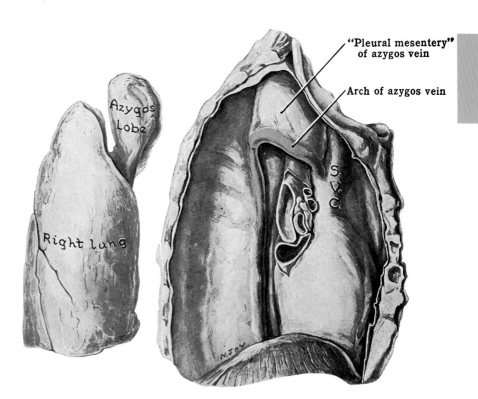

1-88 LOBE OF THE AYZGOS VEIN

Labels in figure 1-88:
- "Pleural mesentery" of azygos vein
- Arch of azygos vein
- Azygos Lobe
- Right lung
- S.V.C.

This lobe, said to be constant in the porpoise, is really part of a bifid apex. It results during embryonic life when the apex of a developing right lung encounters the arch of the azygos vein and is cleft by it, a portion of the apex coming to lie on each side of the venous arch. The venous arch is suspended, so to speak, within a pleural mesentery.

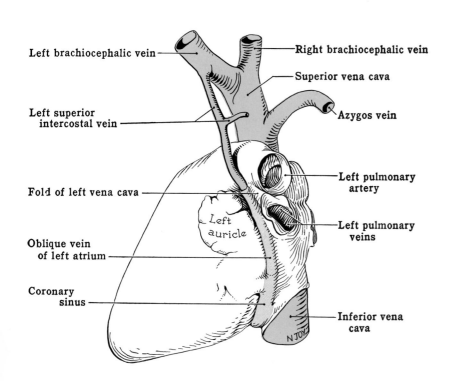

Labels in figure 1-89:
- Left brachiocephalic vein
- Right brachiocephalic vein
- Superior vena cava
- Azygos vein
- Left superior intercostal vein
- Left pulmonary artery
- Fold of left vena cava
- Left auricle
- Left pulmonary veins
- Oblique vein of left atrium
- Coronary sinus
- Inferior vena cava

1-89 PERSISTING LEFT SUPERIOR VENA CAVA

The vessel connecting the left brachiocephalic or innominate vein to the coronary sinus is the primitive left superior vena cava. It has three component segments: (a) an upper, the stem of the left superior intercostal vein; (b) a lower, the oblique vein of the left atrium; and (c) a middle, the vein uniting the intercostal vein to the oblique vein. This channel to the right atrium functioned until the left superior vena cava established connection with the right by means of the left brachiocephalic vein, after which the blood was re-routed, and the middle segment disappeared, although it may persist as a vestigial thread within the fold of the left vena cava (Fig. 1-60).

For persisting left *inferior* vena cava, see Figure 2-112.

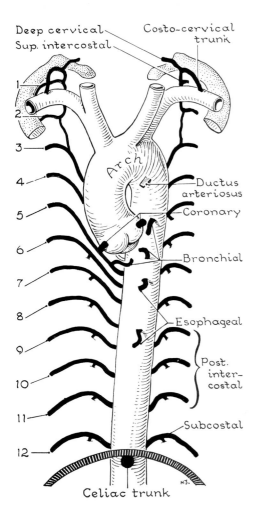

1-90 BRANCHES OF THORACIC AORTA

From the ascending aorta:
  Right coronary
  Left coronary
From the arch of the aorta:
  Brachio-cephalic trunk
    Right subclavian
    Right common carotid
  Left common carotid
  Left subclavian (ductus arteriosus)
From the descending thoracic aorta:
  Visceral branches:
    Esophageal
    Bronchial, left upper and lower
    Right bronchial
    Pericardial
    Mediastinal
  Parietal branches, paired:
    Posterior intercostal (Th. 3 to 11)
    Subcostal (Th. 12)
    Superior phrenic
Note: The right bronchial artery arises from either the upper left bronchial or third right posterior intercostal artery (here the fifth) or the aorta direct. The small arteries to pericardium, tissues in posterior mediastinum, and upper surface of diaphragm are not shown.

(For abdominal aorta and its branches, see Fig. 2-39.)

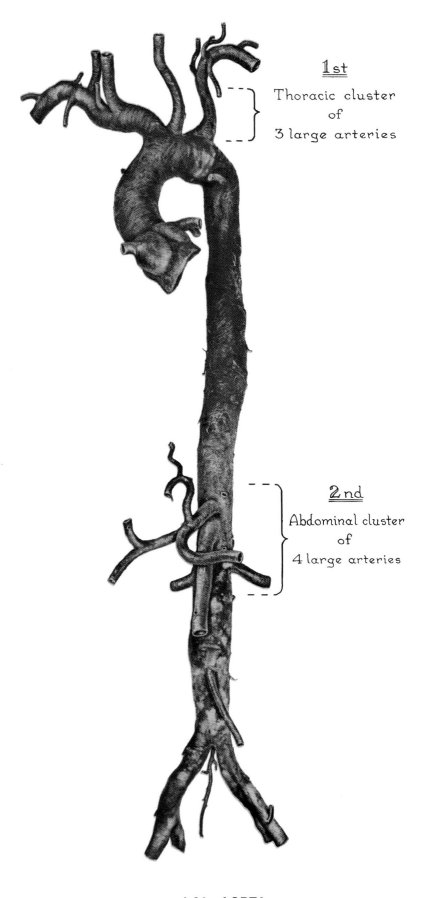

1-91 AORTA

Specimen of an entire aorta showing the two sites (aortic arch and upper limit of abdominal aorta) where the large branches cluster.

The Abdomen

CONTENTS

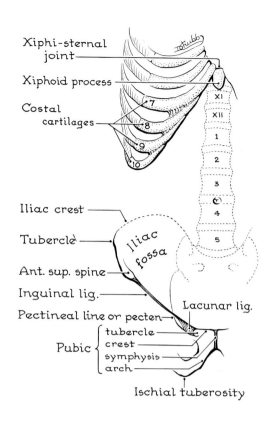

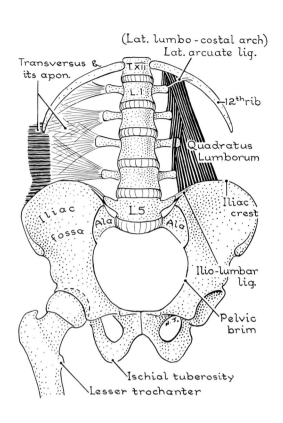

2-1 SKELETON OF ABDOMEN

Xiphi-sternal joint
Xiphoid process
Costal cartilages
Iliac crest
Tubercle
Iliac fossa
Ant. sup. spine
Inguinal lig.
Pectineal line or pecten
Lacunar lig.
Pubic { tubercle, crest, symphysis, arch }
Ischial tuberosity

2-2 POSTERIOR WALL

(Lat. lumbo-costal arch)
Lat. arcuate lig.
Transversus & its apon.
12th rib
Quadratus Lumborum
Iliac fossa
Ala
Ala
Iliac crest
Ilio-lumbar lig.
Pelvic brim
Ischial tuberosity
Lesser trochanter

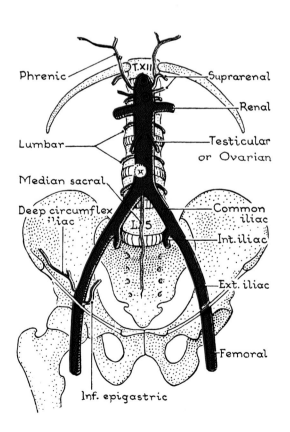

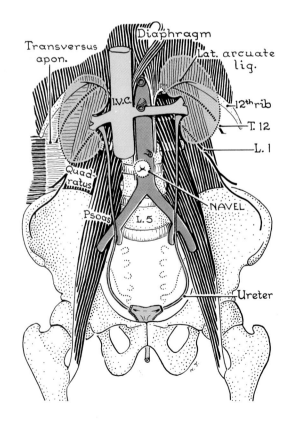

2-3 ABDOMINAL AORTA

Phrenic
Suprarenal
Renal
Lumbar
Testicular or Ovarian
Median sacral
Deep circumflex iliac
Common iliac
Int. iliac
Ext. iliac
Femoral
Inf. epigastric

2-4 URINARY APPARATUS

Transversus apon.
Diaphragm
Lat. arcuate lig.
I.V.C.
12th rib
T. 12
L. 1
Quadratus
NAVEL
Psoas
Ureter

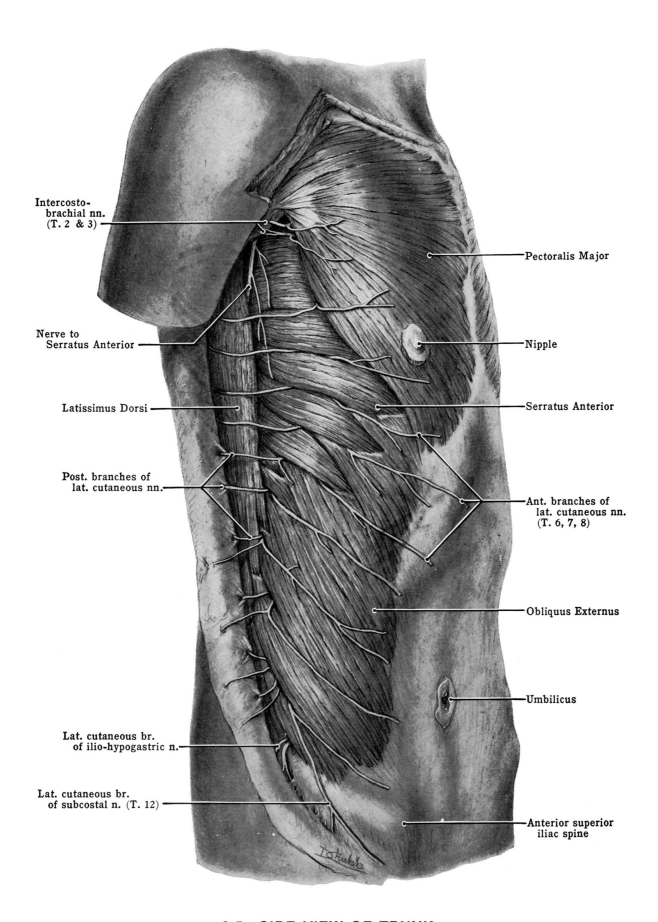

Intercosto-
brachial nn.
(T. 2 & 3)

Nerve to
Serratus Anterior

Latissimus Dorsi

Post. branches of
lat. cutaneous nn.

Lat. cutaneous br.
of ilio-hypogastric n.

Lat. cutaneous br.
of subcostal n. (T. 12)

Pectoralis Major

Nipple

Serratus Anterior

Ant. branches of
lat. cutaneous nn.
(T. 6, 7, 8)

Obliquus Externus

Umbilicus

Anterior superior
iliac spine

2-5 SIDE VIEW OF TRUNK

This superficial dissection shows the external oblique muscle and the lateral
cutaneous nerves.

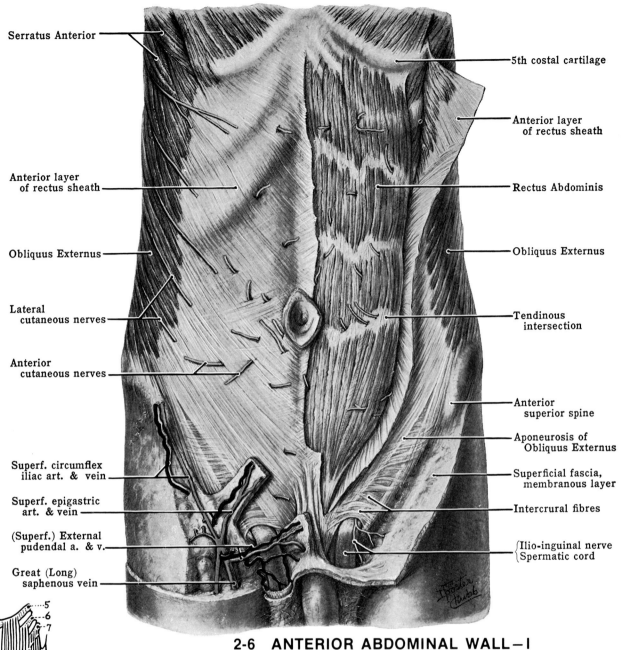

Serratus Anterior

5th costal cartilage

Anterior layer of rectus sheath

Anterior layer of rectus sheath

Rectus Abdominis

Obliquus Externus

Obliquus Externus

Lateral cutaneous nerves

Tendinous intersection

Anterior cutaneous nerves

Anterior superior spine

Aponeurosis of Obliquus Externus

Superf. circumflex iliac art. & vein

Superficial fascia, membranous layer

Superf. epigastric art. & vein

Intercrural fibres

(Superf.) External pudendal a. & v.

{ Ilio-inguinal nerve
  Spermatic cord

Great (Long) saphenous vein

2-6 ANTERIOR ABDOMINAL WALL—I

Anterior layer of Rectus sheath is reflected, on the left side.

Observe:

1. Obliquus Externus is aponeurotic medial to a line that curves upward from point 2.5 cm lateral to the anterior superior iliac spine to the 5th rib. The aponeurosis gives origin to a slip of Pectoralis Major.
2. The lateral border of Rectus Abdominis curving from the pubic tubercle, through the midpoint between the umbilicus and the anterior superior spine, and across the chest margin to the 5th rib.
3. The anterior cutaneous nerves (Th. 7 to 12) piercing Rectus and the anterior layer of its sheath. T10 supplies the region of the umbilicus.
4. In the superficial fascia (of Camper), the three superficial inguinal branches of the femoral artery and the three superficial inguinal tributaries of the great saphenous vein. Of these, the external pudendal artery and vein cross the spermatic cord.
5. The membranous layer of the superficial fascia (of Scarpa) blending with the fascia lata a finger's breadth below the inguinal ligament and thereby forming a gutter that empties into the superficial perineal pouch (of Colles).
6. The spermatic cord and the ilio-inguinal nerve issuing through the superficial (subcutaneous) inguinal ring.

Gall bladder

Linea alba

5
6
7

A.S. S.P.

Rectus Abdominis

2-7 RECTUS ABDOMINIS

Rectus Abdominis is three times as wide cranially, where it is fleshy, as it is caudally, where it is tendinous. Its lateral border crosses the gallbladder at the costal margin.

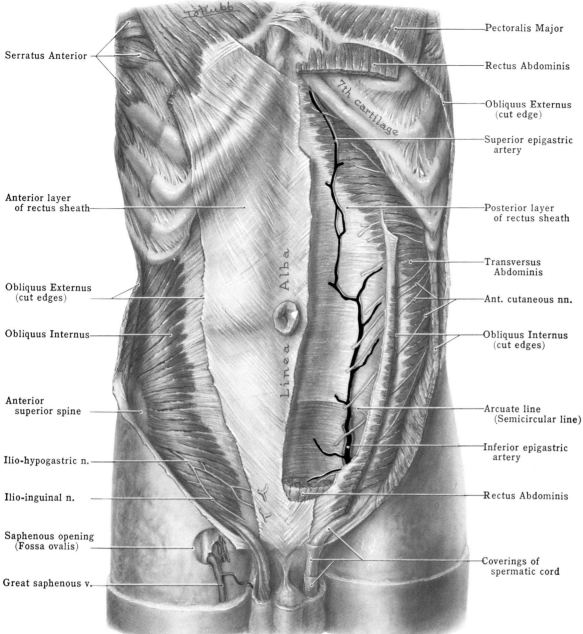

Serratus Anterior

Pectoralis Major

Rectus Abdominis

7th cartilage

Obliquus Externus
(cut edge)

Superior epigastric
artery

Anterior layer
of rectus sheath

Posterior layer
of rectus sheath

Linea Alba

Transversus
Abdominis

Obliquus Externus
(cut edges)

Ant. cutaneous nn.

Obliquus Internus

Obliquus Internus
(cut edges)

Anterior
superior spine

Arcuate line
(Semicircular line)

Inferior epigastric
artery

Ilio-hypogastric n.

Ilio-inguinal n.

Rectus Abdominis

Saphenous opening
(Fossa ovalis)

Coverings of
spermatic cord

Great saphenous v.

2-8 ANTERIOR ABDOMINAL WALL—II

On the right side, most of Obliquus Externus is excised. On the left, Rectus Abdominis is excised and Obliquus Internus divided.

Observe:

1. Obliquus Internus fibers running horizontally at the level of the anterior superior spine; running obliquely upward above this level, and obliquely downward below it.
2. The arcuate line at the level of the anterior superior iliac spine.
3. The anastomosis between the superior and inferior epigastric arteries which indirectly unites the arteries of the upper limb to those of the lower (subclavian to external iliac.)
4. The *inset* to the right shows Rectus Abdominis in action. Refer to Figures 2-6 and 2-7.

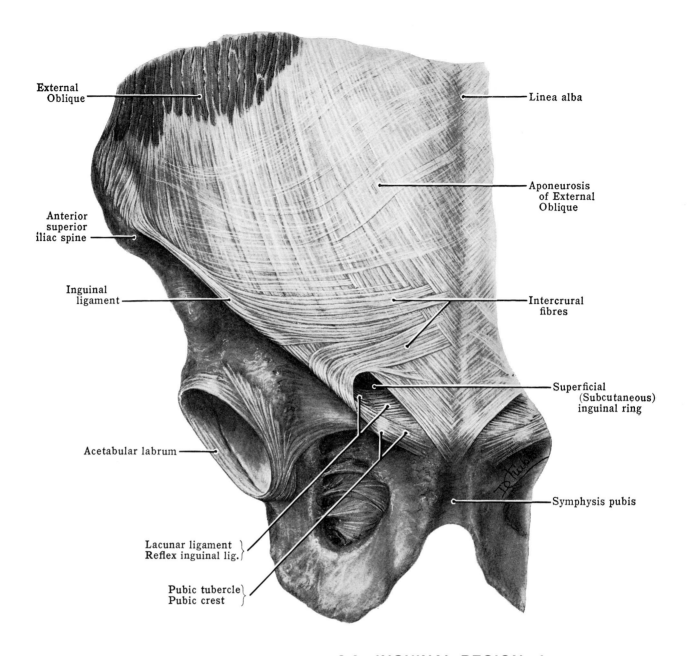

External
Oblique

Linea alba

Aponeurosis
of External
Oblique

Anterior
superior
iliac spine

Inguinal
ligament

Intercrural
fibres

Superficial
(Subcutaneous)
inguinal ring

Acetabular labrum

Symphysis pubis

Lacunar ligament
Reflex inguinal lig.

Pubic tubercle
Pubic crest

2-9   INGUINAL REGION — I

Observe:
1. The linea alba is not an unyielding ligament uniting the sternum to the symphysis pubis, but a line across which fibers decussate in bias and therefore extensible.
2. The intercrural fibers, well developed in this specimen, preventing the crura of the inguinal ring from spreading (see also Fig. 2-6).
3. The superficial inguinal ring is triangular in shape, its central point being above the pubic tubercle, its base being the lateral half of the pubic crest, its lateral crus being the inguinal ligament, and its medial crus being fibers of External Oblique aponeurosis that cross the pubic crest at its midpoint.
4. Behind the ring, some fibers of External Oblique aponeurosis from the opposite side passing to the pubic crest and pecten pubis are called the reflected inguinal ligament.
5. Most fleshy fibers, here shown, of External Oblique sending their tendinous fibers to reinforce the inguinal ligament.

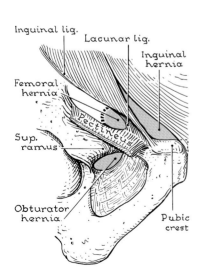

Inguinal lig.

Lacunar lig.

Inguinal
hernia

Femoral
hernia

Sup.
ramus

Pectineus

Obturator
hernia

Pubic
crest

2-10   THREE HERNIA SITES

Relate this diagram to Figure 2-9.

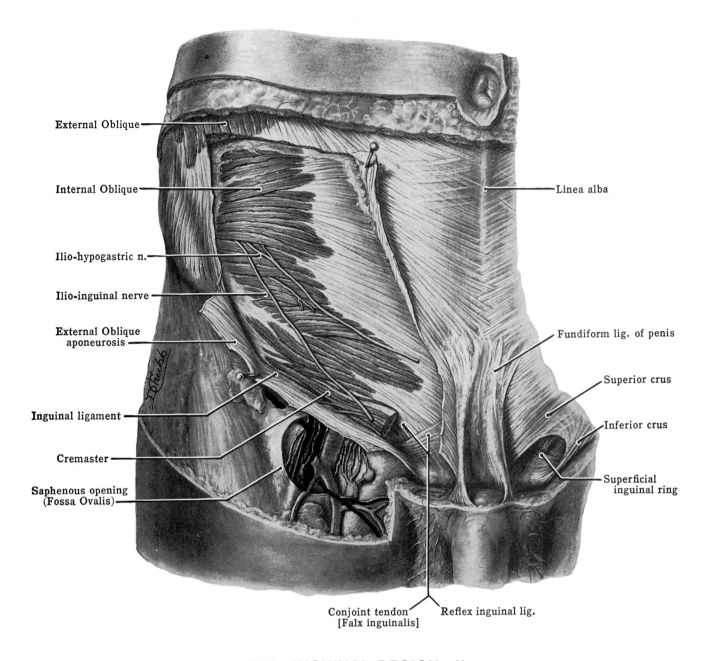

External Oblique

Internal Oblique

Linea alba

Ilio-hypogastric n.

Ilio-inguinal nerve

External Oblique
aponeurosis

Fundiform lig. of penis

Superior crus

Inguinal ligament

Inferior crus

Cremaster

Saphenous opening
(Fossa Ovalis)

Superficial
inguinal ring

Conjoint tendon
[Falx inguinalis]

Reflex inguinal lig.

2-11 INGUINAL REGION—II

External Oblique aponeurosis is partly cut away; spermatic cord is cut short.

Observe:

1. The laminated, fundiform ligament of the penis descending to the junction of the fixed and mobile parts of the organ.
2. The reflex inguinal ligament, which represents External Oblique, lying anterior to the conjoint tendon (falx inguinalis) which represents Internal Oblique and Transversus.
3. The only two structures that course between External and Internal Obliques, namely, the ilio-hypogastric and ilio-inguinal branches of the first lumbar nerve segment. They are sensory from this point to their terminations.
4. The fleshy fibers of Internal Oblique at the level of the anterior superior spine running horizontally; those from the iliac crest passing medio-cranially; and those from the inguinal ligament arching medio-caudally.
5. The Cremaster muscle covering the cord and filling the arched space between conjoint tendon and inguinal ligament.
6. At the level of the navel, the aponeurosis of External Oblique blending with the aponeurosis of Internal Oblique near the lateral border of the Rectus, but in the suprapubic region free as far as the median plane.
7. Numerous lymph vessels (not labeled) streaming cranially around the femoral vessels.

(For femoral sheath see Fig. 4-17.)

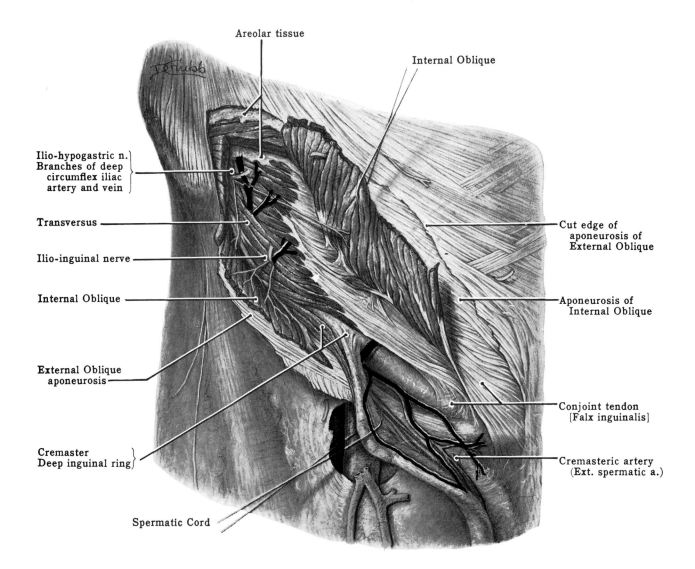

Areolar tissue

Internal Oblique

Ilio-hypogastric n.
Branches of deep
circumflex iliac
artery and vein

Cut edge of
aponeurosis of
External Oblique

Transversus

Ilio-inguinal nerve

Aponeurosis of
Internal Oblique

Internal Oblique

External Oblique
aponeurosis

Conjoint tendon
[Falx inguinalis]

Cremaster
Deep inguinal ring

Cremasteric artery
(Ext. spermatic a.)

Spermatic Cord

2-12 INGUINAL REGION—III

The Internal Oblique is reflected and the spermatic cord is retracted.

Observe:

1. Transversus fibers taking, in this region, the same common medio-caudal direction as the fibers of External Oblique aponeurosis and Internal Oblique.
2. Transversus having a less extensive origin from the inguinal ligament than Internal Oblique.
3. The Internal Oblique portion of the conjoint tendon attached to the pubic crest; Transversus portion extending laterally along the pecten pubis (pectineal line).
4. Conjoint tendon not sharply defined from fascia transversalis, but blending with it.
5. Segment lumbar I, via the ilio-hypogastric and ilio-inguinal nerves, supplying the fibers of Internal Oblique and Transversus that control the conjoint tendon.
6. Fascia transversalis evaginated to form the tubular internal spermatic fascia. The mouth of the tube, called the deep (abdominal) inguinal ring, situated lateral to the inferior epigastric vessels.
7. The cremasteric artery (a branch of the inferior epigastric artery) which in Figure 2-21 is seen to anastomose with the testicular artery and the artery to the deferent duct (vas deferens).
8. The Cremaster muscle arising from the inguinal ligament.

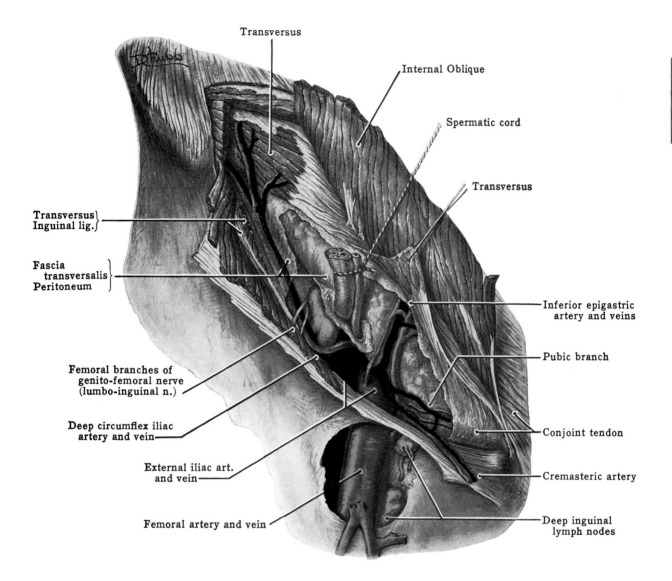

Transversus

Internal Oblique

Spermatic cord

Transversus

Transversus }
Inguinal lig. }

Fascia
transversalis }
Peritoneum }

Inferior epigastric
artery and veins

Pubic branch

Femoral branches of
genito-femoral nerve
(lumbo-inguinal n.)

Deep circumflex iliac
artery and vein

Conjoint tendon

Cremasteric artery

External iliac art.
and vein

Femoral artery and vein

Deep inguinal
lymph nodes

2-13   INGUINAL REGION – IV

The inguinal part of Transversus and fascia transversalis are partly cut away and the spermatic cord is excised.

Observe:

1. The lower limit of the peritoneal sac. It lies some distance above the inguinal ligament laterally but close to it medially.
2. The location of the deep (abdominal) inguinal ring, a finger's breadth above the inguinal ligament at the midpoint between the anterior superior spine and the pubic tubercle.
3. The testicular vessels and the deferent duct (retracted) starting to part company at the deep inguinal ring.
4. The proximity of the external iliac artery and vein to the inguinal canal.
5. The only two branches of the external iliac artery – the deep circumflex iliac and inferior epigastric arteries; note also the cremasteric and pubic branches of the latter.

A new description of the anterior abdominal wall muscles shows that the aponeuroses of the external oblique, internal oblique, and transversus are each *bilaminar*. The decussation that occurs at the linea alba is not only from side to side but from superficial to deep. Because of this interweaving of aponeurotic fibers among the three muscles, they may be considered to act as a *triceps* abdominis. Since each muscle has a right and left fleshy component connected by an intermediate tendon, its aponeurosis, they may also be seen as *digastric* muscles. Above the umbilicus, both anterior and posterior rectus sheaths are trilaminar: in front, the two layers of the external oblique and the superficial layer of the internal oblique; behind, the deep layer of the internal oblique and the two layers of the transversus aponeurosis. Approximately midway between the umbilicus and the symphysis pubis, all three aponeuroses pass in front of the rectus muscle, thus the posterior rectus sheath ends as the arcuate line. Consult Rizk, N.N. (1980) A new description of the anterior abdominal wall in man and mammals. *J. Anat., 131:* 373–385.

2–14   ABDOMINAL WALL MUSCLES

Muscle	Ribs			Lumbar fascia	Linea alba	Inguinal ligament	Iliac crest
	Number	Surface	Interdigitates with				
External oblique	Lower 8	Outer	Serratus anterior Latissimus dorsi	No	Yes	All	Anterior ½ outer lip
Internal oblique	Lower 3	Inferior	In series with internal intercostals	Yes	No	Lateral ⅔	Anterior ⅔ intermediate line
Transversus abdominis	Lower 6	Inner	Diaphragm	Yes	Yes	Lateral ⅓	Anterior ⅔ + inner lip

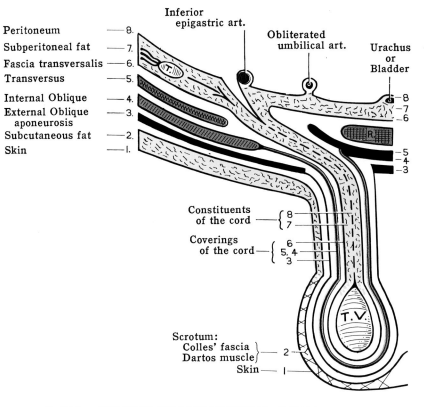

Peritoneum — 8.
Subperitoneal fat — 7.
Fascia transversalis — 6.
Transversus — 5.
Internal Oblique — 4.
External Oblique aponeurosis — 3.
Subcutaneous fat — 2.
Skin — 1.

Inferior epigastric art.

Obliterated umbilical art.

Urachus or Bladder

Constituents of the cord { 8 7
Coverings of the cord { 6 5. 4 3

Scrotum:
Colles' fascia }
Dartos muscle } — 2
Skin — 1

2-15A   SCHEME OF THE INGUINAL CANAL

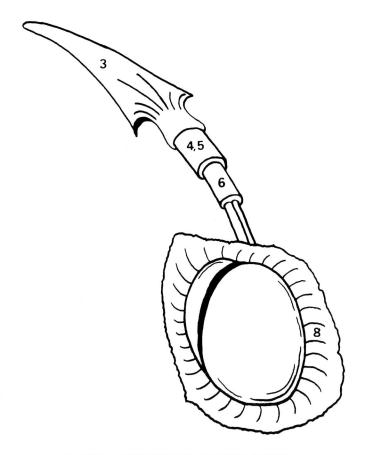

2-15B   COVERINGS OF THE CORD

Figure 2-15A is a schematic horizontal section. The scrotum and testis are assumed to have been raised to the level of the superficial inguinal ring. This shows the eight layers of the abdominal wall and their three evaginations: (a) the scrotum, (b) the coverings of the spermatic cord, and (c) the constituents of the cord.

Figure 2-15B is a diagram showing the coverings of the spermatic cord exposed in layers. The numbers relate to the layers of the abdominal wall as shown in Figure 2-15A.

3. External spermatic fascia, derived from the fascia of the External oblique.

4,5. Cremaster muscle, derived from Internal oblique and Transversus.

6. Internal spermatic fascia, derived from Transversalis.

8. Tunica vaginalis (T.V.), derived from peritoneum.

Note that the deep inguinal ring is *lateral* to the inferior epigastric artery. *Indirect* inguinal hernias pass through this ring, the sac following the course of the spermatic cord. *Direct* inguinal hernias bulge *directly* through the abdominal wall *medial* to the inferior epigastric artery.

See McVay, C.B. (1974) The anatomical basis for inguinal and femoral hernioplasty. *Surg. Gynec. Obst., 139:* 931–945.

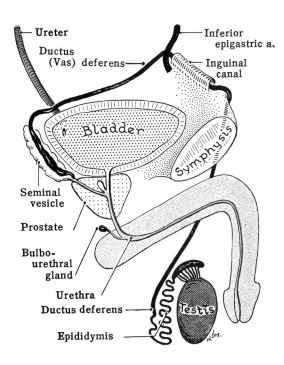

Observe:
1. The *ureters* open into the urinary bladder at the upper angles of the trigone. The *urethra* leaves the bladder at the lower angle of the trigone (Fig. 3-48).
2. Within the prostate gland, the *ejaculatory ducts* from the seminal vesicles enter the urethra (Fig. 3-47).
3. The *efferent ductules* at the upper pole of the testis emptying into the head of the *epididymis* (Fig. 2-20).
4. The *vas deferens* ascends from the scrotum, enters the inguinal canal through the superficial ring, and exits through the deep ring, hooking around the lateral side of the inferior epigastric artery (Fig. 2-13).

2-16 MALE UROGENITAL SYSTEM

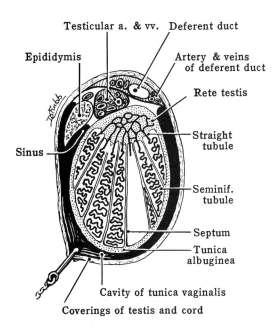

2-17 CROSS-SECTION OF RIGHT TESTIS

Observe:
1. The cavity of the tunica vaginalis testis surrounding the testis in front and at the sides and extending between testis and epididymis as the sinus of the epididymis.
2. The epididymis lying postero-lateral to the testis. It indicates to which side a testis belongs, for it is on the right side of a right testis and on the left side of a left testis.
3. The deferent duct with its fine lumen and thick wall lying postero-medial to the testis.
4. The three groups of longitudinal veins (a) around the testicular artery, (b) medial to the duct with the artery of the duct, and (c) lateral to it.
5. The pyramidal compartments of the seminiferous tubules, shown semidiagrammatically. Each of the 250 compartments contains two or three hairlike seminiferous tubules which join in the mediastinum testis to form a rete.

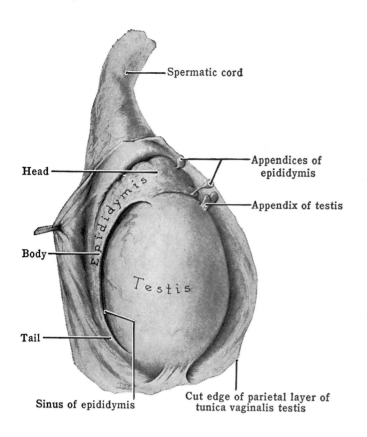

Spermatic cord

Head

Appendices of
epididymis

Appendix of testis

Epididymis

Body

Testis

Tail

Sinus of epididymis

Cut edge of parietal layer of
tunica vaginalis testis

2-18 TESTIS, LATERAL VIEW

The tunica vaginalis testis has been incised longitudinally.

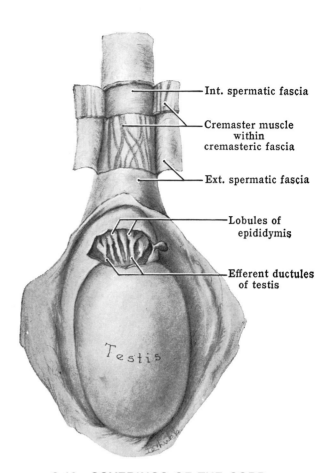

Int. spermatic fascia

Cremaster muscle
within
cremasteric fascia

Ext. spermatic fascia

Lobules of
epididymis

Efferent ductules
of testis

Testis

2-19 COVERINGS OF THE CORD

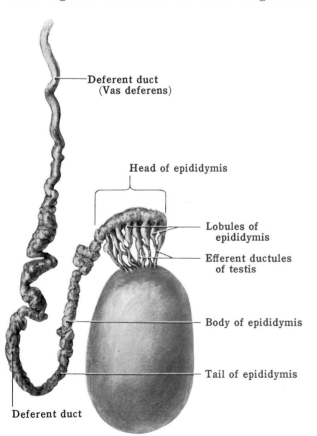

Deferent duct
(Vas deferens)

Head of epididymis

Lobules of
epididymis

Efferent ductules
of testis

Body of epididymis

Tail of epididymis

Deferent duct

2-20 EPIDIDYMIS

Note the eight efferent ductules uniting the epididymis to the upper pole of the testis.

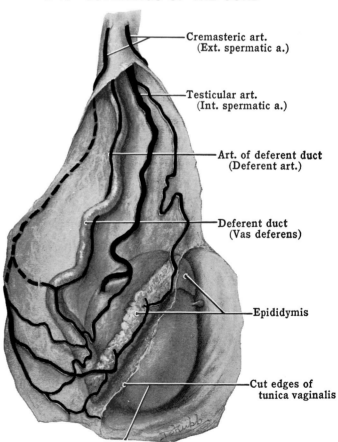

Cremasteric art.
(Ext. spermatic a.)

Testicular art.
(Int. spermatic a.)

Art. of deferent duct
(Deferent art.)

Deferent duct
(Vas deferens)

Epididymis

Cut edges of
tunica vaginalis

2-21 BLOOD SUPPLY OF THE TESTIS

The epididymis is displaced slightly to the lateral side. Note the free anastomosis between the three arteries.

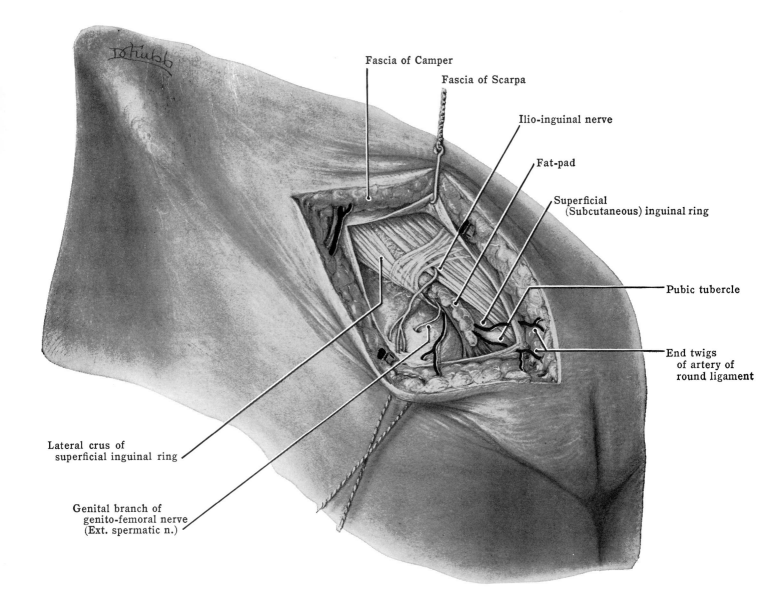

Fascia of Camper

Fascia of Scarpa

Ilio-inguinal nerve

Fat-pad

Superficial
(Subcutaneous) inguinal ring

Pubic tubercle

End twigs
of artery of
round ligament

Lateral crus of
superficial inguinal ring

Genital branch of
genito-femoral nerve
(Ext. spermatic n.)

2-22 FEMALE INGUINAL CANAL—I

PROGRESSIVE DISSECTIONS OF THE FEMALE INGUINAL CANAL

Observe:

1. In Figure 2-22, the superficial inguinal ring, small, and its crura prevented from spreading by the intercrural fibers.

2. Issuing from the superficial inguinal ring: (a) the round ligament of the uterus, (b) a closely applied pad of fat, (c) the genital branch of the genito-femoral nerve, and (d) the artery of the round ligament of the uterus. This artery is homologous with the cremasteric artery in the male, shown in Figure 2-13.

3. The ilio-inguinal nerve, here perforating the medial crus of the superficial inguinal ring.

4. In Figure 2-23, the Cremaster muscle, not extending beyond the ring.

5. In Figure 2-24, the round ligament breaking up into strands as it leaves the inguinal canal and approaches the labium majus.

6. In Figure 2-25, the close relationship of the external iliac artery and vein to the inguinal canal.

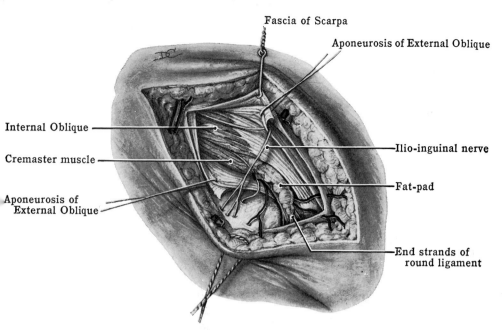

Fascia of Scarpa

Aponeurosis of External Oblique

Internal Oblique

Cremaster muscle

Ilio-inguinal nerve

Fat-pad

Aponeurosis of
External Oblique

End strands of
round ligament

2-23 FEMALE INGUINAL CANAL — II

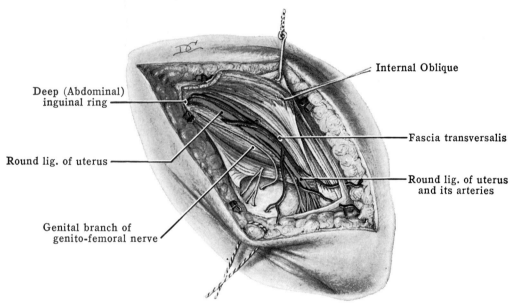

Deep (Abdominal)
inguinal ring

Internal Oblique

Round lig. of uterus

Fascia transversalis

Round lig. of uterus
and its arteries

Genital branch of
genito-femoral nerve

2-24 FEMALE INGUINAL CANAL — III

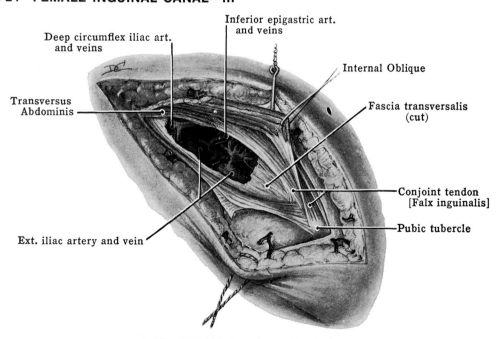

Deep circumflex iliac art.
and veins

Inferior epigastric art.
and veins

Internal Oblique

Transversus
Abdominis

Fascia transversalis
(cut)

Conjoint tendon
[Falx inguinalis]

Ext. iliac artery and vein

Pubic tubercle

2-25 FEMALE INGUINAL CANAL — IV

2-26 DIAGRAM OF THE DIGESTIVE SYSTEM

In the *lower* figure, three organs—spleen, pancreas, and duodenum—are uncovered by removal of the stomach and the transverse colon.

DIGESTIVE SYSTEM

This system extends from the lips to the anus. Here is a list of its chief parts:

ORAL CAVITY
  Vestibule of mouth
    Lips or labia
    Cheeks or buccae
    Gums or gingivae
    Teeth or dentes
  Cavity proper of mouth
    Palate
      Hard
      Soft
    Tongue or lingua
    Salivary glands
      Parotid
      Sublingual
      Submandibular
    Palato-glossal arch
CAVITY OF THE PHARYNX
  Oral part
    Palatine tonsil
  Laryngeal part
    Entrance, or aditus, to larynx
ALIMENTARY CANAL
  Esophagus
    Cervical part
    Thoracic part
    Abdominal part
  Stomach or ventriculus or gaster
    Cardiac orifice
    Incisura cardiaca
    Greater curvature
    Lesser curvature
      Incisura angularis
    Fundus
    Body
    Pyloric part
      Pyloric antrum
      Pyloric canal
      Pylorus
        Pyloric orifice
  Small intestine
    Duodenum
      1st or superior part
      2nd or descending part
      3rd or horizontal part
      4th or ascending part
    Duodeno-jejunal flexure
    Jejunum
    Ileum
    Ileo-cecal orifice and valve
  Large intestine
    Vermiform appendix
    Cecum
    Ascending colon
    Right colic flexure (hepatic flexure)
    Transverse colon
    Left colic flexure (splenic flexure)
    Descending colon
    Sigmoid colon
    Rectum
    Anal canal
    Anus
ASSOCIATED ORGANS
  Liver or hepar
    Gall bladder
      Fundus, body, and neck
    Biliary passages
  Pancreas
    Head, neck, body, tail, and ducts
  Spleen or lien

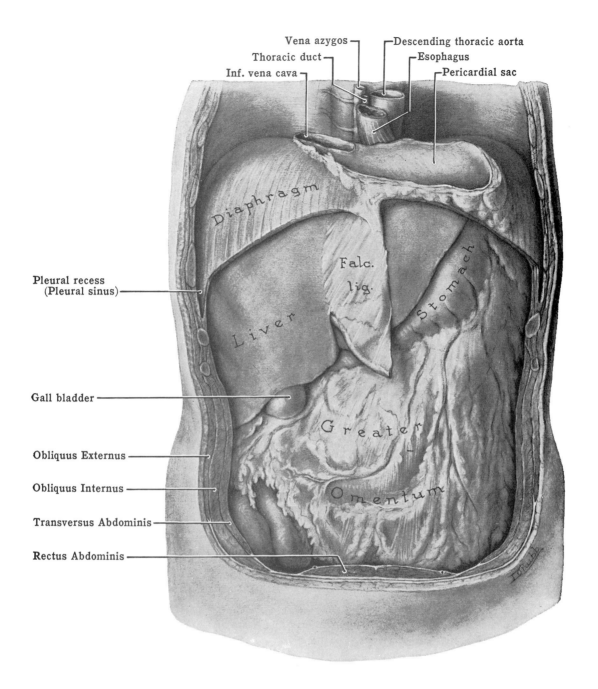

Vena azygos — ┌— Descending thoracic aorta
Thoracic duct — ┌— Esophagus
Inf. vena cava — ┌— Pericardial sac

Diaphragm

Falc. lig.

Stomach

Pleural recess
(Pleural sinus) —

Liver

Gall bladder —

Greater

Obliquus Externus —

Obliquus Internus —

Omentum

Transversus Abdominis —

Rectus Abdominis —

2-27 ABDOMINAL
CONTENTS, UNDISTURBED

The anterior abdominal and thoracic walls are cut away.

Observe:

1. The falciform ligaments, with the ligamentum teres hepatis (round ligament) in its free edge, severed at its attachment to the abdominal wall and
   Diaphragm in the median plane. Its attachment to the liver is its own width
   to the right of the median plane. It resists displacement of the liver to the
   right.
2. The gallbladder projecting below the sharp, inferior border of the liver.
3. The two Recti meeting in the median plane above the pubis.
4. Obliquus Internus, the thickest of the three flat abdominal muscles.
5. The pleural cavities separating the upper abdominal viscera from the body
   wall.
6. Two-thirds of the pericardial sac lying to the left of the median plane; its
   apex, *i.e.*, the lowest and leftmost point, overlying the stomach.

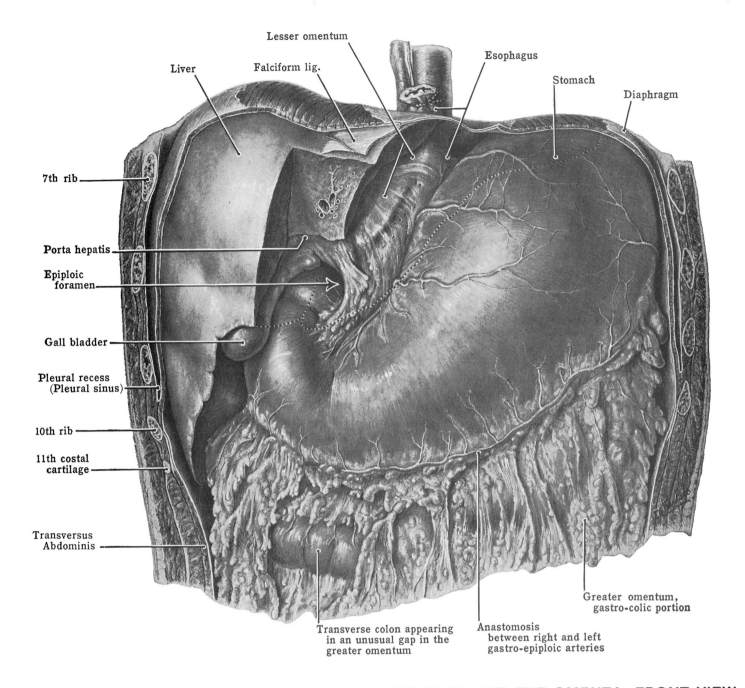

Lesser omentum

Liver

Falciform lig.

Esophagus

Stomach

Diaphragm

7th rib

Porta hepatis

Epiploic foramen

Gall bladder

Pleural recess (Pleural sinus)

10th rib

11th costal cartilage

Transversus Abdominis

Transverse colon appearing in an unusual gap in the greater omentum

Anastomosis between right and left gastro-epiploic arteries

Greater omentum, gastro-colic portion

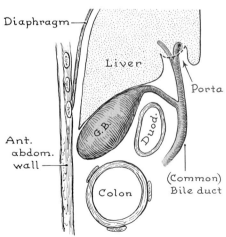

Diaphragm

Liver

Porta

G.B.

Duod.

Ant. abdom. wall

Colon

(Common) Bile duct

2-29 RELATIONS OF GALLBLADDER

2-28 STOMACH AND THE OMENTA, FRONT VIEW

The stomach is inflated with air; the left part of the liver is cut away.

Observe:

1. The pyloric end of the stomach which during development has moved to the right and has come to lie infero-posterior to the gallbladder. The first or superior part of the duodenum—placed behind the *arrowhead*—almost occluding the epiploic foramen (mouth of the lesser sac).

2. The gallbladder, followed cranially, leading to the free margin of the lesser omentum, and hence acting as a guide to the epiploic foramen, which lies behind that free margin.

3. The lesser omentum passing from the lesser curvature of the stomach and first inch of the duodenum to the fissure for the ligamentum venosum and porta hepatis. This omentum, thickened at its free margin but elsewhere like gossamer, much perforated, and the caudate lobe of the liver visible through it.

4. The greater omentum hanging from the greater curvature of the stomach.

5. The right cupola of the diaphragm rising higher than the left cupola.

6. The relative thinness of the Transversus Abdominis. The pleural recess about two fingers breadth above the costal margin in the midlateral line

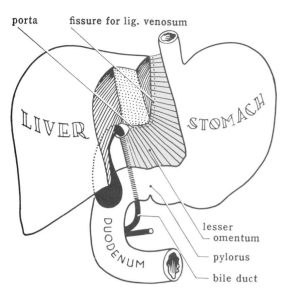

porta fissure for lig. venosum

LIVER STOMACH

DUODENUM

lesser omentum
pylorus
bile duct

2-30 THE LESSER OMENTUM

Two sagittal cuts have been made through the liver—one at the fissure for the ligamentum venosum; the other at the right limit of the porta hepatis. These two cuts have been joined by a coronal cut.

Note:

1. The lesser omentum may be regarded as the "mesentery" of the bile passages, seeing they occupy its free edge (Fig. 2-78).
2. The lesser omentum extends from the lesser curvature of the stomach and first inch of the duodenum to the fissure for the ligamentum venosum and to the porta. The part attached to the body of the stomach passes to the fissure; the part attached to the pyloric part of the stomach and duodenum passes to the porta.

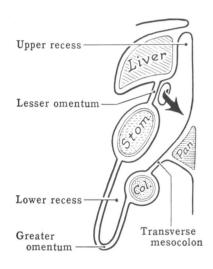

Upper recess
Liver

Lesser omentum
Stom.
Pan.

Col.

Lower recess

Greater omentum
Transverse mesocolon

2-31 VERTICAL EXTENT OF OMENTAL BURSA (LESSER SAC) IN THE MEDIAN PLANE

The *arrow* passes from the greater sac of peritoneum through the epiploic foramen (mouth of the lesser sac) into the omental bursa (lesser sac).

Note the three approaches to the bursa; it may be opened by cutting through (a) the lesser omentum, (b) the greater omentum, or (c) the transverse mesocolon.

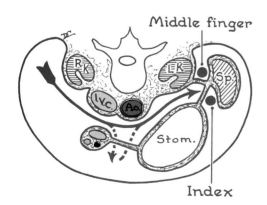

Middle finger

R.K. L.K. Sp.

I.V.C. Ao.

Stom.

Index

2-32 HORIZONTAL EXTENT OF OMENTAL BURSA AT THE LEVEL OF THE EPIPLOIC FORAMEN

Instructions:

1. While standing on the right side of the body, run your right middle finger upward between the left kidney and the spleen and your right index finger upward between the stomach and the spleen. The "pedicle or stalk" of the spleen now lies in the cleft between these two fingers, as though in a clamp. The pedicle has a free lower border (and a free upper border), or you could not grasp it as you are doing. Its linear site of attachment to the spleen is around the hilus.
2. Clamped between your index and middle finger are four layers of peritoneum. Of this you can satisfy yourself by passing your left index finger through the epiploic foramen and across the abdomen, behind the stomach, until it touches the spleen between your two right fingers. If your left index finger will not reach all the way, pass it as far as it will go, tear through the lesser omentum over its tip, withdraw the finger, and reinsert it at the halfway opening just made. The hilus of the spleen, which you are palpating, is situated at the left extremity of the omental bursa.

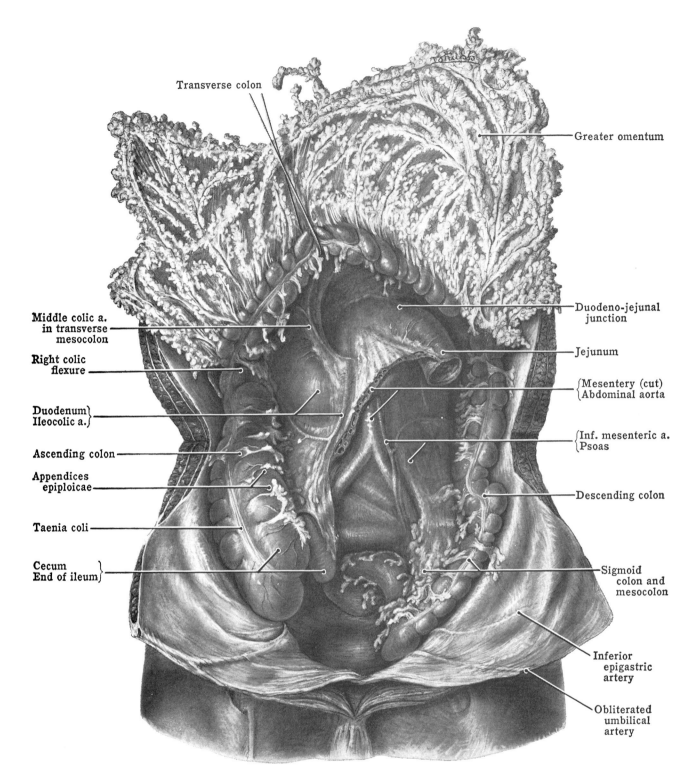

Transverse colon

Greater omentum

Duodeno-jejunal junction

Jejunum

Mesentery (cut) / Abdominal aorta

Inf. mesenteric a. / Psoas

Descending colon

Sigmoid colon and mesocolon

Inferior epigastric artery

Obliterated umbilical artery

Middle colic a. in transverse mesocolon

Right colic flexure

Duodenum / Ileocolic a.

Ascending colon

Appendices epiploicae

Taenia coli

Cecum / End of ileum

2-33 INTESTINES

The greater omentum is thrown upward and with it go the transverse colon and transverse mesocolon. The jejunum and ileum are cut away, excepting their end pieces, and the mesentery is cut short. Examine in conjunction with Figure 2-34 facing.

Observe:

1. The duodeno-jejunal junction, situated to the left of the median plane and immediately below the root of the transverse mesocolon.

2. The first few centimeters of the jejunum descending downward and to the left, anterior to the left kidney. The last few centimeters of the ileum ascending upward and to the right out of the pelvic cavity.

Of these two parallel pieces of gut, the former is much the larger.

3. The large gut forming 3½ sides of a square picture frame around the jejunum and ileum (removed); the missing half-side being between the cecum and the sigmoid colon.

4. The distinguishing features of the large gut: (a) its position around the small gut, (b) the teniae coli or longitudinal muscle bands, (c) the sacculations or haustra, and (d) the appendices epiploicae.

5. The right colic flexure, which lies below the liver, placed at a lower level than the left colic flexure which lies below the spleen.

6. The vermiform appendix had been removed at operation.

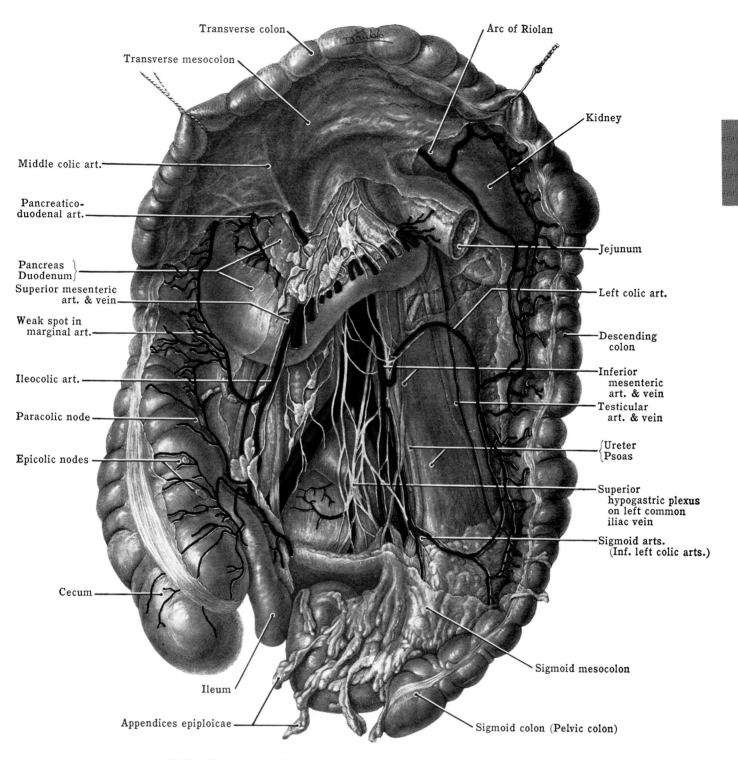

Transverse colon

Transverse mesocolon

Arc of Riolan

Kidney

Middle colic art.

Pancreatico-
duodenal art.

Pancreas }
Duodenum }
Superior mesenteric
art. & vein

Weak spot in
marginal art.

Ileocolic art.

Paracolic node

Epicolic nodes

Cecum

Ileum

Appendices epiploicae

Jejunum

Left colic art.

Descending
colon

Inferior
mesenteric
art. & vein

Testicular
art. & vein

Ureter
Psoas

Superior
hypogastric plexus
on left common
iliac vein

Sigmoid arts.
(Inf. left colic arts.)

Sigmoid mesocolon

Sigmoid colon (Pelvic colon)

2-34   STRUCTURES ON THE POSTERIOR ABDOMINAL WALL

This is the same specimen as Figure 2-33 with much peritoneum removed.

Observe:

1. The duodenum, of large diameter before the site of crossing of the superior mesenteric vessels, and narrow beyond. It is, indeed, much wider than the descending colon.

2. The jejunal and ileal branches (cut) passing from the left side of the superior mesenteric artery; the right colic artery, here as commonly, being a branch of the ileocolic artery.

3. An accessory artery, called the arch of Riolan, which connects the superior mesenteric artery to the left colic artery.

4. On the right side: small (epicolic) lymph nodes on the colon; small (paracolic) nodes beside the colon; nodes along the ileo-colic artery which drain into main nodes ventral to the pancreas.

5. (a) The intestines and the intestinal vessels lying on a plane anterior to (b) that of the testicular vessels, and these in turn lying (c) anterior to the plane of the kidney, its vessels, and the ureter.

6. The right and left ureters asymmetrically placed in this specimen.

7. The superior hypogastric plexus (presacral nerve) lying within the fork of the aorta and ventral to the left common iliac vein, the body of the 5th lumbar vertebra, and the 5th disc.

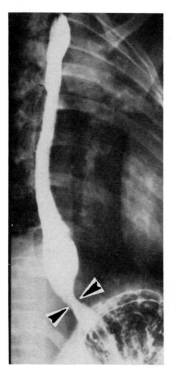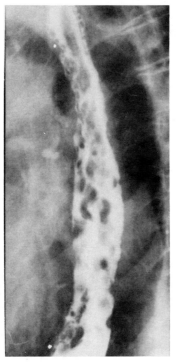

2-35A RADIOGRAPHS OF ESOPHAGUS

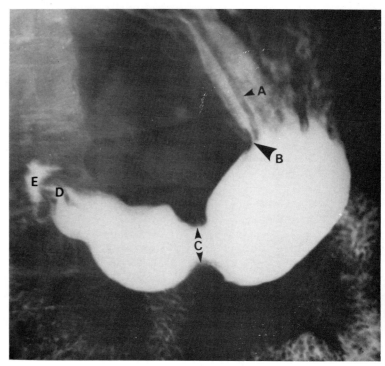

2-35B RADIOGRAPH OF STOMACH

On the *left*, a normal esophagus after swallowing barium. Note the constriction as it passes through the diaphragm (*arrows*) and its short (2 cm) course in the abdomen before entering the stomach.

On the *right*, barium outlines distended veins (esophageal varices) encroaching on the lumen. See Figure 2-58 for explanation.

In this radiograph following a barium meal, observe: *A*, longitudinal ridges of mucous membrane (rugae); *B*, the angular notch; *C*, a peristaltic wave; *D*, pylorus; *E*, duodenal "cap." Note also the feathery appearance of barium in the small intestine.

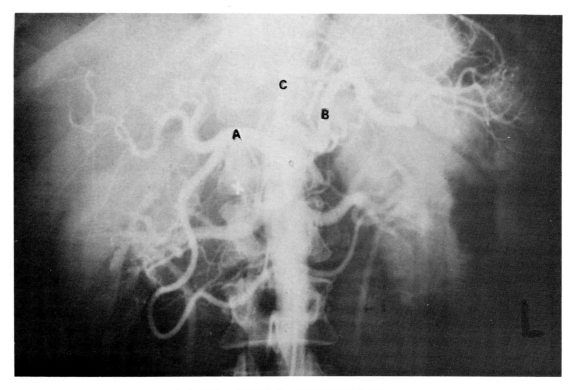

2-35C CELIAC ARTERIOGRAM

Observe: *A*, the hepatic artery. Its right gastroepiploic branch is well shown, following the greater curvature of the stomach. *B*, The tortuous splenic artery. *C*, The left gastric artery.

Consult Figures 2-43 and 2-44.

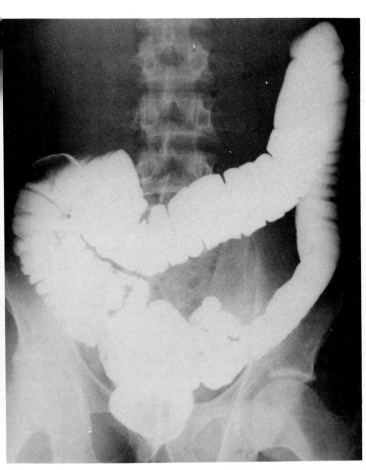 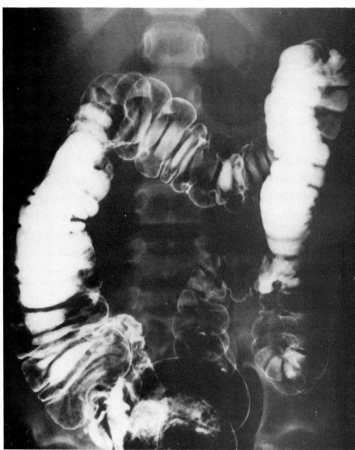

2-36A   RADIOGRAPHS OF LARGE INTESTINE

In the illustration on the *left,* a barium enema has filled the colon. Note the relative levels of the hepatic and splenic flexures. On the *right,* barium coats the walls of a colon distended with air, giving a vivid view of haustra.

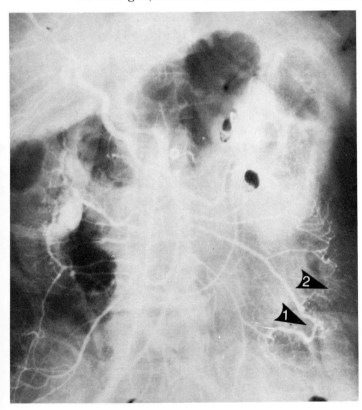

2-36B   SUPERIOR MESENTERIC ARTERIOGRAM

Branches of the celiac trunk are also prominent in this radiograph. Observe in particular examples of anastomotic loops (*1*) and vasa recta (*2*). Consult Figure 2-49.

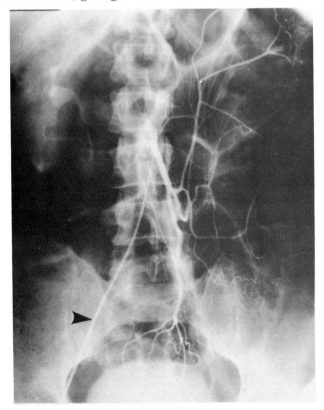

2-36C   INFERIOR MESENTERIC ARTERIOGRAM

Consult Figure 2-50. Observe the left colic artery, sigmoid arteries, and termination as the superior rectal artery. An *arrow* points to an arterial catheter.

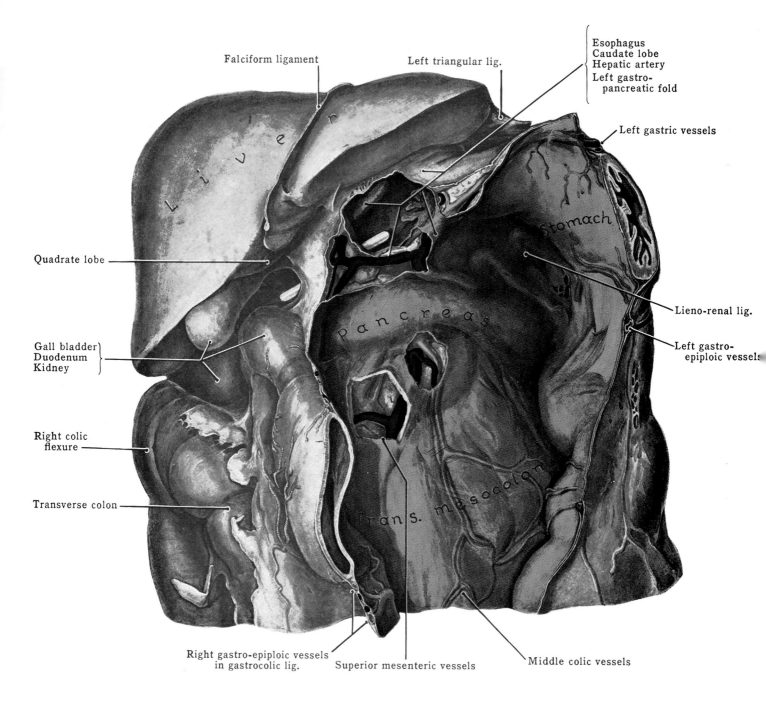

Falciform ligament

Left triangular lig.

Esophagus
Caudate lobe
Hepatic artery
Left gastro-
pancreatic fold

Left gastric vessels

Liver

Stomach

Quadrate lobe

Pancreas

Lieno-renal lig.

Gall bladder
Duodenum
Kidney

Left gastro-
epiploic vessels

Right colic
flexure

Transverse colon

Trans. mesocolon

Right gastro-epiploic vessels
in gastrocolic lig.

Superior mesenteric vessels

Middle colic vessels

2-37 OMENTAL BURSA OPENED—I

The anterior wall of the bursa, consisting of the stomach with its two omenta and the vessels along its two curvatures, has been divided vertically and the two parts have been turned to the left and right.

Observe:

1. The esophagus and the body of the stomach on the left side of the cadaver, and the pyloric part and the superior (first) part of the duodenum on the right side. The first part of the duodenum running backward, upward, and to the right, and having the gallbladder and quadrate lobe in contact with it above.

2. The right kidney forming the posterior wall of the hepato-renal pouch, and the white rod passed from that pouch, through the epiploic foramen, into the omental bursa.

3. The pancreas on the posterior wall of the bursa, molded on the vertebral column, and lying roughly horizontally. Below this, the transverse mesocolon, the root of which has slipped caudally from its attachment to the pancreas.

4. Above the pancreas, the left gastro-pancreatic fold, which is the "mesentery" of the arch of the left gastric vessels, separates the upper recess, into which the caudate lobe projects, from the splenic recess, which lies to the left.

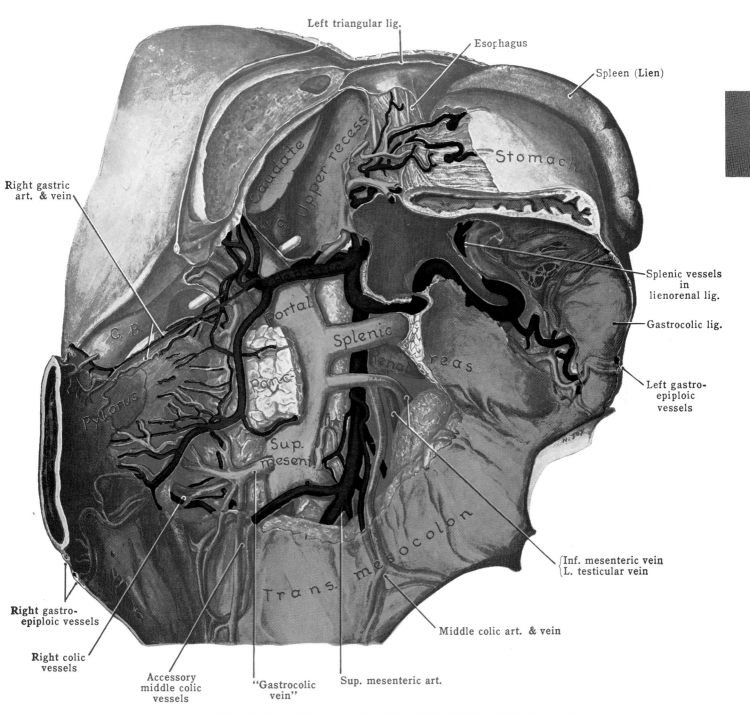

Left triangular lig.

Esophagus

Spleen (Lien)

Caudate

Lv.G.

Upper recess

Stomach

Right gastric
art. & vein

Portal

Splenic

Splenic vessels
in
lienorenal lig.

Pancreas

Gastrocolic lig.

C. B.

Pylorus

Pancreas

Renal

Left gastro-
epiploic
vessels

Sup.
mesent.

Trans. mesocolon

Inf. mesenteric vein
L. testicular vein

Right gastro-
epiploic vessels

Right colic
vessels

Accessory
middle colic
vessels

"Gastrocolic
vein"

Sup. mesenteric art.

Middle colic art. & vein

N. 707

2-38 POSTERIOR WALL OF THE OMENTAL BURSA—II

The peritoneum of the posterior wall has largely been removed and a section of pancreas has been excised.

Observe:

1. The white rod passed through the epiploic foramen.
2. The esophagus and the left gastro-pancreatic fold (Fig. 2-37) bounding the left side of the upper recess. The esophageal branches of the left gastric vessels and the anterior and posterior vagal trunks applied to the esophagus.
3. The celiac trunk (not labeled) from which (a) the left gastric artery arches upward, (b) the splenic artery runs tortuously to the left, and (c) the hepatic artery runs straight to the right to the front of the portal vein where it bifurcates like the letter Y.
4. The right hepatic artery here arising early and passing dorsal to the portal vein (Fig. 2-59).
5. The portal vein formed behind the neck of the pancreas

by the union of the superior mesenteric and splenic veins, with the inferior mesenteric vein joining at or near the angle of union. Here, the left gastric vein joins it. Usually, it is joined by right gastric vein (Fig. 2-80), posterior, superior pancreatico-duodenal vein (Fig. 2-79), and one or two small duodenal or pancreatic veins.

6. The superior mesenteric vein to be joined by the gastro-colic trunk. The middle colic vein accompanying the middle colic artery here ends in the inferior mesenteric vein.

For details, consult Falconer, C. W. A., and Griffiths, E. (1950) The anatomy of the blood vessels in the region of the pancreas. *Brit. J. Surg., 37:* 334; Michels, N. A., *et al.* (1964) The superior mesenteric vein—an anatomic and surgical study of eighty-one subjects. *J. Int. Coll. Surgeons, 41:* 339.

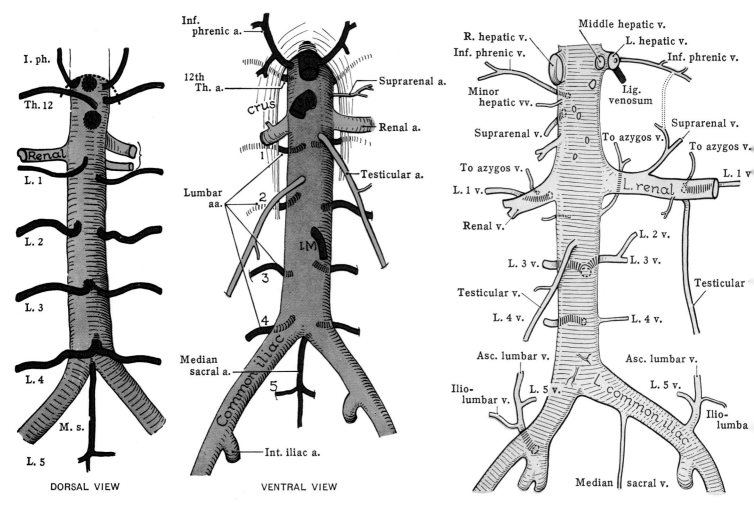

2-39 ABDOMINAL AORTA AND ITS BRANCHES

DORSAL VIEW   VENTRAL VIEW

2-40 INFERIOR VENA CAVA AND ITS TRIBUTARIES

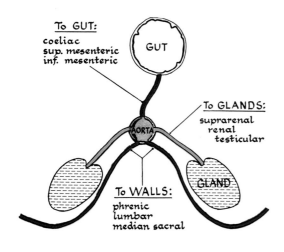

2-41 ABDOMINAL AORTA: THREE TYPES OF BRANCHES

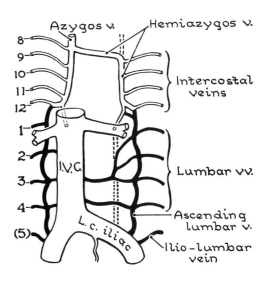

2-42 ASCENDING LUMBAR VEIN

A long channel connecting segmental veins, lying on transverse processes, and terminating in the subcostal vein.

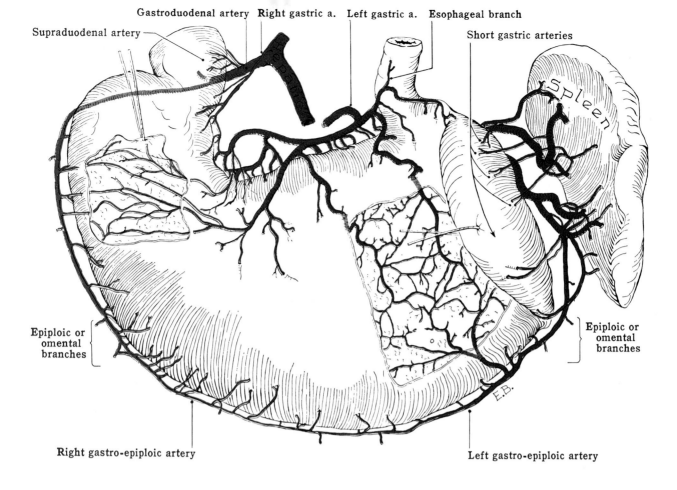

2-43 ARTERIES OF THE STOMACH AND SPLEEN

The serous and muscular coats are removed from two areas of the stomach, thereby revealing the anastomotic networks in the submucous coat.

Observe:

1. The arterial arch on the lesser curvature formed by the larger left gastric artery and the much smaller right gastric artery.
2. The arterial arch on the greater curvature formed equally by the right and the left gastro-epiploic artery. The anastomosis between their two trunks is attenuated; commonly it is absent.
3. The anastomoses between the branches of the two fore-going arterial arches taking place in the submucous coat two-thirds of the distance from lesser to greater curvature.
4. Four or five tenuous short gastric arteries leaving the terminal branches of the splenic artery close to the spleen; and the left gastro-epiploic artery, belonging to the short gastric artery series, arising within 2.5 cm of the hilus of the spleen.

For details, consult Michels, N. A. (1955) *Blood Supply and Anatomy of the Upper Abdominal Organs*. J. B. Lippincott Company, Philadelphia.

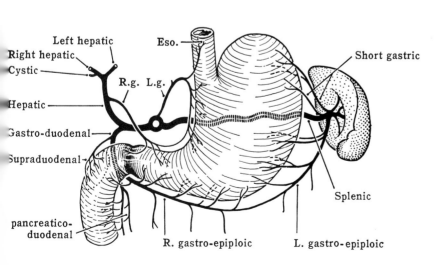

2-44 BRANCHES OF COELIAC TRUNK

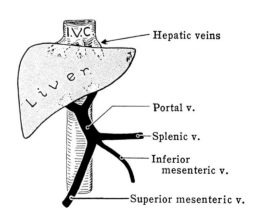

2-45 PORTAL SYSTEM

Blood from the gastrointestinal tract enters the liver via the portal vein and leaves via the hepatic veins to enter the inferior vena cava.

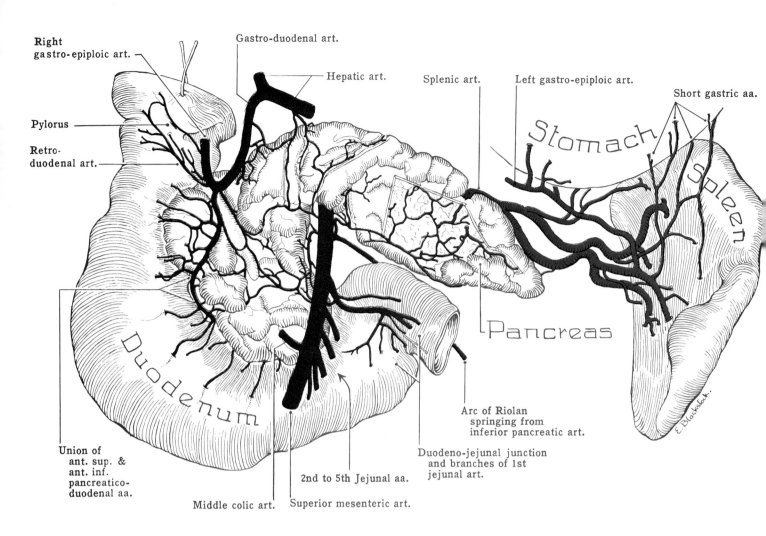

Right gastro-epiploic art.

Gastro-duodenal art.

Hepatic art.

Splenic art.

Left gastro-epiploic art.

Short gastric aa.

Pylorus

Stomach

Spleen

Retro-duodenal art.

Pancreas

Duodenum

Arc of Riolan springing from inferior pancreatic art.

Union of ant. sup. & ant. inf. pancreatico-duodenal aa.

Duodeno-jejunal junction and branches of 1st jejunal art.

Middle colic art.

2nd to 5th Jejunal aa.

Superior mesenteric art.

E. Blackstock.

2-46   BLOOD SUPPLY TO THE PANCREAS, DUODENUM, AND SPLEEN, FRONT VIEW

(See text on facing page)

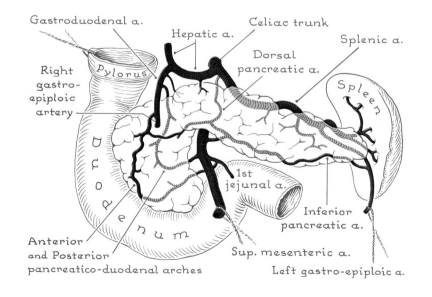

Gastroduodenal a.

Celiac trunk

Hepatic a.

Splenic a.

Dorsal pancreatic a.

Right gastro-epiploic artery

Pylorus

Spleen

Duodenum

1st jejunal a.

Inferior pancreatic a.

Anterior and Posterior pancreatico-duodenal arches

Sup. mesenteric a.

Left gastro-epiploic a.

2-47   BLOOD SUPPLY TO PANCREAS

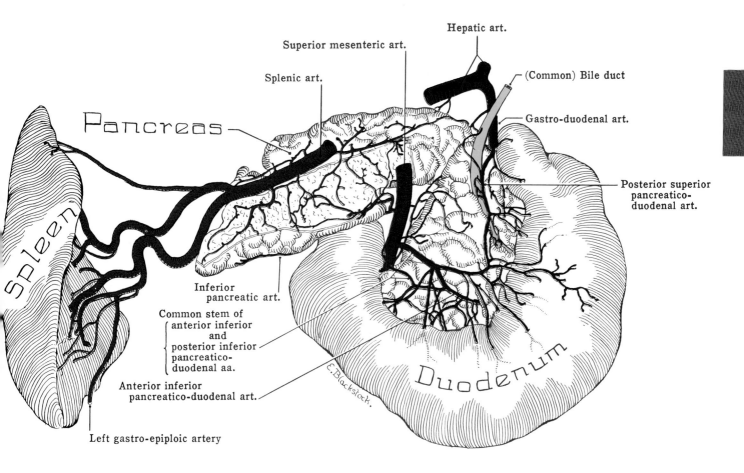

Hepatic art.

Superior mesenteric art.

Splenic art.

(Common) Bile duct

Pancreas

Gastro-duodenal art.

Spleen

Posterior superior
pancreatico-
duodenal art.

Inferior
pancreatic art.

Common stem of
anterior inferior
and
posterior inferior
pancreatico-
duodenal aa.

Anterior inferior
pancreatico-duodenal art.

E. Blackstock.

Duodenum

Left gastro-epiploic artery

2-48   BLOOD SUPPLY TO PANCREAS, DUODENUM, AND SPLEEN, DORSAL VIEW

Two or three short slices have been removed from the pancreas.

Observe:

1. This territory is supplied by the hepatic, splenic, and superior mesenteric arteries.
2. Several retroduodenal branches spring from the right gastro-epiploic artery.
3. The anterior superior pancreatico-duodenal branch of the gastro-duodenal artery and the anterior inferior pancreatico-duodenal branch of the superior mesenteric artery form an arch in front of the head of the pancreas. The posterior superior and posterior inferior branches of the same two arteries form another arch behind the pancreas. The two inferior arteries here, as usual, spring from a common stem. From each arch thus formed straight vessels, called vasa recta duodeni, pass to the anterior and posterior surfaces respectively of the 2nd, 3rd, and 4th parts of the duodenum. The duodeno-jejunal junction supplied by a large branching vessel.
4. The fine network of arteries that pervades the pancreas to be derived from: the common hepatic artery, the gastro-duodenal artery, the pancreatico-duodenal arches, the splenic artery, and also from the superior mesenteric artery.
5. The arc of Riolan, an occasional artery on the posterior abdominal wall, which connects the superior mesenteric artery to a branch of the inferior mesenteric artery.

For more details, consult Michels, N. A. (1962) The anatomic variations of the arterial pancreaticoduodenal arcades, etc. *J. Int. Coll. Surgeons, 37:* 13; Woodbourne, R. T., and Olsen, L. L. (1951). The arteries of the pancreas. *Anat. Rec., 111:* 255.

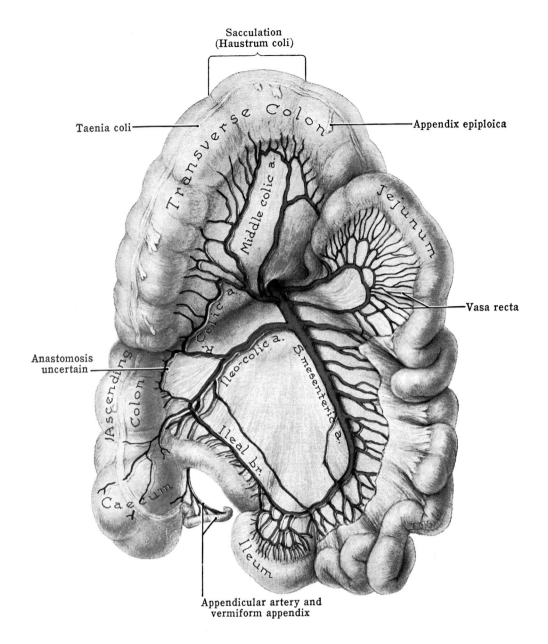

Sacculation
(Haustrum coli)

Taenia coli —

Transverse Colon

Middle colic a.

Jejunum

Appendix epiploica

R. Colic a.

Anastomosis
uncertain —

Ascending Colon

Ileo-colic a.

S. mesenteric a.

Vasa recta

Caecum

Ileal br.

Ileum

Appendicular artery and
vermiform appendix

2-49 SUPERIOR MESENTERIC ARTERY

The peritoneum is in part stripped off.

Observe:

1. The superior mesenteric artery ending by anastomosing with one of its own branches, the ileal
branch of the ileo-colic artery.
2. Its branches:
   a. From its left side, 12 or more jejunal and ileal
   branches. These anastomose to form arcades
   from which vasa recta pass to the small gut.
   b. From its right side, the middle colic, the ileo-
   colic, and commonly, but not here, an independent right colic artery. These anastomose to
   form a marginal artery (labeled in Fig. 2-50)

from which vasa recta pass to the large gut.
   c. The two inferior pancreatico-duodenal arteries
   (not in view, but seen in Fig. 2-87) arise from
   the main artery either directly or in conjunction with the first jejunal branch.
3. Teniae coli, sacculations, and appendices epiploicae which distinguish the large gut from the
smooth walled small gut.

For details, consult Basmajian, J. V. (1954) The marginal anastomoses of the arteries to the large intestine. *Surg. Gynec. Obst., 99:* 614; Basmajian, J. V.
(1955) The main arteries of the large intestine. *Surg.
Gynec. Obst., 101:* 585.

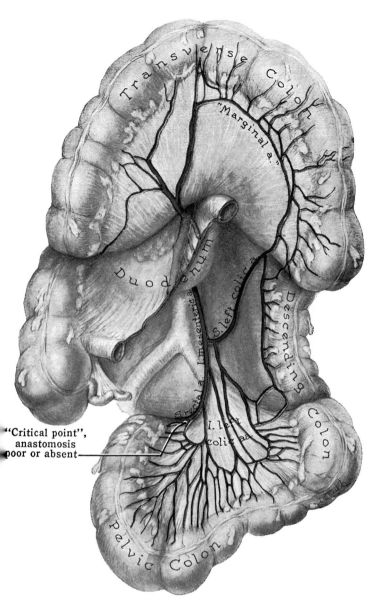

"Critical point", anastomosis poor or absent

2-50 INFERIOR MESENTERIC ARTERY

The mesentery has been cut at its root and discarded with the jejunum and ileum. (*Pelvic Colon* = sigmoid colon.)

Observe:

1. The inferior mesenteric artery arising behind the duodenum, 4 cm above the bifurcation of the aorta. On crossing the left common iliac artery, it becomes the superior rectal (hemorrhoidal) artery.
2. Its branches:
   a. A single (superior) left colic artery and
   b. Several sigmoid arteries (inferior left colic arteries) springing from its left side. In this specimen the two lowest sigmoid arteries spring from the superior rectal artey. The point at which the last artery to the colon leaves the artery to the rectum is known as the "critical point of Sudeck."

For details see Michels, N. A., Siddharth, P., Kornblith, P. L., and Parke, W. W. (1963) The variant blood supply to the small and large intestine—based on four hundred dissections, etc. *J. Int. Coll. Surgeons,* 39: 127; *also* (1965) The variant blood supply to the descending colon, rectosigmoid and rectum. *Dis. Colon Rectum,* 8: 251.

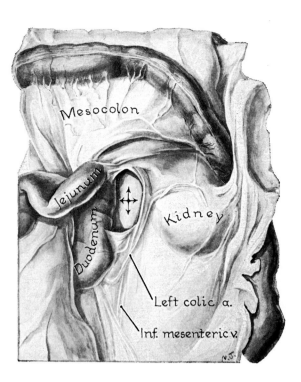

2-51 DUODENAL RECESSES

These occasional peritoneal fossae may pass either upward or downward or to the right or to left. The lower fold is bloodless. When present, the left fold does, and the upper fold may, contain the inferior mesenteric vein. The right recess is retroduodenal.

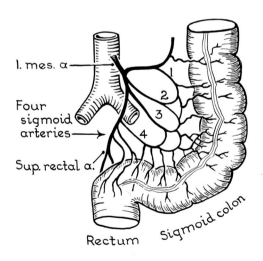

2-52 SIGMOID ARTERIES

A common pattern of the two to four sigmoid arteries.

For details see Goligher, J. C. (1949) The blood supply to the sigmoid colon and rectum. *Brit. J. Surg.,* 37: 157.

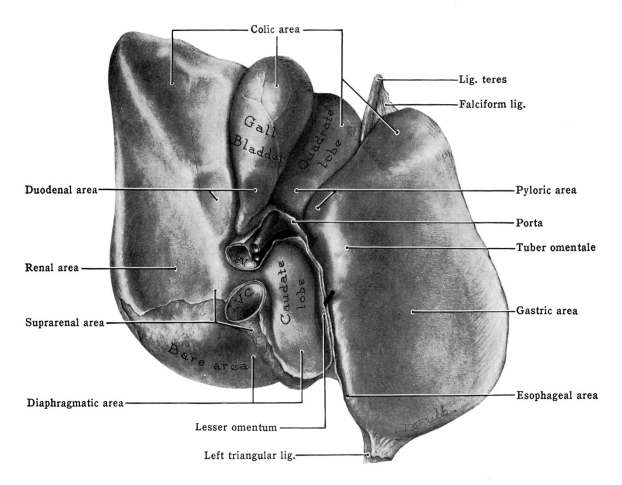

Colic area

Lig. teres

Falciform lig.

Gall Bladder

Quadrate lobe

Duodenal area

Pyloric area

Porta

Tuber omentale

P.V.

Renal area

Caudate lobe

I.V.C.

Suprarenal area

Gastric area

Bare area

Diaphragmatic area

Esophageal area

Lesser omentum

Left triangular lig.

2-53   THE INFERIOR AND POSTERIOR SURFACES OF THE LIVER

You are standing on the right side of the cadaver and facing the head. The attachments of the liver are divided and the sharp, inferior border is raised.

Observe:

1. The visceral areas: (a) for esophagus, stomach, pylorus, and duodenum; (b) for transverse colon; and (c) for right kidney and right suprarenal gland. The gallbladder rests on the transverse colon and on the duodenum (Fig. 2-28).

2. The posterior surface comprising: (a) the bare area, occupied on its left by the inferior vena cava, (b) the caudate lobe, and (c) the groove for the esophagus.

3. The caudate lobe separated from the quadrate lobe by the porta, and joined to the right lobe by the caudate process (not labeled) which is squeezed between the inferior vena cava and the portal vein.

4. The cut edge of the peritoneum. At the right end of the bare area the right triangular ligament (not labeled) bifurcates into the upper and the lower layer of the coronary ligament. The lower layer crosses the renal and suprarenal areas, and, after passing in front of the inferior vena cava, turns cranially as the left layer, or base, of the coronary ligament. Followed to the left this layer of peritoneum forms the upper limit of the upper recess and then turns caudally as the posterior layer of the lesser omentum. This omentum is attached to the fissure for the ligamentum venosum and to the porta. In this specimen it contains an accessory hepatic artery, a branch of the left gastric artery.

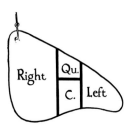

A. The four lobes of the liver: right, quadrate, caudate, and left.

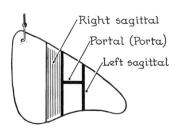

Right sagittal

Portal (Porta)

Left sagittal

B. The H-shaped deep fissures and wide sulci defining the lobes.

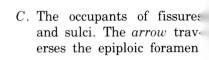

Gall bladder

Bile passages
Hepatic art.
Portal vein.

Umbilical vein (Lig. teres)

Ductus venosus (Lig. venosum)

C.   Inf. vena cava

C. The occupants of fissures and sulci. The *arrow* traverses the epiploic foramen

2-54   DIAGRAMS OF THE LIVER

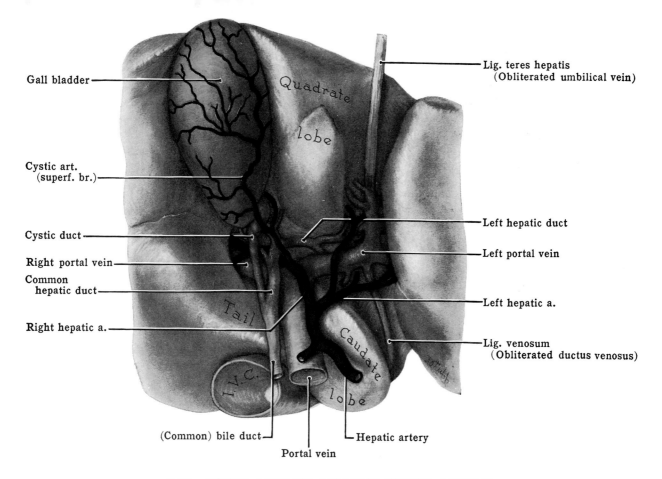

Gall bladder

Cystic art.
(superf. br.)

Cystic duct

Right portal vein

Common
hepatic duct

Right hepatic a.

Quadrate
lobe

Tail

Caudate
lobe

I.V.C.

Lig. teres hepatis
(Obliterated umbilical vein)

Left hepatic duct

Left portal vein

Left hepatic a.

Lig. venosum
(Obliterated ductus venosus)

(Common) bile duct

Portal vein

Hepatic artery

2-55 PORTA HEPATIS AND THE CYSTIC ARTERY

Observe:
1. The tail of the caudate lobe forming the roof of the epiploic foramen, and lying between the upper end of the portal vein, and the inferior vena cava.
2. The relation of structures as they ascend to the porta—duct to the right, artery to the left, vein behind.
3. The order of structures at the porta—duct, artery, vein—from before backward.
4. The left portal vein and left hepatic artery supplying the quadrate and caudate lobes en route to the

left lobe, and accompanied by tributaries of the left hepatic duct.
5. The ligamentum teres hepatis passing to the left portal vein, and the ligamentum venosum arising opposite it and ascending to the inferior vena cava (as in Fig. 2-61).
6. The cystic artery springing from the right hepatic artery and dividing into a superficial and a deep branch which arborize on the respective surfaces of the bladder.
7. The cystic duct, sinuous at its origin.

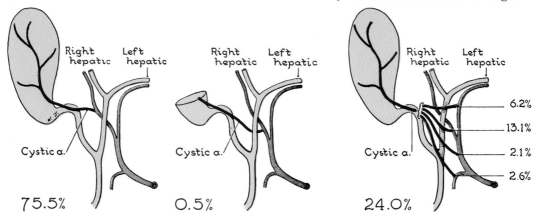

Right
hepatic

Left
hepatic

Cystic a.

75.5%

Right
hepatic

Left
hepatic

Cystic a.

0.5%

Right
hepatic

Left
hepatic

Cystic a.

6.2%

13.1%

2.1%

2.6%

24.0%

2-56 VARIATIONS IN THE ORIGIN AND COURSE OF THE CYSTIC ARTERY

The cystic artery usually arises from the right hepatic artery in the angle between the common hepatic duct and the cystic duct, and has no occasion to cross the common hepatic duct, as in Figures 2-59 and 2-78. When, however, it arises on the left of the bile passages, it almost always crosses anterior to the

passages, as in Figure 2-55, and only rarely does it cross behind.

Diagrams are based on 580 cases by Daseler, E. H., Anson, B. J., Hambley, W. C., and Reimann, A. F. (1947) The cystic artery and constituents of the hepatic pedicle. *Surg. Gynec. Obst., 85:* 47.

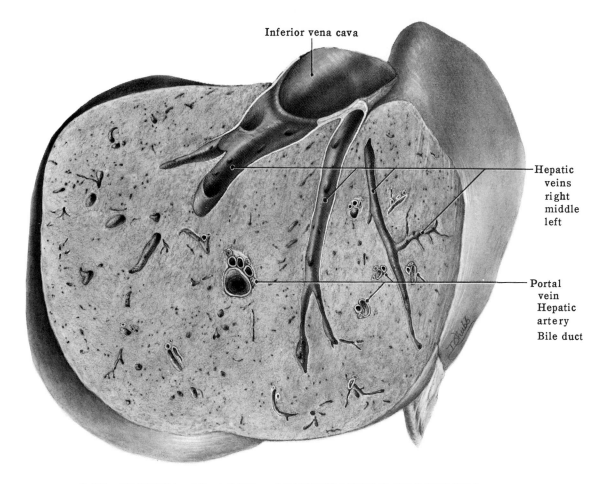

Inferior vena cava

Hepatic
veins
right
middle
left

Portal
vein
Hepatic
artery
Bile duct

2-57 SECTION OF LIVER, APPROXIMATELY HORIZONTAL

Observe:
1. Perivascular fibrous capsules (Glisson's capsule), each containing a branch (or branches) of the portal vein, hepatic artery, bile ductules, and lymph vessels cut across throughout the section (*cf.* Fig. 2-62).

2. Interdigitating with these are branches of the three main hepatic veins which, unaccompanied and having no capsules, converge fanwise on the inferior vena cava (*cf.* Fig. 2-65).

2-58 PORTA-CAVAL SYSTEM

In this diagram, portal tributaries are *blue*, systemic tributaries are *striped*, and communicating veins are *black*. In portal hypertension (as in hepatic cirrhosis) the tiny anastomotic veins become varicose and may rupture. The sites of anastomosis shown are:

a. Between esophageal veins (when dilated these are esophageal varices).
b. Between rectal veins (when dilated these are hemorrhoids).
c. Para-umbilical veins (may produce the "Caput Medusae").
d. Twigs of colic and splenic veins with renal, bare area of liver.

For details see Edwards, E. A. (1951) Functional anatomy of the porta-systemic (portal-caval) communications. *Arch. Int. Med.*, *88:* 137.

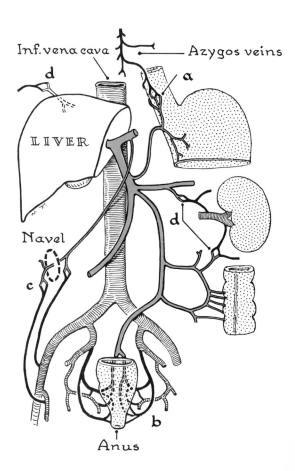

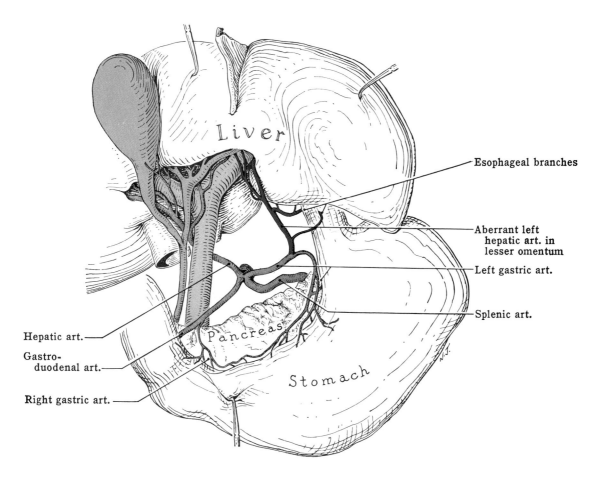

Esophageal branches

Aberrant left hepatic art. in lesser omentum

Left gastric art.

Splenic art.

Hepatic art.

Gastro-duodenal art.

Right gastric art.

A. ABERRANT LEFT HEPATIC ARTERY

The left hepatic artery was entirely replaced by a branch of the left gastric artery, as in this specimen, in 11.5 per cent of 200 cadavers, and in another 11.5 per cent it was partially replaced.

See Michels, N. A. (1966) Newer anatomy of the liver and its variant blood supply and collateral circulation. *Am. J. Surg., 112:* 337.

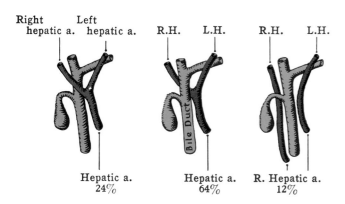

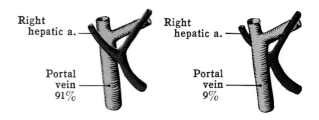

B. RIGHT HEPATIC ARTERY VARIANTS

In a study of 165 cadavers, three patterns were seen: 24 per cent, right hepatic artery crossed ventral to bile passages; 64 per cent, right hepatic artery crossed dorsal to bile passages; 12 per cent, aberrant artery arising from the superior mesenteric artery.

C. RIGHT HEPATIC ARTERY AND PORTAL VEIN

In 165 cadavers, the artery crossed ventral to the portal vein in 91 per cent and dorsal in 9 per cent.

2-59 VARIATIONS IN THE HEPATIC ARTERIES

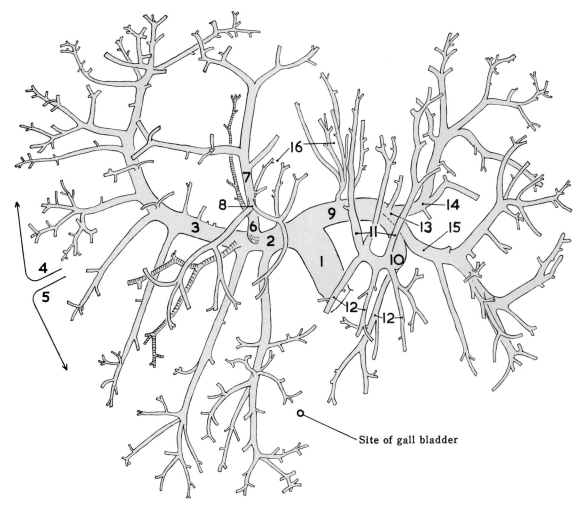

2-60   PORTAL VEIN AND ITS BRANCHES, LEFT ANTERO-INFERIOR VIEW

Figures 2-60 and 2-62 were traced from a projected Kodachrome of a corrosion specimen. Perhaps it is simplest to regard the portal vein as branching and rebranching dichotomously. If branches 10 and 15 were split back to where 14 arises (*broken line*), branch 13 would be created. This figure of the portal vein would then serve equally as a plan of the hepatic artery and of the bile passages, as Figure 2-62 shows. The portal vein (*1*) divides into right (*2*) and left (*9*) portal veins and caudate veins (*16*).

Each portal vein divides into two segmental veins:
    Right (*2*): posterior (*3*) and anterior (*6*).
    Left (*9*): lateral (*13*) and medial (*10*).

Each segmental vein divides into two area veins named:

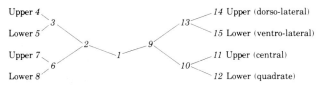

Upper *4*		*14* Upper (dorso-lateral)
	3 *13*	
Lower *5*		*15* Lower (ventro-lateral)
	2 *1* *9*	
Upper *7*		*11* Upper (central)
	6 *10*	
Lower *8*		*12* Lower (quadrate)

See Doehner, G. A. (1968) The hepatic venous system. *Radiology*, 90: 1119–1123.

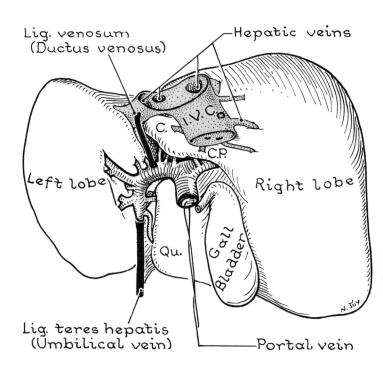

2-61   VEINS OF THE LIVER, POSTERIOR VIEW

Note the branches of the portal vein (*yellow*) entering the liver and 3 large and several small hepatic veins (*blue*) leaving it to join the inferior vena cava. (*C.*, caudate lobe; *C. P.*, caudate process.)

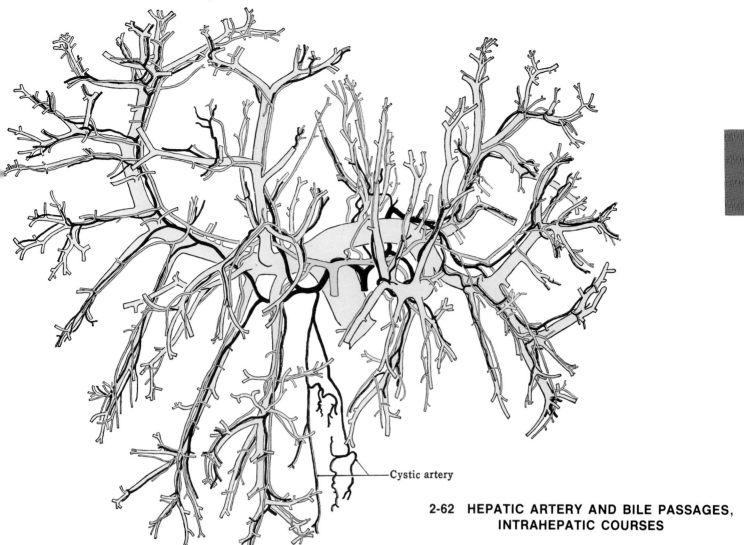

—Cystic artery

2-62 HEPATIC ARTERY AND BILE PASSAGES, INTRAHEPATIC COURSES

2-63 PORTAL STRUCTURES, INTRAHEPATIC COURSES, POSTERIOR VIEW

This is the same specimen as Figure 2-62. Examine with a magnifying lens.

For details see Hjortsjo, C. H. (1951) The topography of the intrahepatic duct systems (and of the portal vein). *Acta Anat., 11:* 599; Hjortsjo, C. H. (1956) The intrahepatic ramifications of the portal vein. *Lunds Universitets Arsskrift, 52:* 20; Healey, J. E., Schroy, P. C., and Sorensen, R. J. (1953) The intrahepatic artery in man. *J. Int. Coll. Surgeons, 20:* 133; Healey, J. E., and Schroy, P. C. (1953) Anatomy of the biliary ducts within the human liver. *Arch. Surg. (Chicago), 66:* 599.

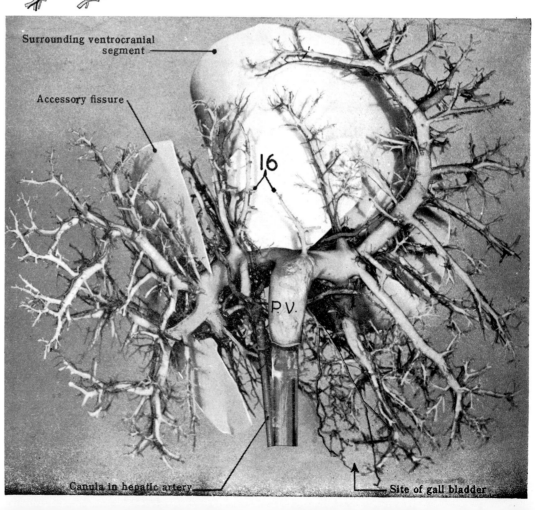

Surrounding ventrocranial segment

Accessory fissure

16

P.V.

Canula in hepatic artery

Site of gall bladder

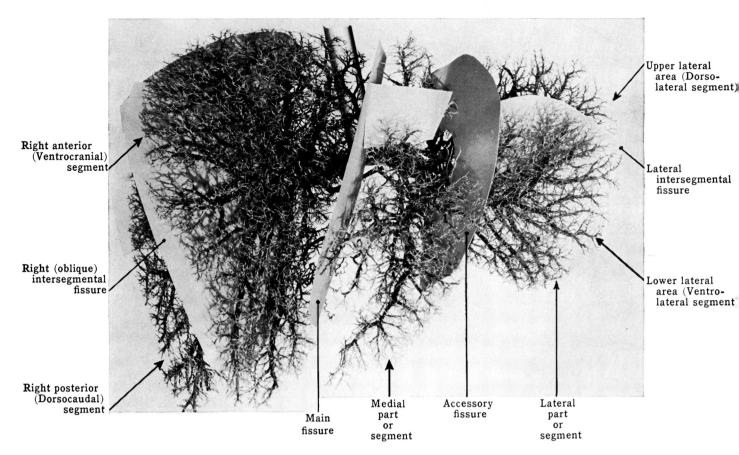

Right anterior
(Ventrocranial)
segment

Right (oblique)
intersegmental
fissure

Right posterior
(Dorsocaudal)
segment

Upper lateral
area (Dorso-
lateral segment)

Lateral
intersegmental
fissure

Lower lateral
area (Ventro-
lateral segment)

Main
fissure

Medial
part
or
segment

Accessory
fissure

Lateral
part
or
segment

2-64 SUBDIVISIONS OF THE LIVER, ANTERIOR VIEW

Corrosion specimen of the portal structures.

The main fissure divides the liver into almost equal halves, "the right and left portal lobes," served by right and left portal veins, hepatic arteries, and bile passages. Its surface markings are the right sagittal sulci and an imaginary line, 3 or 4 cm from the falciform ligamentum.

The right half is divided into an anterior and a posterior segment; the left half is divided into a medial and a lateral

segment by the accessory fissure, whose surface markings are the left sagittal fissures and the attachment of the falciform ligamentum.

Each of these 4 segments is subdivisible into an upper and a lower area (segment), as revealed by the inserted slips of paper and the branches seen in Figures 2-60 and 2-62. Notable are the upper and lower lateral areas.

2-65 LARGE HEPATIC VEINS, ANTERIOR VIEW

These 3 veins are intersegmental, as are pulmonary veins. They occupy the main, the right, and the accessory and lateral intersegmental fissures. (The portal vein is injected white.)

Right hepatic vein

I.V.C.

Middle hepatic vein

Left hepatic vein

Rod supporting portal

Rod supporting inf. vena cava

Site of gall bladder

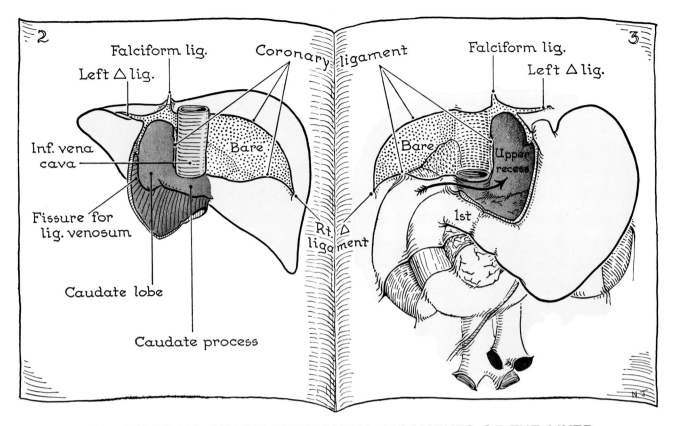

2-66 DIAGRAM OF THE PERITONEAL LIGAMENTS OF THE LIVER

The attachments of the liver are cut through and the liver is turned to the right side of the cadaver, as you would turn the page of a book. Hence, the posterior aspect of the liver is shown on page 2 and its posterior relations on page 3.

Note:1. The *arrow* passing from the greater sac of peritoneum (*yellow*) into the lesser sac (*orange*).
2. The upper recess of the lesser sac, or omental bursa, occupies the median plane of the body. Dorsally lies the descending thoracic aorta, diaphragm intervening (Fig. 2-67). Ventrally is the lesser omentum. On the *left* is esophagus, and on the *right* is inferior vena cava, both of which lie bare on the diaphragm.
3. The inferior vena cava occupies the left or medial limit of the bare area of the liver.
4. The bare area is triangular; hence, the so-called coronary ligament, which surrounds it, is not a corona, but is three-sided. Its left side or base is between the inferior vena cava and the caudate lobe; it is palpated when the

right index is inserted into the upper recess.
5. Its apex is at the right triangular ligament, where the cranial and caudal layers of the coronary ligament meet.
6. The lower or caudal layer of the ligament is reflected from the liver onto: diaphragm, right kidney, and right suprarenal gland. It is called the hepato-renal ligament by the surgeon. Followed medially this layer crosses the inferior vena cava at the epiploic foramen, and turning cranially becomes the left or basal layer of the ligament (Fig. 2-73).
7. The lesser omentum is divided close to the stomach. It is not labeled, nor is the kidney, nor the suprarenal gland.

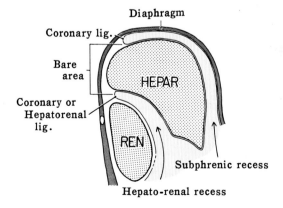

2-67 BARE AREA OF LIVER

A paramedian section through diaphragm, liver, and right kidney to show the bare area of liver to be situated between the dorsal ends of 2 peritoneal recesses or pouches.

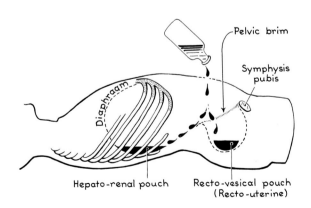

2-68 PERITONEAL CAVITY

This diagram illustrates the location of the two lowest (most dorsal) parts of the peritoneal cavity, when the subject lies recumbent.

2-69 STOMACH, FRONT VIEW

The cardiac notch and the angular notch (incisura angularis) are here well marked. The pyloric antrum is here not demarcated from the tubular pyloric canal.

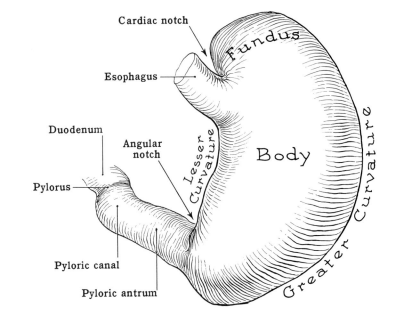

2-70 MUCOUS MEMBRANE OF THE STOMACH

Along the lesser curvature several longitudinal ridges extend from esophagus to pylorus; elsewhere the mucous membrane is rugose when the stomach is empty.

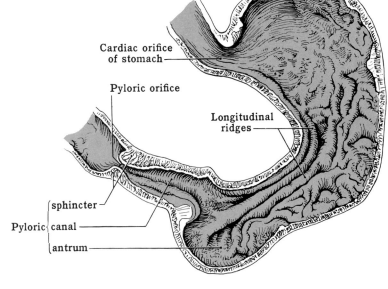

2-71 MUSCULAR COAT OF THE STOMACH, FROM WITHIN

The stomach is cut along the greater curvature and the loosely adherent mucous and submucous coats are dissected off.

The fibers of the innermost or oblique muscle layer form ∩-shaped loops that extend over the fundus and down both surfaces of the stomach as far as the antrum. Their medial limit is at the cardiac notch; hence the fibers of the middle or circular layer become the innermost layer along the lesser curvature and in the pyloric region. At the pylorus they are thickened to form the pyloric sphincter. They are present everywhere except at the fundus. The fibers of the outermost or longitudinal layer (not shown here) are best marked along the curvatures.

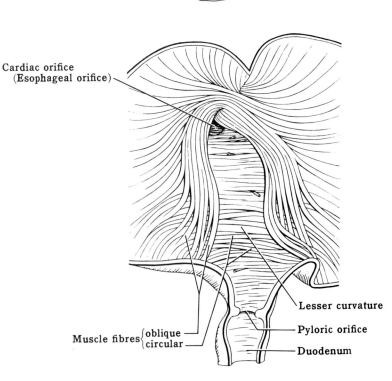

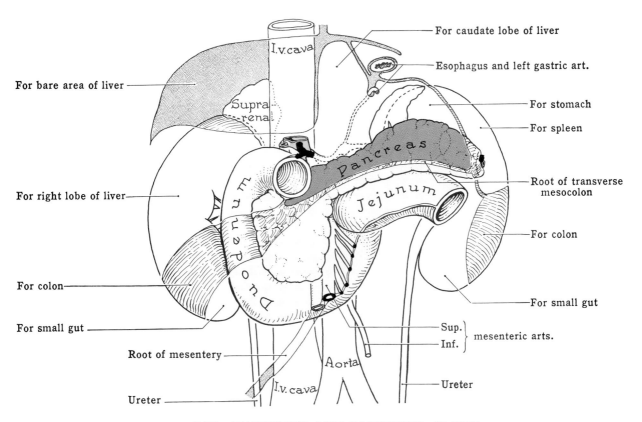

For caudate lobe of liver

I.v.cava

For bare area of liver

Supra-renal

Esophagus and left gastric art.

For stomach

For spleen

Pancreas

Duodenum

Jejunum

Root of transverse mesocolon

For right lobe of liver

For colon

For colon

For small gut

For small gut

Root of mesentery

Sup.
Inf. } mesenteric arts.

Aorta

Ureter

I.v. cava

Ureter

Ureter

2-72 DUODENUM AND PANCREAS, *IN SITU*

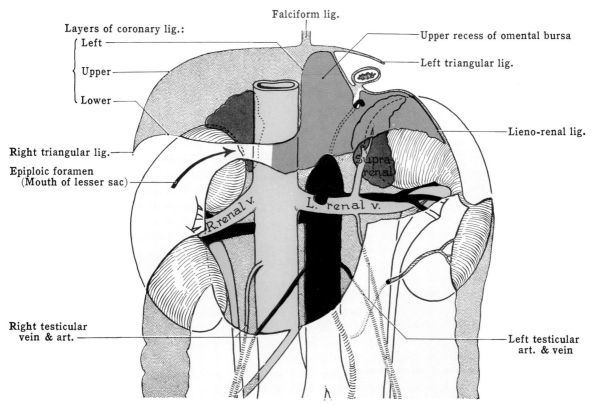

Falciform lig.

Layers of coronary lig.:
{ Left

Upper recess of omental bursa

Upper

Left triangular lig.

Lower

Lieno-renal lig.

Supra-renal

Right triangular lig.

Epiploic foramen
(Mouth of lesser sac)

R. renal v.

L. renal v.

Right testicular
vein & art.

Left testicular
art. & vein

2-73 DUODENUM AND PANCREAS, REMOVED

POSTERIOR ABDOMINAL VISCERA AND THEIR VENTRAL RELATIONS

Observe:

1. The peritoneal covering (*yellow*) of pancreas and duodenum.
2. The colic area of right kidney, second part of duodenum, and head of pancreas; the line of attachment of the transverse mesocolon to the body and tail of the pancreas and the colic area of the left kidney.

3. The ventral relations of the kidneys and suprarenal glands.
4. The right suprarenal gland at the epiploic foramen.
5. The three parts of the coronary ligament attached to the diaphragm except where inferior vena cava, suprarenal gland, and kidney intervene.

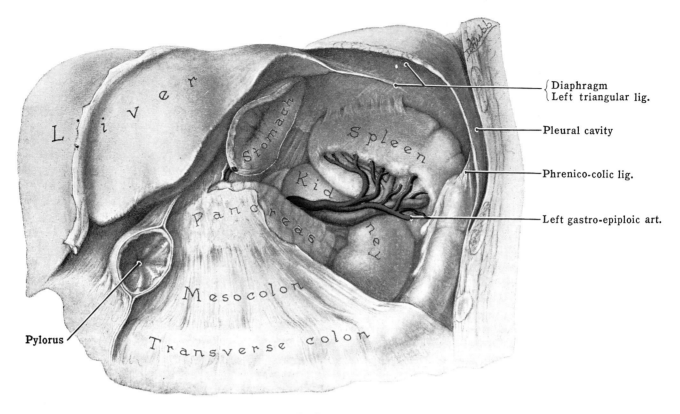

Diaphragm
Left triangular lig.

Pleural cavity

Phrenico-colic lig.

Left gastro-epiploic art.

Liver

Stomach

Spleen

Kidney

Pancreas

Mesocolon

Transverse colon

Pylorus

2-74 STOMACH BED

The stomach is excised. The peritoneum of the omental bursa, or lesser sac, covering the stomach bed is largely removed; so is the peritoneum of the greater sac covering the lower part of the kidney and pancreas.

The pancreas is unusually short; the adhesions binding the spleen to Diaphragm are pathological but not unusual.

Unlabeled are: (a) the branch of the posterior vagal trunk that descends on the right side of the left gastric artery to the celiac plexus (Fig. 2-117); (b) the left suprarenal gland; (c) three short gastric branches of the splenic artery, cut short (see Fig. 2-43); (d) the tuber omentale of the liver and that of the pancreas, which fit into the lesser curvature of the stomach; (e) the lesser omentum attached to the upper border of the pylorus and the greater omentum to the lower border.

Note the pleural cavity separating the spleen and Diaphragm from the thoracic wall.

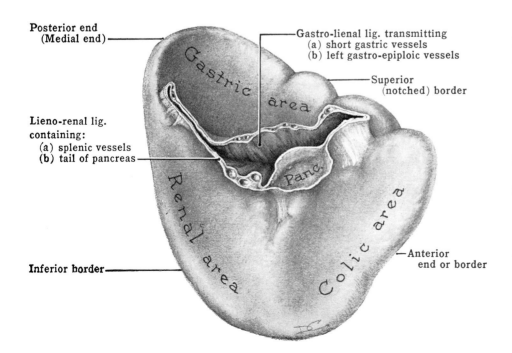

Posterior end
(Medial end)

Gastro-lienal lig. transmitting
(a) short gastric vessels
(b) left gastro-epiploic vessels

Superior
(notched) border

Lieno-renal lig.
containing:
(a) splenic vessels
(b) tail of pancreas

Gastric area

Panc.

Renal area

Colic area

Anterior
end or border

Inferior border

For orientation see Figure 2-74.

Observe:

1. A "circumferential border" comprising the inferior, superior, and anterior borders, and separating the visceral surface from the diaphragmatic surface.

2. The notches characteristic of the superior border.

3. The left limit of the omental bursa at the hilus of the spleen, between the lieno-renal and gastrolienal ligaments.

4. The spleen taking the impressions of the structures in contact with it. The large colic area seen here is reduced in Figure 2-77 — presumably in the one case the colon was full (of gas), and in the other it was empty.

2-75 SPLEEN OR LIEN, VISCERAL SURFACE

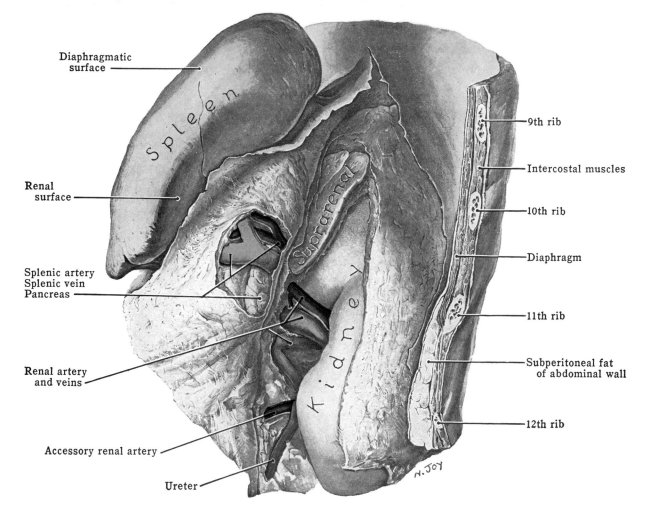

Diaphragmatic surface

Renal surface

Splenic artery
Splenic vein
Pancreas

Renal artery
and veins

Accessory renal artery

Ureter

9th rib

Intercostal muscles

10th rib

Diaphragm

11th rib

Subperitoneal fat
of abdominal wall

12th rib

2-76   REFLEXION OF THE SPLEEN: EXPOSURE OF THE LEFT KIDNEY AND SUPRARENAL GLAND, FRONT VIEW

The spleen and the lieno-renal ligament are turned forward and to the subject's right, like opening a book, taking the splenic vessels and the tail of the pancreas with them. Part of the fatty capsule of the kidney is cut away. Note that this exposure is, in effect, a restoration to the embryological state.

Observe:

1. The exposed left kidney, renal vessels, and ureter, also the left suprarenal gland, separated from the kidney by a thin layer of fat.

2. When *in situ*, the splenic vessels lie ventral to the renal vessels (Fig. 2-74), but two areolar membranes, closely applied to each other, and some fatty capsule are seen to intervene.

3. The proximity to each other of the splenic vein and the left renal vein, a relationship used in the surgical relief of portal hypertension through a splenorenal shunt.

4. The subperitoneal fat of the abdominal wall stops on reaching the diaphragm.

Accessory spleens resemble lymph nodes, about 1 cm in diameter, but they are usually covered with peritoneum, as is the spleen itself. They lie along the course of the splenic artery or its gastro-epiploic branch, but they may be elsewhere. The most common location is at or near the hilus of the spleen, but 1 in 6 are partially (or wholly) embedded in the tail of the pancreas. They occur in about 10 per cent of people.

For details see Curtis, G. M., and Movitz, D. (1946) The surgical significance of the accessory spleen. *Ann. Surg., 123:* 276; Halpert, B., and Gyorkey, F. (1959) Lesions observed in accessory spleens of 311 patients. *Am. J. Clin. Path., 32:* 165.

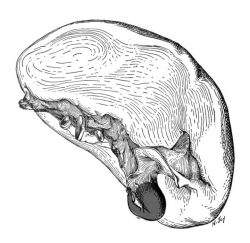

2-77   AN ACCESSORY SPLEEN

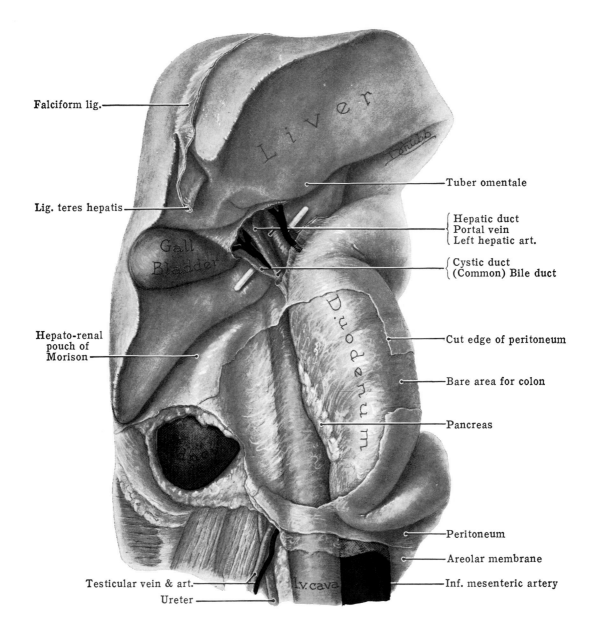

Falciform lig.

Lig. teres hepatis

Hepato-renal pouch of Morison

Testicular vein & art.

Ureter

Liver

Gall Bladder

Duodenum

Tuber omentale

{ Hepatic duct
Portal vein
Left hepatic art.

{ Cystic duct
(Common) Bile duct

Cut edge of peritoneum

Bare area for colon

Pancreas

Peritoneum

Areolar membrane

Inf. mesenteric artery

I.v. cava

2-78 EXPOSURE OF THE (COMMON) BILE DUCT, POSTERIOR ASPECT—I

A white rod is passed through the epiploic foramen. The lesser omentum is removed. The transverse colon is separated from the front of the descending part of the duodenum and thrown down; the peritoneum is cut along the right or convex border of the duodenum, and this part of the duodenum is swung forward like a door on a hinge.

Note:
1. The space opened up is virtually a bursal space, comparable to the retropubic space where two smooth areolar membranes are applied to each other. Here one membrane covers the posterior aspect of the second part of the duodenum and the head of the pancreas; the other covers the aorta, inferior vena cava, renal vessels, and perinephric fat.
2. To find the epiploic foramen either: (a) follow the liver, at the upper limit of the hepato-renal pouch, medially to the caudate process (tail of the caudate lobe), which forms the roof of the foramen and which lies in front of the inferior vena cava, which forms the posterior wall; or (b) follow the gallbladder to the cystic duct, which occupies the free edge of the lesser omentum, which forms the anterior wall of the foramen.
3. Of the three main structures in the anterior wall, the portal vein is posterior, the hepatic artery ascends from the left, and the bile passages descend to the right (Fig. 2-55).
4. In this specimen the right hepatic artery springs from the superior mesenteric artery, a common variant as shown in Figure 2-59.

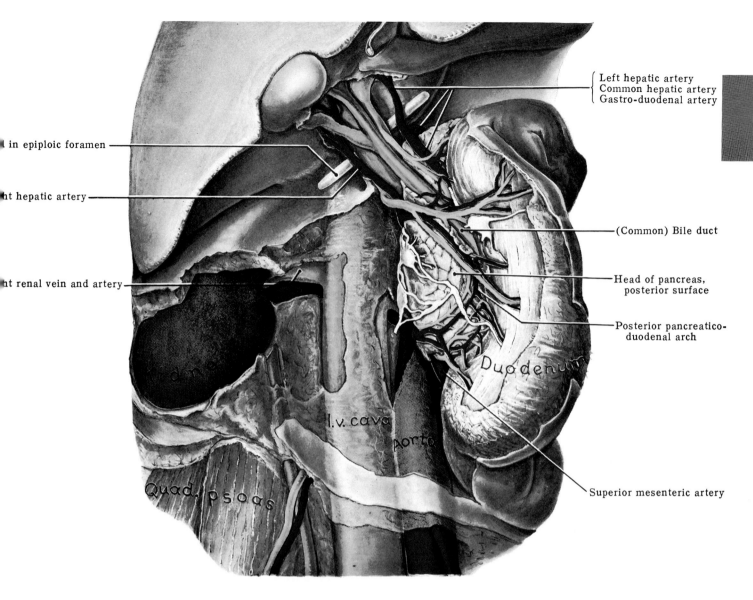

in epiploic foramen

ht hepatic artery

ht renal vein and artery

Left hepatic artery
Common hepatic artery
Gastro-duodenal artery

(Common) Bile duct

Head of pancreas,
posterior surface

Posterior pancreatico-
duodenal arch

Superior mesenteric artery

l. v. cava

Aorta

Duodenum

Quad. psoas

2-79 EXPOSURE OF THE (COMMON) BILE DUCT, POSTERIOR ASPECT—II

In this further dissection of Figure 2-78, the duodenum is swung still further forward and to the left, taking the head of the pancreas with it. In effect, the epiploic foramen has been enlarged caudally. The areolar membrane covering these two organs is largely removed; that covering the great vessels is in part removed.

Observe:

1. The bile duct descending in a groove on the head of the pancreas, and a tongue of that head (reflected) lying behind the end of the duct.

2. The bile duct ends at the level of the hilus of the kidney, the ureter being at this level.

3. The very close relationship of the inferior vena cava to the portal vein; they are separated by the epiploic foramen. A portacaval shunt to divert the portal circulation into the caval system may be done here by an end-to-side anastomosis.

4. Vasa recta, accompanied by veins and lymph vessels, passing from the posterior pancreatico-duodenal arch to the duodenum.

5. Of the two posterior pancreatico-duodenal arteries that form the posterior arch, the inferior arises from the superior mesenteric artery, and the superior here from the right hepatic artery, but usually from the gastro-duodenal artery, as in Figure 2-48.

6. The posterior superior pancreatico-duodenal vein ending in the portal vein.

7. The right renal vein is short; the right renal artery is long—it passes behind the inferior vena cava.

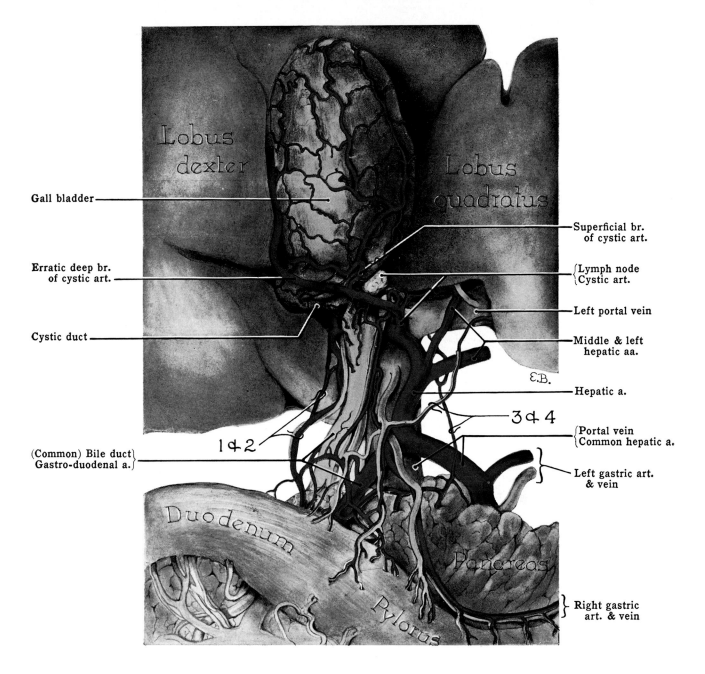

Gall bladder

Erratic deep br. of cystic art.

Cystic duct

(Common) Bile duct
Gastro-duodenal a.

Lobus dexter

Lobus quadratus

Superficial br. of cystic art.

Lymph node
Cystic art.

Left portal vein

Middle & left hepatic aa.

E.B.

Hepatic a.

3 & 4

Portal vein
Common hepatic a.

Left gastric art. & vein

1 & 2

Duodenum

Pancreas

Pylorus

Right gastric art. & vein

2-80 GALLBLADDER, BILE PASSAGES, AND RELATED BLOOD VESSELS

The liver is turned up and the duodenum is pulled down.

Observe:

1. The network of arteries on the gallbladder. Most of the smaller arteries lying on a deeper plane than the larger arteries.
2. A large erratic, deep branch of the cystic artery, crossing superficial to the neck of the gallbladder.
3. The cystic lymph node.
4. Many fine sinuous arterial twigs supplying the bile passages and springing from nearby arteries.
5. The right gastric artery, here arising very low—indeed, from the gastro-duodenal artery.
6. The right gastric vein, here ending high in the portal vein.

7. Several, here 5, anastomotic arteries capable of bringing blood from various gastric and pancreatic arteries to the porta hepatis. Four of these have been retracted, one of them passing to the left lobe.
8. Veins (not all shown), accompanying faithfully most arteries.
9. A middle hepatic artery, seen here, passing to the quadrate lobe is common.

For details see Michels, N. A. (1955) *Blood Supply and Anatomy of the Upper Abdominal Organs*. J. B. Lippincott Company, Philadelphia; Parke, W. W., Michels, N. A., and Ghosh, C. M. (1963) Blood supply of the common bile duct. *Surg. Gynec. Obst., 117*: 47.

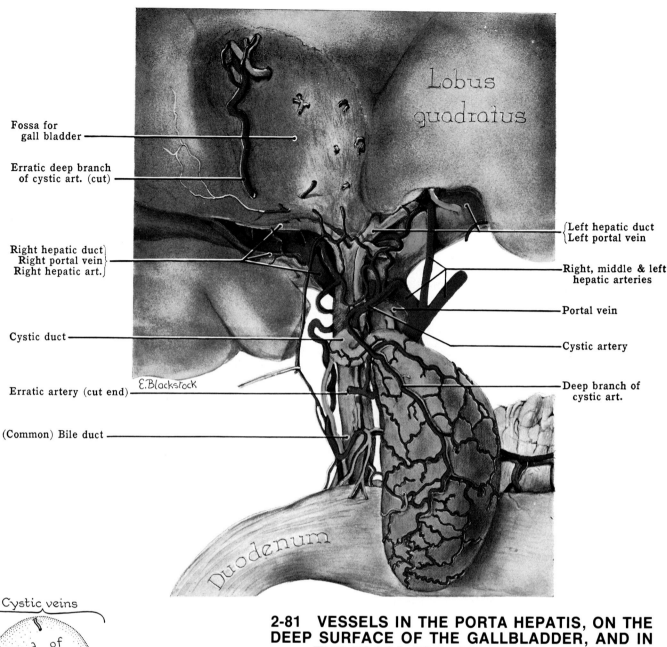

Fossa for gall bladder

Erratic deep branch of cystic art. (cut)

Right hepatic duct
Right portal vein
Right hepatic art.

Cystic duct

Erratic artery (cut end)

E. Blackstock

(Common) Bile duct

Lobus quadratus

Left hepatic duct
Left portal vein

Right, middle & left hepatic arteries

Portal vein

Cystic artery

Deep branch of cystic art.

Duodenum

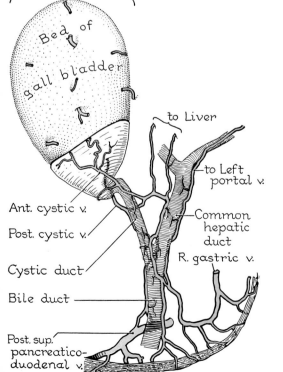

Cystic veins

Bed of gall bladder

to Liver

to Left portal v.

Ant. cystic v.

Post. cystic v.

Common hepatic duct

Cystic duct

R. gastric v.

Bile duct

Post. sup. pancreatico-duodenal v.

2-81 VESSELS IN THE PORTA HEPATIS, ON THE DEEP SURFACE OF THE GALLBLADDER, AND IN THE FOSSA FOR THE GALLBLADDER

The gallbladder is freed from its bed or fossa and turned down.

Observe:

1. The anastomoses seen in the porta hepatis, here involving various branches of the right hepatic artery.
2. A very large anastomotic artery passing behind the origin of the common hepatic duct; several smaller arteries forming a network (rete) in front of the ducts. From the network, twigs passing to the subcapsular arterial plexus.
3. The deep branch of the cystic artery, aided by a branch of the erratic deep cystic artery, ramifying on the deep or attached surface of the gallbladder, anastomosing with twigs of the superficial branch of the cystic artery, and sending twigs into the bed of the gallbladder. (The cut ends of the arterial and venous twigs can be seen.)
4. The erratic artery plunging into the bed.

2-82 VEINS OF THE EXTRAHEPATIC BILE PASSAGES AND GALLBLADDER

Note: The venous twigs draining the passages and the neck of the bladder join veins which, clinging to the passages, connect gastric, duodenal, and pancreatic veins partly to the liver directly and partly via a portal vein. The veins of the fundus and body, here 8, plunge directly into the liver.

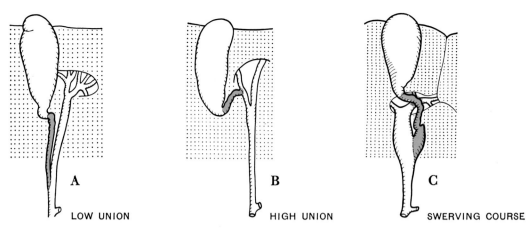

A. VARIATIONS IN THE LENGTH AND COURSE OF THE CYSTIC DUCT

The cystic duct usually lies on the right of the common hepatic duct and joins it just above the first part of the duodenum. Three variations are shown here.

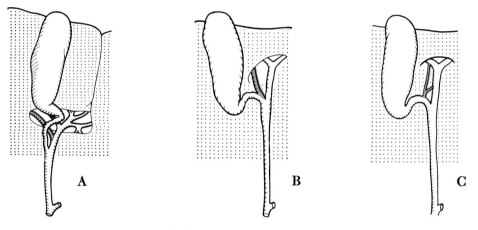

B. VARIETIES OF ACCESSORY HEPATIC DUCTS

"Accessory" ducts are either area or segmental ducts that arise early. They are common and in positions of surgical danger. Of 95 gallbladders and bile passages studied, 7 had accessory ducts. Of these: (A) 4 joined the common hepatic duct near the cystic ducts; (B) 2 joined the cystic duct; and (C) 1 was an anastomosing duct.

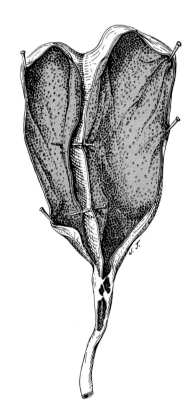

D. DOUBLE GALLBLADDER

See Boyden, E. A. (1962) The accessory gallbladder. *Am. J. Anat., 38:* 177.

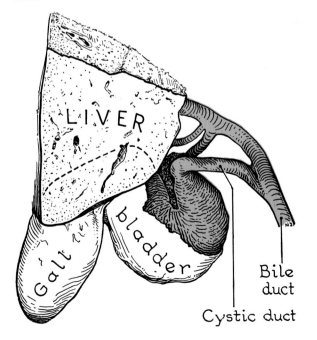

C. FOLDED GALLBLADDER

Radiologically, the body of the gallbladder was found to be folded, or kinked, upon itself in 14.5 per cent of 165 persons. The fundus was folded upon the bladder in 3.5 per cent, as shown here.

See Boyden, E. A. (1935) The "Phrygian cap" in cholecystography: congenital anomaly of the gallbladder. *Am. J. Roentgenol., 33:* 589.

2-83 VARIATIONS IN THE BILE PASSAGES AND GALLBLADDER

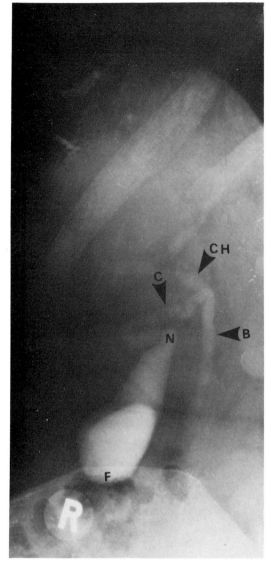

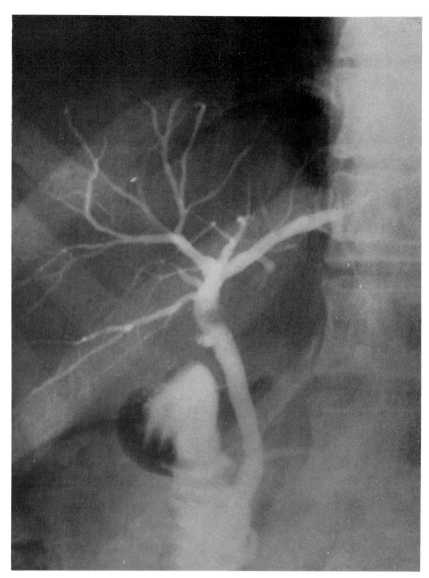

2-84 RADIOGRAPHS OF THE BILIARY PASSAGES

A. On the *left,* a cholecystogram of the gallbladder and biliary passages which are visualized with contrast medium. Observe:

1. The body, fundus (*F*), and neck (*N*) of the gallbladder.
2. The cystic (*C*), common hepatic (*CH*), and bile (*B*) ducts. The spiral valve of the cystic duct gives it a tortuous appearance.

B. On the *right,* following cholecystectomy contrast medium has been injected via a T tube into the bile passages. Consult Figure 2-62 for a dissection of the intrahepatic passages.

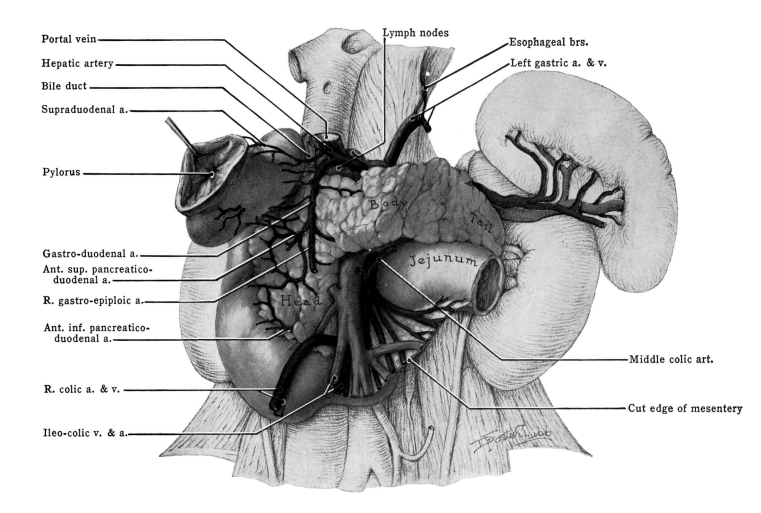

Portal vein —
Hepatic artery —
Bile duct —
Supraduodenal a. —

Pylorus —

Gastro-duodenal a. —
Ant. sup. pancreatico-
duodenal a. —
R. gastro-epiploic a. —
Ant. inf. pancreatico-
duodenal a. —

R. colic a. & v. —

Ileo-colic v. & a. —

Lymph nodes
Esophageal brs.
Left gastric a. & v.

Body
Tail
Head
Jejunum

Middle colic art.
Cut edge of mesentery

2-85   DUODENUM AND PANCREAS *IN SITU*, FRONT VIEW

Observe:

1. The *duodenum* molded around the head of the pancreas. Its 1st or superior part (retracted) overlapping the pancreas and passing backward, upward, and to the right. The remaining parts (2nd, 3rd, and 4th) overlapped by the pancreas. Near the junction of its 3rd and 4th parts the duodenum is crossed by the superior mesenteric vessels, which appear from under cover of the neck of the pancreas, descend in front of the uncinate process, and enter the root of the mesentery. They may, as here, by constricting the duodenum, cause its 1st, 2nd, and 3rd parts to be dilated.

2. The *pancreas*, here very short—its tail usually abuts on the spleen and is blunt (Fig. 2-75). It is arched forward because it crosses the vertebral column and aorta.

3. The *celiac trunk*, which lies behind the upper border of the pancreas, sending (a) the left gastric artery upward on the diaphragm toward the cardiac orifice of the stomach to enter the lesser omentum; (b) the splenic artery to the left and (c) the hepatic artery to the right. The hepatic artery passing onto the front of the portal vein and giving off the gastro-duodenal artery, which descends between the duodenum and pancreas, less than 2.5 cm from the pylorus, and divides into the right gastro-epiploic artery and the superior pancreatico-duodenal artery. Note the single supraduodenal artery and the several retroduodenal arteries supplying the first 2.5 cm of the duodenum.

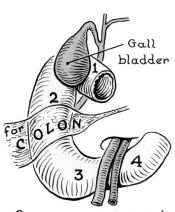

Gall
bladder

1
2
for COLON
3   4

Sup. mesenteric vessels

2-86   VENTRAL CONTACT
RELATIONS OF DUODENUM

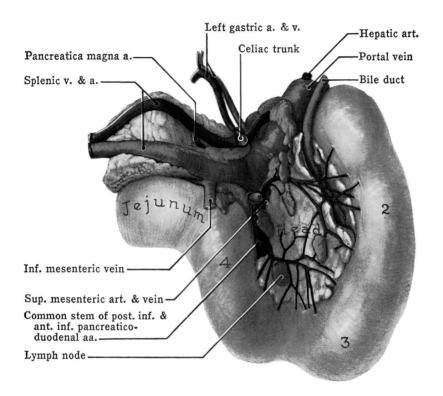

Left gastric a. & v.

Celiac trunk

Pancreatica magna a.

Splenic v. & a.

Hepatic art.

Portal vein

Bile duct

Jejunum

Head

2

Inf. mesenteric vein

Sup. mesenteric art. & vein

Common stem of post. inf. &
ant. inf. pancreatico-
duodenal aa.

Lymph node

4

3

2-87 DUODENUM, PANCREAS, AND BILE DUCT, FROM BEHIND

This is the reverse of the specimen depicted in Figure 2-85.

Observe:

1. Only the end of the 1st or superior part of the duodenum is in view.
2. The bile duct here descending in a long fissure (opened up) in the posterior part of the head of the pancreas; contrast Figure 2-79.
3. Arteries. In the absence of the usual posterior superior pancreatico-duodenal artery, the vasa recta that pass behind the duct springing from an accessory branch of the superior mesenteric artery; compare Figure 2-48.

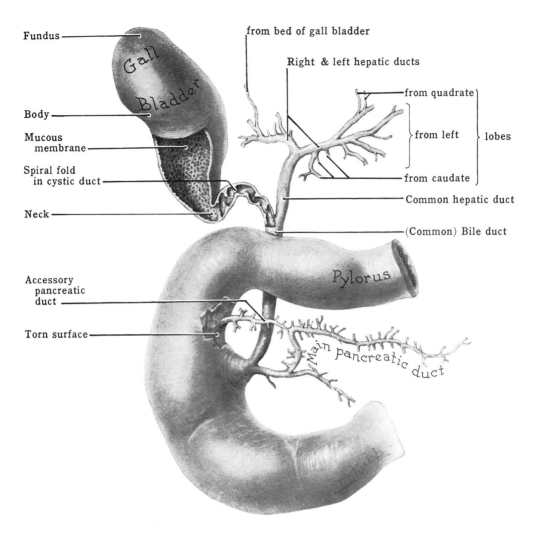

Fundus

Gall Bladder

Body

Mucous membrane

Spiral fold in cystic duct

Neck

Accessory pancreatic duct

Torn surface

from bed of gall bladder

Right & left hepatic ducts

from quadrate

from left — lobes

from caudate

Common hepatic duct

(Common) Bile duct

Pylorus

Main pancreatic duct

2-88 EXTRAHEPATIC BILE PASSAGES AND THE PANCREATIC DUCTS

Note: The right hepatic duct collects from the right lobe of the liver; the left hepatic duct collects from the left, quadrate, and caudate lobes. The common hepatic duct unites with the cystic duct close above the duodenum. The mucous membrane of the gallbladder has a low honeycomb surface. The cystic duct is sinuous; its mucous membrane forms a spiral fold (spiral valve).

The bile duct, after descending behind the 1st part of the duodenum and the accessory pancreatic duct, is joined by the main pancreatic duct; these open on the duodenal papilla (Fig. 2-93).

The main pancreatic duct with its tributaries resembles a herring bone; the common hepatic duct with its tributaries resembles a deciduous tree.

The accessory pancreatic duct retains its anastomosis with the main duct. The pancreas invades the duodenal wall around the accessory duct, and cannot be removed without lacerating the duodenum.

In about 5 per cent of cases the bile and pancreatic ducts open separately on the duodenal papilla.

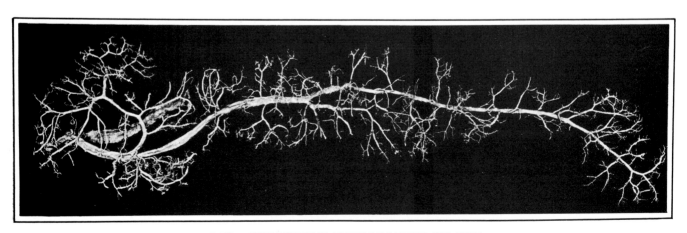

2-89 CORROSION PREPARATION OF THE PANCREATIC DUCTS

The ducts were injected with plastic by retrograde flow, via the bile duct, and then corroded with acid.

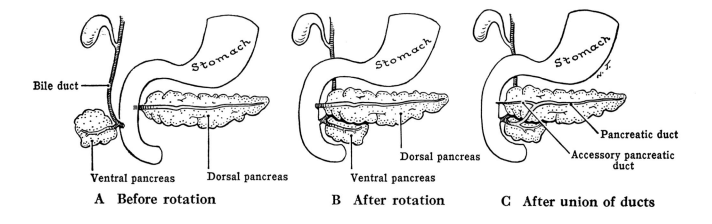

Bile duct

Ventral pancreas   Dorsal pancreas

A Before rotation

Dorsal pancreas

Ventral pancreas

B After rotation

Pancreatic duct

Accessory pancreatic duct

C After union of ducts

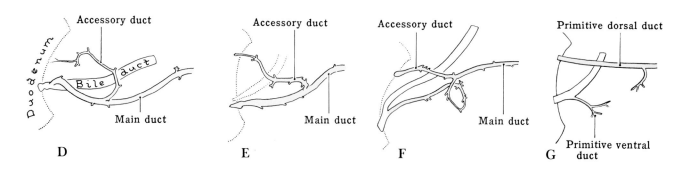

Accessory duct

Bile duct

Main duct

Duodenum

D

Accessory duct

Main duct

E

Accessory duct

Main duct

F

Primitive dorsal duct

Primitive ventral duct

G

2-90 VARIABILITY OF PANCREATIC DUCTS

In seeking a developmental explanation for variability of pancreatic ducts, examine the three upper pictures of stages in the development of the pancreas:

A. A smaller primitive ventral bud arises in common with the bile duct, and a larger primitive dorsal bud arises independently from the duodenum, cranial to this.

B. The 2nd or descending part of the duodenum rotates on its long axis, which brings the ventral bud and the bile duct behind the dorsal bud.

C. A connecting segment unites the dorsal duct to the ventral duct, whereupon the duodenal end of the dorsal duct tends to atrophy and the direction of flow within it is reversed.

Millbourn's radiographic study of 200 cases showed the following:

  44 per cent: The accessory duct has lost its connection with the duodenum (as in *D* above).
  8 per cent: Accessory duct ends blindly at the duodenum.
  9 per cent: A negligibly small opening is retained.
  10 per cent: The accessory duct is large enough to relieve an obstructed main duct (as in *E*).
  20 per cent: The accessory duct could probably substitute for the main duct (as in *F*).
  9 per cent: The primitive dorsal duct persists, usually quite unconnected with the primitive ventral duct (as in G).

For details see Millbourn, E. (1950) On the excretory ducts of the pancreas in man, with special reference to their relations to each other, to the common bile duct and to the duodenum: radiological and anatomical study. *Acta Anat., 9:* 1. Dawson, W., and Langman, J. (1961) An anatomical-radiological study on the pancreatic duct pattern in man. *Anat. Rec., 139:* 59.

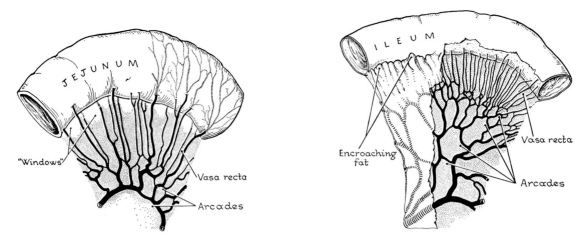

2-91 JEJUNUM CONTRASTED WITH ILEUM

Compare diameter, thickness of wall, number of arterial arcades, long or short vasa recta, presence of translucent (fat-free) areas at the mesenteric border, and fat encroaching on the wall of the gut.

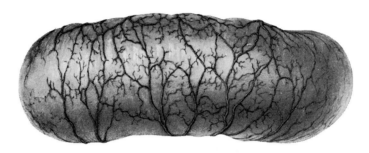

Antimesenteric border of gut

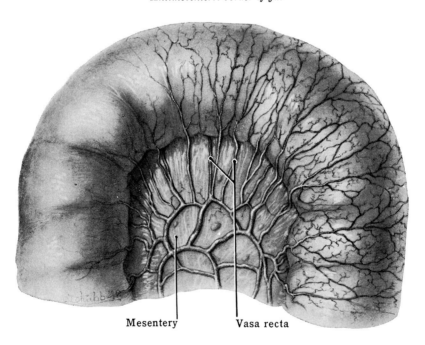

Segment of Intestinum Jejunum with Mesentery and Arteries.

2-92 BLOOD SUPPLY OF THE SMALL INTESTINE

Observe:
1. The series of anastomotic arterial arches in the mesentery.
2. The vasa recta that proceed from the arches to the mesenteric border of the gut and then pass more or less alternately to opposite sides of the gut.
3. The arborizations of the vasa in the wall of the gut and the fine anastomoses effected between adjacent arborizations.
4. The efficient anastomoses across the antimesenteric border.

See: Ross, J. A. (1952) Vascular patterns of small and large intestine compared. *Brit. J. Surg.*, *39:* 330.

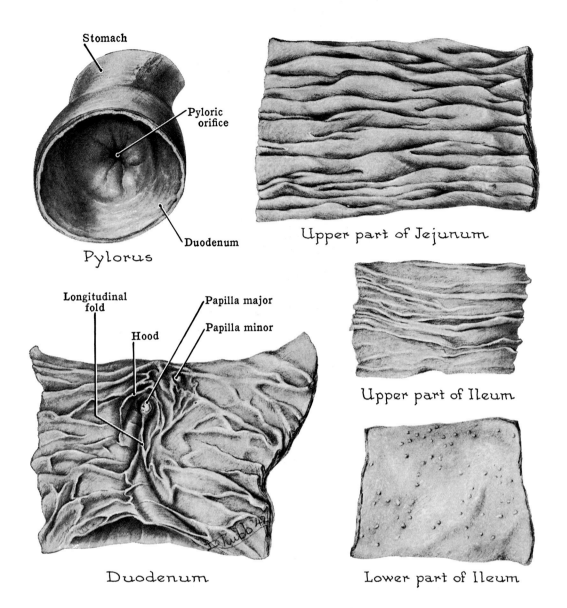

Stomach

Pyloric
orifice

Duodenum

Pylorus

Upper part of Jejunum

Longitudinal
fold

Papilla major

Papilla minor

Hood

Duodenum

Upper part of Ileum

Lower part of Ileum

2-93 INTERIOR OF THE SMALL INTESTINE

Pylorus: The pylorus pouts into the 1st (superior) part of the duodenum and, having a fornix around it, resembles the cervix projecting into the vagina. The first 4 cm of the duodenum have no plicae circulares, but the mucous membrane may be rugose.

Duodenum: The larger duodenal papilla (of Vater) projects into the 2nd (descending) part of the duodenum on the concave border 9 cm from the pylorus (in Fig. 2-88 it is at the apex of a conical evagination). On its tip is the orifice of the pancreatic duct, but usually these 2 ducts open together (see Fig. 2-89). A hood is thrown over the larger papilla, a longitudinal fold descends from it, and the small duodenal papilla,

of the accessory pancreatic duct, lies ¾ to 2 cm antero-superior to it. Plicae circulares are pronounced.

Jejunum, upper part: The plicae circulares are tall, closely packed, and commonly branched.

Ileum, upper part: The plicae circulares are low and becoming sparse. The caliber of the gut is reduced and the wall is thinner.

Ileum, lower part: Plicae are now absent. Solitary lymph nodules stud the wall. (In youth, aggregated lymph nodules, up to 4 cm long and 2 cm wide, are scattered along the antimesenteric border.)

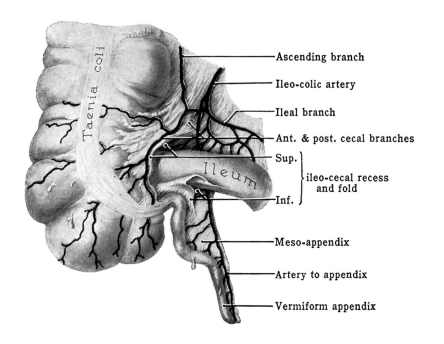

Ascending branch

Ileo-colic artery

Ileal branch

Ant. & post. cecal branches

Sup. ⎫
         ⎬ ileo-cecal recess
Inf. ⎭   and fold

Meso-appendix

Artery to appendix

Vermiform appendix

2-94 ILEO-CECAL REGION

Observe:
1. The appendix in one free border of the mesoappendix and the artery in the other.
2. The anterior tenia coli leading to the appendix. This is a guide to the surgeon.
3. The inferior ileo-cecal (bloodless) fold extending from ileum to meso-appendix.
4. The vascular cecal fold is the official name for the superior ileo-cecal fold.

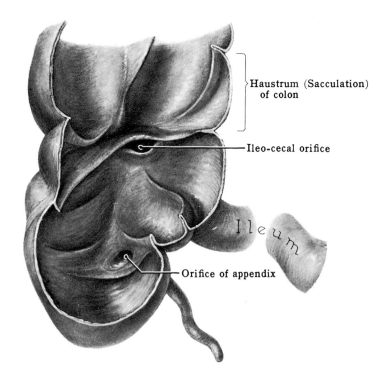

Haustrum (Sacculation) of colon

Ileo-cecal orifice

Orifice of appendix

2-95 INTERIOR OF A DRIED CECUM

This cecum was filled with air until dry, opened, and varnished.
Observe:
1. The ileo-cecal valve guarding the ileo-cecal orifice; its pouting upper lip overhanging the lower lip; and the folds or frenula running horizontally from the commissures of the lips.
2. The slight fold closing the upper part of the orifice of the appendix.

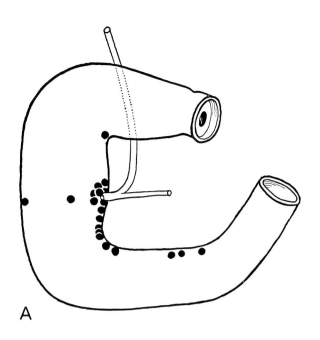

A

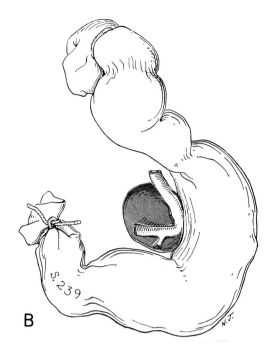

B

2-96 DUODENAL DIVERTICULA

A. Shows the sites of 20 diverticula found in 15 of 133 duodena studied.

B. A specimen of a single diverticulum, anterior to the bile and pancreatic ducts, posterior view.

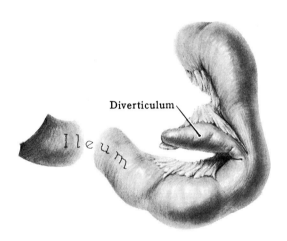

Diverticulum

Ileum

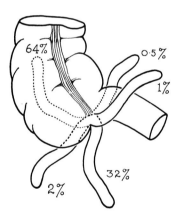

64%   0·5%

1%

2%   32%

2-97 DIVERTICULUM ILEI (MECKEL'S DIVERTICULUM)

Meckel's diverticulum, found in 2 per cent of persons, is the remains of the prenatal vitello-intestinal duct. It projects from the side of the ileum (theoretically, from the antimesenteric border) and it is attached to it by a short peritoneal fold. Usually less than 5 cm in length, it may attain 25 cm. About 72 per cent of diverticula are located within 100 cm of the ileo-cecal orifice and 25 per cent from 100 to 160 cm.

See Jay, G.D. III, Margulis, R.R., McGraw, A.B., and Northrip, R.R. (1950) Meckel's diverticulum: survey of 103 cases, *Arch. Surg. (Chicago), 61:* 158.

2-98 VERMIFORM APPENDIX

This diagram shows the approximate incidence of various locations of the appendix. Like the hands of a clock, the appendix may be long or short, and it may occupy any position consistent with its length.

See Wakeley, C. P. G. (1933) The position of the vermiform appendix as ascertained by the analysis of 10,000 cases. *J. Anat., 67:* 277; Maisel, H. (1960) The position of the human vermiform appendix in fetal and adult age groups. *Anat. Rec., 136:* 385.

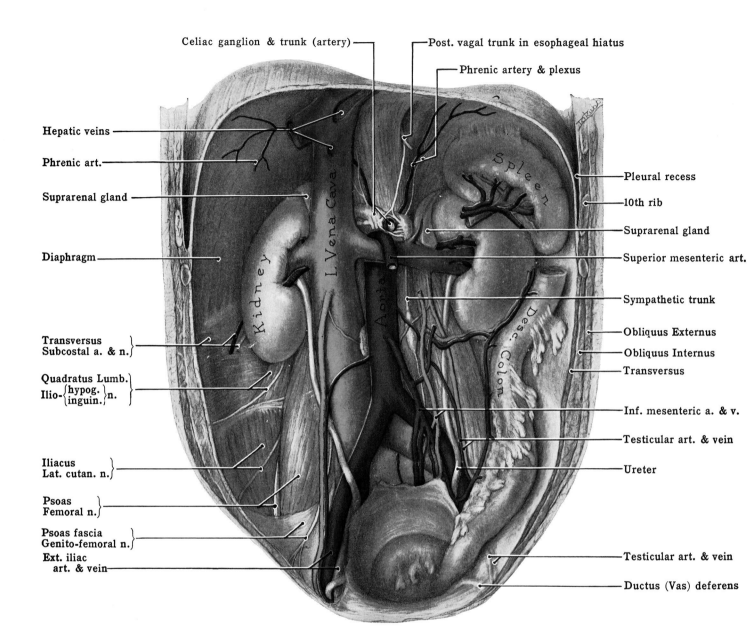

Celiac ganglion & trunk (artery) — | — Post. vagal trunk in esophageal hiatus

— Phrenic artery & plexus

Hepatic veins —

Phrenic art.—

Suprarenal gland —

Diaphragm —

Transversus
Subcostal a. & n.

Quadratus Lumb.
Ilio-{hypog. inguin.} n.

Iliacus
Lat. cutan. n.

Psoas
Femoral n.

Psoas fascia
Genito-femoral n.

Ext. iliac
art. & vein —

— Pleural recess

— 10th rib

— Suprarenal gland

— Superior mesenteric art.

— Sympathetic trunk

— Obliquus Externus

— Obliquus Internus

— Transversus

— Inf. mesenteric a. & v.

— Testicular art. & vein

— Ureter

— Testicular art. & vein

— Ductus (Vas) deferens

2-99 GREAT VESSELS: KIDNEYS: SUPRARENAL GLANDS

Observe that:

The abdominal aorta is shorter than the combined common and external iliac arteries.

The celiac trunk, straddled by crura and the median arcuate ligamentum (Fig. 2-116), appears above the pancreas (Fig. 2-47).

The superior mesenteric artery arises just below the celiac trunk.

The inferior mesenteric artery arises 4 cm above the aortic bifurcation and crosses the left common iliac vessels to become the superior rectal artery.

The kidneys lie ventral to Diaphragm, Transversus aponeurosis, Quadratus Lumborum, and Psoas (Fig. 2-119).

The ureter crosses the external iliac artery just beyond the common iliac bifurcation.

The testicular vessels cross ventral to the ureter and join the ductus deferens at the deep inguinal ring (Fig. 3-38).

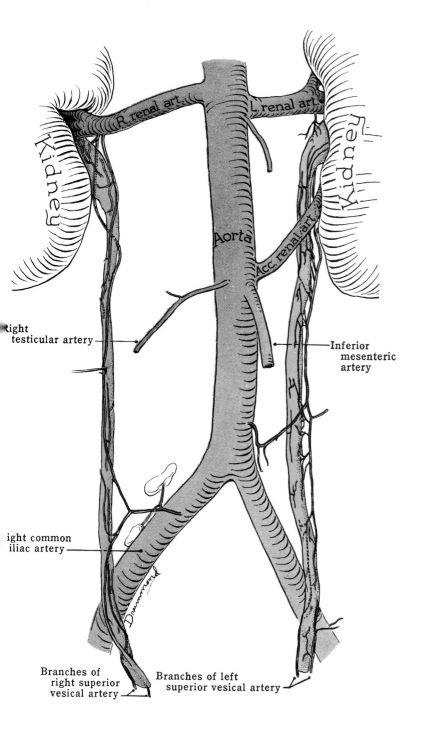

2-100   BLOOD SUPPLY TO THE URETER

The arterial system has been injected with latex by way of the femoral artery.

Observe:

The blood supply to the ureter comes from three main sources; (a) from the renal artery above, (b) from a vesical artery below, and (c) from either the common iliac artery or the aorta.

In this specimen an accessory renal artery also supplies branches. The testicular artery may also contribute a branch.

Branches approach the ureter from its medial side.

These long branches form an excellent anastomatic chain.

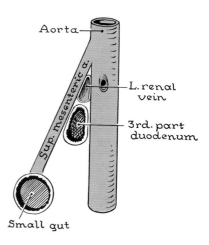

2-101   NUTCRACKERS

Diagram illustrating that the left renal vein and duodenum are compressed between the aorta behind and the superior mesenteric artery suspending the weight of the gut in front.

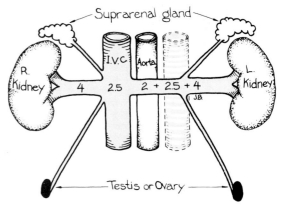

2-102   LEFT RENAL VEIN

Because of the disappearance of the left inferior vena cava, the left renal vein becomes in reality the "vein of the 3 left paired glands," suprarenals, kidneys, and gonads. Thus, it must cross the midline to join the (*right*) inferior vena cava, a journey 4.5 cm longer than that of the right renal vein.

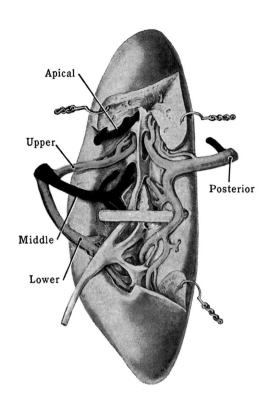

2-103 BRANCHES OF THE RENAL ARTERY WITHIN THE SINUS

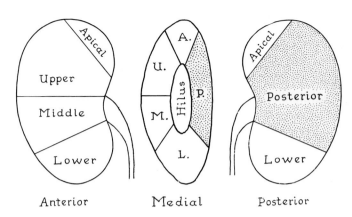

2-104 SEGMENTS OF THE KIDNEY

According to its arterial supply the kidney has 5 segments: (a) apical (superior); (b) superior (antero-superior); (c) middle (antero-inferior); (d) lower or inferior; and (e) posterior.

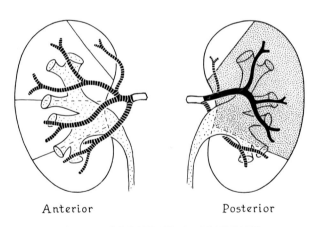

2-105 SEGMENTAL ARTERIES

Only the apical and inferior arteries supply the whole thickness of the kidney. The posterior artery crosses cranial to the renal pelvis to reach its segment.

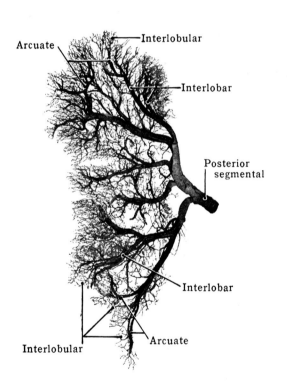

2-106 A SEGMENTAL ARTERY

THE ARTERIES OF THE KIDNEY

Typically, the renal artery divides into 5 branches, each supplying a segment of the kidney. A *segmental* artery provides a *lobar* artery to each pyramid. These divide to provide 2 or 3 *interlobar* arteries which travel between pyramids. Near the junction of medulla and cortex, *arcuate* arteries are given off at right angles to the parent stem. These do not anastomose. From the arcuate arteries (and some from the inter-lobar arteries) *interlobular* arteries pass into the cortex. The afferent arterioles supplying glomeruli are mainly from these interlobular arteries. Although the veins of the kidney anastomose freely, segmental arteries are end arteries. See Graves, F. T. (1971) *The Arterial Anatomy of the Kidney.* John Wright & Sons, Ltd., Bristol.

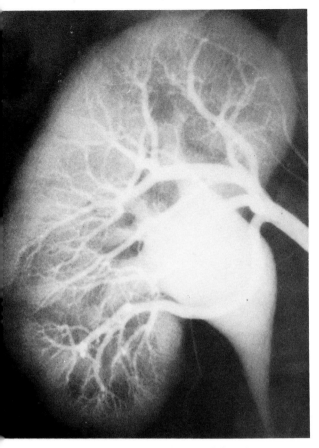

2-106A   RENAL ARTERIOGRAM

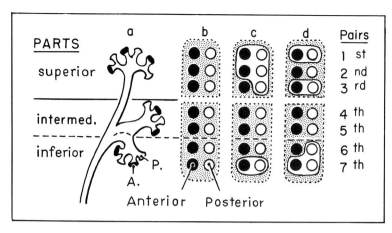

2-107   VARIABILITY

Scheme explaining the variation in the number of pyramids and calices minores.

Of the 7 paired (anterior and posterior) primitive pyramids, each dips into its own calix minor, there being 14 in all. Some of those at the upper and lower ends of the kidney, being crowded, coalesce, hence the maximum number of 14 (see *b*) is almost always reduced to about 9 or 10 compound pyramids and calices minores (see *c* and *d*).

An intermediate (or middle) calix major commonly splits off the lower calix major, as shown.

See Lofgren, F. (1956) *Some Features in the Renal Morphogenesis and Anatomy with Practical Considerations.* Institute of Anatomy, University of Lund, Lund, Sweden.

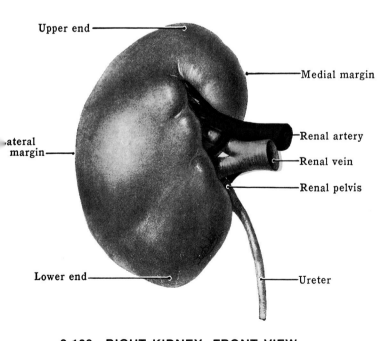

2-108   RIGHT KIDNEY, FRONT VIEW

The order of structures at the hilus (entrance to the renal sinus) from before backward is vein, artery, duct (pelvis or ureter), and a branch of the artery crosses behind the pelvis.

The upper end of the kidney is usually wider than the lower, and closer to the median plane, as in Figure 2-111C.

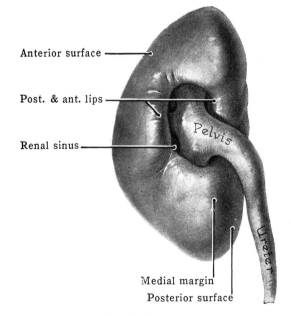

2-109   SINUS OF THE KIDNEY

The renal sinus is a vertical "pocket" on the medial side of the kidney. Tucked into the pocket is the renal pelvis (and the renal vessels.)

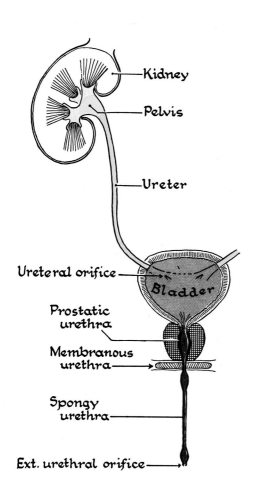

Kidney

Pelvis

Ureter

Ureteral orifice

Bladder

Prostatic urethra

Membranous urethra

Spongy urethra

Ext. urethral orifice

2-110 MALE URINARY SYSTEM

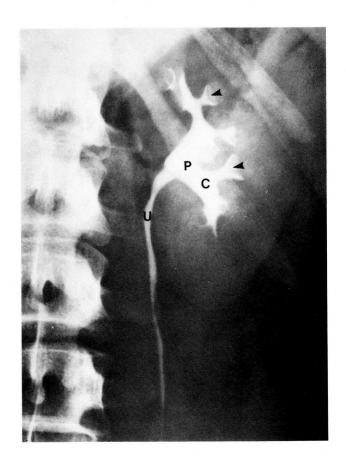

2-111A PYELOGRAM

Radiopaque material outlines the cavities conducting urine. *Arrows* point to two examples of papillae bulging into minor calices which empty into a major calyx (*C*), which in turn opens into the renal pelvis (*P*), which drains into the ureter (*U*).

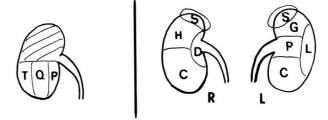

2-111B RELATIONS OF THE KIDNEY

The diagram on the *left* shows *posterior* relations of the kidney: upper half lies on diaphragm and ribs; lower half on three muscles, Transversus, Quadratus lumborum, and Psoas.

The diagram on the *right* shows *anterior* relations of the two kidneys. *S*, suprarenal glands; *H*, liver; *C*, colon; *D*, duodenum; *G*, stomach; *P*, pancreas; *L*, spleen.

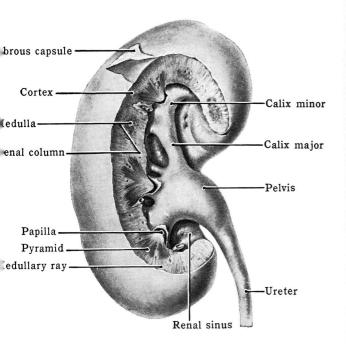

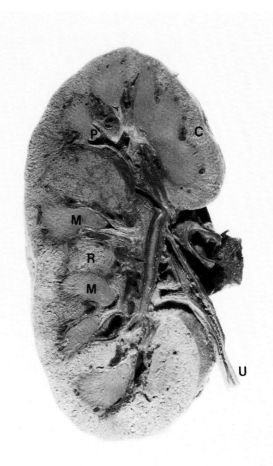

2-111C   EXPOSED CALICES

The anterior lip of the sinus has been cut away to expose the renal pelvis and the calcies. This gives a three-dimensional view to assist in interpreting coronal sections of the kidney such as Figure 2-111D.

2-111D   CORONAL SECTION OF KIDNEY

INTERNAL STRUCTURE OF THE KIDNEY

In Figure 2-111D observe:

Conical *renal pyramids* radiate from the renal sinus toward the surface of the kidney. Their blunted apex (*P*) pouts into a minor calyx into which it discharges urine from the openings of its collecting tubules. The pyramids, which appear striated, form the *medulla (M)* of the kidney and contain loops of Henle and collecting tubules. The renal *cortex (C)* forms the outer one-third of the renal substance, extends between pyramids as *renal columns (R)*, appears rather granular, and contains glomeruli and convoluted tubules. Interlobar arteries travel in the renal columns. The ureter (*U*) drains the renal pelvis which receives 2 or 3 major calices. Each kidney has 7 to 14 minor calices. Verify these observations in Figures 2-111A and 2-111C.

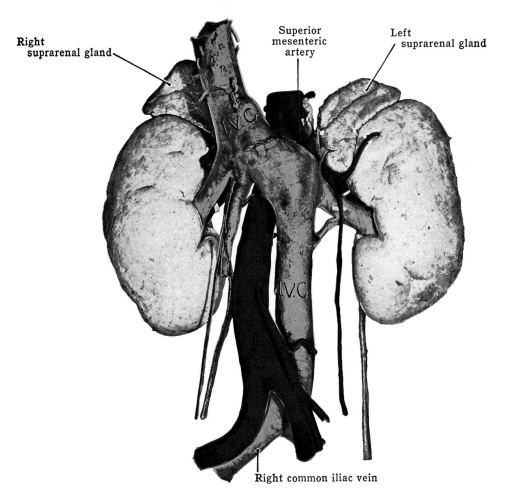

Right
suprarenal gland

Superior
mesenteric
artery

Left
suprarenal gland

I.V.C.

I.V.C.

Right common iliac vein

A. LEFT POSTRENAL INFERIOR VENA CAVA

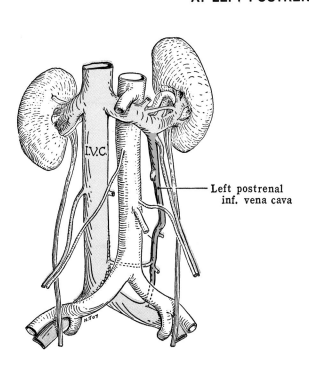

I.V.C.

Left postrenal
inf. vena cava

B. PERSISTING LEFT INFERIOR VENA CAVA

Joins the left renal and the left common iliac veins.
This occasional vessel may be small, as in this specimen, or it may be large.

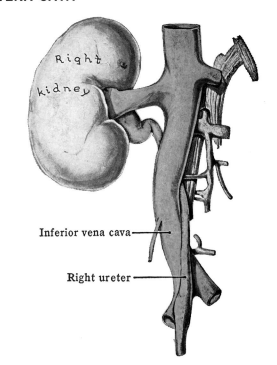

Right
kidney

Inferior vena cava

Right ureter

C. RETROCAVAL URETER

See Pick, J. W., and Anson, B. J. (1940) Retrocaval
ureter. *J. Urol., 43:* 672; Lowsley, O. S. (1946) Postca-
val ureter—operation for its correction. *Surg. Gynec
Obst., 82:* 549.

2-112   ANOMALIES OF THE INFERIOR VENA CAVA

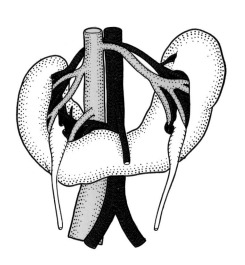

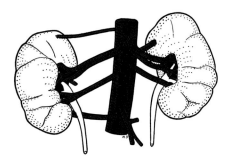

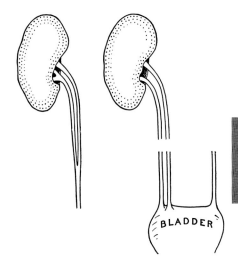

A. HORSESHOE KIDNEY

This occurs in 0.25 per cent of cases. See Yamaguchi, B. (1964) A dissection-case of horseshoe kidney. *Sapporo Med. J.*, 25: 141.

B. MULTIPLE RENAL ARTERIES AND FETAL LOBULATION

About 25 per cent of kidneys receive directly from the aorta 2nd, 3rd, and even 4th branches. These enter either through the renal sinus or at the upper or lower pole.

See Lloyd, L. W. (1935) The renal artery in whites and American Negroes. *Am. J. Phys. Anthropol., 20:* 153.

C. DUPLICATED OR BIFID URETERS

These may be either unilateral or bilateral, and either complete or incomplete. The incidence is less than 1 per cent.

See Campbell, M. (1954) Ureteral reduplication (double ureter). *In Urology*, Vol. 1, p. 309. W. B. Saunders Company, Philadelphia.

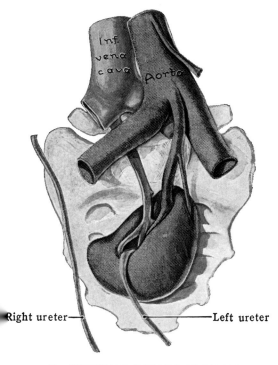

Right ureter——————Left ureter

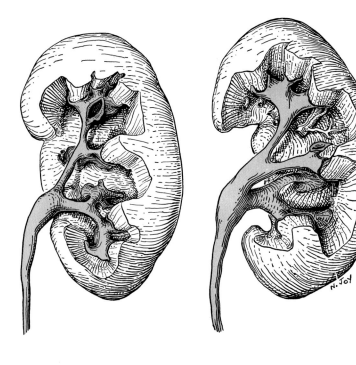

D. ECTOPIC PELVIC KIDNEY

Pelvic kidneys have no fatty capsule and may be unilateral or bilateral. During childbirth, they may both cause obstruction and suffer injury.

See Anderson, G. A., Rice, G. G., and Harris, B. A. (1949) Pregnancy and labor complicated by pelvic ectopic kidney anomalies; a review. *Obst. Gynec. Surv., 4:* 737.

E. BIFID PELVES

The pelves are almost replaced by two long calyces majores, which lie entirely within the sinus (*left*) and partly within and partly without (*right*).

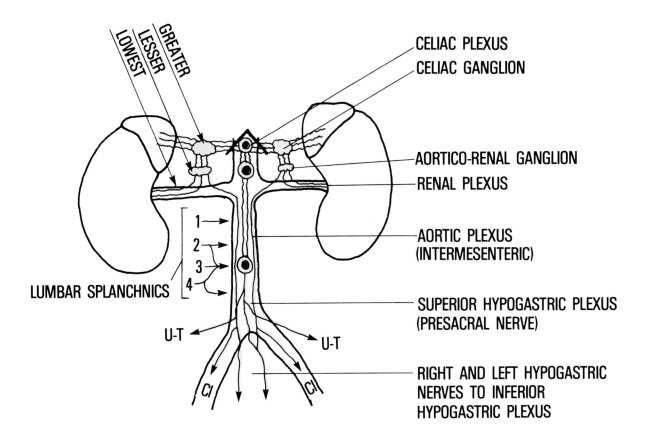

2-114 AUTONOMIC SUPPLY TO ABDOMEN AND PELVIS

Consult Figure 5-50 to review the general plan of the autonomic nervous system. Both sympathetic and parasympathetic fibers are delivered via a complex tangle of nerves around abdominal and pelvic arteries. This network is variable, difficult to dissect, peculiarly named, and described differently by different authors. The above simplified plan is for orientation. Observe:

1. The interconnected plexuses on abdominal arteries.
2. The stems of celiac, superior mesenteric, and inferior mesenteric arteries are surrounded by nerve fibers.
3. Sympathetic fibers here synapse *outside* the sympathetic trunk in ganglia, some of which are small and scattered but two of which are large and named: celiac and aortico-renal.
4. *Arrows* show the source of sympathetic input: greater, lesser, lowest, and (four) lumbar splanchnics. Although paired, they are shown here only on one side.
5. The superior hypogastric plexus supplies ureteric and testicular (*U-T*) plexuses and a plexus on each of the common iliac arteries (*CI*).
6. Parasympathetic supply is not shown here. Branches of the vagus nerve are distributed to foregut and midgut. Pelvic splanchnics join the lower part of the nerve network shown and supply the hindgut and pelvic viscera.

Consult other figures for details in dissected specimens: sympathetic trunk (Figs. 1-19, 2-119, and 3-73), celiac plexus and ganglia (Figs. 2-117 and 2-118), splanchnic nerves (Figs. 1-46, 47, and 3-72), superior hypogastric plexus (Fig. 2-34), and autonomic nerves in the female pelvis (Figs. 3-75 and 3-76).

SPLANCHNIC NERVES

The term "splanchnic" is a common stimulus to student despair. It simply means "visceral." There are three types of nerves called "splanchnic":

Names	Type	Origin
1. Thoracic (greater, lesser, and lowest)	Sympathetic	Branches of thoracic sympathetic ganglia 5–12
2. Lumbar splanchnics	Sympathetic	Branches of the 4 lumbar sympathetic ganglia
3. Pelvic splanchnics	Parasympathetic	Branches of anterior rami of sacral spinal nerves 2, 3, (4)

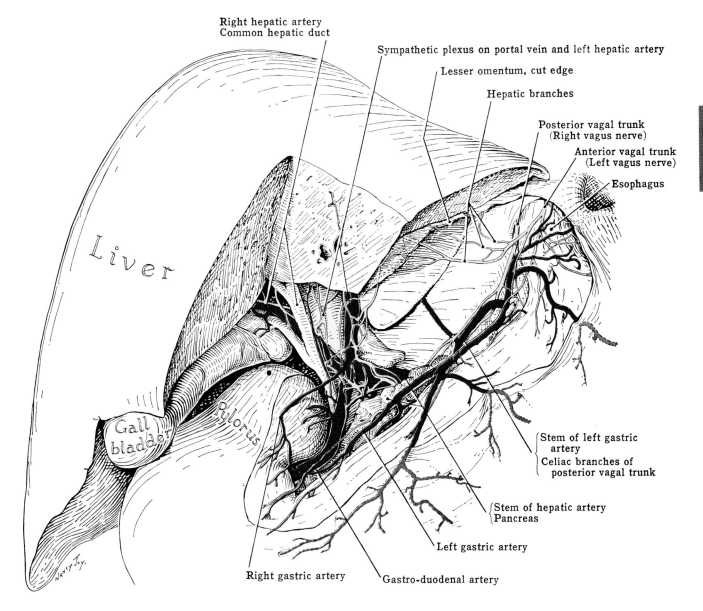

Right hepatic artery
Common hepatic duct

Sympathetic plexus on portal vein and left hepatic artery

Lesser omentum, cut edge

Hepatic branches

Posterior vagal trunk
(Right vagus nerve)

Anterior vagal trunk
(Left vagus nerve)

Esophagus

Liver

Gall bladder

Pylorus

Stem of left gastric artery
Celiac branches of posterior vagal trunk

Stem of hepatic artery
Pancreas

Left gastric artery

Right gastric artery

Gastro-duodenal artery

Nancy Joy.

2-115   VAGUS NERVES WITHIN THE ABDOMEN

(For orientation see Figure 2-29)

First, let us oversimplify by saying that autonomic enervation of the abdominal viscera is of two sorts:

a. *Parasympathetic* via right and left vagus nerves (to the end of the midgut).

b. *Sympathetic* via preganglionic fibers from right and left sympathetic trunks (splanchnic nerves) which synapse in preaortic ganglia.

Both kinds of nerves mingle in the rich tangle of nerve plexuses on the front of the aorta especially around the celiac trunk (celiac plexus). Right and left celiac *ganglia* connect with the celiac *plexus* medially and send large plexuses into the suprarenal glands. Both kinds of fibers are distributed by hitchhiking on the walls of branches of the abdominal aorta to their destinations.

Now, in this figure, observe:

1. The posterior and anterior vagal trunks (formerly right and left vagus nerves) entering on the esophagus and supplying gastric branches.

2. The celiac branch of the posterior vagal trunk leaving to make its contribution to the preaortic plexuses.

3. Hepatic branches of the anterior vagal trunk joined by sympathetic fibers from the celiac plexus.

For details see McCrea, E. A. (1924) The abdominal distribution of the vagus. *J. Anat.*, *59:* 18; Mitchell, G. A. G. (1953) Innervation of the stomach. In *Anatomy of the Autonomic Nervous System*, p. 281. E. & S. Livingstone, Ltd., Edinburgh.

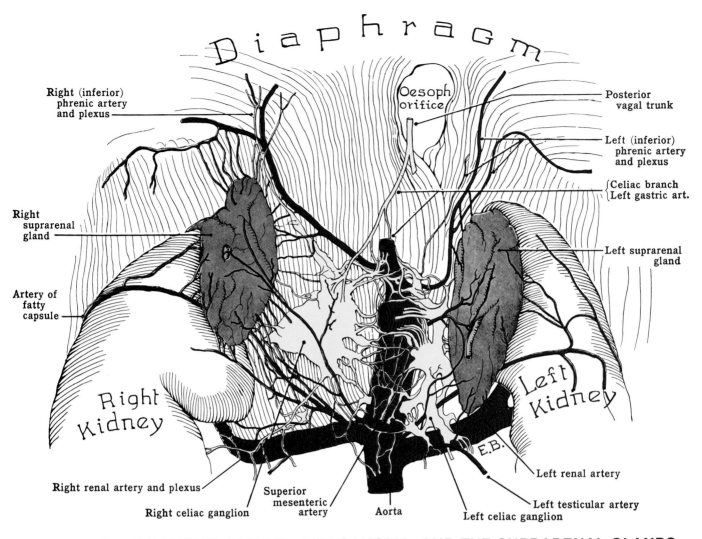

Right (inferior) phrenic artery and plexus

Right suprarenal gland

Artery of fatty capsule

Right Kidney

Right renal artery and plexus

Right celiac ganglion

Superior mesenteric artery

Aorta

Left celiac ganglion

Left testicular artery

Left renal artery

E.B.

Left Kidney

Left suprarenal gland

Celiac branch
Left gastric art.

Left (inferior) phrenic artery and plexus

Posterior vagal trunk

Oesoph orifice

Diaphragm

2-117A   CELIAC TRUNK, PLEXUS, AND GANGLIA, AND THE SUPRARENAL GLANDS

1. The celiac plexus of nerves surrounding the celiac trunk (not labeled) and connecting the right and left celiac ganglia.
2. A stout branch from the posterior vagal trunk descending along the stem of the left gastric artery and conveying vagal (parasympathetic) fibers to the celiac ganglion.
3. Nerves extending along the arteries to the viscera, and down the aorta. The nerves to the suprarenal

glands are mostly preganglionic.
4. The suprarenal glands, abreast of the celiac trunk, 5 cm apart, but not equidistant from the medial plane. (For relations, see Fig. 2-73.)
5. The numerous threadlike arteries from various sources converging on the periphery of the suprarenal gland.
6. Relationship to vertebral levels shown in Figure 2-117B.

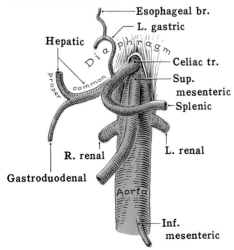

Esophageal br.

L. gastric

Hepatic

Diaphragm

proper

common

Celiac tr.

Sup. mesenteric

Splenic

R. renal

L. renal

Gastroduodenal

Aorta

Inf. mesenteric

2-116   BRANCHES OF AORTA

Nerve plexuses travel with these arteries to viscera.

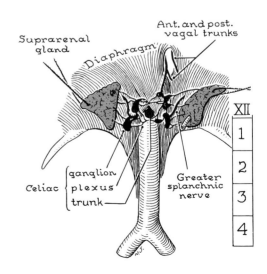

Suprarenal gland

Diaphragm

Ant. and post. vagal trunks

Celiac { ganglion plexus trunk

Greater splanchnic nerve

XII
1
2
3
4

2-117B

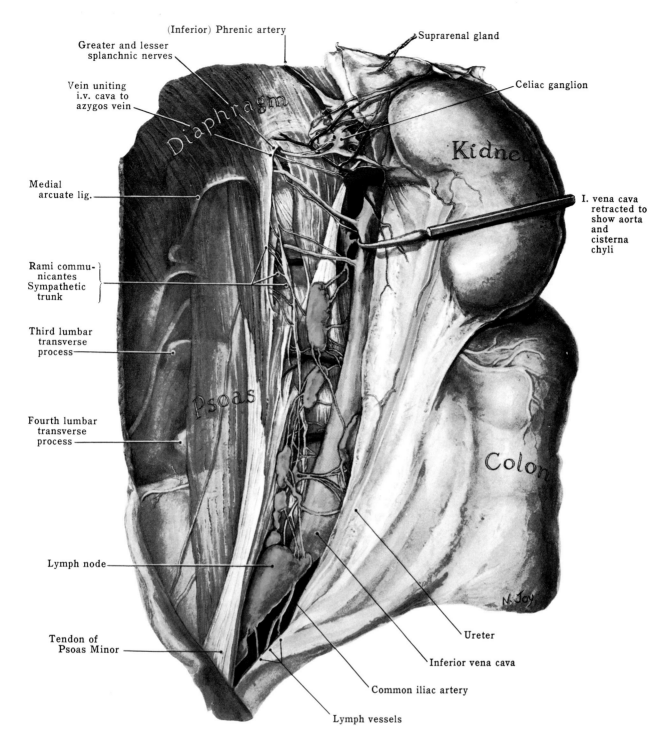

Labels on figure:
(Inferior) Phrenic artery
Greater and lesser splanchnic nerves
Vein uniting i.v. cava to azygos vein
Suprarenal gland
Celiac ganglion
Diaphragm
Medial arcuate lig.
Kidney
I. vena cava retracted to show aorta and cisterna chyli
Rami communicantes
Sympathetic trunk
Third lumbar transverse process
Psoas
Fourth lumbar transverse process
Colon
Lymph node
N. Joy
Ureter
Inferior vena cava
Tendon of Psoas Minor
Common iliac artery
Lymph vessels

2-118—RIGHT CELIAC GANGLION, SPLANCHNIC NERVES, SYMPATHETIC TRUNK

he right suprarenal gland, kidney, ureter, and colon are reflected; e inferior vena cava is pulled medially; and the third and fourth mbar veins are removed.

call that the splanchnic nerves are preganglionic fibers coming om the sympathetic trunk in the thoracic region. The greater anchnic nerve is from ganglia 5 to 9; the lesser from ganglia 10 to

serve:
A wide cleft in the right crus (not labeled) of the diaphragm through which pass both splanchnic nerves, the sympathetic trunk, and a communicating vein. This unusually wide cleft is present on the left side also.
The greater splanchnic nerve ending in the celiac ganglion; the lesser splanchnic nerve here ending in the renal plexus (aortico-renal ganglion). Usually, as in Figure 2–119, the greater splanchnic nerve pierces the crus at the level of the celiac trunk; the lesser

nerve infero-lateral to this; and the sympathetic trunk enters with the Psoas.

3. The sympathetic trunk lying on the bodies of the vertebrae, the lumbar vessels alone intervening, and descending along the anterior border of the Psoas. The trunk is slender where it enters the abdomen; its ganglia are ill-defined: about 11 rami communicantes join it postero-laterally; and about 6 visceral branches, or lumbar splanchnic nerves, leave it antero-medially.

4. The 4 right lumbar arteries and cranial 2 veins (not labeled); the tip of the 3rd lumbar transverse process projecting further than the other transverse processes; and the right lumbar lymph nodes and vessels draining into the cisterna chyli.

5. In this specimen there is a Psoas Minor.

See Wrete, M. (1959) The anatomy of the sympathetic trunks in man. *J. Anat.*, 93: 448.

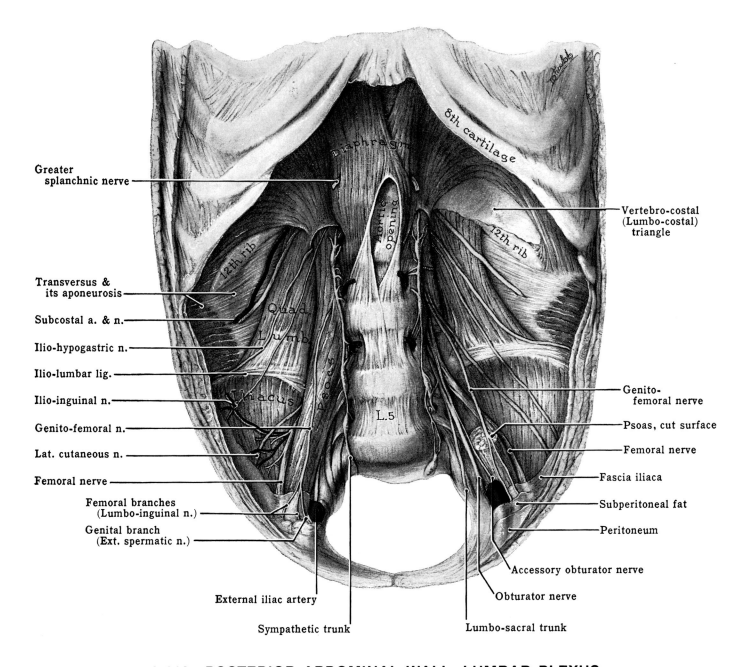

Greater
splanchnic nerve

Transversus &
its aponeurosis

Subcostal a. & n.

Ilio-hypogastric n.

Ilio-lumbar lig.

Ilio-inguinal n.

Genito-femoral n.

Lat. cutaneous n.

Femoral nerve

Femoral branches
(Lumbo-inguinal n.)

Genital branch
(Ext. spermatic n.)

External iliac artery

Sympathetic trunk

8th cartilage

diaphragm

Aortic opening

12th rib

12th rib

Quad. Lumb.

Iliacus

Psoas

L.5

Vertebro-costal
(Lumbo-costal)
triangle

Genito-
femoral nerve

Psoas, cut surface

Femoral nerve

Fascia iliaca

Subperitoneal fat

Peritoneum

Accessory obturator nerve

Obturator nerve

Lumbo-sacral trunk

2-119—POSTERIOR ABDOMINAL WALL: LUMBAR PLEXUS

Muscles. Transversus Abdominis becomes aponeurotic on a line dropped from the tip of the 12th rib. Quadratus Lumborum has an oblique lateral border; its fascia is thickened to form the lateral arcuate ligament above and the ilio-lumbar ligament below. Iliacus lies below the iliac crest. Psoas rises above the crest and extends above the medial arcuate ligament which is thickened Psoas fascia.

Nerves. The subcostal nerve (T12) passes behind the lateral arcuate ligament and runs at some distance below the 12th rib (with its artery). The next four nerves appear at the lateral border of Psoas. Of these, the ilio-hypogastric (T12, L1) takes the characteristic course here shown; the ilio-inguinal (L1) and the lateral cutaneous of the thigh (L2, L3) are varia-

ble; the femoral (L2, L3, L4) descends in the angle between Iliacus and Psoas. The genito-femoral nerve (L1, L2) pierces Psoas and its fascia anteriorly. The obturator nerve (L2, L3, L4) and a branch of L4 that joins with L5 to form the lumbo-sacral trunk appear at the medial border of Psoas, and, crossing the ala of the sacrum, enter the pelvis.

The *sympathetic trunk* enters the abdomen with Psoas from behind the medial arcuate ligament. It descends on vertebral bodies and intervertebral discs, following closely the attached border of Psoas to enter the pelvis. Its rami communicates run dorsally with, or near, the lumbar arteries to join the lumbar nerves.

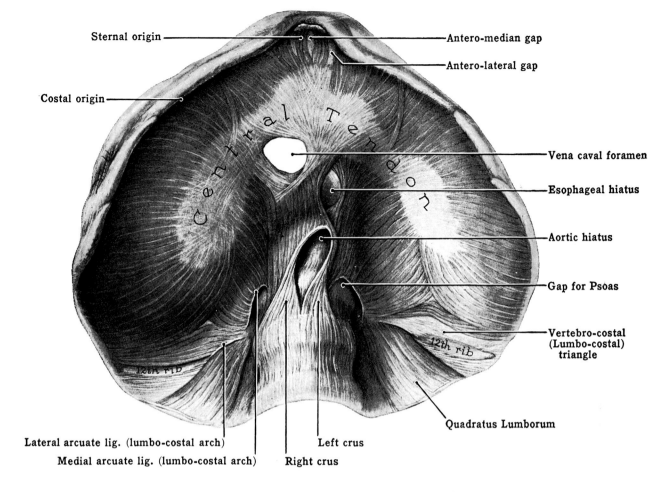

Sternal origin ——— ——— Antero-median gap

——— Antero-lateral gap

Costal origin ———

Central Tendon

——— Vena caval foramen

——— Esophageal hiatus

——— Aortic hiatus

——— Gap for Psoas

——— Vertebro-costal (Lumbo-costal) triangle

12th rib

12th rib

Quadratus Lumborum

Lateral arcuate lig. (lumbo-costal arch)   Left crus

Medial arcuate lig. (lumbo-costal arch)   Right crus

2-120   DIAPHRAGM, VIEWED FROM BELOW

Observe:
1. The trefoil-shaped central tendon which is the aponeurotic insertion of the muscle.
2. The fleshy origins: *In front* from the inner surface of the xiphoid process, *circumferentially* from the lower six costal cartilages, and *behind* from the upper three lumbar vertebral bodies via right and left crura which unite in front of the aortic hiatus

to form the median arcuate ligament. Thickenings of the Psoas and Quadratus lumborum fascia (the medial and lateral arcuate ligaments) also afford origin to the diaphragm.

3. The diaphragm, in this specimen, failing to arise from the left arcuate ligament, hence the vertebro-costal trigone through which a herniation of abdominal contents into the pleural cavity may occur.

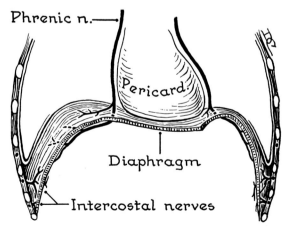

Phrenic n. ———

Pericard.

Diaphragm

Intercostal nerves

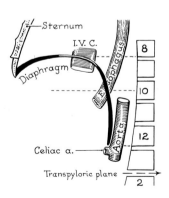

Sternum

I.V.C.

Diaphragm

8

10

12

Celiac a.

Transpyloric plane   2

A. Nerve Supply: Each phrenic nerve (C3, C4, C5) is the sole motor nerve to its own half of the diaphragm. It is also sensory to its own half, including the pleura above and the peritoneum below, but the lower intercostal nerves are sensory to the peripheral fringe.

B. The three large openings in the diaphragm for major structures: 1. I.V.C., most anterior, at T8 level, and to the right of the midline. 2. Esophagus, intermediate, at T10, and to the left. 3. Aorta, most posterior, at T12, and in the midline.

2-121   DIAPHRAGM DIAGRAMS

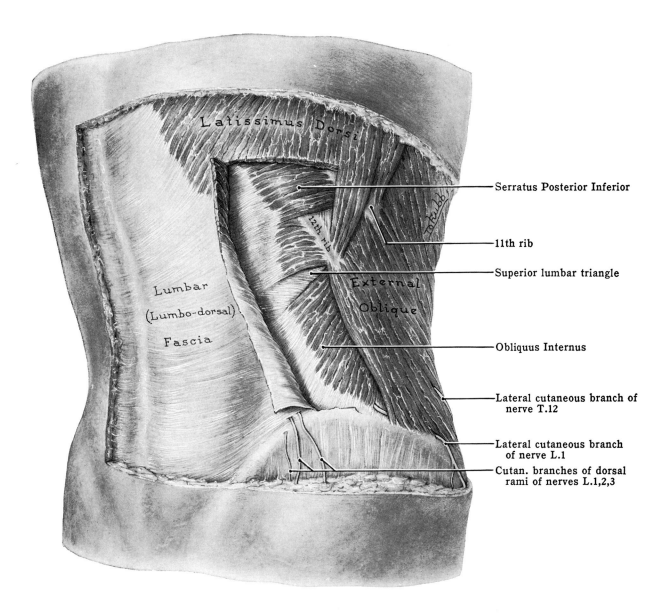

Latissimus Dorsi

Lumbar
(Lumbo-dorsal)
Fascia

12th rib

External
Oblique

Serratus Posterior Inferior

11th rib

Superior lumbar triangle

Obliquus Internus

Lateral cutaneous branch of
nerve T.12

Lateral cutaneous branch
of nerve L.1

Cutan. branches of dorsal
rami of nerves L.1,2,3

2-122 — POSTERIOR ABDOMINAL WALL — I

POSTERO-LATERAL VIEW

Latissimus Dorsi is in part reflected.

Observe:

1. External Oblique having an oblique, free, posterior border which extends from the tip of the 12th rib to the midpoint of the iliac crest.
2. The small, triangular space between External Oblique, Latissimus Dorsi, and the iliac crest. This is the (inferior) lumbar triangle (Fig. 6–30).
3. Internal Oblique extending behind External Oblique. It forms the floor of the lumbar triangle, creeps up on to the lumbar fascia, and has a triangle between it and Serratus Posterior Inferior. This is the "superior lumbar triangle."

(In N.A.P. lumbo-dorsal fascia reads thoraco-lumbar fascia.)

When External Oblique has been incised and turned laterally, and Internal Oblique incised and turned medially, Transversus Abdominis and its posterior aponeurosis are exposed where pierced by the subcostal (T12) and ilio-hypogastric (L1) nerves. These nerves give off motor twigs and lateral cutaneous (iliac) branches, and continue forward between Internal Oblique and Transversus.

2-123 — POSTERIOR ABDOMINAL WALL — II

On dividing the posterior aponeurosis of Transversus between the subcostal and ilio-hypogastric nerves, and lateral to the oblique lateral border of Quadratus Lumborum, the retroperitoneal fat surrounding the kidney is exposed. The renal fascia is within this fat. The portion of fat inside the renal fascia is termed fatty renal capsule (perinephric fat); the fat outside is paranephric fat.

(For Quadratus Lumborum, ventral view, see Fig. 2–119.)

2-124 — POSTERIOR ABDOMINAL WALL — III

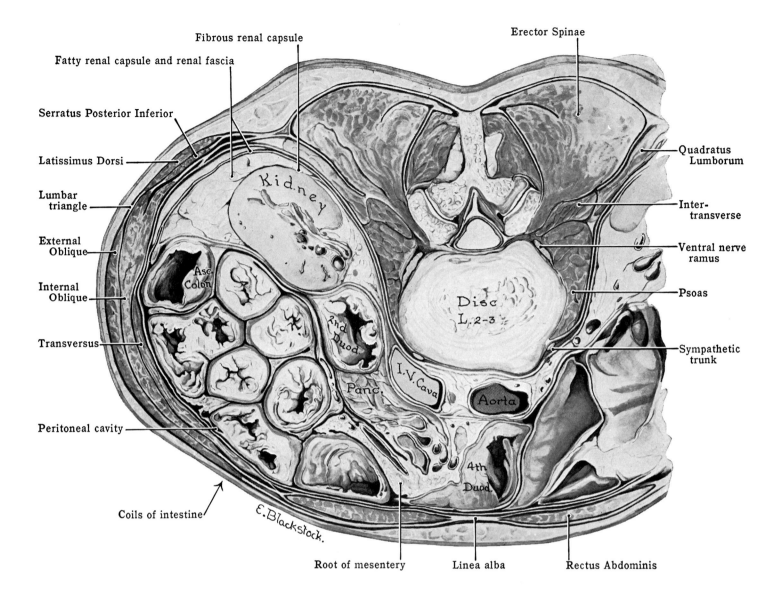

Fibrous renal capsule

Fatty renal capsule and renal fascia

Serratus Posterior Inferior

Latissimus Dorsi

Lumbar triangle

External Oblique

Internal Oblique

Transversus

Peritoneal cavity

Coils of intestine

Erector Spinae

Quadratus Lumborum

Inter-transverse

Ventral nerve ramus

Psoas

Sympathetic trunk

Kidney

Asc. Colon

2nd Duod

Panc.

I.V. Cava

Disc L.2-3

Aorta

4th Duod.

E. Blackstock.

Root of mesentery

Linea alba

Rectus Abdominis

2-125—TRANSVERSE SECTION THROUGH THE ABDOMEN AT L2, L3

Observe:

1. The anterior aspect of the vertebral column, nearer to the ventral surface of the body than to the dorsal surface in this thin recumbent subject.

2. The coils of small intestine, not circular as seen at operation, but mutually compressed.

3. The mucous membrane of the small intestine—which in life is much more moist and succulent than in this hardened specimen—in loose folds which almost completely occlude the lumen.

4. The ascending colon with an appendix epiploica, having no mesentery but being bare posteriorly.

5. The anterior and posterior surfaces of the kidney facing not ventrally and dorsally but ventro-laterally and dorso-medially.

6. The adipose renal capsule (perinephric fat) massed along the borders of the kidney and leaving the anterior surface close to the peritoneum.

7. The kidney projecting lateral to Quadratus Lumborum.

8. The 2nd or descending part of the duodenum overlapping this surface of the kidney.

9. The sympathetic trunk lying along the anterior border of Psoas; and on the right side behind the inferior vena cava.

10. The lipping at the epiphyseal plate of the vertebral body which first begins to show itself before the age of 30 years.

11. The linea alba, bandlike above the navel, though linear below it.

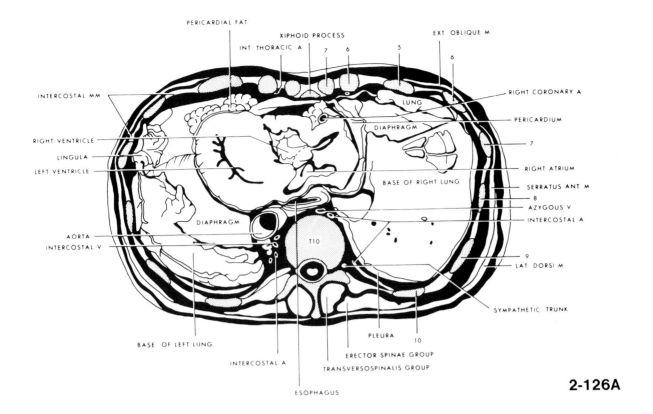

2-126A

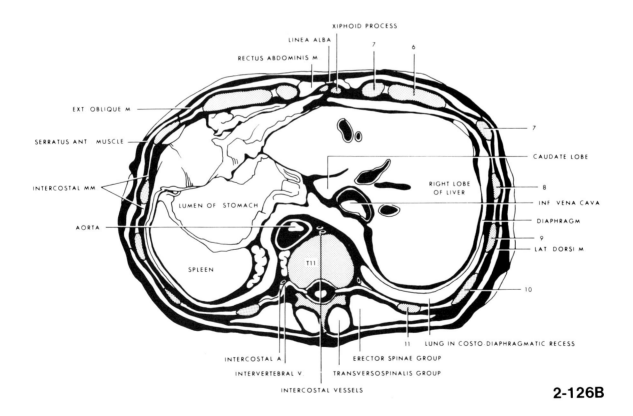

2-126B

HORIZONTAL SECTIONS THROUGH ABDOMEN

Figures 2-126 to 2-129 are a series of eight tracings made from horizontal sections through the body at vertebral levels T10 to L5.

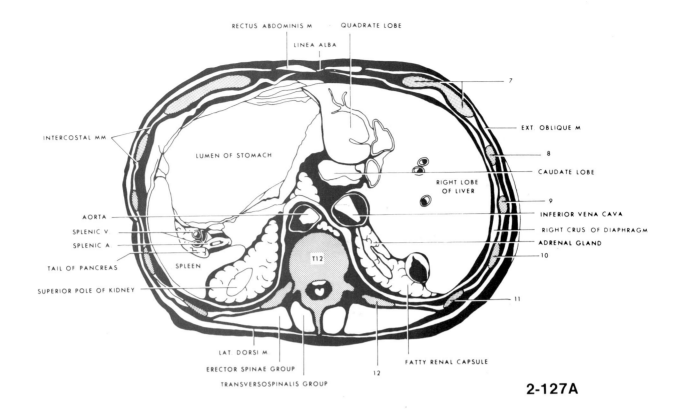

RECTUS ABDOMINIS M
QUADRATE LOBE
LINEA ALBA

7

EXT. OBLIQUE M

INTERCOSTAL MM

8

CAUDATE LOBE

LUMEN OF STOMACH

RIGHT LOBE OF LIVER

9

AORTA

INFERIOR VENA CAVA

SPLENIC V

RIGHT CRUS OF DIAPHRAGM

SPLENIC A

ADRENAL GLAND

TAIL OF PANCREAS

SPLEEN

10

T12

SUPERIOR POLE OF KIDNEY

11

LAT DORSI M

ERECTOR SPINAE GROUP

12

FATTY RENAL CAPSULE

TRANSVERSOSPINALIS GROUP

2-127A

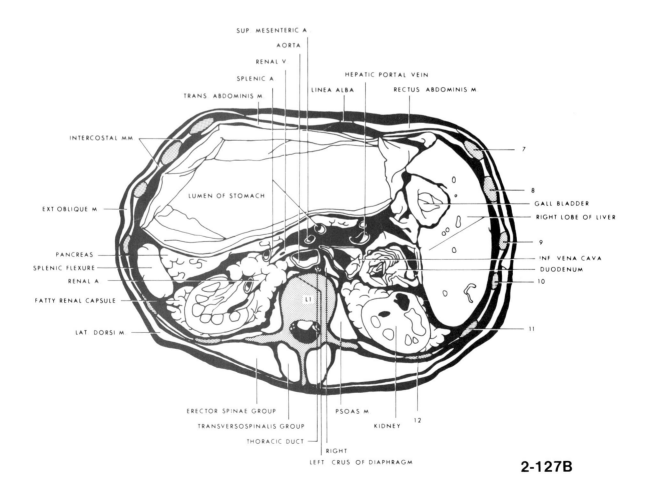

SUP. MESENTERIC A

AORTA

RENAL V

HEPATIC PORTAL VEIN

SPLENIC A

TRANS. ABDOMINIS M

LINEA ALBA

RECTUS ABDOMINIS M

INTERCOSTAL MM

7

8

GALL BLADDER

LUMEN OF STOMACH

EXT OBLIQUE M

RIGHT LOBE OF LIVER

9

PANCREAS

INF VENA CAVA

SPLENIC FLEXURE

DUODENUM

RENAL A

10

FATTY RENAL CAPSULE

L1

LAT. DORSI M

11

ERECTOR SPINAE GROUP

PSOAS M

12

TRANSVERSOSPINALIS GROUP

KIDNEY

THORACIC DUCT

RIGHT

LEFT CRUS OF DIAPHRAGM

2-127B

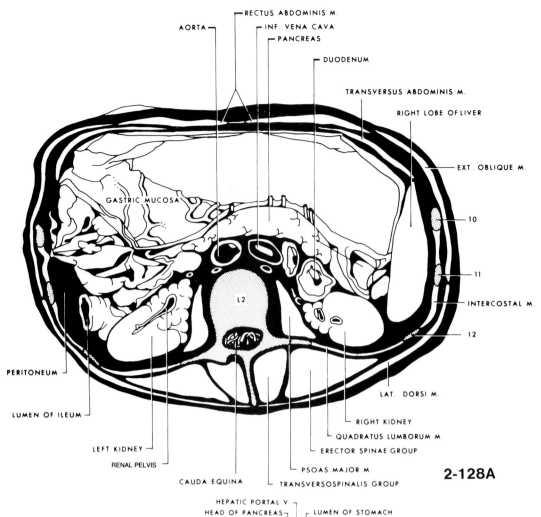

AORTA
RECTUS ABDOMINIS M.
INF. VENA CAVA
PANCREAS
DUODENUM
TRANSVERSUS ABDOMINIS M.
RIGHT LOBE OF LIVER
EXT. OBLIQUE M.
10
11
INTERCOSTAL M.
12
GASTRIC MUCOSA
L2
PERITONEUM
LUMEN OF ILEUM
LAT. DORSI M.
LEFT KIDNEY
RENAL PELVIS
RIGHT KIDNEY
QUADRATUS LUMBORUM M.
ERECTOR SPINAE GROUP
PSOAS MAJOR M
CAUDA EQUINA
TRANSVERSOSPINALIS GROUP

2-128A

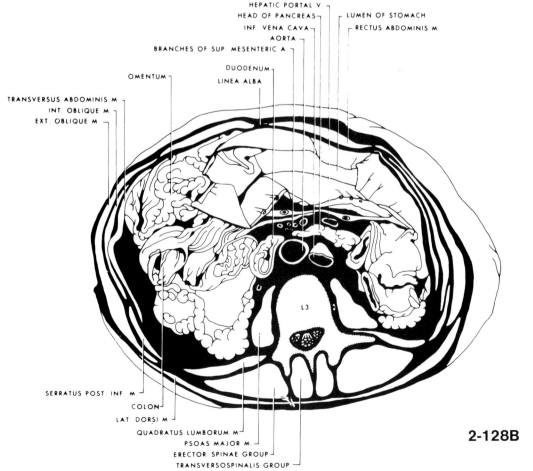

HEPATIC PORTAL V
HEAD OF PANCREAS
INF. VENA CAVA
AORTA
BRANCHES OF SUP. MESENTERIC A
DUODENUM
LINEA ALBA
OMENTUM
TRANSVERSUS ABDOMINIS M
INT. OBLIQUE M
EXT. OBLIQUE M
LUMEN OF STOMACH
RECTUS ABDOMINIS M
L3
SERRATUS POST. INF. M
COLON
LAT. DORSI M
QUADRATUS LUMBORUM M
PSOAS MAJOR M
ERECTOR SPINAE GROUP
TRANSVERSOSPINALIS GROUP

2-128B

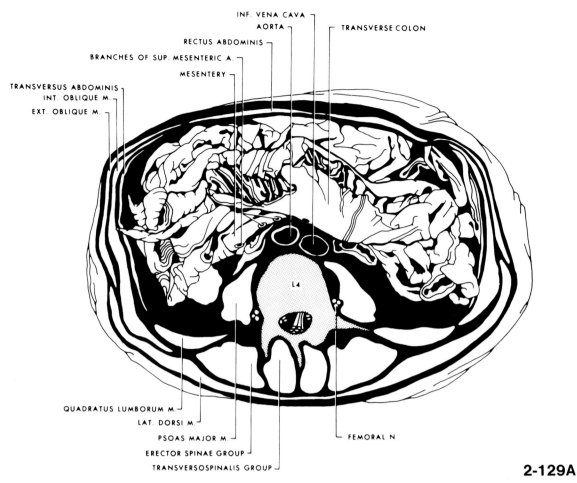

INF. VENA CAVA
AORTA
TRANSVERSE COLON
RECTUS ABDOMINIS
BRANCHES OF SUP. MESENTERIC A.
MESENTERY
TRANSVERSUS ABDOMINIS
INT. OBLIQUE M.
EXT. OBLIQUE M.

L4

QUADRATUS LUMBORUM M.
LAT. DORSI M.
PSOAS MAJOR M.
ERECTOR SPINAE GROUP
TRANSVERSOSPINALIS GROUP

FEMORAL N

2-129A

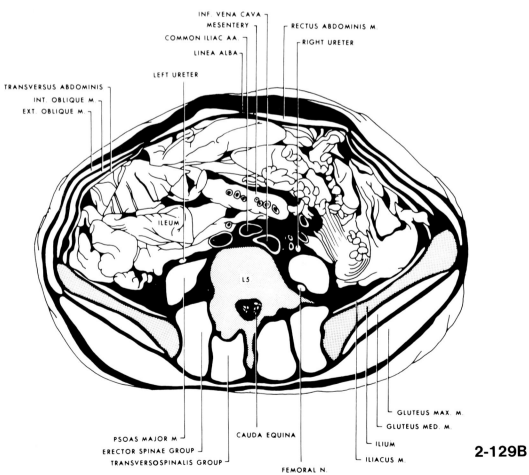

INF. VENA CAVA
MESENTERY
COMMON ILIAC AA.
LINEA ALBA
LEFT URETER
RECTUS ABDOMINIS M.
RIGHT URETER

TRANSVERSUS ABDOMINIS
INT. OBLIQUE M.
EXT. OBLIQUE M.

ILEUM

L5

PSOAS MAJOR M.
ERECTOR SPINAE GROUP
TRANSVERSOSPINALIS GROUP

CAUDA EQUINA

FEMORAL N.

GLUTEUS MAX. M.
GLUTEUS MED. M.
ILIUM
ILIACUS M.

2-129B

Perineum and Pelvis

CONTENTS

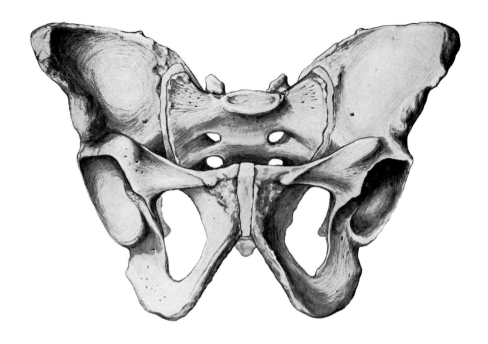

3-1—MALE PELVIS, FROM THE FRONT

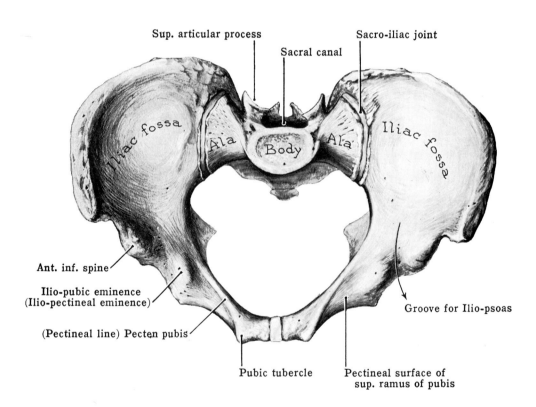

Sup. articular process

Sacral canal

Sacro-iliac joint

Iliac fossa

Ala

Body

Ala

Iliac fossa

Ant. inf. spine

Ilio-pubic eminence
(Ilio-pectineal eminence)

(Pectineal line) Pecten pubis

Groove for Ilio-psoas

Pubic tubercle

Pectineal surface of
sup. ramus of pubis

3-2—MALE PELVIS, FROM ABOVE

The *pelvis major* (false pelvis) is formed by the iliac fossae and the alae of the sacrum. The *pelvis minor* (true pelvis) is formed by the inner surface of ischium, pubis, some ilium, sacrum, and coccyx. The groove between the anterior inferior iliac spine and the ilio-pubic eminence conducts ilio-psoas from the false pelvis to the thigh. Three bones separated by three joints share in the formation of the *pelvic brim*. The bones are: two hip bones and the sacrum. The joints are: the symphysis pubis and the two sacro-iliac joints.

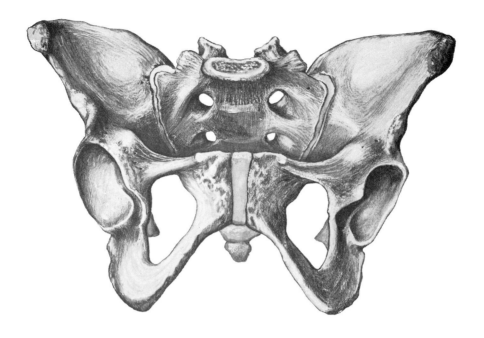

3-3—FEMALE PELVIS, FROM THE FRONT

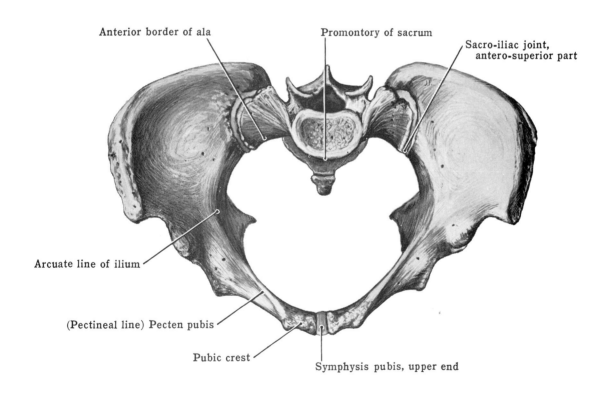

Anterior border of ala   Promontory of sacrum

Sacro-iliac joint,
antero-superior part

Arcuate line of ilium

(Pectineal line) Pecten pubis

Pubic crest

Symphysis pubis, upper end

3-4—FEMALE PELVIS, FROM ABOVE

Compare the following features of the male and female pelvis:
1. The angle of the pubic arch.
2. The relative strength of the attachments of crus penis or clitoridis.
3. The relative breadth of alae to body of sacrum.
4. The shape and dimensions of the pelvic brim.

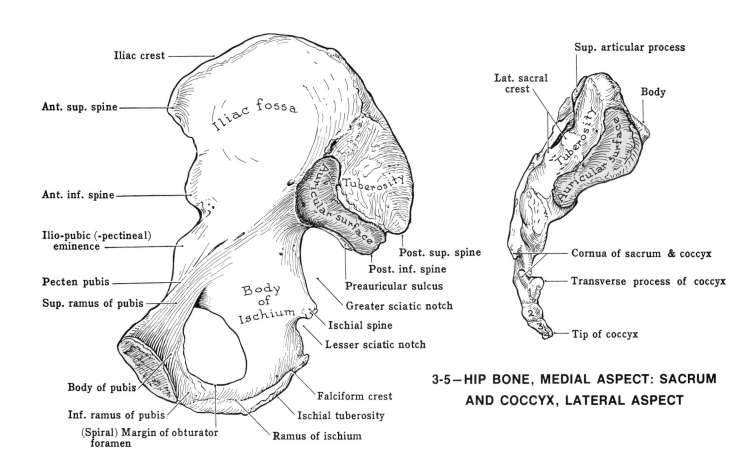

Iliac crest

Ant. sup. spine

Ant. inf. spine

Ilio-pubic (-pectineal) eminence

Pecten pubis

Sup. ramus of pubis

Body of pubis

Inf. ramus of pubis

(Spiral) Margin of obturator foramen

Iliac fossa

Body of Ischium

Auricular surface

Tuberosity

Post. sup. spine

Post. inf. spine

Preauricular sulcus

Greater sciatic notch

Ischial spine

Lesser sciatic notch

Falciform crest

Ischial tuberosity

Ramus of ischium

Sup. articular process

Lat. sacral crest

Body

Auricular surface

Tuberosity

Cornua of sacrum & coccyx

Transverse process of coccyx

Tip of coccyx

3-5 — HIP BONE, MEDIAL ASPECT: SACRUM AND COCCYX, LATERAL ASPECT

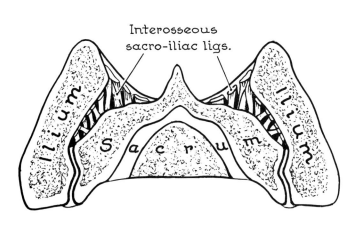

Interosseous sacro-iliac ligs.

Ilium

Ilium

Sacrum

3-6 — SACRO-ILIAC JOINT, TRANSVERSE SECTION

For dorsal sacro-iliac ligaments see Figure 4–36, sacro-tuberous and sacro-spinous ligaments Figures 3–57 and 4–36, and ilio-lumbar ligaments Figure 2–119.

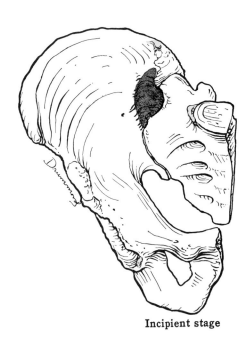

Incipient stage

3-7 — SYNOSTOSIS OF SACRO-ILIAC JOINT

Ossification commonly begins as shown here in *red* in the ventral sacro-iliac ligament and is associated with bony overgrowth at other articular surfaces. In a study of 91 cadavers aged 30 to 84 years, synostosis was present in 8 unilaterally and 3 bilaterally.

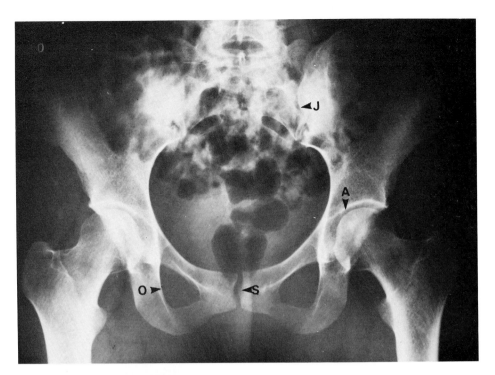

3-8A — RADIOGRAPH OF THE PELVIS

For orientation, see Figure 3–9. Note air in the colon.

Arrows indicate:

    S: symphysis pubis
    J: sacroiliac joint
    A: acetabular fossa
    O: obturator foramen

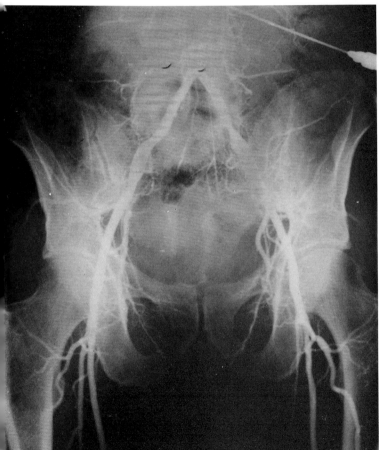

3-8B — ILIAC ARTERIOGRAM

Injection has been made into the aorta in the lumbar region.

Observe:

1. The bifurcation of the aorta into right and left common iliac arteries (in front of L4).
2. The bifurcation of the common iliacs into internal and external iliac arteries (opposite the sacroiliac joint, at the level of the lumbosacral disc).

Consult Figure 4–11 for branches of the external iliac and Figure 3–53 for the internal iliac artery.

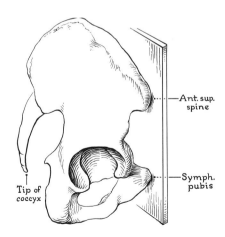

3-9 — ORIENTATION

When the subject is standing, the anterior superior spines and the upper end of the symphysis pubis lie in the same vertical plane. The tip of the coccyx is on a level with the upper half of the body of the pubis.

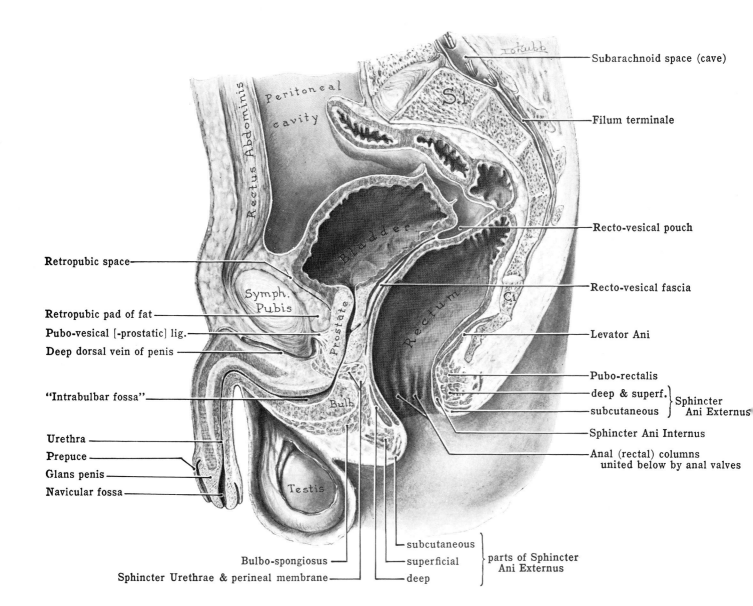

Subarachnoid space (cave)

Filum terminale

Recto-vesical pouch

Recto-vesical fascia

Levator Ani

Pubo-rectalis

deep & superf. } Sphincter
subcutaneous } Ani Externus

Sphincter Ani Internus

Anal (rectal) columns
united below by anal valves

Retropubic space

Retropubic pad of fat

Pubo-vesical [-prostatic] lig.

Deep dorsal vein of penis

"Intrabulbar fossa"

Urethra

Prepuce

Glans penis

Navicular fossa

Rectus Abdominis

Peritoneal cavity

Peritoneal cavity

Bladder

Symph. Pubis

Prostate

Rectum

Bulb

Testis

Bulbo-spongiosus

Sphincter Urethrae & perineal membrane

subcutaneous
superficial
deep
} parts of Sphincter
Ani Externus

3-10—MALE PELVIS, IN MEDIAN SECTION

Observe:
1. The urinary bladder slightly distended and resting on the rectum; the prostatic urethra descending vertically through a somewhat elongated prostate and showing the prostatic utricle opening on to its posterior wall; the short membranous urethra passing through the deep perineal space; the spongy urethra with a low lying dilation in the bulb and another in the glans; and the Bulbo-spongiosus which by contracting empties the urethra.
2. The involuntary Sphincter Ani Internus not descending so far as the voluntary Sphincter Ani Externus, and separated from it by an areolar layer.
3. The two layers of recto-vesical fascia in the median plane between bladder and rectum. On each side it contains the deferent duct, seminal vesicle, and vesical vessels (Figs. 3–39 and 3–41).
4. The peritoneum passing from the abdominal wall above the symphysis to the distended bladder, over the bladder to the bottom of the recto-vesical pouch, and up the anterior aspect of the rectum.
5. The tunica vaginalis (not labeled), opened in order to expose the testis, which here happens to be rotated so that the epididymis is to the front.

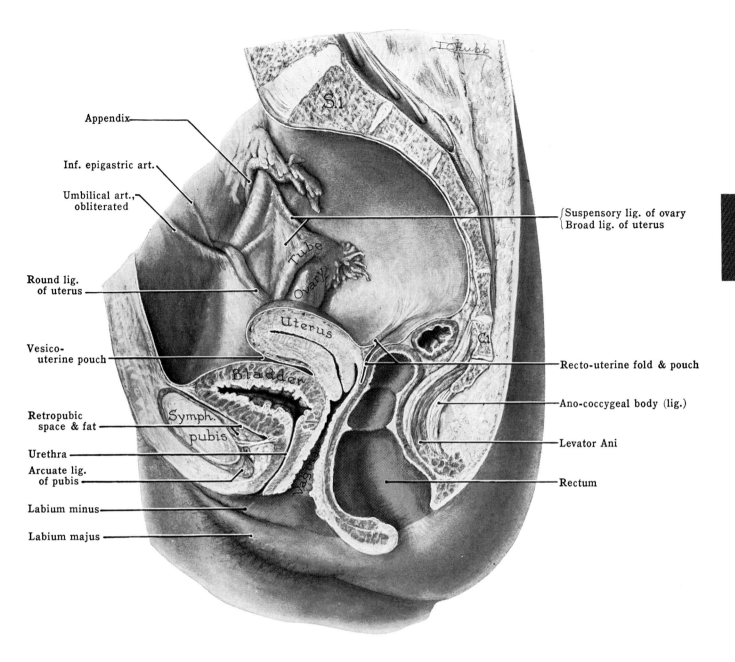

Appendix

Inf. epigastric art.

Umbilical art.,
obliterated

Round lig.
of uterus

Vesico-
uterine pouch

Retropubic
space & fat

Urethra

Arcuate lig.
of pubis

Labium minus

Labium majus

Suspensory lig. of ovary
Broad lig. of uterus

Recto-uterine fold & pouch

Ano-coccygeal body (lig.)

Levator Ani

Rectum

3-11—FEMALE PELVIS, IN MEDIAN SECTION

Plugs of cotton wool in the lower parts of the rectum and vagina had slightly distorted these parts. The uterus was sectioned in its own median plane and depicted as though this coincided with the median plane of the body, which is seldom the case.

Observe:

1. The uterine tube and the ovary, in their virginal positions on the side wall of the pelvis, *i.e.*, in the angle between ureter and the umbilical artery, and medial to the obturator nerve and vessels (Fig. 3-51).
2. The uterus, bent on itself at the junction of body and cervix. The cervix, opening on the anterior wall of the vagina, and having a short, round, anterior lip and a long, thinner, posterior lip.
3. The ostium (external os) of the uterus, at the level of the upper end of the symphysis pubis.
4. The anterior fornix of the vagina, 1 cm or more from the vesico-uterine pouch. The posterior fornix covered with 1 cm or more of the recto-uterine pouch, which is the lowest part of the peritoneal cavity, when the subject is erect.
5. The urethra (3 cm long), the vagina, and the rectum parallel to one another and to the pelvis brim. The uterus, nearly at right angles to them, when the bladder is empty.

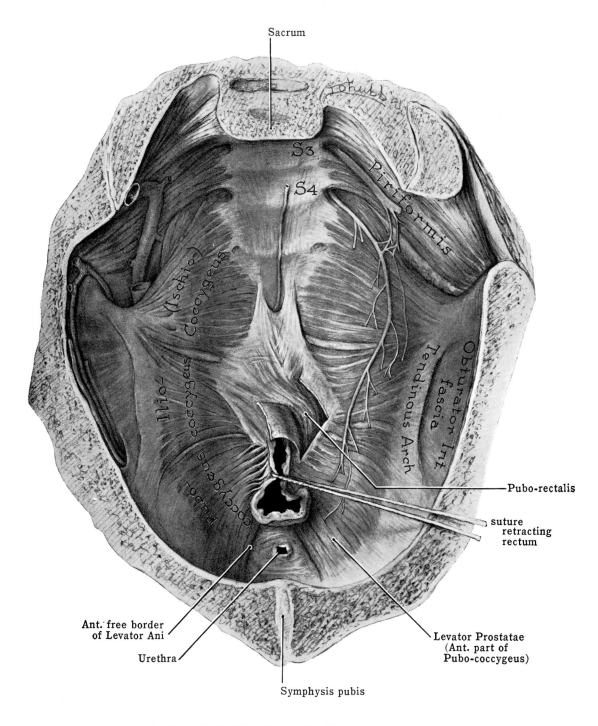

Sacrum

S3

S4

Piriformis

(Ischio)
Coccygeus

Ilio-coccygeus

Pubo-coccygeus

Obturator Int. fascia
Tendinous Arch

Pubo-rectalis

suture
retracting
rectum

Ant. free border
of Levator Ani

Urethra

Symphysis pubis

Levator Prostatae
(Ant. part of
Pubo-coccygeus)

3-12—FLOOR OF THE MALE PELVIS

The pelvic viscera are removed and the bony pelvis is sawn through transversely to demonstrate the Levatores Ani and Coccygei.

Observe:

1. Pubo-coccygeus arising mainly from the pubic bone, Ischio-coccygeus from the ischial spine, and Ilio-coccygeus from the tendinous arch in between. Pubo-coccygeus is strong (Fig. 3-55); Ilio-coccygeus is weak; Ischio-coccygeus is largely transformed into the sacrospinous ligament (Fig. 3-57); Pubo- and Ilio-coccygei together constitute Levator Ani.

2. The anterior free border of Pubo-coccygeus, the posterior free border of Ischio-coccygeus; the clefts at the borders of Ilio-coccygeus closed by areolar membranes.

3. The urethra passing between the anterior borders of Pubo-coccygei. The rectum perforating Pubo-coccygei, thus: (a) the anterior fibers of the muscles of opposite sides meet and unite in the central tendon of the perineum (perineal body) in front of the rectum; (b) the posterior fibers unite behind the rectum in an aponeurosis that extends backward to the anterior sacro-coccygeal ligament; (c) the middle fibers blend with the outer wall of the anal canal and pass between Internal and External Sphincters of the anus (Fig. 3-21); (d) on the right hand, this aponeurosis is reflected to show Pubo-rectalis (Fig. 3-39).

4. Branches of S3 and S4 supplying Levator Ani and Coccygeus. (The pudendal nerve, via its perineal branch (Fig. 3-71) also supplies Levator Ani.)

For a perineal view of Levatores Ani see Figure 3-21.

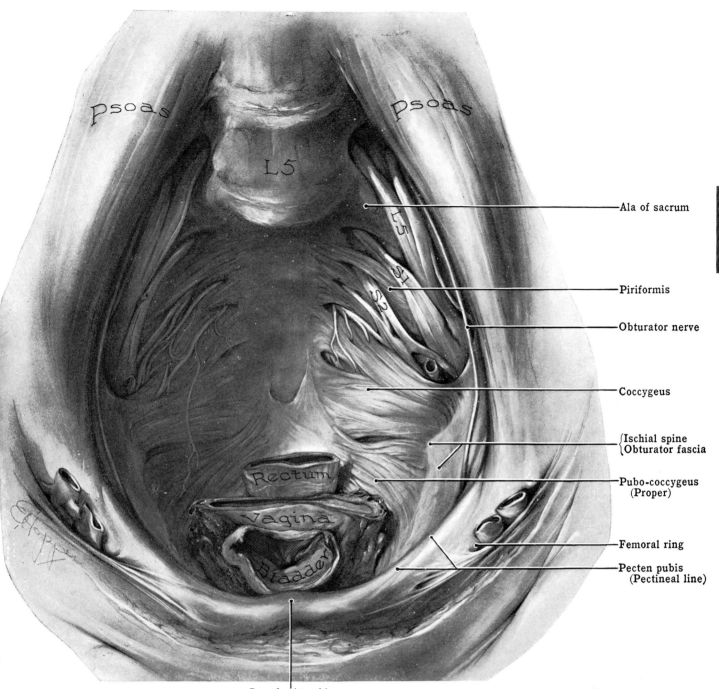

Psoas

Psoas

L5

L5

S1

S2

Ala of sacrum

Piriformis

Obturator nerve

Coccygeus

{Ischial spine
 {Obturator fascia

Pubo-coccygeus
(Proper)

Femoral ring

Pecten pubis
(Pectineal line)

Rectum

Vagina

Bladder

Symphysis pubis

3-13 — FLOOR OF THE FEMALE PELVIS

Observe:
1. The muscles of the pelvic floor.
2. The relative positions of bladder, vagina, and rectum.
3. The obturator nerve, derived from lumbar nerves 2, 3, 4, running along the side wall of the pelvis to enter the thigh through the obturator foramen.
4. The femoral ring, the doorway into the femoral canal, the site of femoral hernia.

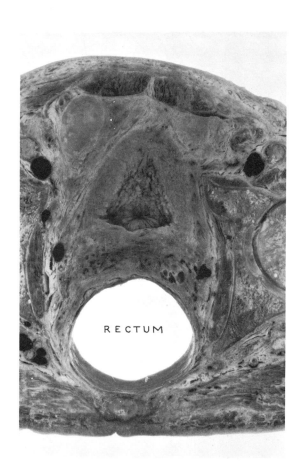

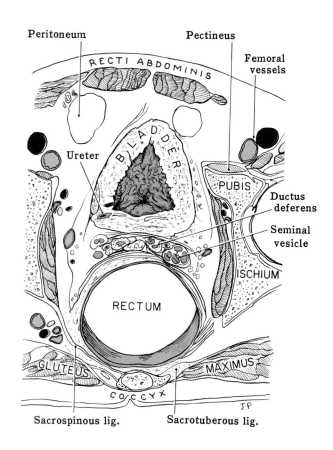

Peritoneum

Pectineus

Femoral vessels

RECTI ABDOMINIS

Ureter

BLADDER

PUBIS

Ductus deferens

Seminal vesicle

ISCHIUM

RECTUM

GLUTEUS

MAXIMUS

COCCYX

J.P.

Sacrospinous lig.

Sacrotuberous lig.

3-14—MALE PELVIS, ON HORIZONTAL SECTION

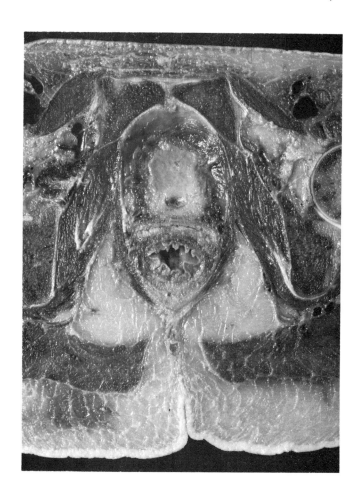

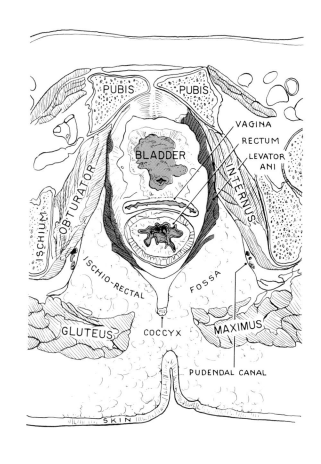

PUBIS

PUBIS

VAGINA

RECTUM

BLADDER

LEVATOR ANI

OBTURATOR

INTERNUS

ISCHIUM

ISCHIO-RECTAL

FOSSA

GLUTEUS

COCCYX

MAXIMUS

PUDENDAL CANAL

SKIN

3-15—FEMALE PELVIS, ON HORIZONTAL SECTION

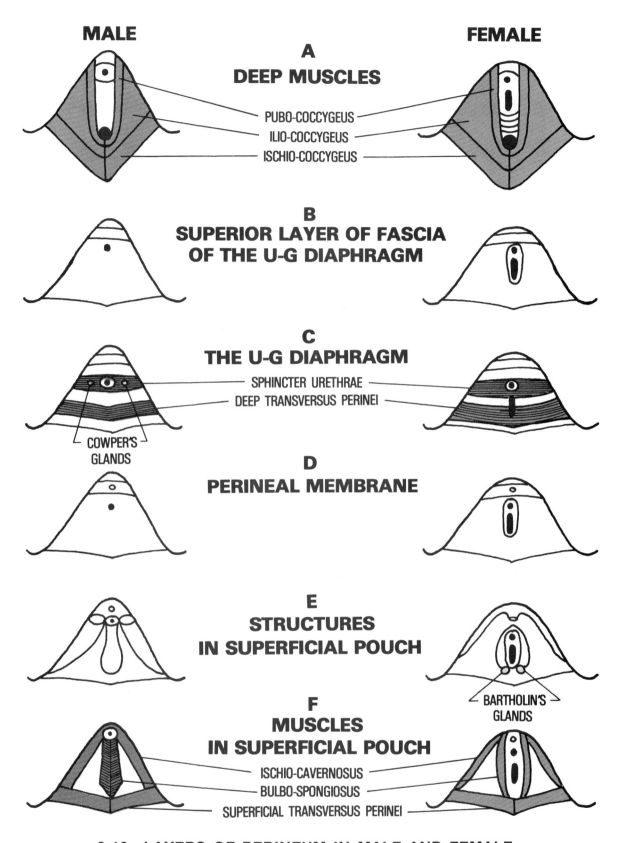

MALE **FEMALE**

A
DEEP MUSCLES

PUBO-COCCYGEUS
ILIO-COCCYGEUS
ISCHIO-COCCYGEUS

B
SUPERIOR LAYER OF FASCIA
OF THE U-G DIAPHRAGM

C
THE U-G DIAPHRAGM

SPHINCTER URETHRAE
DEEP TRANSVERSUS PERINEI

COWPER'S
GLANDS

D
PERINEAL MEMBRANE

E
STRUCTURES
IN SUPERFICIAL POUCH

BARTHOLIN'S
GLANDS

F
MUSCLES
IN SUPERFICIAL POUCH

ISCHIO-CAVERNOSUS
BULBO-SPONGIOSUS
SUPERFICIAL TRANSVERSUS PERINEI

3-16—LAYERS OF PERINEUM IN MALE AND FEMALE

These schematic diagrams show layers of the perineum built up from deep to superficial. In *A* the angle between the two ischio-pubic rami is almost filled by the three coccygeus muscles. The urethra (and vagina in the female) peer through anteriorly, the rectum posteriorly. A superior layer of fascia *B* and an inferior layer of fascia—the perineal membrane *D*—enclose a deep perineal space or pouch *C* containing two muscles and, in the male, the bulboure-thral (Cowper's) glands. The sandwich formed by the two layers of fascia and the contents of the deep pouch comprise the urogenital diaphragm.

Superficial and deep layers of perineal fascia attach to the ischio-pubic ramus and to the posterior margin of the uro-genital diaphragm and so enclose the *superficial* perineal space or pouch which contains the structures shown in *E* and the muscles shown in *F*. In the female, greater vestibu-lar (Bartholin's) glands lie behind the bulb of the vestibule.

Figures 3–31 to 3–36 show a sequential dissection of the female perineum from superficial to deep.

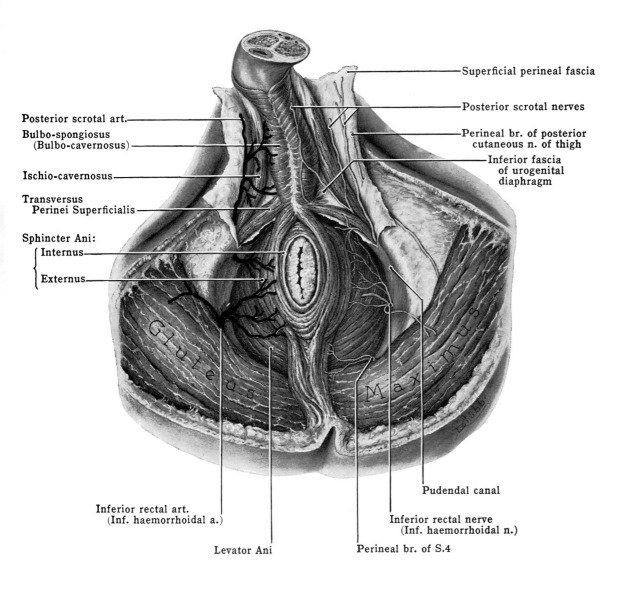

Posterior scrotal art.

Bulbo-spongiosus
(Bulbo-cavernosus)

Ischio-cavernosus

Transversus
Perinei Superficialis

Sphincter Ani:
Internus
Externus

Superficial perineal fascia

Posterior scrotal nerves

Perineal br. of posterior
cutaneous n. of thigh

Inferior fascia
of urogenital
diaphragm

Inferior rectal art.
(Inf. haemorrhoidal a.)

Levator Ani

Pudendal canal

Inferior rectal nerve
(Inf. haemorrhoidal n.)

Perineal br. of S.4

3-17—MALE PERINEUM

Observe in the anal region or triangle:
1. The anal orifice at the center of the anal triangle, surrounded by Sphincter Ani Externus, and an ischio-rectal fossa on each side.
2. The superficial fibers of Sphincter Ani Externus anchoring the anus in front to the perineal body, or central tendon of the perineum, and behind to the coccyx, here to the skin.
3. The ischio-rectal fossa, filled with fat, bounded medially by Levator Ani and Sphincter Ani Externus; laterally by Obturator Internus fascia; behind by Gluteus Maximus overlying the sacro-tuberous ligament; in front by the base of the perineal membrane (Figs. 3–18 and 3–19). The apex or roof is where the medial and lateral walls meet. The base or floor, formed by tough skin and deep fascia, is removed.
4. The inferior rectal nerve leaving the pudendal canal and, with the perineal branch of S.4, supplying Sphincter Ani Externus. Its cutaneous twigs to the anus are removed. The branch turning round Gluteus Maximus is replacing the perforating cutaneous nerve.

Observe in the urogenital region or triangle:
5. The superficial perineal fascia (of Colles) incised in the midline, freed from its attachment to the base of the perineal membrane, and reflected.
6. The cutaneous nerves and artery in the superficial perineal space (pouch).
7. The three paired superficial perineal muscles—Bulbo-spongiosus, Ischio-cavernosus, and Transversus Perinei Superficialis.
8. The exposed triangular portion of the perineal membrane or inferior fascia of urogenital diaphragm.

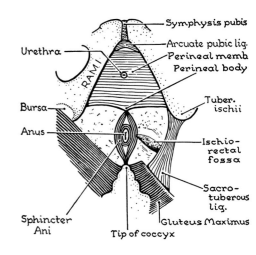

Symphysis pubis

Arcuate pubic lig.

Perineal memb

Perineal body

Urethra

RAMI

Bursa

Anus

Tuber.
ischii

Ischio-
rectal
fossa

Sacro-
tuberous
lig.

Sphincter
Ani

Tip of coccyx

Gluteus Maximus

3-18—PERINEAL REGION

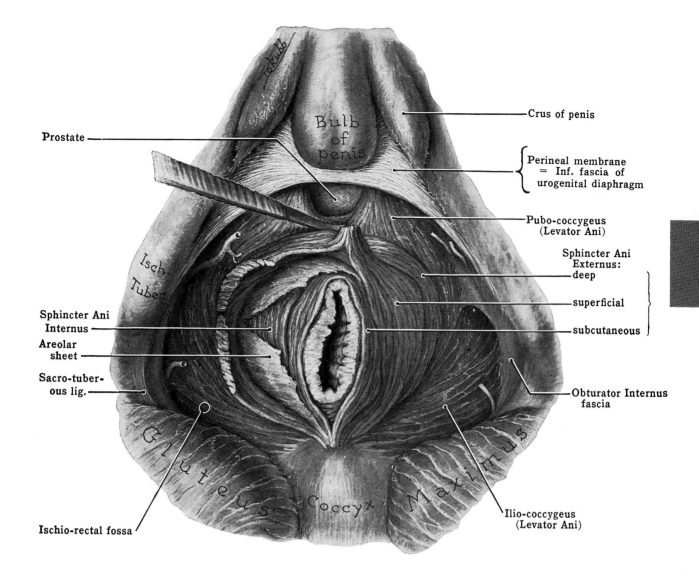

Prostate

Crus of penis

Bulb of penis

Perineal membrane = Inf. fascia of urogenital diaphragm

Pubo-coccygeus (Levator Ani)

 Isch. Tuber.

Sphincter Ani Externus:
deep

superficial

subcutaneous

Sphincter Ani Internus

Areolar sheet

Sacro-tuber-ous lig.

Obturator Internus fascia

Gluteus

Coccyx

Maximus

Ilio-coccygeus (Levator Ani)

Ischio-rectal fossa

3-19—DISSECTION OF SPHINCTER ANI EXTERNUS

Observe:

1. The 3 parts of this voluntary sphincter: (a) subcutaneous—encircling the anal orifice; (b) superficial—anchoring the anus in the median plane, to perineal body in front and to the coccyx behind (Fig. 3–17); and (c) deep—forming a wide encircling band (Fig. 3–39).

2. On the *left* of the figure: the superficial and deep parts of the sphincter are reflected, and the underlying sheet, consisting of areolar tissue, Levator Ani fibers, and outer longitudinal muscular coat of the gut, is cut, in order to reveal the inner circular muscular coat of the gut, which is thickened to form an involuntary Sphincter Ani Internus (Fig. 3–43).

3. The anterior free borders of Levatores Ani meeting in front of the anal canal, and pressed backward in order to expose the prostate.

4. On the *right* of the figure: the remains of the "false roof" of the ischio-rectal fossa, *i.e.*, a layer of fascia that stretches from Obturator Internus fascia to the thin fascia covering Levator Ani.

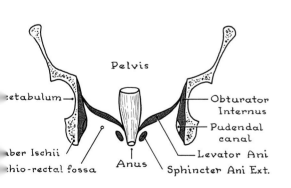

Pelvis

Acetabulum

Obturator Internus

Pudendal canal

Tuber Ischii

Ischio-rectal fossa

Anus

Levator Ani

Sphincter Ani Ext.

3-20—CORONAL SECTION

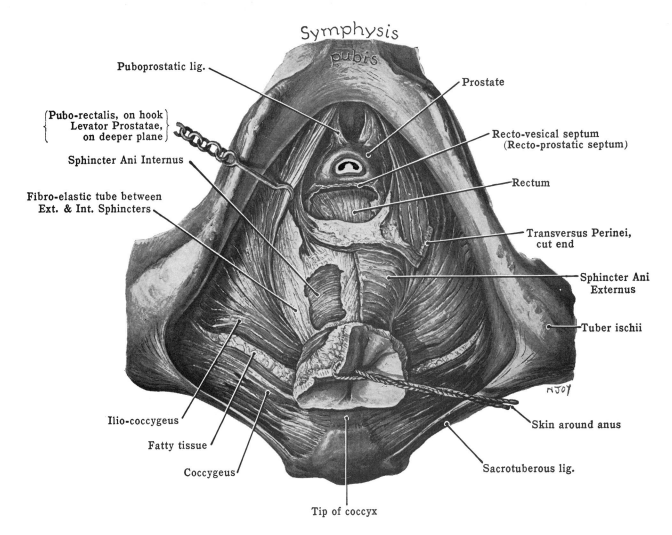

Symphysis
pubis

Puboprostatic lig.

{ Pubo-rectalis, on hook
Levator Prostatae,
on deeper plane }

Sphincter Ani Internus

Fibro-elastic tube between
Ext. & Int. Sphincters

Ilio-coccygeus

Fatty tissue

Coccygeus

Tip of coccyx

Prostate

Recto-vesical septum
(Recto-prostatic septum)

Rectum

Transversus Perinei,
cut end

Sphincter Ani
Externus

Tuber ischii

Skin around anus

Sacrotuberous lig.

3-21—LEVATORES ANI AND COCCYGEI, THE EXPOSURE OF THE PROSTATE

The urogenital diaphragm and its fasciae have been removed.

Observe:

1. The anal canal guarded by two sphincters: Sphincter Ani Internus, a continuation of the circular muscle coat of the gut; it is smooth muscle. Sphincter Ani Externus, extending below Sphincter Internus and having three parts; it is skeletal muscle.
2. Pubo-rectalis, the slinglike muscle uniting with Sphincter Externus behind and at the sides. (It is better shown in Figs. 3-10 and 3-39.)
3. The anal columns united half-way down the Internal Sphincter by semilunar anal valves.
4. The pecten, or smooth zone of simple stratified epithelium, between the anal valves above and the lower border of the Internal Sphincter below. It is transitional between intestinal mucosa above and skin having dermal papillae and appendages below.
5. The longitudinal muscle coat of the rectum and its fascia blending with the Levator Ani and its fasciae to form a fibroelastic tube which descends between the two Sphincters. From this tube septa pass through the Internal Sphincter to the submucous coat, through the External Sphincter to the skin, and, as the anal intermuscular septum, below the Internal Sphincter.

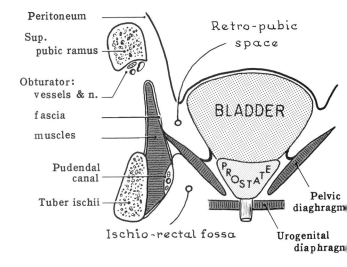

Peritoneum

Sup.
pubic ramus

Obturator:
vessels & n.

fascia

muscles

Pudendal
canal

Tuber ischii

Ischio-rectal fossa

Retro-pubic
space

BLADDER

PROSTATE

Pelvic
diaphragm

Urogenital
diaphragm

3-22—PELVIC DIAPHRAGM

Only the pelvic diaphragm (Levator Ani portion) intervenes between the ischio-rectal fossa and the retro-pubic space. For pelvic views of the Levator Ani see Figures 3–12 (male) and 3–13 (female).

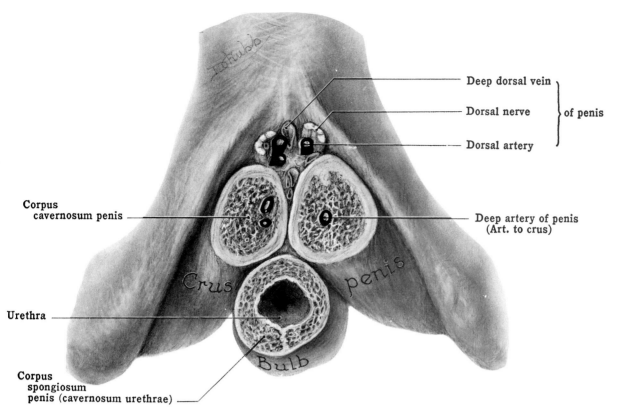

Deep dorsal vein ⎱
Dorsal nerve ⎰ of penis
Dorsal artery

Corpus cavernosum penis

Deep artery of penis (Art. to crus)

Urethra

Corpus spongiosum penis (cavernosum urethrae)

3-23—SECTION ACROSS THE ROOT OF THE PENIS

Observe that the urethra is dilated within the bulb of the penis.

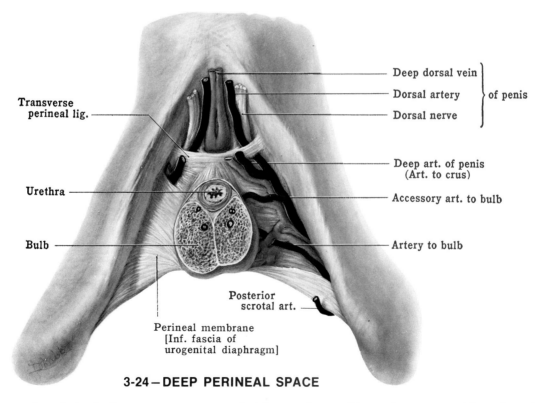

Transverse perineal lig.

Deep dorsal vein ⎱
Dorsal artery ⎰ of penis
Dorsal nerve

Urethra

Deep art. of penis (Art. to crus)

Accessory art. to bulb

Bulb

Artery to bulb

Posterior scrotal art.

Perineal membrane [Inf. fascia of urogenital diaphragm]

3-24—DEEP PERINEAL SPACE

The crura are removed and the bulb is cut shorter than in Figure 3–23 and viewed more from below.

On the *right* of page, the perineal membrane is in part removed and the deep perineal space is thereby opened.

Observe:

1. The fibers of the perineal membrane converging on the bulb and mooring it to the pubic arch.

2. The urethra, still membranous and bound to the dorsum of the bulb.

3. The septum in the bulb indicating its bilateral origin.

4. The artery to the bulb (here double); the artery to the crus, called the deep artery; and the dorsal artery which ends in the glans penis. The deep dorsal vein, originally double, which ends in the prostatic plexus.

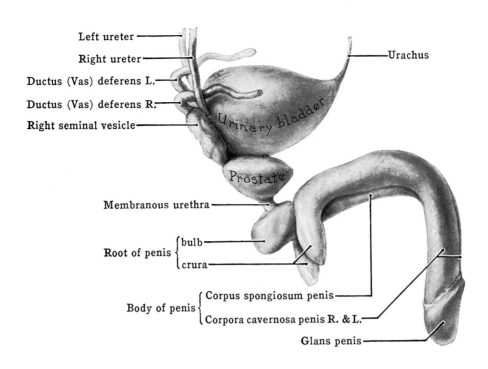

Left ureter

Right ureter

Ductus (Vas) deferens L.

Ductus (Vas) deferens R.

Right seminal vesicle

Urachus

Urinary bladder

Prostate

Membranous urethra

Root of penis { bulb

crura

Body of penis { Corpus spongiosum penis

Corpora cavernosa penis R. & L.

Glans penis

3-25—LOWER PARTS OF MALE GENITAL AND URINARY TRACTS

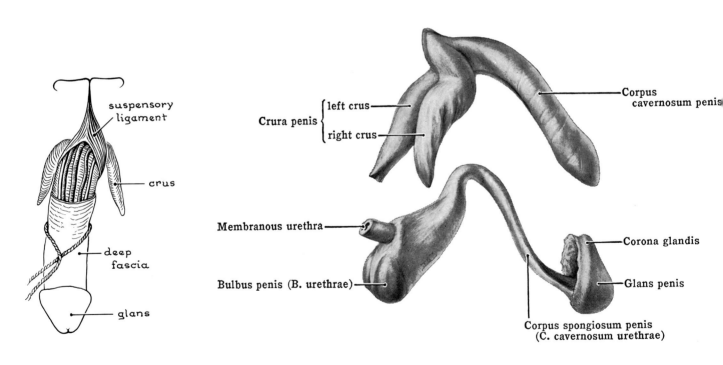

suspensory ligament

crus

deep fascia

glans

Crura penis { left crus

right crus

Corpus cavernosum penis

Membranous urethra

Bulbus penis (B. urethrae)

Corona glandis

Glans penis

Corpus spongiosum penis (C. cavernosum urethrae)

3-26—SUSPENSORY LIGAMENT

A triangular band fixed to the front of the symphysis.

3-27—DISSECTION OF THE PENIS

The corpus spongiosum is separated from the corpora cavernosa penis. The natural flexures are preserved.

Observe:

1. The corpora cavernosa penis bent where that organ is slung by the suspensory ligament of the penis, and grooved by encircling vessels.
2. The corpus spongiosum penis massed (a) below the urethra posteriorly to form the bulb of the penis, and (b) above the urethra anteriorly to form the glans which fits like a cap on the blunt ends of the corpora cavernosa penis.

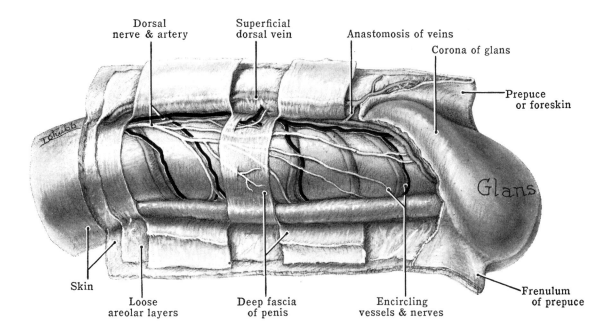

3-28 — PENIS, SIDE VIEW

The three tubular envelopes of the penis are reflected; so also is the prepuce.

Observe:

1. The skin carried forward as the prepuce.
2. The loose, laminated, subcutaneous, areolar tissue (Dartos and Colles' fascia), called the superficial fascia of the penis, carried forward into the prepuce, and containing the superficial dorsal vein. This vein begins in the prepuce, anastomoses with the deep dorsal vein from the glans, and ends in the superficial inguinal veins.
3. The deep fascia penis (Buck's fascia). It ends at the glans penis.
4. Large encircling tributaries of the deep dorsal vein; thread-like companion arteries; numerous oblique nerves.
5. The vessels and nerves at the neck plunging into the glans penis.

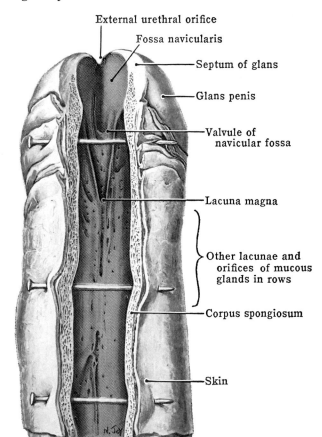

3-29 — INTERIOR OF THE SPONGY URETHRA

A longitudinal incision was made on the urethral surface of the penis and carried through the floor of the urethra, so the view is of the dorsal surface of the interior of the urethra. The prepuce is retracted.

3-30 – CLITORIS

This miniature penis comprises two corpora cavernosa which are bent, suspended by a suspensory ligament, and capped by a glans as in the male (Fig. 3-27).

The male bulb and body of the corpus spongiosum (corpus cavernosum urethrae) are represented in the female by the bulbs of the vestibule and the commissure of the bulbs (pars intermedia). These, however, are not regarded as part of the clitoris and are not traversed by the urethra.

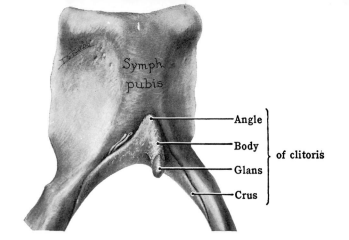

Symph. pubis

Angle
Body
Glans } of clitoris
Crus

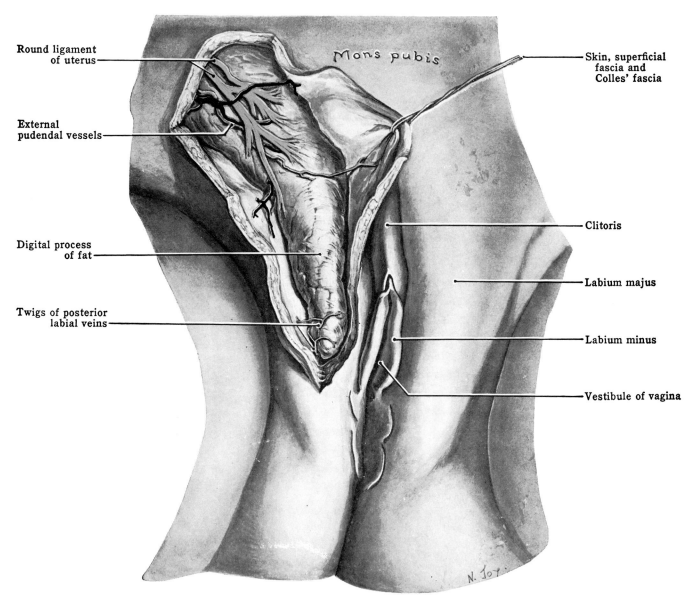

Round ligament of uterus

External pudendal vessels

Digital process of fat

Twigs of posterior labial veins

Mons pubis

Skin, superficial fascia and Colles' fascia

Clitoris

Labium majus

Labium minus

Vestibule of vagina

N. Joy.

3-31 – FEMALE PERINEUM – I

Observe:

1. A long digital, or fingerlike, process of fat lying deep to the subcutaneous fatty tissue and descending far into the labium majus.
2. The thin transparent veil of areolar tissue that encloses the process and through which its lobulated nature is apparent.
3. The round ligament of the uterus (ligamentum teres uteri) ending as a branching band of fascia which spreads out in front of the digital process; and the (superficial) external pudendal vessels crossing the process.
4. The vestibule of the vagina is the region between the labia minora and external to the hymen.

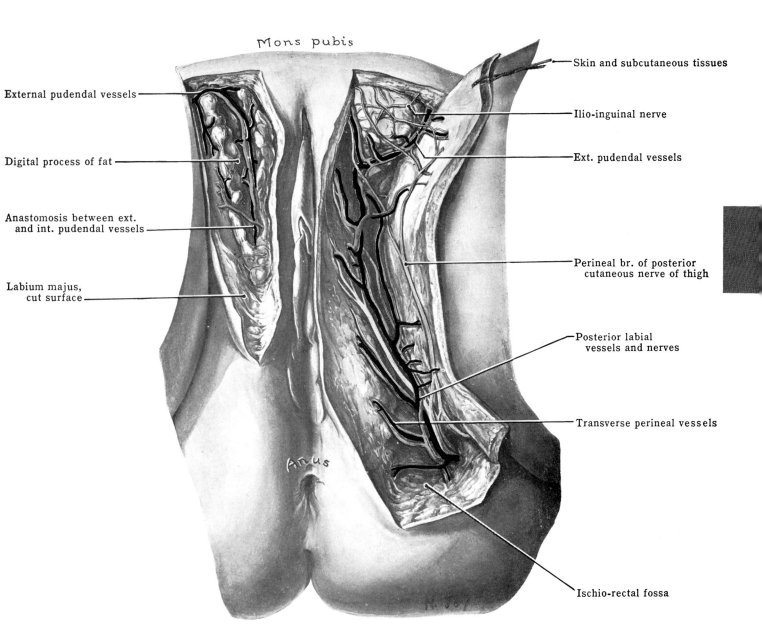

Mons pubis

Skin and subcutaneous tissues

External pudendal vessels

Ilio-inguinal nerve

Digital process of fat

Ext. pudendal vessels

Anastomosis between ext. and int. pudendal vessels

Perineal br. of posterior cutaneous nerve of thigh

Labium majus, cut surface

Posterior labial vessels and nerves

Transverse perineal vessels

Anus

Ischio-rectal fossa

3-32—FEMALE PERINEUM—II

Observe:

1. On the *left* of the page, the lobulated digital process of fat opened up, and anastomotic vessels, which unite the external to the internal pudendal vessels, running through its long axis like a core.

2. On the *right* of the page, the digital process of fat is largely removed. The posterior labial vessels and nerves (S2, S3), joined by the perineal branch of the posterior cutaneous nerve of the thigh (S1, S2, S3) running forward almost to the mons pubis, the vessels there anastomosing with the external pudendal vessels, and the nerves meeting the ilio-inguinal nerve (L1).

3. Note that there is here a hiatus in the numerical sequence of the nerve segments, accounted for by the fact that L2, L3, L4, L5, and S1 are drawn down into the lower limb with the result that L1 is succeeded by S2 (*cf.* the similar hiatus in the pectoral region between C3 and C4 and T2. See Figs. 6–14 and 5–41).

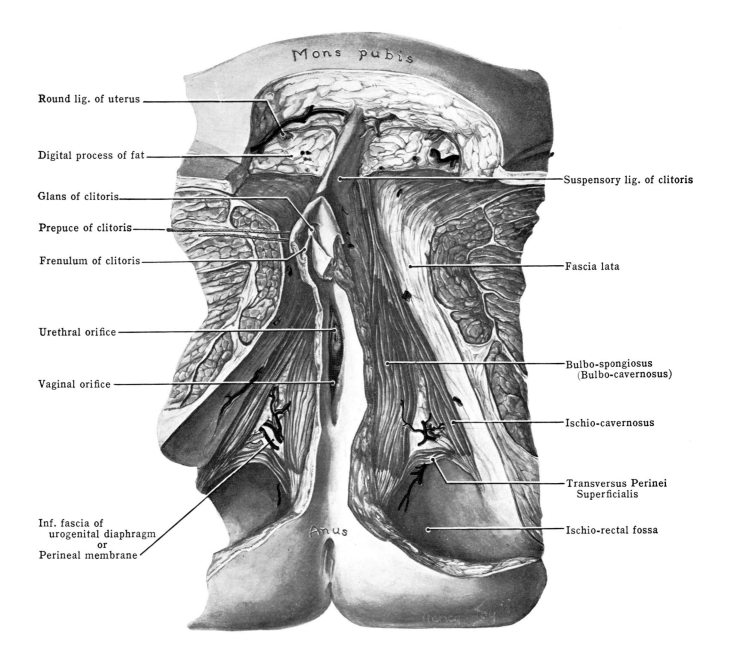

Mons pubis

Round lig. of uterus

Digital process of fat

Glans of clitoris

Prepuce of clitoris

Frenulum of clitoris

Urethral orifice

Vaginal orifice

Inf. fascia of
urogenital diaphragm
or
Perineal membrane

Anus

Suspensory lig. of clitoris

Fascia lata

Bulbo-spongiosus
(Bulbo-cavernosus)

Ischio-cavernosus

Transversus Perinei
Superficialis

Ischio-rectal fossa

3-33—FEMALE PERINEUM—III

Observe:
1. The thickness of the superficial fatty tissue at the mons and the encapsuled digital process of fat deep to this. The suspensory ligament of the clitoris descending from the linea alba and symphysis pubis.
2. The prepuce of the clitoris, thrown like a hood over the clitoris, and the anterior ends of the labia minora uniting to form the frenulum of the clitoris.
3. The 3 muscles on each side: Bulbo-spongiosus, Ischio-cavernosus, and Transversus Perinei Superficialis, which when slightly separated reveal the perineal membrane. The Bulbo-spongiosus overlies the bulb of the vestibule. In the male, the muscles of the two sides are united by a median raphe (Fig. 3–17); in the female, the orifice of the vagina separates the two.
4. The pinpoint orifices of the right and the left paraurethral duct below the urethral orifice.

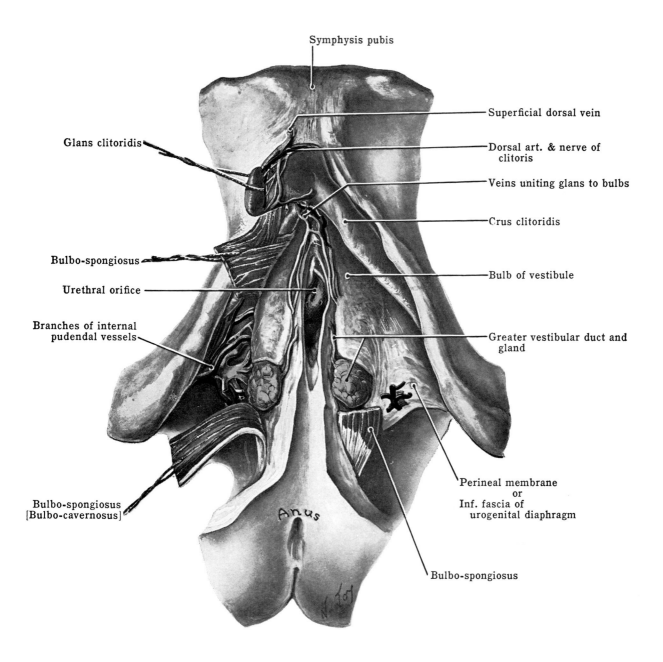

Symphysis pubis

Glans clitoridis

Bulbo-spongiosus

Urethral orifice

Branches of internal
pudendal vessels

Bulbo-spongiosus
[Bulbo-cavernosus]

Anus

Superficial dorsal vein

Dorsal art. & nerve of
clitoris

Veins uniting glans to bulbs

Crus clitoridis

Bulb of vestibule

Greater vestibular duct and
gland

Perineal membrane
or
Inf. fascia of
urogenital diaphragm

Bulbo-spongiosus

3-34 — FEMALE PERINEUM — IV

On the right side the perineal membrane is removed.

Observe:

1. The right Bulbo-spongiosus, divided and reflected on the right side, and largely excised on the left side.
2. The glans clitoridis, pulled over to the right side and the dorsal vessels and nerve of the clitoris running to it, as in the male (Fig. 3–28).
3. The bulb of the vestibule (paired, right and left) one on each side of the vestibule of the vagina and therein differing from the bulb of the penis, which is unpaired (Fig. 3–23).

4. Veins (pars intermedia) connecting the bulbs of the vestibule to the glans of the clitoris.
5. The greater vestibular gland situated at the posterior blunt end of the bulb and like it covered with Bulbo-spongiosus, and having a long duct (about $^3/_4$ inch) which opens into the vestibule (*cf.* the bulbo-urethral glands and ducts.)
6. The perineal membrane to which the bulb is seen to be fastened, on the left side. On the right side the membrane is cut away, thereby revealing the vessels of the bulb and the dorsal nerve and vessels of the clitoris within the deep perineal space.

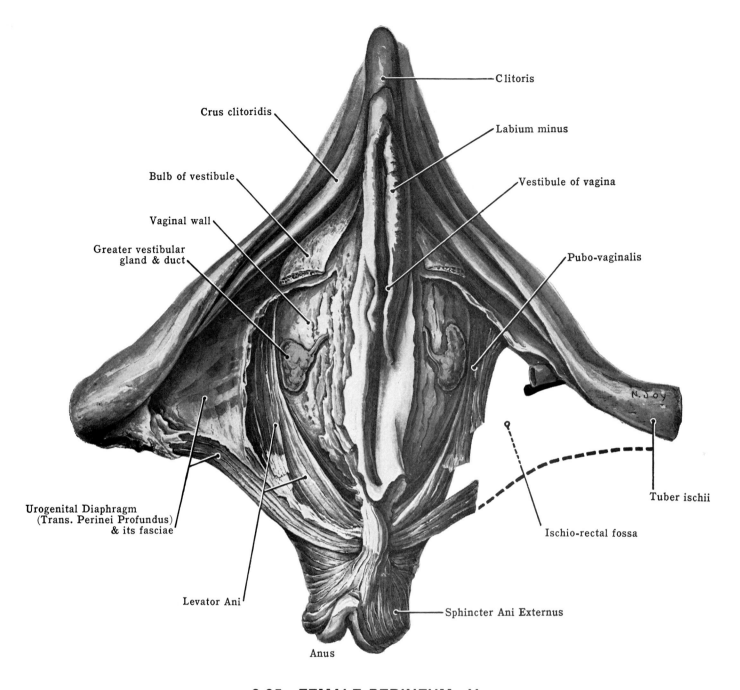

Clitoris

Crus clitoridis

Labium minus

Bulb of vestibule

Vestibule of vagina

Vaginal wall

Greater vestibular
gland & duct

Pubo-vaginalis

Urogenital Diaphragm
(Trans. Perinei Profundus)
& its fasciae

Tuber ischii

Ischio-rectal fossa

Levator Ani

Sphincter Ani Externus

Anus

3-35—FEMALE PERINEUM—V

THE UROGENITAL DIAPHRAGM

Observe:

1. The bulbs of the vestibule cut short. The urogenital diaphragm and its fascia partly cut away on the right side and extensively cut away on the left.
2. The urogenital diaphragm—a sheet of striated muscle, mainly Deep Transversus Perinei, placed between a superior and an inferior sheet of fascia, and having 2 parts: (a) a posterior which is a strong fleshy band that meets its fellow in the perineal body (central tendon of perineum) and (b) an anterior which is more areolar than fleshy.
3. Medially, the diaphragm and its fasciae have been raised from the sloping inferior surface of Levator

Ani and detached from the sloping outer wall of vagina with which it fuses.

4. The anterior parts of the Levatores Ani (Pubovaginales) meeting behind the vaginal orifice. The greater vestibular glands and the bulbs of the vestibule applied to the sides of the vagina, medial to the anterior borders of the Levatores Ani.
5. The laminated nature of this part of the wall of the vagina. It is laminated because several layers of fascia fuse with it and lose their identity in it, namely, the superficial perineal fascia, the diaphragm and its fasciae, and the fasciae of the Levatores Ani.

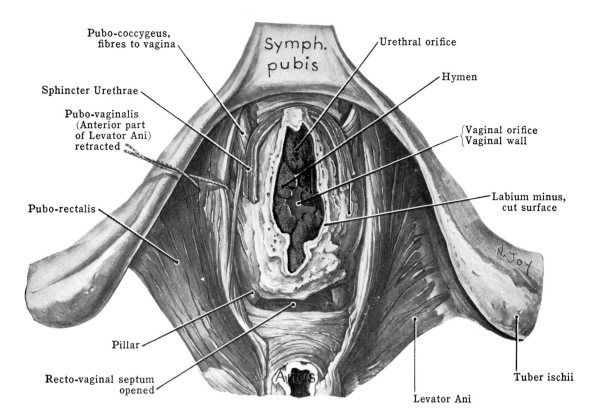

Pubo-coccygeus, fibres to vagina

Sphincter Urethrae

Pubo-vaginalis (Anterior part of Levator Ani) retracted

Pubo-rectalis

Pillar

Recto-vaginal septum opened

Symph. pubis

Urethral orifice

Hymen

Vaginal orifice / Vaginal wall

Labium minus, cut surface

Anus

Tuber ischii

Levator Ani

N. Joy

3-36 — FEMALE PERINEUM — VI

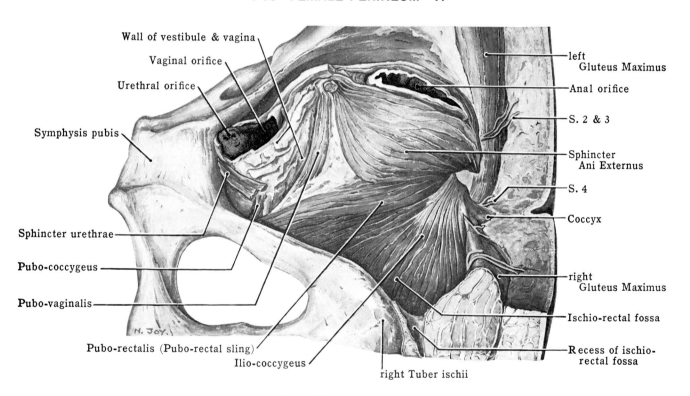

Wall of vestibule & vagina

Vaginal orifice

Urethral orifice

Symphysis pubis

Sphincter urethrae

Pubo-coccygeus

Pubo-vaginalis

Pubo-rectalis (Pubo-rectal sling)

Ilio-coccygeus

right Tuber ischii

N. Joy.

left Gluteus Maximus

Anal orifice

S. 2 & 3

Sphincter Ani Externus

S. 4

Coccyx

right Gluteus Maximus

Ischio-rectal fossa

Recess of ischio-rectal fossa

3-37 — THE LEVATOR ANI

(The front view (*above*) and the obliquely tilted side view (*below*) are of different specimens.)

Observe:
1. The Sphincter Urethrae, of striated muscle, like a saddle, resting on the urethra and straddling the vagina.
2. The labia minora, cut short, bounding the vestibule of the vagina, and the hymen, separating the vestibule from the cavity of the vagina.
3. The subdivisions of Levator Ani: Pubo-coccygeus (including Pubo-coccygeus proper, Pubo-vaginalis, and Pubo-rectalis) and Ilio-coccygeus.
4. The recto-vaginal (areolar) septum easily converted into

a space, and limited on each side by a "fascial pillar" between rectal and vaginal fasciae.
5. The thick anterior pillar of Pubo-coccygeus Proper extending backward to the vagina. Pubo-vaginal fibers of opposite sides meeting behind the vaginal orifice. Pubo-rectal fibers of opposite sides meeting in the ano-rectal junction to form the "pubo-rectal sling". Ilio-coccygeus meeting its fellow in an aponeurosis between rectum and coccyx.
6. The vaginal wall laminated, as in Figure 3–35.

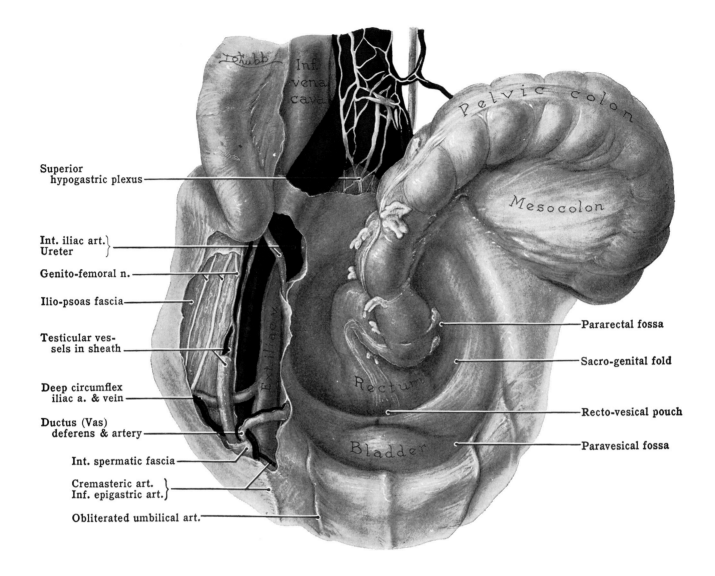

Superior hypogastric plexus

Int. iliac art.
Ureter

Genito-femoral n.

Ilio-psoas fascia

Testicular vessels in sheath

Deep circumflex iliac a. & vein

Ductus (Vas) deferens & artery

Int. spermatic fascia

Cremasteric art.
Inf. epigastric art.

Obliterated umbilical art.

Inf. vena cava

Pelvic colon

Mesocolon

Pararectal fossa

Sacro-genital fold

Rectum

Recto-vesical pouch

Bladder

Paravesical fossa

3-38—MALE PELVIS AND SURROUNDINGS, ANTERO-SUPERIOR VIEW

Observe:

1. One limb of the inverted V-shaped root of the sigmoid (pelvic) mesocolon ascending near the external iliac vessels; the other descending to the third piece of the sacrum. At the apex is the mouth of the intersigmoid recess, and behind the mouth lies the left ureter.

2. The teniae coli forming two wide bands; one in front of the rectum, the other behind.

3. The crescentic fold of peritoneum called the sacro-genital fold.

4. The superior hypogastric plexus (presacral nerve) lying in the fork of the aorta and in front of the left common iliac vein.

5. The ureter adhering to the peritoneum, crossing the external iliac vessels, and descending in front of the internal iliac artery. The ductus deferens and its artery also adhering to peritoneum, crossing the external iliac vessels, and then hooking round the inferior epigastric artery to join the other constituents of the spermatic cord.

6. The genito-femoral nerve on the Psoas fascia. Its two lateral (femoral) branches become cutaneous; its medial (genital) branch supplies Cremaster and becomes cutaneous.

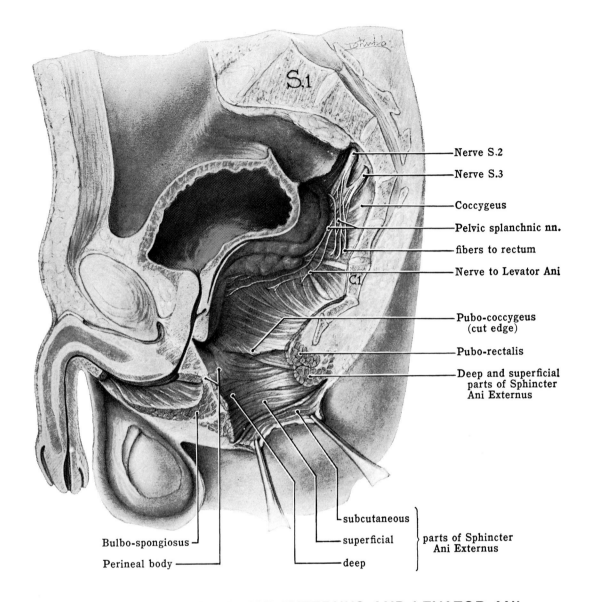

Nerve S.2

Nerve S.3

Coccygeus

Pelvic splanchnic nn.

fibers to rectum

Nerve to Levator Ani

Pubo-coccygeus
(cut edge)

Pubo-rectalis

Deep and superficial
parts of Sphincter
Ani Externus

Bulbo-spongiosus

Perineal body

subcutaneous

superficial

deep

parts of Sphincter
Ani Externus

3-39—SPHINCTER ANI EXTERNUS AND LEVATOR ANI

This is Figure 3-10 from which the rectum, anal canal, and bulb of the penis are removed.

Observe:

1. The subcutaneous fibers of Sphincter Ani Externus held reflected with forceps; the superficial fibers mingling posteriorly with deep fibers; and deep fibers mingling posteriorly with Pubo-rectalis (inferior fibers of Pubo-coccygeus) which forms a sling that occupies the angle between the rectum and the anal canal (Fig. 3-21).
2. Pubo-coccygeus divided to allow of the removal of the anal canal to which it is in part attached (Fig. 3-12).
3. The ampulla of the deferent duct and seminal vesicle curving to fit the cyclindrical rectum.

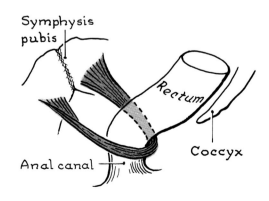

Symphysis
pubis

Rectum

Coccyx

Anal canal

3-40—PUBO-RECTALIS

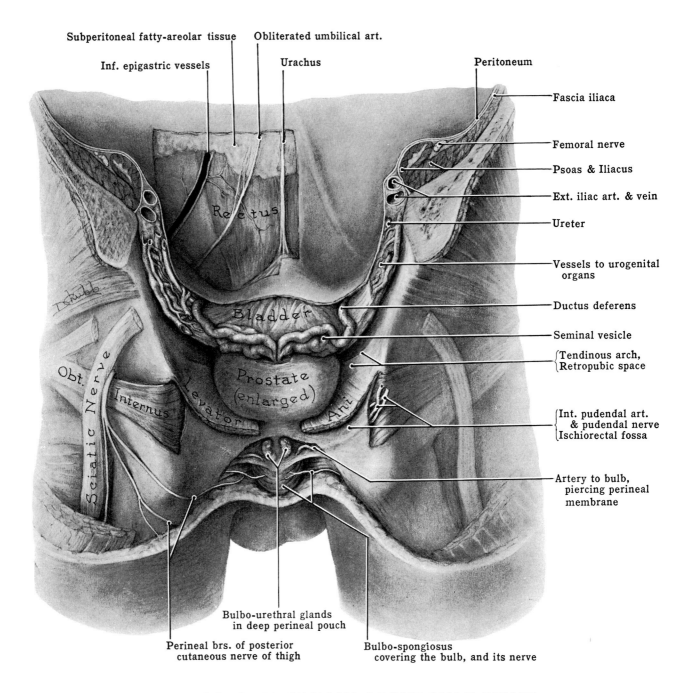

Subperitoneal fatty-areolar tissue   Obliterated umbilical art.

Inf. epigastric vessels   Urachus   Peritoneum

Fascia iliaca

Femoral nerve

Psoas & Iliacus

Ext. iliac art. & vein

Ureter

Vessels to urogenital organs

Ductus deferens

Seminal vesicle

Tendinous arch, Retropubic space

Int. pudendal art. & pudendal nerve / Ischiorectal fossa

Artery to bulb, piercing perineal membrane

Rectus

Ischium

Bladder

Obt. Nerve

Internus

Levator Ani

Prostate (enlarged)

Sciatic Nerve

Bulbo-urethral glands in deep perineal pouch

Perineal brs. of posterior cutaneous nerve of thigh

Bulbo-spongiosus covering the bulb, and its nerve

3-41—CORONAL SECTION OF THE MALE PELVIS

The section is through the pelvis just in front of the rectum. The view is of the anterior portion from behind.

Observe:

1. The inferior epigastric artery and venae comitantes entering the Rectus sheath, while the obliterated umbilical artery and the urachus, like the bladder, remain subperitoneal.

Terminology. The peritoneal folds having these 3 structures in their free edges are called: lateral, medial, and median umbilical folds.

2. The femoral nerve lying between Psoas and Iliacus outside the Psoas fascia, which is attached to the pelvic brim, while the external iliac artery and vein lie inside.

3. The ductus deferens and ureter, both subperitoneal. Near the bladder, the ureter accompanying a leash of vesical vessels enclosed in recto-vesical fascia.

4. Levator Ani and its fascial coverings separating the retropubic (prevesical) space from the ischiorectal fossa.

5. The free, anterior borders of Levatores Ani. They are the width of the handle of a scalpel apart.

6. The bulbo-urethral glands (Cowper's) and the artery to the bulb lying above the perineal membrane (inferior fascia of the urogenital diaphragm), *i.e.*, in the deep perineal space.

7. Obturator Internus making a right-angled turn as it escapes from its osseo-fascial pocket (Fig. 4–46).

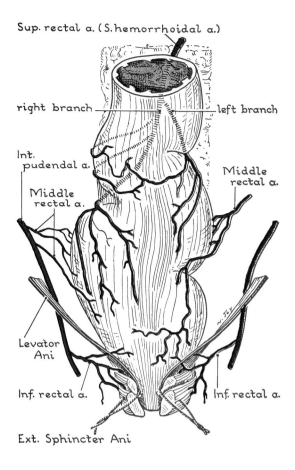

Sup. rectal a. (S. hemorrhoidal a.)

right branch — — left branch

Int. pudendal a.

Middle rectal a.

Middle rectal a.

Levator Ani

Inf. rectal a. — Inf. rectal a.

Ext. Sphincter Ani

3-42 ARTERIES OF THE RECTUM AND ANAL CANAL, FRONT VIEW

The Levatores Ani are semidiagrammatic.

Observe:

1. The branches of the right and left divisions of the superior rectal artery obliquely encircling the rectum, much as the branches of the dorsal arteries of the penis encircle the penis (Fig. 3–28).
2. The middle rectal arteries (branches of the internal iliac arteries) are usually small. In this specimen the right artery is small, but the left one is large and partly replaces the left division of the superior rectal artery.
3. The inferior rectal arteries (branches of the internal pudendal arteries) are largely expended on the anal canal.
4. Similarly, venous drainage is via superior, middle, and inferior rectal veins. There is a rich anastomosis between all three. Since the superior rectal vein drains into the portal system while middle and inferior veins drain into the systemic, this is an important area of porta-caval anastomosis.

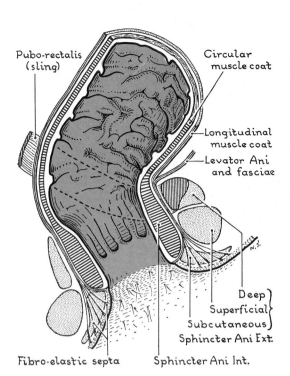

Pubo-rectalis (sling)

Circular muscle coat

Longitudinal muscle coat

Levator Ani and fasciae

Deep
Superficial
Subcutaneous
Sphincter Ani Ext.

Fibro-elastic septa   Sphincter Ani Int.

3-43 ANAL SPHINCTERS

In this median section observe:
1. The Sphincter Ani Internus is a thickening of the inner circular muscular coat of the canal.
2. The three parts of Sphincter Ani Externus. Its deep part is associated with the Pubo-rectalis posteriorly.

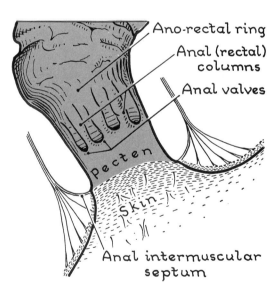

Ano-rectal ring

Anal (rectal) columns

Anal valves

Pecten

Skin

Anal intermuscular septum

3-44 ANAL CANAL

Observe:
1. Anal columns: vertical folds of mucosa containing twigs of superior rectal artery and vein. Varicosity of these veins produces internal hemorrhoids.
2. The lower ends of anal columns joined by crescentic folds, anal valves. Together they form the sinuous pectinate line.
3. Varicosity of the veins draining the end of the anal canal—inferior rectal veins—produces external hemorrhoids.

See Wilde, F. R. (1949) The anal intermuscular septum. *Brit. J. Surg.*, 36: 279.

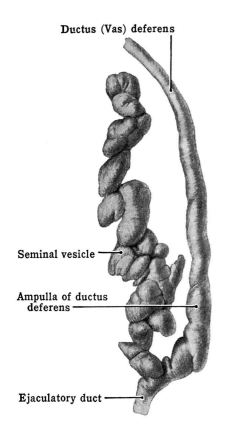

Ductus (Vas) deferens

Seminal vesicle

Ampulla of ductus deferens

Ejaculatory duct

3-45 SEMINAL VESICLE, UNRAVELED

The vesicle is a tortuous tube with numerous outpouchings. The lower end of the ampulla of the ductus deferens has similar outpouchings.

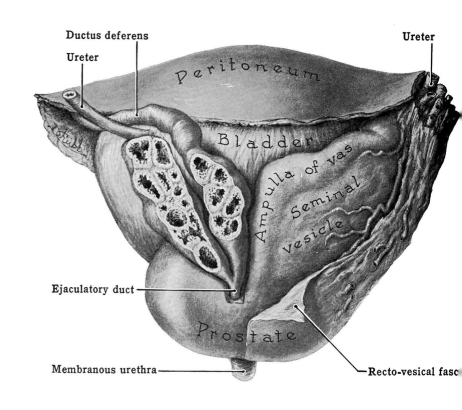

Ductus deferens

Ureter

Ureter

Peritoneum

Bladder

Ampulla of vas

Seminal vesicle

Ejaculatory duct

Prostate

Membranous urethra

Recto-vesical fasc

3-46 BLADDER, DEFERENT DUCTS, SEMINAL VESICLES, AND PROSTATE, FROM BEHIND

The left vesicle and ampulla are dissected free and sliced open. (For cross-section, see Fig. 3-14.)

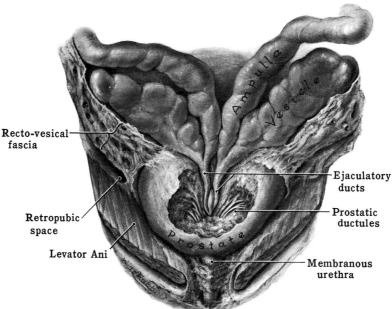

Recto-vesical fascia

Retropubic space

Levator Ani

Ampulla

Vesicle

Ejaculatory ducts

Prostatic ductules

Prostate

Membranous urethra

3-47 PROSTATE, DISSECTED, FROM BEHIND

Observe:

1. The right and the left ejaculatory duct, each formed where the duct of a seminal vesicle joins the ampullary end of a deferent duct.
2. The prostatic utricle (uterus masculinus), lying in between the ends of the two ejaculatory ducts, and all three flattened from side to side. All three opened into the prostatic urethra (Fig. 3-48).
3. The prostatic ductules, in all about 63 in number and mostly opening on to the prostatic sinus (Fig. 3-48).

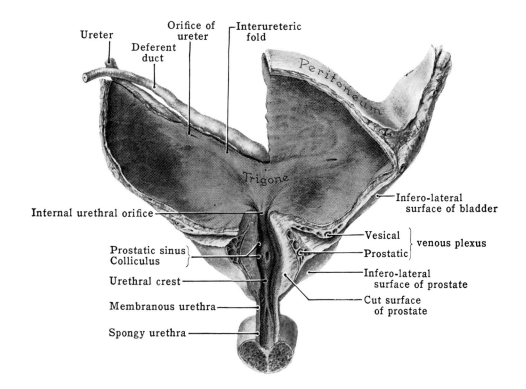

3-48 INTERIOR OF THE MALE URINARY BLADDER AND PROSTATIC URETHRA

The anterior parts of the bladder, prostate, and urethra were cut away. The knife was then carried through the posterior wall of the bladder, at the upper border of the right ureter and interureteric fold, which unites the two ureters along the upper limit of the smooth trigone.

Observe:
1. The right ureter not joining the bladder wall but traversing it obliquely as far as its slitlike orifice, which is situated 2 to 4 cm from the left orifice.
2. The mucous membrane, smooth over the trigone, but rugose elsewhere, especially when the bladder is empty.
3. The slight fullness behind the internal urethral orifice, which, when exaggerated, becomes the uvula vesicae.
4. The mouth of the prostatic utricle (not labeled) at the summit of the colliculus, on the urethral crest, and the orifice of an ejaculatory duct on each side of the utricle.
5. The urethral crest extending rather higher than usual and bifurcating rather lower than usual.
6. The prostatic fascia enclosing a venous plexus.

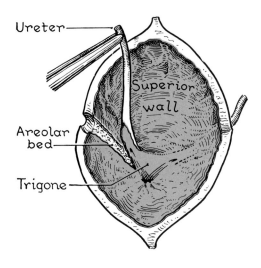

3-49 THE TRIGONE

Sketch of the ureter being raised from its areolar bed as it passes obliquely through the muscular coat of the bladder to blend with the trigonal muscle.

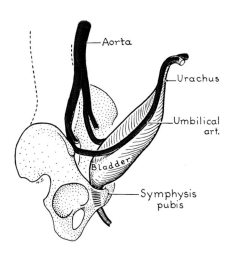

3-50 BLADDER AT BIRTH

At birth the bladder is abdominal in position and fusiform in shape and the two umbilical arteries lie along its sides.

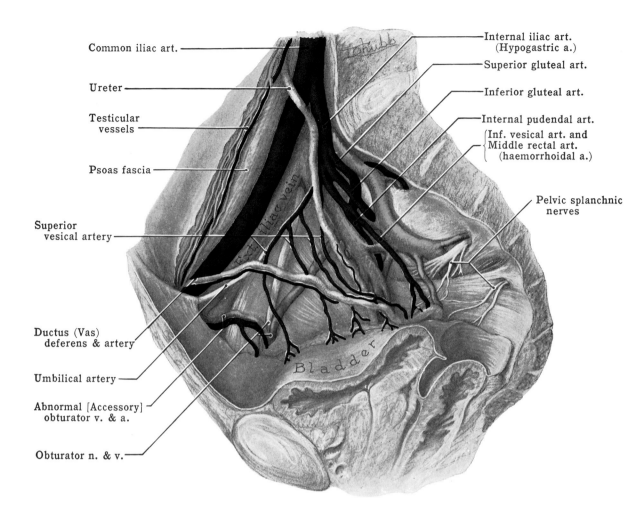

Common iliac art.

Ureter

Testicular
vessels

Psoas fascia

Superior
vesical artery

Ductus (Vas)
deferens & artery

Umbilical artery

Abnormal [Accessory]
obturator v. & a.

Obturator n. & v.

Internal iliac art.
(Hypogastric a.)

Superior gluteal art.

Inferior gluteal art.

Internal pudendal art.

Inf. vesical art. and
Middle rectal art.
(haemorrhoidal a.)

Pelvic splanchnic
nerves

Bladder

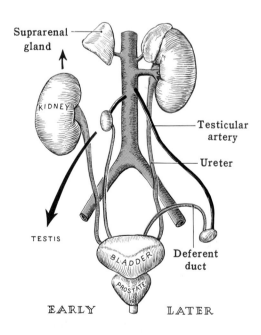

Suprarenal
gland

KIDNEY

TESTIS

BLADDER

PROSTATE

Testicular
artery

Ureter

Deferent
duct

EARLY   LATER

3-52 THREE PAIRED GLANDS

This diagram explains developmentally
how the ureter comes to be crossed by testicular vessels in the abdomen and the
deferent duct in the pelvis.

3-51 SIDE WALL OF THE MALE PELVIS

Observe:

1. The ureter and ductus deferens running a strictly subperitoneal course across the external iliac vessels, umbilical artery,
obturator nerve and vessels, and each receiving a branch
from a vesical artery.

The ureter crosses the exterior iliac artery at its origin (at
common iliac bifurcation); the ductus crosses it at its termination (at the deep inguinal ring). See also Figure 2-99.

2. The umbilical artery, obliterated beyond the origin of the last
superior vesical artery and creating a peritoneal fold.

3. The obturator artery here springing from the inferior epigastric artery; i.e., the artery is "abnormal." There are here both
a normal and an "abnormal" obturator vein.

4. The veins forming an open network through which the arteries are threaded.

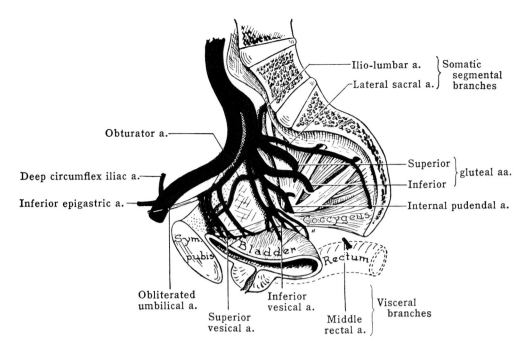

3-53   ILIAC ARTERIES AND THEIR BRANCHES, SIDE VIEW

Observe:

1. The common iliac artery having two terminal branches, but no collateral branches.
2. The external iliac artery having two collateral branches, and ending as the femoral artery.
3. The internal iliac (hypogastric) artery ending as an anterior and a posterior division. From these, branches arise variably. Commonly, as here, the ilio-lumbar artery, lateral sacral artery, and superior gluteal artery spring from the posterior division; the others spring from the anterior division.
4. Of the 10 branches of the internal iliac artery, the obliterated umbilical, which in the fetus passed to the placenta, and 3 others are visceral; 2 supply the 5th lumbar and the sacral segments and are somatic segmental; 3 enter the gluteal region; and 1 passes to the front of the thigh.

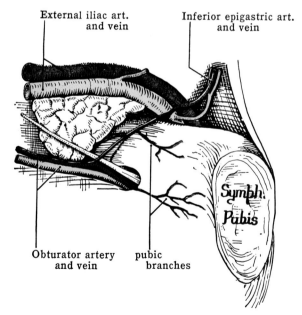

A Normal

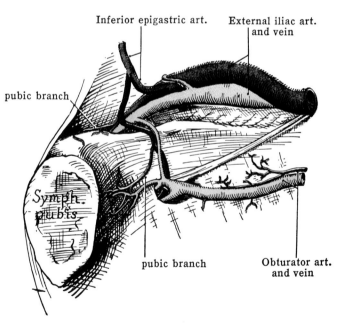

B Abnormal [Accessory]

3-54   NORMAL AND ABNORMAL (ACCESSORY) OBTURATOR ARTERIES

A. The pubic branch of the obturator artery anastomoses behind the body of the pubis with the pubic branch of the inferior epigastric artery.

B. The obturator artery arises from the inferior epigastric via the pubic anastomoses.

In a study of 283 limbs, the obturator artery arose from the internal iliac in 70 per cent, from the inferior epigastric in 25.4 per cent, and nearly equally from both in 4.6 per cent.

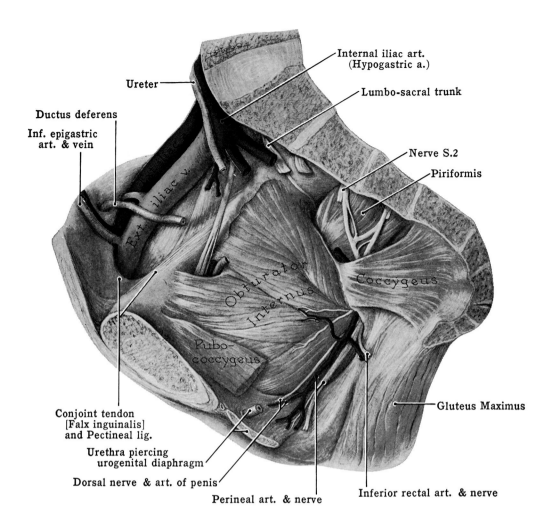

Internal iliac art.
(Hypogastric a.)

Lumbo-sacral trunk

Ureter

Ductus deferens

Inf. epigastric
art. & vein

Nerve S.2

Piriformis

Obturator Internus

Coccygeus

Pubo-coccygeus

Gluteus Maximus

Conjoint tendon
[Falx inguinalis]
and Pectineal lig.

Urethra piercing
urogenital diaphragm

Dorsal nerve & art. of penis

Perineal art. & nerve

Inferior rectal art. & nerve

3-55 MUSCLES OF THE PELVIS MINOR (MALE SPECIMEN)

Observe:

1. Obturator Internus padding the side wall of the pelvis and escaping through the lesser sciatic foramen; its nerve is seen. Piriformis padding the posterior wall and escaping through the greater sciatic foramen. Coccygeus concealing the sacro-spinous ligament. Pubo-coccygeus, which is the chief and strongest part of Levator Ani, springing from the body of the pubis.

2. The obturator nerve, artery, and vein escaping through the obturator foramen. The internal pudendal artery and the pudendal nerve making an exit through the greater foramen, and a re-entry through the lesser foramen, and taking a forward course (in the pudendal canal) within the Obturator Internus fascia to the urogenital diaphragm.

3. In the pelvis major, the deferent duct and the ureter descending across the external iliac artery and vein, the Psoas fascia, and the pelvic brim to enter the pelvis minor or true pelvis.

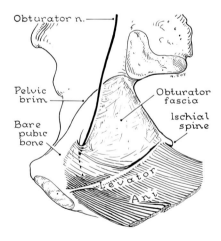

Obturator n.

Pelvic
brim

Bare
pubic
bone

Obturator
fascia

Ischial
spine

Levator Ani

3-56 ORIGIN OF LEVATOR ANI

From the body of the pubis, the ischial spine, and the tendinous arch in between. Compare with Figure 3–12.

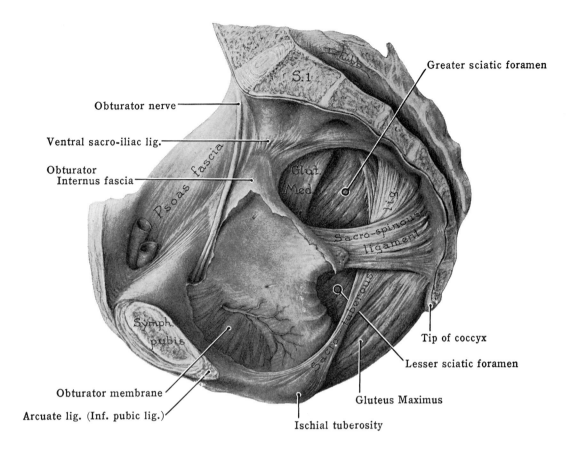

Obturator nerve

Ventral sacro-iliac lig.

Obturator
Internus fascia

Greater sciatic foramen

S.1

Glut. Med.

Psoas fascia

Sacro-spinous ligament

Symph. pubis

Sacro-tuberous lig.

Tip of coccyx

Lesser sciatic foramen

Obturator membrane

Arcuate lig. (Inf. pubic lig.)

Gluteus Maximus

Ischial tuberosity

3-57 WALLS OF THE PELVIS MINOR (FEMALE SPECIMEN)

In front is the pubis; behind are the sacrum and coccyx. Postero-laterally, the coccyx and lower part of the sacrum are fastened to the ischial tuberosity by the sacro-tuberous ligament and to the ischial spine by the sacro-spinous ligament, while the upper part of the sacrum is joined by the ilium by the ventral sacro-iliac ligament. Anterior to the sacro-tuberous ligament are the greater and lesser sciatic foramina, the one being above, and the other below, the sacro-spinous ligament.

Antero-laterally, the fascia covering Obturator Internus is snipped away and Obturator Internus removed from its osseo-fascial pocket, thereby exposing the ischium and obturator membrane. The mouth of this pocket is the lesser sciatic foramen. Through it Obturator Internus escapes from the pelvis, and the grooves made by its tendon are conspicuous.

Obturator Internus fascia is attached along the line of the obturator nerve above; to the sacro-tuberous ligamentum below; and to the posterior border of the body of the ischium behind.

Consult for other ligaments: Figure 3–7 (interosseous), Figure 4–36 (dorsal), and Figure 2–119 (ilio-lumbar).

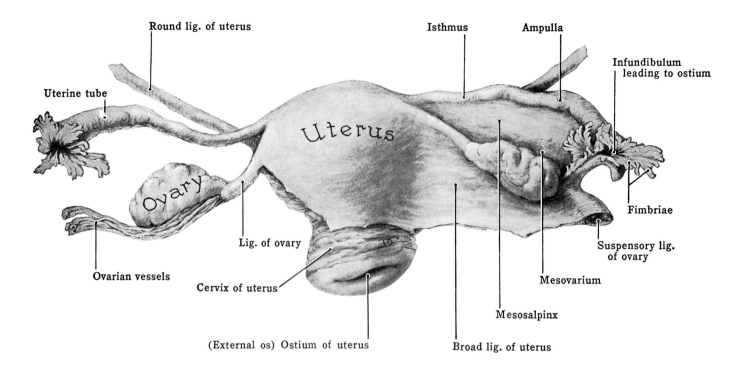

3-58 UTERUS AND ITS APPURTENANCES, FROM BEHIND

On the *left side:* The broad ligament of the uterus is removed, thereby setting free: uterine tube, round ligament of uterus, and ligament of ovary. These three are attached to the side of the uterus close together, at the junction of its fundus and body.

On the *right side:* The "mesentery" of the uterus and tube is called the broad ligament. The ovary is at-

tached (a) to the broad ligament by a "mesentery" of its own, called the mesovarium; (b) to the uterus by the ligament of the ovary; and (c) near the pelvic brim, by the suspensory ligament of the ovary, which transmits the ovarian vessels. The part of the broad ligament above the level of the mesovarium is called the mesosalpinx.

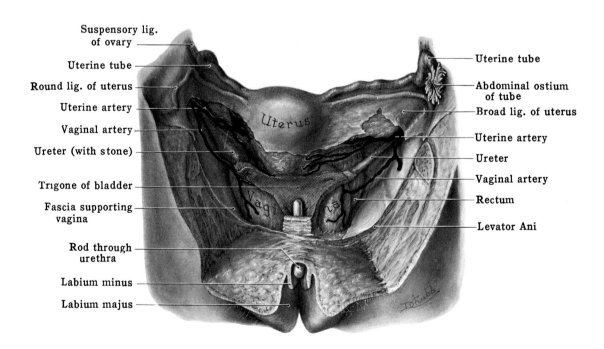

3-59 UTERUS IN ITS SETTING, VIEWED FROM THE FRONT

The pubic bones and the bladder, trigone excepted, are removed. Note: (a) Ureters, trigone of bladder, and urethra in relation to the asymmetrically placed uterus and vagina; (b) ostium of the left uterine tube here happens to face forward; and (c) a calculus (stone) in the dilated right ureter.

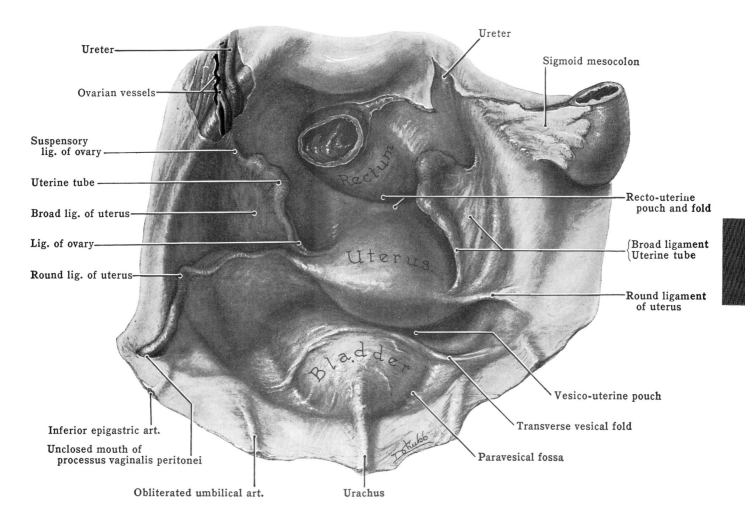

Ureter

Ovarian vessels

Suspensory lig. of ovary

Uterine tube

Broad lig. of uterus

Lig. of ovary

Round lig. of uterus

Inferior epigastric art.

Unclosed mouth of processus vaginalis peritonei

Obliterated umbilical art.

Ureter

Sigmoid mesocolon

Rectum

Uterus

Recto-uterine pouch and **fold**

{Broad ligament
{Uterine tube

Round ligament of uterus

Bladder

Vesico-uterine pouch

Transverse vesical fold

Urachus

Paravesical fossa

3-60 FEMALE TRUE PELVIS, FROM ABOVE

Observe:
1. The pear-shaped uterus asymmetrically placed, as usual; here leaning to the left.
2. The right round ligament (ligamentum teres uteri); here longer than the left and having an acquired "mesentery." The round ligament of the female takes the same subperitoneal course as the deferent duct of the male (Fig. 3–51).
3. The free edge of the medial four-fifths of the broad ligament (ligamentum latum uteri) occupied by the uterine tube (fallopian tube). The lateral one-fifth, occupied by the ovarian vessels, is the suspensory ligament of the ovary.
4. The ovarian vessels crossing the external iliac vessels very close to the ureter; the left ureter crossing at the apex of the inverted V-shaped root of the sigmoid mesocolon.

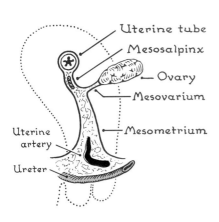

Uterine tube

Mesosalpinx

Ovary

Mesovarium

Mesometrium

Uterine artery

Ureter

3-61 BROAD LIGAMENT

A paramedian section to show "mesenteries" with prefix "meso-." *Salpinx* is Greek for trumpet or tube, *metro* for uterus.

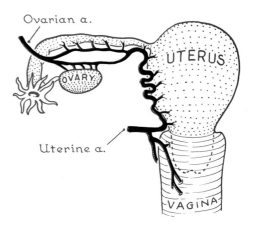

Ovarian a.

UTERUS

OVARY

Uterine a.

VAGINA

3-62 BLOOD SUPPLY

The ovarian and uterine arteries anastomose in the broad ligament.

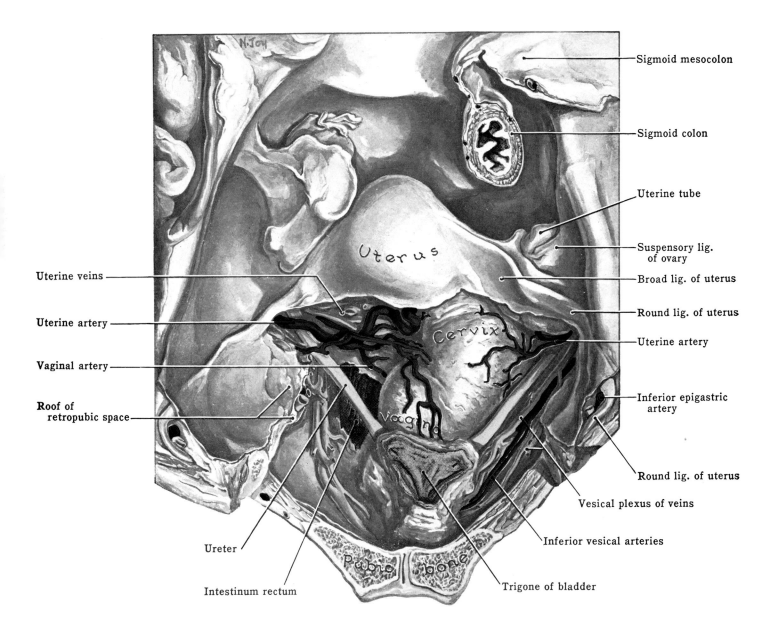

Labels on figure:
- Sigmoid mesocolon
- Sigmoid colon
- Uterine tube
- Suspensory lig. of ovary
- Broad lig. of uterus
- Round lig. of uterus
- Uterine artery
- Inferior epigastric artery
- Round lig. of uterus
- Vesical plexus of veins
- Inferior vesical arteries
- Trigone of bladder
- Uterine veins
- Uterine artery
- Vaginal artery
- Roof of retropubic space
- Ureter
- Intestinum rectum
- Uterus
- Cervix
- Vagina
- Pubic bone

3-63 FEMALE GENITAL ORGANS, FROM ABOVE

Part of the pubic bones and the whole of the bladder, excepting the trigone, are removed, and with them parts of the broad ligaments.

Observe:

1. The uterus asymmetrically placed, here leaning to the right, and also the vagina.
2. As a result, one ureter — in this instance the left — crosses the lateral fornix of the vagina and is close to the cervix of the uterus, the other being correspondingly farther away.
3. The uterine artery, lying with its veins in the base of the broad ligament and running up the side of the uterus.
4. A large vaginal artery, a branch of the uterine artery, supplying the cervix and the anterior surface of the vagina. The vaginal artery arising from the internal iliac artery and supplying the posterior surface of the vagina.
5. The rectal fascia (not labeled) intervening between the foregoing arteries and the rectum.

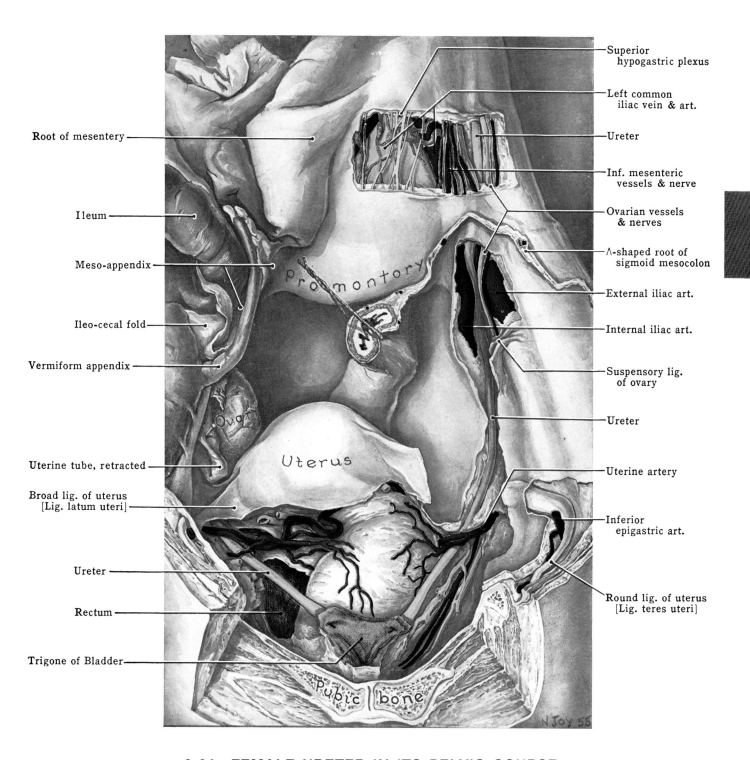

Root of mesentery

Ileum

Meso-appendix

Ileo-cecal fold

Vermiform appendix

Uterine tube, retracted

Broad lig. of uterus
[Lig. latum uteri]

Ureter

Rectum

Trigone of Bladder

Superior
hypogastric plexus

Left common
iliac vein & art.

Ureter

Inf. mesenteric
vessels & nerve

Ovarian vessels
& nerves

∧-shaped root of
sigmoid mesocolon

External iliac art.

Internal iliac art.

Suspensory lig.
of ovary

Ureter

Uterine artery

Inferior
epigastric art.

Round lig. of uterus
[Lig. teres uteri]

Promontory

Ovary

Uterus

Pubic bone

N Joy 55

3-64   FEMALE URETER IN ITS PELVIC COURSE

Observe:

1. The superior hypogastric plexus and some lymph vessels anterior to the left common iliac vein.

2. The left ureter, which has just been crossed by the ovarian vessels and nerves, and which is about to be crossed by sigmoid branches of the inferior mesenteric vessels. The apex of the "∧"-shaped root of the sigmoid mesocolon situated in front of the left ureter, and acting as a guide to it. The ureter, crossing the external iliac artery—at the bifurcation of the common iliac artery and close behind the ovarian vessels—and descending in front of the internal iliac artery. Its subperitoneal course from where it enters the pelvis to where it passes deep to the broad ligament and is there crossed by the uterine artery.

3. The vermiform appendix in one of its rarer positions—postileal. The ileo-cecal fold extending from the end of the ileum to the meso-appendix.

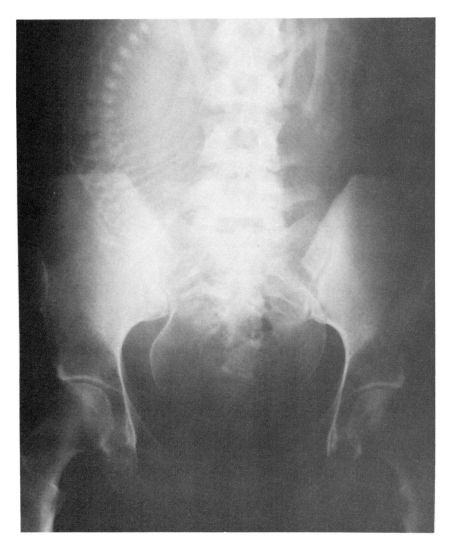

3-65A RADIOGRAPH OF FETUS

A. In this antero-posterior view of a fetus *in utero* the head is framed by the pelvic brim; the incompletely ossified vertebral column arches on the mother's right side. At the top of the radiograph, the flexed lower limb is visible.

B. Radiopaque contrast medium has been injected into the uterus through the external os. The triangular uterine cavity is clearly outlined. (Consult Fig. 3-65*C.*) Contrast medium has traveled through the uterine tubes (indicated by *small arrows*) to the infundibulum and leaks into the peritoneal cavity on both sides.

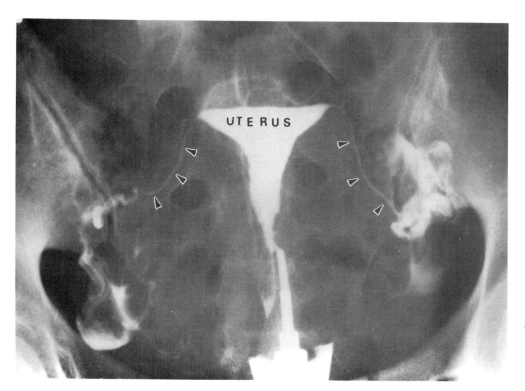

3-65B HYSTEROSALPINGOGRAM

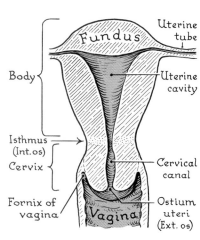

3-65C PARTS OF THE UTERUS

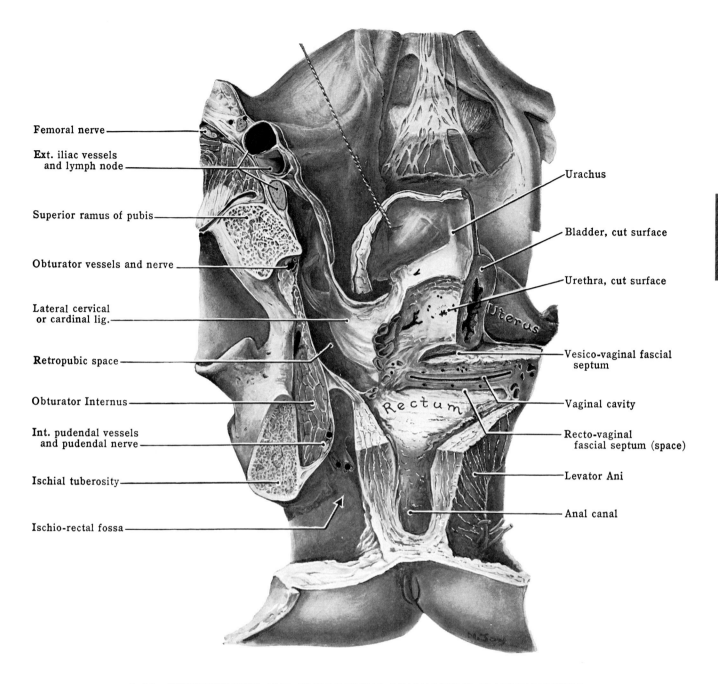

Femoral nerve

Ext. iliac vessels and lymph node

Superior ramus of pubis

Obturator vessels and nerve

Lateral cervical or cardinal lig.

Retropubic space

Obturator Internus

Int. pudendal vessels and pudendal nerve

Ischial tuberosity

Ischio-rectal fossa

Urachus

Bladder, cut surface

Urethra, cut surface

Uterus

Vesico-vaginal fascial septum

Vaginal cavity

Recto-vaginal fascial septum (space)

Levator Ani

Anal canal

Rectum

3-66 SUSPENSORY AND SUPPORTING MECHANISM OF THE VAGINA

This is approximately a coronal section of the pelvis. The neck of the bladder and the vagina are cut across transversely; the bladder is divided sagitally and is rotated backward.

Observe:
1. The partition that separates the retropubic space from the ischiorectal fossa to be formed merely by the thin origin of Levator Ani from the Obturator Internus fascia, and its areolar coverings.
2. The rectum supporting the posterior wall of the vagina; the posterior wall supporting the anterior wall; and the anterior wall supporting the bladder.
3. The dense areolar tissue within which the vesico-vaginal plexus of veins passes postero-superiorly to join the internal iliac veins. This acts as a suspensory ligament for the cervix and vagina. It is called the cardinal ligament.
4. The cardinal ligament blending with the fascia that encapsules the vagina and with adjacent fasciae.
5. Anteriorly, the fascia encapsuling the vagina blending with the vesical fascia and intimately adherent to the urethra; and posteriorly, loosely attached to the rectal fascia.

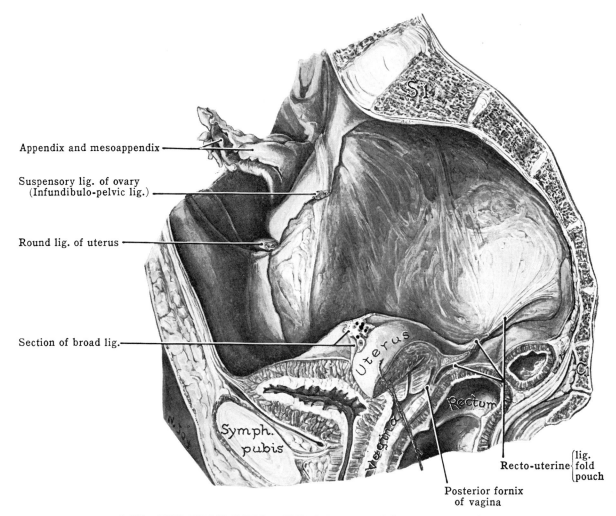

Appendix and mesoappendix

Suspensory lig. of ovary
(Infundibulo-pelvic lig.)

Round lig. of uterus

Section of broad lig.

Uterus

Vagina

Rectum

Symph. pubis

Recto-uterine {lig. fold pouch}

Posterior fornix
of vagina

3-67   BED FROM WHICH FIGURE 3-68 HAS BEEN REMOVED

Observe:

1. Divided at the pelvic brim are: the round ligament and the suspensory ligament with the contained ovarian vessels and nerves.

2. Divided at the side of the uterus are: the broad ligament with the uterine tube in its free margin the round ligament anteriorly, the ligament of the ovary posteriorly, and several branches of the uterine vessels.

3. The structures shown in Figure 3-68, thus set free, peeled off the side wall of the pelvis leaving the subperitoneal fatty-areolar tissue (tela subserosa) exposed.

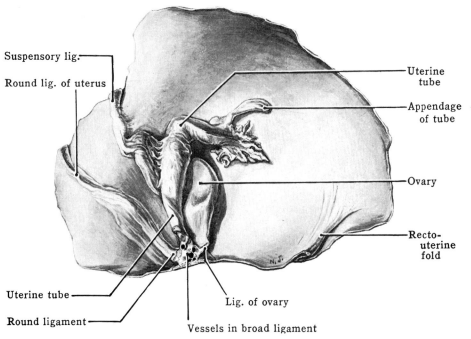

Suspensory lig.

Round lig. of uterus

Uterine tube

Appendage of tube

Ovary

Recto-uterine fold

Uterine tube

Round ligament

Lig. of ovary

Vessels in broad ligament

3-68   BROAD LIGAMENT AND RELATED STRUCTURES, REMOVED FROM FIGURE 3-67

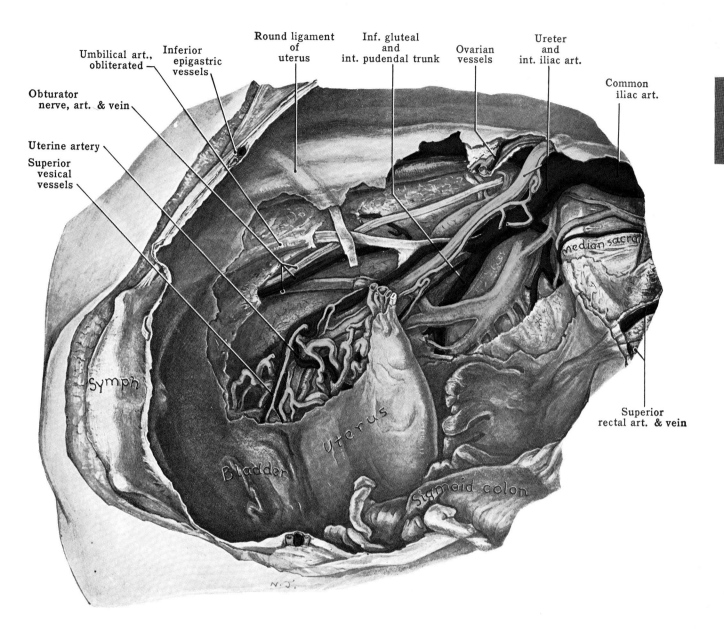

Obturator
nerve, art. & vein

Uterine artery
Superior
vesical
vessels

Umbilical art.,
obliterated

Inferior
epigastric
vessels

Round ligament
of
uterus

Inf. gluteal
and
int. pudendal trunk

Ovarian
vessels

Ureter
and
int. iliac art.

Common
iliac art.

median sacra

Symph

Bladder

Uterus

Sigmoid colon

Superior
rectal art. & vein

3-69 BLOOD VESSELS ON THE SIDE WALL OF THE FEMALE PELVIS

This figure is viewed from the left side of a supine cadaver.

In this older subject the uterus is retroverted. The inferior gluteal and internal pudendal arteries here spring from a common trunk, not separately as in Figure 3–51. The uterine plexus of veins communicates with the superior rectal veins, so providing in the female an additional area of porta-caval anastomosis.

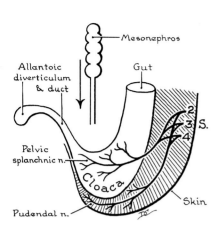

3-70 THE CLOACA

1. The embryonic cloaca about to receive the mesonephric duct (paired) from which springs the ureter (paired). The nerve supply to the whole region.

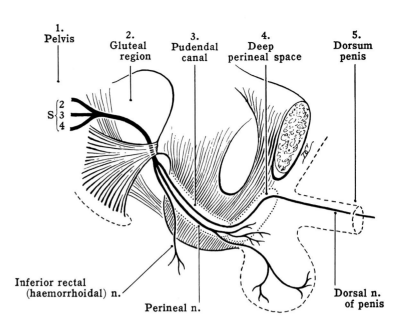

3-71 DIAGRAM OF THE PUDENDAL NERVE

Note:
a. The 5 regions in which it runs, and
b. The 3 divisions into which it divides.

3-72 DIAGRAM OF THE NERVE SUPPLY TO THE BLADDER AND URETHRA

(*Broken lines* indicate afferent fibers.)

Parasympathetic: The pelvic splanchnic nerves (nervi erigentes: S2, S3, S4) are the motor nerves to the bladder; when they are stimulated the bladder empties, the blood vessels dilate, and the penis becomes erect. They transmit also the sensory nerves of the bladder.

Sympathetic, through the superior hypogastric plexus (presacral nerve: Th. lower, L1, L2, L3), is motor to a continuous muscle sheet comprising the ureteric musculature, the trigonal muscle, and the muscle of the urethral crest. It also supplies the muscle of the epididymis, ductus deferens, seminal vesicle, and prostate. When the plexus is stimulated, the seminal fluid is ejaculated into the urethra but is hindered from entering the bladder perhaps by the muscle sheet which is drawn toward the internal urethral orifice. The sympathetic is also vaso-constrictor and to some slight extent it is sensory to the trigonal region.

The pudendal nerve is motor to the sphincter urethrae and sensory to the glans penis and the urethra.

It would seem that the sympathetic supply to the bladder has a vaso-constrictor and a sexual effect and that as regards micturition it is not antagonistic to the parasympathetic supply.

See Langworthy, O. R., Kolb, L. C., and Lewis, L. G. (1940) In *Physiology of Micturition.* The Williams & Wilkins Co., Baltimore.

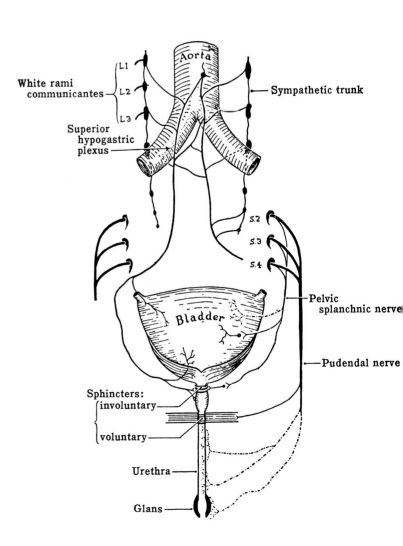

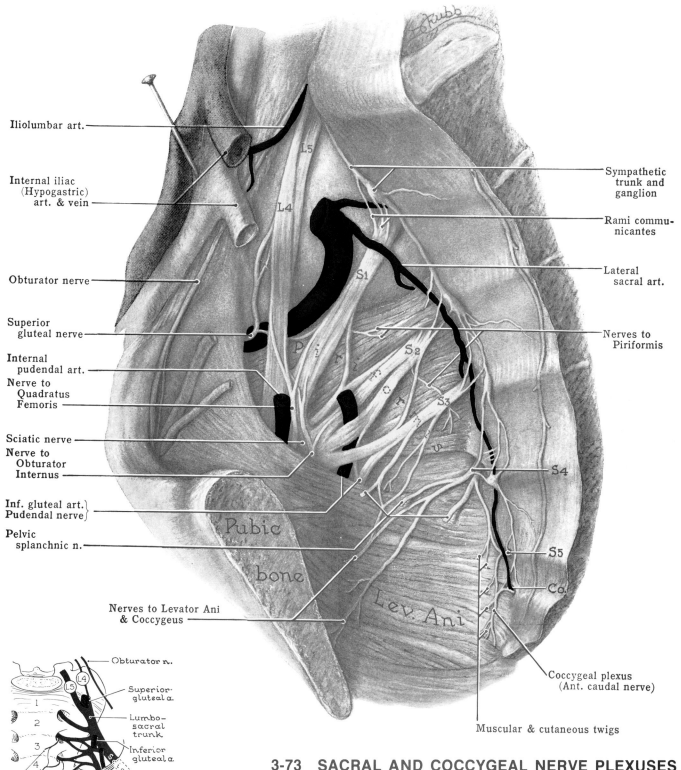

Iliolumbar art.

Internal iliac
(Hypogastric)
art. & vein

Obturator nerve

Superior
gluteal nerve

Internal
pudendal art.
Nerve to
Quadratus
Femoris

Sciatic nerve
Nerve to
Obturator
Internus

Inf. gluteal art.
Pudendal nerve

Pelvic
splanchnic n.

Nerves to Levator Ani
& Coccygeus

Sympathetic
trunk and
ganglion

Rami communi-
cantes

Lateral
sacral art.

Nerves to
Piriformis

Coccygeal plexus
(Ant. caudal nerve)

Muscular & cutaneous twigs

L5
L4
S1
S2
S3
S4
S5
Co.

Pubic
bone

Lev. Ani

3-73 SACRAL AND COCCYGEAL NERVE PLEXUSES

Obturator n.

Superior
gluteal a.

Lumbo-
sacral
trunk

Inferior
gluteal a.

Int. pudendal n. and a.
Anterior caudal n.
Pelvic splanchnic n.

L4
L5

3-74 SACRAL PLEXUS

e sacral plexus is pierced by the supe-
r and inferior gluteal arteries which
n dorsally, whereas the pudendal ar-
y continues downward and forward
ward the ischial spine.

Observe:
1. Either the sympathetic trunk or its ganglia sending gray rami communicantes to each sacral nerve and the coccygeal nerve.
2. The branch from L4 joining L5 to form the lumbo-sacral trunk.
3. The roots of S1, S2 supplying Piriformis; S3, S4 supplying Coccygeus and
Levator Ani; S(2), S3, S4 each contributing a branch to the formation of the
pelvic splanchnic nerve.
4. The sciatic nerve springing from segments L4, L5, S1, S2, S3; the pudendal
nerve from S2, S3, S(4); the coccygeal plexus from S4, S5, Co.
5. The ilio-lumbar artery accompanying nerve L5; the branches of the lateral
sacral artery accompanying the sacral nerves; the superior gluteal artery
passing backward between L5 and S1 — its position is not constant.

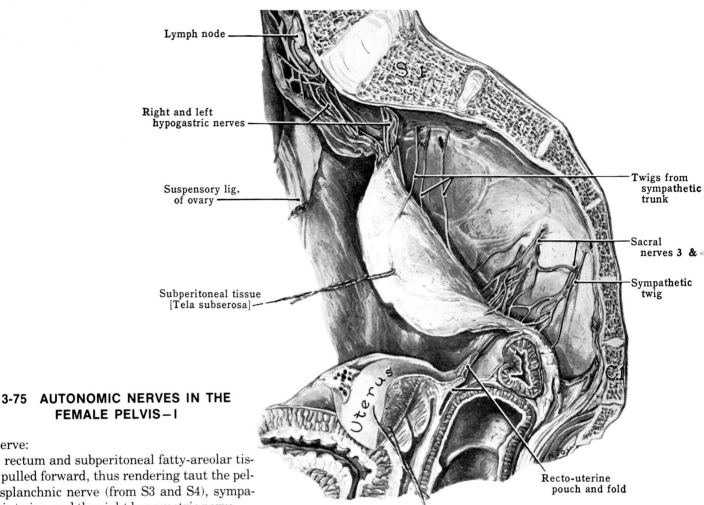

Lymph node

Right and left
hypogastric nerves

Suspensory lig.
of ovary

Subperitoneal tissue
[Tela subserosa]

S 1

Twigs from
sympathetic
trunk

Sacral
nerves 3 &

Sympathetic
twig

Uterus

Recto-uterine
pouch and fold

3-75 AUTONOMIC NERVES IN THE FEMALE PELVIS—I

Observe:
The rectum and subperitoneal fatty-areolar tissue pulled forward, thus rendering taut the pelvic splanchnic nerve (from S3 and S4), sympathetic twigs, and the right hypogastric nerve.

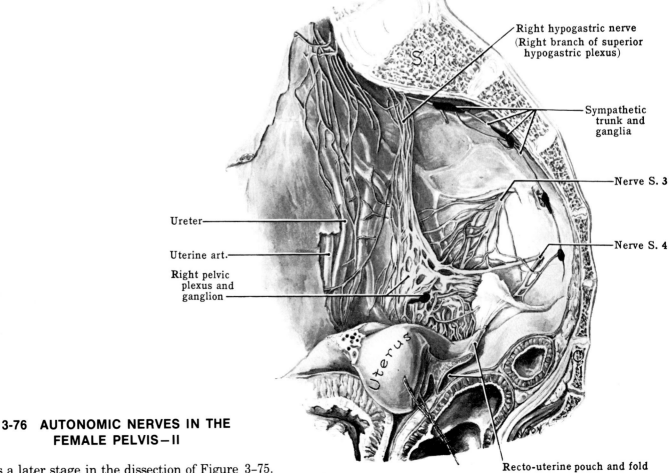

S. 1

Right hypogastric nerve
(Right branch of superior
hypogastric plexus)

Sympathetic
trunk and
ganglia

Nerve S. 3

Nerve S. 4

Ureter

Uterine art.

Right pelvic
plexus and
ganglion

Uterus

3-76 AUTONOMIC NERVES IN THE FEMALE PELVIS—II

This is a later stage in the dissection of Figure 3-75.

Recto-uterine pouch and fold

Lower Limb

CONTENTS

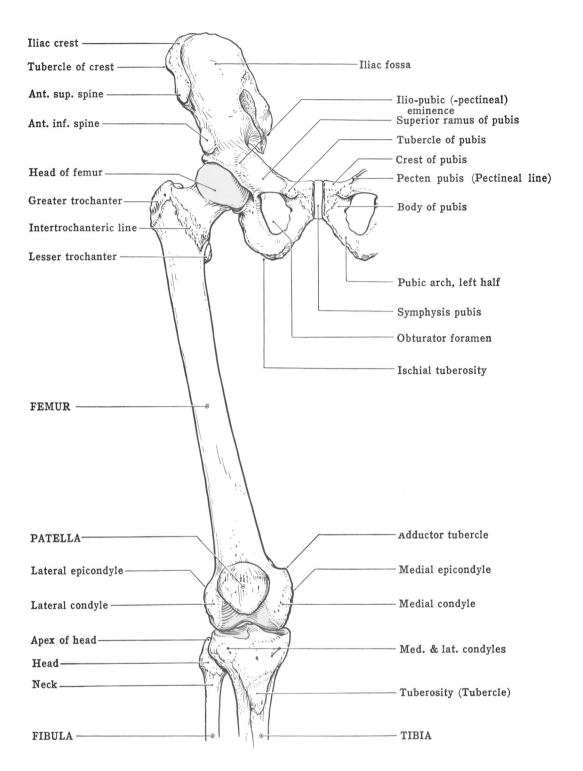

Iliac crest

Tubercle of crest

Ant. sup. spine

Ant. inf. spine

Head of femur

Greater trochanter

Intertrochanteric line

Lesser trochanter

FEMUR

PATELLA

Lateral epicondyle

Lateral condyle

Apex of head

Head

Neck

FIBULA

Iliac fossa

Ilio-pubic (-pectineal) eminence

Superior ramus of pubis

Tubercle of pubis

Crest of pubis

Pecten pubis (Pectineal line)

Body of pubis

Pubic arch, left half

Symphysis pubis

Obturator foramen

Ischial tuberosity

Adductor tubercle

Medial epicondyle

Medial condyle

Med. & lat. condyles

Tuberosity (Tubercle)

TIBIA

4-1   BONES OF THE LOWER LIMB, FRONT VIEW

For bones of the leg, see Figure 4-70.

For bones of the foot, dorsal aspect, see Figures 4-103, 4-104, 4-106, and 4-120.

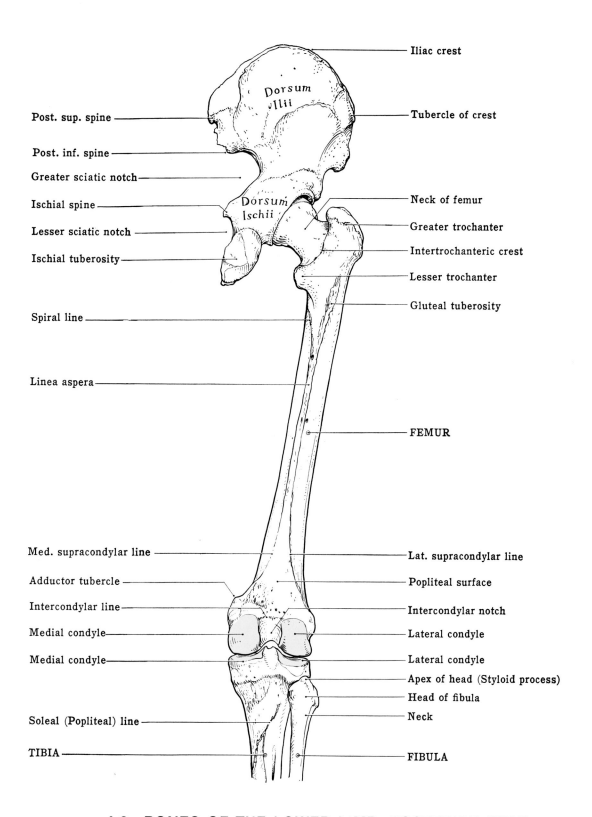

Iliac crest

Dorsum Ilii

Post. sup. spine

Tubercle of crest

Post. inf. spine

Greater sciatic notch

Ischial spine

Dorsum Ischii

Neck of femur

Greater trochanter

Lesser sciatic notch

Intertrochanteric crest

Ischial tuberosity

Lesser trochanter

Gluteal tuberosity

Spiral line

Linea aspera

FEMUR

Med. supracondylar line

Lat. supracondylar line

Adductor tubercle

Popliteal surface

Intercondylar line

Intercondylar notch

Medial condyle

Lateral condyle

Medial condyle

Lateral condyle

Apex of head (Styloid process)

Head of fibula

Neck

Soleal (Popliteal) line

TIBIA

FIBULA

4-2 BONES OF THE LOWER LIMB, POSTERIOR VIEW

For bones of the leg, posterior view, see Figure 4–80.

For bones of the foot, plantar aspect, see Figure 4–107.

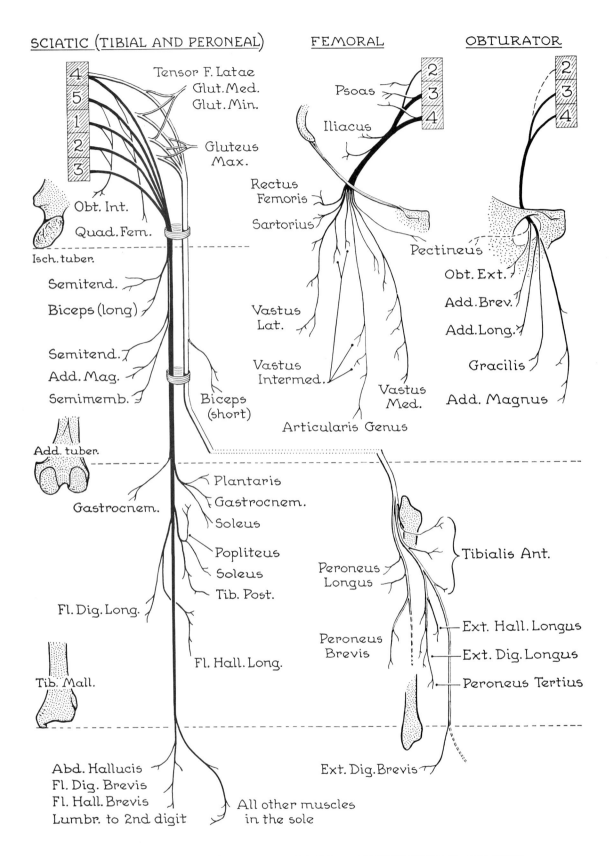

SCIATIC (TIBIAL AND PERONEAL)   **FEMORAL**   **OBTURATOR**

4-3 SCHEME OF THE MOTOR DISTRIBUTION OF THE NERVES OF THE LOWER LIMB

Levels at which the motor branches leave the stems of the main nerves are shown with reference to: ischial tuberosity, adductor tubercle, and tibial malleolus.

4-4 A LIST OF THE MUSCLES OF THE LOWER LIMB

Ilio-psoas
  Iliacus
  Psoas Major
Psoas Minor
Gluteus Maximus
Gluteus Medius
Gluteus Minimus
Tensor Fasciae Latae
Piriformis
Obturator Internus
Gemellus Superior
Gemellus Inferior
Quadratus Femoris
Sartorius
Quadriceps Femoris
  Rectus Femoris
  Vastus Lateralis
  Vastus Intermedius
  Vastus Medialis
Articularis Genus
Pectineus
Gracilis
Adductor Longus
Adductor Brevis
Adductor Magnus
Obturator Externus
Biceps Femoris
  Long head
  Short head
Semitendinosus
Semimembranosus
Tibialis Anterior
Extensor Digitorum Longus
Peroneus Tertius
Extensor Hallucis Longus
Peroneus Brevis
Peroneus Longus
Gastrocnemius
  Lateral head
  Medial head
Soleus
Plantaris
Popliteus
Tibialis Posterior
Flexor Digitorum Longus
Flexor Hallucis Longus
Extensor Hallucis Brevis
Extensor Digitorum Brevis
Abductor Hallucis
Flexor Hallucis Brevis
Adductor Hallucis
  Oblique head
  Transverse head
Abductor Digiti Minimi (V)
  (Abductor Ossis Metatarsi Quinti)
Flexor Digiti Minimi Brevis
Flexor Digitorum Brevis
Flexor Digitorum Accessorius
Lumbricales
Interossei
  Dorsal
  Plantar

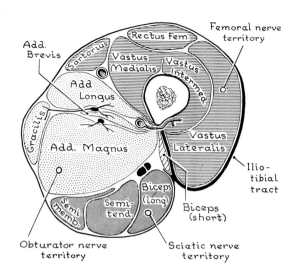

4-5   CROSS-SECTION OF THIGH

This diagram shows that the muscles of the thigh are in three groups, each with its own nerve supply and primary function:
1. Anterior: femoral nerve: extend the leg at the knee.
2. Medial: Obturator nerve: adductors.
3. Posterior: Sciatic nerve: flex the leg at the knee.

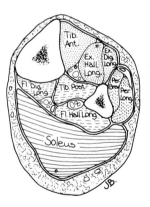

4-6   CROSS-SECTION OF LEG

This diagram shows that the muscles of the leg are in three groups each with its own nerve supply:
1. Anterior: Deep Peroneal nerve: extend the toes.
2. Lateral: Superficial Peroneal nerve: evert the foot.
3. Posterior: Tibial nerve. The superficial group act in plantar flexion; the deep group flex the toes.

For details, see Figure 4-71.

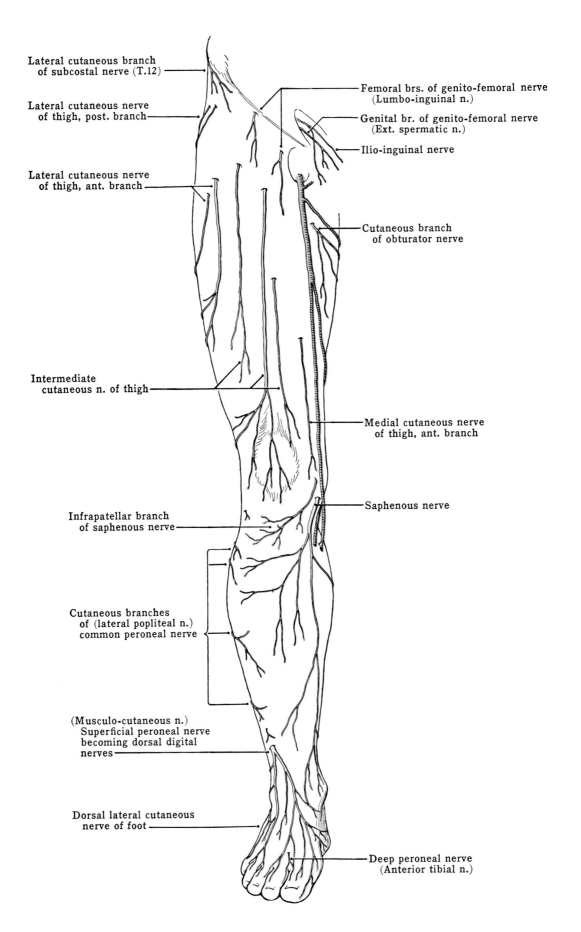

Lateral cutaneous branch
of subcostal nerve (T.12)

Lateral cutaneous nerve
of thigh, post. branch

Lateral cutaneous nerve
of thigh, ant. branch

Intermediate
cutaneous n. of thigh

Infrapatellar branch
of saphenous nerve

Cutaneous branches
of (lateral popliteal n.)
common peroneal nerve

(Musculo-cutaneous n.)
Superficial peroneal nerve
becoming dorsal digital
nerves

Dorsal lateral cutaneous
nerve of foot

Femoral brs. of genito-femoral nerve
(Lumbo-inguinal n.)

Genital br. of genito-femoral nerve
(Ext. spermatic n.)

Ilio-inguinal nerve

Cutaneous branch
of obturator nerve

Medial cutaneous nerve
of thigh, ant. branch

Saphenous nerve

Deep peroneal nerve
(Anterior tibial n.)

**4-7 CUTANEOUS NERVES OF THE LOWER LIMB,
FRONT VIEW**

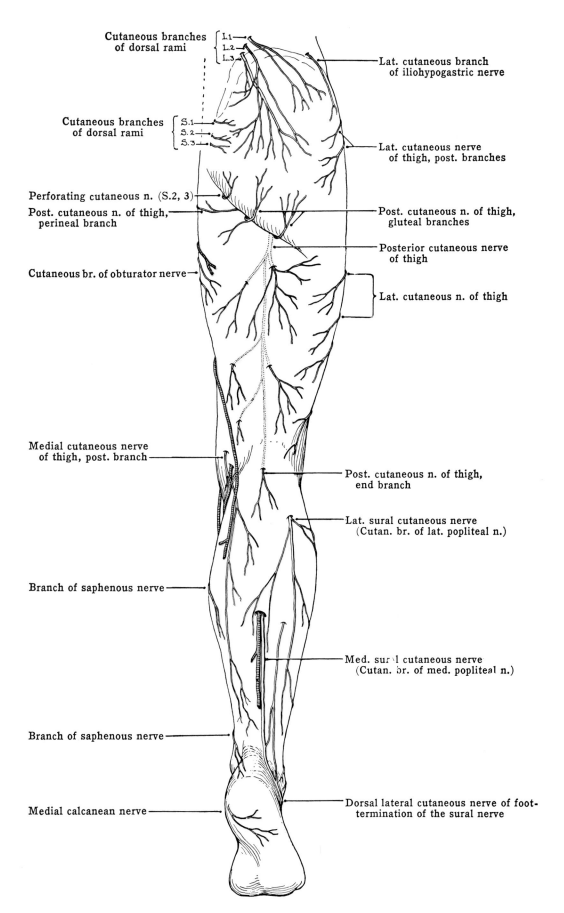

Cutaneous branches of dorsal rami {L.1 L.2 L.3}

Lat. cutaneous branch of iliohypogastric nerve

Cutaneous branches of dorsal rami {S.1 S.2 S.3}

Lat. cutaneous nerve of thigh, post. branches

Perforating cutaneous n. (S.2, 3)

Post. cutaneous n. of thigh, perineal branch

Post. cutaneous n. of thigh, gluteal branches

Posterior cutaneous nerve of thigh

Cutaneous br. of obturator nerve

Lat. cutaneous n. of thigh

Medial cutaneous nerve of thigh, post. branch

Post. cutaneous n. of thigh, end branch

Lat. sural cutaneous nerve (Cutan. br. of lat. popliteal n.)

Branch of saphenous nerve

Med. sural cutaneous nerve (Cutan. br. of med. popliteal n.)

Branch of saphenous nerve

Medial calcanean nerve

Dorsal lateral cutaneous nerve of foot—termination of the sural nerve

4-8 CUTANEOUS NERVES OF THE LOWER LIMB, BACK VIEW

Note: *Sura* is Latin for the calf. The medial sural cutaneous nerve is here joined close above the ankle by a communicating branch (not labeled) of the lateral sural cutaneous nerve to form the sural nerve. The level of the junction is variable, being here very low, in Figure 4–52 very high, and in Figure 4–9 in between.

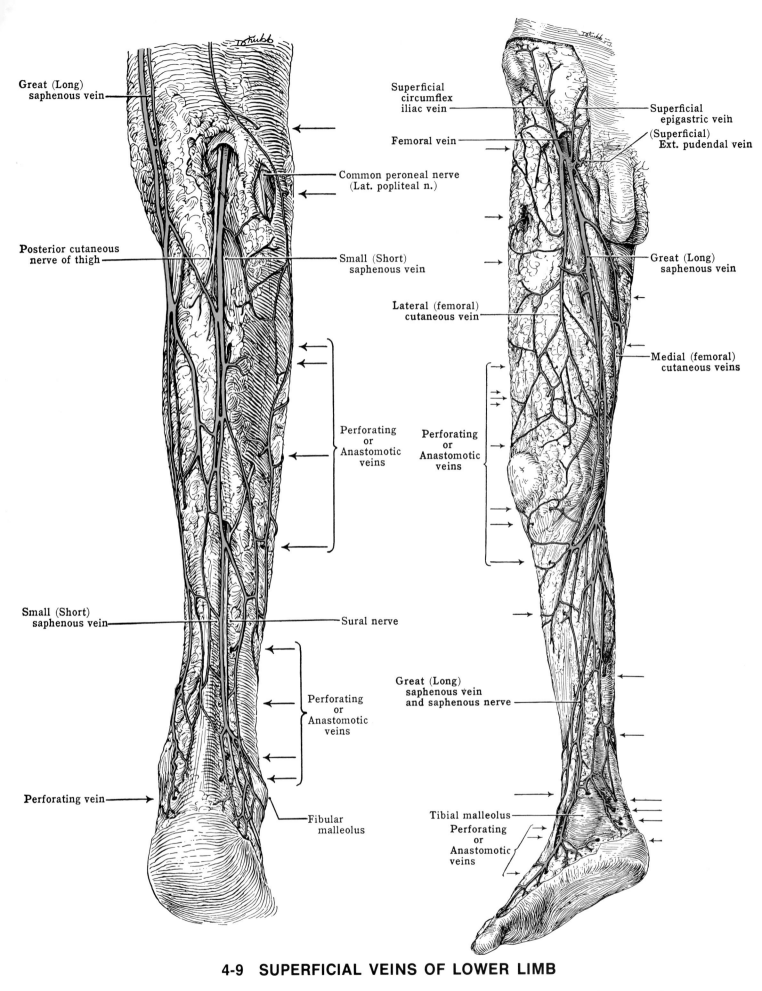

4-9 SUPERFICIAL VEINS OF LOWER LIMB

Posterior and antero-medial views. The *arrows* indicate where anastomotic veins perforate the deep fascia and bring the superficial and deep veins into communication with each other.

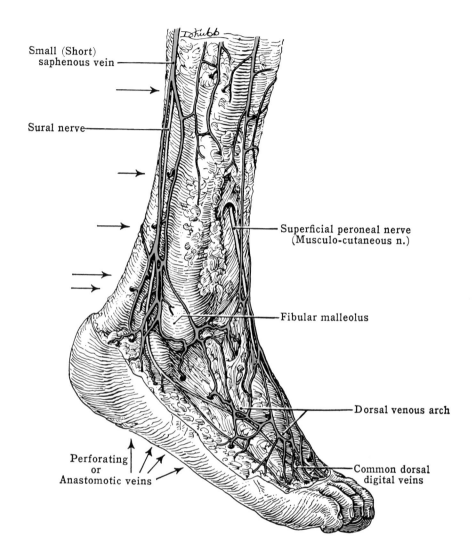

Small (Short) saphenous vein

Sural nerve

Superficial peroneal nerve (Musculo-cutaneous n.)

Fibular malleolus

Dorsal venous arch

Perforating or Anastomotic veins

Common dorsal digital veins

4-10 SUPERFICIAL VEINS OF THE ANKLE AND DORSUM OF THE FOOT, ANTERO-LATERAL VIEW

Figures 4-9 and 4-10 show the superficial veins of the lower limb which lie in the subcutaneous fat. When these veins become enlarged and tortuous, their valves become incompetent and they are termed varicose veins.

See Conrad, M. C. (1971) *Functional Anatomy of the Circulation to the Lower Extremities.* Year Book Medical Publishers, Chicago.

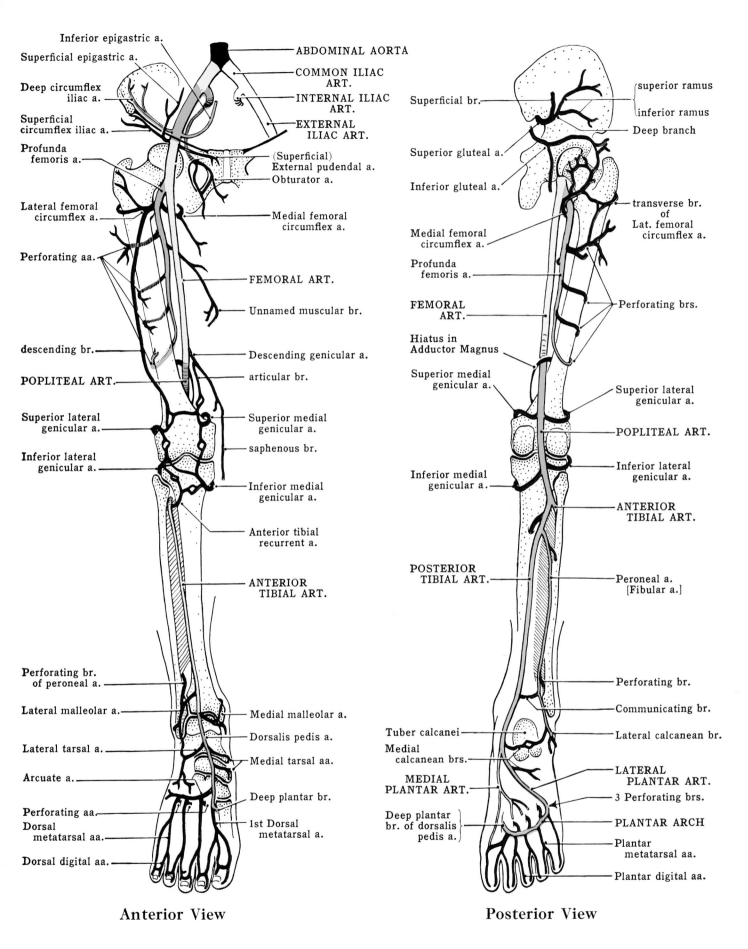

Inferior epigastric a.
Superficial epigastric a.
Deep circumflex iliac a.
Superficial circumflex iliac a.
Profunda femoris a.
Lateral femoral circumflex a.
Perforating aa.
descending br.
POPLITEAL ART.
Superior lateral genicular a.
Inferior lateral genicular a.
Perforating br. of peroneal a.
Lateral malleolar a.
Lateral tarsal a.
Arcuate a.
Perforating aa.
Dorsal metatarsal aa.
Dorsal digital aa.

ABDOMINAL AORTA
COMMON ILIAC ART.
INTERNAL ILIAC ART.
EXTERNAL ILIAC ART.
(Superficial) External pudendal a.
Obturator a.
Medial femoral circumflex a.
FEMORAL ART.
Unnamed muscular br.
Descending genicular a.
articular br.
Superior medial genicular a.
saphenous br.
Inferior medial genicular a.
Anterior tibial recurrent a.
ANTERIOR TIBIAL ART.
Medial malleolar a.
Dorsalis pedis a.
Medial tarsal aa.
Deep plantar br.
1st Dorsal metatarsal a.

Anterior View

Superficial br.
Superior gluteal a.
Inferior gluteal a.
Medial femoral circumflex a.
Profunda femoris a.
FEMORAL ART.
Hiatus in Adductor Magnus
Superior medial genicular a.
Inferior medial genicular a.
POSTERIOR TIBIAL ART.
Tuber calcanei
Medial calcanean brs.
MEDIAL PLANTAR ART.
Deep plantar br. of dorsalis pedis a.

superior ramus
inferior ramus
Deep branch
transverse br. of Lat. femoral circumflex a.
Perforating brs.
Superior lateral genicular a.
POPLITEAL ART.
Inferior lateral genicular a.
ANTERIOR TIBIAL ART.
Peroneal a. [Fibular a.]
Perforating br.
Communicating br.
Lateral calcanean br.
LATERAL PLANTAR ART.
3 Perforating brs.
PLANTAR ARCH
Plantar metatarsal aa.
Plantar digital aa.

Posterior View

4-11 A DIAGRAM OF THE ARTERIES OF THE LOWER LIMB

A LIST OF THE NAMED ARTERIES OF THE LOWER LIMB

Internal iliac artery
  Obturator artery
    Acetabular branch
    Anterior branch
    Posterior branch
  Superior gluteal artery
    Superficial branch
    Deep branch
      Superior ramus
      Inferior ramus
  Inferior gluteal artery
    Branch to sciatic nerve
External iliac artery
Femoral artery
  Superficial epigastric artery
  Superficial circumflex iliac artery
  (Superficial) external pudendal artery
    Anterior scrotal (or labial) branches
    Inguinal branches
  Profunda femoris artery
    Medial femoral circumflex artery
      Acetabular branch
      Ascending branch
      Transverse branch
    Lateral femoral circumflex artery
      Ascending branch
      Transverse branch
      Descending branch
    Perforating arteries
  Descending genicular artery
    Articular branch
    Saphenous branch
  Popliteal artery
    Superior lateral genicular artery
    Superior medial genicular artery
    Middle genicular artery
    Sural arteries
    Inferior lateral genicular artery
    Inferior medial genicular artery
  Anterior tibial artery
    Anterior tibial recurrent artery
    Lateral malleolar artery
    Medial malleolar artery
  Dorsalis pedis artery
    Lateral tarsal artery
    Medial tarsal arteries
    Arcuate artery
      Dorsal metatarsal arteries
        Dorsal digital arteries
    Deep plantar branch
    1st dorsal metatarsal artery
  Posterior tibial artery
    Circumflex fibular branch
    Peroneal (fibular) artery
      Perforating branch
      Communicating branch
      Lateral malleolar artery
      Lateral calcanean artery
    Medial malleolar artery
    Medial calcanean branches
  Medial plantar artery
  Lateral plantar artery
    Plantar arch
      Plantar metatarsal arteries
      Perforating branches
        Plantar digital arteries

ARTERIES OF THE LOWER LIMB

Use Figure 4-11 as a preliminary tour guide of the arterial supply of the lower limb. Observe that the *common iliac* artery bifurcates into a smaller *internal iliac* and a larger *external iliac* artery anterior to the sacroiliac joint at the level of the lumbosacral intervertebral disc. (In the fetus, the relative size of the two iliac arteries is reversed since the umbilical arteries arise from the internal iliac. See Fig. 10-1C.)

Internal Iliac: Most branches supply the pelvic viscera and the perineum. Only three branches are shown here: The *obturator* artery passes through the obturator foramen and divides into anterior and posterior branches which anastomose with each other and with adjacent arteries. The *anterior* branch supplies the adductor group of thigh muscles; the *posterior* branch supplies hamstring muscles and sends an *acetabular* branch to the head of the femur accompanying the ligament of the head (Fig. 4-47). The *superior gluteal* artery leaves the pelvis through the greater sciatic foramen and divides into superficial and deep branches. The former supplies Gluteus maximus. The deep branch divides into a superior ramus which anastomoses with arteries of the region, and an inferior ramus which supplies Gluteus medius and minimus. The inferior gluteal artery leaves the pelvis through the lower part of the greater sciatic foramen and supplies the buttock, upper part of the thigh, and the sciatic nerve. The gluteal arteries are shown in more detail in Figures 4-33 and 4-34.

External Iliac (*Blue*): Two named branches are shown: *inferior epigastric* going to the anterior abdominal wall, and *deep circumflex iliac* to the lateral abdominal wall.

Femoral Artery (*Yellow*): The external iliac artery changes its name as it passes behind the inguinal ligament and almost immediately gives off three superficial branches. It travels first in the femoral triangle (Fig. 4-20) then in the adductor canal (Fig. 4-25), supplying thigh muscles of the flexor and adductor groups. The major branch is the *profunda femoris* artery which supplies the thigh through *medial and lateral femoral circumflex* branches and through *perforating* arteries which pass through the Adductor magnus on their way to the back of the thigh (Fig. 4-34). Both femoral and lateral circumflex arteries have *descending genicular* branches which contribute to the complex anastomoses around the knee (Figs. 4-54 and 4-55).

Popliteal Artery (*Green*): As the femoral artery passes through the hiatus in Adductor magnus, another change of name occurs. Its branches supply skin, muscles, and the knee joint (Fig. 4-53). At the inferior angle of the popliteal fossa it divides into anterior and posterior tibial arteries, on either side of the interosseous membrane.

Anterior Tibial Artery (*Yellow*): Supplies the anterior compartment of the leg and the ankle (Fig. 4-75). It continues as the *dorsalis pedis* artery.

Posterior Tibial Artery (*Blue*): Supplies the posterior compartment of the leg and terminates as *medial and lateral plantar* arteries (Fig. 4-84). Its major branch is the *peroneal* artery (*green*) which supplies the posterior and lateral compartments of the leg (Fig. 4-86).

Arterial Supply of the Foot (Figs. 4-76 and 4-91 to 4-94): Observe the dorsal arterial arch formed by the dorsalis pedis artery which also contributes to the plantar arch formed by the *medial and lateral plantar* arteries.

Pulses: Pulsation of the *femoral* artery can be felt just distal to the inguinal ligament midway between the anterior superior spine and the pubic tubercle (Fig. 4-20). *Dorsalis pedis* is palpated on the dorsum of the foot midway between the two malleoli, just lateral to the tendon of extensor hallucis longus (Fig. 4-77). *Posterior tibial* artery is felt just behind the two tendons which use the medial malleolus as a pulley (Fig. 4-87).

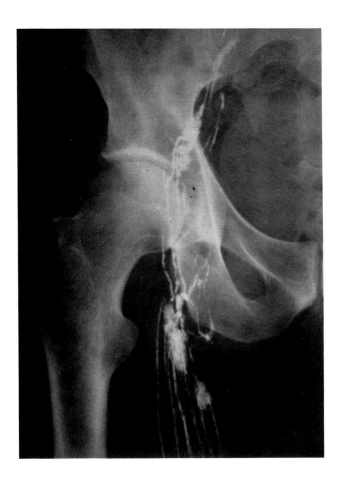

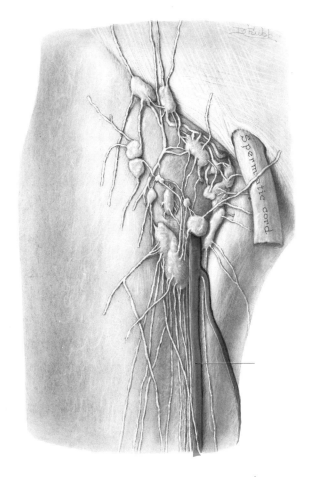

4-12A   INJECTION   **4-12B   DISSECTION**

INGUINAL LYMPH NODES

Observe:
1. The arrangement of the nodes: (a) a proximal chain parallel to the inguinal ligament (superficial inguinal nodes); (b) a distal chain on the sides of the great saphenous vein (superficial subinguinal nodes); and, above this, a chain of two or three nodes on the medial side of the femoral vein (deep inguinal nodes), one being below the femoral canal and one or two within it.
2. The free anastomosis between the lymph vessels. About two dozen efferent vessels leave these nodes and, passing deep to the inguinal ligament, enter the external iliac nodes. Of these, less than half traverse the femoral canal; the others ascend alongside the femoral artery and vein, some being inside the femoral sheath and some outside it (Fig. 2-11).

Note:
1. These nodes receive the superficial and deep lymph vessels of the lower limb; the superficial vessels of the lower part of the abdominal wall; the vessels of the penis, including the glans penis and spongy urethra, and of the scrotum (but not of the testis or ovary); the vessels of the vulva, lower part of the vagina, and some vessels from the uterus that run with the round ligament; and vessels of the lower part of the anal canal.
2. Because of their origin, lymph from the ovary or testis drains through vessels following ovarian and testicular arteries to para-aortic and preaortic nodes which are considerably less accessible than those in the inguinal region.

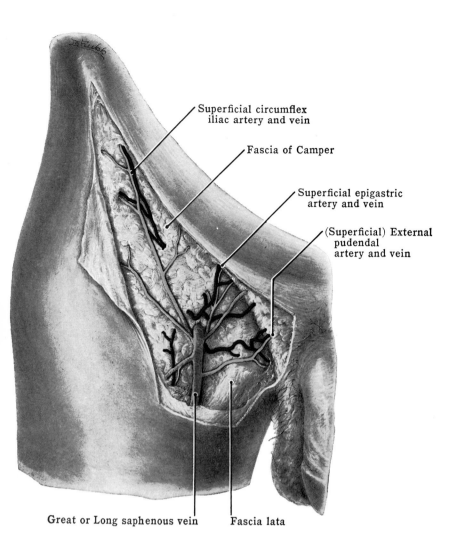

Superficial circumflex
iliac artery and vein

Fascia of Camper

Superficial epigastric
artery and vein

(Superficial) External
pudendal
artery and vein

The arteries are branches of the femoral artery but the veins are tributaries of the great saphenous vein.

Great or Long saphenous vein   Fascia lata

4-13 SUPERFICIAL INGUINAL ARTERIES AND VEINS

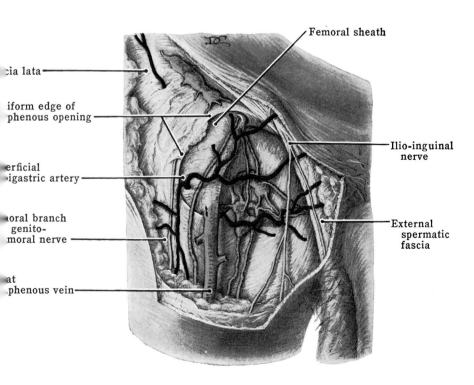

Femoral sheath

cia lata

iform edge of
phenous opening

erficial
igastric artery

oral branch
genito-
moral nerve

at
phenous vein

Ilio-inguinal
nerve

External
spermatic
fascia

Observe:
1. The oval shape of this hiatus in the fascia lata through which the great saphenous vein passes to join the femoral vein, which lies within the femoral sheath.
2. The sharp superior and inferior free margins, or cornua, of the opening and the less sharply definable lateral margin or falciform edge.
3. The great saphenous vein hooking over the inferior cornu. Occasionally it joins the femoral vein 1 or even 2 cm above this cornu.

4-14 SAPHENOUS OPENING (FOSSA OVALIS)

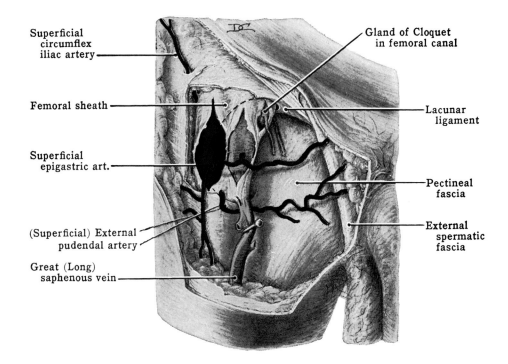

Superficial circumflex iliac artery

Femoral sheath

Superficial epigastric art.

(Superficial) External pudendal artery

Great (Long) saphenous vein

Gland of Cloquet in femoral canal

Lacunar ligament

Pectineal fascia

External spermatic fascia

4-15 FEMORAL SHEATH, CANAL AND RING

Observe:
1. The falciform edge of the saphenous opening or hiatus cut away.
2. The superior cornu of the opening passing toward the pubic tubercle and blending with the inguinal ligament and, in this specimen, with the lacunar ligament also.
3. The medial border of the opening formed by the fascia covering Pectineus and as such passing laterally behind the femoral sheath.

4. The 3 compartments of the sheath, each incised longitudinally: (a) the lateral one for the artery, (b) the middle one for vein, and (c) the medial one, called the femoral canal, for lymph vessels.
5. The proximal end of the femoral canal, called the femoral ring, bounded: medially by the lacunar ligament, anteriorly by the inguinal ligament, posteriorly by Pectineus and its fascia (Fig. 4–42), and laterally by femoral vein.

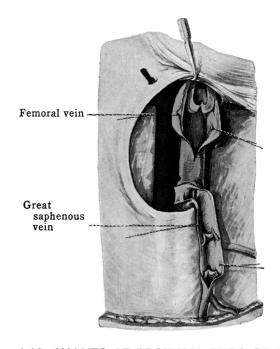

Femoral vein

Great saphenous vein

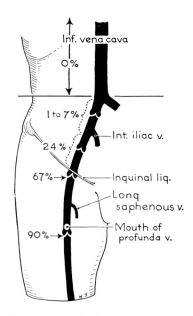

Inf. vena cava

0%

1 to 7%

Int. iliac v.

24%

67%

Inguinal lig.

Long saphenous v.

Mouth of profunda v.

90%

Percentage Incidence of Valves

4-16 VALVES AT PROXIMAL ENDS OF THE FEMORAL AND GREAT SAPHENOUS VEINS

Between the mouth of the great saphenous vein and the heart there were no valves in 21 per cent of 506 limbs, 1 valve in 71 per cent, 2 in 7 per cent, and 3 in 1 per cent. Valves in the common iliac vein are rarely competent, but two-thirds of those in the external iliac vein are competent. For most proximal valve, see Figure 4–46.

See Basmajian, J. V. (1952) The distribution of valves in the femoral, external iliac and common iliac veins and their relationship to varicose veins. *Surg. Gynec. Obst., 95: 537.*

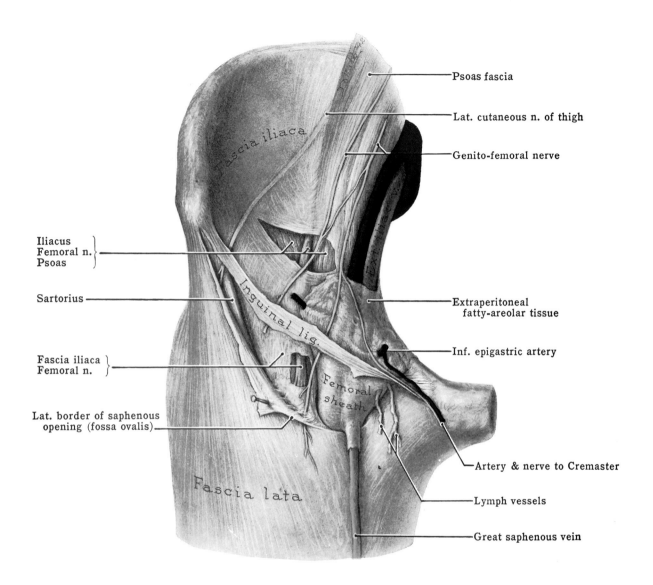

- Psoas fascia
- Lat. cutaneous n. of thigh
- Genito-femoral nerve

Iliacus
Femoral n.
Psoas

Sartorius

Fascia iliaca
Femoral n.

- Extraperitoneal fatty-areolar tissue
- Inf. epigastric artery

Lat. border of saphenous opening (fossa ovalis)

Fascia lata

- Artery & nerve to Cremaster
- Lymph vessels
- Great saphenous vein

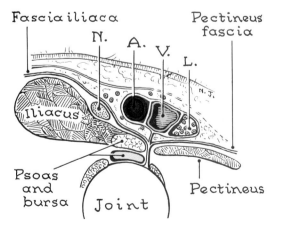

Fascia iliaca   Pectineus fascia

N. A. V. L.

Iliacus

Psoas and bursa   Joint   Pectineus

4-18 RELATION OF FEMORAL STRUCTURES

The tough Psoas tendon separates the femoral artery from the hip joint.

4-17 FEMORAL SHEATH

The three flat muscles of the abdominal wall are cut away from the upper border of the inguinal ligament, and the fascia lata from the lower border. The falciform margin of the saphenous opening or hiatus in the fascia lata is cut and reflected. The inferior epigastric artery is pulled medially. The deep circumflex iliac artery is not labeled.

Observe:

1. The fascia iliaca, continuous medially with the Psoas fascia and carried downward in front of Iliacus into the thigh. As it passes behind the inguinal ligament it adheres to it.
2. The extraperitoneal fatty-areolar tissue, which lines the abdominal cavity and in which the external iliac vessels run, carried downward around these vessels into the thigh as delicate funnel-shaped sac, called the femoral sheath. This is loosely adherent to the inguinal ligament in front and to the pecten pubis behind.
3. The femoral sheath containing the femoral artery, vein, and lymph vessels; the femoral nerve, being behind the fascia iliaca, is outside the sheath.
4. The falciform margin of the saphenous opening (part of the fascia lata) lying in front of the femoral sheath. Medially the fascia passes behind the sheath as the pectineal fascia.
5. The genito-femoral nerve pierces the Psoas fascia high up in one or two branches; the lateral cutaneous nerve of the thigh pierces the fascia iliaca at a variable point, commonly, as here, low down near the anterior superior iliac spine.

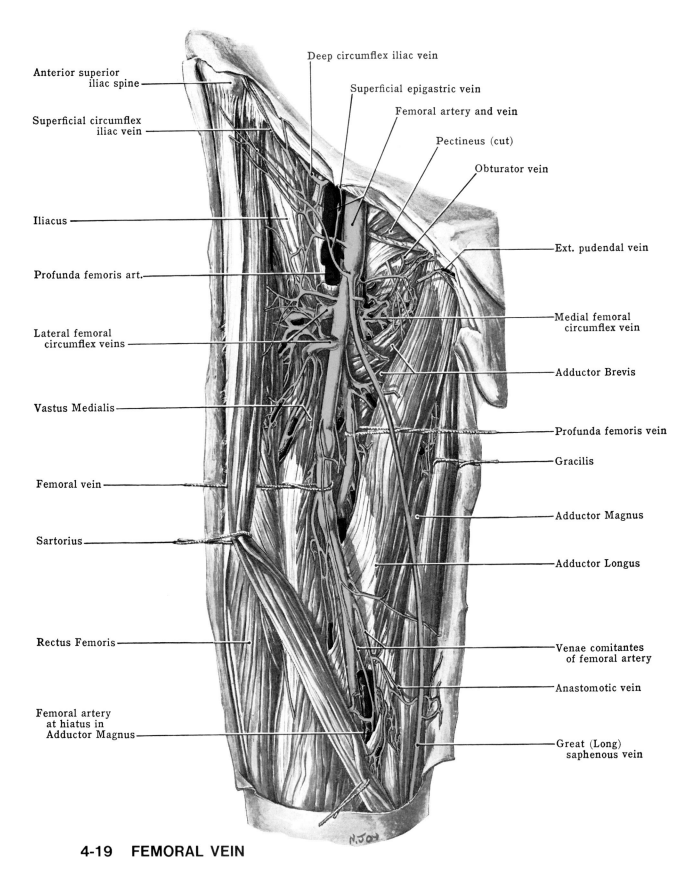

Anterior superior iliac spine

Superficial circumflex iliac vein

Iliacus

Profunda femoris art.

Lateral femoral circumflex veins

Vastus Medialis

Femoral vein

Sartorius

Rectus Femoris

Femoral artery at hiatus in Adductor Magnus

Deep circumflex iliac vein

Superficial epigastric vein

Femoral artery and vein

Pectineus (cut)

Obturator vein

Ext. pudendal vein

Medial femoral circumflex vein

Adductor Brevis

Profunda femoris vein

Gracilis

Adductor Magnus

Adductor Longus

Venae comitantes of femoral artery

Anastomotic vein

Great (Long) saphenous vein

4-19   FEMORAL VEIN

The veins are injected with latex. Only the stumps of the arteries remain. (The limb is rotated laterally at the hip joint through the force of gravity.)

Observe:

1. In this specimen the profunda femoris *artery* arises from the femoral artery about 4 cm below the inguinal ligament. The profunda femoris *vein* joins the femoral vein about 8 cm below the ligament.

2. The large size of the lateral femoral circumflex vein, here double

and, like the medial femoral circumflex, ending in the femoral vein.

3. The many long, slender, paired venae comitantes that accompany the various arteries.

4. The superficial circumflex iliac vein, in this specimen receiving the superficial epigastric vein, communicating with the vein proximal and distal to it, and ending both in the great saphenous and in the femoral vein.

5. The great saphenous vein communicating with a vena comitans of the femoral artery. The medial femoral circumflex vein communicating with the obturator vein.

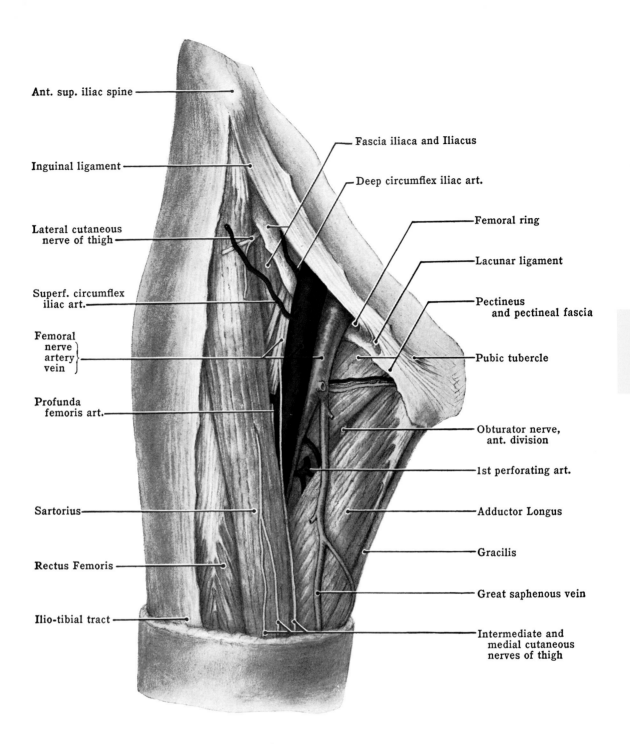

Ant. sup. iliac spine

Inguinal ligament

Lateral cutaneous nerve of thigh

Superf. circumflex iliac art.

Femoral { nerve artery vein }

Profunda femoris art.

Sartorius

Rectus Femoris

Ilio-tibial tract

Fascia iliaca and Iliacus

Deep circumflex iliac art.

Femoral ring

Lacunar ligament

Pectineus and pectineal fascia

Pubic tubercle

Obturator nerve, ant. division

1st perforating art.

Adductor Longus

Gracilis

Great saphenous vein

Intermediate and medial cutaneous nerves of thigh

4-20 FEMORAL TRIANGLE

Observe:
1. The boundaries of the triangle: the inguinal ligament, which curves gently from anterior superior spine to pubic tubercle, being the base; the medial border of Sartorius being the lateral side; the lateral border of Adductor Longus being the medial side; and the point where the two converging sides meet distally being the apex. (Some authors regard the medial border of Adductor Longus as the medial side of the triangle.)
2. The femoral artery and vein lying in front of the fascia covering Ilio-psoas and Pectineus, and the femoral nerve lying behind.
3. The artery appearing midway between the anterior superior spine and the pubic tubercle, and disappearing where the medial border of Sartorius crosses the lateral border of Adductor Longus—in short, from the midpoint of the base of the triangle to the apex.
4. That when the adjacent borders of Pectineus and Adductor Longus are not contiguous, as here, a glimpse is had of the anterior branch of the obturator nerve lying in front of Adductor Brevis.

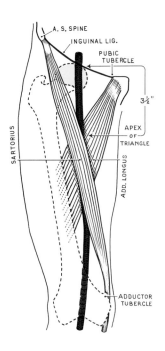

4-21A   BOUNDARIES OF TRIANGLE

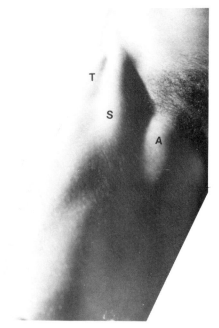

4-21B   FEMORAL TRIANGLE

Note:
1. The trough between Sartorius (*S*) and Adductor longus (*A*).
2. The deep groove between Sartorius and Tensor fascia lata (*T*).

4-21C   SARTORIUS

The longest muscle in the body, its origin on the anterior superior spine, its insertion on the medial side of the tibia (See Fig. 4-65*A*), and like other flexors of the hip it is supplied by the femoral nerve.

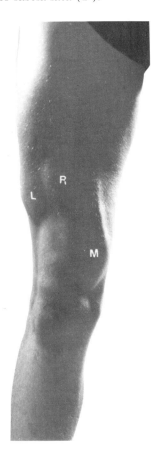

4-21D   ANTERIOR THIGH MUSCLES

Vastus medialis (*M*), Vastus lateralis (*L*), and Rectus femoris (*R*), all inserting (with Vastus intermedius) on the patella and thence via the ligamentum patellae to the tibial tuberosity. All supplied, of course, by the femoral nerve.

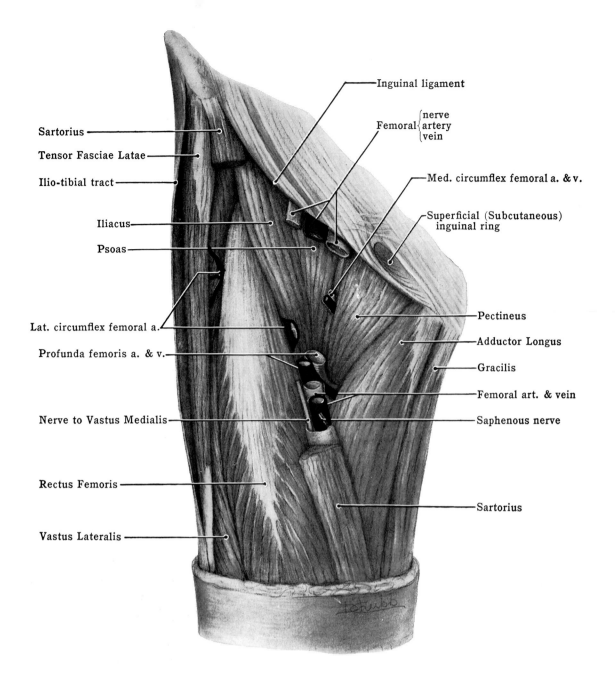

Sartorius

Tensor Fasciae Latae

Ilio-tibial tract

Iliacus

Psoas

Lat. circumflex femoral a.

Profunda femoris a. & v.

Nerve to Vastus Medialis

Rectus Femoris

Vastus Lateralis

Inguinal ligament

Femoral { nerve / artery / vein }

Med. circumflex femoral a. & v.

Superficial (Subcutaneous) inguinal ring

Pectineus

Adductor Longus

Gracilis

Femoral art. & vein

Saphenous nerve

Sartorius

4-22 FLOOR OF THE FEMORAL TRIANGLE

Sections are removed from Sartorius and from the femoral vessels and nerve.

Observe:

1. The floor of the triangle is a trough with sloping lateral and medial walls. This is notably so, if Adductor Longus is included with Pectineus in the medial wall; Ilio-psoas (medial border of Rectus Femoris) and Sartorius form the lateral wall.
2. The trough is shallow at the base and deep at the apex.
3. At the apex four vessels, one in front of the other, and two nerves pass into the adductor canal of Hunter (subsartorial canal).

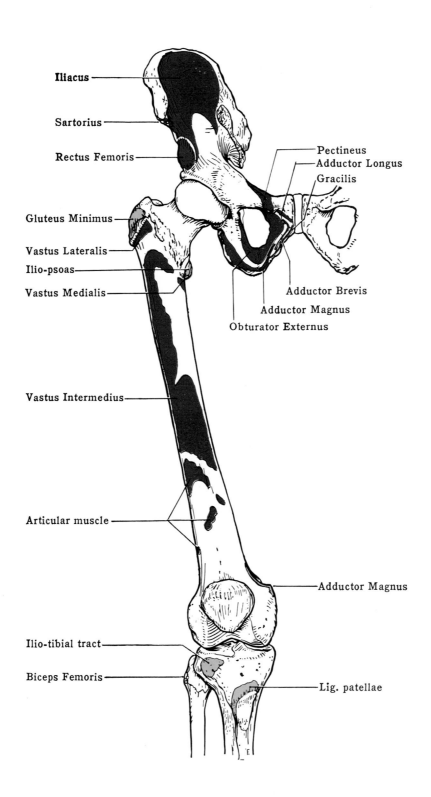

Iliacus

Sartorius

Rectus Femoris

Pectineus
Adductor Longus
Gracilis

Gluteus Minimus

Vastus Lateralis

Ilio-psoas

Vastus Medialis

Adductor Brevis

Adductor Magnus

Obturator Externus

Vastus Intermedius

Articular muscle

Adductor Magnus

Ilio-tibial tract

Biceps Femoris

Lig. patellae

4-23 BONES OF THE LOWER LIMB SHOWING ATTACHMENTS OF MUSCLES, ANTERIOR VIEW

For tibia and fibula, see Figure 4-70

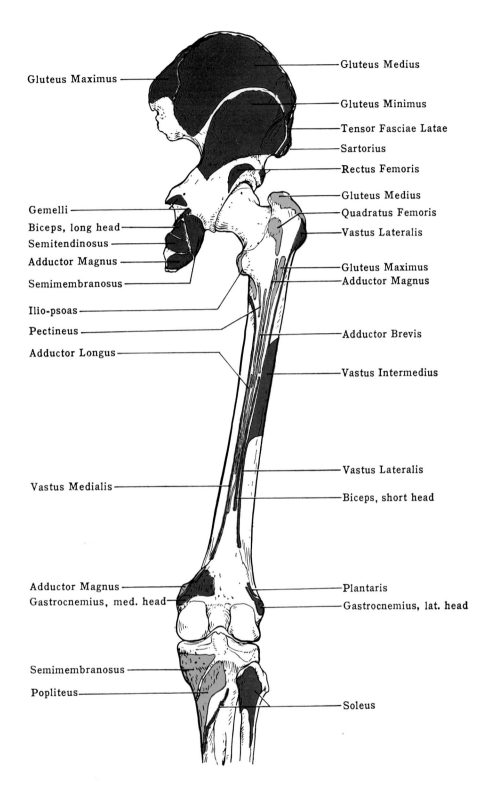

Gluteus Maximus

Gluteus Medius

Gluteus Minimus

Tensor Fasciae Latae

Sartorius

Rectus Femoris

Gluteus Medius

Quadratus Femoris

Vastus Lateralis

Gemelli

Biceps, long head

Semitendinosus

Adductor Magnus

Semimembranosus

Gluteus Maximus

Adductor Magnus

Ilio-psoas

Pectineus

Adductor Longus

Adductor Brevis

Vastus Intermedius

Vastus Lateralis

Biceps, short head

Vastus Medialis

Adductor Magnus

Gastrocnemius, med. head

Plantaris

Gastrocnemius, lat. head

Semimembranosus

Popliteus

Soleus

**4-24 BONES OF THE LOWER LIMB SHOWING
ATTACHMENTS OF MUSCLES, POSTERIOR VIEW**

For tibia and fibula, posterior aspect, see Figure 4-81.

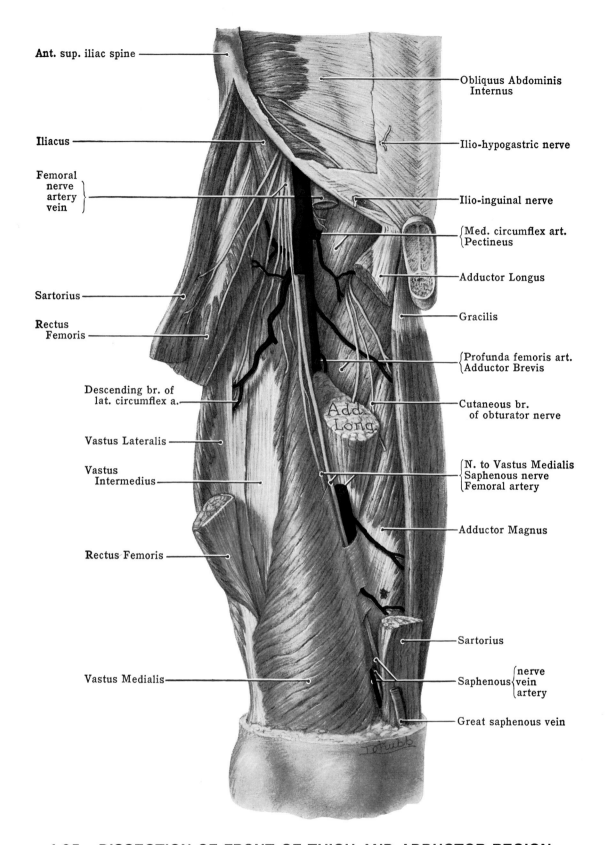

4-25   DISSECTION OF FRONT OF THIGH AND ADDUCTOR REGION

(The limb is rotated laterally.)

Observe:

1. The femoral nerve breaking up into a leash of nerves on entering the thigh.

2. The femoral artery lying between two motor territories, namely, that of the obturator nerve which is medial and that of the femoral nerve which is lateral. No motor nerve crosses in front of the femoral artery, but the twig to Pectineus is seen crossing behind it.

3. The nerve to Vastus Medialis and the saphenous nerve accompanying the femoral artery into the adductor canal. The saphenous nerve and artery and their companion anastomotic vein emerging from the lower end of the canal. They become superficial between Sartorius and Gracilis.

4. The profunda femoris artery arising 4 cm below the inguinal ligament, lying behind the femoral artery, and disappearing behind Adductor Longus.

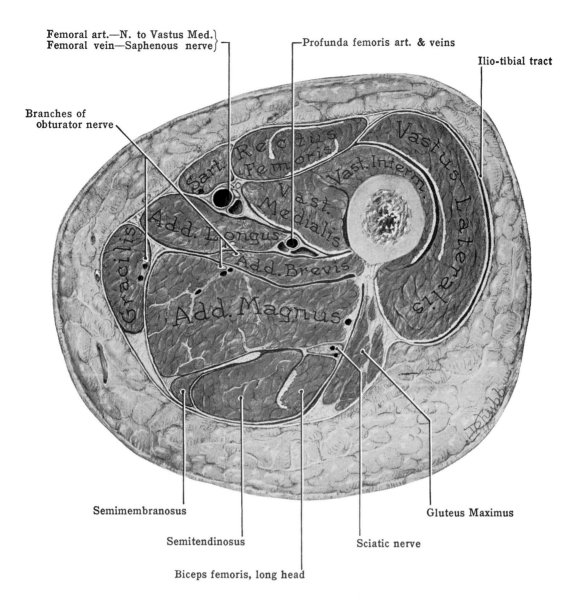

Femoral art.—N. to Vastus Med.
Femoral vein—Saphenous nerve

Profunda femoris art. & veins

Ilio-tibial tract

Branches of
obturator nerve

Sart. Rectus Femoris

Vast. Interm.

Vastus Lateralis

Vast. Medialis

Add. Longus

Gracilis

Add. Brevis

Add. Magnus

Semimembranosus

Semitendinosus

Biceps femoris, long head

Sciatic nerve

Gluteus Maximus

4-26   CROSS-SECTION THROUGH THE THIGH, FEMALE

Observe:

Level of section: (a) through the insertion of Gluteus Maximus,
therefore, above the origin of short head of Biceps; (b) below
insertion of Pectineus, above that of Adductor Longus and,
therefore, through that of Adductor Brevis; (c) through the
adductor canal and, therefore, below the apex of the femoral
triangle, about 10 to 15 cm down the femur.

Gracilis abutting against the free, medial borders of the Adductor muscles.

Adductor Longus intervening between the femoral and the
profunda femoris vessels.

Adductor Brevis intervening between the anterior and the
posterior division of the obturator nerve.

The aponeurosis of semimembranosus not dissimilar from the
sciatic nerve and mistakable for it.

Vastus Intermedius arising from the anterior and lateral surfaces of the shaft of the femur; Vastus Medialis covering the
medial surface but not arising from it; that is to say, Vastus
Intermedius alone arises from the surfaces, the other muscles
are relegated to the linea aspera or to its upward and downward
extensions. Proximally, Vastus Lateralis is large.

The considerable amount of subcutaneous fat present in the
female.

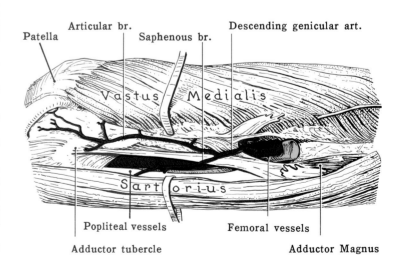

Articular br.

Descending genicular art.

Patella

Saphenous br.

Vastus Medialis

Sartorius

Popliteal vessels

Adductor tubercle

Femoral vessels

Adductor Magnus

4-27   ORIGIN OF POPLITEAL ARTERY

The femoral artery becomes the popliteal artery at the (tendinous) opening in Adductor Magnus (Adductor hiatus).

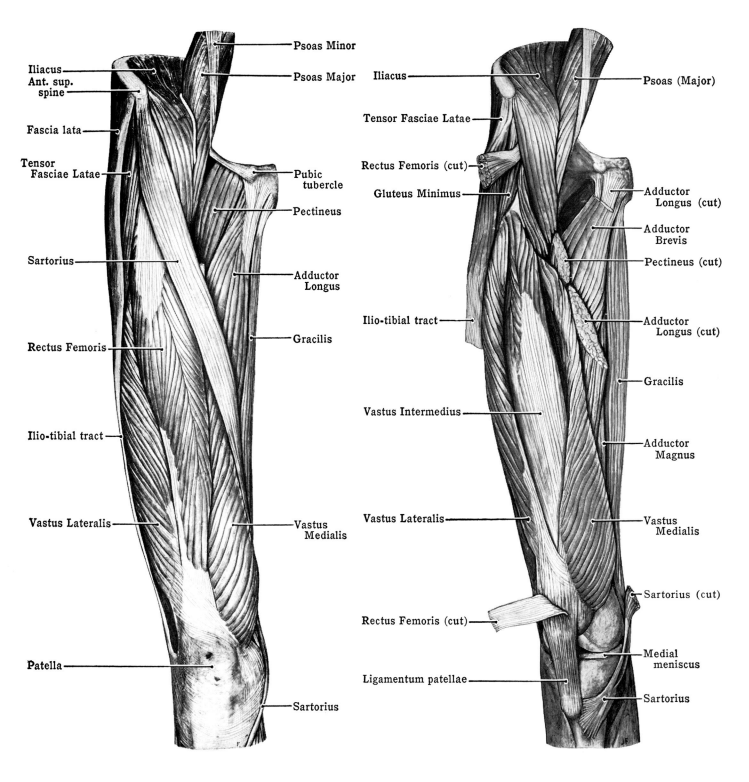

Iliacus
Ant. sup. spine

Fascia lata

Tensor Fasciae Latae

Sartorius

Rectus Femoris

Ilio-tibial tract

Vastus Lateralis

Patella

Psoas Minor

Psoas Major

Pubic tubercle

Pectineus

Adductor Longus

Gracilis

Vastus Medialis

Sartorius

Iliacus

Tensor Fasciae Latae

Rectus Femoris (cut)

Gluteus Minimus

Ilio-tibial tract

Vastus Intermedius

Vastus Lateralis

Rectus Femoris (cut)

Ligamentum patellae

Psoas (Major)

Adductor Longus (cut)

Adductor Brevis

Pectineus (cut)

Adductor Longus (cut)

Gracilis

Adductor Magnus

Vastus Medialis

Sartorius (cut)

Medial meniscus

Sartorius

4-28A   MUSCLES, FRONT OF THIGH—I   **4-28B   MUSCLES, FRONT OF THIGH—II**

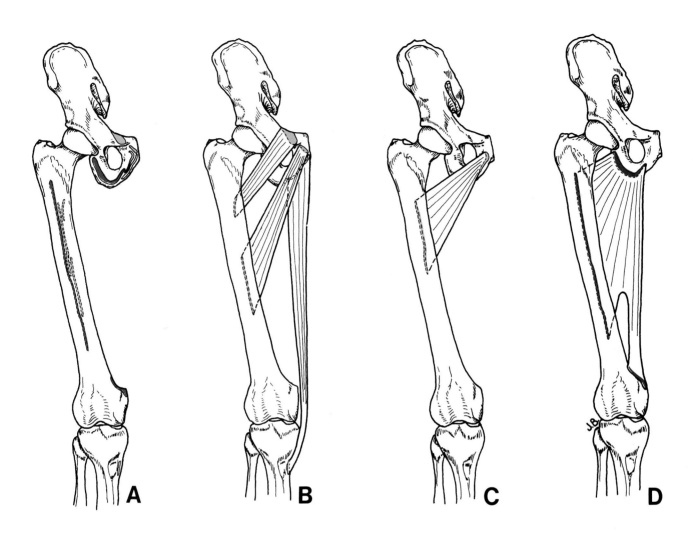

4-29   THE ADDUCTOR GROUP

Each of the adductor group of thigh muscles has a linear attachment to the linea aspera on the posterior surface of the femur.

A. The insertions of the adductor muscles are seen from in front as through a transparent femur. Most medial (*green*) are Pectineus and Adductor longus, Adductor brevis (*blue*) is intermediate, and Adductor magnus (*red*) is most lateral, but most extensive and swings medially to reach the adductor tubercle.

B. The anterior group: Pectineus, Adductor longus, and Gracilis. The latter alone avoids the femur and inserts on the medial side of the proximal femur.

C. Adductor brevis attaches to the intermediate area of the linea aspera.

D. Adductor magnus is deepest, most lateral on the femur, and has the most extensive origin and insertion. Its aponeurosis is punctured by perforating arteries and through the wide hiatus in its insertion passes the femoral artery.

All are adductors of the thigh; their attachments disclose their other actions: Pectineus flexes the thigh, Gracilis flexes the leg and rotates it medially. All contribute to normal gait and posture.

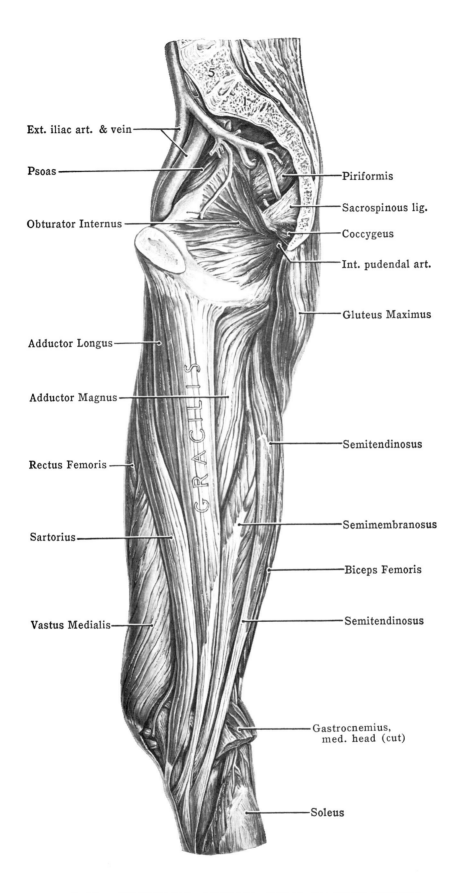

Ext. iliac art. & vein

Psoas

Obturator Internus

Adductor Longus

Adductor Magnus

Rectus Femoris

Sartorius

Vastus Medialis

GRACILIS

Piriformis

Sacrospinous lig.

Coccygeus

Int. pudendal art.

Gluteus Maximus

Semitendinosus

Semimembranosus

Biceps Femoris

Semitendinosus

Gastrocnemius, med. head (cut)

Soleus

4-30   MUSCLES, MEDIAL SIDE OF THE THIGH

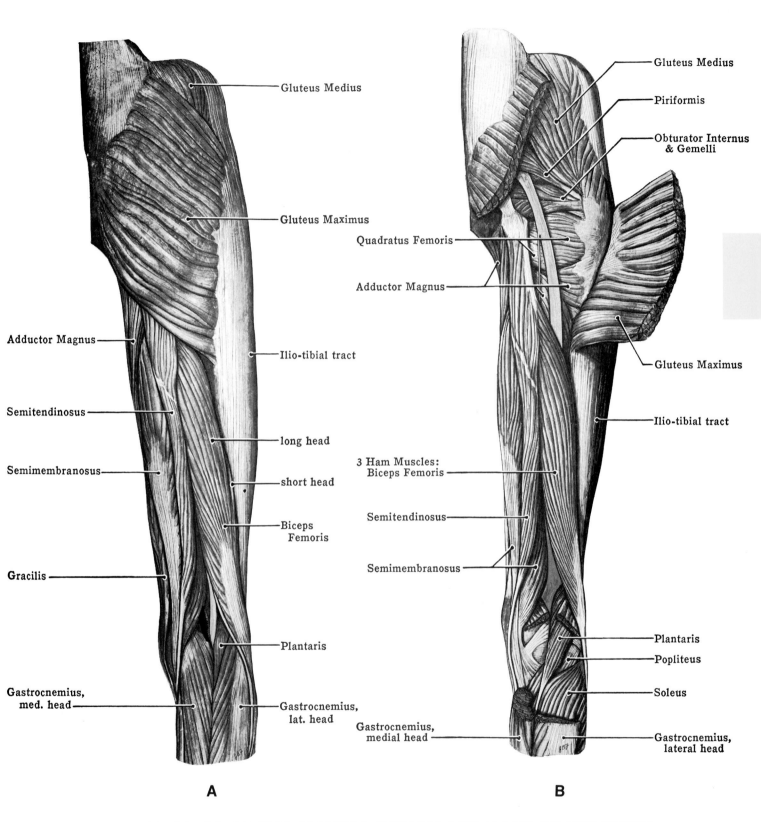

Gluteus Medius

Gluteus Maximus

Adductor Magnus

Semitendinosus

Semimembranosus

Gracilis

Gastrocnemius, med. head

Ilio-tibial tract

long head

short head

Biceps Femoris

Plantaris

Gastrocnemius, lat. head

A

Gluteus Medius

Piriformis

Obturator Internus & Gemelli

Quadratus Femoris

Adductor Magnus

Gluteus Maximus

Ilio-tibial tract

3 Ham Muscles: Biceps Femoris

Semitendinosus

Semimembranosus

Plantaris

Popliteus

Soleus

Gastrocnemius, medial head

Gastrocnemius, lateral head

B

4-31 MUSCLES OF THE GLUTEAL REGION AND BACK OF THE THIGH

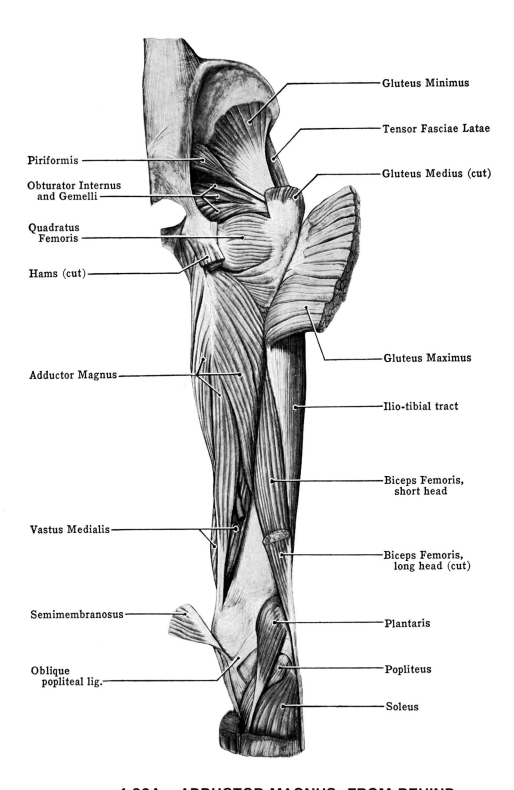

Gluteus Minimus

Tensor Fasciae Latae

Gluteus Medius (cut)

Piriformis

Obturator Internus
and Gemelli

Quadratus
Femoris

Hams (cut)

Gluteus Maximus

Adductor Magnus

Ilio-tibial tract

Biceps Femoris,
short head

Vastus Medialis

Biceps Femoris,
long head (cut)

Semimembranosus

Plantaris

Oblique
popliteal lig.

Popliteus

Soleus

4-32A   ADDUCTOR MAGNUS, FROM BEHIND

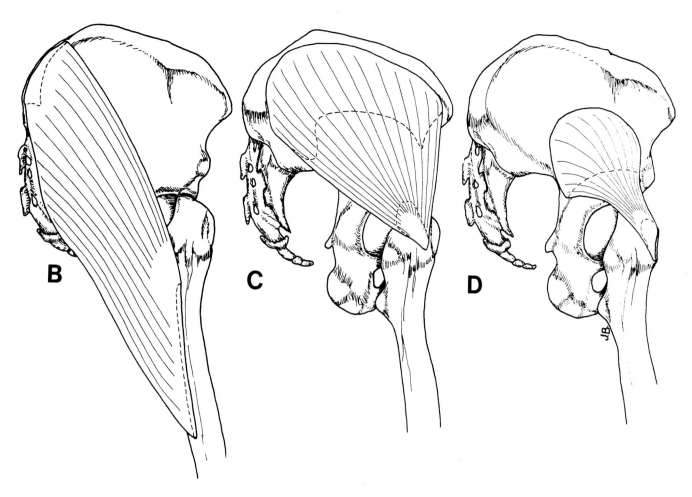

4-32B, C, D   THE GLUTEI

These oblique views of the hip region show the bony attachments of the three gluteal muscles. Gluteus maximus (*B*) arises from ilium, sacrum, and coccyx as well as from the aponeurosis of the erector spinae and the sacrotuberous ligament. As well as its attachment to the femur, shown here, it has a strong attachment to the iliotibial tract. Gluteus medius (*C*) arises from the ilium and inserts on the greater trochanter. Gluteus minimus (*D*) passes from the ilium to the *front* of the greater trochanter. As may be determined by their attachments, all three are extensors and abductors of the hip. In addition, maximus is a strong lateral rotator of the thigh, and minimus (passing in front of the joint) is a medial rotator. Medius and minimus are crucial in maintaining stability at the hip when the opposite leg is raised in walking.

In *E*, the subject in the *upper* photograph is lying prone with glutei relaxed. In the *lower* picture he is raising his trunk from the table. With fixed femur, the glutei are extending the hip joint by rotating the pelvis.

See Stern, J. T. (1972) Anatomical and functional specializations of the human Gluteus Maximus. *Am. J. Phys. Anthrop.*, *36*: 315–340.

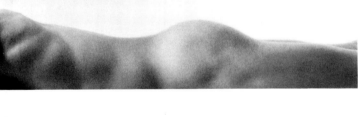

4-32E   GLUTEUS MAXIMUS

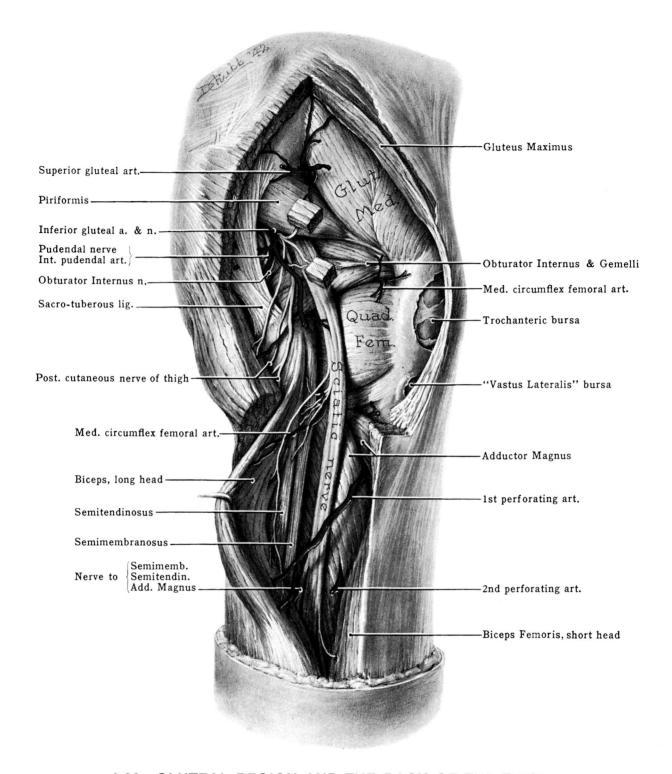

Superior gluteal art.

Piriformis

Inferior gluteal a. & n.

Pudendal nerve
Int. pudendal art.

Obturator Internus n.

Sacro-tuberous lig.

Post. cutaneous nerve of thigh

Med. circumflex femoral art.

Biceps, long head

Semitendinosus

Semimembranosus

Nerve to { Semimemb.
Semitendin.
Add. Magnus }

Gluteus Maximus

Obturator Internus & Gemelli

Med. circumflex femoral art.

Trochanteric bursa

"Vastus Lateralis" bursa

Adductor Magnus

1st perforating art.

2nd perforating art.

Biceps Femoris, short head

4-33 GLUTEAL REGION AND THE BACK OF THE THIGH—I

Gluteus Maximus is split, both above and below, in the direction of its fibers, and the middle part excised, but two cubes remain to identify its nerve.

Observe:

1. Gluteus Maximus is the only muscle to cover the greater trochanter. It is aponeurotic where it plays on the trochanter and the aponeurosis of Vastus Lateralis, and has underlying bursae.
2. The nerve (inferior gluteal nerve) enters Gluteus Maximus in two chief branches near its center.
3. Above Piriformis is Gluteus Medius, which covers Gluteus Minimus (Fig. 4-34).

4. The sciatic nerve appears below Piriformis and crosses in turn: dorsum ischii, Obturator Internus and Gemelli, Quadratus Femoris, and Adductor Magnus. Its branches spring from its medial side at variable levels to supply the hams and part of Adductor Magnus. Only the branch to Biceps (short head) springs (usually low down) from its lateral side, which, therefore, is the safer side to dissect on.
5. The richness of vessels and nerves is disquieting when you recall that the gluteal region is a common site for intramuscular injections. A safe approach is described in Von Hochstetter, A., *et al.* (1959) *Intragluteal Injection*. Georg Thieme-Verlag, Stuttgart.

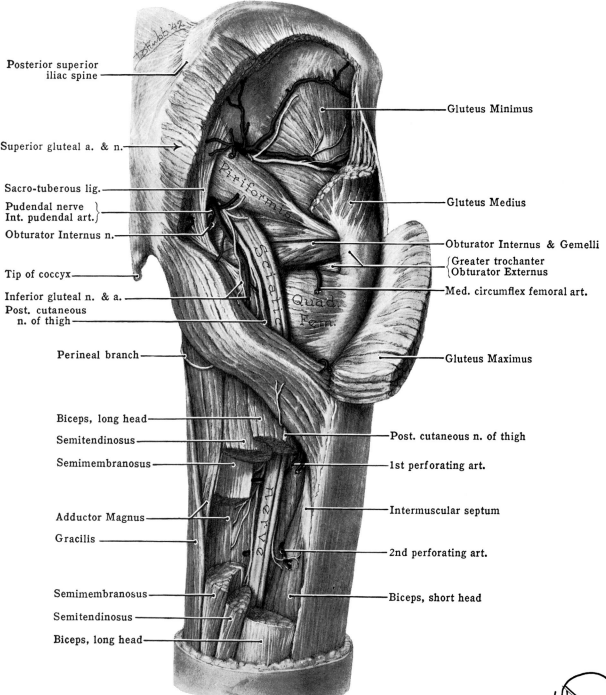

Posterior superior
iliac spine

Superior gluteal a. & n.

Sacro-tuberous lig.
Pudendal nerve
Int. pudendal art.
Obturator Internus n.

Tip of coccyx

Inferior gluteal n. & a.
Post. cutaneous
n. of thigh

Perineal branch

Biceps, long head
Semitendinosus
Semimembranosus

Adductor Magnus
Gracilis

Semimembranosus
Semitendinosus
Biceps, long head

Gluteus Minimus

Gluteus Medius

Obturator Internus & Gemelli
Greater trochanter
Obturator Externus
Med. circumflex femoral art.

Gluteus Maximus

Post. cutaneous n. of thigh
1st perforating art.

Intermuscular septum

2nd perforating art.

Biceps, short head

4-34 GLUTEAL REGION AND THE BACK OF THE THIGH—II

The upper three-quarters of Gluteus Maximus is reflected, and parts of Gluteus
Medius and the three ham muscles are excised.

Observe:
. The superior gluteal vessels and nerve appearing above Piriformis; all other
  vessels and nerves appearing below it.
. There are no nerves and no vessels of importance lateral to the sciatic nerve.
. The horizontal groove (the natal fold) crossing the lower border of Gluteus
  Maximus indicates the upper limit of the sleevelike deep fascia of the thigh.
. Gluteus Maximus consisting of bundles of parallel fibers. It is rhomboidal
  and the deep fascia covering it is thin. Gluteus Medius arising in part from
  the covering deep fascia which, therefore, is strong and thick. It is fan-
  shaped.
. The sciatic nerve is accessible deep in the angle between the lower border of
  Gluteus Maximus and the lateral border of the long head of Biceps.

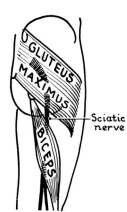

4-35 SCIATIC NERVE

Exaggerated to show that the sciatic
nerve is most readily accessible in
the angle deep to Gluteus Maximus
and long head of Biceps.

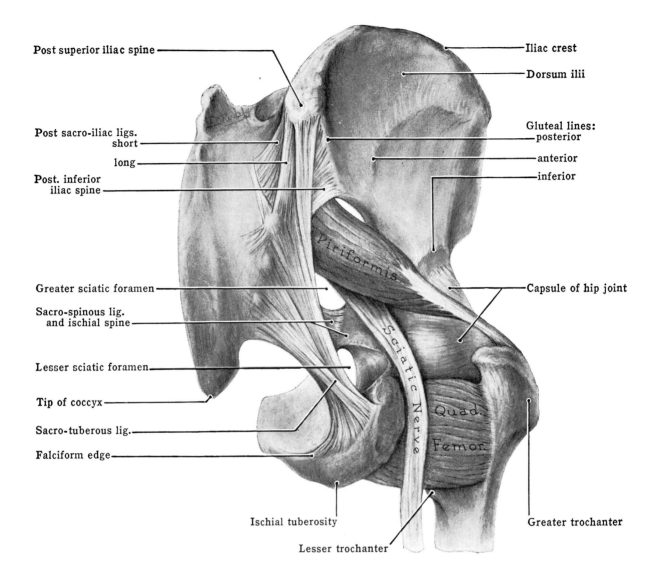

Post superior iliac spine

Post sacro-iliac ligs. short

long

Post. inferior iliac spine

Greater sciatic foramen

Sacro-spinous lig. and ischial spine

Lesser sciatic foramen

Tip of coccyx

Sacro-tuberous lig.

Falciform edge

Iliac crest

Dorsum ilii

Gluteal lines: posterior

anterior

inferior

Capsule of hip joint

Piriformis

Sciatic Nerve

Quad. Femor.

Greater trochanter

Ischial tuberosity

Lesser trochanter

4-36 BONY AND LIGAMENTOUS PARTS OF GLUTEAL REGION

Observe:
1. The tip of the coccyx lies above the level of the ischial tuberosity and below that of the ischial spine.
2. The lower border of Piriformis is defined by joining the midpoint between the tip of the coccyx and the posterior superior iliac spine to the top of the greater trochanter.
3. The lower border of Quadratus Femoris is level with the lower end of the ischial tuberosity and it crosses the lesser trochanter.
4. The lateral border of the sciatic nerve lies midway between the lateral surface of the greater trochanter and the medial surface of the ischial tuberosity, provided the body is in the anatomical posture—toes pointing forward.

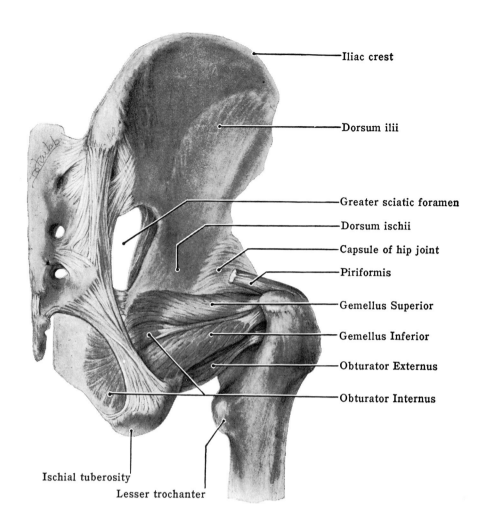

Iliac crest

Dorsum ilii

Greater sciatic foramen

Dorsum ischii

Capsule of hip joint

Piriformis

Gemellus Superior

Gemellus Inferior

Obturator Externus

Obturator Internus

Ischial tuberosity

Lesser trochanter

4-37 OBTURATOR MUSCLES FROM BEHIND

Observe:
1. Obturator Internus and Gemelli fill the gap between Piriformis above and Quadratus Femoris below. (For origin within the pelvis see Fig. 3-55.)
2. Obturator Externus passing obliquely, below neck of femur, to its insertion. (For origin see Figs. 4-42 and 4-39.)
3. That the lower end of the ischial tuberosity is on the level of the lesser trochanter.

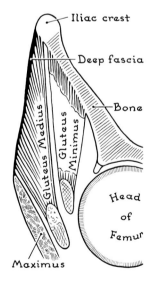

Iliac crest

Deep fascia

Bone

Gluteus Medius

Gluteus Minimus

Head of Femur

Maximus

4-38 GLUTEUS MEDIUS

The most anterior part of Gluteus Medius has but little bone available to it (Fig. 4-41), so it uses extensively, as an aponeurosis, the deep fascia covering it.

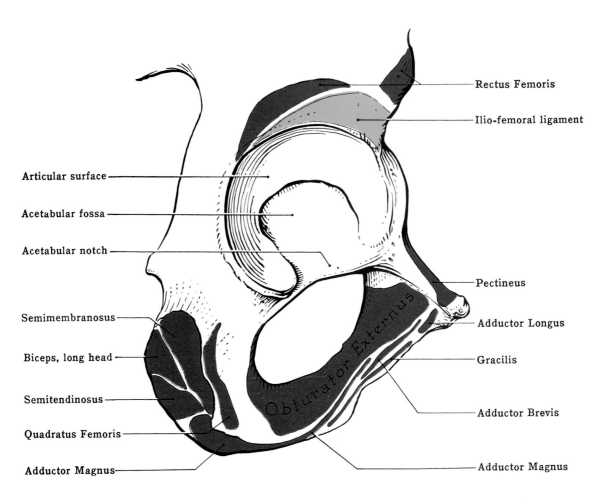

Rectus Femoris

Ilio-femoral ligament

Articular surface

Acetabular fossa

Acetabular notch

Pectineus

Semimembranosus

Adductor Longus

Biceps, long head

Gracilis

Obturator Externus

Semitendinosus

Adductor Brevis

Quadratus Femoris

Adductor Magnus

Adductor Magnus

4-39 ACETABULAR REGION: ORIGINS OF NEIGHBORING MUSCLES

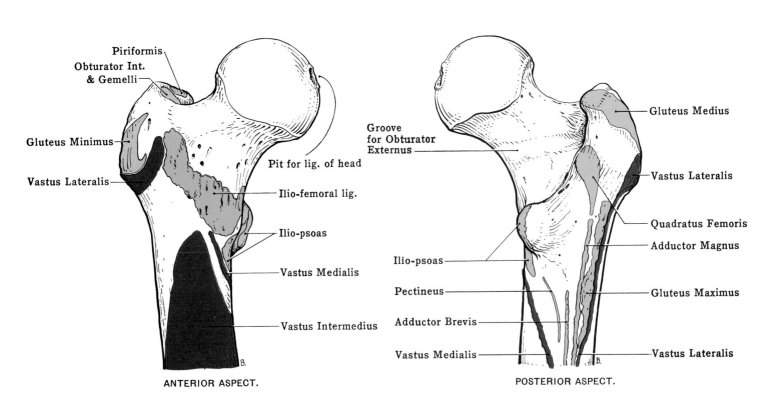

Piriformis
Obturator Int.
& Gemelli

Gluteus Medius

Gluteus Minimus

Pit for lig. of head

Groove
for Obturator
Externus

Vastus Lateralis

Vastus Lateralis

Ilio-femoral lig.

Quadratus Femoris

Ilio-psoas

Adductor Magnus

Ilio-psoas

Vastus Medialis

Pectineus

Gluteus Maximus

Adductor Brevis

Vastus Intermedius

Vastus Medialis

Vastus Lateralis

ANTERIOR ASPECT.

POSTERIOR ASPECT.

4-40 UPPER END OF FEMUR SHOWING ATTACHMENTS OF MUSCLES

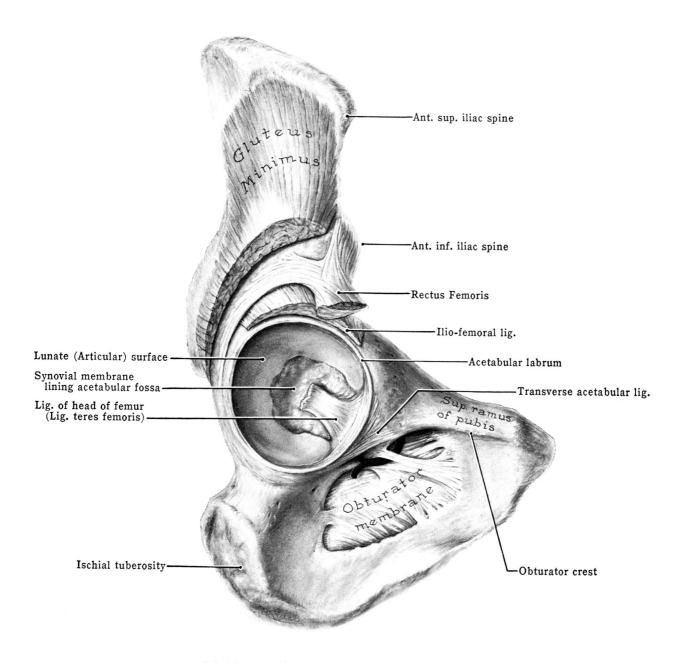

Ant. sup. iliac spine

Ant. inf. iliac spine

Rectus Femoris

Ilio-femoral lig.

Acetabular labrum

Transverse acetabular lig.

Sup. ramus of pubis

Obturator membrane

Obturator crest

Gluteus Minimus

Lunate (Articular) surface

Synovial membrane lining acetabular fossa

Lig. of head of femur (Lig. teres femoris)

Ischial tuberosity

4-41 SOCKET FOR THE HEAD OF THE FEMUR

Observe:
1. The transverse acetabular ligament (whose fibers decussate like the limbs of a St. Andrew's cross) converting the acetabular notch into the acetabular foramen.
2. The acetabular labrum, attached to the acetabular rim and to the transverse ligament. It forms a complete ring around the head of the femur beyond its equator.
3. The articular or lunate surface.
4. The synovial membrane attached to the margin of the articular cartilage and covering the pad of fat and the vessels in the acetabular fossa.
5. The ligament of the head of the femur, which is a hollow cone of synovial membrane compressed between the head of the femur and its socket. It resembles a collapsed bell tent. It envelops ligamentous fibers; these are attached above to the pit on the head of the femur, and below to the transverse ligament and the margins of the acetabular notch. Through it passes the artery to the head of the femur (Fig. 4-47).
6. Gluteus Minimus which, being a medial rotator as well as an abductor of the hip joint, is thickest in front. It covers the two heads of Rectus femoris which in turn cover, and are nearly co-extensive with, the attachment of the ilio-femoral ligament (Fig. 4-39).

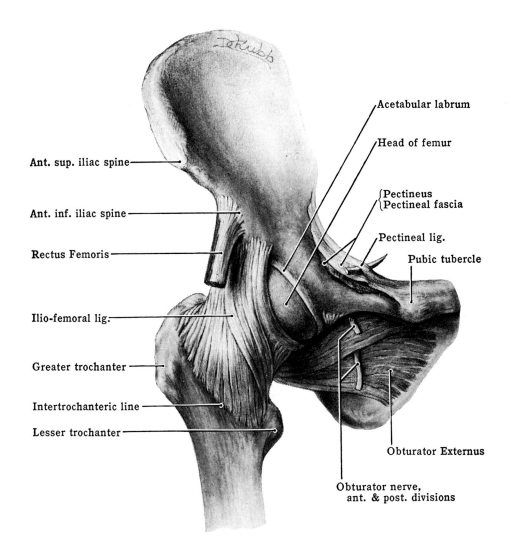

Ant. sup. iliac spine

Ant. inf. iliac spine

Rectus Femoris

Ilio-femoral lig.

Greater trochanter

Intertrochanteric line

Lesser trochanter

Acetabular labrum

Head of femur

{ Pectineus
{ Pectineal fascia

Pectineal lig.

Pubic tubercle

Obturator Externus

Obturator nerve,
ant. & post. divisions

4-42 HIP JOINT, FROM THE FRONT

Observe:
1. The head of the femur exposed just medial to the ilio-femoral ligament and facing not only upward and medially, but also forward. Here, at the side of the Psoas bursa, the capsule is weak or, as in this specimen, partially deficient, but it is guarded by the Psoas tendon.
2. The ilio-femoral ligament, shaped like an inverted Y, attached above deep to Rectus Femoris, and so directed as to become taut on medial rotation of the femur (Figs. 4-39 and 4-40).
3. Obturator Externus crossing obliquely below the neck of the femur (see Fig. 4-37).
4. The thinness of Pectineus; and its fascia blending with the pectineal ligament (Cooper's ligament) along the pecten pubis (pectineal line).

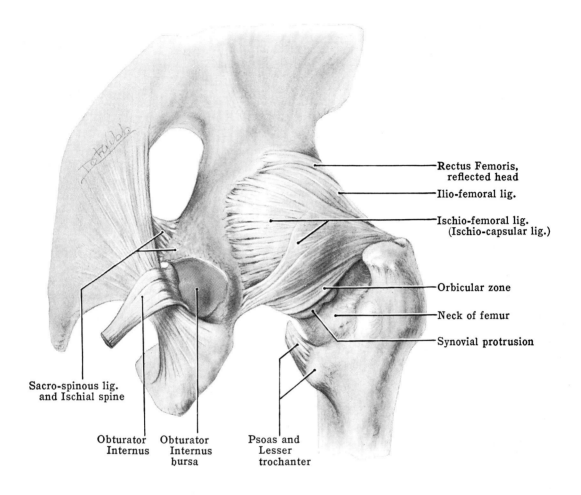

Rectus Femoris,
reflected head

Ilio-femoral lig.

Ischio-femoral lig.
(Ischio-capsular lig.)

Orbicular zone

Neck of femur

Synovial protrusion

Sacro-spinous lig.
and Ischial spine

Obturator
Internus

Obturator
Internus
bursa

Psoas and
Lesser
trochanter

4-43 HIP JOINT, FROM BEHIND

Observe:
1. The fibers of the capsule so directed spirally as to become taut during
extension and medial rotation of the femur.
2. The fibers crossing the neck posteriorly, but not attached to it; indeed, the
synovial membrane protrudes below the fibrous capsule and there forms a
bursa for the tendon of Obturator Externus.

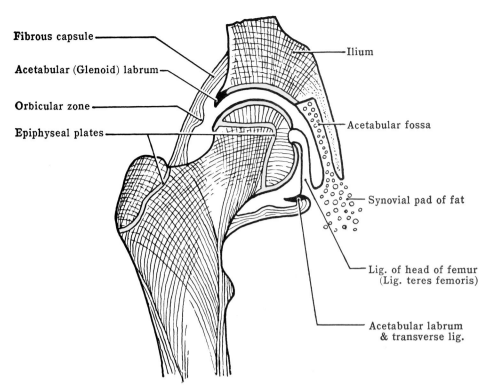

Fibrous capsule

Acetabular (Glenoid) labrum

Orbicular zone

Epiphyseal plates

Ilium

Acetabular fossa

Synovial pad of fat

Lig. of head of femur
(Lig. teres femoris)

Acetabular labrum
& transverse lig.

Observe:
1. The bony trabeculae of the ilium projected into the head of the femur as lines of pressure and the trabeculae that cross these as lines of tension.
2. The epiphysis of the head of the femur, entirely within the capsule of the joint.
3. The ligament of the head of the femur, as a synovial tube that is fixed above at the pit (fovea) on the head of the femur and open below at the acetabular foramen where it is continuous with the synovial membrane covering the fat in the acetabular fossa and also with the synovial membrane covering the transverse ligament (Fig. 4-41).
4. The ligament of the head obviously becomes taut during adduction of the hip joint, as when crossing the legs.
5. The fluid fat below the joint can be sucked into the acetabular fossa during flexion.

4-44 HIP JOINT ON CORONAL SECTION

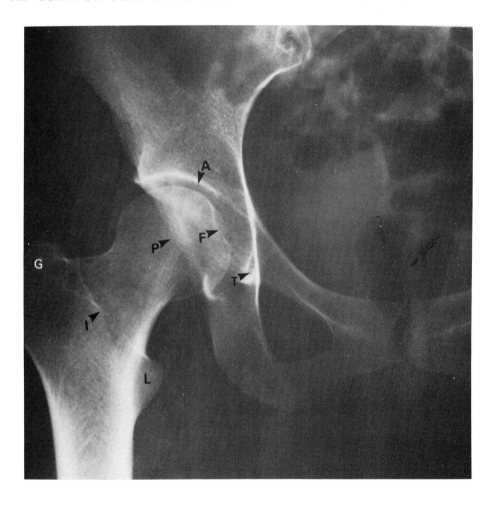

4-45 RADIOGRAPH OF HIP

Observe:

On the femur: greater (G) and lesser (L) trochanters, the intertrochanteric crest (I), and the pit or fovea (F) for the ligament of the head.

On the pelvis: the roof (A) and posterior rim (P) of the acetabulum, and the "teardrop" appearance (T) caused by the superimposition of structures at the lower margin of the acetabulum.

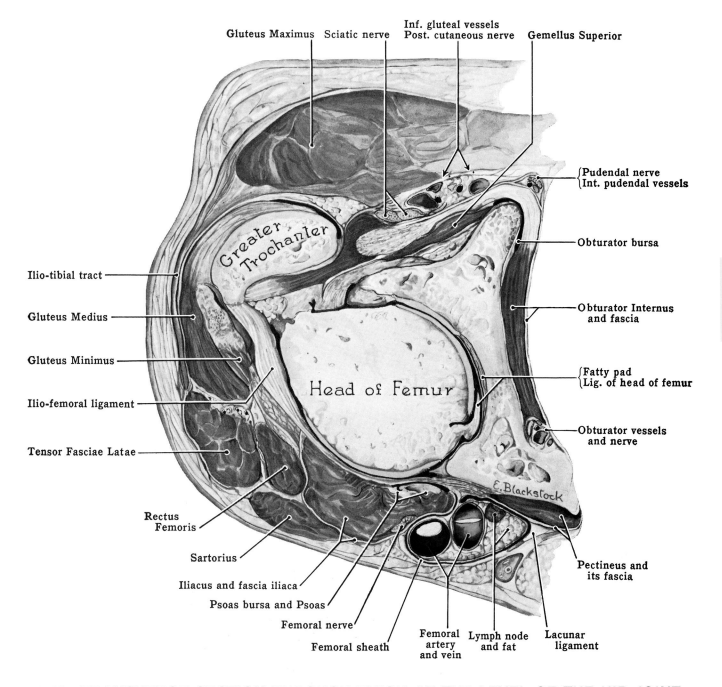

Gluteus Maximus Sciatic nerve Inf. gluteal vessels / Post. cutaneous nerve Gemellus Superior

Pudendal nerve / Int. pudendal vessels

Obturator bursa

Obturator Internus and fascia

Fatty pad / Lig. of head of femur

Obturator vessels and nerve

Ilio-tibial tract

Gluteus Medius

Gluteus Minimus

Ilio-femoral ligament

Tensor Fasciae Latae

Greater Trochanter

Head of Femur

E. Blackstock

Rectus Femoris

Sartorius

Iliacus and fascia iliaca

Psoas bursa and Psoas

Femoral nerve

Femoral sheath

Femoral artery and vein

Lymph node and fat

Lacunar ligament

Pectineus and its fascia

4-46 TRANSVERSE SECTION THROUGH THIGH AT THE LEVEL OF THE HIP JOINT

Observe:
1. The articular cartilage spread unevenly over the head of the femur.
2. The fibrous capsule of the joint is very thick where forming the ilio-femoral ligament, and thin dorsal to the Psoas tendon, the Psoas bursa here intervening.
3. The femoral sheath, which encloses the femoral artery, vein, lymph node, lymph vessels, and fat, to be free except posteriorly where, between Psoas and Pectineus, it is attached to the capsule of the hip joint.
4. The femoral artery separated from the joint by the tough Psoas tendon; the vein at the interval beween Psoas and Pectineus; the lymph node anterior to Pectineus. The femoral nerve lying between Iliacus and fascia Iliaca.
5. The two cusps of the valve in the femoral vein so placed that pressure on the skin surface closes the valve. (For data on valves, see Fig. 4-16.)
6. The sciatic nerve descending between Gluteus Maximus and the short lateral rotators of the femur.

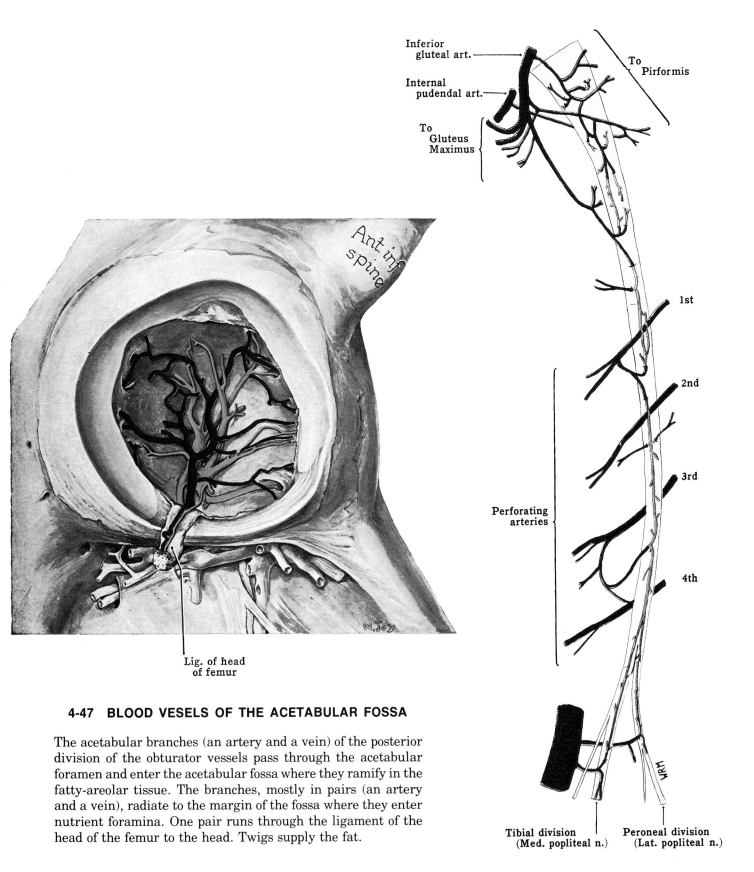

4-47 BLOOD VESELS OF THE ACETABULAR FOSSA

The acetabular branches (an artery and a vein) of the posterior division of the obturator vessels pass through the acetabular foramen and enter the acetabular fossa where they ramify in the fatty-areolar tissue. The branches, mostly in pairs (an artery and a vein), radiate to the margin of the fossa where they enter nutrient foramina. One pair runs through the ligament of the head of the femur to the head. Twigs supply the fat.

4-48 BLOOD SUPPLY TO THE SCIATIC NERVE

Note the continuous anastomotic chain of arteries.

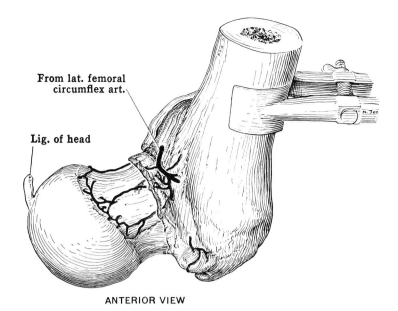

From lat. femoral
circumflex art.

Lig. of head

ANTERIOR VIEW

From med. femoral
circumflex art.

POSTERO-SUPERIOR VIEW

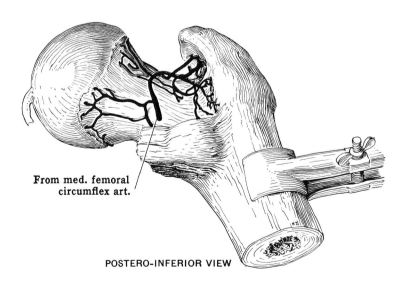

From med. femoral
circumflex art.

POSTERO-INFERIOR VIEW

4-49 BLOOD SUPPLY TO THE HEAD OF THE FEMUR

Note: The head receives 3 sets of arteries: (a) the main set of 3 or 4 ascends in the synovial retinacula on the postero-superior and postero-inferior parts of the neck, to perforate just distal to the head, bend at 45° toward its center, and anastomose freely with (b) terminal branches of the medullary artery of the shaft, and in 80 per cent of cases with (c) the artery of the ligament of the head. This last artery enters the head only when the center of ossification has extended to the pit for the ligament of the head (12th to 14th year). This anastomosis persists even in advanced age, but in 20 per cent it is never established. The blood supply is in danger in fractures of the neck of the femur.

See Wolcott, W. E. (1943) The evolution of the circulation in the developing femoral head and neck. *Surg. Gynec. Obst.*, 77: 61; Trueta, J., and Harrison, M. H. M. (1953) Normal vascular anatomy of the femoral head in adult man. *J. Bone Joint Surg.*, 35B: 442.

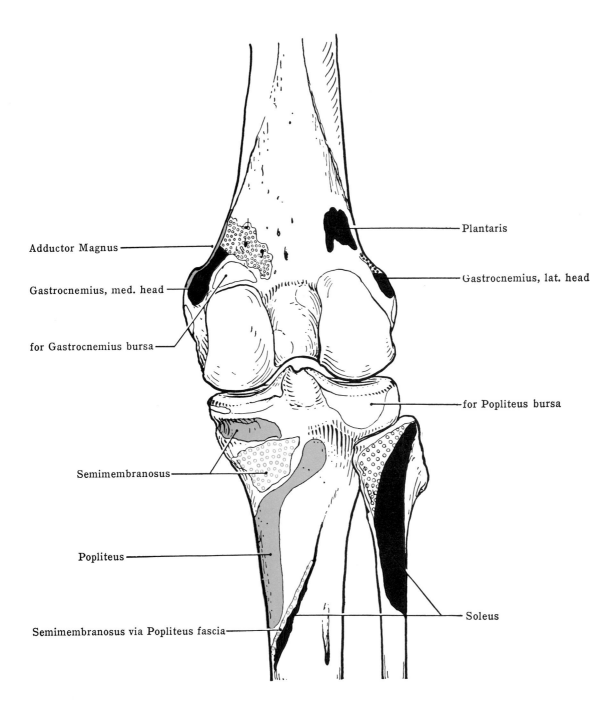

Adductor Magnus

Gastrocnemius, med. head

for Gastrocnemius bursa

Semimembranosus

Popliteus

Semimembranosus via Popliteus fascia

Plantaris

Gastrocnemius, lat. head

for Popliteus bursa

Soleus

**4-50 BONES OF THE KNEE JOINT SHOWING
ATTACHMENTS OF MUSCLES, FROM BEHIND**

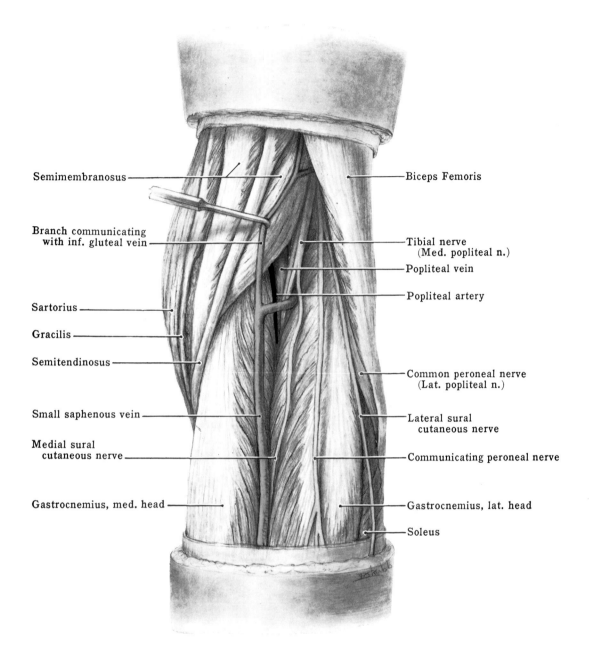

Semimembranosus

Branch communicating
with inf. gluteal vein

Sartorius

Gracilis

Semitendinosus

Small saphenous vein

Medial sural
cutaneous nerve

Gastrocnemius, med. head

Biceps Femoris

Tibial nerve
(Med. popliteal n.)

Popliteal vein

Popliteal artery

Common peroneal nerve
(Lat. popliteal n.)

Lateral sural
cutaneous nerve

Communicating peroneal nerve

Gastrocnemius, lat. head

Soleus

4-51 SUPERFICIAL DISSECTION OF THE POPLITEAL FOSSA

Observe:
1. The two heads of Gastrocnemius, embraced on the medial side by Semimembranosus, which is overlaid by Semitendinosus, and on the lateral side by Biceps. The result is the lozenge-shaped popliteal fossa.
2. The small saphenous vein running between the two heads of Gastrocnemius. Deep to this vein is the medial sural cutaneous nerve which, followed proximally, leads to the tibial nerve. The tibial nerve is superficial to the popliteal vein, which in turn is superficial to the popliteal artery.
3. The common peroneal nerve following the posterior border of Biceps, and here giving off two cutaneous branches. (See cutaneous nerves, Fig. 4-8.)

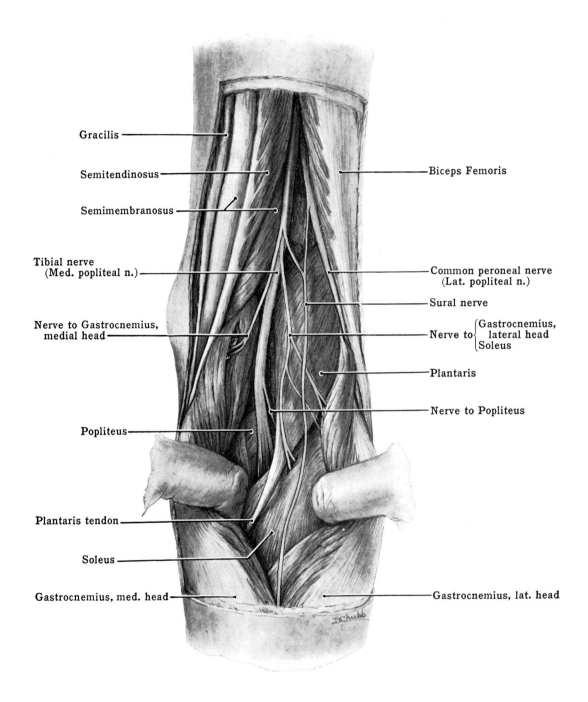

Gracilis

Semitendinosus

Semimembranosus

Tibial nerve
(Med. popliteal n.)

Nerve to Gastrocnemius,
medial head

Popliteus

Plantaris tendon

Soleus

Gastrocnemius, med. head

Biceps Femoris

Common peroneal nerve
(Lat. popliteal n.)

Sural nerve

Nerve to {Gastrocnemius, lateral head / Soleus}

Plantaris

Nerve to Popliteus

Gastrocnemius, lat. head

4-52   NERVES OF THE POPLITEAL FOSSA

The two heads of Gastrocnemius are pulled forcibly apart.

Observe:

1. A cutaneous branch of the tibial nerve joining a cutaneous branch of the common peroneal nerve to form the sural nerve. Here the junction is very high; usually it is 5 to 8 cm above the ankle.

2. All motor branches in this region springing from the tibial nerve, one branch coming from its medial side, the others from its lateral side. Hence, it is safer to dissect on the medial side.

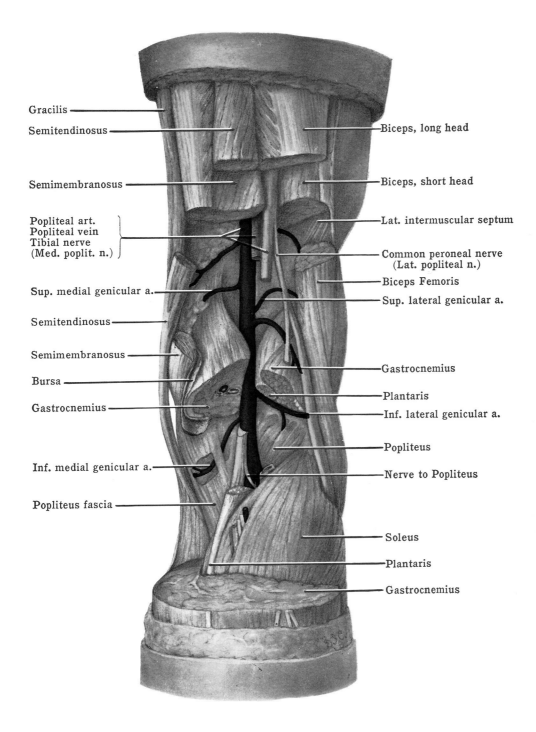

Gracilis

Semitendinosus

Semimembranosus

Popliteal art.
Popliteal vein
Tibial nerve
(Med. poplit. n.)

Sup. medial genicular a.

Semitendinosus

Semimembranosus

Bursa

Gastrocnemius

Inf. medial genicular a.

Popliteus fascia

Biceps, long head

Biceps, short head

Lat. intermuscular septum

Common peroneal nerve
(Lat. popliteal n.)

Biceps Femoris

Sup. lateral genicular a.

Gastrocnemius

Plantaris

Inf. lateral genicular a.

Popliteus

Nerve to Popliteus

Soleus

Plantaris

Gastrocnemius

4-53   STEP DISSECTION OF THE POPLITEAL FOSSA

Observe:

1. The thickness of the various muscles.
2. The popliteal artery lying on the floor of the fossa (*i.e.*, femur, capsule of joint, Popliteus fascia), much fat intervening, and giving off genicular branches which also lie on the floor, and ending by bifurcating into the anterior and the posterior tibial artery at the upper border of Soleus.

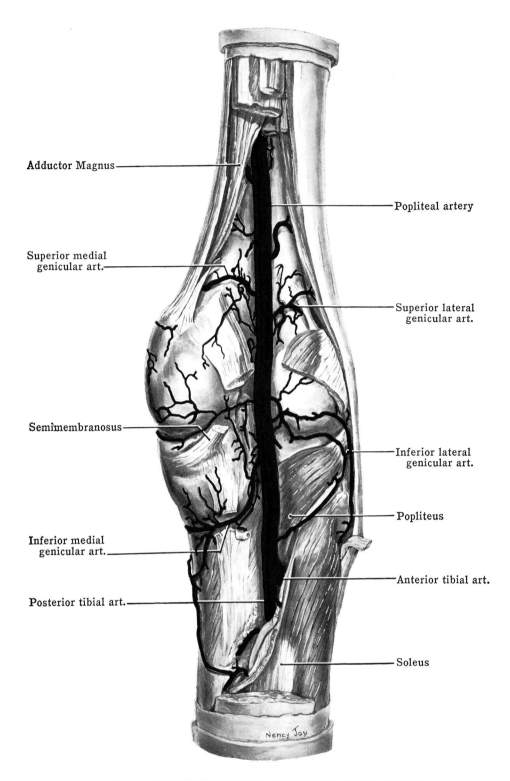

Adductor Magnus

Popliteal artery

Superior medial
genicular art.

Superior lateral
genicular art.

Semimembranosus

Inferior lateral
genicular art.

Popliteus

Inferior medial
genicular art.

Anterior tibial art.

Posterior tibial art.

Soleus

Nancy Joy

4-54 ANASTOMOSES AROUND THE KNEE, POSTERIOR VIEW

Observe:
1. The popliteal artery (injected with latex) throughout its course, from the hiatus in Adductor Magnus proximally to the lower border of Popliteus distally, where it bifurcates into the anterior and posterior tibial arteries.
2. The 3 ventral relations of the artery: (a) femur (fat intervening), (b) capsule of the joint, and (c) Popliteus (covered with popliteus fascia).
3. The 4 named genicular branches that hug the skeletal plane, nothing intervening except the popliteus tendon which the inferior lateral genicular artery must cross. (The median genicular artery is not in view.)
4. An unnamed genicular artery arising on each side.

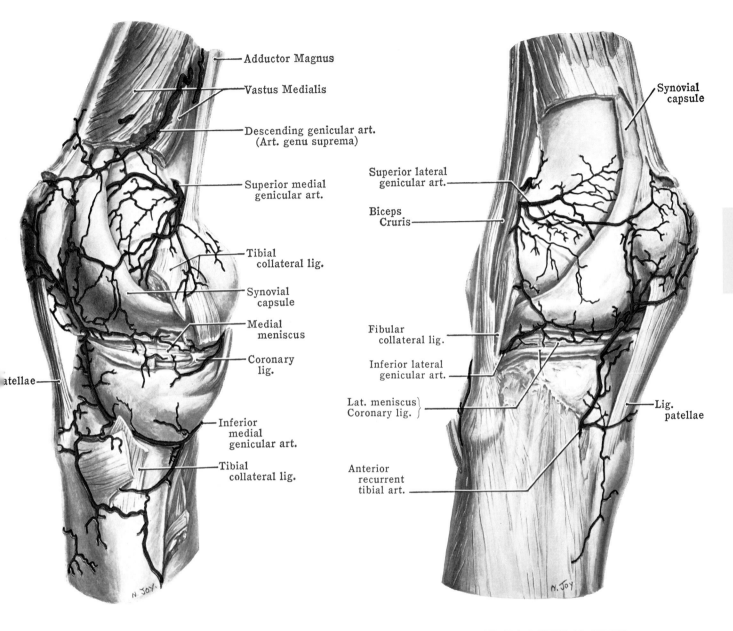

Adductor Magnus

Vastus Medialis

Descending genicular art.
(Art. genu suprema)

Superior medial
genicular art.

Tibial
collateral lig.

Synovial
capsule

Medial
meniscus

Coronary
lig.

Inferior
medial
genicular art.

Tibial
collateral lig.

atellae

Synovial
capsule

Superior lateral
genicular art.

Biceps
Cruris

Fibular
collateral lig.

Inferior lateral
genicular art.

Lat. meniscus}
Coronary lig. }

Anterior
recurrent
tibial art.

Lig.
patellae

A. ANTERO-MEDIAL VIEW

B. ANTERO-LATERAL VIEW

4-55 ANASTOMOSES AROUND THE KNEE

Observe:
1. Two named genicular branches of the popliteal artery; on each side, a superior and an inferior.
2. Three supplementary arteries: (a) descending genicular branch of the femoral artery, supero-medially; (b) descending branch of lateral femoral circumflex artery, supero-laterally (Fig 4-11); and (c) anterior recurrent branch of anterior tibial artery, infero-laterally (Fig. 4-75).
3. The inferior lateral genicular artery running along the lateral meniscus; an unnamed artery running similarly along the medial meniscus.

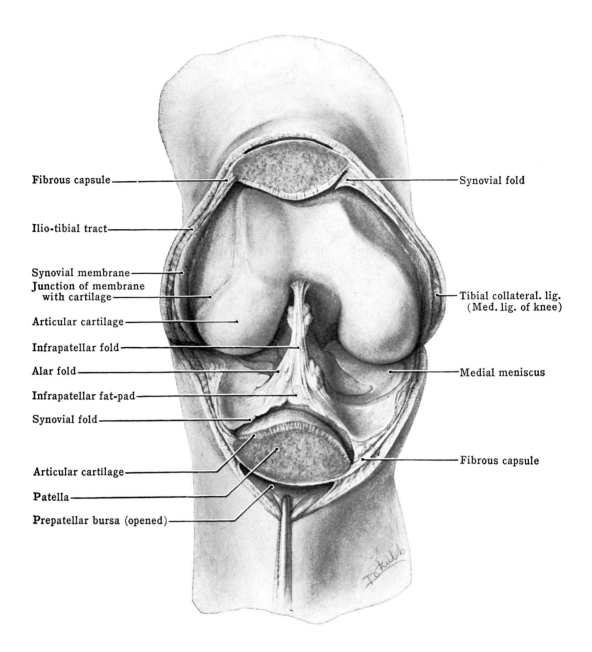

Fibrous capsule

Ilio-tibial tract

Synovial membrane
Junction of membrane
with cartilage

Articular cartilage

Infrapatellar fold

Alar fold

Infrapatellar fat-pad

Synovial fold

Articular cartilage

Patella

Prepatellar bursa (opened)

Synovial fold

Tibial collateral. lig.
(Med. lig. of knee)

Medial meniscus

Fibrous capsule

4-56 KNEE JOINT, OPENED FROM THE FRONT

The patella is sawn through; the skin and joint capsule are cut through; and the joint is flexed.

Observe:

1. The articular cartilage of the patella, not of uniform thickness but spread unevenly, as on other bones.
2. The infrapatellar synovial fold resembling a partially collapsed bell-tent whose apex is attached to the intercondylar notch and whose base is below the patella (cf. ligament of the head of femur, Fig. 4-44). The infrapatellar pad of fat is continued into the tent.
3. A fracture of the patella would bring the prepatellar bursa into the communication with the joint cavity.
4. Articular cartilage and synovial membrane continuous with each other on the side of the condyle, as in other joints.

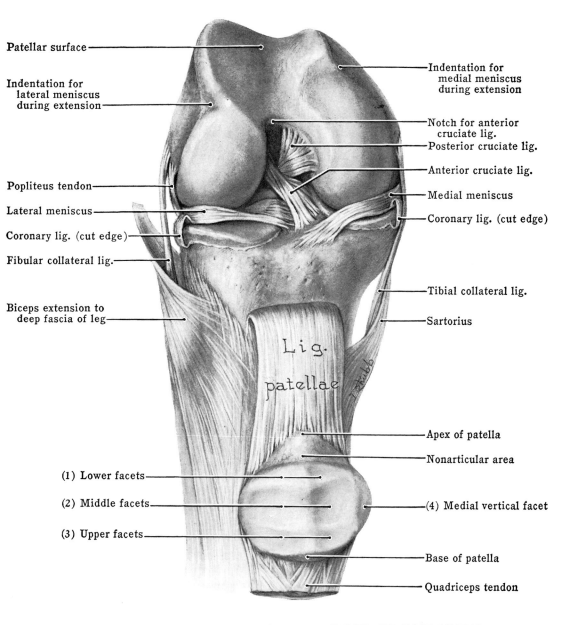

Patellar surface

Indentation for
lateral meniscus
during extension

Popliteus tendon

Lateral meniscus

Coronary lig. (cut edge)

Fibular collateral lig.

Biceps extension to
deep fascia of leg

(1) Lower facets

(2) Middle facets

(3) Upper facets

Lig.
patellae

Indentation for
medial meniscus
during extension

Notch for anterior
cruciate lig.

Posterior cruciate lig.

Anterior cruciate lig.

Medial meniscus

Coronary lig. (cut edge)

Tibial collateral lig.

Sartorius

Apex of patella

Nonarticular area

(4) Medial vertical facet

Base of patella

Quadriceps tendon

4-57 LIGAMENTS OF THE KNEE JOINT, FRONT VIEW

The patella is thrown down and the joint is fixed.

Observe:

1. The indentations on the sides of the femoral condyles at the junction of the patellar and tibial articular areas. The lateral tibial articular area, shorter than the medial one.
2. The subsidiary notch, at the antero-lateral part of the intercondylar notch, for the reception of the anterior cruciate ligament on full extension.
3. The three paired facets on the posterior surface of the patella for articulation with the patellar surface of the femur successively during (1) extension, (2) slight flexion, (3) flexion; and the most medial facet on the patella (4) for articulation during full flexion with the crescentic facet that skirts the medial margin of the intercondylar notch of the femur.

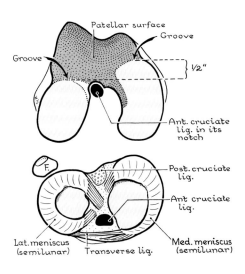

Patellar surface

Groove

Groove

½″

Ant. cruciate
lig. in its
notch

Post. cruciate
lig.

Ant cruciate
lig.

Lat. meniscus
(semilunar)

Transverse lig.

Med. meniscus
(semilunar)

4-58 ARTICULAR SURFACES OF KNEE JOINT

In each illustration one-half of the femur is removed with the proximal part of the corresponding cruciate ligament.

Observe:
1. The posterior cruciate ligament prevents forward sliding of the femur, particularly when the knee is flexed.
2. The anterior cruciate ligament prevents backward sliding of the femur and hyperextension of the knee and limits medial rotation of the femur when the foot is on the ground—*i.e.*, when the leg is fixed.

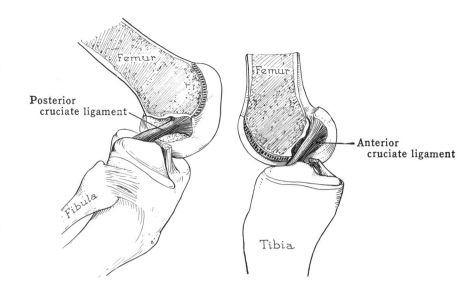

4-59 CRUCIATE LIGAMENTS

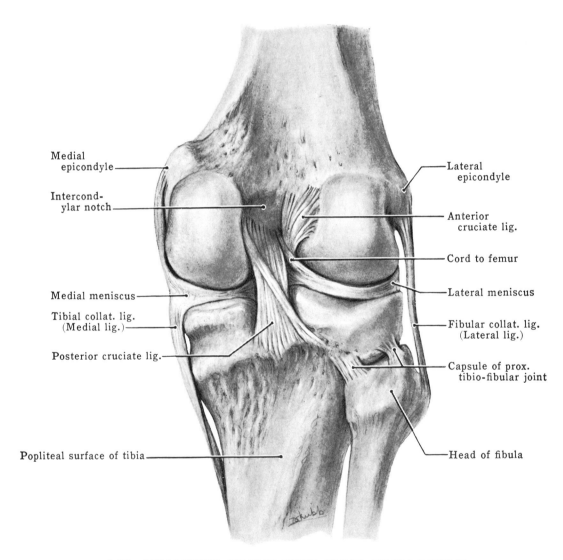

4-60 LIGAMENTS OF THE KNEE JOINT, FROM BEHIND

Observe:
1. The bandlike medial ligament attached to the medial meniscus (semilunar cartilage). The cordlike lateral ligament separated from the lateral meniscus by the width of the Popliteus tendon (removed).

2. The posterior cruciate ligament joined by a cord from the lateral meniscus and passing to the fore part of the medial condyle of the femur. The anterior cruciate ligament attached to the hinder part of the lateral condyle.

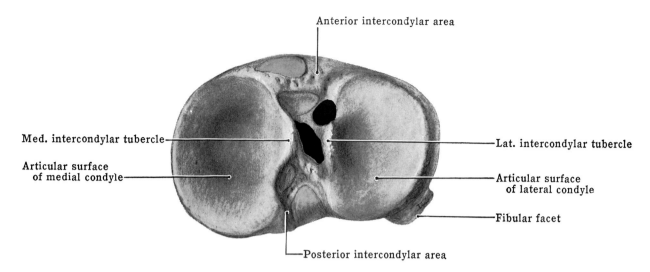

Anterior intercondylar area

Med. intercondylar tubercle

Articular surface
of medial condyle

Lat. intercondylar tubercle

Articular surface
of lateral condyle

Fibular facet

Posterior intercondylar area

Superior Aspect of the Proximal End of the Tibia

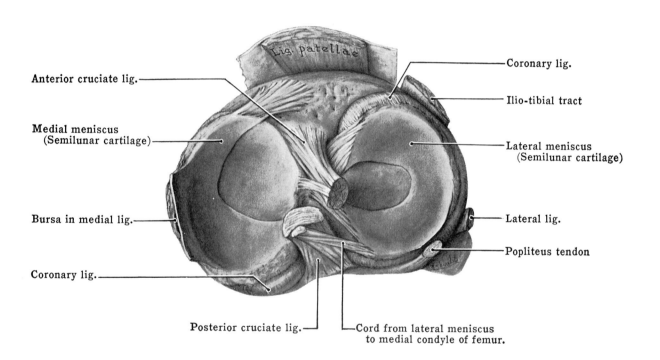

Lig. patellae

Anterior cruciate lig.

Coronary lig.

Ilio-tibial tract

Medial meniscus
(Semilunar cartilage)

Lateral meniscus
(Semilunar cartilage)

Bursa in medial lig.

Lateral lig.

Popliteus tendon

Coronary lig.

Posterior cruciate lig.

Cord from lateral meniscus
to medial condyle of femur.

4-61   CRUCIATE LIGAMENTS AND THE MENISCI (SEMILUNAR CARTILAGES)

e sites of attachment of the cruciate ligaments are colored
low; those of the medial meniscus, *blue*; and those of the lateral
niscus, *red*.

the tibial condyles, the lateral is flatter, shorter from front to
k, and more circular; the medial is concave, longer from front to
k, and more oval.

e menisci are cartilaginous and tough where compressed be-
en femur and tibia, but ligamentous and pliable at their attach-
nts – as in the case with other intra-articular fibro-cartilages.

e menisci conform to the shapes of the surfaces on which they
t. Since the horns of the lateral meniscus are attached close

together and its coronary ligament is slack, this meniscus can slide
forward and backward on the (flat) condyle; since the horns of the
medial meniscus are attached far apart, its movements on the
(concave) condyle are restricted. The medial meniscus is commonly
trapped, injured, and torn.

Note the bursa between the long and short parts of the medial
ligament of the knee.

See Robichon, J., and Romero, C. (1968) The functional anatomy of
the knee joint. *Can. J. Surg., 11:* 36; Kennedy, J. C., Weinberg, H.
W., and Wilson, A. S. (1974) The anatomy and function of the
anterior cruciate ligament. *J. Bone Joint Surg., 56A:* 223.

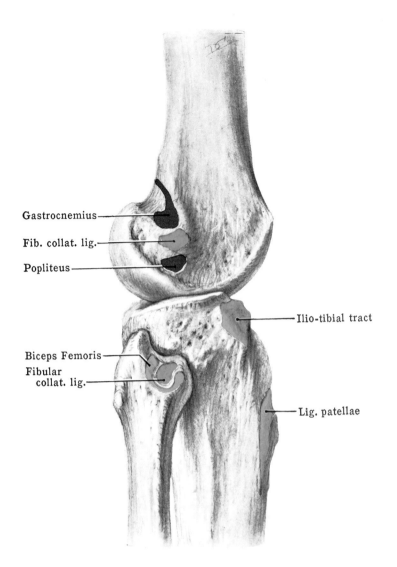

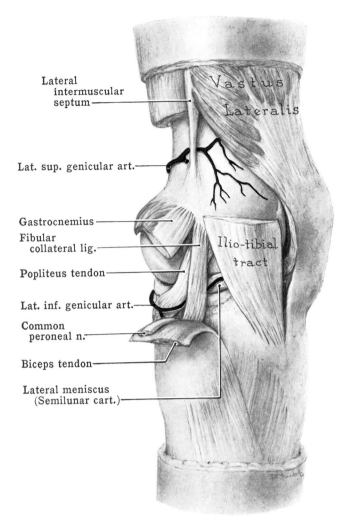

Gastrocnemius

Fib. collat. lig.

Popliteus

Ilio-tibial tract

Biceps Femoris
Fibular
collat. lig.

Lig. patellae

Lateral
intermuscular
septum

Vastus
Lateralis

Lat. sup. genicular art.

Gastrocnemius

Fibular
collateral lig.

Ilio-tibial
tract

Popliteus tendon

Lat. inf. genicular art.

Common
peroneal n.

Biceps tendon

Lateral meniscus
(Semilunar cart.)

**4-62A   BONES OF THE KNEE JOINT: ATTACHMENTS
OF MUSCLES AND LIGAMENTS, LATERAL VIEW**

**4-62B   DISSECTION OF THE KNEE,
LATERAL ASPECT**

Observe:
1. The ilio-tibial tract intervening between the skin and the synovial membrane
   and which, by virtue of its toughness, protects this exposed aspect of the joint.
2. The 3 structures that arise from the lateral epicondyle and that are uncovered by reflecting Biceps. Of these, Gastrocnemius is postero-superior; Popliteus is antero-inferior; the fibular collateral ligament is in between, and it
   crosses superficial to Popliteus.
3. The lateral inferior genicular artery coursing along the lateral meniscus.
4. The proximal Tibio-fibular joint, commonly neglected, has important functions: (a) dissipation of torsional stresses applied at the ankle, (b) dissipation
   of lateral tibial bending moments, and (c) tensile weight bearing.

See Ogden, J. A. (1974) The anatomy and function of the proximal tibiofibular
joint. *Clin. Orthop., No. 101:* 186.

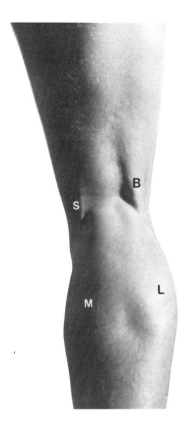

4-63A POPLITEAL FOSSA

The hamstring muscles diverge, Biceps (*B*) to the fibula; the two Semi-muscles (*S*) to the tibia. Medial (*M*) and Lateral (*L*) heads of gastrocnemius are grasped at their origins by the hamstrings as they near their insertion.

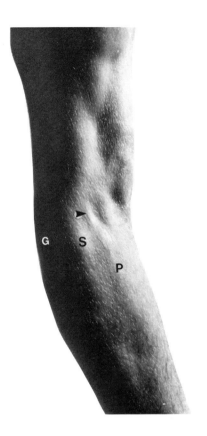

4-63B LATERAL SIDE OF THE KNEE

The *arrow* points to the tendon of Biceps femoris inserting on the head of the fibula. The bulges of the lateral head of Gastrocnemius (*G*), Soleus (*S*), and Peroneus longus (*P*) can be seen.

4-63C THE PROXIMAL TIBIOFIBULAR JOINT

The proximal Tibiofibular joint exists in two basic forms: 1. *Horizontal*, with two almost flat surfaces articulating posterior to the lateral edge of the tibia, and 2. *Oblique*, with joint surfaces inclined at an angle greater than 20 degrees. Generally, the greater the angle, the smaller the surface area of the joint. Rotation at this joint occurs during dorsiflexion of the ankle, especially in horizontal joints. In knee flexion the fibula moves anteriorly; in extension, posteriorly. See Ogden, J. A. (1974) The anatomy and function of the proximal tibiofibular joint. *Clin. Orthop.*, *101*: 186–191.

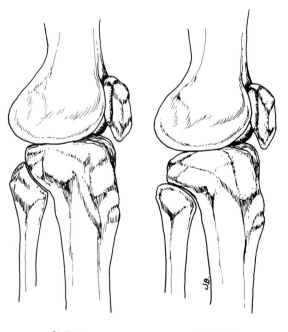

Oblique   Horizontal

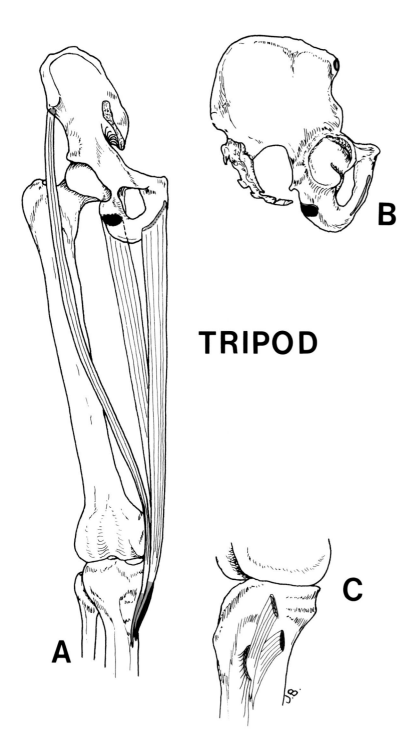

TRIPOD

A

B

C

A. These three muscles: Sartorius (*Green*), Gracilis (*Blue*), and Semitendinosus (*Red*) form an inverted "tripod" with its base separated at the hip bone and its three legs converging to an apex on the medial side of the proximal end of the tibia.

B. Each has its origin on a different bone: Sartorius on the ilium; Gracilis on the pubis; Semitendinosus on the ischium. Each has a different nerve supply: Sartorius, femoral; Gracilis, obturator; Semitendinosus, sciatic. Each belongs to a different muscle group: Sartorius is an anterior (flexor) thigh muscle, Gracilis is in the adductor compartment of the thigh, and Semitendinosus is a hamstring (extensor) muscle. Examining their attachments, it can be seen that all flex the knee but Sartorius is a lateral rotator and abductor while Gracilis is a medial rotator and adductor.

C. At their insertion to the tibia all three tendons become thin aponeuroses. A sharp knife and a steady hand are required to separate them from each other as they near their linear attachment. In addition, the upper fibers of Sartorius curve backward above the insertion of Gracilis. See Figure 4-65A.

4-64   MUSCULAR TRIPOD

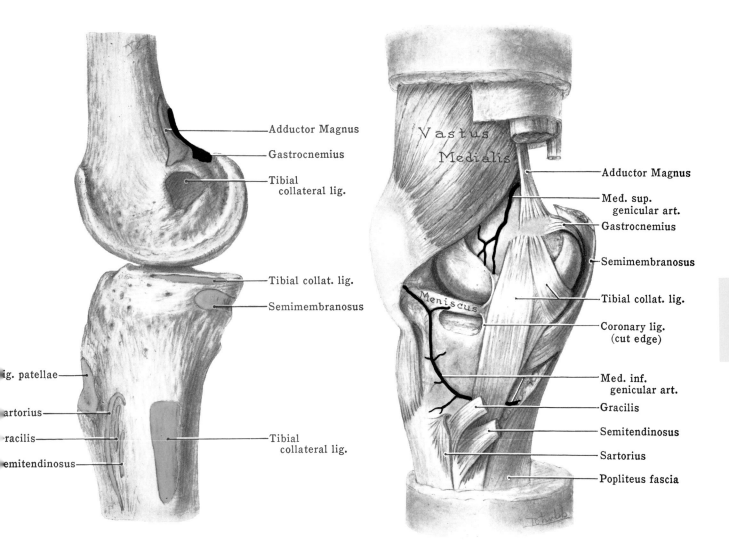

— Adductor Magnus

— Gastrocnemius

— Tibial
   collateral lig.

— Tibial collat. lig.

— Semimembranosus

ig. patellae —

artorius —

racilis —

emitendinosus —

— Tibial
   collateral lig.

Vastus
Medialis

Meniscus

— Adductor Magnus

— Med. sup.
   genicular art.
— Gastrocnemius

— Semimembranosus

— Tibial collat. lig.

— Coronary lig.
   (cut edge)

— Med. inf.
   genicular art.
— Gracilis

— Semitendinosus

— Sartorius

— Popliteus fascia

4-65A   BONES OF THE KNEE
MUSCLE AND LIGAMENT ATTACHMENTS,
MEDIAL VIEW

4-65B   DISSECTION OF THE KNEE,
MEDIAL VIEW

Note that the bandlike part of the tibial collateral ligament attaches to the medial epicondyle of the femur, is almost in line with the tendon of Adductor magnus, bridges over the insertion of Semimembranosus, crosses the medial inferior genicular artery, and is crossed by three tendons (Sartorius, Gracilis, Semitendinosus) on their way to their insertion. See Figure 4-64.

4-66   ARTICULARIS GENU

Articularis genu, deep to Vastus intermedius, consists of a few fibers arising from the anterior surface of the femur and inserting into the synovial capsule of the knee joint which it retracts during extension.

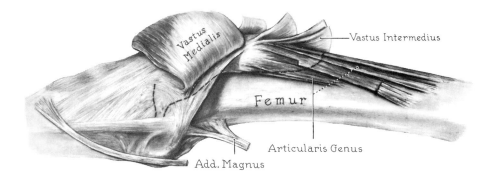

Vastus
Medialis

— Vastus Intermedius

Femur

Articularis Genus

Add. Magnus

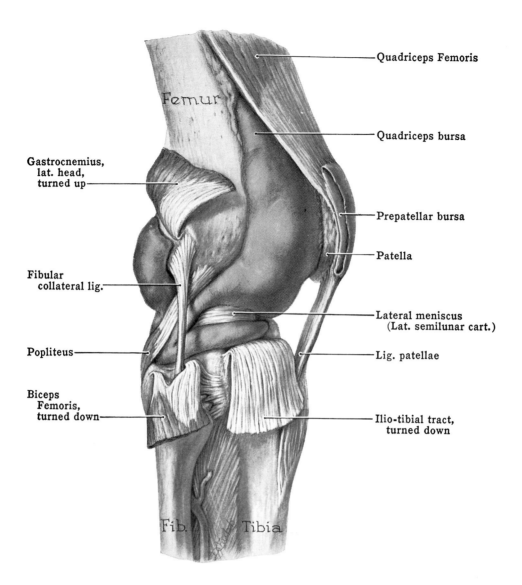

Quadriceps Femoris

Quadriceps bursa

Gastrocnemius, lat. head, turned up

Prepatellar bursa

Patella

Fibular collateral lig.

Popliteus

Lateral meniscus (Lat. semilunar cart.)

Lig. patellae

Biceps Femoris, turned down

Ilio-tibial tract, turned down

Femur

Fib.   Tibia

4-67   DISTENDED KNEE JOINT, LATERAL VIEW

Latex was injected into the joint cavity and fixed with acetic acid; the distended synovial capsule was exposed and cleaned. Gastrocnemius is thrown up; Biceps and the ilio-tibial tract are thrown down. The latex, in this specimen, flowed into the proximal tibio-fibular joint cavity.

Observe:

1. The extent of the synovial capsule:
    a. Superiorly, it rises about 2 fingers' breadth above the patella and here rests on a layer of fat which allows it to glide freely in movements of the joint. This upper part, called the suprapatellar (Quadriceps Femoris) bursa, is obviously not a frictional bursa.
    b. Posteriorly, it rises as high as the origin of Gastrocnemius.
    c. Laterally, it curves below the lateral femoral epicondyle where popliteus tendon and the fibular collateral ligament are attached.
    d. Inferiorly, it bulges below the lateral meniscus, overlapping about 1/3 inch of the tibia. The coronary ligament is removed to show this.
2. Biceps and ilio-tibial tract protecting the joint laterally.
3. The prepatellar bursa, here more extensive than usual, more than covering the patella.

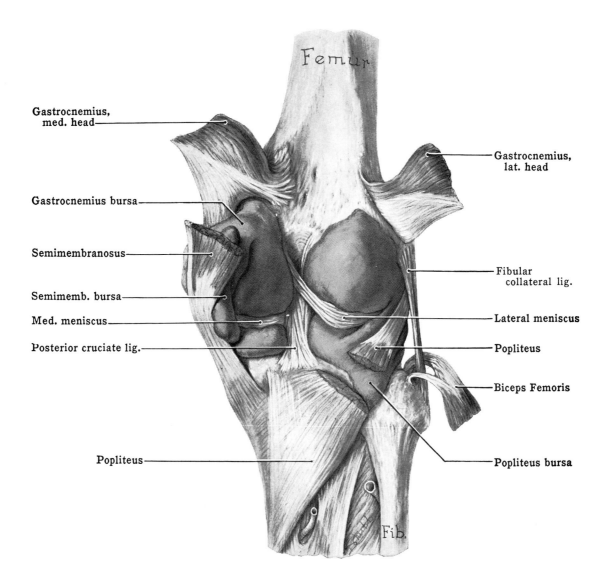

Femur

Gastrocnemius, med. head

Gastrocnemius bursa

Semimembranosus

Semimemb. bursa

Med. meniscus

Posterior cruciate lig.

Popliteus

Gastrocnemius, lat. head

Fibular collateral lig.

Lateral meniscus

Popliteus

Biceps Femoris

Popliteus bursa

Fib.

4-68 DISTENDED KNEE JOINT, POSTERIOR VIEW

Both heads of Gastrocnemius are thrown up, Biceps is thrown down, and a section is removed from Popliteus.

Observe:
1. The posterior cruciate ligament exposed from behind without opening the synovial capsule (articular cavity).
2. The origins of Gastrocnemius limiting the extent to which the synovial capsule can rise.
3. Semimembranosus bursa here communicating with Gastrocnemius bursa, which in turn communicates with the synovial cavity as in Figure 4-53.
4. The Popliteus tendon separated from the lateral meniscus, the upper end of the tibia, and the proximal tibio-fibular joint by an elongated bursa. This Popliteus bursa communicates with the synovial cavity of the knee joint both above and below the meniscus and in this specimen it also communicates with the proximal tibio-fibular synovial cavity, as revealed by Figure 4-67.

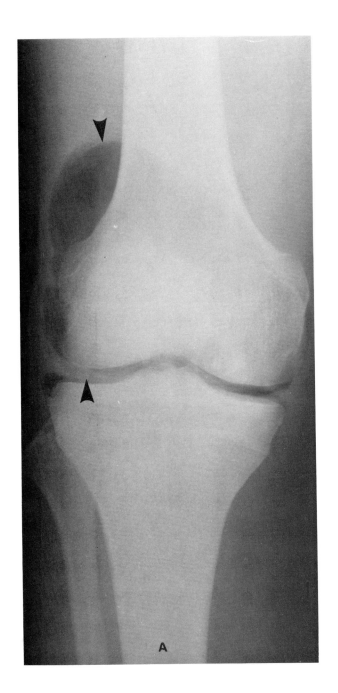
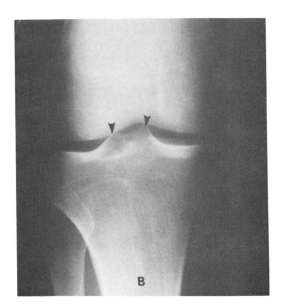
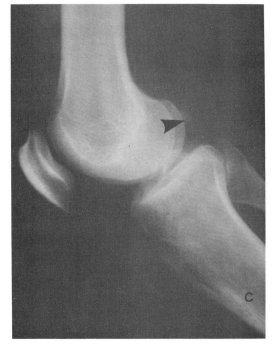

4-69 RADIOGRAPHS OF KNEE

Three radiographs of the knee region.

A. In this AP view, air has been injected into the joint cavity. Being less opaque, it appears *black* in the x-ray. The *upper arrow* points to the highest margin of the Quadriceps bursa. The *lower arrow* draws attention to the lateral meniscus outlined with air. Consult Figure 4-67.

B. In this AP view, *arrows* point to the lateral and medial intercondylar tubercles. See Figure 4-61.

C. A lateral view of the flexed knee. The *arrows* points to a fabella, a sesamoid bone in the lateral head of Gastrocnemius.

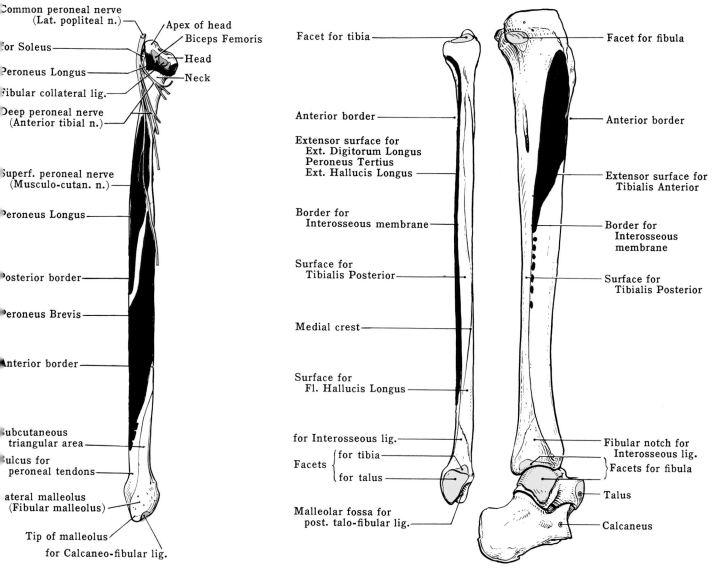

Common peroneal nerve
(Lat. popliteal n.)
Apex of head
Biceps Femoris
for Soleus
Head
Peroneus Longus
Neck
Fibular collateral lig.
Deep peroneal nerve
(Anterior tibial n.)

Superf. peroneal nerve
(Musculo-cutan. n.)

Peroneus Longus

Posterior border

Peroneus Brevis

Anterior border

Subcutaneous
triangular area
Sulcus for
peroneal tendons
Lateral malleolus
(Fibular malleolus)
Tip of malleolus
for Calcaneo-fibular lig.

Facet for tibia
Facet for fibula

Anterior border
Anterior border

Extensor surface for
Ext. Digitorum Longus
Peroneus Tertius
Ext. Hallucis Longus
Extensor surface for
Tibialis Anterior

Border for
Interosseous membrane
Border for
Interosseous
membrane

Surface for
Tibialis Posterior
Surface for
Tibialis Posterior

Medial crest

Surface for
Fl. Hallucis Longus

for Interosseous lig.
Fibular notch for
Interosseous lig.
Facets { for tibia
for talus
Facets for fibula
Talus

Malleolar fossa for
post. talo-fibular lig.
Calcaneus

A. LATERAL SURFACE OF FIBULA **B. TIBIA AND FIBULA, OPPOSED ASPECTS**

4-70 BONES OF THE LEG

Observe:

1. The lateral (or peroneal) surface of the fibula describing a quarter of a spiral, its distal end being grooved and facing posteriorly. This allows the lateral malleolus to act as a pulley for the long and short peroneal tendons.
2. The common peroneal nerve and its terminal branches having long contact with the fibula.
3. On the fibula, 2 small articular facets for the tibia, one proximally and one distally. Below the latter a large triangular facet for articulation with almost the entire depth of the lateral surface of the body of the talus.
4. The extensor surface of the fibula narrow below and almost linear above.

5. Interosseous borders of the tibia and fibula, for the attachment of the interosseous membrane, separating the anterior or extensor surface from the posterior or flexor surface. Each of these borders widening below into a triangular area for the interosseous ligament. During walking, the fibula moves *downward* and laterally, stretching the interosseous membrane.

See Weinert, C. R., Jr., McMaster, J. H., and Ferguson, R. J. (1973) Dynamic function of the human fibula. *Am. J. Anat. 138:* 145.

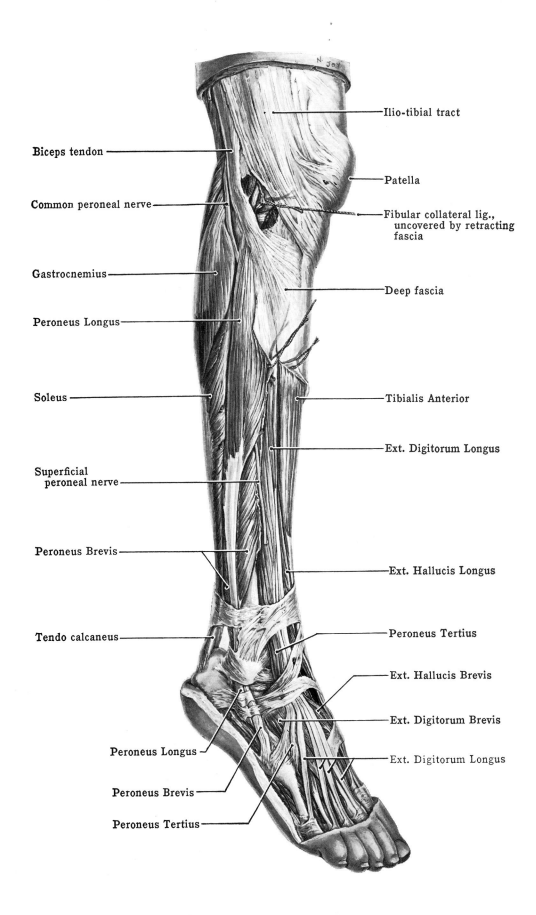

Ilio-tibial tract

Biceps tendon

Common peroneal nerve

Gastrocnemius

Peroneus Longus

Soleus

Superficial
peroneal nerve

Peroneus Brevis

Tendo calcaneus

Peroneus Longus

Peroneus Brevis

Peroneus Tertius

Patella

Fibular collateral lig.,
uncovered by retracting
fascia

Deep fascia

Tibialis Anterior

Ext. Digitorum Longus

Ext. Hallucis Longus

Peroneus Tertius

Ext. Hallucis Brevis

Ext. Digitorum Brevis

Ext. Digitorum Longus

**4-71A   MUSCLES OF THE LEG AND FOOT,
ANTERO-LATERAL VIEW**

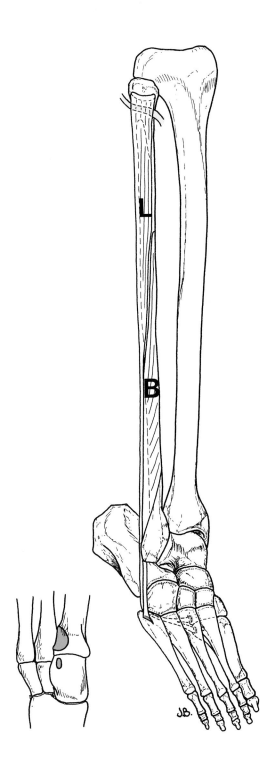

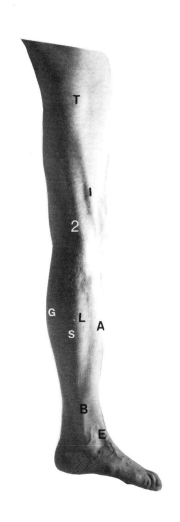

4-71C LATERAL VIEW OF LEG

T = Tensor fascia lata
2 = Biceps
S = Soleus
I = Iliotibial tract
G = Gastrocnemius
A = Tibialis Anterior
E = Extensor digitorum longus
Peroneus longus (L) and brevis (B).

4-71B PERONEAL MUSCLES

This oblique lateral view of the leg shows the attachment of the two peroneal muscles. Both attach to two-thirds of the fibula: Peroneus longus to the upper two-thirds; Peroneus brevis to the lower two-thirds. Where they overlap, Peroneus brevis is in front. Peroneus brevis inserts on the proximal end of the 5th metatarsal; Peroneus longus enters the foot by hooking around the cuboid and traveling medially. The *inset* shows the plantar surface of the foot with the insertion of Peroneus longus to the lateral side of the base of the 1st metatarsal and medial cuneiform. Note at the proximal end of the fibula: the common peroneal nerve in contact with the neck of the fibula deep to Peroneus longus. Here, it is vulnerable to injury with serious implications as it supplies the extensor group of muscles whose loss results in foot drop.

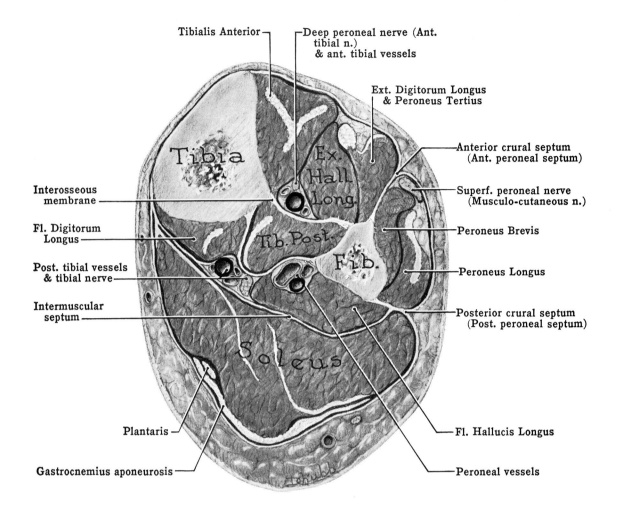

Tibialis Anterior

Deep peroneal nerve (Ant. tibial n.) & ant. tibial vessels

Ext. Digitorum Longus & Peroneus Tertius

Tibia

Ex. Hall. Long.

Anterior crural septum (Ant. peroneal septum)

Interosseous membrane

Superf. peroneal nerve (Musculo-cutaneous n.)

Tib. Post.

Fib.

Peroneus Brevis

Fl. Digitorum Longus

Post. tibial vessels & tibial nerve

Peroneus Longus

Intermuscular septum

Posterior crural septum (Post. peroneal septum)

Soleus

Plantaris

Fl. Hallucis Longus

Gastrocnemius aponeurosis

Peroneal vessels

4-72   CROSS-SECTION THROUGH THE LEG, MALE

Observe:
1. This section is through the lower part of the middle third of the leg: Gastrocnemius is aponeurotic and Peroneus longus and brevis are both attaching to the fibula.
2. The anterior tibio-fibular compartment, bounded by tibia, interosseous membrane, fibula, anterior intermuscular crural septum, and deep fascia, and containing the anterior tibial vessels and deep peroneal nerve. The unyielding walls of this compartment may lead to catastrophe—necrosis of the muscles—if pressure increases in the compartment following injury or ischemia. See Waddell, J. P. (1977) Anterior tibial compartment syndrome. *C.M.A. Journal, 116:* 653.
3. The peroneal compartment (*peroneal* is the Greek equivalent of the latin *fibular*) bounded by fibula, anterior and posterior intermuscular crural septa, and the deep fascia, and containing the superficial peroneal nerve.
4. The posterior tibio-fibular compartment bounded by tibia, interosseous membrane, fibula, posterior intermuscular crural septum, and deep fascia. This compartment is subdivided by two coronal septa into three subcompartments: 1st, or deepest, contains Tibialis Posterior; the 2nd, or intermediate, contains Flexor Hallucis Longus, Flexor Digitorum Longus, and posterior tibial vessels and tibial nerve; and the 3rd, or most superficial, contains Soleus, Gastrocnemius, and Plantaris.

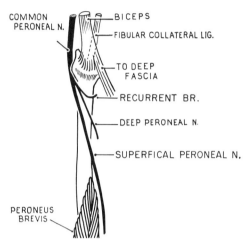

COMMON PERONEAL N.

BICEPS

FIBULAR COLLATERAL LIG.

TO DEEP FASCIA

RECURRENT BR.

DEEP PERONEAL N.

SUPERFICAL PERONEAL N.

PERONEUS BREVIS

4-74   COMMON PERONEAL NERVE

The exposed position of the common peroneal nerve. It is applied to the back of the head of the fibula (a film of Soleus intervening); its branches are applied directly to the neck and body of the fibula for 8 to 10 cm deep to Peroneus Longus, as shown in Figure 4-75.

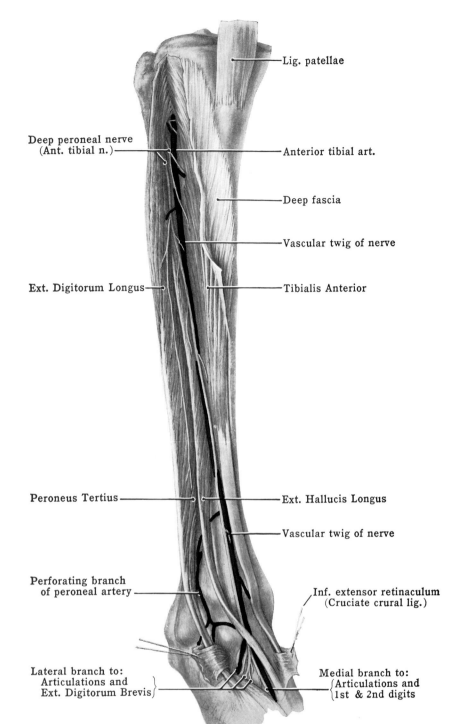

Lig. patellae

Deep peroneal nerve (Ant. tibial n.)

Anterior tibial art.

Deep fascia

Vascular twig of nerve

Ext. Digitorum Longus

Tibialis Anterior

Peroneus Tertius

Ext. Hallucis Longus

Vascular twig of nerve

Perforating branch of peroneal artery

Inf. extensor retinaculum (Cruciate crural lig.)

Lateral branch to: Articulations and Ext. Digitorum Brevis

Medial branch to: Articulations and 1st & 2nd digits

4-73   FRONT OF THE LEG

The muscles are separated in order to display the artery and nerve.

Observe:

1. Tibialis Anterior, arising in part from the deep fascia which, therefore, is strong, has longitudinal fibers, and makes sharp the upper part of the anterior border of the tibia.

2. Peroneus Tertius is merely the lower part of Extensor Digitorum Longus. Extensor Hallucis Longus extends farther proximally than usual. These three muscles are unipennate and arise from the fibula.

3. The vascular and articular branches of the deep peroneal nerve.

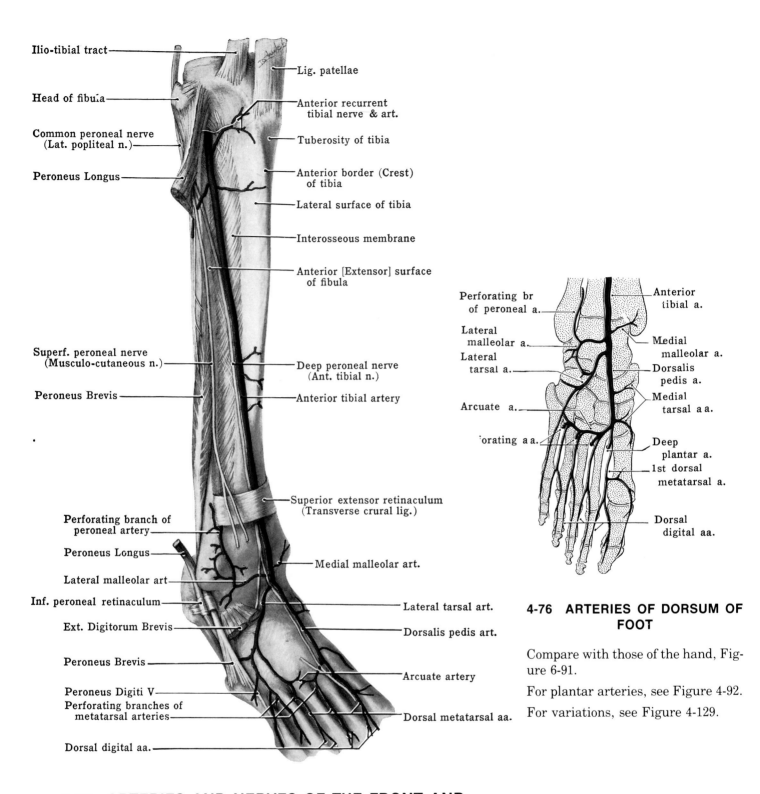

Ilio-tibial tract

Head of fibula

Common peroneal nerve
(Lat. popliteal n.)

Peroneus Longus

Superf. peroneal nerve
(Musculo-cutaneous n.)

Peroneus Brevis

Perforating branch of
peroneal artery

Peroneus Longus

Lateral malleolar art.

Inf. peroneal retinaculum

Ext. Digitorum Brevis

Peroneus Brevis

Peroneus Digiti V

Perforating branches of
metatarsal arteries

Dorsal digital aa.

Lig. patellae

Anterior recurrent
tibial nerve & art.

Tuberosity of tibia

Anterior border (Crest)
of tibia

Lateral surface of tibia

Interosseous membrane

Anterior [Extensor] surface
of fibula

Deep peroneal nerve
(Ant. tibial n.)

Anterior tibial artery

Superior extensor retinaculum
(Transverse crural lig.)

Medial malleolar art.

Lateral tarsal art.

Dorsalis pedis art.

Arcuate artery

Dorsal metatarsal aa.

Perforating br
of peroneal a.

Lateral
malleolar a.

Lateral
tarsal a.

Arcuate a.

Perforating a a.

Anterior
tibial a.

Medial
malleolar a.

Dorsalis
pedis a.

Medial
tarsal a a.

Deep
plantar a.

1st dorsal
metatarsal a.

Dorsal
digital aa.

4-76 ARTERIES OF DORSUM OF FOOT

Compare with those of the hand, Figure 6-91.

For plantar arteries, see Figure 4-92.

For variations, see Figure 4-129.

4-75 ARTERIES AND NERVES OF THE FRONT AND DORSUM OF FOOT

The anterior crural muscles are removed and Peroneus Longus is excised.

Observe:

1. The anterior tibial artery entering the region in contact with the medial side of the neck of the fibula; the deep peroneal nerve in contact with the lateral side. Hence, the nerve approaches the artery from the lateral side (Fig. 4-57).

2. The artery and nerve and their named branches lying strictly on the skeletal plane and undisturbed by the removal of the muscles.

3. The superficial peroneal nerve following the anterior border of Peroneus Brevis which guides it to the surface a variable distance above the triangular subcutaneous area of the fibula.

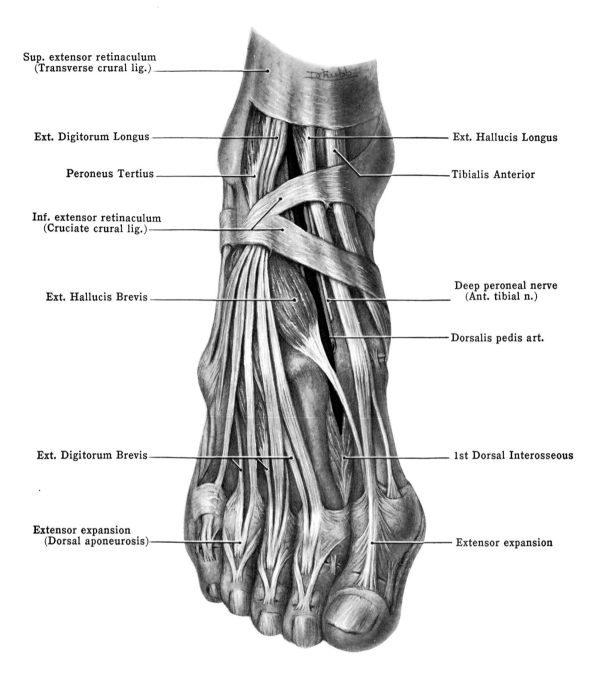

Sup. extensor retinaculum
(Transverse crural lig.)

Ext. Digitorum Longus

Peroneus Tertius

Inf. extensor retinaculum
(Cruciate crural lig.)

Ext. Hallucis Brevis

Ext. Digitorum Brevis

Extensor expansion
(Dorsal aponeurosis)

Ext. Hallucis Longus

Tibialis Anterior

Deep peroneal nerve
(Ant. tibial n.)

Dorsalis pedis art.

1st Dorsal Interosseous

Extensor expansion

4-77 DORSUM OF THE FOOT, FRONT VIEW

The vessels and nerves are cut short.

Observe:

1. At the ankle, the vessels and nerve lying midway between the malleoli and
   having two tendons on each side.
2. On the dorsum of the foot, the artery crossed by Extensor Hallucis Brevis and
   disappearing between the two heads of the 1st Dorsal Interosseous (*cf.* the
   radial artery on the dorsum of the hand, Figs. 6-91 and 6-93).
3. The inferior extensor retinaculum restraining the tendons from bowstringing
   forward and also from bowstringing medially; *i.e.*, it restrains them in two
   planes.

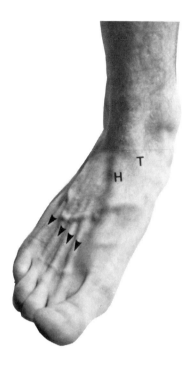

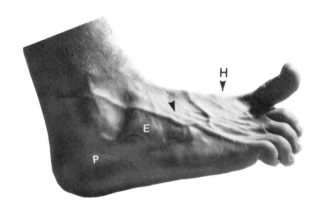

4-78A   DORSAL TENDONS

Arrows indicate individual tendons of Extensor digitorum longus. Follow the tendon of the Extensor hallucis longus (*H*) from its exit through the extensor retinaculum to its insertion into the base of the distal phalanx of the great toe. The tendon of Tibialis anterior (*T*) disappears as it moves to its insertion on the medial cuneiform and base of the first metatarsal (Fig. 4-95).

4-78B   LATERAL VIEW OF FOOT

Observe the soft swelling of the fleshy belly of Extensor digitorum brevis (*E*). An *arrow* points the location of the synovial sheath distal to which the tendons of Extensor digitorum longus fan out to reach the digits. The tendon of Extensor hallucis longus (*H*) is prominent. At *P*, the tendons of Peroneus longus and brevis are hooking around the lateral malleolus.

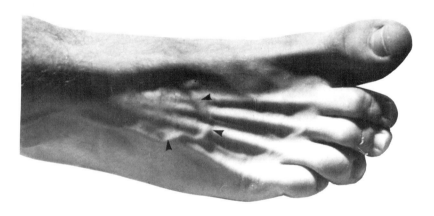

4-78C   DORSUM OF THE FOOT

Arrows point to the dorsal venous arch. The tendons of the Extensor hallucis longus and Extensor digitorum longus are easily followed.

The ankle, subtalar, and calcaneo-cuboid joints are exposed in order to reveal their positions.

Observe:

1. The calcaneo-fibular ligament attached to the fibular malleolus in front of its tip, thereby allowing that tip to overlap the Peronei tendons and so prevent them from slipping forward.

2. The inferior peroneal retinaculum attached to the lateral surface of the calcaneus, and in line with the inferior extensor retinaculum which is attached to the superior surface.

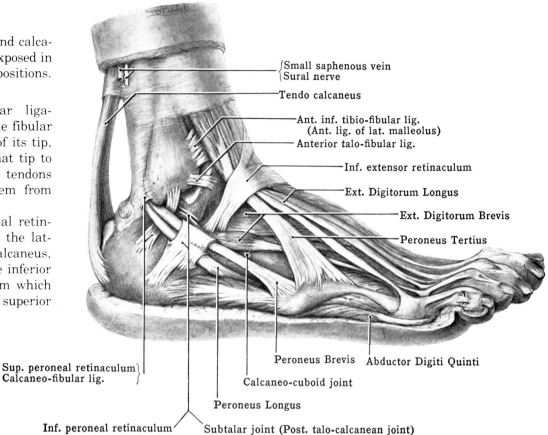

Small saphenous vein
Sural nerve
Tendo calcaneus
Ant. inf. tibio-fibular lig. (Ant. lig. of lat. malleolus)
Anterior talo-fibular lig.
Inf. extensor retinaculum
Ext. Digitorum Longus
Ext. Digitorum Brevis
Peroneus Tertius
Peroneus Brevis   Abductor Digiti Quinti
Calcaneo-cuboid joint
Peroneus Longus
Subtalar joint (Post. talo-calcanean joint)
Inf. peroneal retinaculum
Sup. peroneal retinaculum
Calcaneo-fibular lig.

4-78D   DORSUM OF THE FOOT, LATERAL VIEW

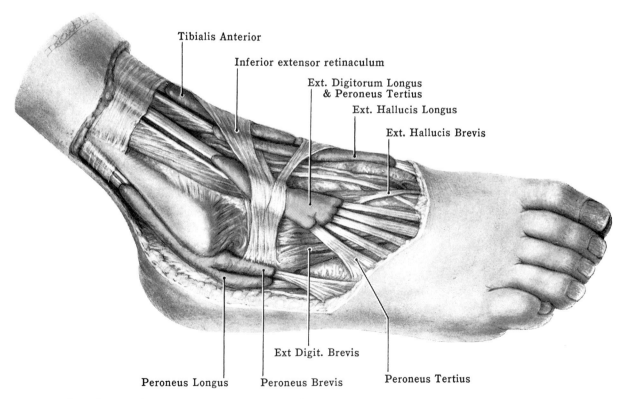

Tibialis Anterior
Inferior extensor retinaculum
Ext. Digitorum Longus & Peroneus Tertius
Ext. Hallucis Longus
Ext. Hallucis Brevis
Peroneus Longus   Peroneus Brevis   Peroneus Tertius
Ext Digit. Brevis

4-79   SYNOVIAL SHEATHS OF THE TENDONS AT THE ANKLE, ANTERO-LATERAL VIEW

The tendons of Peroneus Longus and Peroneus Brevis are enclosed in a common synovial sheath behind the fibular malleolus and this sheath splits into two, one for each tendon, behind the peroneal trochlea.

The tendon of Peroneus Longus has a second sheath (not in view) which accompanies it across the sole of the foot.

In almost half of 131 feet studied, the two sheaths of the Longus were demonstrated by injection to be in continuity on their deep, or frictional, surface, although not on their superficial surface.

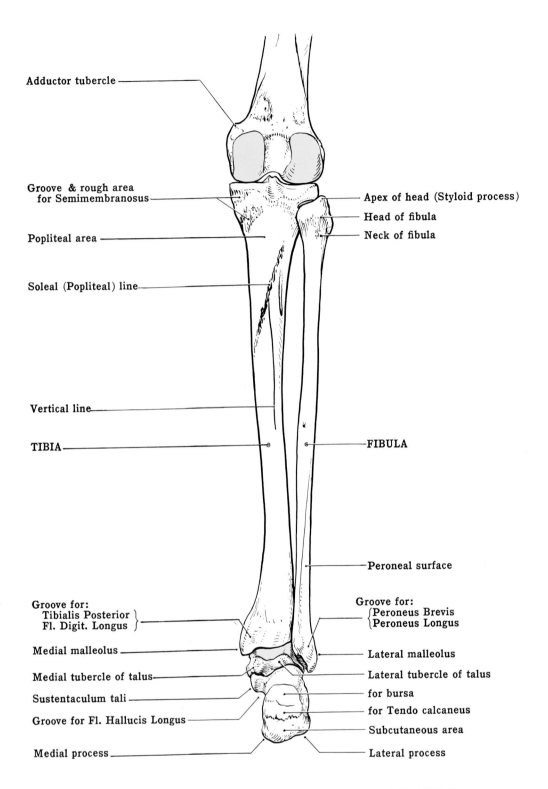

Adductor tubercle

Groove & rough area
  for Semimembranosus

Popliteal area

Soleal (Popliteal) line

Vertical line

TIBIA

Apex of head (Styloid process)
Head of fibula
Neck of fibula

FIBULA

Peroneal surface

Groove for:
  Tibialis Posterior
  Fl. Digit. Longus

Medial malleolus

Medial tubercle of talus

Sustentaculum tali

Groove for Fl. Hallucis Longus

Medial process

Groove for:
  Peroneus Brevis
  Peroneus Longus

Lateral malleolus

Lateral tubercle of talus

for bursa

for Tendo calcaneus

Subcutaneous area

Lateral process

4-80 BONES OF THE LEG, POSTERIOR VIEW

For anterior view, see also Figure 4-70

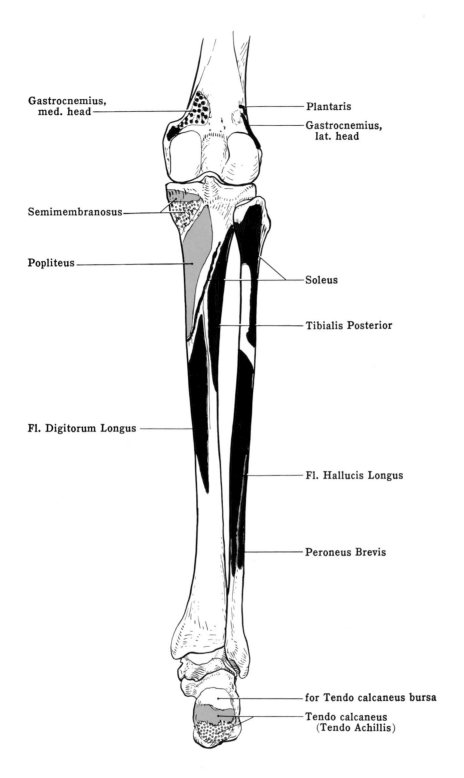

Gastrocnemius, med. head

Plantaris

Gastrocnemius, lat. head

Semimembranosus

Popliteus

Soleus

Tibialis Posterior

Fl. Digitorum Longus

Fl. Hallucis Longus

Peroneus Brevis

for Tendo calcaneus bursa

Tendo calcaneus (Tendo Achillis)

4-81 BONES OF THE LEG SHOWING ATTACHMENTS OF MUSCLES, POSTERIOR VIEW

For plantar aspect of bones of the foot, see Figure 4-107.

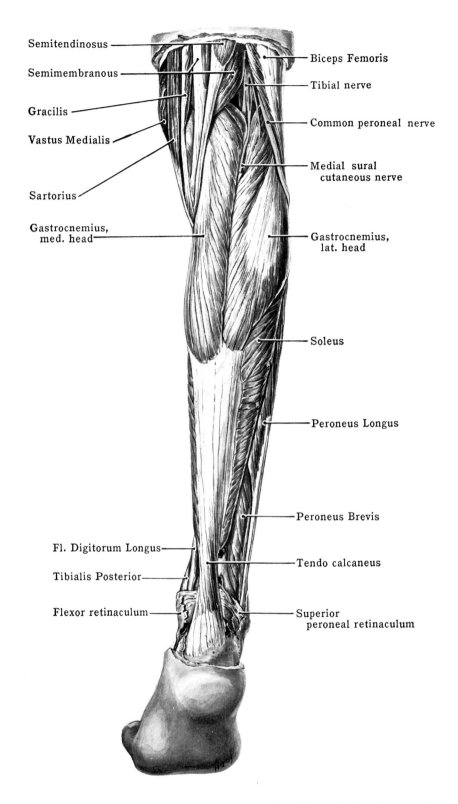

Semitendinosus

Semimembranous

Gracilis

Vastus Medialis

Sartorius

Gastrocnemius, med. head

Biceps Femoris

Tibial nerve

Common peroneal nerve

Medial sural cutaneous nerve

Gastrocnemius, lat. head

Soleus

Peroneus Longus

Peroneus Brevis

Fl. Digitorum Longus

Tibialis Posterior

Flexor retinaculum

Tendo calcaneus

Superior peroneal retinaculum

4-82A   MUSCLES OF THE LEG, POSTERIOR VIEW—I

4-82B POPLITEAL FOSSA BOUNDARIES

The four muscles bounding the diamond shaped popliteal fossa resemble an upper outside caliper, Semimembranosus (*SM*) and Biceps (*B*), grasping a lower inside caliper, medial and lateral heads of Gastrocnemius (*M* and *L*). Note that Semitendinosus (*ST*) is posterior (superficial) to the Semimembranosus and is distinguished by the length of its tendon.

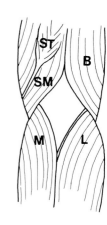

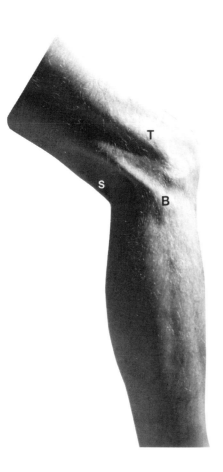

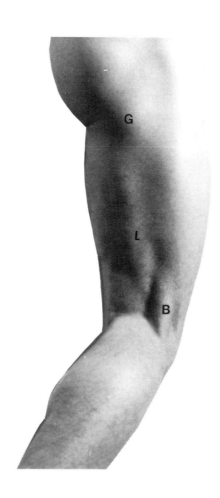

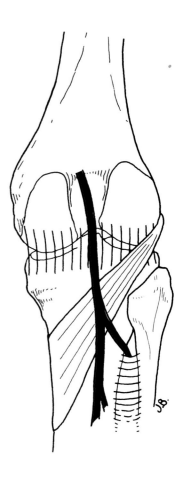

4-82C LATERAL VIEW OF KNEE

Observe the strongly marked posterior edge of the iliotibial tract (a thickening of the fascia lata) which attaches to the lateral condyle of the tibia. The Biceps femoris tendon is inserting on the head of the fibula. Semimembranosus (*S*) is forming the medial boundary of the popliteal fossa.

4-82D BICEPS FEMORIS

The long head of Biceps (*L*) is traveling laterally from its origin on the ischial tuberosity to its insertion (*B*) with the short head. Proximally, it emerges from the lower edge of Gluteus maximus (*G*).

4-82E POPLITEAL ARTERY

In this posterior view of the skeletal plane of the popliteal fossa, the popliteal artery successively lies on the bone, joint capsule, and Popliteus muscle before dividing into anterior and posterior tibial arteries.

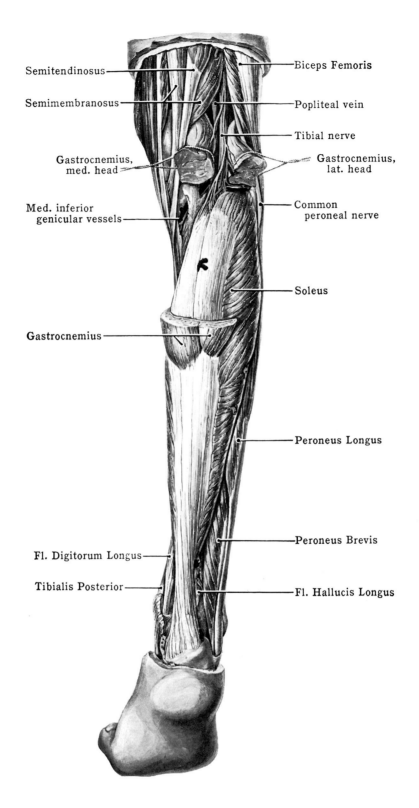

Semitendinosus

Semimembranosus

Gastrocnemius, med. head

Med. inferior genicular vessels

Gastrocnemius

Fl. Digitorum Longus

Tibialis Posterior

Biceps Femoris

Popliteal vein

Tibial nerve

Gastrocnemius, lat. head

Common peroneal nerve

Soleus

Peroneus Longus

Peroneus Brevis

Fl. Hallucis Longus

4-83 MUSCLES OF THE LEG, POSTERIOR VIEW—II

The fleshy bellies of Gastrocnemius are largely excised, and the origin of Soleus is thereby exposed. Plantaris is absent from this specimen.

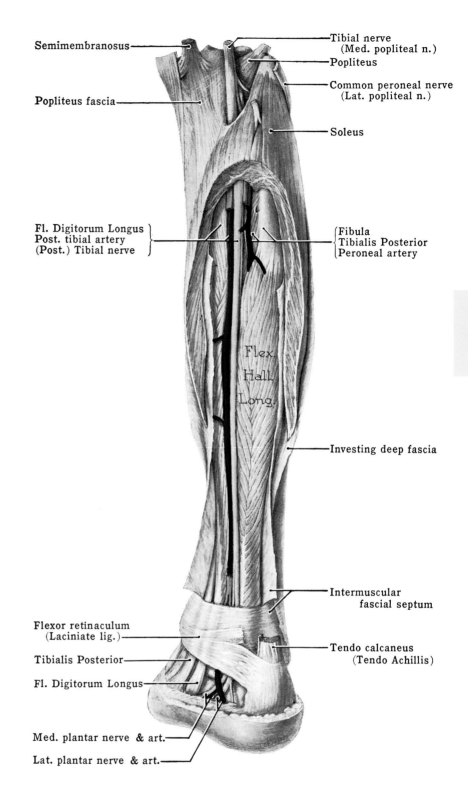

Semimembranosus

Tibial nerve
(Med. popliteal n.)

Popliteus

Common peroneal nerve
(Lat. popliteal n.)

Popliteus fascia

Soleus

Fl. Digitorum Longus
Post. tibial artery
(Post.) Tibial nerve

Fibula
Tibialis Posterior
Peroneal artery

Flex.
Hall.
Long.

Investing deep fascia

Intermuscular
fascial septum

Flexor retinaculum
(Laciniate lig.)

Tendo calcaneus
(Tendo Achillis)

Tibialis Posterior

Fl. Digitorum Longus

Med. plantar nerve & art.

Lat. plantar nerve & art.

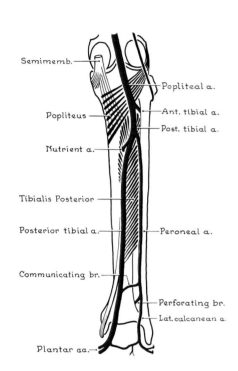

Semimemb.

Popliteal a.

Ant. tibial a.

Post. tibial a.

Popliteus

Nutrient a.

Tibialis Posterior

Posterior tibial a.

Peroneal a.

Communicating br.

Perforating br.

Lat. calcanean a.

Plantar aa.

4-85   ARTERIES OF BACK OF LEG

For variations see Figure 4-129.

4-84   BACK OF THE LEG, DEEP STRUCTURES—I

Tendo-calcaneus is divided; Gastrocnemius and a horseshoe-shaped section of Soleus are removed.

Observe:

1. The bipennate structure of the large Flexor Hallucis Longus and of the smaller Flexor Digitorum Longus.
2. The posterior tibial artery and the tibial nerve descending between these two muscles, on a layer of fascia that covers Tibialis Posterior. (For cross-section, see Fig. 4-72.)
3. The tough, intermuscular fascial septum deep to Soleus and tendo-calcaneus that acts as a restraining anklet at the ankle and there blends medially with the weaker investing deep fascia to form the flexor retinaculum.

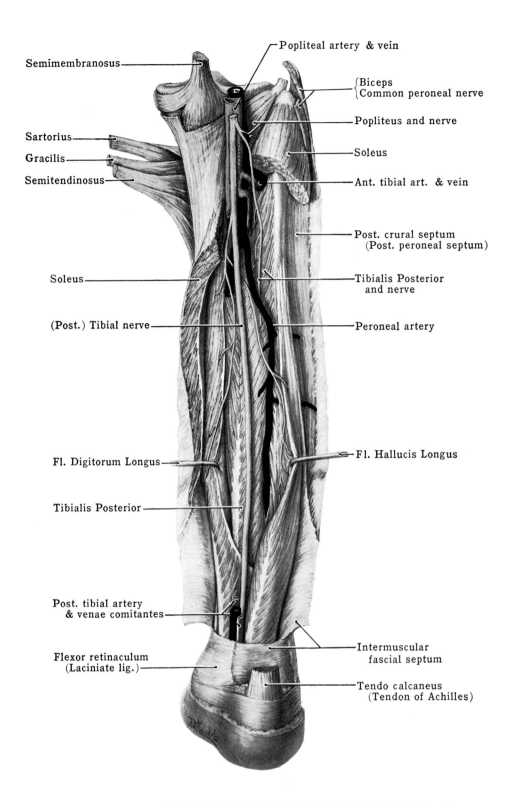

Semimembranosus

Popliteal artery & vein

Biceps
Common peroneal nerve

Popliteus and nerve

Sartorius

Gracilis

Semitendinosus

Soleus

Ant. tibial art. & vein

Post. crural septum
(Post. peroneal septum)

Soleus

Tibialis Posterior
and nerve

(Post.) Tibial nerve

Peroneal artery

Fl. Digitorum Longus

Fl. Hallucis Longus

Tibialis Posterior

Post. tibial artery
& venae comitantes

Intermuscular
fascial septum

Flexor retinaculum
(Laciniate lig.)

Tendo calcaneus
(Tendon of Achilles)

4-86   BACK OF THE LEG, DEEP STRUCTURES—II

Soleus is largely cut away; the two long digital flexors are pulled apart; the posterior tibial artery is partly excised.

Observe:

1. Tibialis Posterior, bipennate and powerful, lying deep to the two long digital flexors.
2. The peroneal artery overlapped by Flexor Hallucis Longus.
3. The nerve to Tibialis Posterior arising in conjunction with the nerve to Popliteus, and the nerve to Flexor Digitorum Longus arising in conjunction with the nerve to Flexor Hallucis Longus.
4. In the popliteal fossa the nerve is superficial to the artery; at the ankle the artery is superficial to the nerve.

The posterior part of Abductor Hallucis is excised.

Observe:

1. The posterior tibial artery and the tibial nerve lying between Flexor Digitorum Longus and Flexor Hallucis Longus; separated from the tibial malleolus by the width of two tendons (Tibialis Posterior and Flexor Digitorum Longus); and dividing into medial and lateral plantar branches on the surface of the osseo-fibrous tunnel of Flexor Hallucis Longus (Figs. 4-89 and 4-109).

2. Tibialis Posterior and Flexor Digitorum Longus occupying separate and individual *osseofibrous* tunnels behind the medial malleolus, which is their pulley.

3. The medial and lateral plantar nerves lying within the fork of the medial and lateral plantar arteries.

4. The deep veins of the foot emerging to join the great saphenous vein.

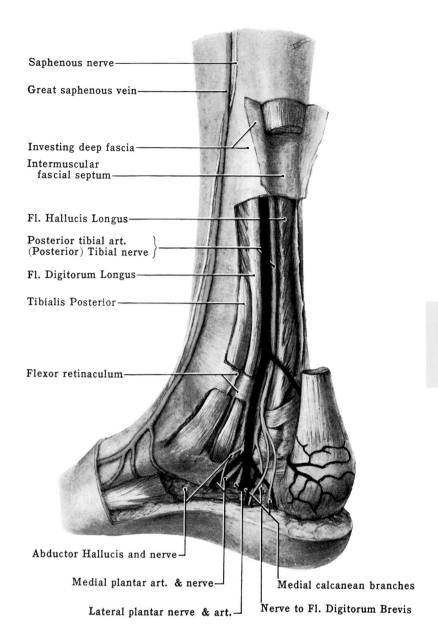

4-87 ANKLE AND HEEL, MEDIAL VIEW

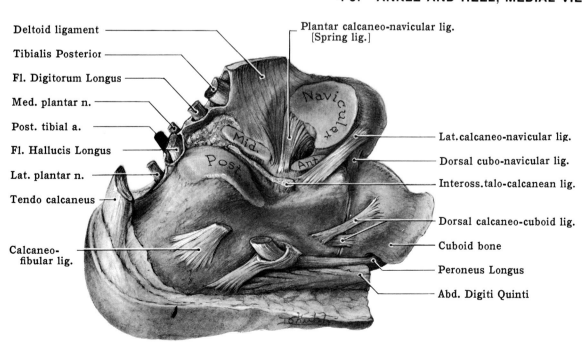

4-88 STRUCTURES ON MEDIAL SIDE OF THE ANKLE, LATERAL VIEW

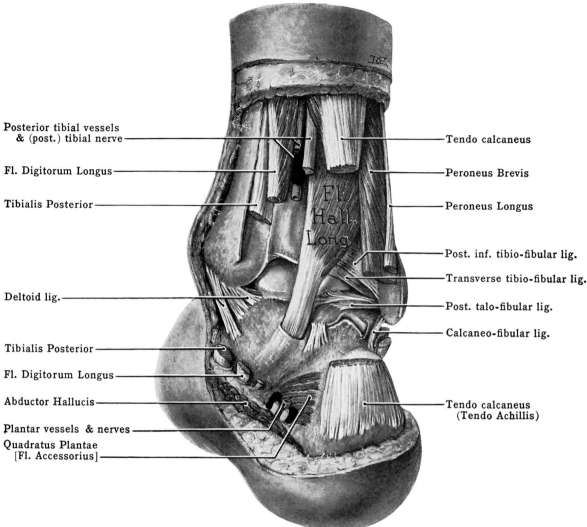

Posterior tibial vessels & (post.) tibial nerve

Fl. Digitorum Longus

Tibialis Posterior

Deltoid lig.

Tibialis Posterior

Fl. Digitorum Longus

Abductor Hallucis

Plantar vessels & nerves

Quadratus Plantae [Fl. Accessorius]

Tendo calcaneus

Peroneus Brevis

Peroneus Longus

Post. inf. tibio-fibular lig.

Transverse tibio-fibular lig.

Post. talo-fibular lig.

Calcaneo-fibular lig.

Tendo calcaneus (Tendo Achillis)

4-89 ANKLE AND HEEL, POSTERIOR VIEW

Observe:

1. Flexor Hallucis Longus placed midway between the two malleoli; and having the two tendons (Flexor Digitorum Longus and Tibialis Posterior) that groove the tibial malleolus medial to it and the two tendons (Peronei Longus and Brevis) that groove the fibular melleolus lateral to it.

2. The entrance to the sole "porta pedis" lying deep to Abductor Hallucis. The plantar vessels and nerves, the two long digital flexors, and part of Tibialis Posterior enter here. Quadratus Plantae serves as a soft pad for the vessels and nerves.

3. The posterior tibial artery and the tibial nerve lying medial to Flexor Hallucis Longus above and, after bifurcating, postero-lateral to it below. The crossing takes place where the long flexor is within its osseofibrous tunnel.

4. The strongest parts of the ligaments of the ankle are those that prevent forward displacement of the leg bones, namely, the posterior part of the deltoid (posterior tibio-talar), posterior talo-fibular, tibio-calcanean, and calcaneo-fibular.

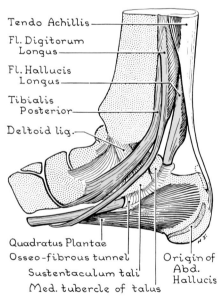

Tendo Achillis

Fl. Digitorum Longus

Fl. Hallucis Longus

Tibialis Posterior

Deltoid lig.

Quadratus Plantae

Osseo-fibrous tunnel

Sustentaculum tali

Med. tubercle of talus

Origin of Abd. Hallucis

4-90 A PULLEY

Only Flexor Hallucis Longus uses the sustentaculum as a pulley.

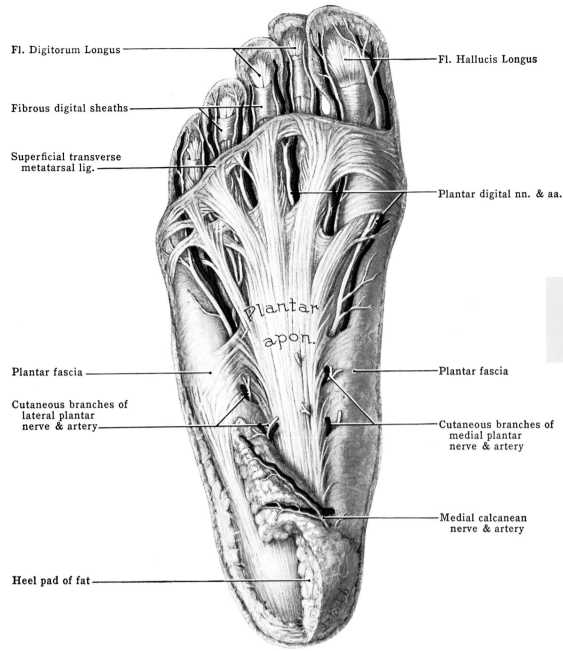

Fl. Digitorum Longus

Fibrous digital sheaths

Superficial transverse metatarsal lig.

Plantar fascia

Cutaneous branches of lateral plantar nerve & artery

Heel pad of fat

Fl. Hallucis Longus

Plantar digital nn. & aa.

Plantar apon.

Plantar fascia

Cutaneous branches of medial plantar nerve & artery

Medial calcanean nerve & artery

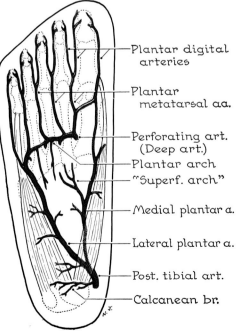

Plantar digital arteries

Plantar metatarsal aa.

Perforating art. (Deep art.)

Plantar arch

"Superf. arch"

Medial plantar a.

Lateral plantar a.

Post. tibial art.

Calcanean br.

4-91 SUPERFICIAL DISSECTION OF PLANTAR ASPECT, OR SOLE, OF THE FOOT

The plantar aponeurosis, the medial and lateral parts of the plantar fascia, and the digital vessels and nerves should be compared and contrasted with the corresponding structures in the palm (Figs. 6-66, 6-78, and 6-79).

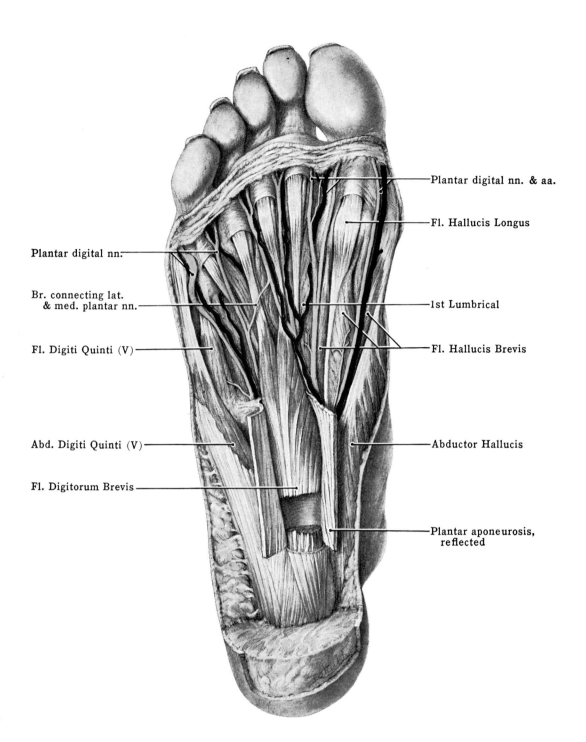

Plantar digital nn. & aa.

Fl. Hallucis Longus

Plantar digital nn.

Br. connecting lat. & med. plantar nn.

1st Lumbrical

Fl. Digiti Quinti (V)

Fl. Hallucis Brevis

Abd. Digiti Quinti (V)

Abductor Hallucis

Fl. Digitorum Brevis

Plantar aponeurosis, reflected

4-93 THE FIRST LAYER OF PLANTAR MUSCLES, DIGITAL NERVES AND ARTERIES

Observe:

1. The muscles of this layer are: Abductor Digiti Quinti, Flexor Digitorum Brevis, and Abductor Hallucis.

2. The plantar aponeurosis and fascia are reflected or removed, and a section is taken from Flexor Digitorum Brevis in order to show the fibrous box encasing it.

3. The lateral and medial plantar digital nerves, like the corresponding palmar digital branches of the ulnar and median nerves, supply $1\frac{1}{2}$ and $3\frac{1}{2}$ digits respectively and are united by a connecting (communicating) branch (*cf.* Fig. 6-79).

4. The lateral nerve to the little toe is here thickened. Flexor Digitorum Brevis here, as commonly, fails to send a tendon to the little toe.

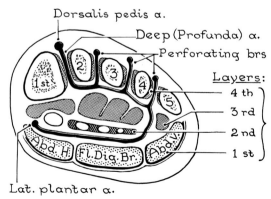

Dorsalis pedis a.

Deep (Profunda) a.

Perforating brs

Layers:
4 th
3 rd
2 nd
1 st

Lat. plantar a.

4-94 CROSS-SECTION NEAR METATARSAL BASES

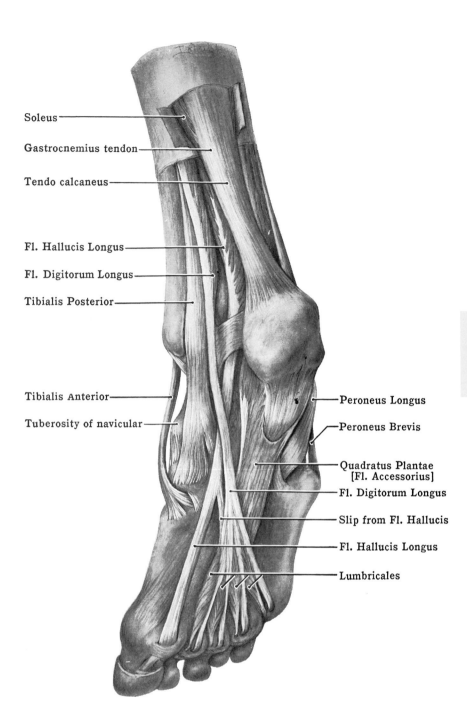

Soleus

Gastrocnemius tendon

Tendo calcaneus

Fl. Hallucis Longus

Fl. Digitorum Longus

Tibialis Posterior

Tibialis Anterior

Tuberosity of navicular

Peroneus Longus

Peroneus Brevis

Quadratus Plantae [Fl. Accessorius]

Fl. Digitorum Longus

Slip from Fl. Hallucis

Fl. Hallucis Longus

Lumbricales

4-95 SECOND LAYER OF PLANTAR MUSCLES

Observe:

1. The muscles of this layer are: Flexor Hallucis Longus, Flexor Digitorum Longus, four Lumbricals, and Quadratus Plantae.
2. Flexor Digitorum Longus crossing superficial to Tibialis Posterior behind the medial malleolus and superficial to Flexor Hallucis Longus abreast of the tuberosity of the navicular bone.
3. The four Lumbricals passing to the hallux side of the toes just as, in the hand, they pass to the pollex side of the fingers (Fig. 6-69).
4. Flexor Hallucis Longus sending a strong tendinous slip to Flexor Digitorum Longus.

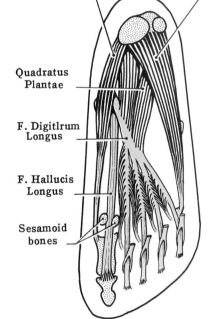

Abd. Hallucis

Abd. Digiti V.

Quadratus Plantae

F. Digitlrum Longus

F. Hallucis Longus

Sesamoid bones

4-96 SECOND LAYER FRAMED BY ABDUCTORS

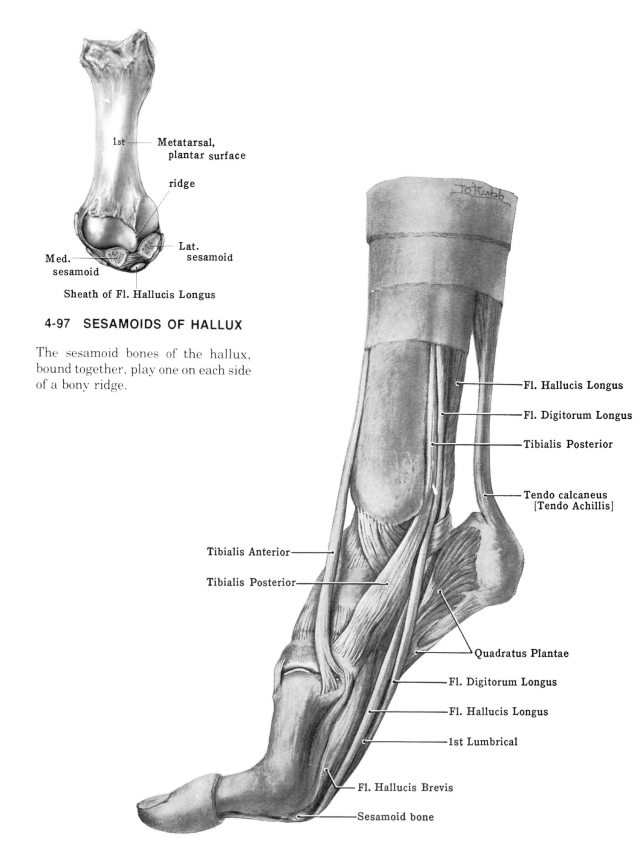

1st —— Metatarsal,
plantar surface

ridge

Lat.
sesamoid

Med. —
sesamoid

Sheath of Fl. Hallucis Longus

4-97 SESAMOIDS OF HALLUX

The sesamoid bones of the hallux,
bound together, play one on each side
of a bony ridge.

Fl. Hallucis Longus

Fl. Digitorum Longus

Tibialis Posterior

Tendo calcaneus
[Tendo Achillis]

Tibialis Anterior

Tibialis Posterior

Quadratus Plantae

Fl. Digitorum Longus

Fl. Hallucis Longus

1st Lumbrical

Fl. Hallucis Brevis

Sesamoid bone

4-98 FOOT RAISED AS IN WALKING, MEDIAL VIEW

Note:
1. The heel is raised but the toes remain applied to the ground.
2. The sesamoid bones act as a footstool for the first metatarsal, giving it increased height.
3. Quadratus Plantae lines the concave medial surface of the calcaneus between Flexor Hallucis Longus
   tendon and tuber calcanei.
4. By inserting into Flexor digitorum longus, Quadratus plantae acts as a guy line modifying the
   oblique pull of the flexor tendons.
5. The Flexor hallucis longus uses three pulleys: a groove on the back of the distal end of the tibia, a
   groove on the back of the talus, and a groove beneath the sustentaculum tali.

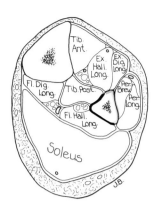

4-99A   CROSS SECTION OF LEG

Note:
1. The fibula's irregular shape and less than helpful anatomical descriptions of its surfaces and borders confounds an understanding of muscle attachments. Note that its anterior surface (*blue*) is narrow and gives linear origin to 3 muscles of the anterior group: Extensor digitorum longus, Extensor hallucis longus, and Peroneus tertius. The *lateral* surface (*green*) provides origin for the two muscles of the lateral group: Peroneus longus and brevis. The so-called *posterior* surface (*red*) is divided obliquely by a crest into a medial and a posterior part. At the junction of *blue* and *red* is the *interosseous* border to which the interosseous membrane attaches. At the junction of *blue* and *green* is the *anterior* border to which the anterior crural septum attaches, separating the anterior and lateral groups of muscles. At the junction of *red* and *green* is the posterior border to which the posterior crural septum attaches, separating the lateral and posterior groups of muscles.
2. The three deep muscles of the posterior group vie with each other for space. The central (and deepest), Tibialis posterior, arises from the interosseous membrane, the lateral part of the posterior surface of the tibia, and the concave medial part of the "posterior" surface of the fibula. Medial to it on the tibia is Flexor digitorum longus. Flexor hallucis longus arises from the posterior surface of the fibula and the fascia covering Tibialis posterior.

4-99B   LONG TOE FLEXORS

Note:
1. It is surprising to discover that the flexor of the great toe has a lateral (fibular) origin while the flexor of the lateral four toes has a medial (tibial) origin.
2. As their tendons pass behind the medial malleolus (and Tibialis posterior), Flexor digitorum is in front, but in the sole it moves superficially to reach a lateral position in relation to Flexor hallucis.
3. Both attach to the base of the distal phalanx of their respective toes and so flex the distal interphalangeal joints. However, in spanning many joints they also contribute to flexion of metacarpo-phalangeal joints and to plantar flexion.
4. It is evident that both contribute to inversion of the foot.

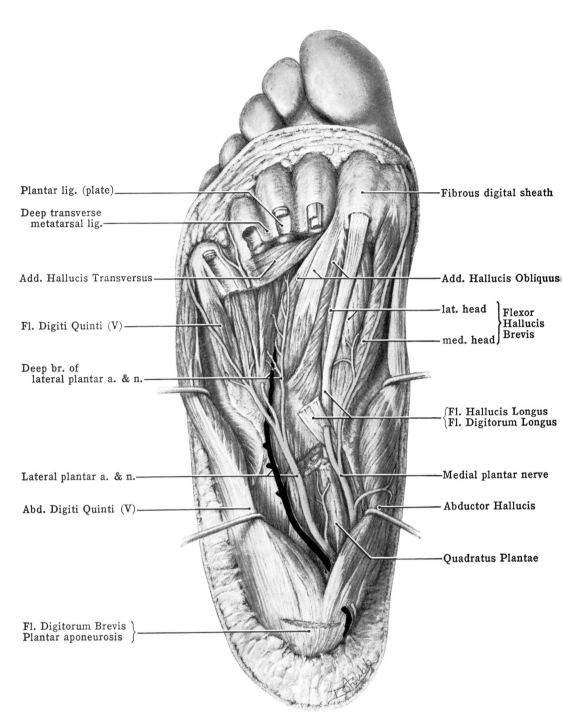

Plantar lig. (plate)

Deep transverse metatarsal lig.

Add. Hallucis Transversus

Fl. Digiti Quinti (V)

Deep br. of lateral plantar a. & n.

Lateral plantar a. & n.

Abd. Digiti Quinti (V)

Fl. Digitorum Brevis }
Plantar aponeurosis }

Fibrous digital sheath

Add. Hallucis Obliquus

lat. head } Flexor
          } Hallucis
med. head } Brevis

Fl. Hallucis Longus
Fl. Digitorum Longus

Medial plantar nerve

Abductor Hallucis

Quadratus Plantae

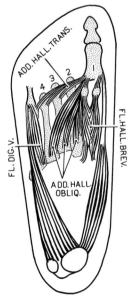

4-100   THIRD LAYER OF PLANTAR MUSCLES

Observe:
1. The muscles of this layer are: Flexor Digiti V, Adductor Hallucis Transversus, and Flexor Hallucis Brevis which form three sides of a square in the anterior half of the sole, largely filled by Abductor Hallucis Obliquus.
2. Of the 1st layer, Adductor Digiti V and Abductor Hallucis are pulled aside and Flexor Digitorum Brevis is cut short. Of the 2nd layer, Flexor Digitorum Longus and Lumbricales are excised and Quadratus Plantae is cut long.
3. The lateral Interossei are seen in the floor of the square.
4. The lateral plantar nerve and artery course laterally between muscles of the 1st and 2nd layers; their deep branches then course medially between muscles of the 3rd and 4th layers.

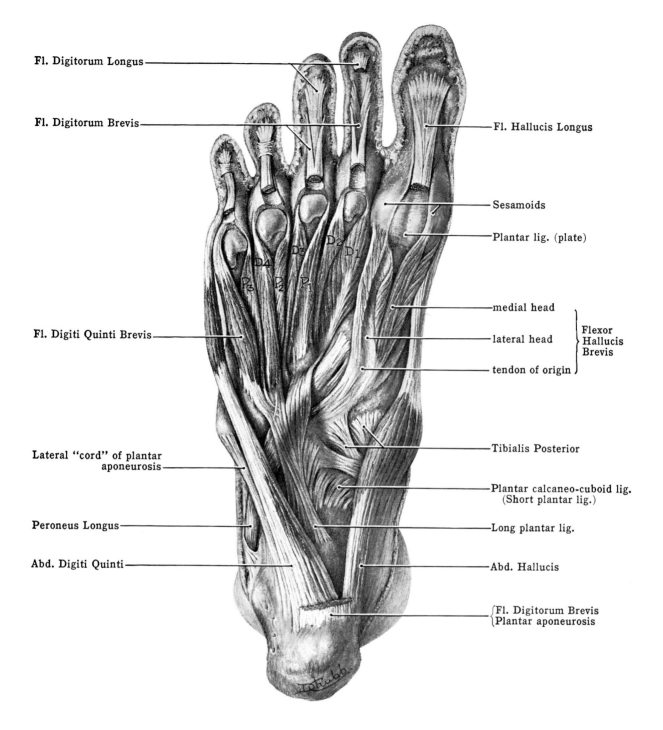

Fl. Digitorum Longus

Fl. Digitorum Brevis

Fl. Digiti Quinti Brevis

Lateral "cord" of plantar aponeurosis

Peroneus Longus

Abd. Digiti Quinti

Fl. Hallucis Longus

Sesamoids

Plantar lig. (plate)

medial head
lateral head   Flexor Hallucis Brevis
tendon of origin

Tibialis Posterior

Plantar calcaneo-cuboid lig. (Short plantar lig.)

Long plantar lig.

Abd. Hallucis

{ Fl. Digitorum Brevis
{ Plantar aponeurosis

4-102 FOURTH LAYER OF PLANTAR MUSCLES

Observe:
1. The muscles of this layer are: (a) three Plantar and four Dorsal Interossei in the anterior half of the foot and (b) the tendon of Peroneus Longus and of Tibialis Posterior in the posterior half.
2. Of the first three layers, Abductor and Flexor Brevis of the fifth toe and Abductor and Flexor Brevis of the big toe remain for purposes of orientation.
3. Plantar Interossei adduct the three lateral toes toward an axial line that passes through the 2nd metatarsal bone and second toe, whereas Dorsal Interossei abduct from this line.

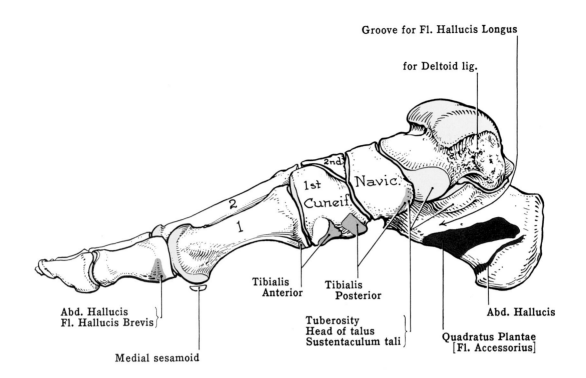

Groove for Fl. Hallucis Longus

for Deltoid lig.

2nd

1st

Navic.

Cuneif.

2

1

Tibialis Anterior

Tibialis Posterior

Abd. Hallucis Fl. Hallucis Brevis }

Medial sesamoid

Tuberosity Head of talus Sustentaculum tali }

Abd. Hallucis

Quadratus Plantae [Fl. Accessorius]

4-103 BONES OF THE FOOT, MEDIAL ASPECT

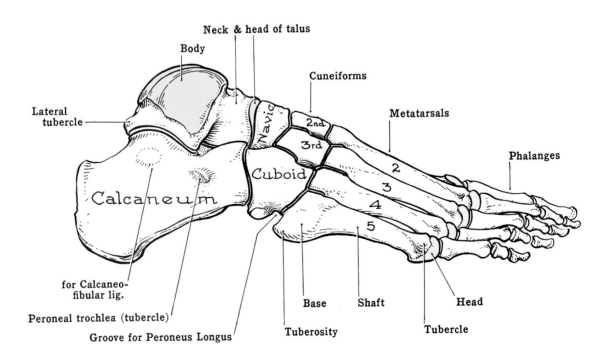

Neck & head of talus

Body

Cuneiforms

Lateral tubercle

Navic.

2nd

3rd

Metatarsals

Cuboid

2

Phalanges

Calcaneum

3

4

5

for Calcaneo-fibular lig.

Peroneal trochlea (tubercle)

Groove for Peroneus Longus

Base

Shaft

Head

Tuberosity

Tubercle

4-104 BONES OF THE FOOT, LATERAL ASPECT

Note terminology: The trochlea of the talus is the part of the body of the talus that articulates with the ankle socket. It has an upper, a medial malleolar, and a lateral malleolar part.

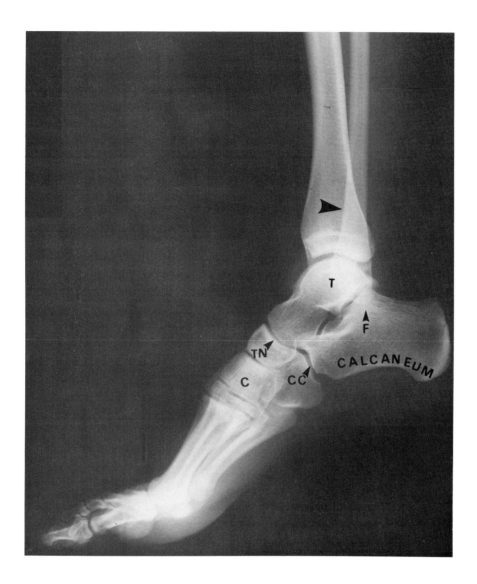

4-105 LATERAL RADIOGRAPH OF FOOT AND ANKLE

This radiograph was taken with the foot in a walking position similar to that of Figure 4-98.

Observe:

1. The *large arrow above* points to the edge of the triangular area where tibia and fibula are superimposed on each other.
2. The *small arrow* (F) reminds us of how far the fibula extends distally.
3. The talus (T) participates in the talo-navicular joint (TN) and the calcaneum in calcaneo-cuboid (CC) joint.
4. The cuneiforms (C) and the proximal ends of the metatarsals are superimposed upon each other.

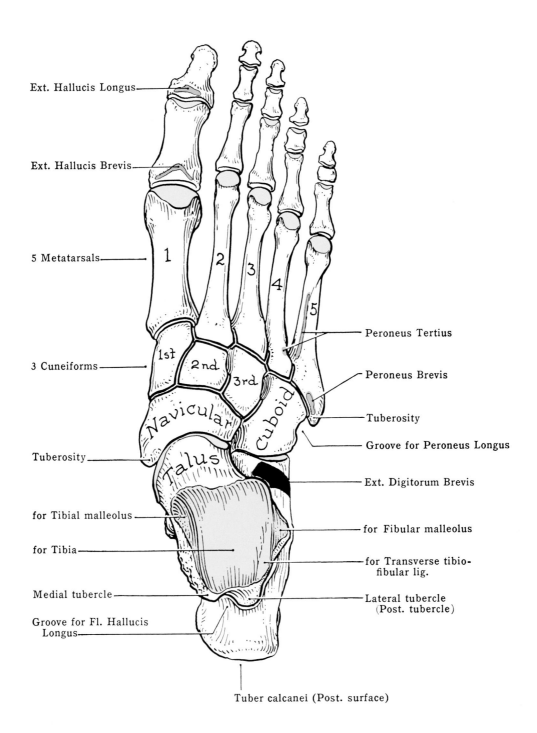

Ext. Hallucis Longus

Ext. Hallucis Brevis

5 Metatarsals

1 2 3 4 5

3 Cuneiforms

1st 2nd 3rd

Navicular Cuboid

Talus

Peroneus Tertius

Peroneus Brevis

Tuberosity

Groove for Peroneus Longus

Ext. Digitorum Brevis

Tuberosity

for Tibial malleolus

for Fibular malleolus

for Tibia

for Transverse tibio-fibular lig.

Medial tubercle

Lateral tubercle (Post. tubercle)

Groove for Fl. Hallucis Longus

Tuber calcanei (Post. surface)

4-106 BONES OF THE FOOT, DORSAL ASPECT

For upper surface of calcaneus, see Figure 4-121.

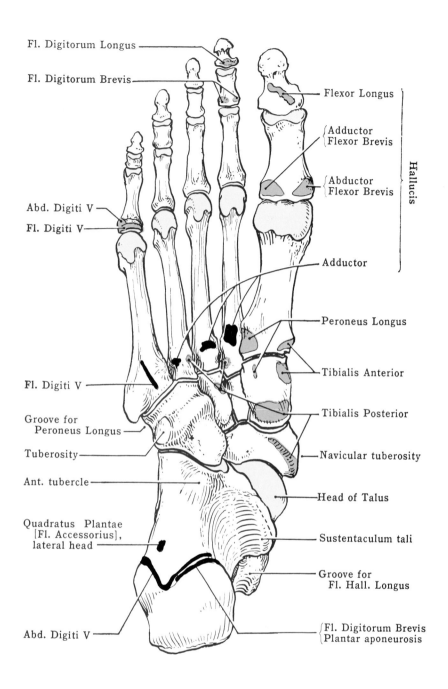

Fl. Digitorum Longus

Fl. Digitorum Brevis

Flexor Longus

{ Adductor
Flexor Brevis

{ Abductor
Flexor Brevis

Hallucis

Abd. Digiti V

Fl. Digiti V

Adductor

Peroneus Longus

Tibialis Anterior

Fl. Digiti V

Tibialis Posterior

Groove for
Peroneus Longus

Navicular tuberosity

Tuberosity

Ant. tubercle

Head of Talus

Quadratus Plantae
[Fl. Accessorius],
lateral head

Sustentaculum tali

Groove for
Fl. Hall. Longus

Abd. Digiti V

{ Fl. Digitorum Brevis
Plantar aponeurosis

4-107 BONES OF THE FOOT, PLANTAR ASPECT

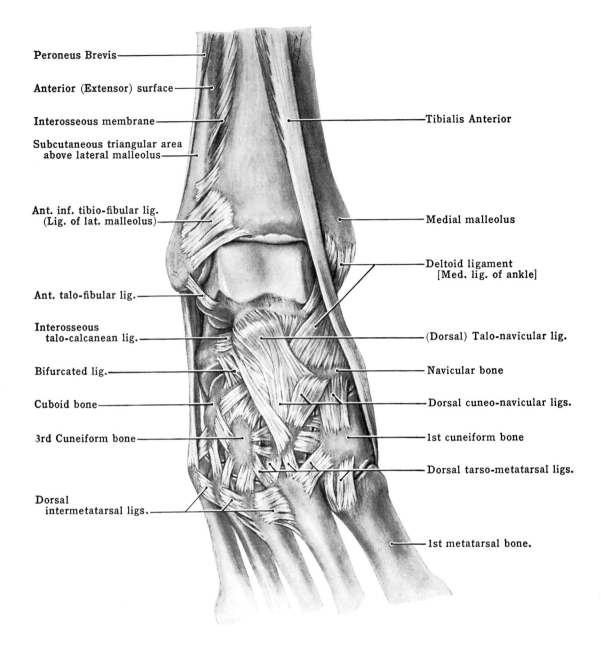

Peroneus Brevis

Anterior (Extensor) surface

Interosseous membrane

Subcutaneous triangular area above lateral malleolus

Ant. inf. tibio-fibular lig. (Lig. of lat. malleolus)

Ant. talo-fibular lig.

Interosseous talo-calcanean lig.

Bifurcated lig.

Cuboid bone

3rd Cuneiform bone

Dorsal intermetatarsal ligs.

Tibialis Anterior

Medial malleolus

Deltoid ligament [Med. lig. of ankle]

(Dorsal) Talo-navicular lig.

Navicular bone

Dorsal cuneo-navicular ligs.

1st cuneiform bone

Dorsal tarso-metatarsal ligs.

1st metatarsal bone.

4-108 ANKLE JOINT AND THE JOINTS OF THE FOOT, DORSAL VIEW

The ankle joint is extended (plantar-flexed); its anterior capsular fibers are removed.

Observe:

1. The fibers of the membrane and ligaments uniting the fibula to the tibia are so directed as to resist the downward pull of (eight) muscles, but allow the fibula to be forced upward.
2. The anterior talo-fibular ligament is but a weak band, easily torn (Fig. 4-114).
3. The dorsal ligaments of the foot resist the same thrusts as the plantar ligaments, and, therefore, are identically disposed, as reference to Figure 4-118 shows. The plantar ligaments, however, act also as tie beams for the arches of the foot and, therefore, are stronger.
4. Tibialis Anterior clinging to the skeleton throughout its entire course, as does Tibialis Posterior (Figs. 4-86 and 4-117).

Observe:

1. The body of the talus on section, wedge-shaped and grasped by the malleoli, which are bound to it by the deltoid and the posterior talo-fibular ligament and are thereby prevented from sliding forward.
2. Several synovial folds projecting into the joint.
3. Flexor Hallucis Longus, within its fibro-osseous sheath, lying between medial and lateral tubercles of talus.

Two tendons each within a separate sheath (fibrous and synovial) behind the medial malleolus, and 2 within a common sheath behind the lateral malleolus.

4. Because of the intervening fibrous sheath, the posterior tibial vessels and the tibial nerve are not disturbed by the excursions of Flexor Hallucis Longus.
5. The small inconstant bursa superficial to tendo Achillis and the large constant bursa deep to it and containing a long synovial fold.
6. The anterior tibial artery and its companion nerve at the midpoint of the front of the ankle, with 2 tendons medial to it and 2 lateral to it (Fig. 4-77).
7. The intermuscular fascial septum, shown and described in Figure 4-84.

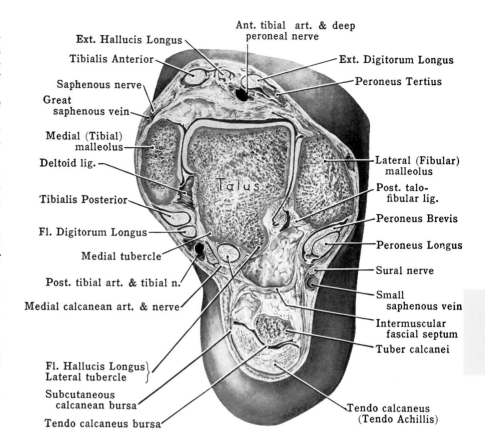

4-109 HORIZONTAL SECTION THROUGH ANKLE JOINT

Observe:

1. Tibia resting on the talus, and talus resting on the calcaneus. Between the calcaneus and the skin several large and many small encapsuled cushions of fat.
2. The fibular malleolus descending much farther than tibial malleolus.
3. The weak interosseous tibio-fibular ligament.
4. The interosseous band between talus and calcaneus separating the subtalar or posterior talo-calcanean joint from the talo-calcaneo-navicular joint. (For clarification see Fig. 4-113.)
5. Sustentaculum tali acting as a pulley for Flexor Hallucis Longus and giving attachment to the calcaneo-tibial band of the deltoid ligament. Tibialis Posterior rubbing on the band and Flexor Digitorum Longus on the sustentaculum.

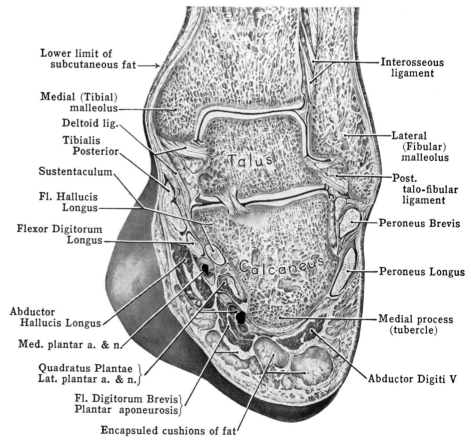

4-110 VERTICAL SECTION THROUGH ANKLE REGION

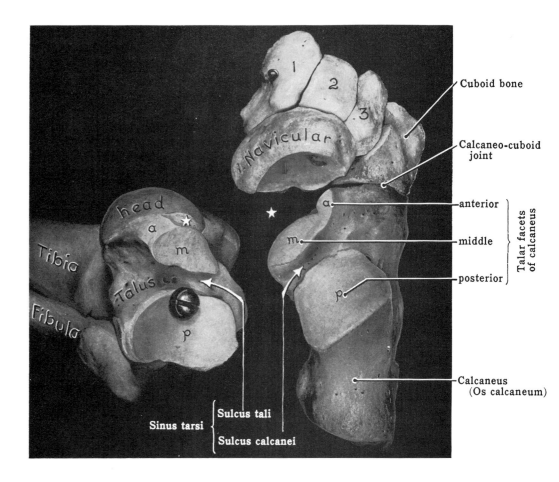

Cuboid bone

Calcaneo-cuboid joint

anterior

middle — Talar facets of calcaneus

posterior

Calcaneus (Os calcaneum)

Sinus tarsi { Sulcus tali / Sulcus calcanei

4-112 JOINTS OF INVERSION AND EVERSION

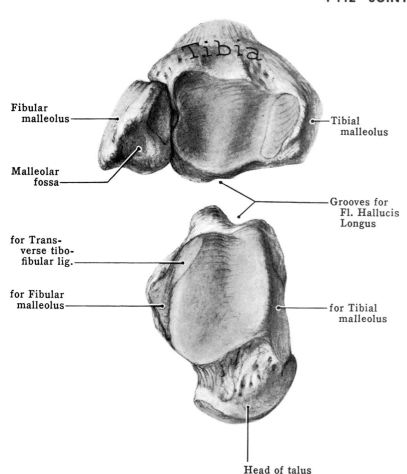

Fibular malleolus

Malleolar fossa

for Transverse tibofibular lig.

for Fibular malleolus

Tibia

Tibial malleolus

Grooves for Fl. Hallucis Longus

for Tibial malleolus

Head of talus

Observe:

1. The ankle joint has been immobilized by nailing together tibia, fibula, and talus, thereby making a single rigid unit of these 3 bones. The remaining bones of the foot—all but talus—have been wired together into another unit. Movements between these 2 units constitute inversion and eversion of the foot.
2. The talus takes part in 3 joints: (a) "supratalar joint," *i.e.*, the ankle joint; (b) "infratalar joints," the posterior talo-calcanean (the subtalar joint) and anterior talo-calcanean; (c) "pretalar joint," *i.e.*, talo-navicular.
3. At the supratalar joint only movements of flexion and extension are normally permitted—they are here eliminated by a nail. At the infratalar and pretalar joints movements of inversion and eversion take place.
4. The 2 parts of the infratalar joint are separated from each other by the sulcus tali and the sulcus calcanei, which, when the talus and calcaneus are in articulation, become the tarsal sinus or tunnel.
5. The convex posterior talar facet of the calcaneus, the concave middle and anterior talar facets, and the concave talar facet of the navicular all have their counterpart on the talus. The *white star* (∗) is at the site of the spring ligament. The middle talar facet is the cartilage-covered upper surface of the sustentaculum tali.
6. The calcaneo-cuboid joint is accessory to the foregoing joints.

4-111 ARTICULAR SURFACES OF THE ANKLE JOINT

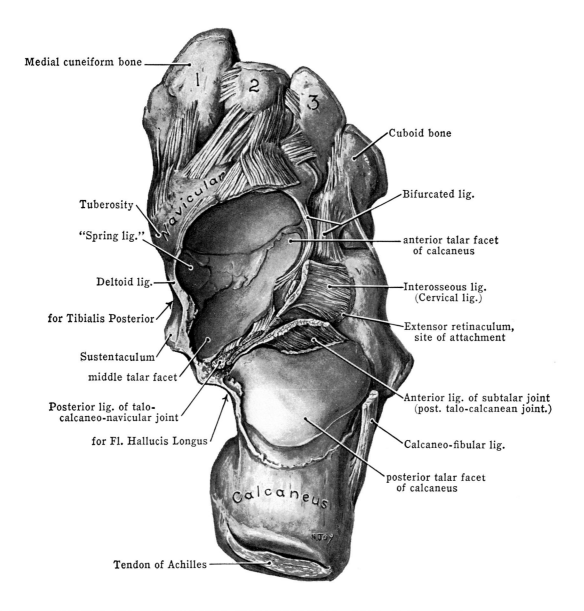

Medial cuneiform bone

Cuboid bone

Tuberosity

"Spring lig."

Deltoid lig.

for Tibialis Posterior

Sustentaculum

middle talar facet

Posterior lig. of talo-calcaneo-navicular joint

for Fl. Hallucis Longus

Bifurcated lig.

anterior talar facet of calcaneus

Interosseous lig. (Cervical lig.)

Extensor retinaculum, site of attachment

Anterior lig. of subtalar joint (post. talo-calcanean joint.)

Calcaneo-fibular lig.

posterior talar facet of calcaneus

Tendon of Achilles

4-113 JOINTS OF INVERSION (SUPINATION) AND EVERSION (PRONATION)

This specimen was prepared by sawing through the body of the talus and, after discarding it, by nibbling away the neck and head of the talus, thereby fully exposing the structures in the tarsal sinus. For plantar apsect of these joints see Figures 4-117 and 4-118.

Observe:

1. The convex, posterior talar facet separated from the concave, middle, and anterior facets by the ligamentous structures within the tarsal sinus.

2. At the wide lateral end of the sinus: (a) the strong interosseous talo-calcanean ligament; and (b) in *blue*, the attachments of the extensor retinaculum, which extends medially between the posterior ligament of the anterior talo-calcanean joint and the anterior ligament of the posterior talo-calcanean, or subtalar, joint.

3. The subtalar joint has a synovial cavity to itself, whereas the talo-navicular joint and the anterior talo-calcanean joint share a common synovial cavity, hence the collective title, talo-calcaneo-navicular joint.

4. The angular space between the navicular bone and the middle talar facet, on the sustentaculum tali, to be bridged by the plantar calcaneo-navicular (spring) ligament, the central part of which is fibro-cartilaginous.

5. The socket for the head of the talus to be deepened medially by the part of the deltoid ligament that is attached to the spring ligament, and laterally by the calcaneo-navicular part of the bifurcated ligament.

6. Various synovial folds overlying the margins of the articular cartilage, some of these containing fat.

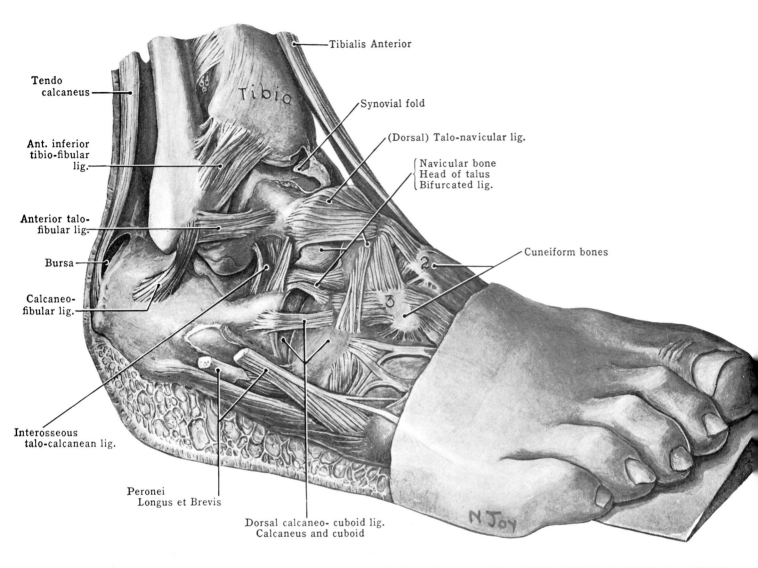

Tendo calcaneus

Tibialis Anterior

Tibia

Synovial fold

Ant. inferior tibio-fibular lig.

(Dorsal) Talo-navicular lig.

Navicular bone
Head of talus
Bifurcated lig.

Anterior talo-fibular lig.

Bursa

Cuneiform bones

Calcaneo-fibular lig.

Interosseous talo-calcanean lig.

Peronei Longus et Brevis

Dorsal calcaneo- cuboid lig.
Calcaneus and cuboid

4-114 ANKLE JOINT AND THE JOINTS OF INVERSION AND EVERSION, LATERAL VIEW

The joints of inversion and eversion are: (a) the subtalar (posterior talo-calcanean) joint, (b) the talo-calcaneo-navicular (*i.e.*, combined anterior talo-calcanean and talo-navicular) joint, and (c) the transverse tarsal (*i.e.*, combined calcaneo-cuboid and talo navicular) joint. Hence, the talo-navicular joint is twice involved.

The foot has been inverted in order to demonstrate: (a) the areas of articular surface uncovered, and (b) the ligaments rendered taut, on inverting the foot.

Observe:
1. The uncovered articular parts are: (a) posterior talar facet of calcaneus, (b) anterior surface of the calcaneus, and (c) head of the talus, all of which are palpable. Since inversion of the foot is commonly associated with plantar flexion of the ankle joint, (d) the upper and side parts of the trochlea of the talus are commonly uncovered too.
2. The ligaments that resist further inversion.
3. The anterior talo-fibular ligament and the dorsal calcaneo-cuboid ligament are weak and easily torn; the bifurcated and talo-navicular ligaments are under strain. The strong calcaneo-fibular ligament, not attached to the tip of the malleolus but to a facet in front of the tip. Hence, the projecting tip, being free, helps to retain the peroneal tendons.

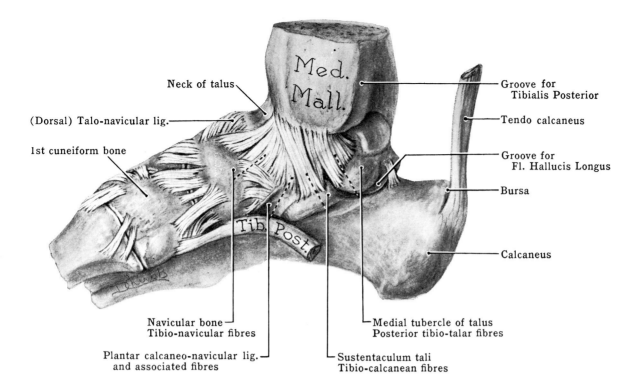

Neck of talus

(Dorsal) Talo-navicular lig.

1st cuneiform bone

Med. Mall.

Tib. Post.

Groove for
Tibialis Posterior

Tendo calcaneus

Groove for
Fl. Hallucis Longus

Bursa

Calcaneus

Navicular bone
Tibio-navicular fibres

Plantar calcaneo-navicular lig.
and associated fibres

Medial tubercle of talus
Posterior tibio-talar fibres

Sustentaculum tali
Tibio-calcanean fibres

4-115   LIGAMENTS OF THE ANKLE JOINT AND FOOT, MEDIAL VIEW

Observe:
1. Tibialis Posterior displaced from its bed—malleolus, deltoid ligament, and plantar calcaneo-navicular (spring) ligament.
2. The bed of Flexor Hallucis Longus—the groove between the two tubercles of the talus and the continuation of that groove beneath the sustentaculum tali.
3. The chief parts of the medial or deltoid ligament which is attached above to the medial malleolus of the tibia and below to talus, navicular, and calcaneus.

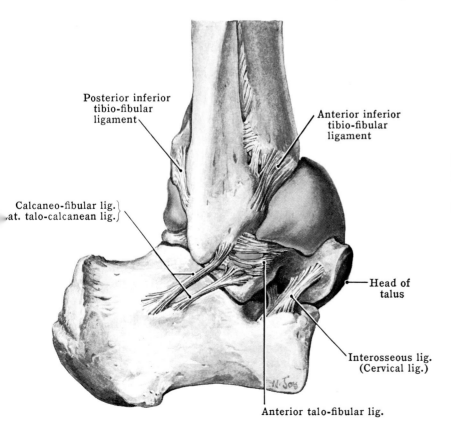

Posterior inferior
tibio-fibular
ligament

Anterior inferior
tibio-fibular
ligament

Calcaneo-fibular lig.
at. talo-calcanean lig.

Head of
talus

Interosseous lig.
(Cervical lig.)

Anterior talo-fibular lig.

N. Joy

Observe:
1. The far forward extension of the synovial cavity on the neck of the talus
2. Posteriorly and laterally, the closeness of the synovial cavity of the ankle joint to that of the subtalar (posterior talo-calcanean) joint.

4-116   A DISTENDED ANKLE JOINT

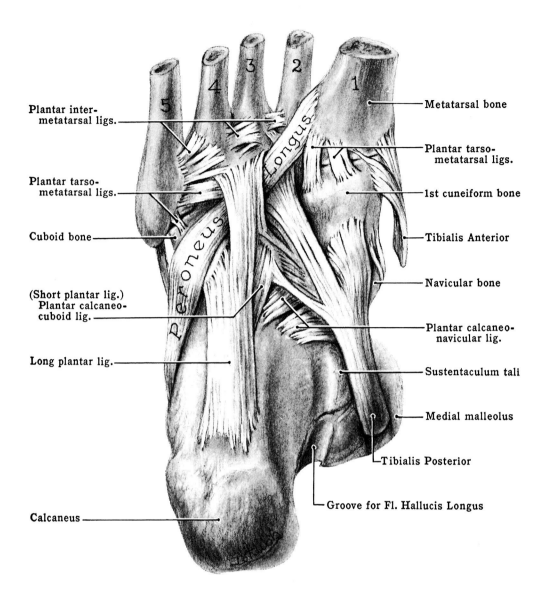

Plantar inter-metatarsal ligs.

Plantar tarso-metatarsal ligs.

Cuboid bone

(Short plantar lig.)
Plantar calcaneo-cuboid lig.

Long plantar lig.

Calcaneus

Peroneus Longus

5 4 3 2 1

Metatarsal bone

Plantar tarso-metatarsal ligs.

1st cuneiform bone

Tibialis Anterior

Navicular bone

Plantar calcaneo-navicular lig.

Sustentaculum tali

Medial malleolus

Tibialis Posterior

Groove for Fl. Hallucis Longus

4-117 PLANTAR LIGAMENTS—I

Observe:
1. The insertions of three long tendons: Peroneus Longus, Tibialis Anterior, and Tibialis Posterior.
2. The tendon of Peroneus Longus crossing the sole in the groove in front of the ridge of the cuboid; bridged by some fibers of the long plantar ligament; and inserted into the base of the 1st metatarsal. Usually, like Tibialis Anterior, it is also inserted into the 1st cuneiform. It is an evertor (pronator) of the foot (Fig. 4-107).
3. Slips of the tendon of Tibialis Posterior extending like the fingers of an open hand to grasp the bones anterior to the transverse tarsal joint (*i.e.*, the five small tarsal bones and several metatarsal bones, Fig. 4-120). It is an invertor (supinator) of the foot.

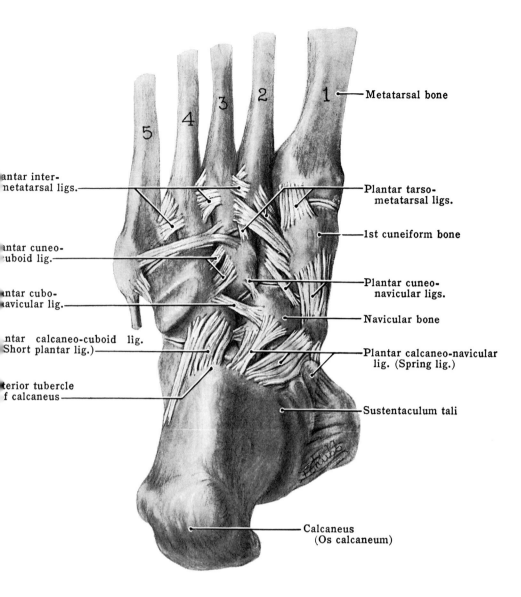

Metatarsal bone

Plantar inter-
metatarsal ligs.

Plantar cuneo-
cuboid lig.

Plantar cubo-
navicular lig.

Plantar calcaneo-cuboid lig.
(Short plantar lig.)

Anterior tubercle
of calcaneus

Plantar tarso-
metatarsal ligs.

1st cuneiform bone

Plantar cuneo-
navicular ligs.

Navicular bone

Plantar calcaneo-navicular
lig. (Spring lig.)

Sustentaculum tali

Calcaneus
(Os calcaneum)

4-118 PLANTAR LIGAMENTS—II

observe:

The plantar calcaneo-cuboid (short plantar) ligament and the plantar calcaneo-navicular (spring)
ligament are the inferior ligaments of the transverse tarsal joint. Having a common purpose, they
have a common direction.

The ligaments in the fore part of the foot diverge
backward from each side of the long axis of the 3rd
metatarsal and 3rd cuneiform. Hence a backward
thrust to the 1st metatarsal, as when rising on the
big toe in walking, is transmitted directly to the
navicular and talus by the 1st cuneiform, and
indirectly by the 2nd metatarsal and 2nd cuneiform and also by the 3rd metatarsal and 3rd cuneiform.

A backward thrust to the 4th and 5th metatarsals
is transmitted directly to the cuboid and calcaneus. That these four bones (i.e., the bones of the
lateral longitudinal arch of the foot) are not displaced backward is to the credit of the adjoining
ligaments. For dorsal aspect see Figure 4-113.

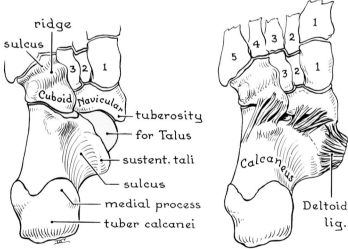

4-119 SUPPORT FOR HEAD OF TALUS

Seen from below, the head of the talus is supported by
plantar calcaneo-navicular ligaments and by the tendon of Tibialis Posterior (Fig. 4-117). For upper and
side views, see Figures 4-113 and 4-114.

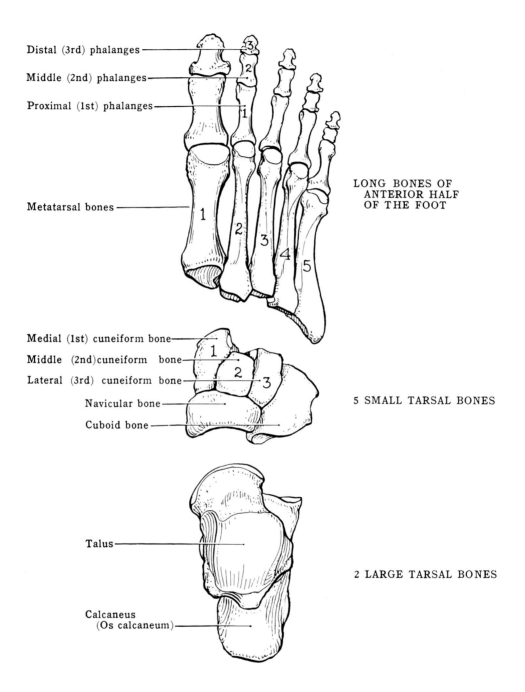

Distal (3rd) phalanges

Middle (2nd) phalanges

Proximal (1st) phalanges

Metatarsal bones

LONG BONES OF
ANTERIOR HALF
OF THE FOOT

Medial (1st) cuneiform bone

Middle (2nd) cuneiform bone

Lateral (3rd) cuneiform bone

Navicular bone

Cuboid bone

5 SMALL TARSAL BONES

Talus

2 LARGE TARSAL BONES

Calcaneus
(Os calcaneum)

4-120 BONES OF THE FOOT, DORSAL ASPECT

Note that the bones are divisible, at the transverse tarsal and tarso-metatarsal joints, into three sections—anterior, middle, and posterior.

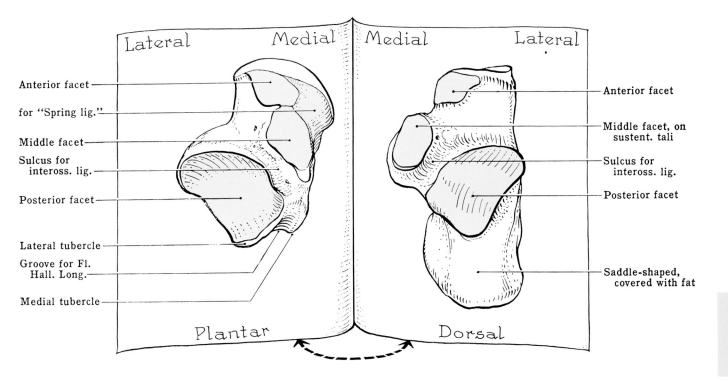

4-121 BONY SURFACES OF THE TALO-CALCANEAN JOINTS

The under or plantar surface of the talus and the upper or dorsal surface of the calcaneus are displayed as pages in a book.

The joints are gliding joints: hence apposed or corresponding facets are not exact counterparts of each other, one being more extensive than the other.

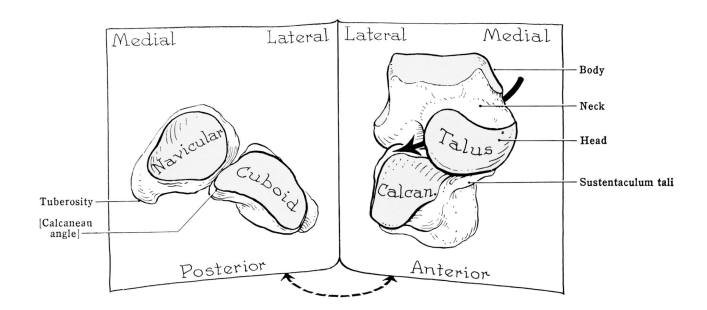

4-122 BONY SURFACES OF THE TRANSVERSE TARSAL JOINT

This joint includes the talo-navicular and the calcaneo-cuboid. The posterior surfaces of the navicular and cuboid bones and the anterior surfaces of the talus and calcaneus are displayed as pages in a book. The *black arrow* traverses the tarsal sinus (tunnel) which lodges the interosseous talo-calcanean ligament (Fig. 4-113).

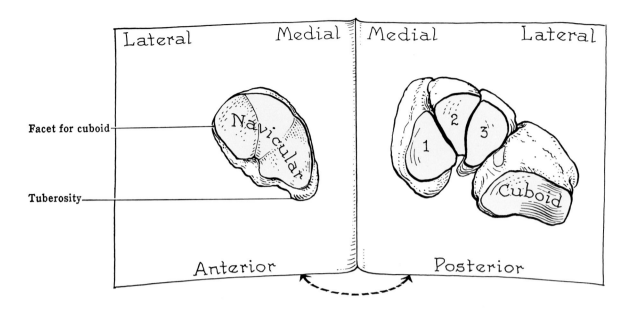

4-123 BONY SURFACES OF THE CUNEO-NAVICULAR AND CUBO-NAVICULAR JOINTS

The anterior surface of the navicular bone, the posterior surfaces of the three cuneiform bones, and the medial and posterior surfaces of the cuboid bone are displayed as pages in a book.

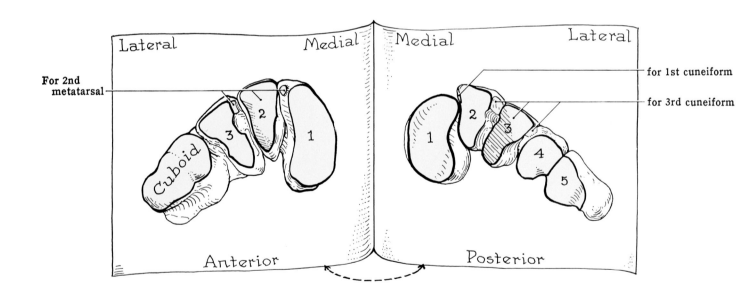

4-124 BONY SURFACES OF THE TARSO-METATARSAL JOINTS

The anterior surfaces of the cuboid and 3 cuneiform bones and the posterior surfaces of the bases of the 5 metatarsal bones are displayed as pages in a book.

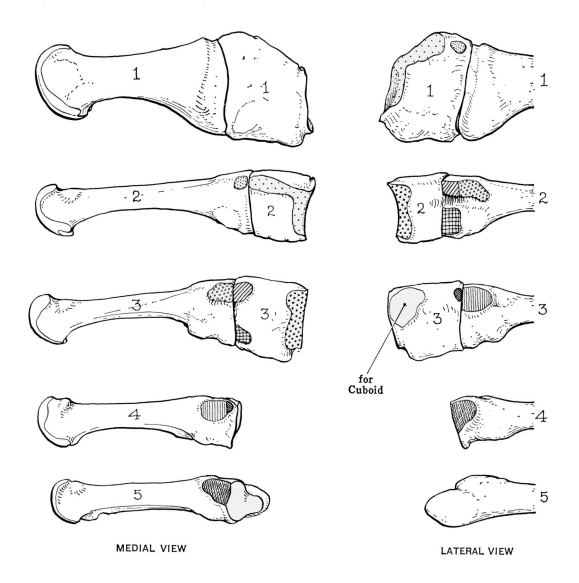

MEDIAL VIEW   LATERAL VIEW

for
Cuboid

4-125 BONY SURFACES OF THE INTERCUNEIFORM AND INTERMETATARSAL JOINTS

The medial and lateral surfaces of the 3 cuneiform and 5 metatarsal bones are displayed.

The joints are gliding joints; hence apposed or corresponding facets are not exact counterparts of each other, one being more extensive than the other.

For purposes of identification, corresponding facets have been marked with corresponding designs.

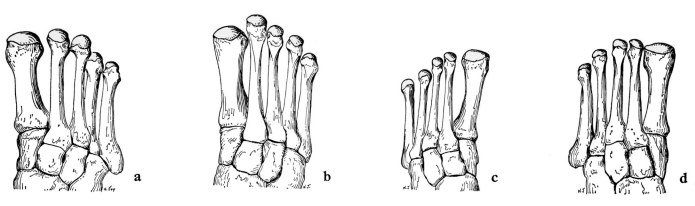

4-126 LONG AND SHORT 1ST METATARSAL BONES

The extent of the forward projection of the 1st, 2nd, and 3rd metatarsal bones varies. In *a* the 1st and 2nd metatarsals project equally far; in *b* the 2nd projects much farther than either the 1st or 3rd; in *c* the 1st is the most projecting; and in *d* the 2nd and 3rd project equally far.

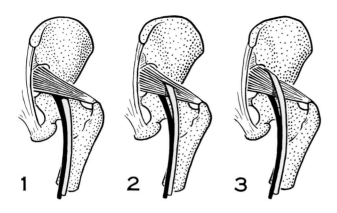

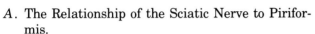

A. The Relationship of the Sciatic Nerve to Piriformis.

Of 640 limbs studied, (*1*) 87.3 per cent: both tibial and peroneal divisions passed *below* Piriformis: (*2*) 12.2 per cent: the peroneal division passed *through* Piriformis: (*3*) 0.5 per cent: the peroneal division passed *above*.

C. Bipartite Patella, Posterior View.
Occasionally the superlateral angle of the patella ossifies independently and remains discrete.

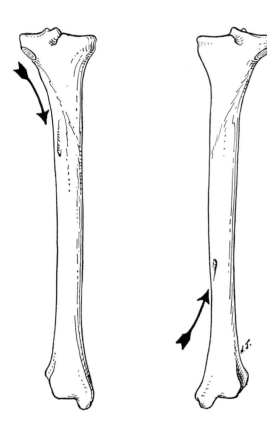

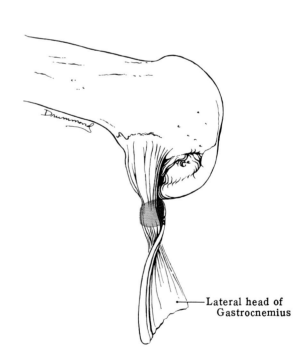

—Lateral head of Gastrocnemius

B. The nutrient canals of long bones are directed away from the more actively growing ends of the bones, but, in this pair of tibiae taken from the same subject, they take opposite directions.

D. A sesamoid bone (fabella) in the lateral head of Gastrocnemius was present in 21.6 per cent of 116 limbs.

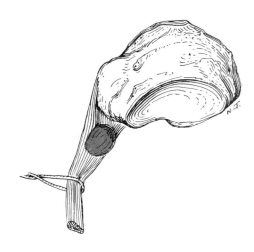

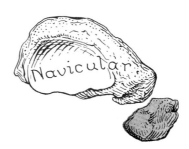

A. A sesamoid bone in Tibialis Posterior tendon was found in 80 or 23 per cent of 348 adult feet; 42 were paired and 38 unpaired.

B. Accessory Navicular Bone (Tibiale Externum). The tuberosity of the navicular bone, to which Tibialis Posterior is largely inserted, appeared as an independent bone in 4.1 per cent of 3619 men examined by x-ray.

See Harris, R. I., and Beath, T. (1947) *Army Foot Survey*, vol. 1, p. 52. National Research of Canada, Ottawa.

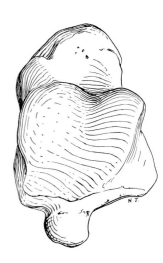

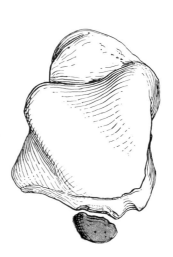

C. A bipartite medial cuneiform bone, in dorsal and plantar halves, is not common.

D. Os Trigonum. The lateral (posterior) tubercle of the talus has a separate center of ossification, which appears between the ages of 7 and 13 years. When this fails to fuse with the body of the talus, as in the left bone of this pair, it is called an os trigonum. It was found in 7.7 per cent of 558 adult feet; 22 were paired, 21 were unpaired.

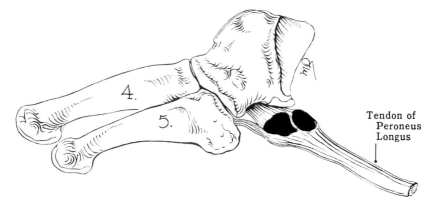

Tendon of
Peroneus
Longus

E. A sesamoid bone in Peroneus Longus tendon was found in 26 per cent of 92 feet. In this specimen it is bipartite, and Peroneus Longus has an additional attachment to the 5th metatarsal bone.

4-128 ANOMALIES OF THE TARSAL BONES

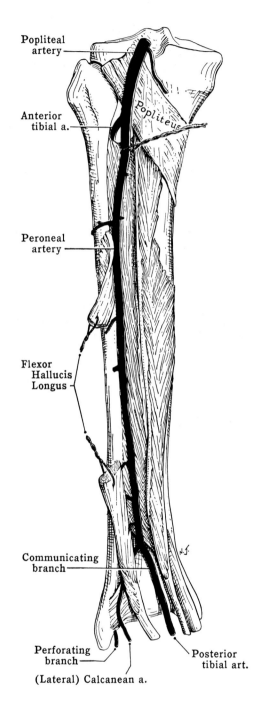

Popliteal artery

Anterior tibial a.

Popliteus

Peroneal artery

Flexor Hallucis Longus

Communicating branch

Perforating branch

(Lateral) Calcanean a.

Posterior tibial art.

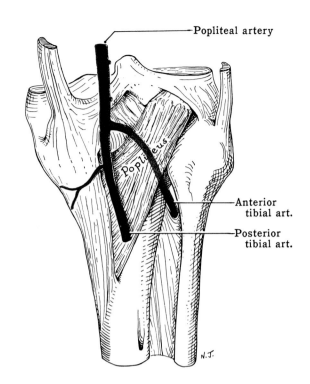

Popliteal artery

Popliteus

Anterior tibial art.

Posterior tibial art.

B. High division of popliteal artery with the anterior tibial artery descending anterior to Popliteus occurs in about 2 per cent of limbs.

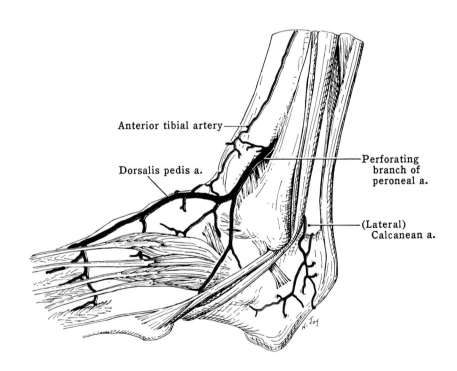

Anterior tibial artery

Dorsalis pedis a.

Perforating branch of peroneal a.

(Lateral) Calcanean a.

A. Absence of posterior tibial artery (*left side*), with compensatory enlargement of the peroneal artery, occurs in about 5 per cent of limbs.

The reverse condition, *i.e.,* absence of peroneal artery with enlargement of the posterior tibial artery, probably never occurs.

C. The dorsalis pedis artery (*left side*) as a continuation of the perforating branch of the peroneal artery. When this occurs (3.7 per cent of 592 limbs) the anterior tibial artery either fails to reach the ankle or is a very slender vessel.

The Back

CONTENTS

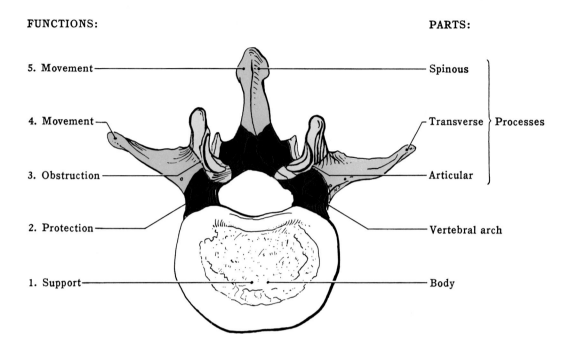

5. Movement ———————————————————— Spinous

4. Movement ——   — Transverse ⎫ Processes

3. Obstruction ——————   — Articular ⎭

2. Protection ——————   — Vertebral arch

1. Support ————————————   — Body

5-1 FUNCTIONS OF CONSTITUENT PARTS OF A VERTEBRA

A typical vertebra comprises the following parts:

1. A columnar body, situated anteriorly or ventrally. Its function, like that of the femur and tibia, is to support weight. Like them and other long bones, it is narrow about its middle and expanded at both ends. These ends also are articular and during growth have epiphyses.

2. A vertebral arch, placed behind the body. With the body this arch encloses the vertebral foramen. Collectively, the vertebral foramina constitute the vertebral canal wherein lodges the spinal cord. The function of a vertebral arch is to afford protection to the cord much as the bones of the vault of the skull afford protection to the brain.

3. Three processes—2 transverse and 1 spinous. These project from the vertebral arch like spokes from a capstan. They afford attachment to muscles. Indeed, they are the levers that help to move the vertebrae.

4. Four articular processes—2 superior and 2 inferior. These project (cranially and caudally) respectively from the arch and come into apposition with the corresponding processes of the vertebrae above and below. Their function is to restrict movements to certain directions, or at least to decree in what directions movements may be permitted, and they prevent the vertebrae from slipping forward. When one rises from the flexed position, they bear weight temporarily. The lower articular processes of the 5th lumbar vertebra bear weight even in the erect posture (Fig. 5-16).

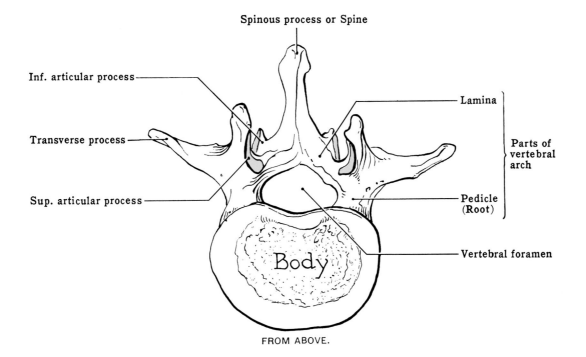

Spinous process or Spine

Inf. articular process

Transverse process

Sup. articular process

Lamina

Parts of vertebral arch

Pedicle (Root)

Vertebral foramen

Body

FROM ABOVE.

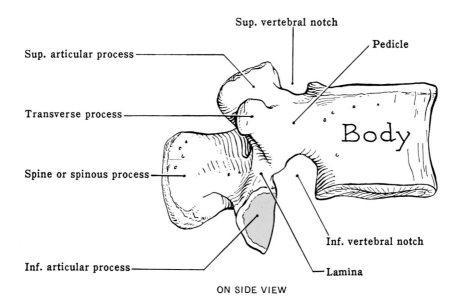

Sup. vertebral notch

Pedicle

Sup. articular process

Transverse process

Spine or spinous process

Body

Inf. vertebral notch

Inf. articular process

Lamina

ON SIDE VIEW

5-2 A VERTEBRA

Observe in the 2nd lumbar vertebra:
1. The vertebral arch. It consists of two stout, rounded pedicles, one on each side which spring from the body and which are united posteriorly by two flat plates or laminae.
2. A small notch above the pedicle and a larger one below it, called the superior and the inferior vertebral notch. When two vertebrae are in articulation, the two adjacent vertebral notches become an intervertebral foramen for the transmission of a spinal nerve and its accompanying intervertebral vessels.
3. Obviously, each articular process has an articular facet—the two terms are not synonymous.

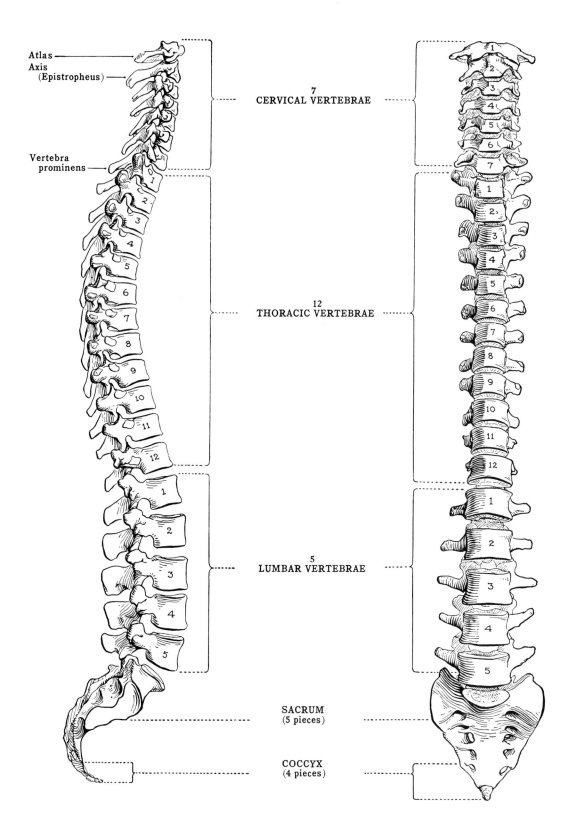

Atlas

Axis
(Epistropheus)

Vertebra
prominens

CERVICAL VERTEBRAE
7

THORACIC VERTEBRAE
12

LUMBAR VERTEBRAE
5

SACRUM
(5 pieces)

COCCYX
(4 pieces)

A. SIDE VIEW　　　　　　　　**B. FRONT VIEW**

5-3   VERTEBRAL COLUMN

Observe:

1. The vertebral column comprises 24 separate pre-sacral vertebrae and 2 composite vertebrae — the sacrum and the coccyx. Of the 24 pre-sacral vertebrae, 12 support ribs and therefore are thoracic; of the other 12, 7 are in the neck and 5 are in the lumbar region.

2. Vertebrae lying behind bony cavities (the thoracic vertebrae behind the thoracic cavity, and the sacrum and coccyx behind the pelvic cavity) are concave forward, but elsewhere (in the cervical and lumbar regions), by way of compensation, they are convex forward.

3. The transverse processes of the atlas spread widely; those of C7 spread almost as far; those of C2 to C6 much less. The spread diminishes progressively from T1 to T12. In the lumbar region it is greatest at L3.

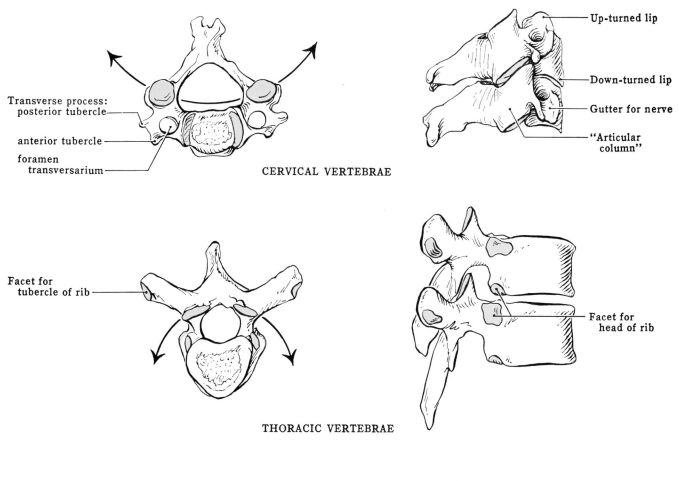

CERVICAL VERTEBRAE

- Transverse process: posterior tubercle
- anterior tubercle
- foramen transversarium

- Up-turned lip
- Down-turned lip
- Gutter for nerve
- "Articular column"

- Facet for tubercle of rib

THORACIC VERTEBRAE

- Facet for head of rib

Processes:
- mammillary
- accessory
- transverse

LUMBAR VERTEBRAE

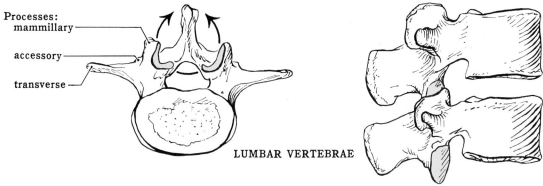

5-5   DISTINGUISHING FEATURES AND MOVEMENTS

Observe:

. The most distinctive feature of *cervical* vertebrae is the presence of foramina transversaria. All *thoracic* vertebrae have facets for articulation with the heads of ribs. The absence of these two features is distinctive of *lumbar* vertebrae.

. The *bodies* of cervical and lumbar vertebrae are greater in the transverse diameter than in the antero-posterior, and the vertebral foramina are triangular. In thoracic vertebrae the two diameters are about equal and the foramen is circular. Further, the superior surface of the body of a cervical vertebra ends at each side in an upturned lip; hence it is concave from side to side. The inferior surface ends anteriorly in a downturned lip. The superior and inferior surfaces of thoracic and lumbar bodies are flat.

. A *transverse process* in the cervical region points laterally, downward and forward, and ends in two tubercles with a gutter between them. In the thoracic region it points laterally, backward and upward, has a facet for the tubercle of a rib, and is

stout. In the lumbar region it points laterally, and is long and slender.

4. *Spinous processes* are bifid if cervical, spinelike if thoracic, and oblong if lumbar.

5. *Articular processes* in the cervical region collectively form a cylinder which is in part weight-bearing; it is cut obliquely into segments. In the thoracic and lumbar regions, the superior articular facets lie behind the pedicles, and the inferior facets are in front of the laminae. *Superior articular facets* in the cervical region face mainly upward; in the thoracic region mainly backward; in the lumbar region mainly medially. The change in direction is gradual from cervical to thoracic, but from thoracic to lumbar it is abrupt.

6. *Movements:* In all three regions the articular processes permit flexion and extension and side to side movement. Cervical vertebrae allow one to look sideways up. Thoracic vertebrae allow medial and lateral rotation, but lumbar vertebrae do not.

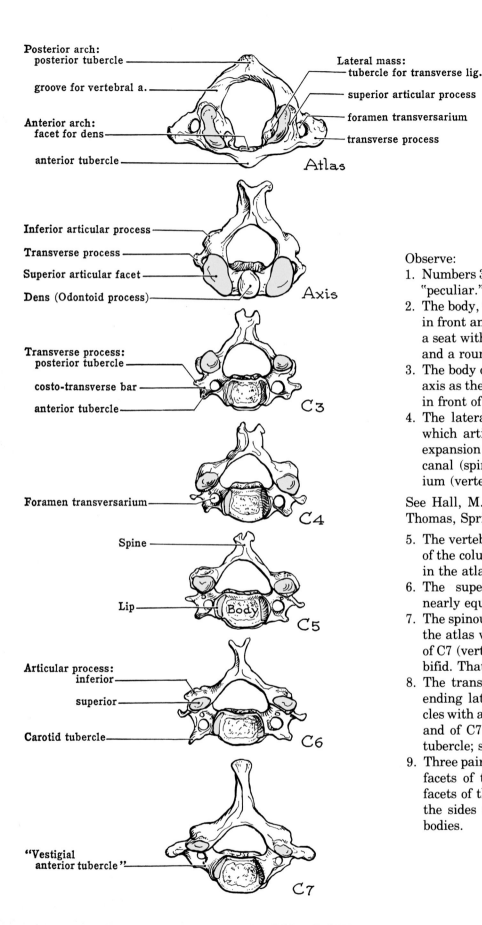

Posterior arch:
  posterior tubercle

groove for vertebral a.

Anterior arch:
  facet for dens

anterior tubercle

Lateral mass:
  tubercle for transverse lig.
  superior articular process
  foramen transversarium
  transverse process

Atlas

Inferior articular process

Transverse process

Superior articular facet

Dens (Odontoid process)

Axis

Transverse process:
  posterior tubercle

costo-transverse bar

anterior tubercle

C3

Foramen transversarium

C4

Spine

Lip

Body

C5

Articular process:
  inferior
  superior

Carotid tubercle

C6

"Vestigial
  anterior tubercle"

C7

Observe:
1. Numbers 3, 4, 5, and 6 are "typical"; 1, 2, and 7 are "peculiar."
2. The body, tranversely elongated, is of equal depth in front and behind. Its upper surface, resembling a seat with upturned side arms which bear facets, and a rounded front but no back.
3. The body of the atlas is missing: it is joined to the axis as the dens. An anterior arch on the atlas lies in front of the dens and articulates with it.
4. The lateral lips on the upper surface of a body which articulate with the body above. Arthritic expansion of this joint encroaches on the vertebral canal (spinal cord) and the foramen transversarium (vertebral artery).

See Hall, M. C. (1965) *Luschka's Joint*, Charles C Thomas, Springfield IL.

5. The vertebral foramen in this most mobile section of the column is large and triangular. It is largest in the atlas.
6. The superior and inferior vertebral notches, nearly equal in depth.
7. The spinous process, short and bifid, except that of the atlas which is reduced to a tubercle, and that of C7 (vertebra prominens) which is long and non-bifid. That of the axis is massive.
8. The transverse processes, short, perforated, and ending laterally in anterior and posterior tubercles with a gutter between them. Those of the atlas and of C7 are long and have but one (posterior) tubercle; so has the axis, but it is short.
9. Three paired articular facets; namely, the superior facets of the axis and the inferior and superior facets of the atlas are in series with the facets at the sides of the upper and lower surfaces of the bodies.

5-6   CERVICAL VERTEBRAE, FROM ABOVE

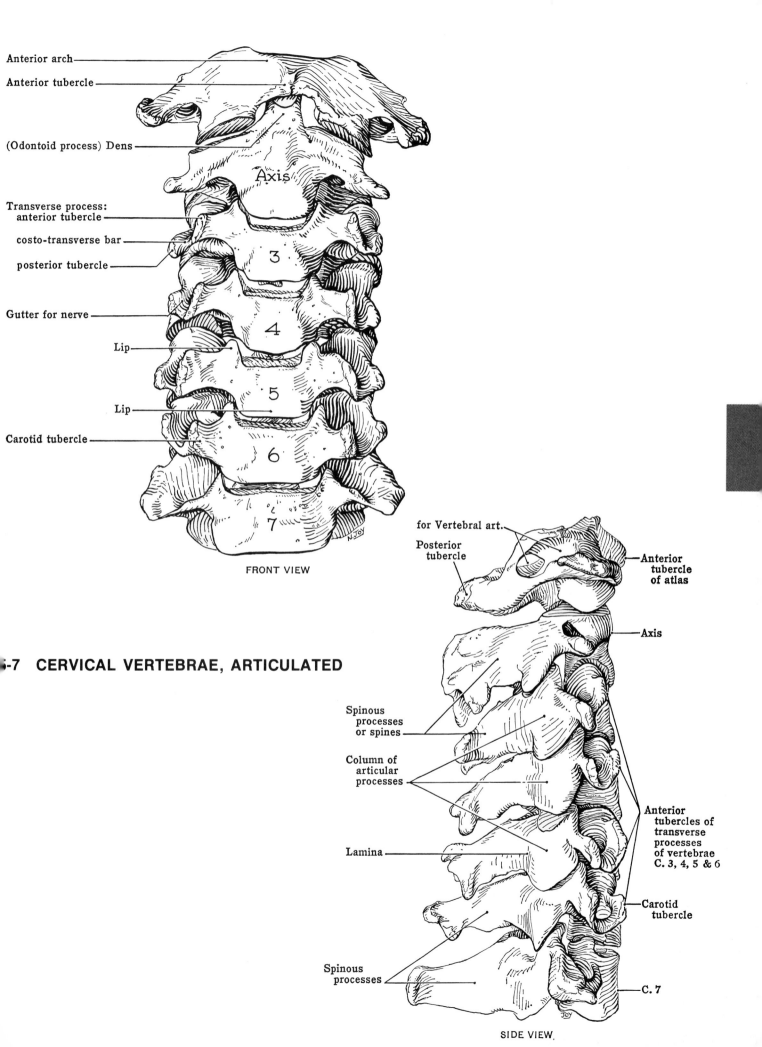

Anterior arch

Anterior tubercle

(Odontoid process) Dens

Axis

Transverse process:
anterior tubercle

costo-transverse bar

posterior tubercle

3

Gutter for nerve

4

Lip

5

Lip

Carotid tubercle

6

7

FRONT VIEW

7 CERVICAL VERTEBRAE, ARTICULATED

for Vertebral art.

Posterior tubercle

Anterior tubercle of atlas

Axis

Spinous processes or spines

Column of articular processes

Anterior tubercles of transverse processes of vertebrae C. 3, 4, 5 & 6

Lamina

Carotid tubercle

Spinous processes

C. 7

SIDE VIEW

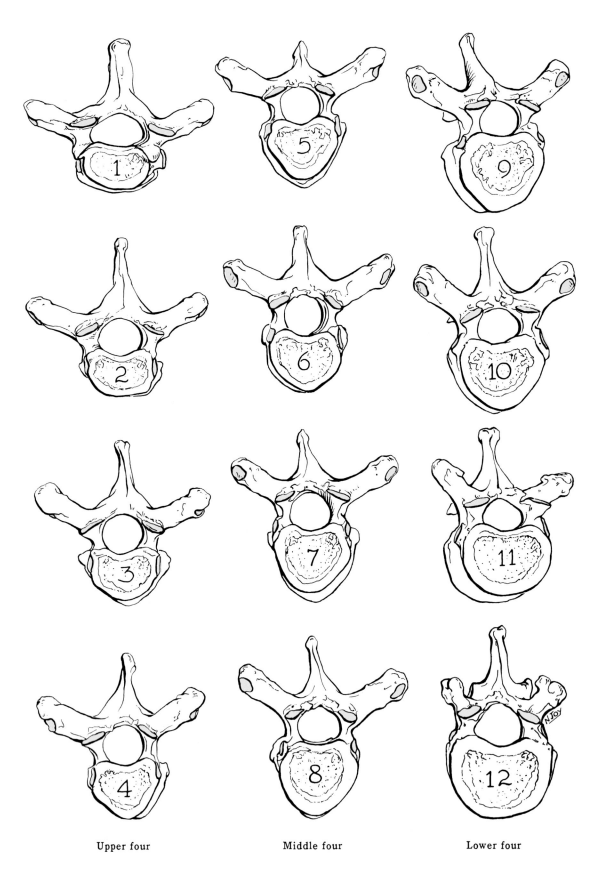

Upper four   Middle four   Lower four

5-8 THORACIC VERTEBRAE, FROM ABOVE

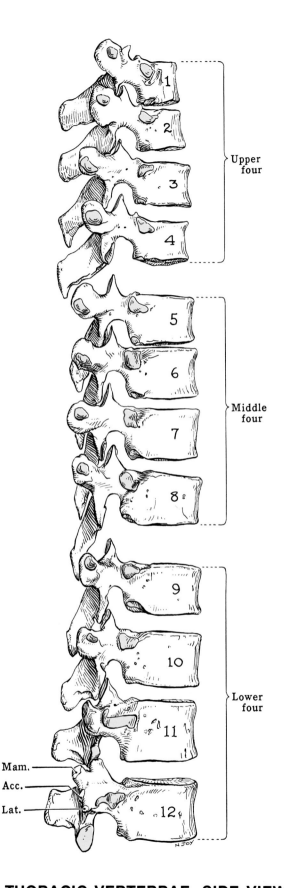

Observe:

1. The middle four are typically thoracic; the upper four have some cervical features; and the lower four some lumbar features.

2. The body, deeper dorsally than ventrally, with flat upper and lower surfaces. The surface area (weight-bearing surface) increasing from T1 to T12. The triangular shape of the middle 4 which have almost equal transverse and antero-posterior diameters. The transverse diameter increases toward the cervical and lumbar ends of the series.

3. The rib facet at the upper postero-lateral angle of the body encroaching on the lower postero-lateral angle of the body above, except for the facets of (T10), T11, and T12 which are on the pedicles.

4. The superior vertebral notch present on T1 only.

5. The vertebral foramen, circular and smaller than a finger ring, and becoming triangular toward the cervical and lumbar ends (Fig. 5-10).

6. The spines of the middle 4, which are long, overlapping, and nearly vertical. Those of 1,2 and 11,12 are nearly horizontal, and those of 3,4 and 9,10 are oblique.

7. The stretch of the transverse processes diminishes progressively from T1 to T12. T1 to T10 have rib facets on their transverse processes. These are concave and placed anteriorly on T1 to T7, flat and superiorly placed on T8 to T10.

8. The cervical features of T1—possession of superior vertebral notches, and upturned side lips on the body.

9. The lumbar features of T12—the lateral direction of the inferior articular processes; possession of mamillary, accessory, and lateral tubercles.

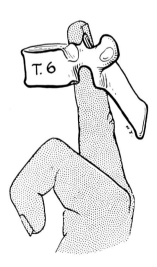

9 THORACIC VERTEBRAE, SIDE VIEW

5-10 SIZE OF VERTEBRAL FORAMEN

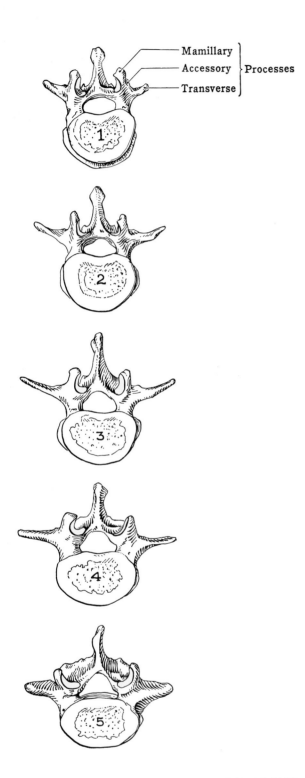

Mamillary ⎤
Accessory ⎬ Processes
Transverse ⎦

Observe:
1. The kidney-shaped bodies, greater in transverse than in antero-posterior diameter. Bodies L1 and L2 are deeper behind; L4 and L5 deeper in front; L3 transitional, being sometimes deeper behind and sometimes deeper in front.
2. The vertebral foramina, small and triangular, and having pinched lateral angles in L5.
3. The slight superior vertebral notches.
4. The large, oblong, and horizontal spinous processes.
5. The long, slender, horizontal transverse processes. That of L3 projects farthest; that of L5 spreads forward onto the body, is conical, and its apex has an upward tilt The mamillary process (for the origin of Multifidus) on the superior articular process. The accessory process (for insertion of Longissimus) on the transverse process.
6. The superior articular processes, facing each other and grasping the inferior processes of the vertebra above. The inferior articular processes, close together in L1, but far apart in L5 and facing more anteriorly.

5-11 LUMBAR VERTEBRAE, FROM ABOVE

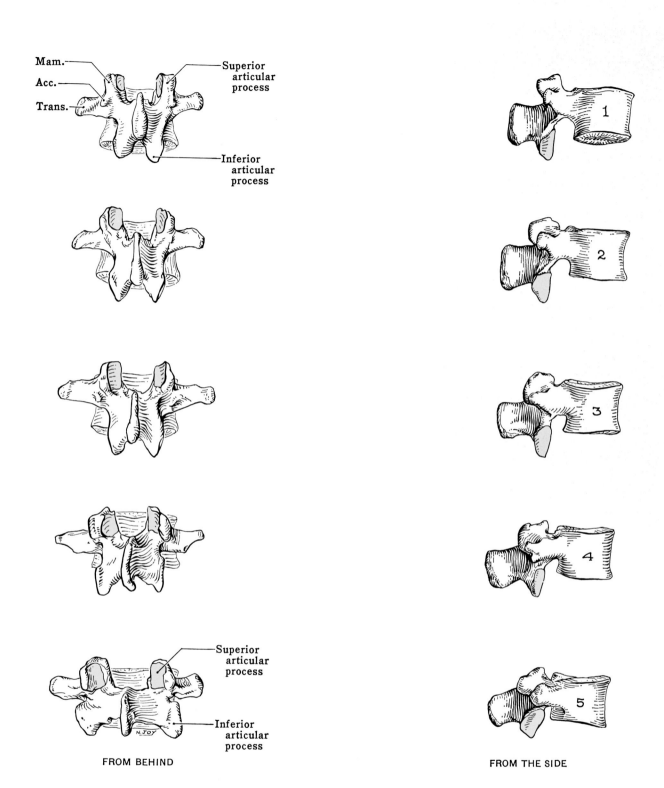

Mam.
Acc.
Trans.

Superior
articular
process

Inferior
articular
process

Superior
articular
process

Inferior
articular
process

FROM BEHIND

1

2

3

4

5

FROM THE SIDE

5-12 LUMBAR VERTEBRAE

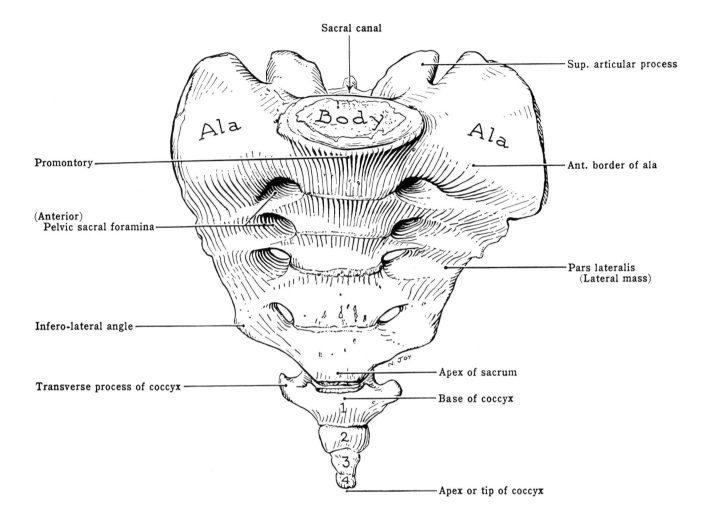

Sacral canal

Sup. articular process

Promontory

Ant. border of ala

(Anterior)
Pelvic sacral foramina

Pars lateralis
(Lateral mass)

Infero-lateral angle

Transverse process of coccyx

Apex of sacrum

Base of coccyx

Apex or tip of coccyx

5-13   SACRUM AND COCCYX, PELVIC SURFACE AND BASE

Observe:
1. The 5 sacral bodies, demarcated by 4 transverse lines which end laterally in 4 pairs of pelvic (anterior) sacral foramina.
2. The foramina of the two sides, approximately equidistant throughout. Their margins are rounded laterally but sharp elsewhere, indicating the courses of the emerging nerves.
3. The coccyx has 4 pieces. The 1st piece bears a pair of transverse processes and a pair of cornua; the other 3 pieces are nodular.

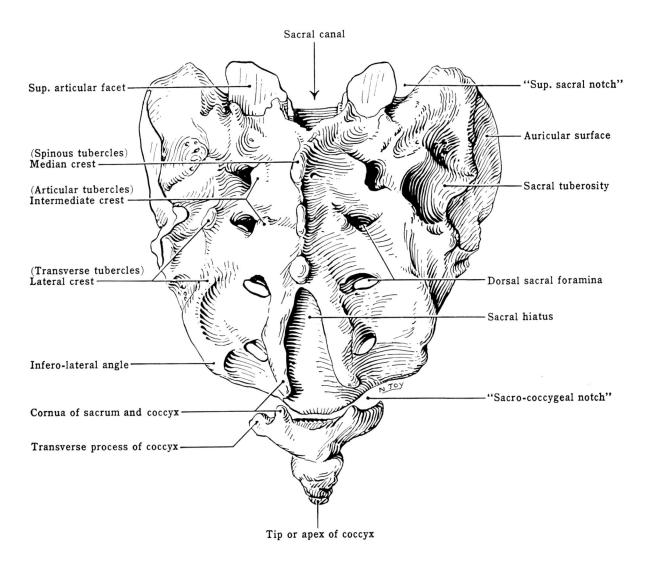

Sacral canal

Sup. articular facet

"Sup. sacral notch"

Auricular surface

(Spinous tubercles)
Median crest

(Articular tubercles)
Intermediate crest

Sacral tuberosity

(Transverse tubercles)
Lateral crest

Dorsal sacral foramina

Sacral hiatus

Infero-lateral angle

N TOY

"Sacro-coccygeal notch"

Cornua of sacrum and coccyx

Transverse process of coccyx

Tip or apex of coccyx

5-14 SACRUM AND COCCYX, DORSAL SURFACE

Observe:
1. The absence of the 4th and 5th sacral spines and laminae.
2. The superior articular processes, the intermediate crest, and the sacral and coccygeal cornua are serially homologous. So, likewise, are the "superior sacral notch," the 4 dorsal sacral foramina, and the "sacro-coccygeal notch."
3. A straight probe can be passed through a lower dorsal foramen, across the sacral canal, and through a pelvic foramen.

For side view see Figure 3-5.

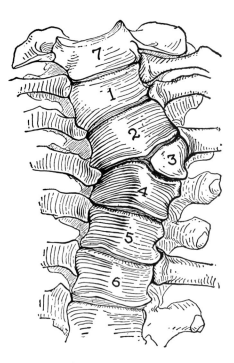

A. HEMIVERTEBRA

The entire right half of the 3rd thoracic vertebra and the corresponding rib are absent. The left lamina and the spine are fused throughout with those of T4, and the left intervertebral foramen is reduced in size. Observe the associated scoliosis (lateral curvature).

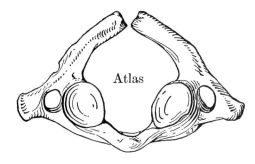

B.   UNFUSED ARCH

Of the three vertebral components at birth, the centrum has fused to the right and left halves of the vertebral arch, but the arch has not fused in the midline posteriorly.

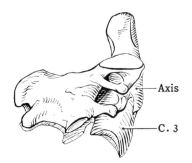

C. SYNOSTOSIS OF AXIS AND C3

Congenital synostosis of two vertebrae is relatively common, especially these two.

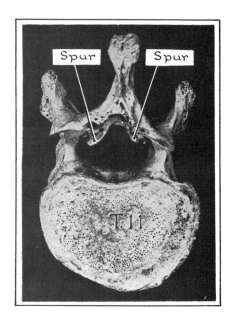

D. OSSIFYING LIGAMENTA FLAVA

Sharp, bony spurs commonly grow from the laminae caudally into the ligamenta flava, thereby reducing the lengths of these elastic bands. Hence, when the vertebral column is flexed, they are likely to be torn. Restricted to the thoracic and lumbar regions, most common and largest on T11, they diminish in size and frequency cranially to T1 and caudally to L5.

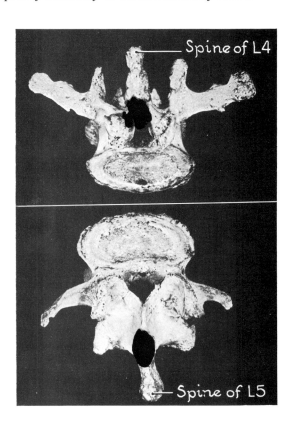

E. "KISSING SPINES"

Contact facets, either horizontal or obliquely overlapping, between lumbar spines (1 and 2), 2 and 3, 3 and 4, and 4 and 5 are commonly found (Fig. 5-20).

5-15   ANOMALIES OF THE VERTEBRAE

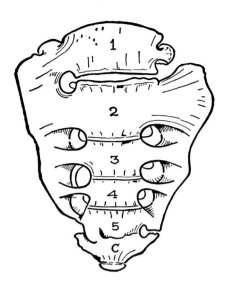

A. TRANSITIONAL LUMBO-SACRAL VERTEBRA

In this instance, the 1st sacral vertebra is partly free (lumbarized). Not uncommonly the 5th lumbar vertebra is partly fused to the sacrum (sacralized).

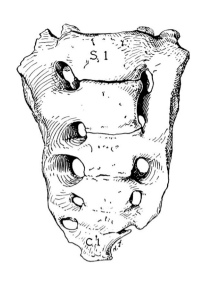

B. MALDEVELOPED SACRUM

The left side of the sacrum is imperfectly developed.

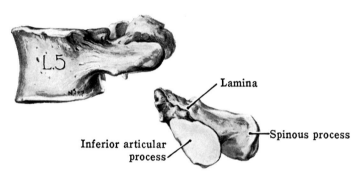

SIDE VIEW

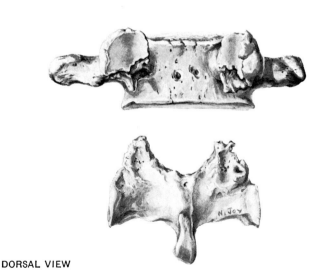

DORSAL VIEW

C. SPONDYLOLYSIS OF L5

In this case, the 5th lumbar vertebra is bipartite; the oblique defect is through the pars interarticularis. The two elements are held together by fibrous tissue. Separation of the two elements is spondylolisthesis.

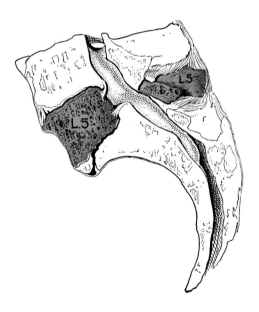

D. SPONDYLOLISTHESIS

The anterior element of a bipartite L5 has slipped forward.

See Stewart, T. D. (1953) The age incidence of neural-arch defects in Alaskan natives considered from the standpoint of etiology. *J. Bone Joint Surg.*, 35A: 937.

5-16   ANOMALIES OF THE VERTEBRAE

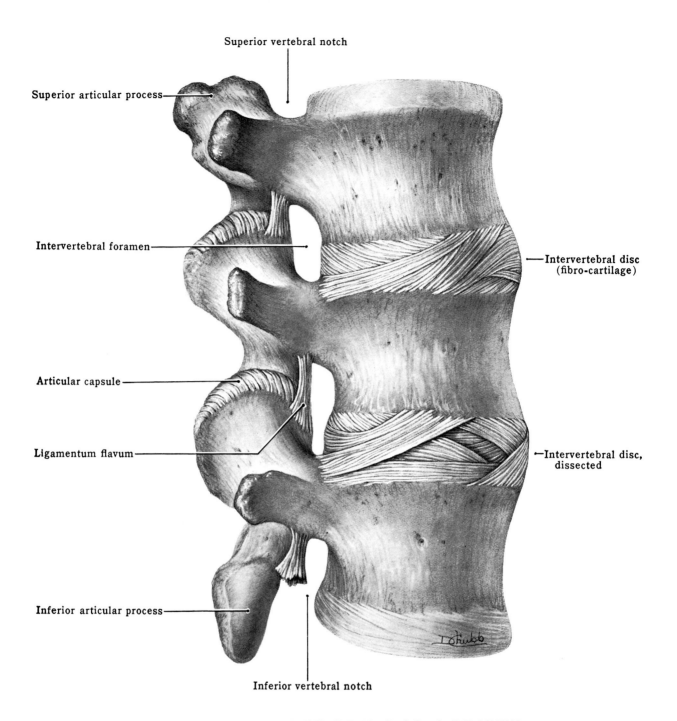

Superior vertebral notch

Superior articular process

Intervertebral foramen

Articular capsule

Ligamentum flavum

Inferior articular process

Intervertebral disc (fibro-cartilage)

Intervertebral disc, dissected

Inferior vertebral notch

5-17 AN INTERVERTEBRAL DISC, SIDE VIEW

Sections have been removed from the superficial layers of the lower disc in order to show the directions of the fibers.

Observe:
1. The anulus fibrosus, resembling the flat muscles of the abdominal wall in being arranged in layers of parallel fibers which criss-cross those of the next layer.
2. An intervertebral foramen, resulting from the apposition of a superior and an inferior vertebral notch, bounded above and below by pedicles, in front by an intervertebral disc and parts of the two bodies united by that disc, and behind by a capsular ligament and parts of the two articular processes united by that capsular ligament. Further, the anterior part of the capsule is strengthened by the lateral border of the ligamentum flavum.
3. The vulnerability of a spinal nerve to the pressure of an extruded nucleus pulposus through a torn annulus fibrosus. The most common site of a disc lesion is between L5 and S1.

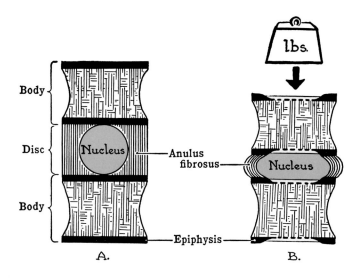

Body

Disc

Body

Nucleus

Anulus
fibrosus

Epiphysis

lbs.

Nucleus

Anulus
fibrosus

Epiphysis

A.

B.

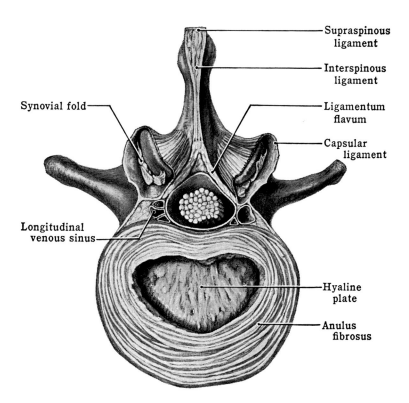

Supraspinous
ligament

Interspinous
ligament

Ligamentum
flavum

Capsular
ligament

Synovial fold

Longitudinal
venous sinus

Hyaline
plate

Anulus
fibrosus

5-19 AN INTERVERTEBRAL DISC AND LIGAMENTS, ON CROSS-SECTION

The nucleus pulposus has been scooped out and the cartilaginous epiphyseal plate exposed.

Observe:

1. The rings of the anulus fibrosus, least numerous dorsally.
2. The continuity of the following ligaments—capsular, flavum, interspinous, and supraspinous.
3. The synovial fold, containing a pad of fat, such as is present in all synovial joints.
4. The longitudinal vertebral venous sinuses which extend extradurally throughout the length of the vertebral canal (Fig. 5-23).
5. The cauda equina of the spinal cord (not labeled) lying free within the subarachnoid space.

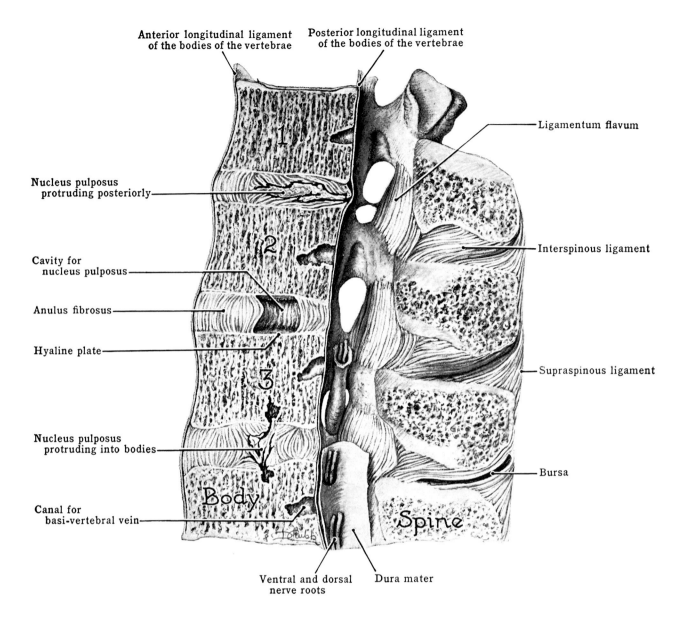

Anterior longitudinal ligament of the bodies of the vertebrae

Posterior longitudinal ligament of the bodies of the vertebrae

Ligamentum flavum

Nucleus pulposus protruding posteriorly

Cavity for nucleus pulposus

Anulus fibrosus

Hyaline plate

Interspinous ligament

Supraspinous ligament

Nucleus pulposus protruding into bodies

Bursa

Canal for basi-vertebral vein

Ventral and dorsal nerve roots

Dura mater

5-20   AN INTERVERTEBRAL DISC AND LIGAMENTS, ON MEDIAN SECTION

Observe:
1. The nucleus pulposus of the normal disc between the 2nd and 3rd vertebrae has been scooped out from the enclosing anulus fibrosus.
2. The ligamentum flavum, extending from the upper border and adjacent part of the posterior aspect of one lamina to the lower border and adjacent part of the anterior aspect of the lamina above, and extending laterally to the intervertebral foramen which it bounds posteriorly.
3. The interspinous ligament, uniting obliquely the upper and lower borders of two adjacent spines. Elastic fibers are sparce. The supraspinous ligament extended as far caudally as L3 in 22 of 100 specimens, to L4 in 73, to L5 in 5, and to the sacrum never. Many of the fibers shown above are not ligamentous, but the fibrous attachments of thoracolumbar fascia, Longissimus, and Multifidus (Figs. 5-27 to 5-29).

See Rissanen, P. M. (1960) The surgical anatomy and pathology of the supraspinous and interspinous ligaments, etc. *Acta Orthop. Scandinav., Suppl. 46.*

4. The adventitious bursa between the 3rd and 4th lumbar spines, acquired presumably as the result of habitual hyperextension which brings the lumbar spines into contact, as in Figure 5-15.
5. Two degenerative changes: (a) The pulp of the disc between the 1st and 2nd vertebrae has herniated backward through the anulus, and (b) the pulp of the disc between the 3rd and 4th vertebrae has herniated through the cartilaginous epiphyseal plates into the bodies of the vertebrae above and below.

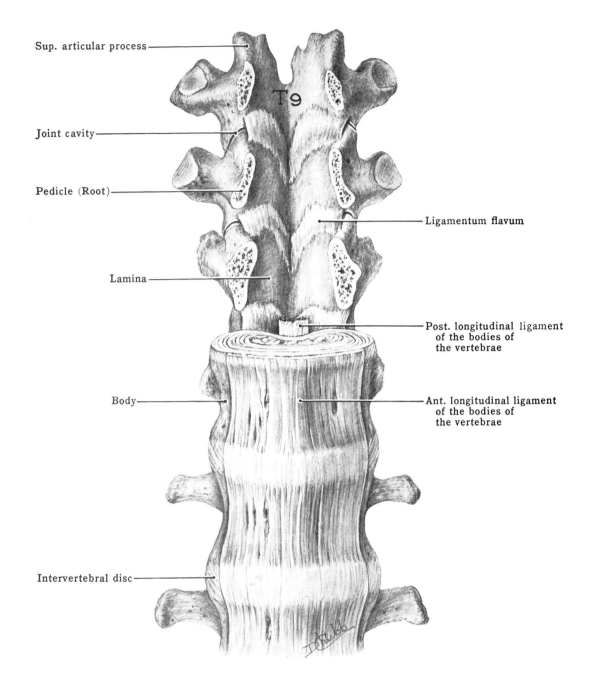

Sup. articular process

Joint cavity

Pedicle (Root)

Lamina

Body

Intervertebral disc

T9

Ligamentum flavum

Post. longitudinal ligament
of the bodies of
the vertebrae

Ant. longitudinal ligament
of the bodies of
the vertebrae

5-21   ANTERIOR LONGITUDINAL LIGAMENT AND THE LIGAMENTA FLAVA, ANTERIOR VIEW

The pedicles of the 9th, 10th, and 11th thoracic vertebrae have been sawn through and their bodies discarded.

Note:
1. The anterior and posterior longitudinal ligaments are ligaments of the bodies; the ligamenta flava are ligaments of the vertebral arches.
2. The anterior longitudinal ligaments are broad, strong, fibrous bands. They are attached to the intervertebral discs and to the adjacent parts of the fronts of the bodies. They have foramina for veins and arteries passing from and to the bodies.
3. The ligamenta flava, composed of yellow or elastic fibers, extend between adjacent laminae. Those of opposite sides meet and blend in the median plane. They extend laterally to the articular processes where they blend with the anterior fibers of the capsule of the joint. Being elastic, they tend, at all times, to restore the vertebral column to the extended or erect position. Above, they are in series with the posterior atlanto-axial and posterior atlanto-occipital membranes.

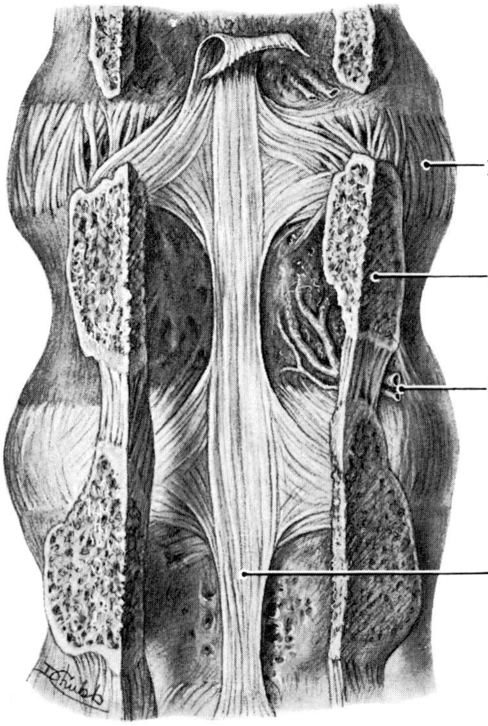

Intervertebral disc
Pedicle of vertebral arch
Intervertebral vessels
Posterior longitudinal lig. of vertebral bodies

5-22 POSTERIOR LONGITUDINAL LIGAMENT, POSTERIOR VIEW

The vertebral arches have been sawn through and largely removed.

Observe:

1. This taut but somewhat flimsy band passing from disc to disc, spanning the posterior surfaces of the bodies of the vertebrae, and rendering smooth the anterior wall of the vertebral canal.
2. The diamond shape taken by the ligament behind each disc, where it both gives and receives fibers.
3. Between the ligament and a vertebral body, a plexus of veins which receives the basivertebral vein from the body (Fig. 5-20), communicates with the longitudinal vertebral venous sinus on each side, and drains by way of the intervertebral veins.

The ligament extends to the sacrum below; it becomes the strong membrana tectoria above (Fig. 5-39).

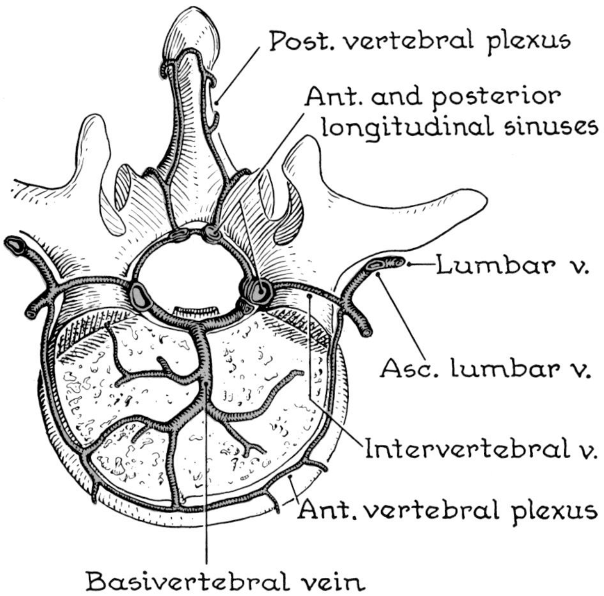

Post. vertebral plexus
Ant. and posterior longitudinal sinuses
Lumbar v.
Asc. lumbar v.
Intervertebral v.
Ant. vertebral plexus
Basivertebral vein

5-23 VERTEBRAL VENOUS PLEXUSES

There is an Internal and an External Plexus, communicating with each other and with both segmental systemic veins and the portal system.

The Internal Plexus: The vertebral canal contains a plexus of thin walled, valveless veins which surround like a basketwork the dura mater of the spinal cord and the posterior longitudinal ligament. Anterior and posterior longitudinal channels (anterior and posterior longitudinal venous sinuses) can be discerned in this plexus. Above, this plexus communicates through the foramen magnum with the occipital and basilar sinuses. At each spinal segment the plexus receives veins from the spinal cord and a basivertebral vein from the body of a vertebra. The plexus in turn is drained by intervertebral veins which pass through the intervertebral and sacral foramina to the vertebral, intercostal, lumbar, and lateral sacral veins.

The External Plexus: Through the body of each vertebra come veins which form a meager anterior vertebral plexus, and through the ligamenta flava pass veins which form a well marked posterior vertebral plexus. In the cervical region, these plexuses communicate freely with the occipital and profunda cervicis veins which receive from the sigmoid sinus the mastoid and condyloid emissary veins. In the thoracic, lumbar, and pelvic regions the azygos (or hemiazygos), the ascending lumbar, and the lateral sacral veins respectively further link segment to segment.

Infection and tumors may spread from systemic and portal areas (*e.g.*, prostate, breast) to the vertebral venous system and lodge in vertebrae, spinal cord, brain, or skull.

See Batson, O. V. (1957) The vertebral vein system. *Am. J. Roentgenol., 78:* 195.

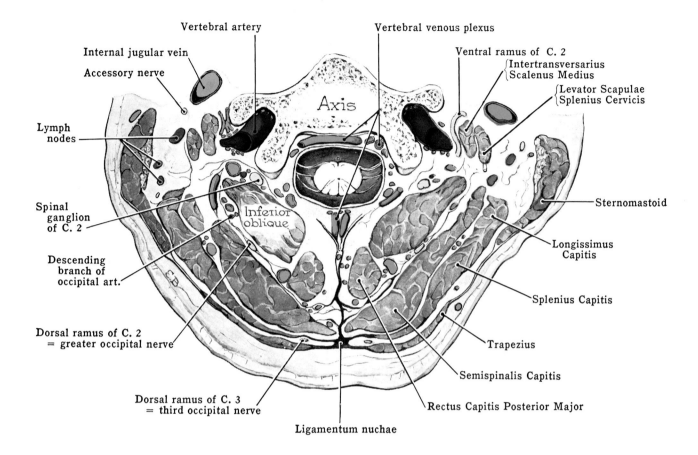

Vertebral artery

Vertebral venous plexus

Internal jugular vein

Ventral ramus of C. 2
{Intertransversarius
\Scalenus Medius

Accessory nerve

{Levator Scapulae
\Splenius Cervicis

Axis

Lymph nodes

Sternomastoid

Spinal ganglion of C. 2

Inferior oblique

Longissimus Capitis

Descending branch of occipital art.

Splenius Capitis

Dorsal ramus of C. 2 = greater occipital nerve

Trapezius

Semispinalis Capitis

Dorsal ramus of C. 3 = third occipital nerve

Rectus Capitis Posterior Major

Ligamentum nuchae

5-24 CROSS-SECTION OF THE NUCHAL REGION, AT THE LEVEL OF THE AXIS

The section, clearly, passes above the level of the spine and laminae of the axis, for Obliquous Inferior and Rectus Capitis Major are present, whereas Semispinalis Cervicis and Multifidus are not. It passes below the posterior arch of the atlas, for Obliquus Superior and Rectus Capitis Minor do not appear.

Observe:
1. Trapezius, Splenius, and Semispinalis Capitis forming a covering or roof for the suboccipital region.
2. The two muscles that ascend from the spine of the axis divided, namely, Inferior Oblique and Rectus Capitis Posterior Major.
3. Many anastomosing veins: (a) those around the vertebral artery unite, before leaving the 6th cervical transverse foramen to form the vertebral vein (Fig. 9-83); (b) the vertebral venous plexus, which followed cranially communicates through the foramen magnum with the basilar and occipital venous sinuses.
4. The ventral ramus of C2 passing forward lateral to the vertebral artery and the dorsal ramus ascending behind Inferior Oblique.
5. The spinal cord having plenty of room at this high level.

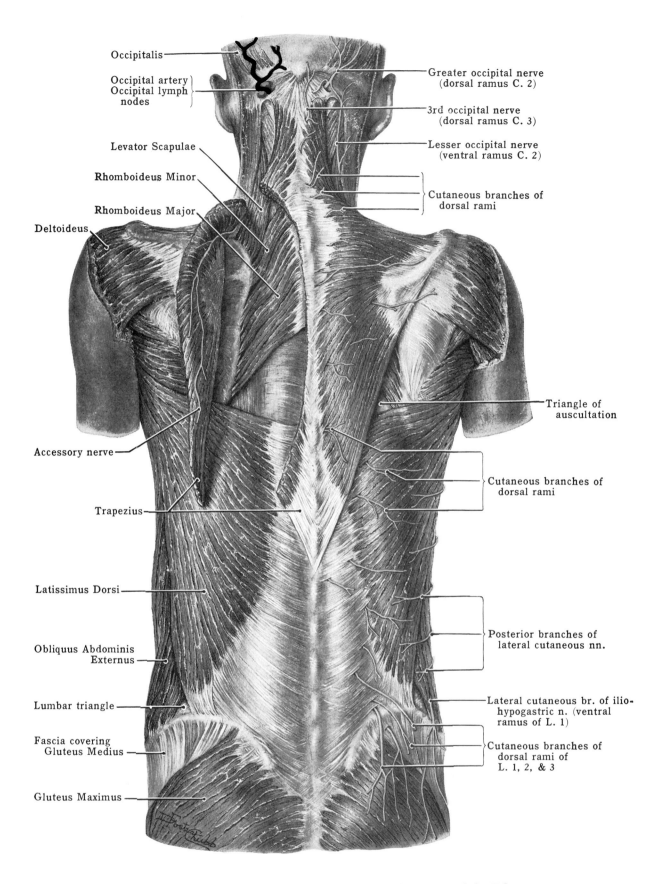

Occipitalis

Occipital artery
Occipital lymph
nodes

Levator Scapulae

Rhomboideus Minor

Rhomboideus Major

Deltoideus

Accessory nerve

Trapezius

Latissimus Dorsi

Obliquus Abdominis
Externus

Lumbar triangle

Fascia covering
Gluteus Medius

Gluteus Maximus

Greater occipital nerve
(dorsal ramus C. 2)

3rd occipital nerve
(dorsal ramus C. 3)

Lesser occipital nerve
(ventral ramus C. 2)

Cutaneous branches of
dorsal rami

Triangle of
auscultation

Cutaneous branches of
dorsal rami

Posterior branches of
lateral cutaneous nn.

Lateral cutaneous br. of ilio-
hypogastric n. (ventral
ramus of L. 1)

Cutaneous branches of
dorsal rami of
L. 1, 2, & 3

5-25A THE BACK—I: SUPERFICIAL MUSCLES

On the left side, Trapezius is reflected.

Note two layers:

1. Trapezius and Latissimus Dorsi.
2. Levator Scapulae and Rhomboidei Minor and Major. These muscles help to attach the upper limb to the trunk.

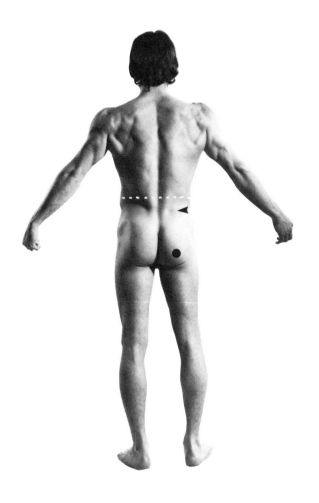

5-25B   THE BACK

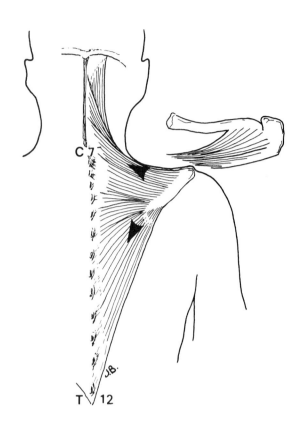

5-25C   TRAPEZIUS MUSCLE

Note:
1. The deep median furrow separating the lateral bulges of the erector spinae group of muscles.
2. The *arrow* points to the dimple overlying the posterior superior iliac spine.
3. A line (or towel) joining the highest points of the iliac crests is a guide to the fourth lumbar spine for the purpose of inserting a needle safely into the subarachnoid space below the termination of the spinal cord.
4. The *gluteal fold* is horizontal, crosses the oblique lower border of Gluteus maximus, marks the upper limit of the back of the thigh, marks the posterior flexion crease of the hip joint, and disappears when the hip is flexed.
5. The ischial tuberosity can be palpated 5 cm lateral to the midline and 5 cm above the gluteal fold.

Note:
1. The attachments of Trapezius:
   Origin: medial one-third of the superior nuchal line, the inion, the ligamentum nuchae, spinous processes of C7 to T12, and the intervening supra-spinous ligaments.
   Insertion (see *inset*): lateral one-third of clavicle, medial margin of the acromion, superior lip of the crest of the scapular spine and its tubercle.
2. The important role of Trapezius in forward rotation of the scapula when raising the arm: upper fibers raise the acromial end and lower fibers pull down the medial end of the scapular spine.
3. Trapezius shares with Sternomastoid the honor of being supplied by a cranial nerve: the spinal part of the accessory nerve (XI).

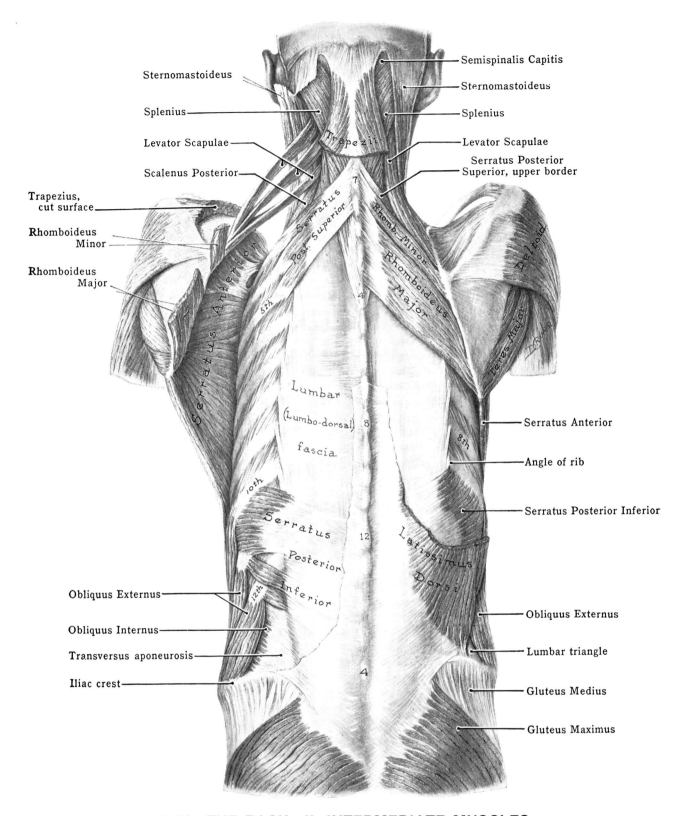

Semispinalis Capitis

Sternomastoideus

Sternomastoideus

Splenius

Splenius

Levator Scapulae

Levator Scapulae

Scalenus Posterior

Serratus Posterior
Superior, upper border

Trapezius,
cut surface

Serratus
Post. Superior

Trapezii

7

Rhomb. Minor

Rhomboideus
Major

Deltoid

Rhomboideus
Minor

Rhomboideus
Major

Serratus Anterior

Teres Major

Teres Minor

Lumbar
(Lumbo-dorsal)
fascia

5th

4

8

8th

Serratus Anterior

Angle of rib

10th

Serratus Posterior Inferior

Serratus

Posterior

Inferior

12

Latissimus

Dorsi

12th

Obliquus Externus

Obliquus Internus

Transversus aponeurosis

Iliac crest

4

Obliquus Externus

Lumbar triangle

Gluteus Medius

Gluteus Maximus

5-26 THE BACK — II: INTERMEDIATE MUSCLES

Trapezius and Latissimus Dorsi are largely cut away on both sides.

Observe:

1. On the *right side:* Levator Scapulae and Rhomboidei, *in situ.* Serratus Superior, rising above Rhomboideus Minor. It is likely to be divided when Rhomboidei are severed.
2. On the *left side:* Rhomboidei, severed and allowing the vertebral border of the scapula to part from the thoracic wall. The 3 (usually 4) digitations of Levator Scapulae.
3. Serrati Posteriores Superior and Inferior—the 3rd or intermediate layer of muscles—bridging the deep muscles, passing from spines to ribs, and sloping in opposite directions. These are muscles of inspiration.
4. The thoraco-lumbar (lumbo-dorsal) fascia, extending laterally to the angles of the ribs, becoming thin superiorly, passing deep to Serratus Superior, and reinforced inferiorly by Latissimus Dorsi and Serratus Inferior.

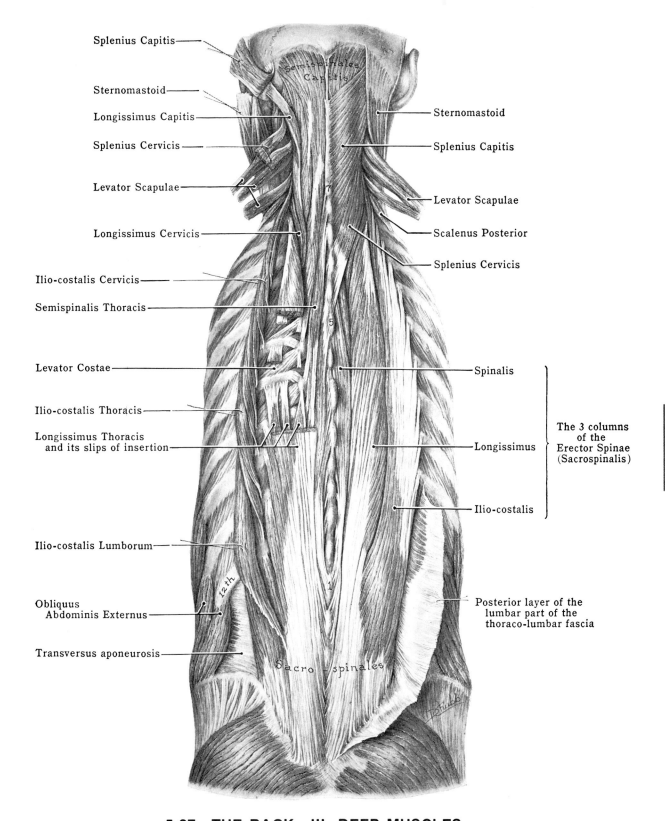

Splenius Capitis

Sternomastoid

Longissimus Capitis

Splenius Cervicis

Levator Scapulae

Longissimus Cervicis

Ilio-costalis Cervicis

Semispinalis Thoracis

Levator Costae

Ilio-costalis Thoracis

Longissimus Thoracis
and its slips of insertion

Ilio-costalis Lumborum

Obliquus
Abdominis Externus

Transversus aponeurosis

Sternomastoid

Splenius Capitis

Levator Scapulae

Scalenus Posterior

Splenius Cervicis

Spinalis

Longissimus

Ilio-costalis

The 3 columns
of the
Erector Spinae
(Sacrospinalis)

Posterior layer of the
lumbar part of the
thoraco-lumbar fascia

5-27 THE BACK—III: DEEP MUSCLES

Observe:

1. Splenius Capitis and Cervicis—the 4th layer of muscles—*in situ* on the *right side;* reflected on the *left side* and attached (Splenius Capitis) to the mastoid process deep to Sterno-mastoid and (Splenius Cervicis) to 1st, 2nd, (and 3rd) transverse processes deep to Levator Scapulae.

2. Erector Spinae—the 5th layer of muscles—*in situ* on the *right side,* lying between the spines medially and the angles of the ribs laterally, and splitting into 3 columns—lateral, middle, and medial. On the *left side,* the lateral column is everted, a section is taken from the middle column, the medial column is *in situ.* Only the middle or Longissimus column extends to the skull, and is there inserted into the mastoid process deep to Splenius Capitis.

3. Semispinalis (thoracis, cervicis, and capitis) *in situ* and belonging to the 6th or transverso-spinalis group of muscles.

4. For suboccipital muscles see Figures 5-32 to 5-36.

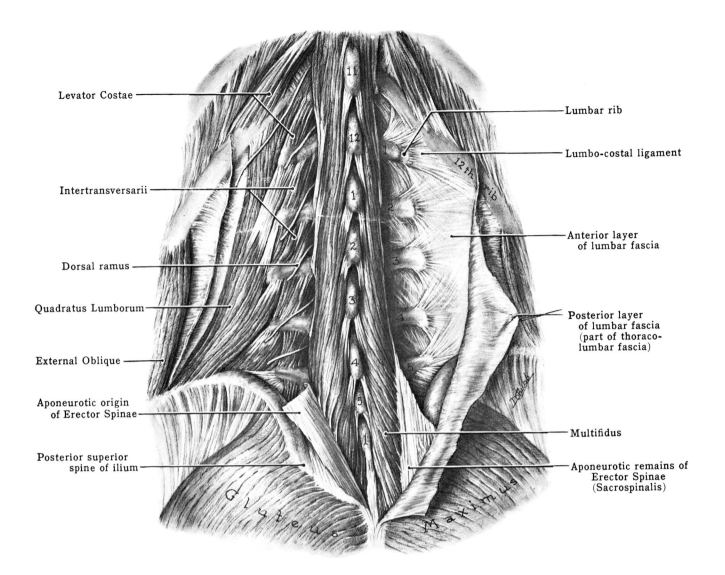

Levator Costae

Intertransversarii

Dorsal ramus

Quadratus Lumborum

External Oblique

Aponeurotic origin
of Erector Spinae

Posterior superior
spine of ilium

Lumbar rib

Lumbo-costal ligament

Anterior layer
of lumbar fascia

Posterior layer
of lumbar fascia
(part of thoraco-
lumbar fascia)

Multifidus

Aponeurotic remains of
Erector Spinae
(Sacrospinalis)

5-28 THE BACK—IV: MULTIFIDUS, QUADRATUS LUMBORUM, LUMBAR FASCIA

Semispinalis, Multifidus, and Rotatores constitute the transverso-spinalis group of deep muscles. In general, their bundles pass obliquely upward and medially, from transverse processes to spines, in successively deeper layers. The bundles of Semispinalis span about 5 interspaces, those of Multifidus about 3, and those of Rotatores, 1 or 2.

a. Semispinalis extends from the lower thoracic region to the skull (Figs. 5-27 and 5-33).

b. Multifidus extends from the sacrum to the spine of the axis (Fig. 5-36).

c. Rotatores are well developed only in the thoracic region (Fig. 5-31).

Observe:

1. Multifidus in the lumbo-sacral region, arising from the aponeurosis of Erector Spinae, from dorsum sacri (not seen), and from mammillary processes (Fig. 5-11), and inserted into spinous processes several segments higher up.

2. On the *right side,* after removal of Erector Spinae, the anterior layer of lumbar fascia (thoraco-lumbar) attached in a fan-shaped manner to the tips of transverse processes. Also, a short lumbar rib.

3. On the *left side,* after removal of the anterior layer of lumbar fascia, the lateral border of Quadratus Lumborum is oblique, and the medial border is in continuity with the Intertransversarii.

See Morris, J. M., Benner, G., and Lucas, D. B. (1962) An electromyographic study of the intrinsic muscles of the back in man. *J. Anat., 96:* 509.

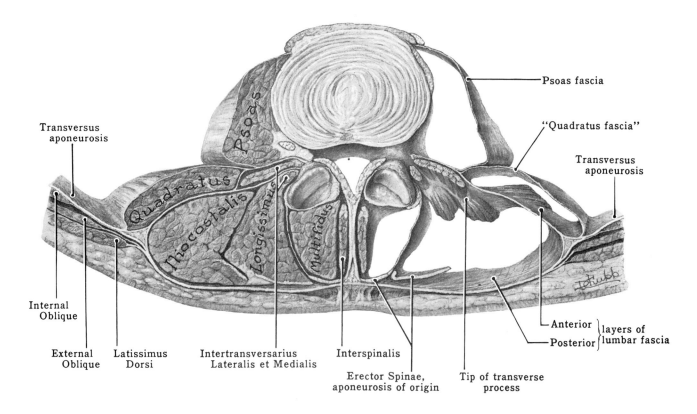

Transversus aponeurosis

Psoas fascia

"Quadratus fascia"

Transversus aponeurosis

Internal Oblique

External Oblique

Latissimus Dorsi

Intertransversarius Lateralis et Medialis

Interspinalis

Erector Spinae, aponeurosis of origin

Tip of transverse process

Anterior
Posterior } layers of lumbar fascia

5-29 MUSCLES OF THE BACK, ON CROSS-SECTION

On the *left side,* the muscles are seen within their sheaths or compartments. On the *right side,* the empty sheaths are shown.

Observe:

1. The posterior aponeurosis of Transversus Abdominis, splitting into two strong sheets — the anterior and the posterior layer of the lumbar fascia (being part of the thoraco-lumbar fascia) which enclose the deep muscles of the back.
2. The posterior layer, reinforced by Latissimus Dorsi and at a higher level (Fig. 5-26) by Serratus Posterior Inferior.
3. The weak areolar layer covering Quadratus Lumborum and that covering Psoas.
4. The ends of Intertransversarius, Longissimus, and Quadratus Lumborum, attached to a transverse process.

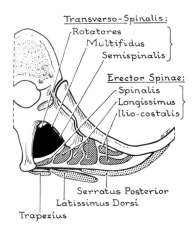

Transverso-Spinalis:
Rotatores
Multifidus
Semispinalis

Erector Spinae:
Spinalis
Longissimus
Ilio-costalis

Serratus Posterior
Latissimus Dorsi
Trapezius

5-30 BACK MUSCLES

This cross-section shows Erector Spinae in three columns and Transverso-spinalis in three layers.

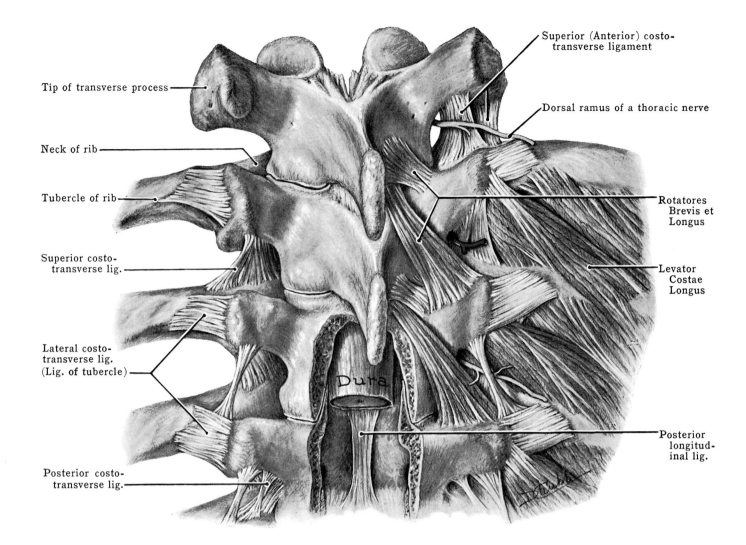

Tip of transverse process

Neck of rib

Tubercle of rib

Superior costo-transverse lig.

Lateral costo-transverse lig. (Lig. of tubercle)

Posterior costo-transverse lig.

Superior (Anterior) costo-transverse ligament

Dorsal ramus of a thoracic nerve

Rotatores Brevis et Longus

Levator Costae Longus

Posterior longitud-inal lig.

Dura

5-31A   ROTATORES AND THE COSTO-TRANSVERSE LIGAMENTS

Note:

1. Of the 3 layers of Transverso-spinalis or oblique muscles of the back—Semispinales, Multifidus, Rotatores—the Rotatores are the deepest and shortest. They pass from the root of one transverse process to the junction of the transverse process and lamina of the vertebra next above. Some (Rotatores Longi) pass to the vertebra two above.

2. Similarly, the Levatores Costarum pass from the tip of one transverse process to the rib next below. Some (Levatores Longi) pass to the rib 2 below.

3. Of the 3 sets of costo-transverse ligaments—superior, lateral, and medial—

   a. The superior ligament splits laterally into two sheets between which the medial border of a Levator Costae and of an External Intercostal are received. The dorsal ramus of a thoracic nerve passes behind this ligament and the ventral ramus (intercostal nerve) passes in front.

   b. The lateral ligament is strong and, if there were no joint cavity between transverse process and rib, it would be continuous with the medial ligament (ligament of the neck, Fig. 1-50).

   c. The medial ligament passes between the front of a transverse process and the back of the neck of its own rib. It is called "the" costo-transverse ligament. (A few fibers of little account, lying postero-medial to the superior ligament, constitute a posterior costo-transverse ligament.)

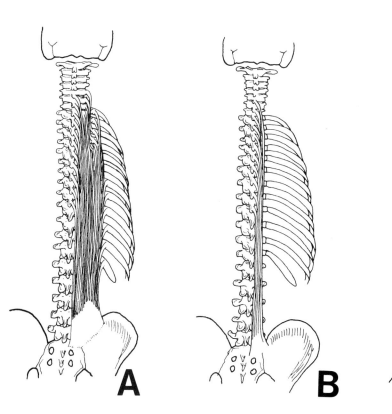
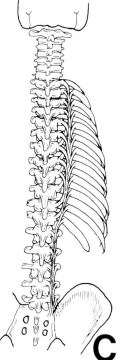
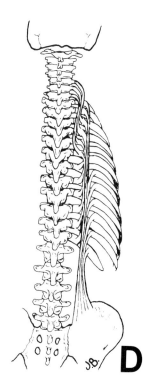

5-31B   ERECTOR SPINAE GROUP

5-31C   SIDE VIEW OF NECK

Flexing the neck renders even more prominent the spinous process of C7, *vertebra prominens*, the first visible spine. Above this, the ligamentum nuchae attaches to the spinous processes. Also well shown is the palpable (and visible) spinous process of scapula, acromion, and clavicle. Note the lateral end of the clavicle higher than the acromion at their articulation.

These muscles are complicated, inconstant, and confusing. They have a common inferior attachment to a strong tendon which attaches to sacrum, ilium, and lumbar spines. They are sandwiched by thoracolumbar fascia.

A. The *Erector Spinae* group of back muscles, extending one hand's breadth on either side of the midline, and divisible into three longitudinal columns (Fig. 5-27).

B. *Spinalis*, thinnest and most medial, running from lower to higher spinous processes, inconstantly extending as high as the neck or even the skull.

C. *Longissimus*, the intermediate column, inserting by twin slips into ribs and transverse processes. Not shown here are its extensions to neck and head, Longissimus cervicis and capitis, the latter inserting into the mastoid part of the temporal bone.

D. *Iliocostocervicalis*, the most lateral, consisting of three overlapping relays. From lateral to medial: *Iliocostalis lumborum*, which inserts on the inferior border of the lower six ribs; *Iliocostalis thoracis*, which runs from the upper border of the lower six ribs to the posterior angles of the upper six ribs; and *Iliocostalis cervicis*, which runs from the posterior angles of ribs 3 to 6 and inserts on the posterior tubercles of lower cervical vertebrae.

See Morris, J.M., Benner, G., and Lucas, D.B. (1962) An electromyographic study of the intrinsic muscles of the back in man. *J. Anat.*, 96: 509–520.

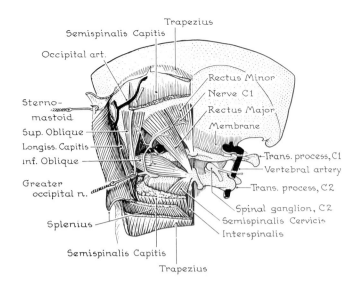

Note that the Suboccipital Region contains four pair of structures:

2 straight muscles: Rectus major and minor.

2 oblique muscles: Superior oblique and Inferior oblique.

2 nerves (Posterior Primary Rami):
C1 Suboccipital (Motor)
C2 Greater Occipital (Sensory)

2 arteries: Occipital and Vertebral

5-32   DIAGRAM OF THE SUBOCCIPITAL REGION

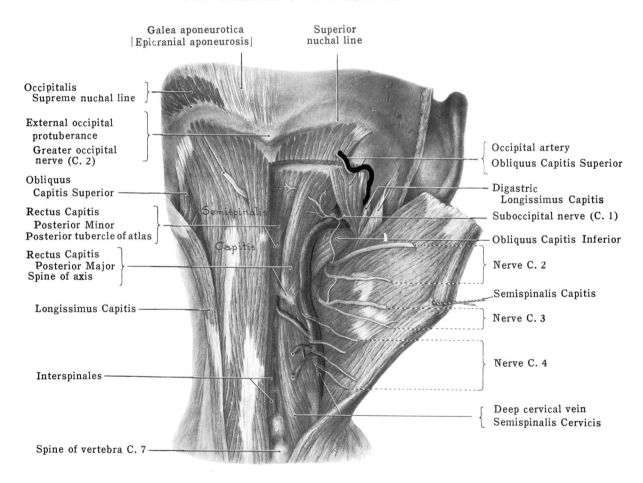

5-33— SUBOCCIPITAL REGION—I

Trapezius, Sternomastoid, and Splenius are removed.

Observe:

1. Semispinalis Capitis, the great extensor of the head and neck, forming the posterior wall of the suboccipital region, pierced by the greater occipital nerve (C2, posterior ramus), and having free medial and lateral borders at this high level. The right Semispinalis is divided and turned laterally.

2. The greater occipital nerve, when followed caudally, leading to the lower border of Obliquus Capitis Inferior around which it turns and to which it is the guide.

3. This border of Obliquus Inferior, followed medially, leading to spine of axis, and, followed laterally, to the transverse process of the atlas.

4. Five muscles (all paired) attached to the spine of the axis: Obliquus Capitis Inferior, Rectus Capitis Posterior Major, Semispinalis Cervicis which largely conceals Multifidus, and Interspinalis.

5. Occipital veins emerging through suboccipital triangle to join the deep cervical vein and, with it, the suboccipital nerve (C1, posterior ramus).

6. The suboccipital triangle bounded by 3 muscles: Obliquus Inferior, Obliquus Superior, and Rectus Major.

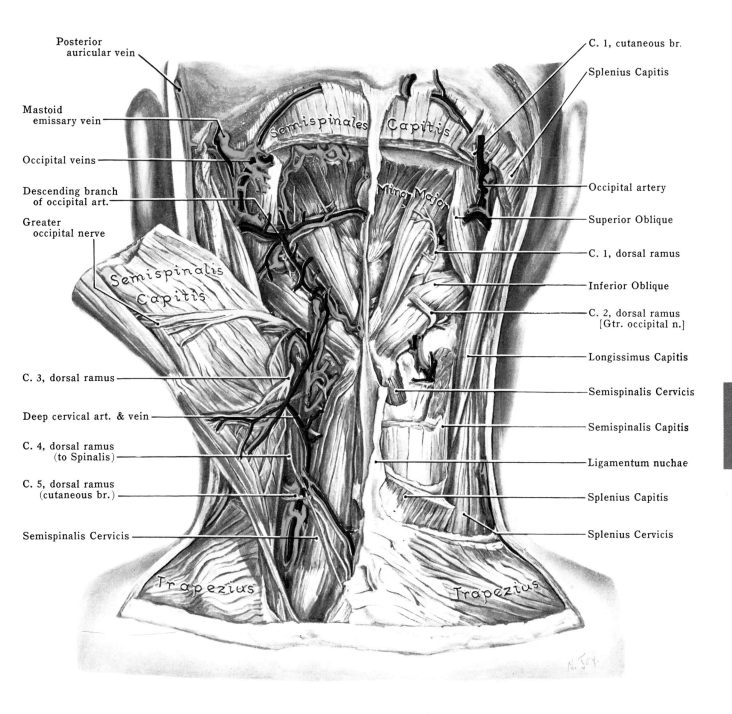

Posterior auricular vein

Mastoid emissary vein

Occipital veins

Descending branch of occipital art.

Greater occipital nerve

Semispinalis Capitis

C. 3, dorsal ramus

Deep cervical art. & vein

C. 4, dorsal ramus (to Spinalis)

C. 5, dorsal ramus (cutaneous br.)

Semispinalis Cervicis

Semispinales Capitis

Minor Major

Trapezius

C. 1, cutaneous br.

Splenius Capitis

Occipital artery

Superior Oblique

C. 1, dorsal ramus

Inferior Oblique

C. 2, dorsal ramus [Gtr. occipital n.]

Longissimus Capitis

Semispinalis Cervicis

Semispinalis Capitis

Ligamentum nuchae

Splenius Capitis

Splenius Cervicis

Trapezius

5-34 SUBOCCIPITAL REGION – II

Observe:

1. The ligamentum nuchae, which represents the cervical part of the supraspinous ligament, is a median, thin, fibrous partition attached to the spines of the cervical vertebrae and the external occipital crest. Its posterior border gives origin to Trapezius and extends upward to the inion or external occipital protuberance.

2. Rectus Capitis Posterior Minor (paired), the only muscle attached to the posterior tubercle of the atlas, which accordingly is upturned. (The atlas has no spine.)

3. The suboccipital nerve (C1, posterior ramus) supplying the 3 muscles bounding the suboccipital triangle, also Rectus Capitis Minor, and communicating with the greater occipital nerve.

4. The 1st cervical nerve here delivering a cutaneous branch, which is unusual.

5. The descending branch of the occipital artery anastomosing with the deep cervical artery, a branch of the subclavian.

6. The posterior vertebral venous plexus. This plexus is largely embedded in fascia, is usually empty and therefore inconsipcuous, and hence is removed unnoticed with the fascia unless specially injected, as here, or engorged with blood.

7. Longissimus Capitis being the only section of Erector Spinae to reach the skull.

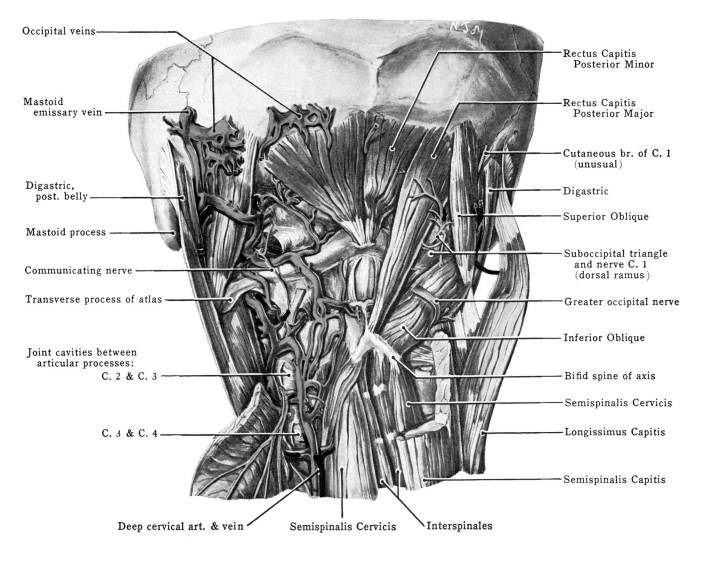

Occipital veins

Mastoid emissary vein

Digastric, post. belly

Mastoid process

Communicating nerve

Transverse process of atlas

Joint cavities between articular processes:
C. 2 & C. 3

C. 3 & C. 4

Deep cervical art. & vein   Semispinalis Cervicis   Interspinales

Rectus Capitis Posterior Minor

Rectus Capitis Posterior Major

Cutaneous br. of C. 1 (unusual)

Digastric

Superior Oblique

Suboccipital triangle and nerve C. 1 (dorsal ramus)

Greater occipital nerve

Inferior Oblique

Bifid spine of axis

Semispinalis Cervicis

Longissimus Capitis

Semispinalis Capitis

5-35 SUBOCCIPITAL REGION—III

On the *left side*, Rectus Capitis Posterior Major and Obliquus Capitis Inferior are removed.

Observe:

1. Rectus Capitis Posterior Major ascending from the spine of the axis to the occipital bone. Obliquus Capitis Superior ascending from the tip of the transverse process of the atlas to the occipital bone. Obliquus Capitis Inferior passing from spine of axis to tip of transverse process of atlas.

2. The foregoing 3 muscles (Inferior Oblique, Rectus Major, and Superior Oblique) forming the sides of the suboccipital triangle, which lies within the suboccipital region, whose lower limit is the axis.

3. Rectus Capitis Posterior Minor arising from the posterior tubercle of the atlas and therefore lying on a deeper plane than the Posterior Major, which arises from a spine.

4. The posterior arch of the atlas forming the floor of the suboccipital triangle. The posterior atlanto-occipital membrane (not labeled) passing from that arch to the margin of the foramen magnum above, and the posterior atlanto-axial membrane passing to the lamina of the axis below. The vertebral artery lying on the arch; the suboccipital nerve (dorsal ramus of C1) appearing between arch and artery and supplying the two straight muscles (minor and major) and the two oblique muscles (superior and inferior).

5. The gaps in these membranes through which pass nerve C1, the vertebral artery, the veins accompanying this artery, and nerve C2.

6. A branch connecting the dorsal rami of nerves C1 and C2 behind the posterior arch of the atlas.

In Figure 9-39 the ventral rami of these nerves are seen in communication.

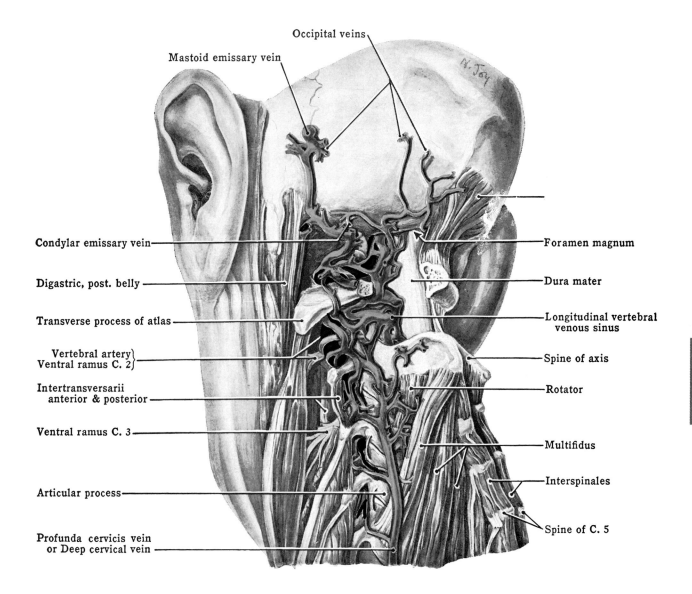

Occipital veins

Mastoid emissary vein

Condylar emissary vein

Digastric, post. belly

Transverse process of atlas

Vertebral artery
Ventral ramus C. 2

Intertransversarii
anterior & posterior

Ventral ramus C. 3

Articular process

Profunda cervicis vein
or Deep cervical vein

Foramen magnum

Dura mater

Longitudinal vertebral
venous sinus

Spine of axis

Rotator

Multifidus

Interspinales

Spine of C. 5

5-36 SUBOCCIPITAL REGION—IV

In this postero-lateral view, the posterior arch of the atlas, the atlanto-occipital, and atlanto-axial membranes have been removed.

Observe:

1. The vertebral venous system of veins and its numerous intercommunications and connections, *e.g.,* through the foramen magnum and the mastoid foramen and condylar canal with the intracranial venous sinuses; between the laminae and through the intervertebral foramina with the longitudinal vertebral venous sinuses (Figs. 5-23 and 5-24); communicating with the veins of the scalp above, with the veins around the vertebral artery and, via the deep cervical vein, with the brachio-cephalic vein below.

2. The Interspinales and the Multifidi extending up to, but not above, the spine of the axis.

Observe in this median section:
The layers of tissue encountered in the median plane. From behind forward they are: (a) dura mater, (b) posterior longitudinal ligament, continued upward as the membrana tectoria, stretching from axis to occipital bone, (c) transverse ligament of the atlas and the upper and lower longitudinal bands, which, like the membrana, stretch from axis to occipital bone and (d) the ligamentum apicis dentis (not labeled), which is a vestigial filament stretching from the apex of the dens to the occipital bone.

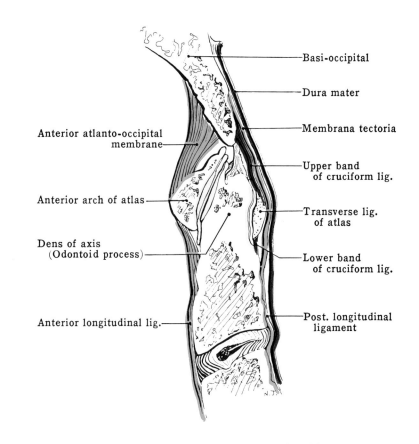

- Basi-occipital
- Dura mater
- Membrana tectoria
- Anterior atlanto-occipital membrane
- Upper band of cruciform lig.
- Anterior arch of atlas
- Transverse lig. of atlas
- Dens of axis (Odontoid process)
- Lower band of cruciform lig.
- Post. longitudinal ligament
- Anterior longitudinal lig.

5-37   LIGAMENTS OF THE ATLANTO-AXIAL AND ATLANTO-OCCIPITAL JOINTS

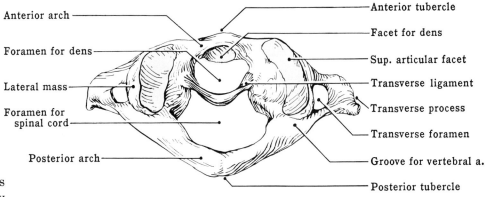

- Anterior arch
- Anterior tubercle
- Facet for dens
- Foramen for dens
- Sup. articular facet
- Lateral mass
- Transverse ligament
- Foramen for spinal cord
- Transverse process
- Transverse foramen
- Posterior arch
- Groove for vertebral a.
- Posterior tubercle

Observe
The large vertebral foramen of the atlas which is divided into two foramina by the transverse ligament of the atlas. In the larger, posterior foramen the spinal cord lies loosely. In the smaller, anterior foramen the dens of the axis fits tightly. It articulates in front with the anterior arch of the atlas and behind with the transverse ligament which, like the annular ligament of the radius, forms an arc of a circle.

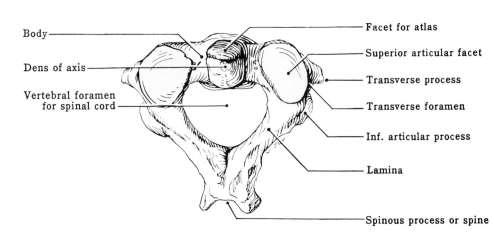

- Body
- Facet for atlas
- Superior articular facet
- Dens of axis
- Transverse process
- Vertebral foramen for spinal cord
- Transverse foramen
- Inf. articular process
- Lamina
- Spinous process or spine

5-38   ATLAS AND ITS TRANSVERSE LIGAMENT AND THE AXIS, FROM ABOVE

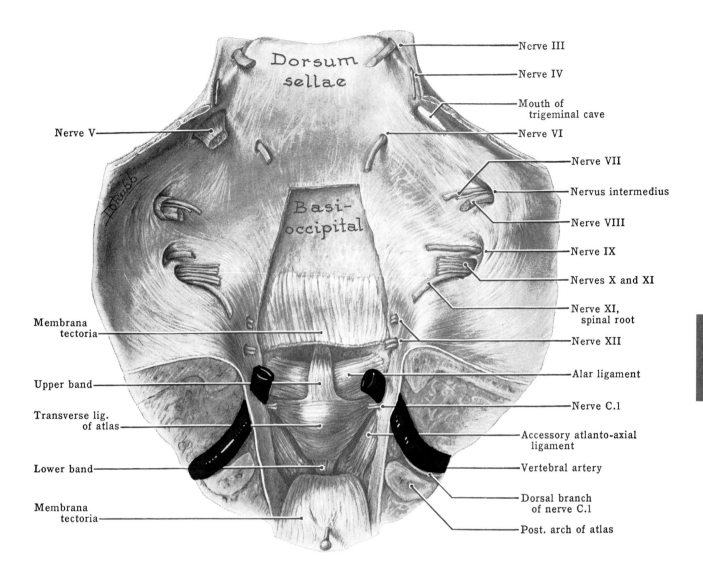

Labels on the figure:

Dorsum sellae

Basi-occipital

Nerve V

Membrana tectoria

Upper band

Transverse lig. of atlas

Lower band

Membrana tectoria

Nerve III

Nerve IV

Mouth of trigeminal cave

Nerve VI

Nerve VII

Nervus intermedius

Nerve VIII

Nerve IX

Nerves X and XI

Nerve XI, spinal root

Nerve XII

Alar ligament

Nerve C.1

Accessory atlanto-axial ligament

Vertebral artery

Dorsal branch of nerve C.1

Post. arch of atlas

5-39 CRANIO-VERTEBRAL JOINTS, DORSAL VIEW

Observe:
1. The bow-shaped transverse ligament of the atlas, which, by the addition of an upper and a lower longitudinal band, becomes a cruciform ligament.
2. The alar or check ligaments passing from the sides of the apex of the dens postero-laterally, above the transverse ligament, to the medial sides of the occipital condyles.
3. The sites where the last 10 pairs of cranial nerves and the first pair of cervical nerves pass through the dura, noting: (a) they are in numerical sequence, cranio-caudally, and (b) nerves III, IV, and VI, which supply the muscles of the eye, and XII, which supplies the muscles of the tongue, are nearly in vertical line with each other and with the ventral or motor root of C1.

5-40A NERVOUS SYSTEM

A photograph of a specimen (prepared by Mr. B. S. Jadon) in which the brain, spinal cord, and major nerves have been removed from a cadaver. It may be explored with a magnifying lens.

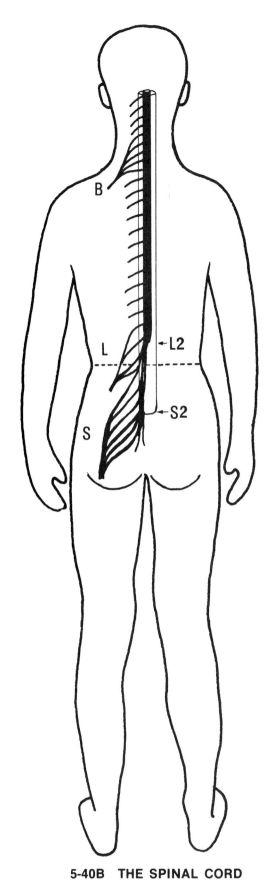

5-40B THE SPINAL CORD

A schematic outline of the spinal cord superimposed on a human figure. Observe:
1. Brachial (*B*), lumbar (*L*), and sacral (*S*) plexuses. Locate these in Figure 5-40*A*.
2. The spinal cord terminates as the conus medularis at the level of the second lumbar vertebra in the adult. For variations see Figure 5-49.
3. The subarachnoid space terminates at the level of S2. Between L2 and S2 the space contains the cauda equina, consisting of the lower nerve roots and the filum terminale. See Figure 5-48.
4. A *dotted line* joining the highest part of the iliac crests is a guide to the space between lumbar vertebrae 3 and 4 where a needle may enter the subarachnoid space without endangering the spinal cord.

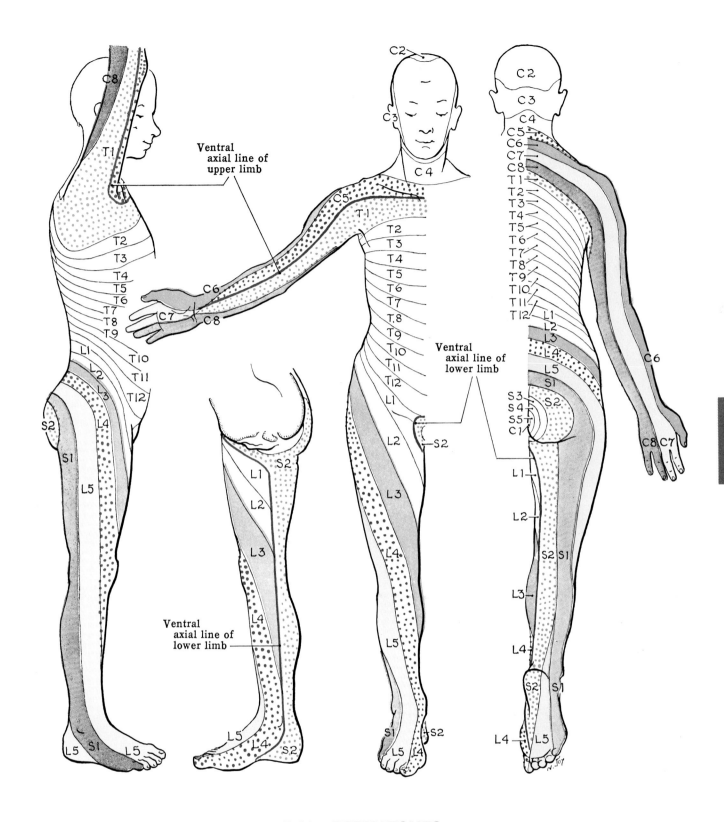

5-41 DERMATOMES

A dermatome is an area of skin supplied by the dorsal (sensory) root of a spinal nerve. In the head and trunk each segment is horizontally disposed with some surprises. C1 has no sensory component and there is a hiatus for the nerve roots of the brachial plexus. The dermatomes of the limbs from the 5th cervical to the 1st thoracic and from the 3rd lumbar to the 2nd sacral extend as a series of bands from the mid-dorsal line of the trunk into the limbs, as illustrated. Note that there is considerable overlapping of contiguous dermatomes; that is to say, each segmental nerve overlaps the territories of its neighbors. As a result, no anesthesia results unless two or more consecutive dorsal roots have lost their functions.

See Fender, F.A. (1939) Foerster's scheme of the dermatomes. *Arch. Neurol. Psychiat., 41*: 688; Keegan, J.J., and Garrett, F.D. (1948) The segmental distribution of the cutaneous nerves in the limbs of man. *Anat. Rec., 102*: 409.

For clinical application, see Barrows, H.S. (1980) *Guide to Neurological Assessment.* J.B. Lippincott, Philadelphia.

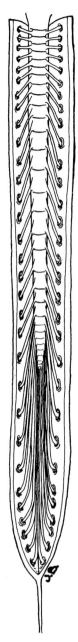

5-42A SUBARACHNOID SPACE

Earlier in development, spinal cord and spinal canal were more equal in length. The canal has grown longer and so spinal nerves have an increasingly longer course to reach the intervertebral foramen at the correct level for their exit. Descending the cord, the spinal nerves become increasingly oblique in the courses. The spinal cord proper terminates at the level of L2 and remaining spinal nerves seeking their portal of exit form the *cauda equina*. At the level of S2 the subarachnoid space ceases, thus spinal taps are done between L2 and S2. See Figures 5-48 and 5-49.

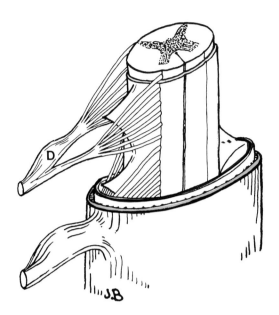

5-42B FORMATION OF SPINAL NERVES

Observe:

1. The cut edges of the three meningeal coverings of the cord have been colored for identification: *dura* mater (*blue*), *arachnoid* mater (*red*), and *pia* mater (*yellow*).
2. Cerebrospinal fluid circulates between pia and arachnoid, the *subarachnoid space.*
3. On each side, two rows of rootlets attach to the cord. The dorsal filaments carry *sensory* information to the central nervous system; the ventral row conveys *motor* enervation to muscles.
4. A number of rootlets combine to form at each segment *dorsal and ventral roots.*
5. The swollen area on the dorsal root, the *dorsal root ganglion (D)* contains cell bodies of sensory neurons.
6. Dorsal and ventral roots unite to form a spinal nerve.
7. Dura (and arachnoid) continues as a sheath around nerves leaving the spinal cord.
8. A row of *denticulate ligaments* continuous with the pia mater separate the rows of dorsal and ventral rootlets (Fig. 5-45).

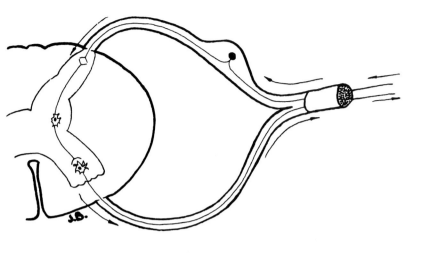

5-43   A REFLEX ARC

This diagram shows a simple, three-neuron reflex arc. The circuit may involve two, three, or more neurons. For a reflex to occur the following components must be intact:

1. A sensory *receptor*.
2. The first degree *sensory neuron* whose cell body resides in the dorsal root ganglion.
3. Transmission through the gray matter, in this case via an *interneuron*.
4. A *lower motor neuron* whose cell body is in the anterior horn of gray matter in the spinal cord.
5. A functioning *neuromuscular* junction and *muscle fiber*.

Interruption of any of these elements results in decreased or absent reflexes.

INTERIOR OF SPINAL CORD

This simple description of the interior of the spinal cord serves to link the gross anatomy portrayed in this atlas with neuroanatomy. All simplifications tend to distort life.

1. The spinal cord, seen in cross-section, consists of a central H-shaped core of gray matter (largely cell bodies) and peripheral white matter consisting of ascending and descending pathways connecting the peripheral nervous system with the brain.
2. Sensory (ascending) pathways consist of a chain of three neurons: (a) conducts the impulse from sensory ending to *spinal cord*; (b) from cord to *thalamus*; and (c) to sensory area of the opposite side of the *brain*. Three ascending columns in the cord convey different types of sensation: (a) *pain and temperature* sensation crosses immediately and ascends in the lateral spinothalamic tract of the cord; (b) *touch and pressure* sensation ascends a few

segments, crosses to the opposite side, and then travels in the anterior spinothalamic tract of the cord; (c) *proprioception, fine touch, and vibration* sense ascend in dorsal columns of white matter and cross to the opposite side in the brain stem.

3. Motor (descending) pathways consist of two neurons: (a) the *upper motor neuron* which begins in the motor cortex of the opposite side, crosses in the pyramid, descends in the corticospinal tract of the spinal cord, and synapses with (b) the *lower motor neuron* whose cell body lives in the anterior horn of gray matter in the spinal cord. Interruption of the lower motor neuron ("a lower motor neuron lesion") results in loss of reflexes. An "upper motor neuron lesion" does not interrupt the reflex arc.
4. There are other pathways in the cord, particularly those which connect the periphery with the cerebellum.

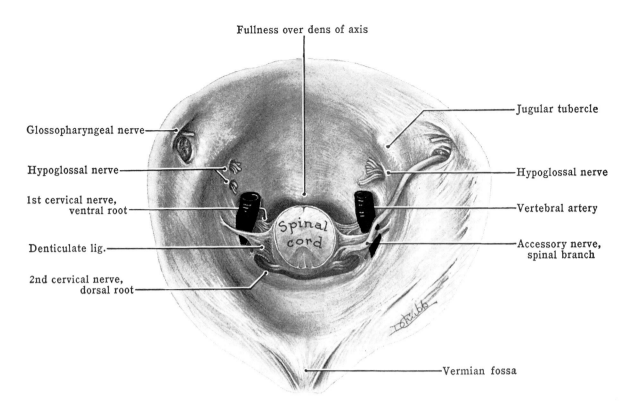

- Fullness over dens of axis
- Glossopharyngeal nerve
- Hypoglossal nerve
- 1st cervical nerve, ventral root
- Denticulate lig.
- 2nd cervical nerve, dorsal root
- Jugular tubercle
- Hypoglossal nerve
- Vertebral artery
- Accessory nerve, spinal branch
- Vermian fossa
- Spinal cord

5-44   STRUCTURES SEEN THROUGH THE FORAMEN MAGNUM, FROM ABOVE

Observe:
1. In front, the fullness over the transverse ligament of the atlas (Fig. 5-38), which curves tightly behind the dens of the axis.
2. Passing through the foramen magnum within the meninges: (a) the spinal cord or medulla, (b) the vertebral arteries, (c) the spinal roots of the accessory nerves (XI), and (d) the highest tooth of the ligamentum denticulatum of each side. (For lower teeth, Figs. 5-45 and 5-48.)
3. The hypoglossal nerves (XII) leaving the dura mater through two openings which are close together on the right side and separated on the left side.
4. In this specimen the first cervical nerve has no posterior (sensory) root.

Observe:
1. The denticulate ligament, running like a band along each side of the spinal cord and, by means of strong toothlike processes, anchoring the cord to the dura between successive nerve roots. (For highest tooth, see Fig. 5-44).
2. The ventral nerve roots, lying in front of the denticulate ligament and the dorsal nerve roots lying behind it.
3. The ventral and the dorsal root of each nerve, leaving the dura by a separate opening. The lowest right dorsal root in this specimen leaves by three openings.
4. The fila of the various dorsal roots, having a linear attachment to the cord.
5. One filum of the lowest left dorsal root, deserting its own root and joining the root above.

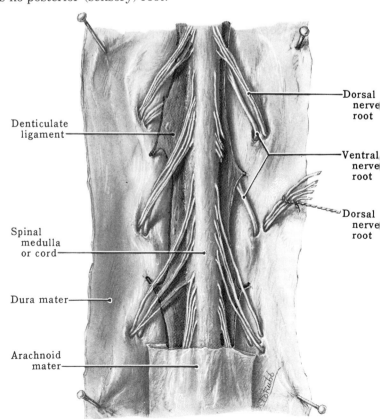

- Denticulate ligament
- Spinal medulla or cord
- Dura mater
- Arachnoid mater
- Dorsal nerve root
- Ventral nerve root
- Dorsal nerve root

5-45   SPINAL CORD WITHIN ITS MEMBRANES, FROM BEHIN

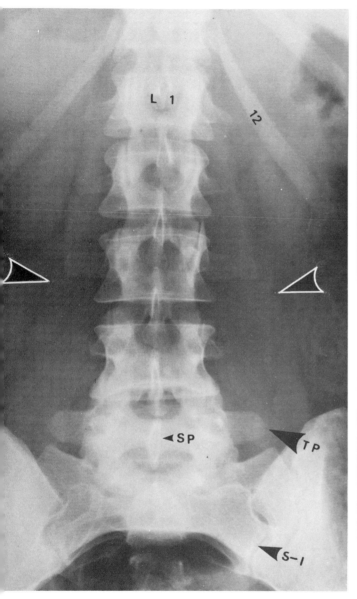

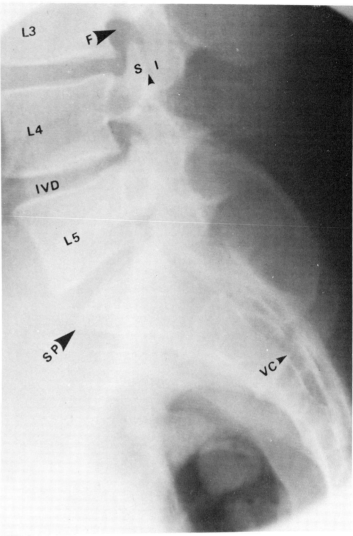

5-46   RADIOGRAPHS OF LUMBO-SACRAL SPINE

A. In this antero-posterior view of the spine observe:
1. The articulation of the last (12) "floating" rib with the last thoracic vertebra.
2. The bodies and processes of the 5 lumbar vertebrae. The spinous process (SP) and transverse process (TP) of L5 are labeled.
3. The sinuous sacro-iliac joint (S-I).
4. *Large arrows* point to the lateral margin of right and left Psoas muscles.

B. In this lateral view of the lumbo-sacral region observe:
1. The last 3 lumbar vertebrae.
2. The spaces for intervertebral discs. The space between L4 and L5 is marked (IVD).
3. The angulation at the lumbo-sacral junction producing the sacral promontory (SP).
4. An *arrow* points to the joint between the superior articular process of L4 (S) and the inferior articular process of L3 (I).
5. A *small arrow* points to the anterior margin of the vertebral canal (VC); a *large arrow* points to an intervertebral foramen (F).

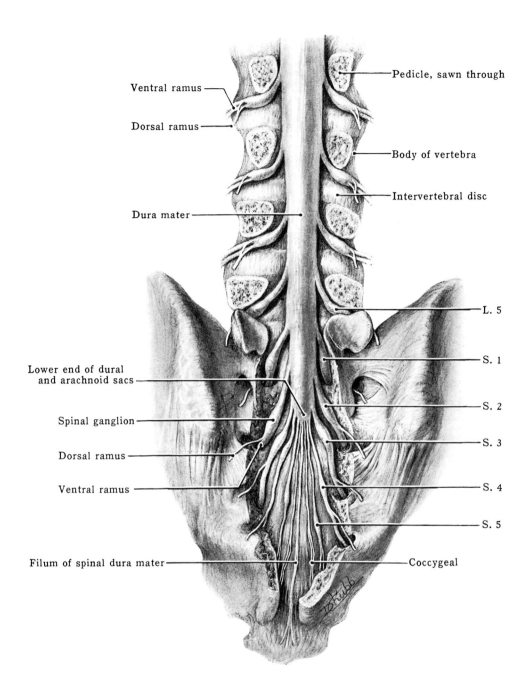

Ventral ramus

Dorsal ramus

Dura mater

Lower end of dural
and arachnoid sacs

Spinal ganglion

Dorsal ramus

Ventral ramus

Filum of spinal dura mater

Pedicle, sawn through

Body of vertebra

Intervertebral disc

L. 5

S. 1

S. 2

S. 3

S. 4

S. 5

Coccygeal

5-47 LOWER END OF THE DURAL SAC FROM BEHIND — I

The posterior parts of the lumbar and sacral vertebrae are sawn and nibbled
away.

Observe:

1. The lower limit of the dural (and the contained arachnoid) sac, at the level of
   the posterior superior iliac spine (=body of 2nd sacral vertebra), and the
   continuation of the dura as the filum of spinal dura mater.
2. The lumbar spinal ganglia in the intervertebral foramina; the sacral spinal
   ganglia, somewhat asymmetrical, within the sacral canal.
3. The dorsal nerve rami, smaller than ventral rami, and having both efferent
   and afferent components.
4. The superior articular processes of the sacrum, asymmetrical.

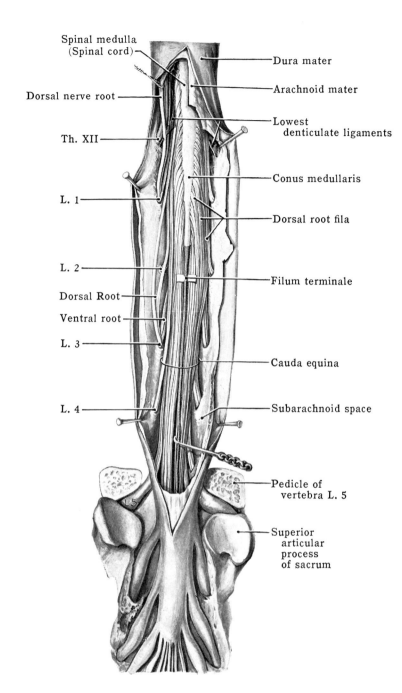

Spinal medulla (Spinal cord)

Dorsal nerve root

Th. XII

L. 1

L. 2

Dorsal Root

Ventral root

L. 3

L. 4

Dura mater

Arachnoid mater

Lowest denticulate ligaments

Conus medullaris

Dorsal root fila

Filum terminale

Cauda equina

Subarachnoid space

Pedicle of vertebra L. 5

Superior articular process of sacrum

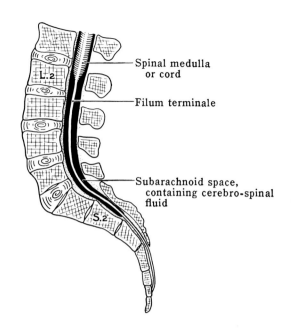

Spinal medulla or cord

Filum terminale

Subarachnoid space, containing cerebro-spinal fluid

L. 2

S. 2

5-49 SPINAL CORD *IN SITU*

Note:
1. The spinal cord ends at the level of the disc between the 1st and 2nd lumbar vertebrae.
2. The subarachnoid space ends at the level of the disc between the 1st and 2nd sacral vertebrae, but it may be lower (Figs. 3-10 and 3-11).
3. Variations: 95 per cent of cords end within the limits of the bodies of vertebrae L1 and L2, whereas 3 per cent end behind the lower half of vertebra T12, and 2 per cent behind vertebra L3.

See Jit, I., and Charnakia, V. M. (1959) The vertebral level of the termination of the spinal cord. *J. Anat. Soc. India, 8:* 93; Reimann, A. F., and Anson, B. J. (1944) Vertebral level of termination of the spinal cord with a report of a case of a sacral cord. *Anat. Rec., 88:* 127.

5-48   LOWER END OF THE DURAL SAC, FROM BEHIND—II

Observe:
1. The lowest tooth or dens of the denticulate ligament, variable in level and asymmetrical.
2. A radicular branch of a spinal vein accompanying the dorsal root of nerve L1. There are only four or five radicular veins and arteries on each side of the cord to accompany the 31 pairs of spinal nerves.

See Suh, T. H., and Alexander, L. (1939) Vascular system of the human spinal cord (radicular vessels). *Arch. Neurol. Psychiat., 41:* 659.

3. The conus medullaris, or conical lower end of the spinal medulla, continued as a glistening thread, the filum terminale, which descends with the dorsal and ventral nerve roots. These constitute the cauda equina.
4. The hook retracting the elongated nerve roots, which surround a tubular space.
5. The subarachnoid space enclosed by arachnoid mater. The subdural space, which is the potential space between the dural and arachnoid maters.

THORACO—LUMBAR OUTFLOW

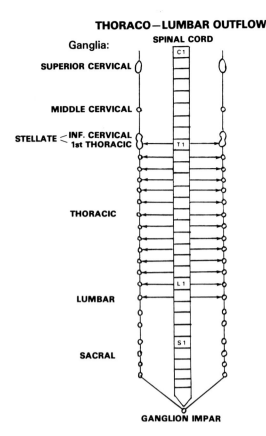

SPINAL CORD

Ganglia:

SUPERIOR CERVICAL

MIDDLE CERVICAL

STELLATE ⟨ INF. CERVICAL / 1st THORACIC

THORACIC

LUMBAR

SACRAL

GANGLION IMPAR

5-50A   SYMPATHETIC TRUNK

The sympathetic trunks run from the base of the skull to the front of the coccyx where they meet at the ganglion impar. Ganglia, containing cell bodies, vary in number but this diagram shows the usual distribution. Input, via *white* rami communicantes, is provided by spinal nerves T1 to L2(3).

See: Wrete, M. (1959) The anatomy of the sympathetic trunks in man. *J. Anat., 93:* 448.

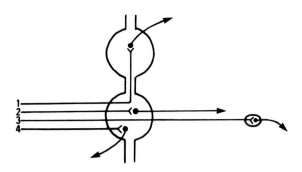

5-50B   SYMPATHETIC GANGLIA

A nerve fiber entering a ganglion of the sympathetic trunk (as a *white* ramus) may follow one of four routes. It may:

1. Travel up (or down) the trunk to synapse at another level and supply a segment without its own sympathetic input (above T1 and below L2).
2. Synapse in the ganglion and exit as a *gray* ramus to supply viscera.
3. Pass through the ganglion to a "prevertebral" ganglion such as the celiac where it synapses.
4. Synapse in the ganglion and rejoin its own segmental nerve as a *gray* ramus.

CRANIO—SACRAL OUTFLOW

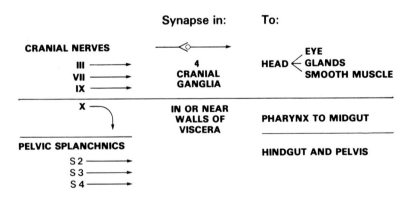

	Synapse in:	To:
CRANIAL NERVES		
III	4 CRANIAL GANGLIA	HEAD ⟨ EYE / GLANDS / SMOOTH MUSCLE
VII		
IX		
X	IN OR NEAR WALLS OF VISCERA	PHARYNX TO MIDGUT
PELVIC SPLANCHNICS		HINDGUT AND PELVIS
S 2		
S 3		
S 4		

5-50C   PARASYMPATHETIC SYSTEM

Parasympathetic fibers are derived from the cranio-sacral outflow. The head is supplied by fibers in cranial nerves III, VII, and IX which synapse in four cranial ganglia (ciliary, pterygopalatine, submandibular, and otic). The vagus nerve (X) in its wanderings provides parasympathetic supply to the viscera of neck, thorax, foregut, and midgut. The pelvic splanchnic (visceral) nerves take over to supply hindgut and pelvic viscera. Below the head, parasympathetic fibers synapse in or near the walls of viscera.

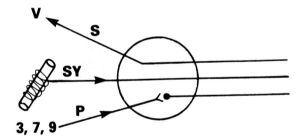

V   S

SY

P

3, 7, 9

5-50D   CRANIAL GANGLIA

Note:

1. Each of the four cranial ganglia receives three types of fibers: *S*, sensory branches of the trigeminal (*V*) nerve; *SY*, sympathetic fibers "hitchhiking" on the walls of the nearest artery; *P*, parasympathetic fibers.
2. Only parasympathetic fibers synapse in the ganglion; other fibers are just passing through to be distributed in a mixed nerve.
3. Parasympathetic supply comes from three cranial nerves: oculomotor (III) to the ciliary ganglion, facial (VII) to the pterygopalatine and submandibular ganglia, and glossopharyngeal (IX) to the otic ganglion.

Consult Figures 7-54 and 7-64 (ciliary ganglion), 7-94 and 7-109 (pterygopalatine), 9-23 (submandibular), and 7-89 (otic).

CONTENTS

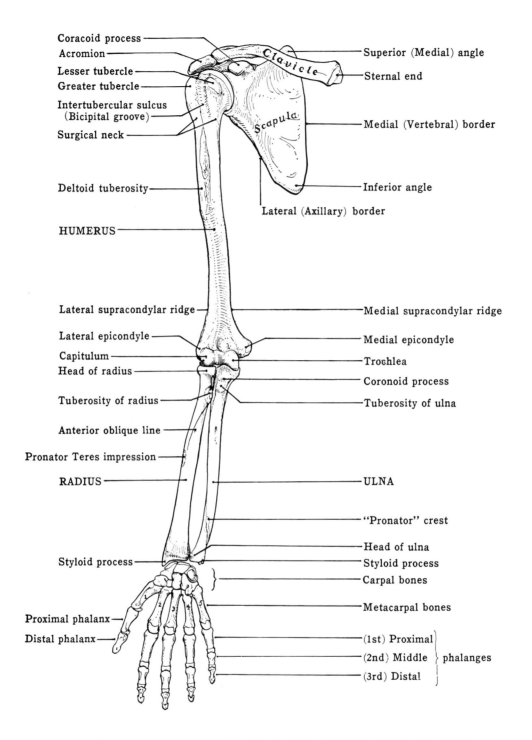

Coracoid process
Acromion
Lesser tubercle
Greater tubercle
Intertubercular sulcus (Bicipital groove)
Surgical neck
Deltoid tuberosity
HUMERUS
Lateral supracondylar ridge
Lateral epicondyle
Capitulum
Head of radius
Tuberosity of radius
Anterior oblique line
Pronator Teres impression
RADIUS
Styloid process
Proximal phalanx
Distal phalanx

Clavicle
Scapula

Superior (Medial) angle
Sternal end
Medial (Vertebral) border
Inferior angle
Lateral (Axillary) border
Medial supracondylar ridge
Medial epicondyle
Trochlea
Coronoid process
Tuberosity of ulna
ULNA
"Pronator" crest
Head of ulna
Styloid process
Carpal bones
Metacarpal bones
(1st) Proximal
(2nd) Middle   } phalanges
(3rd) Distal

6-1 BONES OF THE UPPER LIMB, FROM THE FRONT

For bones of the hand, see Figures 6-97 and 6-114.

For muscle attachments, see Figures 6-10, 6-35, and 6-65.

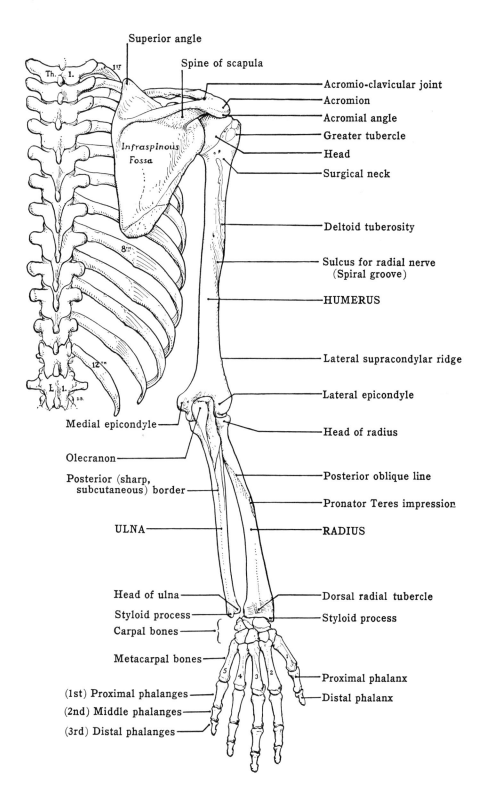

Superior angle

Spine of scapula

Acromio-clavicular joint

Acromion

Acromial angle

Greater tubercle

Head

Surgical neck

Infraspinous Fossa

Deltoid tuberosity

Sulcus for radial nerve (Spiral groove)

HUMERUS

Lateral supracondylar ridge

Lateral epicondyle

Medial epicondyle

Head of radius

Olecranon

Posterior oblique line

Posterior (sharp, subcutaneous) border

Pronator Teres impression

ULNA

RADIUS

Head of ulna

Dorsal radial tubercle

Styloid process

Styloid process

Carpal bones

Metacarpal bones

Proximal phalanx

Distal phalanx

(1st) Proximal phalanges

(2nd) Middle phalanges

(3rd) Distal phalanges

6-2 BONES OF THE UPPER LIMB, FROM BEHIND

For bones of the hand, see Figure 6-97.

For muscle attachments, see Figures 6-36 and 6-90.

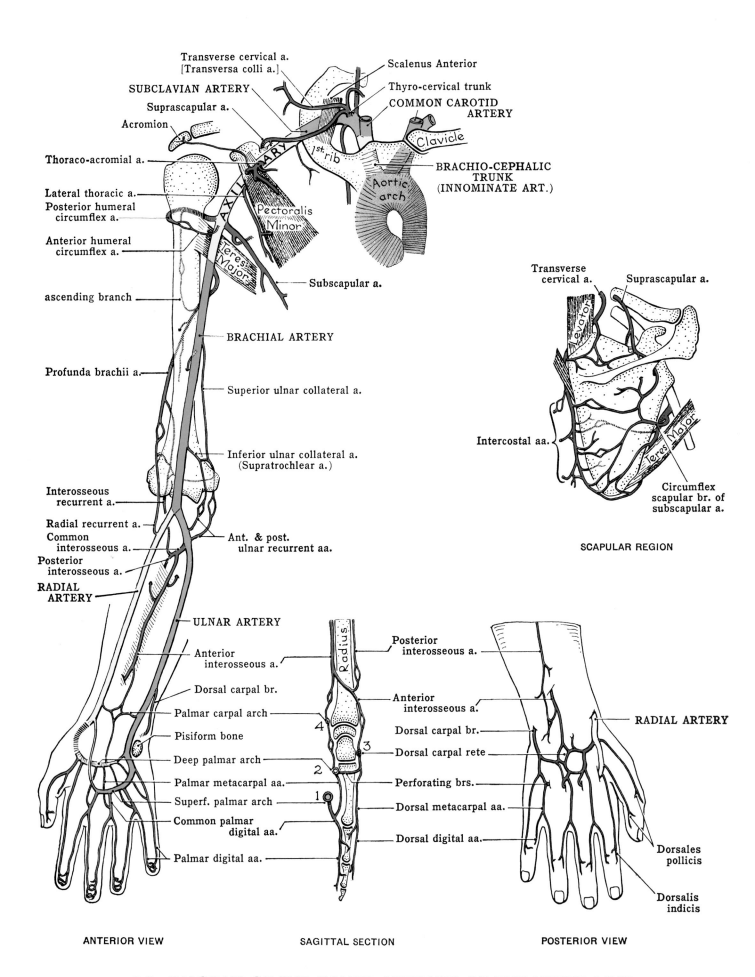

Transverse cervical a.
[Transversa colli a.]

Scalenus Anterior

SUBCLAVIAN ARTERY

Thyro-cervical trunk

Suprascapular a.

COMMON CAROTID ARTERY

Acromion

Clavicle

Thoraco-acromial a.

BRACHIO-CEPHALIC TRUNK (INNOMINATE ART.)

1st rib

Aortic arch

Lateral thoracic a.
Posterior humeral circumflex a.

Pectoralis Minor

Anterior humeral circumflex a.

Teres Major

ascending branch

Subscapular a.

BRACHIAL ARTERY

Profunda brachii a.

Superior ulnar collateral a.

Transverse cervical a.

Suprascapular a.

Levator

Inferior ulnar collateral a.
(Supratrochlear a.)

Intercostal aa.

Teres Major

Interosseous recurrent a.

Circumflex scapular br. of subscapular a.

Radial recurrent a.
Common interosseous a.

Ant. & post. ulnar recurrent aa.

Posterior interosseous a.

SCAPULAR REGION

RADIAL ARTERY

ULNAR ARTERY

Radius

Posterior interosseous a.

Anterior interosseous a.

Anterior interosseous a.

Dorsal carpal br.

4

Palmar carpal arch

Dorsal carpal br.

Pisiform bone

3

Dorsal carpal rete

RADIAL ARTERY

Deep palmar arch

2

Palmar metacarpal aa.

Perforating brs.

Superf. palmar arch

1

Dorsal metacarpal aa.

Common palmar digital aa.

Dorsal digital aa.

Dorsales pollicis

Palmar digital aa.

Dorsalis indicis

ANTERIOR VIEW

SAGITTAL SECTION

POSTERIOR VIEW

6-3   DIAGRAM OF THE NAMED ARTERIES OF THE UPPER LIMB

LIST OF NAMED ARTERIES
OF THE UPPER LIMB

SUBCLAVIAN ARTERY
 Thyro-cervical trunk
  Transverse cervical (colli) artery
   Superficial branch
   Deep branch
  Suprascapular artery
   Acromial branch
AXILLARY ARTERY
 Thoracica suprema (superior thoracic artery)
 Thoraco-acromial artery (acromio-thoracic artery)
  Pectoral branch
  Deltoid branch
  Acromial branch
 Lateral thoracic artery
  Lateral mammary branches
 Subscapular artery
  Circumflex scapular artery
  Thoraco-dorsal artery
 Posterior humeral circumflex artery
  Acromial branch
  Descending branch
 Anterior humeral circumflex artery
BRACHIAL ARTERY
 Profunda brachii artery
  Ascending branch
  Nutrient branch to humerus
  "Terminal descending branches"
  Middle collateral artery
  Radial collateral artery
 Superior ulnar collateral artery
 Nutrient branch to humerus
 Inferior ulnar collateral artery
ULNAR ARTERY
 Anterior ulnar recurrent artery
 Posterior ulnar recurrent artery
 Common interosseous artery
  Anterior interosseous artery
   Median artery
   Nutrient branch to radius
   Nutrient branch to ulna
  Posterior interosseous artery
   Interosseous recurrent artery
 Palmar carpal branch
 Dorsal carpal branch and rete
 Deep palmar branch
 Superficial palmar arch
  Common palmar digital arteries
  Proper palmar digital arteries
RADIAL ARTERY
 Radial recurrent artery
 Palmar carpal branch
 Superficial palmar branch
 Dorsal carpal branch
  Dorsal metacarpal arteries
  Dorsal digital arteries
 Princeps policis artery
 Radialis indicis artery
 Deep palmar arch
  Palmar metacarpal arteries
   Perforating branches

ARTERIES OF THE UPPER LIMB

The stem artery of the upper limb is the subclavian artery. The right subclavian artery springs from the brachio-cephalic trunk (innominate artery) behind the right sterno-clavicular joint, whereas the left subclavian artery springs directly from the aortic arch and ascends behind the left sterno-clavicular joint. From this point onward the arteries of the two sides are symmetrical.

The *subclavian* artery arches over the apex of the lung and pleura, rising about 2.5 cm above the clavicle, and leaves the root of the neck at the lateral border of the 1st rib to enter the axilla, as the *axillary* artery. The axillary artery leaves the axilla at the lower border of the Teres Major to enter the arm or brachium as the *brachial artery.* About 2.5 cm below the crease of the elbow the brachial artery bifurcates into the *radial* and *ulnar* arteries, which traverse the forearm or antebrachium and enter the palm where each ends as an arterial arch. The ulnar artery forms the superficial palmar arch, which descends to the level of the web of the thumb where it is completed by a slender branch of the radial artery—commonly its superficial palmar branch. The radial artery, after crossing the floor of the "snuff box" (the hollow at the root of the thumb), to reach the dorsum of the hand, passes through the 1st intermetacarpal space and so enters the palm. It there forms the deep palmar arch, which lies 1 cm proximal to the superficial palmar arch. The deep palmar arch is completed by a slender artery, the deep palmar branch of the ulnar artery.

Subdivisions: The subclavian artery is divided into 3 unequal parts by the Scalenus Anterior, and the axillary artery is similarly divided by the Pectoralis Minor.

Relationship to bone: The axillary artery passes within a finger's breadth of the tip of the coracoid process. The brachial artery lies medial to the humerus proximally and anterior to it distally. At the wrist the ulnar artery (and nerve) is sheltered from injury by the pisiform bone; its deep palmar branch curves round the lower border of the hamate bone. The radial artery is identified at the wrist by its pulse, which beats against the lower end of the radius.

Branches: The limbs are organs of locomotion and prehension, and the muscles are the motors that move them. Hence, most of the blood delivered to the limbs is for the supply of the muscles and little for the skin, fasciae, bones, and joint. The branches of the arteries are numerous, and they are for the most part muscular and nameless. Most of the named branches in the limbs are muscular too, but their claim to recognition depends mainly on their size and their surgically important anastomoses.

Anastomoses: (a) It should be safe to tie either the subclavian or the axillary artery between the thyro-cervical trunk and the subscapular artery because of the anastomoses around the scapula; (b) it should be safe to tie the brachial artery distal to the inferior ulnar collateral artery because of the anastomoses around the elbow; (c) and it should be safe to tie either the radial or the ulnar artery in the forearm because these arteries are united by 2 palmar arches and by 2 carpal arches.

Arches: The superficial palmar arch lies deep only to skin and palmar aponeurosis—it is well-named. The three other arches lie on the skeletal plane, that is to say, on bones, joints, ligaments, or interosseous membranes. Their order of magnitude (1, 2, 3, 4) is indicated. The deep palmar arch not only unites the radial and ulnar arteries, but it anastomoses with the interosseous arteries of the forearm, of the dorsum of the hand, and of the digits.

Caliber: The caliber of the axillary artery diminishes considerably just beyond the origin of the subscapular and posterior circumflex arteries, which are large arteries. The caliber of the ulnar artery is reduced to that of the radial artery beyond the origin of the common interosseous artery.

The anterior interior interosseous artery does duty for the posterior interosseous artery in the distal half of the back of the forearm.

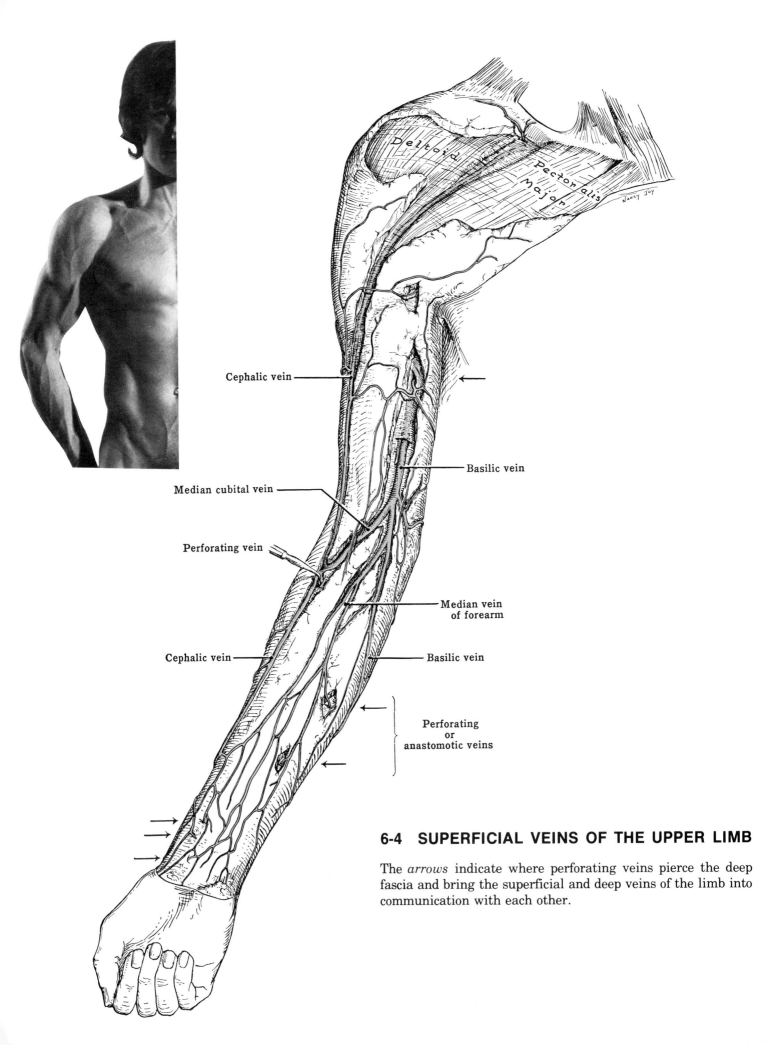

Cephalic vein

Median cubital vein

Perforating vein

Cephalic vein

Deltoid

Pectoralis Major

Nancy Joy

Basilic vein

Median vein
of forearm

Basilic vein

Perforating
or
anastomotic veins

6-4 SUPERFICIAL VEINS OF THE UPPER LIMB

The *arrows* indicate where perforating veins pierce the deep
fascia and bring the superficial and deep veins of the limb into
communication with each other.

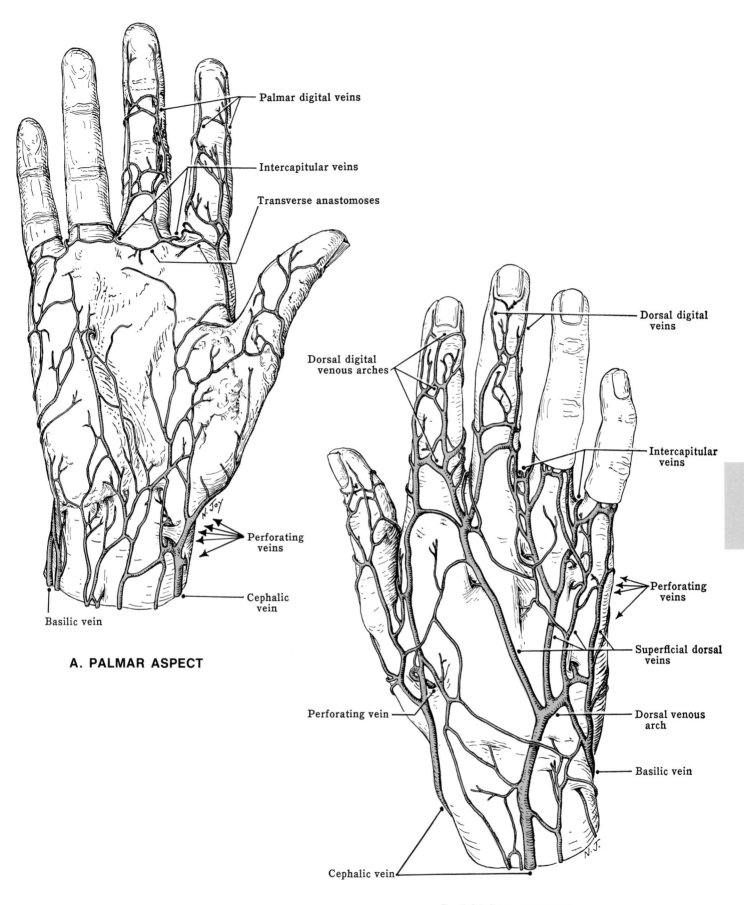

Palmar digital veins

Intercapitular veins

Transverse anastomoses

Perforating veins

Cephalic vein

Basilic vein

A. PALMAR ASPECT

Dorsal digital veins

Dorsal digital venous arches

Intercapitular veins

Perforating veins

Superficial dorsal veins

Perforating vein

Dorsal venous arch

Basilic vein

Cephalic vein

B. DORSAL ASPECT

6-5 SUPERFICIAL VEINS OF THE HAND

For obvious mechanical reasons the palmar veins are few and small, and the dorsal veins are large.

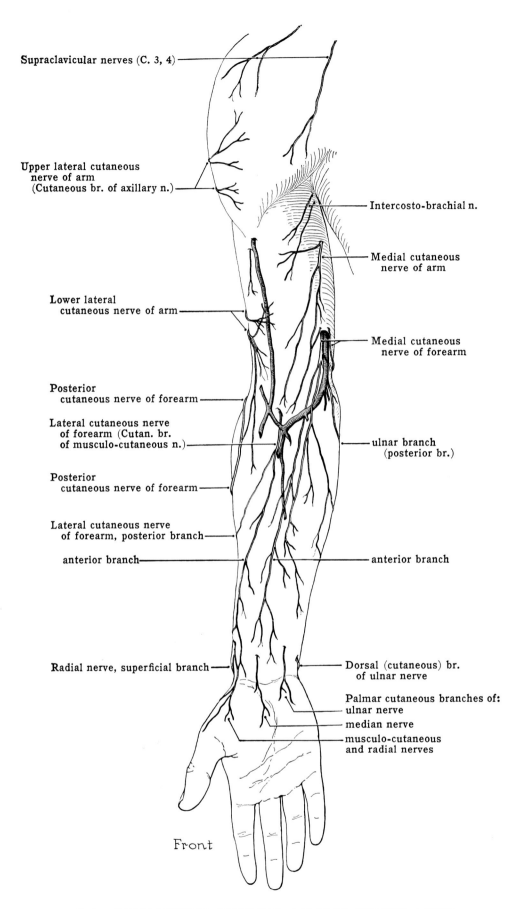

Supraclavicular nerves (C. 3, 4)

Upper lateral cutaneous
nerve of arm
(Cutaneous br. of axillary n.)

Intercosto-brachial n.

Medial cutaneous
nerve of arm

Lower lateral
cutaneous nerve of arm

Medial cutaneous
nerve of forearm

Posterior
cutaneous nerve of forearm

Lateral cutaneous nerve
of forearm (Cutan. br.
of musculo-cutaneous n.)

ulnar branch
(posterior br.)

Posterior
cutaneous nerve of forearm

Lateral cutaneous nerve
of forearm, posterior branch

anterior branch

anterior branch

Radial nerve, superficial branch

Dorsal (cutaneous) br.
of ulnar nerve

Palmar cutaneous branches of:
ulnar nerve
median nerve
musculo-cutaneous
and radial nerves

Front

6-6 CUTANEOUS NERVES OF THE UPPER LIMB

Of the 5 terminal branches of the brachial plexus (Fig. 6-24) — musculo-cutaneous, median, ulnar, radial, and axillary nerves — the first 4 contribute cutaneous branches to the hand. See Figures 6-14 (pectoral region), 6-30 (back), 6-52 (elbow), and 6-99 (hand).

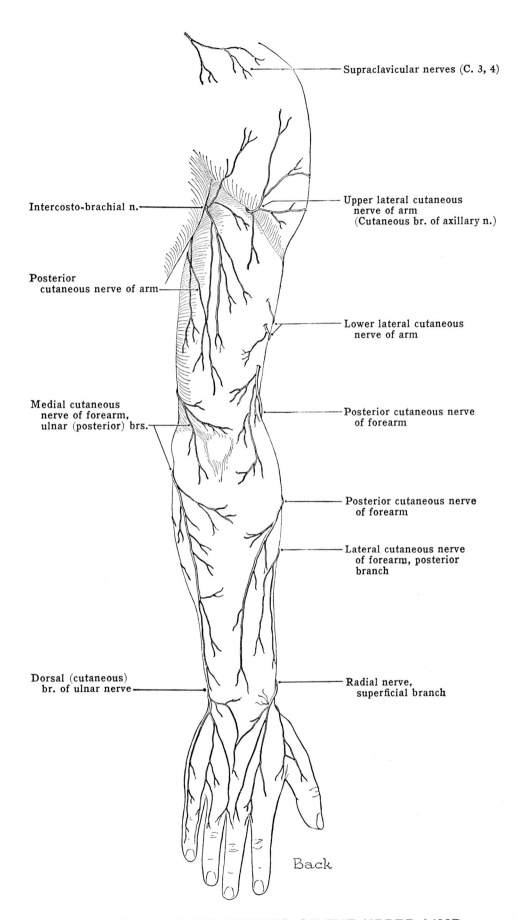

Supraclavicular nerves (C. 3, 4)

Intercosto-brachial n.

Upper lateral cutaneous
nerve of arm
(Cutaneous br. of axillary n.)

Posterior
cutaneous nerve of arm

Lower lateral cutaneous
nerve of arm

Medial cutaneous
nerve of forearm,
ulnar (posterior) brs.

Posterior cutaneous nerve
of forearm

Posterior cutaneous nerve
of forearm

Lateral cutaneous nerve
of forearm, posterior
branch

Dorsal (cutaneous)
br. of ulnar nerve

Radial nerve,
superficial branch

Back

6-7 CUTANEOUS NERVES OF THE UPPER LIMB

The posterior cord of the plexus is represented by 5 cutaneous nerves. Of these (a) one, the upper lateral cutaneous nerve of the arm, is a branch of the axillary nerve, (b) whereas 4 are branches of the radial nerve. They are: the posterior cutaneous nerve of the arm, the lower lateral cutaneous nerve of the arm, the posterior cutaneous nerve of the forearm, and the superficial branch of the radial nerve.

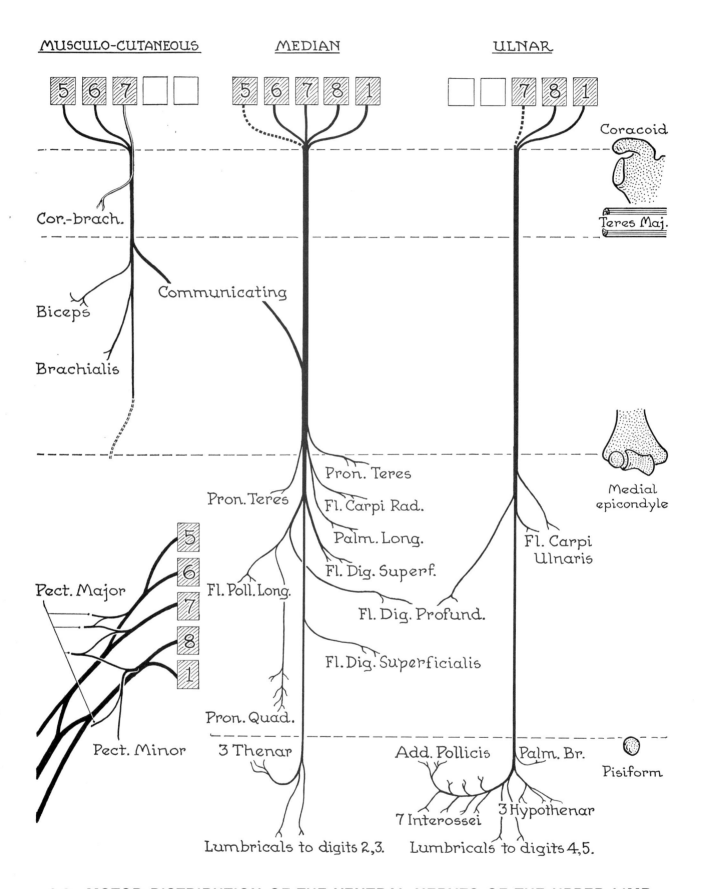

6-8 MOTOR DISTRIBUTION OF THE VENTRAL NERVES OF THE UPPER LIMB

The average levels at which the motor branches leave the stems of the main nerves are shown with reference to the lower border of the axilla (Teres Major), elbow joint (medial epicondyle), and wrist (pisiform bone).

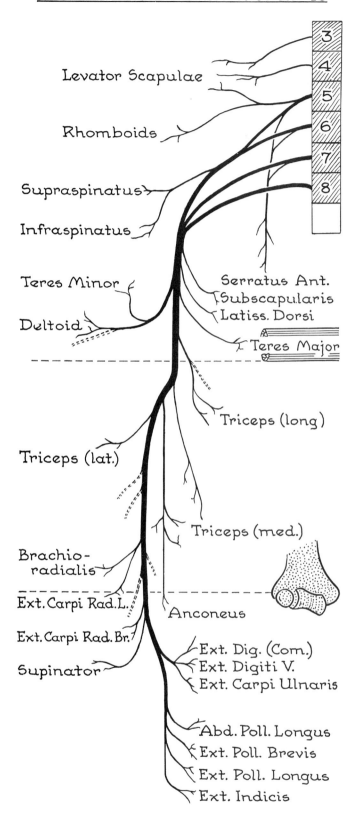

A LIST OF THE MUSCLES OF UPPER LIMB

Trapezius
Latissimus Dorsi
Levator Scapulae
Rhomboideus Major
Rhomboideus Minor
Pectoralis Major
   Clavicular part
   Sternocostal part
   Abdominal part
Pectoralis Minor
Subclavius
Serratus Anterior
Deltoideus
Supraspinatus
Infraspinatus
Teres Minor
Teres Major
Subscapularis
Biceps Brachii
   Long head
   Short head
   Bicipital aponeurosis
Coraco-brachialis
Brachialis
Triceps
   Long head
   Lateral head
   Medial head
   Tricipital aponeurosis
Anconeus
Pronator Teres
Flexor Carpi Radialis
Palmaris Longus
Flexor Carpi Ulnaris
   Humeral head
   Ulnar head
Flexor Digitorum Superficialis
   Humero-ulnar head
   Radial head
Flexor Digitorum Profundus
Flexor Pollicis Longus
Pronator Quadratus
Brachio-radialis
Extensor Carpi Radialis Longus
Extensor Digitorum Communis
Extensor Digiti Minimi (V)
Extensor Carpi Ulnaris
Supinator
Abductor Pollicis Longus
Extensor Pollicis Brevis
Extensor Indicis
Palmaris Brevis
Abductor Pollicis Brevis
Flexor Pollicis Brevis
Opponens Pollicis
Adductor Pollicis
Abductor Digiti Minimi (V)
Flexor Digiti Minimi (V)
Opponens Digiti Minimi (V)
Lumbricales
Interossei
   Palmar
   Dorsal

6-9 MOTOR DISTRIBUTION OF THE DORSAL NERVES OF THE UPPER LIMB

The average levels of origin of the motor branches are shown as in Figure 6-8. There being no fleshy fibers on the dorsum of the hand, there are no motor nerves.

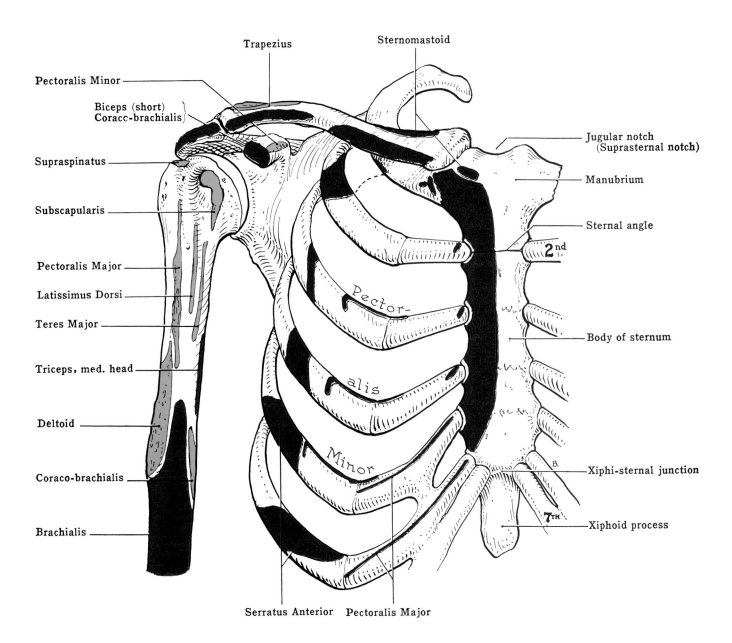

Trapezius

Sternomastoid

Pectoralis Minor

Biceps (short)
Coraco-brachialis

Supraspinatus

Subscapularis

Pectoralis Major

Latissimus Dorsi

Teres Major

Triceps, med. head

Deltoid

Coraco-brachialis

Brachialis

Jugular notch
(Suprasternal notch)

Manubrium

Sternal angle

2nd

Body of sternum

Xiphi-sternal junction

7TH

Xiphoid process

Pector-alis Minor

Serratus Anterior   Pectoralis Major

6-10   BONES OF THE PECTORAL REGION AND AXILLA
SHOWING ATTACHMENTS OF MUSCLES

Observe:

1. The following muscles attached in line with each other:
   Horizontally, on the clavicle: (a) Trapezius and Sternomastoid; (b) Deltoid and clavicular head of Pectoralis Major.
   Longitudinally, on the humerus: (c) Supraspinatus, Pectoralis Major and anterior part of Deltoid; and (d) Subscapularis and Latissimus Dorsi and Teres Major.
2. Pectoralis Major has a crescentic origin from the clavicle, sternum, and the 5th and (or) 6th costal cartilages.
3. Pectoralis Minor here arising from the 3rd, 4th, and 5th ribs. It commonly arises also from either the 2nd or the 6th rib.

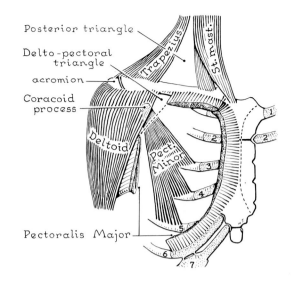

6-11B MUSCLES OF THE REGION

6-11A ANTERIOR CHEST

Note:
1. The clavicle forms a "no man's land" between the neck and the shoulder and pectoral regions: it is subcutaneous (except for Platysma) and can be palpated throughout.
2. Trapezius (*T*) and Sternomastoid (*S*) attach to the upper surface of the lateral and medial thirds of the clavicle, exposing the posterior triangle of the neck.
3. Deltoid (*D*) and clavicular head of Pectoralis major (*C*) fail to meet on the clavicle, exposing the delto-pectoral triangle.
4. The *black dot* marks the sternal angle at the junction of manubrium and sternum, a landmark to the second rib.

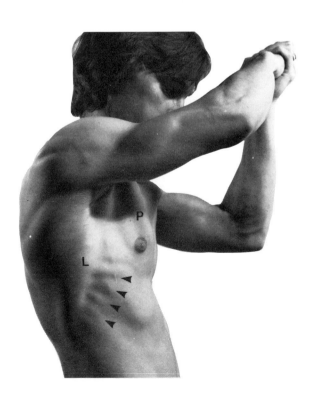

6-11C LATERAL CHEST

Observe:
1. *Arrows* point to digitations of Serratus anterior (Fig. 6-28).
2. Two large muscles of the axillary walls: Pectoralis major (*P*) of the anterior wall passing to its insertion on the *lateral* lip of the bicipital groove; Latissimus dorsi (*L*) of the posterior wall passing to its insertion on the *medial* lip of the bicipital groove in front of Teres major. For bony attachments see Figure 6-35.

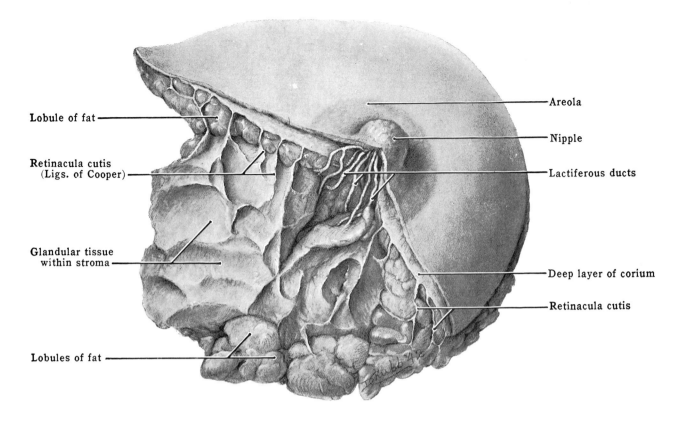

Lobule of fat

Retinacula cutis
(Ligs. of Cooper)

Glandular tissue
within stroma

Lobules of fat

Areola

Nipple

Lactiferous ducts

Deep layer of corium

Retinacula cutis

6-12 MAMMARY GLAND OF THE FEMALE

With the rounded handle of the scalpel, collections of superficial fat were
scooped out of their compartments on the surface of the glandular tissue. The
glandular tissue was incised in order to allow the ducts to be traced. A fringe of
the deeper layer of the skin projects.

Observe:
1. The nipple rising from the center of the pigmented areola.
2. The 7 of the 15 to 20 lactiferous ducts that are displayed. Traced in retrograde
   direction, they run at first dorsally in the long axis of the nipple, enveloped in
   an areolar cuff, and then spread radially and branch to the glandular tissue.
3. The glandular tissue within a dense (fibro-) areolar stroma from which septa
   (retinacula cutis) extend to the deeper layers of the skin, imprisoning lobules
   of superficial fat.

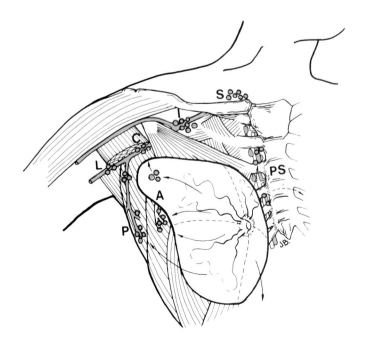

6-13 LYMPHATIC DRAINAGE

Drainage of lymph from the upper limb and
breast passes through nodes arranged irregularly
in groups:

Anterior (*A*) or pectoral, along the lower border
of Pectoralis minor; *Posterior* (*P*), with subscapular veins; *Lateral* (*L*), along the axillary vein;
Infraclavicular (*I*) or apical, between the clavicle and Pectoralis minor; *Central* (*C*), at the
base of the axilla; and *Supraclavicular* (*S*). Most
of the breast is drained through this system to
the subclavian lymph trunk. The medial part of
the breast drains to the *Parasternal* (*PS*) nodes
which follow the internal thoracic vessels and are
much less accessible surgically. When blockage
of the lymphatic system occurs, as in cancer,
drainage may go to the opposite breast and its
nodes or down the anterior abdominal wall to
inguinal nodes.

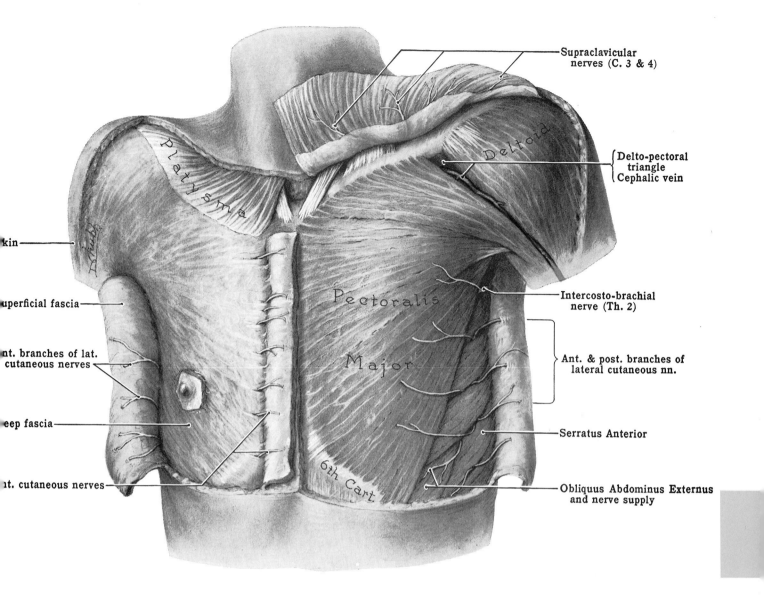

Supraclavicular
nerves (C. 3 & 4)

Delto-pectoral
triangle
Cephalic vein

Intercosto-brachial
nerve (Th. 2)

Ant. & post. branches of
lateral cutaneous nn.

Serratus Anterior

Obliquus Abdominus Externus
and nerve supply

Platysma

Deltoid

Pectoralis

Major

6th Cart.

kin

uperficial fascia

nt. branches of lat.
cutaneous nerves

eep fascia

at. cutaneous nerves

6-14 SUPERFICIAL DISSECTION OF THE PECTORAL REGION

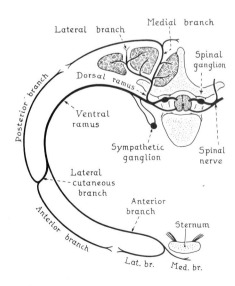

Lateral branch

Medial branch

Posterior branch

Dorsal ramus

Spinal ganglion

Ventral ramus

Sympathetic ganglion

Spinal nerve

Lateral cutaneous branch

Anterior branch

Anterior branch

Sternum

Lat. br.   Med. br.

6-15 SEGMENTAL NERVE

This diagram shows the source of anterior and lateral cutaneous nerves.

Platysma, which descends to the 2nd or 3rd rib, is cut short on the *left side* of the picture; it, together with the supraclavicular nerves, is thrown up on the *right side*.

Observe:
1. The deep fasica covering Pectoralis Major is filmy.
2. The intermuscular bony strip running along the clavicle is both subcutaneous and subplatysmal. Platysma is shown intact in Figure 9-3.
3. The two heads of Pectoralis Major meet at the sternoclavicular joint.
4. The cephalic vein passing through the delto-pectoral triangle.

Note: The brachial plexus (C5, C6, C7, C8, and Th1) does not supply cutaneous branches to the pectoral region, hence the break in the numerical sequence — *i.e.*, branches of supraclavicular nerves C3 and C4 meet those of Th2.

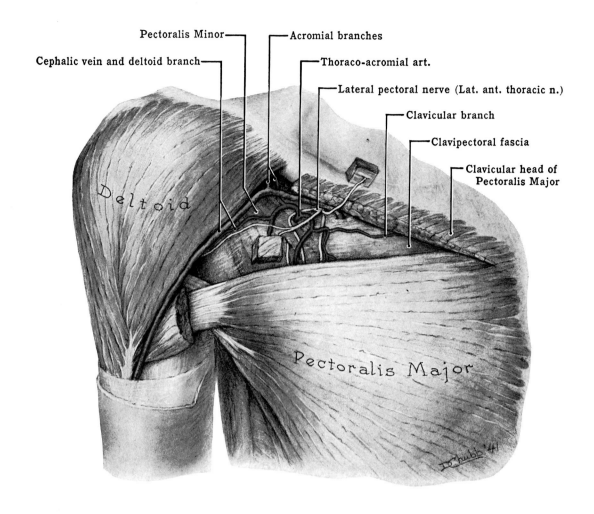

Pectoralis Minor — Acromial branches

Cephalic vein and deltoid branch — Thoraco-acromial art.

Lateral pectoral nerve (Lat. ant. thoracic n.)

Clavicular branch

Clavipectoral fascia

Clavicular head of Pectoralis Major

Deltoid

Pectoralis Major

6-16 CLAVIPECTORAL FASCIA (CORACO-CLAVICULAR FASCIA)

The clavicular head of Pectoralis Major is excised except for 2 cubes which remain to identify its nerves. The thoraco-acromial veins, which join the cephalic vein, are removed.

Observe:

1. The part of the clavipectoral fascia above Pectoralis Minor — the costocoracoid membrane (Fig. 6-17) — pierced by the lateral pectoral nerve and its companion vessels.
2. The part of the fascia enclosing Pectoralis Minor. Here muscle and fascia are pierced by medial pectoral nerve (see Fig. 6-20), thoraco-acromial artery, and cephalic vein.
3. The trilaminar insertion of Pectoralis Major.
4. The course of the cephalic vein through the delto-pectoral triangle and costo-coracoid membrane.

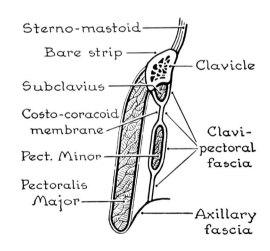

Sterno-mastoid

Bare strip

Clavicle

Subclavius

Costo-coracoid membrane

Clavi-pectoral fascia

Pect. Minor

Pectoralis Major

Axillary fascia

6-17 ANTERIOR WALL OF AXILLA

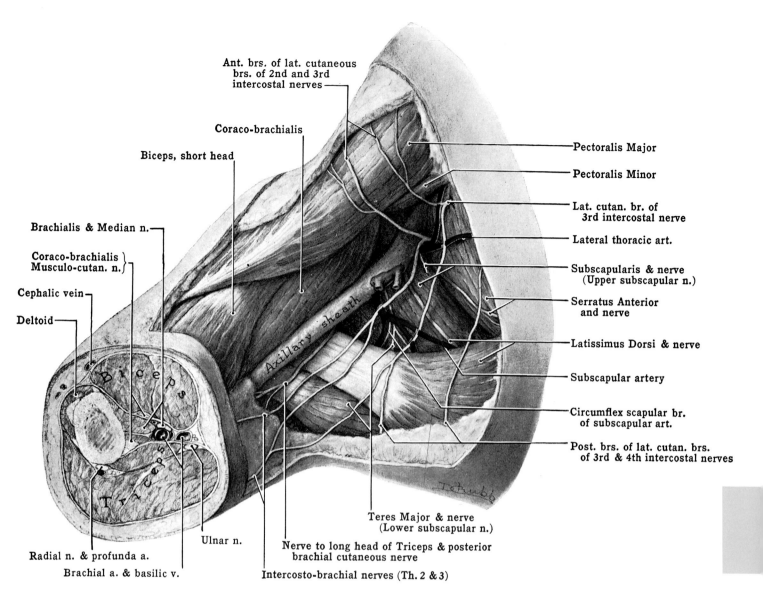

Ant. brs. of lat. cutaneous
brs. of 2nd and 3rd
intercostal nerves

Coraco-brachialis

Biceps, short head

Brachialis & Median n.

Coraco-brachialis ⎫
Musculo-cutan. n. ⎭

Cephalic vein

Deltoid

Pectoralis Major

Pectoralis Minor

Lat. cutan. br. of
3rd intercostal nerve

Lateral thoracic art.

Subscapularis & nerve
(Upper subscapular n.)

Serratus Anterior
and nerve

Latissimus Dorsi & nerve

Subscapular artery

Circumflex scapular br.
of subscapular art.

Post. brs. of lat. cutan. brs.
of 3rd & 4th intercostal nerves

Axillary sheath

Biceps

Triceps

Radial n. & profunda a.

Brachial a. & basilic v.

Ulnar n.

Teres Major & nerve
(Lower subscapular n.)

Nerve to long head of Triceps & posterior
brachial cutaneous nerve

Intercosto-brachial nerves (Th. 2 & 3)

6-18 AXILLA, FROM BELOW. CROSS-SECTION OF THE ARM

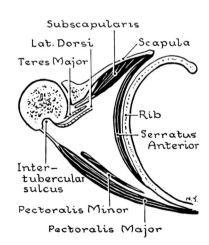

Subscapularis

Lat. Dorsi

Teres Major

Scapula

Rib

Serratus
Anterior

Inter-
tubercular
sulcus

Pectoralis Minor

Pectoralis Major

6-19 WALLS OF AXILLA,
CROSS-SECTION

Observe:
1. The three muscular walls of the axilla:
   a. Anterior wall: Pectoralis Major, Pectoralis Minor, and Subclavius, but only the lower borders of Pectorales are in view.
   b. Posterior wall: Subscapularis, Latissimus Dorsi, and Teres Major.
   c. Medial wall: Serratus Anterior

The lateral or bony wall—intertubercular sulcus (bicipital groove of the humerus)—is concealed by Biceps and Coraco-brachialis.

2. The axillary sheath and the cutaneous nerves crossing Latissimus Dorsi. The most lateral of these nerves is also the sole nerve supply of long head of Triceps (Fig. 6-23).
3. The axillary sheath transmits the great nerves and vessels of the limb. It is a neuro-vascular bundle (see Fig. 6-20).

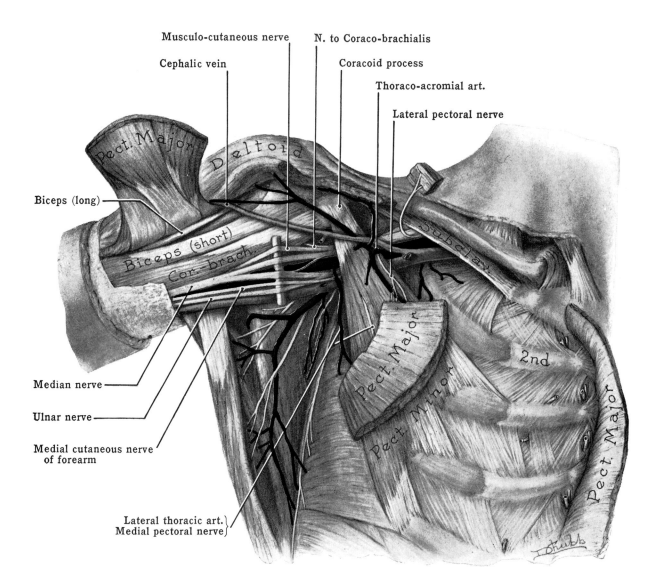

Musculo-cutaneous nerve

Cephalic vein

N. to Coraco-brachialis

Coracoid process

Thoraco-acromial art.

Lateral pectoral nerve

Biceps (long)

Median nerve

Ulnar nerve

Medial cutaneous nerve
of forearm

Lateral thoracic art.
Medial pectoral nerve

6-20 ANTERIOR STRUCTURES OF THE AXILLA

Pectoralis Major is reflected and the clavi-pectoral fascia removed.

Observe:

1. Subclavius and Pectoralis Minor, the two deep muscles of the anterior wall.
2. The axillary artery passing behind Pectoralis Minor, a finger's breadth from the tip of the coracoid process, and having the lateral cord on its lateral side and the medial cord on its medial side.
3. The axillary vein lying medial to the axillary artery.
4. The median nerve, followed proximally, leading by its lateral root to the lateral cord and the musculo-cutaneous nerve, and by its medial root to the medial cord and the ulnar nerve. (These 4 nerves and the medial cutaneous nerve of the forearm are raised on a stick.)
5. The nerve to Coraco-brachialis arising within the axilla.
6. The cube of muscle above the clavicle is cut from the clavicular head of Pectoralis Major.

Note: The lateral root of the median nerve may be in several strands.

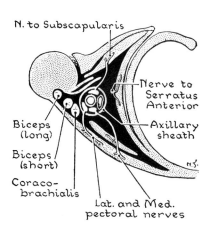

N. to Subscapularis

Nerve to Serratus Anterior

Axillary sheath

Biceps (long)

Biceps (short)

Coraco-brachialis

Lat. and Med. pectoral nerves

**6-21A   CONTENTS OF AXILLA,
CROSS-SECTION**

6-21B   THE AXILLA

Observe:

1. Muscles of the anterior and posterior walls of the axilla, Pectoralis major (*P*) and Latissimus dorsi (*L*) converging on the narrow lateral wall.
2. The belly of Biceps emerging from the axilla where its tendon has occupied a groove in the lateral wall, just passing through.
3. *Arrows* indicating digitations of Serratus anterior whose upper fibers clothe the convex medial wall of the axilla.

4. As the arm is abducted the floor of the axilla becomes increasingly concave because of the attachment of the clavipectoral fascia to the axillary fascia (Fig. 66-17).

Recall that the blunted apex of the axilla is the triangular doorway into the upper limb shown in Figure 9-2B.

6-21C   POSTERIOR WALL MUSCLES

Observe:

1. Teres major (*TM*) and Latissimus dorsi (*L*) moving toward their insertion on the medial lip of the bicipital groove, Latissimus dorsi moving to the more anterior position.
2. The long head of the Triceps (*T*) emerging from the cleft between Deltoid (*D*) and Teres major.

Note that Latissimus dorsi, being a *posterior* axillary wall, will be enervated by *posterior* divisions of the brachial plexus: the thoracodorsal nerve from the posterior cord, C6, 7, (8). Because of Latissimus dorsi's role in forced aspiration, muscle and nerve can be tested by grasping the posterior axillary fold and asking the patient to cough.

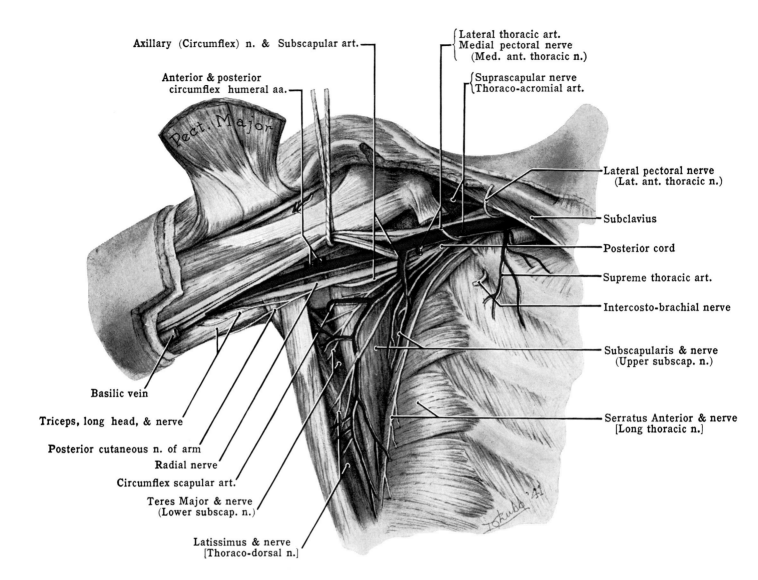

Axillary (Circumflex) n. & Subscapular art.

Anterior & posterior circumflex humeral aa.

Pect. Major

Lateral thoracic art.
Medial pectoral nerve
(Med. ant. thoracic n.)

Suprascapular nerve
Thoraco-acromial art.

Lateral pectoral nerve
(Lat. ant. thoracic n.)

Subclavius

Posterior cord

Supreme thoracic art.

Intercosto-brachial nerve

Subscapularis & nerve
(Upper subscap. n.)

Serratus Anterior & nerve
[Long thoracic n.]

Basilic vein

Triceps, long head, & nerve

Posterior cutaneous n. of arm

Radial nerve

Circumflex scapular art.

Teres Major & nerve
(Lower subscap. n.)

Latissimus & nerve
[Thoraco-dorsal n.]

6-22 POSTERIOR AND MEDIAL WALLS OF THE AXILLA

Pectoralis Minor is excised; the lateral and medial cords are retracted; the axillary vein is removed.

Observe:

1. The posterior cord and its 2 terminal branches (the radial and axillary nerves), lying behind the axillary artery.
2. The nerves to the 3 posterior muscles. Of these:
   a. Nerve to Latissimus Dorsi enters the deep surface of its muscle 1 cm from its free border at a point midway between the chest and the abducted arm.
   b. Upper nerve to Subscapularis lies parallel to (a) but above it (Fig. 6-23).
   c. Lower nerve to Subscapularis and to Teres Major lies parallel to (a) but below it.
3. Nerve to Serratus Anterior clinging to its muscle throughout; high up some fat may intervene.
4. Suprascapular nerve passing toward the root of the coracoid process.
5. Subscapular artery, the largest branch of the axillary artery. Here arising high; usually it arises at the lower border of Subscapularis.
6. Posterior circumflex humeral artery accompanying the axillary nerve through the quadrangular space (Fig. 6-23).

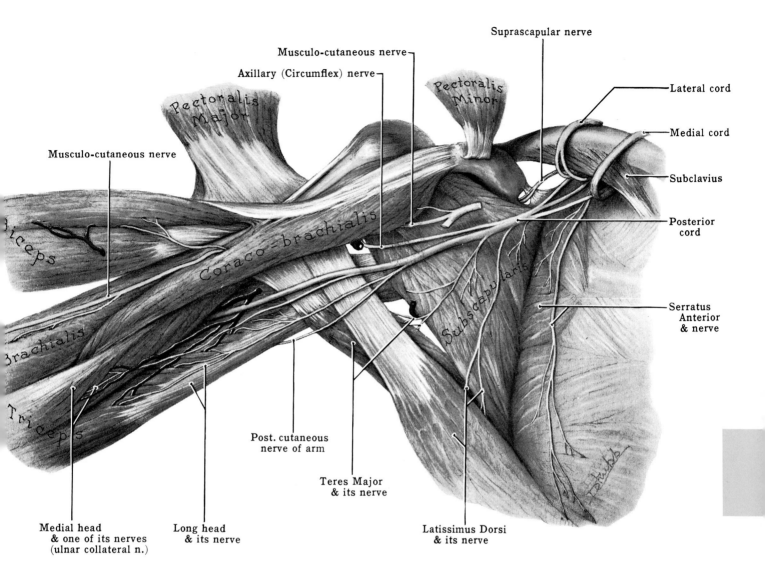

Suprascapular nerve

Musculo-cutaneous nerve

Axillary (Circumflex) nerve

Musculo-cutaneous nerve

Lateral cord

Medial cord

Subclavius

Posterior cord

Serratus Anterior & nerve

Subscapularis

Coraco-brachialis

Biceps

Brachialis

Triceps

Post. cutaneous nerve of arm

Teres Major & its nerve

Latissimus Dorsi & its nerve

Medial head & one of its nerves (ulnar collateral n.)

Long head & its nerve

6-23 POSTERIOR WALL OF THE AXILLA. MUSCULO-CUTANEOUS NERVE, POSTERIOR CORD

Pectoralis Major and Minor are thrown laterally; the lateral and medial cords are thrown upward; the arteries, veins, and median and ulnar nerves are removed.

Observe:

1. Coraco-brachialis arising with the short head of Biceps from the tip of the coracoid process and inserted half-way down the humerus (Fig. 6-35).
2. The musculo-cutaneous nerve piercing Coraco-brachialis, and supplying it, Biceps, and Brachialis before becoming cutaneous.
3. The posterior cord of the plexus formed by the union of the 3 posterior divisions, supplying the 3 muscles of the posterior wall of the axilla, and soon ending as the radial and axillary nerves.
4. The radial nerve giving off, in the axilla, the nerve to the long head of Triceps and a cutaneous branch, and, in this specimen, the ulnar collateral branch to the medial head of Triceps. It then enters the spiral groove of the humerus with the profunda brachii artery.
5. The axillary nerve traversing the quadrangular space with the posterior circumflex humeral artery. The circumflex scapular artery traversing the triangular space.

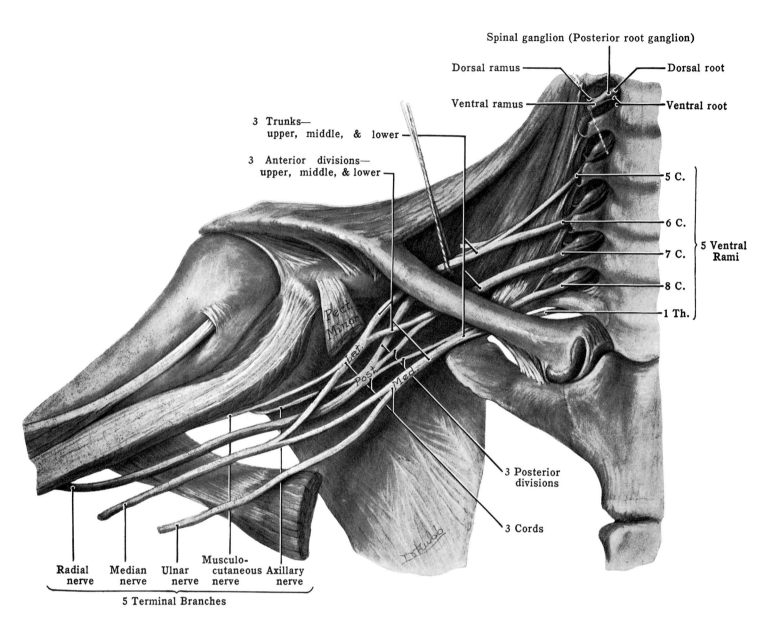

Spinal ganglion (Posterior root ganglion)

Dorsal ramus —

Dorsal root

Ventral ramus —

Ventral root

3 Trunks—
upper, middle, & lower

3 Anterior divisions—
upper, middle, & lower

5 C.

6 C.

7 C.

8 C.

1 Th.

5 Ventral
Rami

Pect.
Minor

Lat.

Post.

Med.

3 Posterior
divisions

3 Cords

Radial
nerve

Median
nerve

Ulnar
nerve

Musculo-
cutaneous
nerve

Axillary
nerve

5 Terminal Branches

6-24 BRACHIAL PLEXUS, LIGAMENTS OF THE CLAVICLE

Observe:

1. A dorsal (sensory) root of a spinal nerve is larger than a ventral (motor) root.
2. The 2 roots uniting beyond the ganglion to form a very short mixed spinal nerve.
3. The mixed nerve at once dividing into a small dorsal ramus and a large ventral ramus.
4. The 5 ventral *rami* forming the brachial plexus. (Of these the middle ramus, C7, is the largest.)
5. These 5 rami uniting to form the 3 *trunks* of the plexus.
6. Each trunk dividing into 2 *divisions*, an anterior and a posterior.
7. These 6 divisions becoming 3 *cords* – a lateral, a medial, and a posterior.
8. The 3 cords lying behind Pectoralis Minor.

Note: The ligaments of the clavicle and the structures of the sterno-clavicular joint are depicted but not labeled (Fig. 6-43).

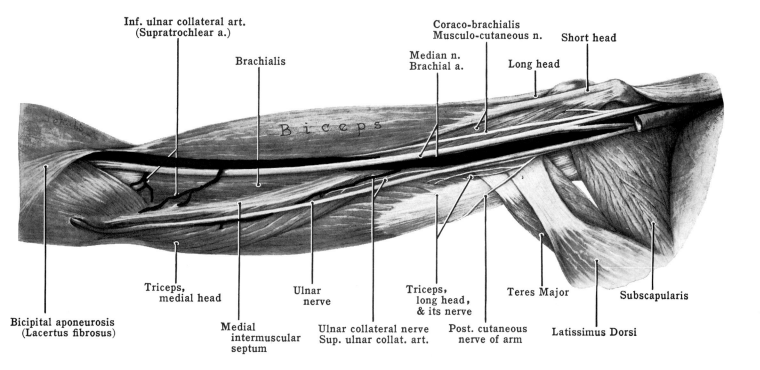

Inf. ulnar collateral art.
(Supratrochlear a.)

Brachialis

Coraco-brachialis
Musculo-cutaneous n.

Median n.
Brachial a.

Short head

Long head

B i c e p s

Bicipital aponeurosis
(Lacertus fibrosus)

Triceps,
medial head

Ulnar
nerve

Medial
intermuscular
septum

Ulnar collateral nerve
Sup. ulnar collat. art.

Triceps,
long head,
& its nerve

Post. cutaneous
nerve of arm

Teres Major

Latissimus Dorsi

Subscapularis

6-25   BRACHIUM OR ARM, MEDIAL VIEW

Observe:

1. Three muscles: Biceps, Brachialis and Coraco-brachialis occupying the front of the brachium or arm; Triceps Brachii occupying the back.

2. The medial intermuscular septum separating these two muscle groups in the distal two-thirds of the arm.

3. The great artery of the limb passing a finger's breadth medial to the tip of the coracoid process, and applied to the medial side of Coraco-brachialis proximally and to the front of Brachialis distally.

4. In the axilla, the lateral and medial cords of the brachial plexus and their end branches making an M-shaped display about the front of the artery.

5. The median nerve applied to the artery throughout and crossing it from lateral to medial side, usually crossing superficially (Fig. 6-119).

6. The ulnar nerve applied to the medial side of the artery as far as the middle of the arm, then passing behind the medial septum, and descending subfascially, on the medial head of Triceps, to the back of the medial epicondyle, where it is palpable (Fig. 6-62).

7. The superior ulnar collateral artery and the ulnar collateral branch of the radial nerve (to medial head of Triceps) accompanying the ulnar nerve.

8. The musculo-cutaneous nerve supplying Coraco-brachialis, following the lateral side of the brachial artery, and disappearing between Biceps and Brachialis accompanied by an arterial branch to these two muscles. More commonly it pierces Coraco-brachialis as in Figure 6-23.

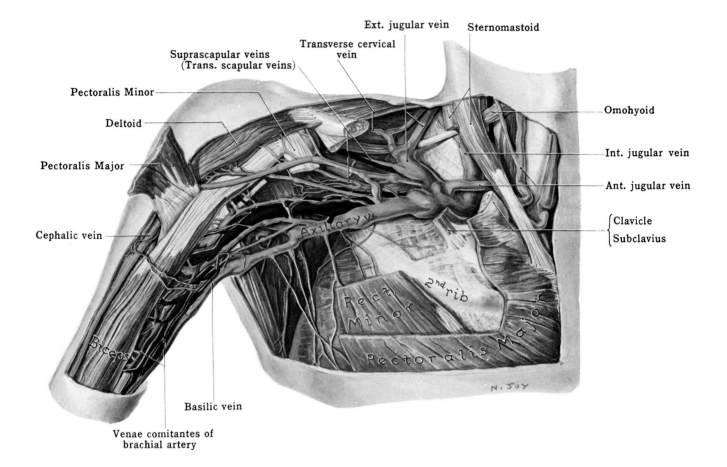

6-26 VEINS OF THE AXILLA

Observe:

1. The basilic vein becoming the axillary vein at the lower border of Teres Major; the axillary vein becoming the subclavian vein at the 1st rib; and the subclavian joining the internal jugular to become the brachiocephalic vein (innominate vein) behind the sternal end of the clavicle. Over 40 venous valves are shown. Note 1 in the basilic vein, 3 in the axillary, and 1 in the subclavian where it rests upon the 1st rib, the last valve on the road to the heart.

2. Venae comitantes of the brachial artery uniting and joining the axillary vein near the middle of the axilla.

3. The cephalic vein here bifurcating to end both in the axillary and in the external jugular vein.

4. The profunda brachii, posterior humeral circumflex, and circumflex scapular venae comitantes united and, as one large vein, joining the axillary vein.

5. Several subscapular veins, tenuous because the latex injection failed to force their valves.

6. Three suprascapular veins: one from below the suprascapular ligament to the axillary vein, and two from above the ligament to the external jugular vein.

7. Anastomotic veins are in view; obviously those on the dorsum of the scapula joining circumflex scapular and suprascapular veins cannot be seen.

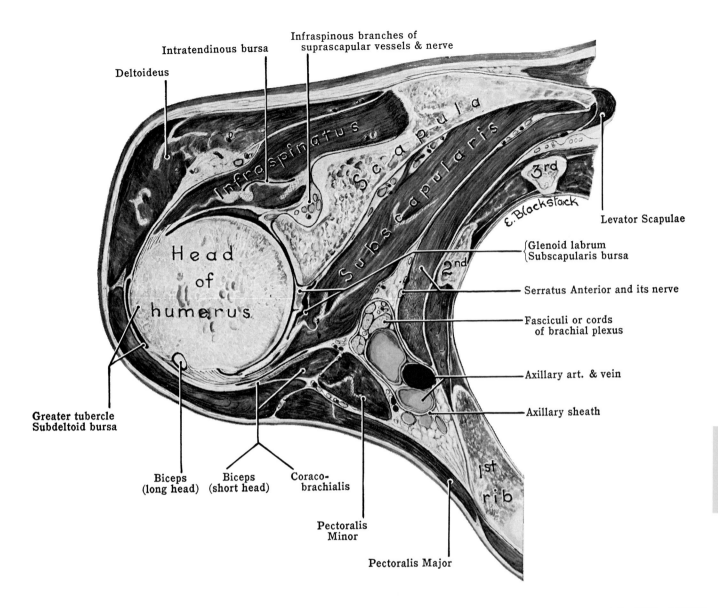

Deltoideus

Intratendinous bursa

Infraspinous branches of
suprascapular vessels & nerve

Head
of
humerus

Greater tubercle
Subdeltoid bursa

Biceps
(long head)

Biceps
(short head)

Coraco-
brachialis

Pectoralis
Minor

Pectoralis Major

E. Blackstock

Levator Scapulae

Glenoid labrum
Subscapularis bursa

Serratus Anterior and its nerve

Fasciculi or cords
of brachial plexus

Axillary art. & vein

Axillary sheath

6-27 CROSS-SECTION THROUGH SHOULDER JOINT AND THE AXILLA, NEAR ITS APEX

Observe:

1. The long head (tendon) of Biceps, in its synovial sheath, facing straight forward. The short head of Biceps and the Coraco-brachialis and Pectoralis Minor evidently sectioned close below their attachment to the coracoid process.

2. The cartilage of the head of the humerus, extensive and very thin peripherally.

3. The glenoid cavity of the scapula, small and with glenoid labrum attached loosely in front but firmly behind where it is part-origin of the long head of Biceps (Fig. 6-46).

4. Delicate synovial folds projecting from the labra between head and glenoid cavity.

5. The fibrous capsule, thin posteriorly and partly fused with the tendon of Infraspinatus; thicker anteriorly (glenohumeral ligaments).

6. Bursae: (a) Subdeltoid bursa, Between Deltoid and greater tubercle (continuous above with subacromial bursa, Fig. 6-42); (b) Subscapular bursa, between Subscapularis tendon and the scapula (Fig. 6-43) (communicating with the joint cavity); (c) Coraco-brachialis bursa, between Coracobrachialis and Subscapularis; (d) an Intra-infraspinatus bursa.

7. Walls of the axilla near its apex, formed by Subscapularis and scapula posteriorly; Serratus Anterior, ribs and Intercostales medially; and Pectoralis Major and Minor anteriorly. There is no bone in the anterior wall.

8. The axillary sheath of areolar tissue (Fig. 6-18), delicate and enclosing the axillary artery and vein and the 3 cords or fasciculi of the brachial plexus (and here an extra vein) to form a neurovascular bundle, surrounded with axillary fat.

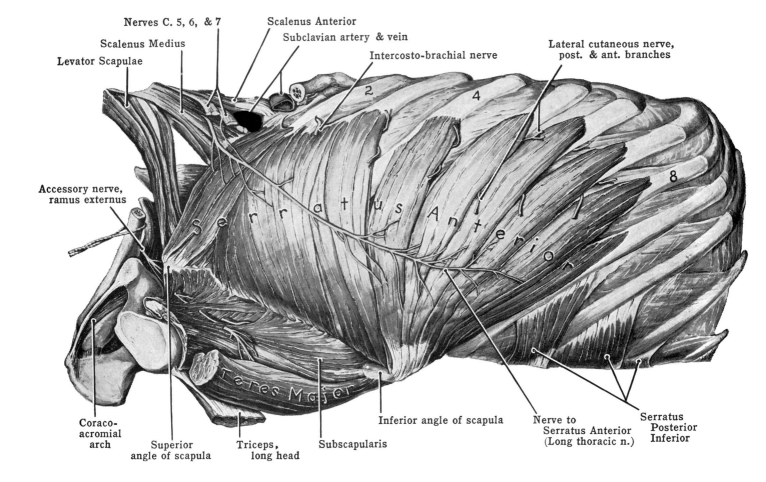

Nerves C. 5, 6, & 7

Scalenus Medius

Levator Scapulae

Scalenus Anterior

Subclavian artery & vein

Intercosto-brachial nerve

Lateral cutaneous nerve, post. & ant. branches

Accessory nerve, ramus externus

Coraco-acromial arch

Superior angle of scapula

Triceps, long head

Subscapularis

Inferior angle of scapula

Nerve to Serratus Anterior (Long thoracic n.)

Serratus Posterior Inferior

6-28 SERRATUS ANTERIOR, SIDE VIEW, SUPINE POSITION

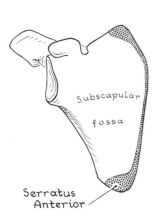

6-29 INSERTION OF SERRATUS ANTERIOR

Observe:

1. Serratus Anterior, which forms the medial wall of the axilla, having an extensive fleshy origin from the upper 8 (here 9) ribs far forward, and an insertion into the whole length of the medial border of the scapula. The fibers from the 1st rib and from the arch between the 1st and 2nd ribs converging on the superior angle, those from the 2nd and 3rd ribs diverging to spread thinly along the medial border; and the remainder (from 4th to 9th ribs), which form the bulk of the muscle, converging on the inferior angle and therefore having a tendinous insertion. (For bony attachments see Figs. 6-10 and 6-35.)

2. The nerve to Serratus Anterior, arising from C5, C6, and C7 and applied to the whole length of the muscle. The fibers from C5 and C6 piercing Scalenus Medius and appearing lateral to the brachial plexus; those from C7 descending dorsal to the plexus.

3. Teres Major applied to the lateral border of Subscapularis. The nerve to Teres Major helping to supply Subscapularis.

4. The brachial plexus and the subclavian artery appearing between Scalenus Anterior and Scalenus Medius; the subclavian vein separated from the artery by Scalenus Anterior.

5. The term "Serratus Anterior" implies the existence of a "Serratus Posterior," and the inferior part of this is seen attached to the lower 4 ribs (Fig. 5-26).

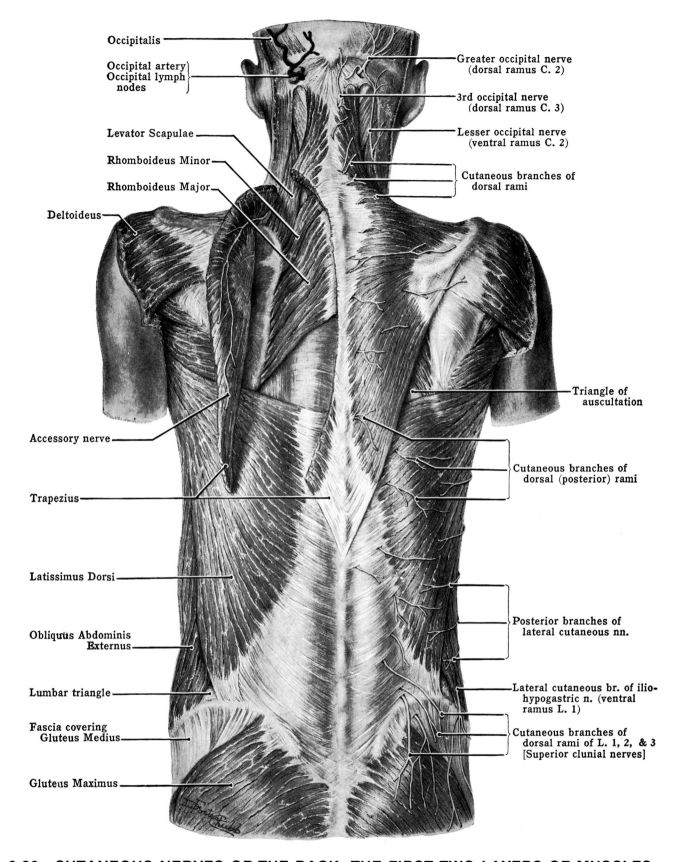

Occipitalis

Occipital artery
Occipital lymph
nodes

Levator Scapulae

Rhomboideus Minor

Rhomboideus Major

Deltoideus

Accessory nerve

Trapezius

Latissimus Dorsi

Obliquus Abdominis
Externus

Lumbar triangle

Fascia covering
Gluteus Medius

Gluteus Maximus

Greater occipital nerve
(dorsal ramus C. 2)

3rd occipital nerve
(dorsal ramus C. 3)

Lesser occipital nerve
(ventral ramus C. 2)

Cutaneous branches of
dorsal rami

Triangle of
auscultation

Cutaneous branches of
dorsal (posterior) rami

Posterior branches of
lateral cutaneous nn.

Lateral cutaneous br. of ilio-
hypogastric n. (ventral
ramus L. 1)

Cutaneous branches of
dorsal rami of L. 1, 2, & 3
[Superior clunial nerves]

6-30 CUTANEOUS NERVES OF THE BACK, THE FIRST TWO LAYERS OF MUSCLES

Trapezius is severed and reflected on the *left side*.

Observe:

1. The cutaneous branches of the dorsal (posterior) nerve rami.
2. Trapezius and Latissimus Dorsi of the 1st layer. Levator Scapulae and
   Rhomboidei to the 2nd layer.
3. Two triangles: (a) the triangle of auscultation where the thoracic wall is
   poorly covered, and (b) the lumbar triangle.

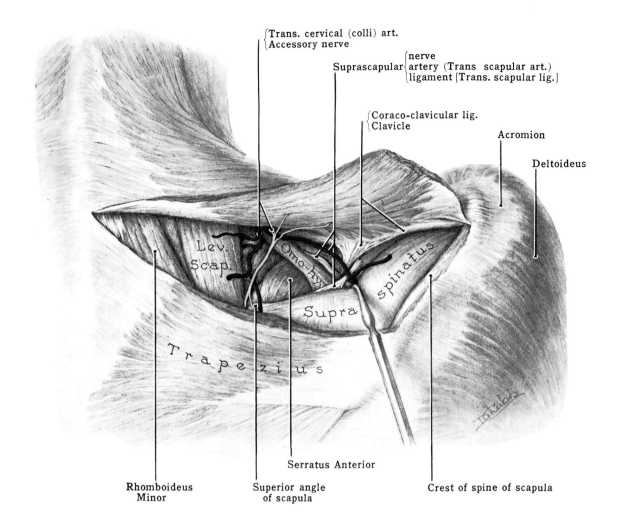

Trans. cervical (colli) art.
Accessory nerve

Suprascapular { nerve
artery (Trans scapular art.)
ligament [Trans. scapular lig.]

Coraco-clavicular lig.
Clavicle

Acromion

Deltoideus

Lev. Scap.

Omo-hy.

Supra

spinatus

Trapezius

Serratus Anterior

Rhomboideus
Minor

Superior angle
of scapula

Crest of spine of scapula

6-31 SUPRASCAPULAR REGION

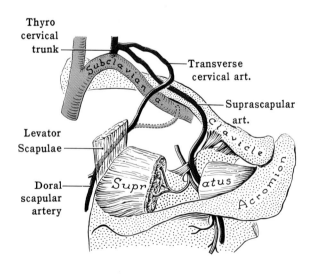

Thyro
cervical
trunk

Subclavian a.

Transverse
cervical art.

Clavicle

Suprascapular
art.

Levator
Scapulae

Supra

atus

Acromion

Doral
scapular
artery

6-32 SUPRASCAPULAR AND DORSAL
SCAPULAR ARTERIES

This diagram shows the source and course of these
two arteries. The dorsal scapular artery has alterna-
tive origins. See Figure 9-38.

The horizontal fibers of Trapezius, at the level of the superior
angle of the scapula, are separated and the incision is carried
laterally along the crest of the spine of the scapula.

Observe:

1. The accessory nerve crossing the superior angle of the scapula.
2. The transverse cervical artery split by Levator Scapulae into
a superficial and a deep branch, one following *the* accessory
nerve, the other (not shown here) the dorsal scapular nerve
(nerve to Rhomboids).
3. The fingertip, placed on the superior angle, may trace a U-
shaped course—laterally along the sharp superior border of
the scapula and along the sharp suprascapular ligament to
the coracoid process, thence cranially along the conoid ligament to the clavicle, and thence medially behind the smooth
posterior surface of the clavicle.
4. The suprascapular artery running behind the clavicle, and
therefore having a retroclavicular course, before crossing
above the suprascapular ligament.
5. The suprascapular nerve crossing below the ligament.

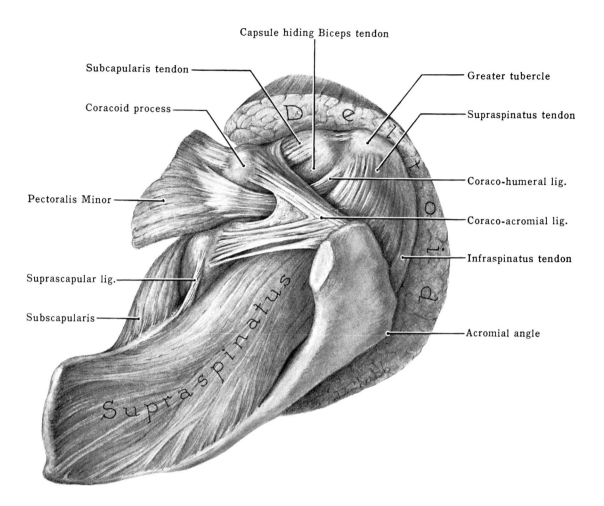

Capsule hiding Biceps tendon

Subcapularis tendon

Coracoid process

Pectoralis Minor

Suprascapular lig.

Subscapularis

Greater tubercle

Supraspinatus tendon

Coraco-humeral lig.

Coraco-acromial lig.

Infraspinatus tendon

Acromial angle

6-33 SUPRASPINOUS AND SUBDELTOID REGIONS

Observe:
1. The clavicular facet on the acromion. It is small, oval, and obliquely set.
2. The triangular coraco-acromial ligament arching from the lateral border of the horizontal part of the coracoid process to the acromion between the facet and the tip.
3. Part of Pectoralis Minor tendon here (as commonly) dividing this ligament into two limbs and continuing, as the anterior part of the coraco-humeral ligament, to the greater tubercle (tuberosity) of the humerus.
4. Supraspinatus passing under the coraco-acromial arch, and then lying between Deltoid above and the capsule of the shoulder joint below. Supraspinatus and the middle fibers of Deltoid are the abductors of the joint. Although the middle part of the Deltoid is thin, it is multipennate and powerful. For subacromial and subdeltoid bursae see Figures 6-42 and 6-47.

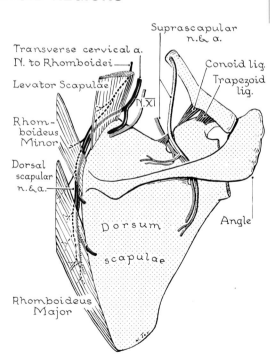

Transverse cervical a.
N. to Rhomboidei
Levator Scapulae
Rhomboideus Minor
Dorsal scapular n. & a.
Rhomboideus Major

Suprascapular n. & a.
Conoid lig.
Trapezoid lig.
Dorsum scapulae
Angle

6-34 MEDIAL BORDER OF SCAPULA

Three muscles and the dorsal scapular nerve (nerve to Rhomboids) and artery (deep branch of transverse cervical artery).

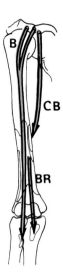

6-35A   THREE MUSCLES

Showing the route from origin
to insertion of the three muscles of the anterior compartment of the arm: Biceps (*B*),
Coraco-brachialis (*CB*), and
Brachialis (*BR*). All are supplied by the Musculocuta-
neous nerve.

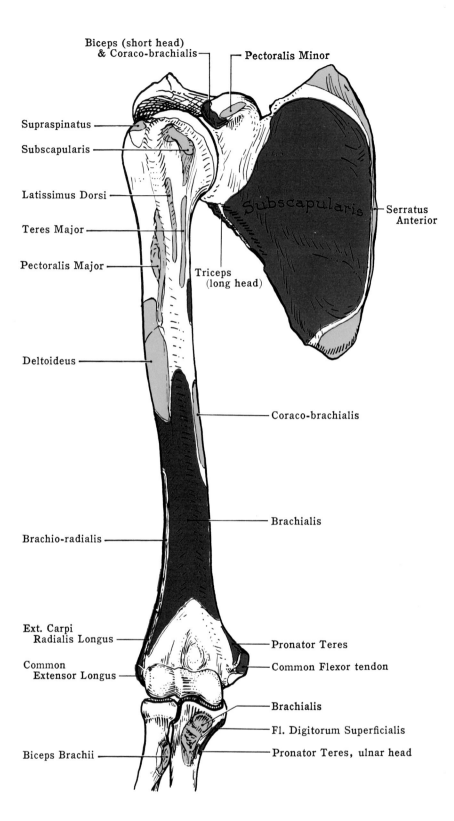

Biceps (short head)
& Coraco-brachialis
Pectoralis Minor

Supraspinatus

Subscapularis

Latissimus Dorsi

Teres Major

Pectoralis Major

Deltoideus

Brachio-radialis

Ext. Carpi
Radialis Longus

Common
Extensor Longus

Biceps Brachii

Subscapularis

Serratus
Anterior

Triceps
(long head)

Coraco-brachialis

Brachialis

Pronator Teres

Common Flexor tendon

Brachialis

Fl. Digitorum Superficialis

Pronator Teres, ulnar head

6-35B   BONES OF THE UPPER LIMB SHOWING
ATTACHMENTS OF MUSCLES, ANTERIOR VIEW

For anterior view of bones of the forearm see Figure 6-65.

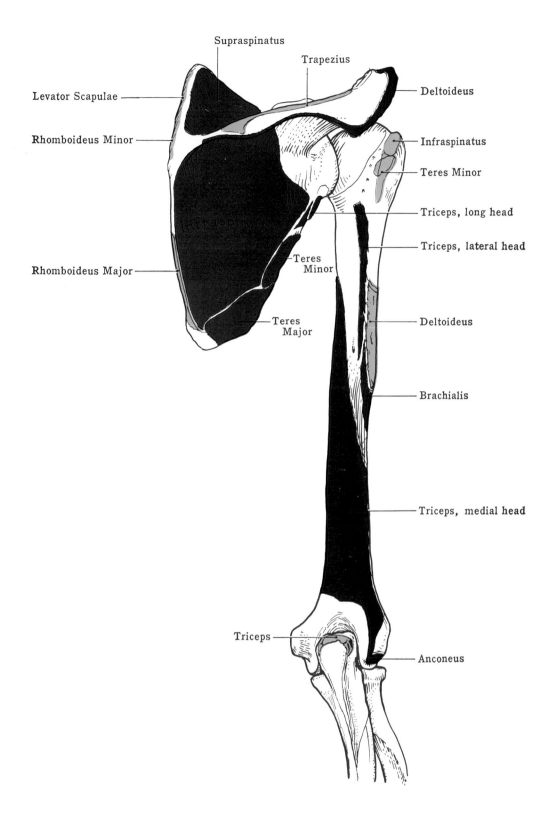

Supraspinatus

Trapezius

Levator Scapulae

Deltoideus

Rhomboideus Minor

Infraspinatus

Teres Minor

Infraspinatus

Triceps, long head

Teres
Minor

Triceps, lateral head

Rhomboideus Major

Deltoideus

Teres
Major

Brachialis

Triceps, medial **head**

Triceps

Anconeus

**6-36 BONES OF THE UPPER LIMB SHOWING
ATTACHMENTS OF MUSCLES, POSTERIOR VIEW**

For posterior view of bones of the forearm see Figure 6-90.

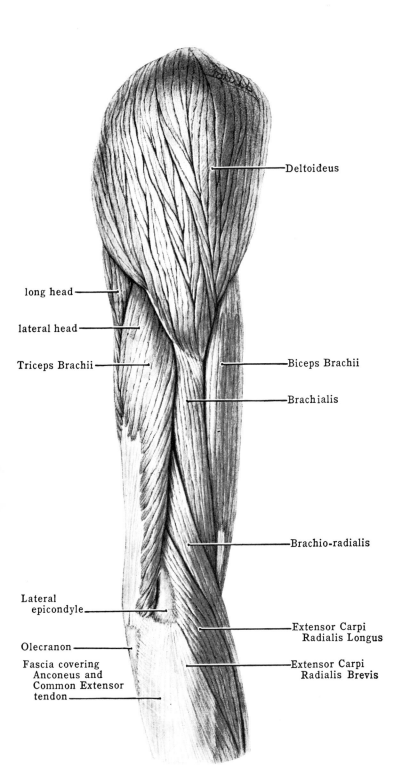

Deltoideus

long head

lateral head

Triceps Brachii

Biceps Brachii

Brachialis

Brachio-radialis

Lateral
epicondyle

Olecranon

Fascia covering
Anconeus and
Common Extensor
tendon

Extensor Carpi
Radialis Longus

Extensor Carpi
Radialis Brevis

6-37A MUSCLES OF THE ARM, LATERAL VIEW

6-37B DELTOIDEUS

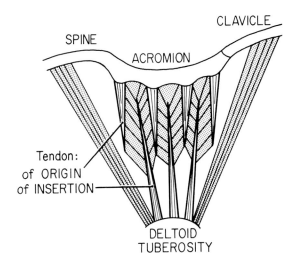

SPINE

CLAVICLE

ACROMION

Tendon:
of ORIGIN
of INSERTION

DELTOID
TUBEROSITY

6-37C INTERNAL STRUCTURE OF DELTOID

Note: The multipennate structure of the middle
part of Deltoid, and the more parallel arrangement of the fibers of the anterior and posterior
parts.

6-38A ARM MUSCLES, ANTERIOR VIEW

6-38B ARM MUSCLES, POSTERIOR VIEW

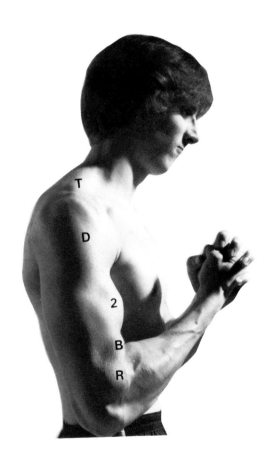

6-38C ARM MUSCLES, LATERAL VIEW

THE MUSCLES OF THE ARM

T, Trapezius	*D,* Deltoid
2, Biceps	*3,* Triceps
B, Brachialis	*R,* Brachioradialis

Note the *arrow* pointing to the delto-pectoral triangle.

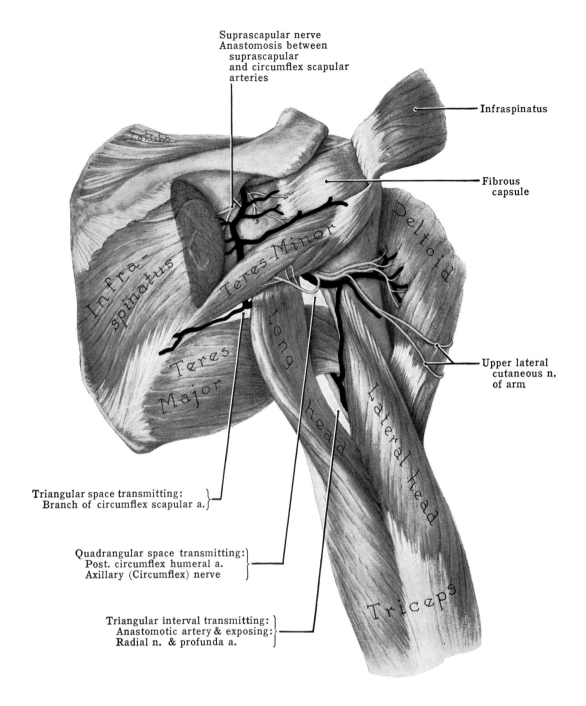

Suprascapular nerve
Anastomosis between suprascapular and circumflex scapular arteries

Infraspinatus

Fibrous capsule

Upper lateral cutaneous n. of arm

Triangular space transmitting:
Branch of circumflex scapular a.

Quadrangular space transmitting:
Post. circumflex humeral a.
Axillary (Circumflex) nerve

Triangular interval transmitting:
Anastomotic artery & exposing:
Radial n. & profunda a.

6-39 DORSAL SCAPULAR AND SUBDELTOID REGIONS

Observe:

1. The thickness of Infraspinatus, which, aided by Teres Minor and posterior fibers of Deltoid rotates the humerus laterally.

2. Long head of Triceps passing between Teres Minor (a lateral rotator) and Teres Major (a medial rotator).

3. Long head of Triceps separating the quadrangular space from the triangular space.

4. Teres Major separating the quadrangular space from another triangular space.

5. The arterial anastomoses behind the scapula and behind the Teres Major.

6. The distribution of the suprascapular and axillary nerves. Each comes from C5 and C6 (Fig. 6-9); each supplies two muscles (Supraspinatus and Infraspinatus—Teres Minor and Deltoid); each supplies the shoulder joint (Figs. 6-39 and 6-48); but only one has cutaneous branches.

Observe: The *long* head (*L*), most medial, coming from the infraglenoid tubercle. The *lateral* head (*LA*) arising from a ridge on the humerus and the *medial* head (*M*) from much of the posterior surface of the humerus. The medial head is attached to the deep surface of the triceps tendon which inserts into the upper aspect of the olecranon and into the deep fascia of the forearm (Fig. 6-62).

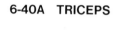

6-40A   TRICEPS

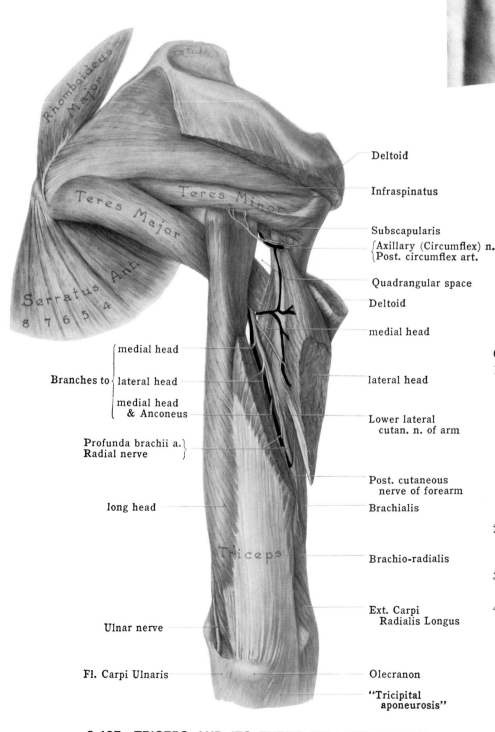

Deltoid

Infraspinatus

Subscapularis
{ Axillary (Circumflex) n.
{ Post. circumflex art.

Quadrangular space

Deltoid

medial head

lateral head

Lower lateral
cutan. n. of arm

Post. cutaneous
nerve of forearm

Brachialis

Brachio-radialis

Ext. Carpi
Radialis Longus

Olecranon

"Tricipital
aponeurosis"

Branches to { medial head
{ lateral head
{ medial head
  & Anconeus

Profunda brachii a.
Radial nerve

long head

Ulnar nerve

Fl. Carpi Ulnaris

Observe:
1. The lateral head of Triceps is divided and displaced. The long and lateral heads lie side by side behind the medial head. The radial nerve, passing in the plane between, is shown supplying the lateral and medial heads of Triceps and Anconeus. The *radial nerve* travels in the gap between the origins of the lateral and medial heads.
2. The *axillary nerve*, passing through the quadrangular space, supplies Deltoid and Teres minor.
3. The *ulnar nerve*, following the medial border of Triceps.
4. Teres Major, Rhomboideus Major, and Serratus Anterior mainly inserted into the inferior angle of the scapula, which is the end of a lever. Their pinwheel arrangement discloses their action in rotation of the scapula.

6-40B   TRICEPS AND ITS THREE RELATED NERVES

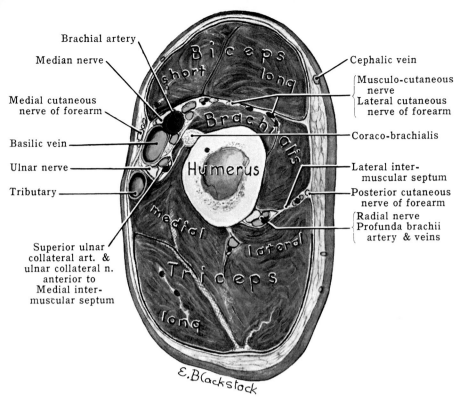

Brachial artery
Median nerve
Medial cutaneous nerve of forearm
Basilic vein
Ulnar nerve
Tributary
Superior ulnar collateral art. & ulnar collateral n. anterior to Medial intermuscular septum

Cephalic vein
Musculo-cutaneous nerve
Lateral cutaneous nerve of forearm
Coraco-brachialis
Lateral intermuscular septum
Posterior cutaneous nerve of forearm
Radial nerve
Profunda brachii artery & veins

E. Blackstock

6-41 CROSS-SECTION THROUGH THE ARM, BELOW ITS MIDPOINT

Observe:

1. The shaft of the humerus is nearly circular, its cortex being thickest here.
2. The 3 heads of Triceps, in the posterior compartment of the arm, *i.e.*, behind the medial and lateral intermuscular septa.
3. The radial nerve and its companion vessels in contact with the bone.
4. The 2 heads of Biceps, the Brachialis, and the insertion of Coraco-brachialis in the anterior compartment of the arm, *i.e.*, in front of the medial and lateral intermuscular septa.
5. The musculocutaneous nerve and its companion vessels in the septum between Biceps and Brachialis.
6. The median nerve crossing to the medial side of the brachial artery and its venae comitantes; the ulnar nerve moving posteriorly onto the side of Triceps. The basilic vein (here as 2 vessels) has pierced the deep fascia.
7. The skin and subcutaneous tissues, thicker postero-laterally where exposed to injury than antero-medially where protected.

Observe:

1. The bursa has been injected with yellow latex.
2. The term "subacromial bursa" is usually understood to include the subdeltoid bursa, for the two bursae are usually combined.
3. Superficial to the bursa are parts of Deltoid, acromion, and coraco-acromial ligament and the acromioclavicular joint.
4. Deep to the bursa are the greater tubercle of the humerus and Supraspinatus tendon (not in view) (Figs. 6-47 and 6-48).
5. The bursa may extend more widely under the acromion, and it may through attrition (Fig. 6-123) communicate with the shoulder joint and also with the acromioclavicular joint.
6. The acromial branches of the thoraco-acromial and suprascapular arteries are seen contributing to the acromial rete (network). The branches of the circumflex artery are destroyed.

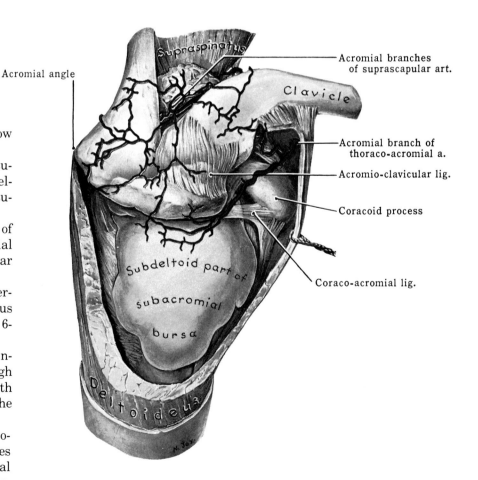

Acromial angle

Supraspinatus
Clavicle
Acromial branches of suprascapular art.
Acromial branch of thoraco-acromial a.
Acromio-clavicular lig.
Coracoid process
Subdeltoid part of subacromial bursa
Coraco-acromial lig.
Deltoideus

N. Joy

6-42 SUBACROMIAL BURSA, SUPEROLATERAL VIEW

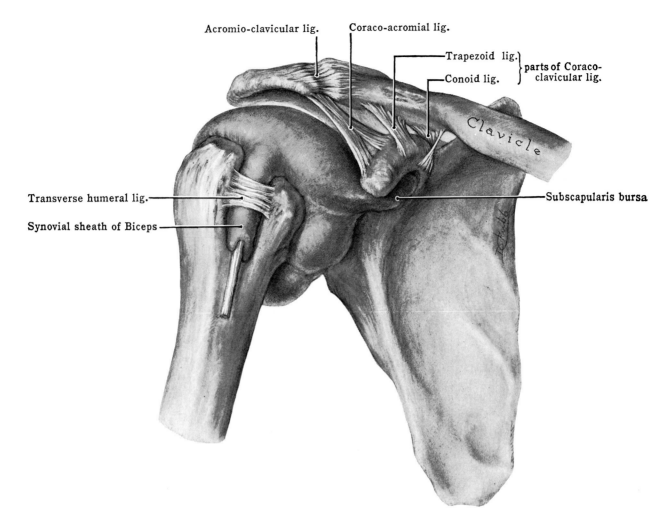

Acromio-clavicular lig. Coraco-acromial lig.

Trapezoid lig.
Conoid lig. } parts of Coraco-
clavicular lig.

Clavicle

Transverse humeral lig.

Synovial sheath of Biceps

Subscapularis bursa

6-43 SYNOVIAL CAPSULE OF THE SHOULDER JOINT, LIGAMENTS AT THE LATERAL END OF THE CLAVICLE

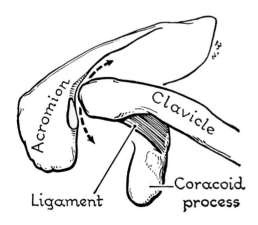

Acromion

Clavicle

Ligament

Coracoid process

6-44 CORACO-CLAVICULAR LIGAMENT

Demonstrating that, so long as the coraco-clavicular ligament is intact, the acromion cannot be driven under the clavicle. This ligament, however, does not prevent protraction and retraction of the acromion.

Observe:
1. The capsule cannot extend onto the lesser and greater tubercles of the humerus, because the 4 short muscles (Subscapularis, Supraspinatus, Infraspinatus, and Teres Minor) are inserted there, but it can and does extend inferiorly onto the surgical neck.
2. The capsule has two prolongations: (a) where it forms a synovial sheath for the tendon of the long head of Biceps in its osseo-fibrous tunnel, and (b) below the coracoid process where it forms a bursa between Subscapularis tendon and the margin of the glenoid cavity.
3. The conoid and trapezoid ligaments are so directed that the clavicle shall hold the scapula laterally.

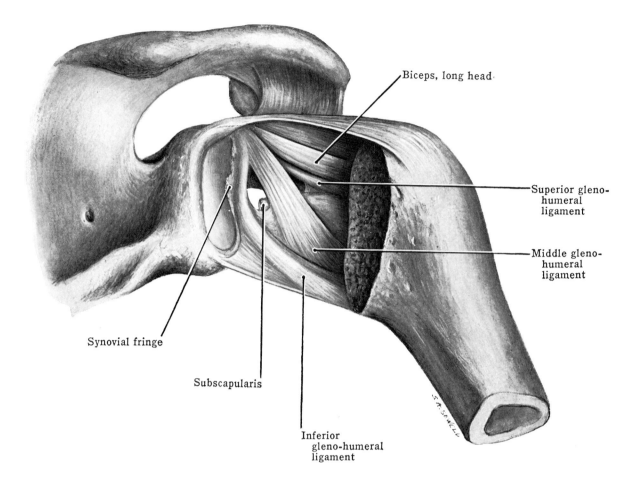

Biceps, long head

Superior gleno-
humeral
ligament

Middle gleno-
humeral
ligament

Synovial fringe

Subscapularis

Inferior
gleno-humeral
ligament

6-45 INTERIOR OF THE SHOULDER JOINT

Exposed from behind by cutting away the posterior part of the capsule and sawing off the head of the humerus.

Observe:

1. The 3 thickenings of the anterior part of the fibrous capsule, called the gleno-humeral ligaments, which are visible from within the joint, but not from without.

2. How these 3 ligaments and the long tendon of Biceps converge on the supraglenoid tubercle.

3. The slender superior ligament parallel to the Biceps tendon; the middle ligament free medially due to the fact that the Subscapularis bursa communicates with the joint cavity both above this ligament and below it; the inferior ligament contributing largely to the anterior lip of the glenoid labrum, much as the Biceps contributes to the posterior lip (see Fig. 6-46).

4. The synovial fringe, constantly present, that overlies the anterior part of the glenoid cavity.

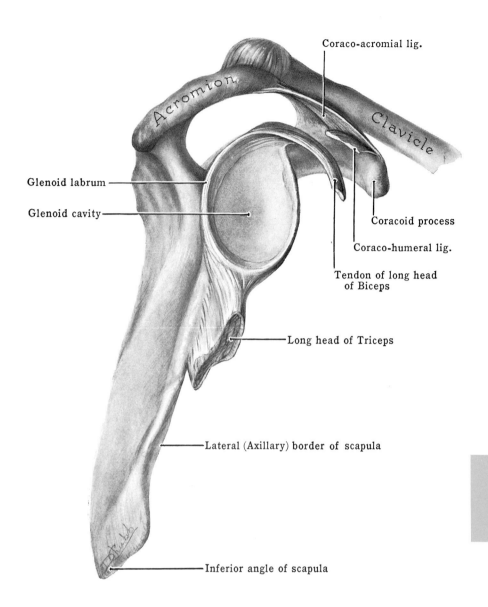

Coraco-acromial lig.

Acromion

Clavicle

Glenoid labrum

Glenoid cavity

Coracoid process

Coraco-humeral lig.

Tendon of long head of Biceps

Long head of Triceps

Lateral (Axillary) border of scapula

Inferior angle of scapula

6-46 GLENOID CAVITY, LATERAL VIEW

Observe:
1. The cavity overhung by the resilient coraco-acromial arch (*i.e.*, coracoid process, coraco-acromial ligament, and acromion), which prevents upward displacement of the head of the humerus.
2. The long head of Triceps arising just below the glenoid cavity.
3. The long head of Biceps arising just above the glenoid cavity. Proximally it is continued as the posterior lip of the glenoid labrum; distally it curves across the front of the head of the humerus, not above it.
4. The orientation of the scapula ensures that should the head of the humerus be dislocated downward it would pass onto the costal surface of the scapula.

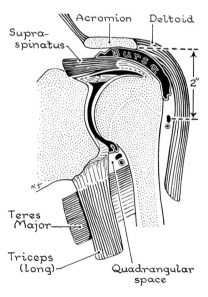

Acromion Deltoid

Supra-spinatus

Bursa

2″

Teres Major

Triceps (long)

Quadrangular space

6-47 CORONAL SECTION OF SHOULDER REGION

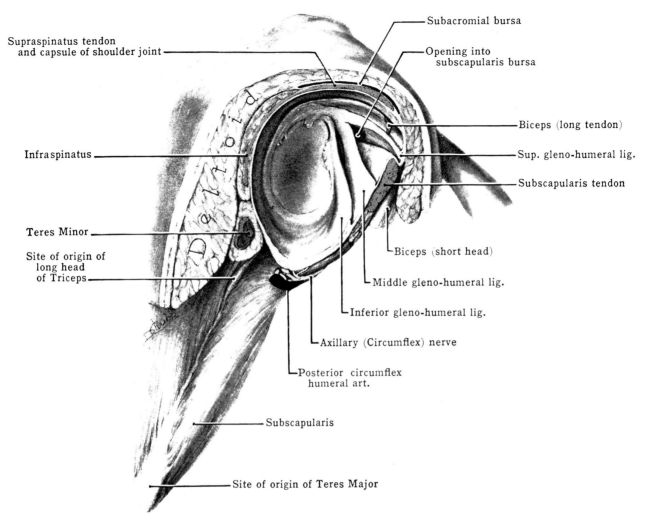

Subacromial bursa

Supraspinatus tendon and capsule of shoulder joint

Opening into subscapularis bursa

Biceps (long tendon)

Infraspinatus

Sup. gleno-humeral lig.

Subscapularis tendon

Teres Minor

Site of origin of long head of Triceps

Biceps (short head)

Middle gleno-humeral lig.

Inferior gleno-humeral lig.

Axillary (Circumflex) nerve

Posterior circumflex humeral art.

Subscapularis

Site of origin of Teres Major

6-48A A HUMERUS VIEW OF THE GLENOID CAVITY

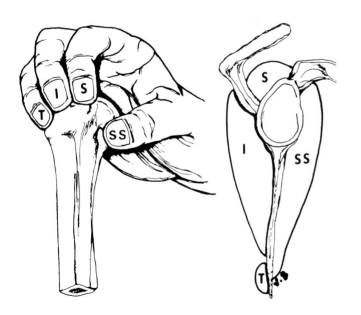

6-48B   THE ROTATOR CUFF

1. The fibrous capsule of the joint thickened in front by the three gleno-humeral ligaments which converge from the humerus to be attached with the long tendon of Biceps to the supraglenoid tubercle.
2. The subacromial bursa between the acromion and Deltoid above and the tendon of Supraspinatus below.
3. The four short muscles which form the "Rotator Cuff": Subscapularis (*SS*), Supraspinatus (*S*), Infraspinatus (*I*), and Teres Minor (*T*). They surround and span the joint, blend with the capsule, and grasp their four points of attachment to the humerus as does the hand in Figure 6-48*B*. Thus they maintain the integrity of the joint by acting as "ligaments under control" as well as moving the humerus in the directions you would expect from their attachments.

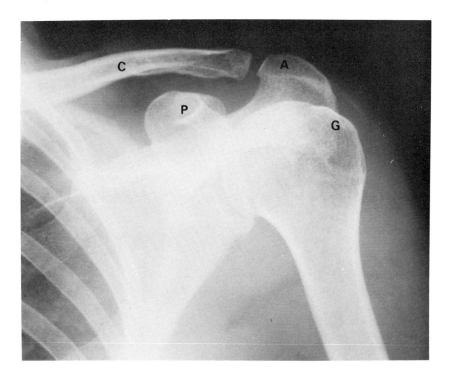

6-49A RADIOGRAPH OF SHOULDER

See Figures 6-35 and 6-43. *C*, clavicle; *A*, acromion; *P*, coracoid process; *G*, greater tubercle of the humerus.

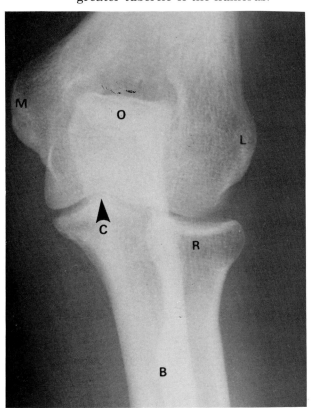 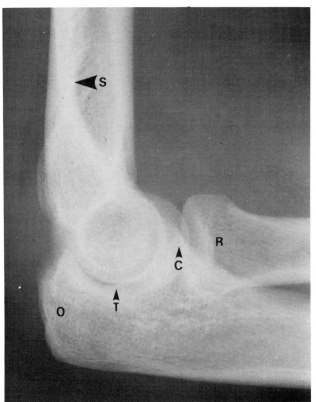

6-49B RADIOGRAPHS OF THE ELBOW

Anteroposterior projection on the *left*, lateral projection on the *right*.

Observe:

On the humerus: medial (*M*) and lateral (*L*) epicondyles, and supracondylar ridges (*S*).

On the ulna: olecranon (*O*), coronoid process (*C*), and trochlear notch (*T*).

On the radius: head (*R*) and radial tuberosity (*B*).

See Figure 6-50.

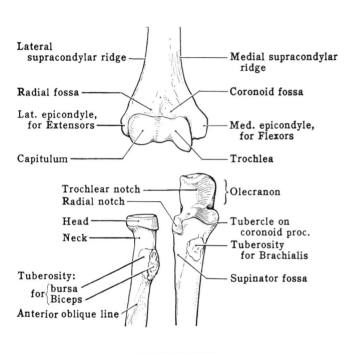

Lateral supracondylar ridge

Medial supracondylar ridge

Radial fossa

Coronoid fossa

Lat. epicondyle, for Extensors

Med. epicondyle, for Flexors

Capitulum

Trochlea

Trochlear notch

Radial notch

Olecranon

Head

Tubercle on coronoid proc.

Neck

Tuberosity for Brachialis

Tuberosity: for {bursa, Biceps}

Supinator fossa

Anterior oblique line

ANTERIOR VIEW

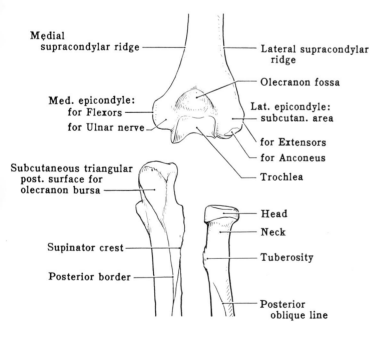

Medial supracondylar ridge

Lateral supracondylar ridge

Olecranon fossa

Med. epicondyle:
for Flexors
for Ulnar nerve

Lat. epicondyle:
subcutan. area

for Extensors

for Anconeus

Trochlea

Subcutaneous triangular post. surface for olecranon bursa

Head

Neck

Supinator crest

Tuberosity

Posterior border

Posterior oblique line

POSTERIOR VIEW

6-50   BONES OF THE ELBOW REGION

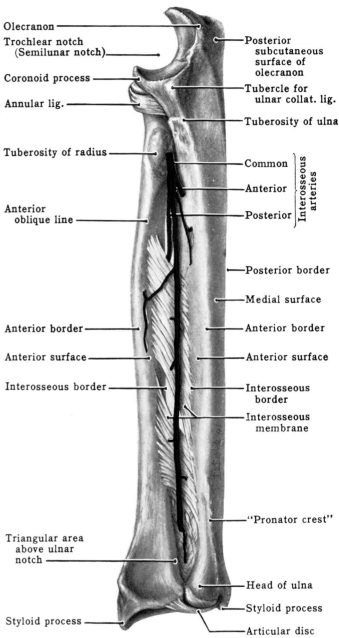

Olecranon

Posterior subcutaneous surface of olecranon

Trochlear notch (Semilunar notch)

Coronoid process

Tubercle for ulnar collat. lig.

Annular lig.

Tuberosity of ulna

Tuberosity of radius

Common

Anterior

Posterior

} Interosseous arteries

Anterior oblique line

Posterior border

Medial surface

Anterior border

Anterior surface

Interosseous border

Anterior border

Anterior surface

Interosseous border

Interosseous membrane

"Pronator crest"

Triangular area above ulnar notch

Head of ulna

Styloid process

Styloid process

Articular disc

6-51   RADIO-ULNAR LIGAMENTS, INTEROSSEOUS ARTERIES

The ligament of the proximal radio-ulnar joint is the annular ligament; that of the distal joint is the articular disc; that of the middle joint is the interosseous membrane. The general direction of the fibers of the membrane is such that an upward thrust to the hand, and therefore received by the radius, is transmitted to the ulna. The membrane is attached to the interosseous borders of the radius and ulna, but it also spreads onto their surfaces.

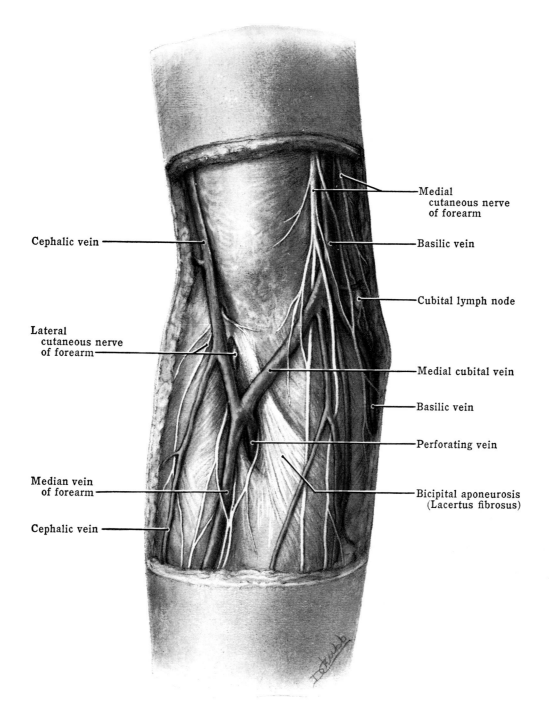

Cephalic vein

Lateral
cutaneous nerve
of forearm

Median vein
of forearm

Cephalic vein

Medial
cutaneous nerve
of forearm

Basilic vein

Cubital lymph node

Medial cubital vein

Basilic vein

Perforating vein

Bicipital aponeurosis
(Lacertus fibrosus)

6-52   FRONT OF ELBOW—I
SUPERFICIAL STRUCTURES

Observe:

1. In the forearm, the superficial veins—cephalic, median, basilic, and their connecting channels—making a (variable) M-shaped pattern.
2. The median cubital vein separated from the brachial artery (Fig. 6-53) only by the bicipital aponeurosis.
3. A perforating vein, lateral to the bicipital aponeurosis, connecting the deep veins to the median cubital vein.
4. In the arm, the cephalic and basilic veins occupying the bicipital furrows, one on each side of Biceps.

5. In the lateral bicipital furrow, the lateral cutaneous nerve of the forearm appearing 2 to 5 cm above the elbow crease, and piercing the deep fascia 2 to 5 cm below it.
6. In the medial bicipital furrow, the medial cutaneous nerve of the forearm becoming cutaneous about the midpoint of the arm.
7. Cutaneous nerves mainly on a deeper plane than superficial veins.
8. The most distal superficial lymph node of the upper limb lying about 2 to 5 cm above the medial epicondyle.

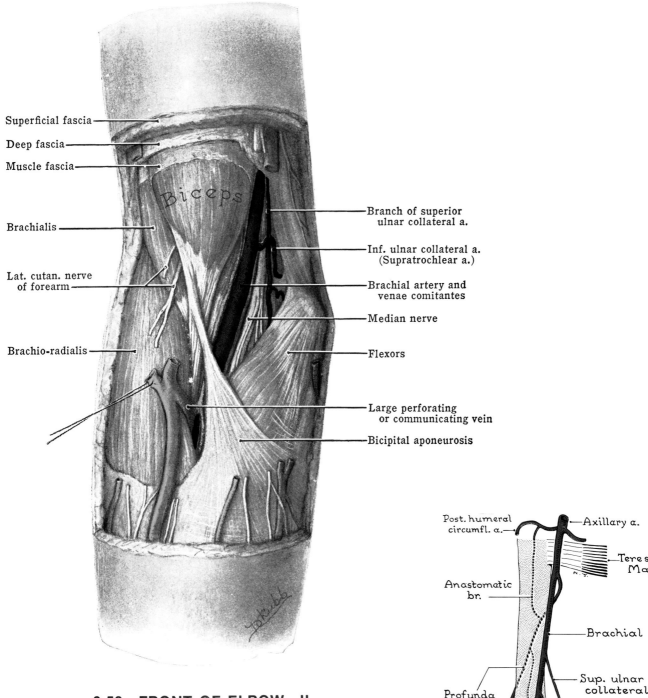

Superficial fascia

Deep fascia

Muscle fascia

Biceps

Brachialis

Lat. cutan. nerve
of forearm

Brachio-radialis

Branch of superior
ulnar collateral a.

Inf. ulnar collateral a.
(Supratrochlear a.)

Brachial artery and
venae comitantes

Median nerve

Flexors

Large perforating
or communicating **vein**

Bicipital aponeurosis

6-53 FRONT OF ELBOW—II
CUBITAL FOSSA

The cubital fossa is the triangular space below the elbow crease. It
is bounded laterally by the extensor muscles (represented by Brachio-radialis) and medially by the flexor muscles (represented by
Pronator Teres). The apex is where these two muscles meet distally.

Observe:

1. The large perforating vein piercing the deep fascia at the apex
   of the fossa.
2. The 3 chief contents—Biceps tendon, brachial artery, and median nerve.
3. Biceps tendon, on approaching its insertion (Fig. 6-35), rotating
   through a right angle, and the bicipital aponeurosis springing
   from the tendon.

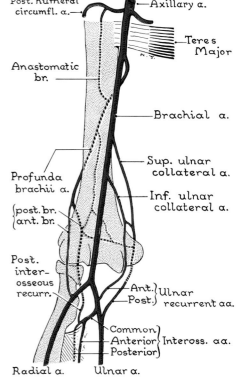

Post. humeral
circumfl. a.

Axillary a.

Teres
Major

Anastomatic
br.

Brachial a.

Profunda
brachii a.

Sup. ulnar
collateral a.

{post. br.
{ant. br.

Inf. ulnar
collateral a.

Post.
inter-
osseous
recurr.

Ant.}Ulnar
Post.}recurrent aa.

Common}
Anterior}Inteross. aa.
Posterior}

Radial a.   Ulnar a.

6-54 ANASTOMOSES
OF ELBOW REGION

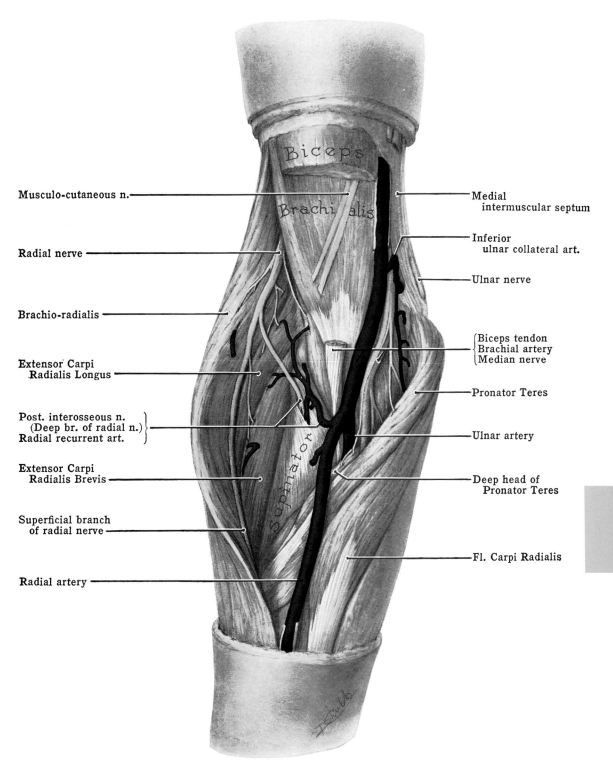

Musculo-cutaneous n.

Radial nerve

Brachio-radialis

Extensor Carpi Radialis Longus

Post. interosseous n. (Deep br. of radial n.) Radial recurrent art.

Extensor Carpi Radialis Brevis

Superficial branch of radial nerve

Radial artery

Medial intermuscular septum

Inferior ulnar collateral art.

Ulnar nerve

Biceps tendon Brachial artery Median nerve

Pronator Teres

Ulnar artery

Deep head of Pronator Teres

Fl. Carpi Radialis

6-55 FRONT OF ELBOW—III

DEEP STRUCTURES

Observe:
1. Part of Biceps is excised and the cubital fossa is opened widely, exposing the floor of the fossa: Brachialis and Supinator.
2. The brachial artery lying between Biceps tendon and median nerve, and dividing into 2 nearly equal branches—the ulnar and radial arteries.
3. The median nerve supplying flexor muscles; hence its motor branches arise from its medial side, the twig to the deep head of Pronator Teres excepted.
4. The radial nerve supplying extensor muscles; hence its motor branches arise from its lateral side, the twig to Brachialis excepted. (The radial nerve has been displaced laterally, so its lateral branches appear in the drawing to run medially.)
5. The posterior interosseous nerve piercing Supinator.

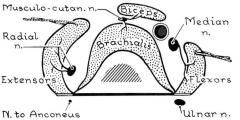

6-56 MOTOR NERVES OF ELBOW REGION

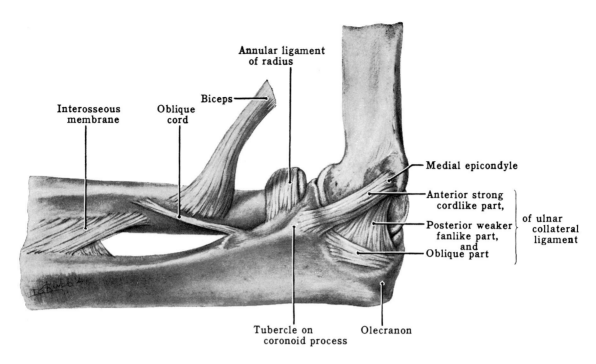

**6-57 ULNAR COLLATERAL LIGAMENT OF THE ELBOW
(MEDIAL LIGAMENT)**

The anterior part is a strong, round cord, taut in extension of the joint; the posterior part is a weak fan, taut in flexion of the joint; the oblique fibers merely deepen the socket for the trochlea of the humerus. Flexor Digitorum Superficialis arises from the cord and from an area of bone at each end of the cord. It also arises from the anterior oblique line of the radius (Fig. 6-51).

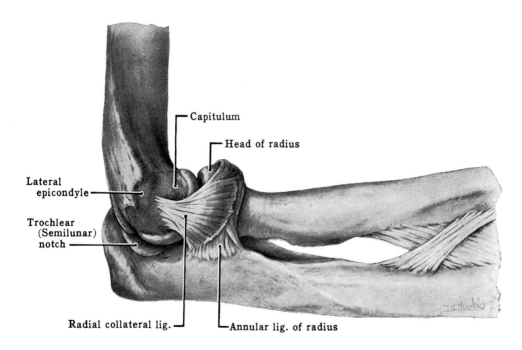

**6-58 RADIAL COLLATERAL LIGAMENT OF THE ELBOW
(LATERAL LIGAMENT)**

The fan-shaped lateral ligament is attached to the annular ligament of radius, but its superficial fibers are continued onward to the radius, as Supinator; hence the ragged edge. Supinator also arises from the annular ligament and the supinator crest of the ulna, which limits the supinator fossa posteriorly.

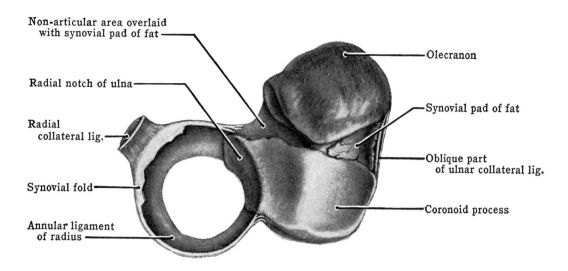

Non-articular area overlaid with synovial pad of fat

Radial notch of ulna

Radial collateral lig.

Synovial fold

Annular ligament of radius

Olecranon

Synovial pad of fat

Oblique part of ulnar collateral lig.

Coronoid process

6-59   SOCKET FOR HEAD OF RADIUS AND TROCHLEA OF HUMERUS

The annular ligament keeps the head of the radius applied to the radial notch of the ulna, and with it forms a cup-shaped socket (i.e., wide above, narrow below). The annular ligament is bound to the humerus by the radial collateral ligament of the elbow.

A common injury in childhood is displacement of the head of the radius following traction on a pronated forearm. Part of the annular ligament becomes trapped between the radial head and the capitulum.

A crescentic synovial fold occupies the angular space between the head of the radius and the capitulum of the humerus. Similarly, synovial folds containing fat occupy the 2 angular nonarticular areas between the coronoid process and the olecranon.

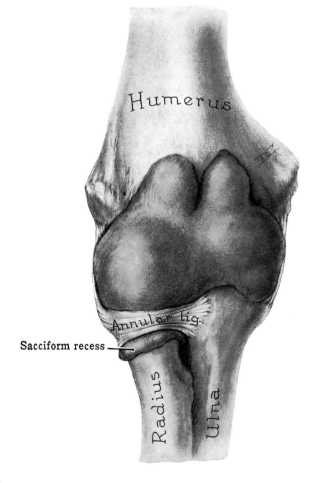

Humerus

Annular lig.

Sacciform recess

Radius   Ulna

6-60   ARTICULAR CAVITY OF THE ELBOW AND PROXIMAL RADIO-ULNAR JOINTS

The cavity was distended with wax. The fibrous capsule has been removed; the synovial capsule remains.

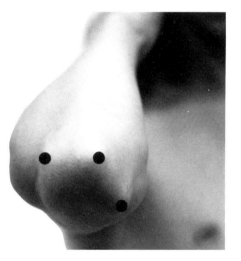

6-61   ELBOW LANDMARKS

Markers have been placed on the medial and lateral epicondyles of the humerus and on the tip of the olecranon process. With elbow flexed, these three points form a triangle. With elbow extended, these three points are on a straight line. These relationships are useful in assessing damage to the elbow region.

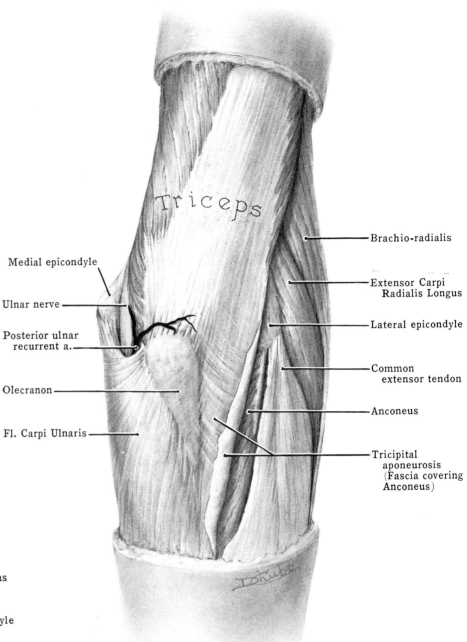

Triceps

Medial epicondyle

Ulnar nerve

Posterior ulnar
recurrent a.

Olecranon

Fl. Carpi Ulnaris

Brachio-radialis

Extensor Carpi
Radialis Longus

Lateral epicondyle

Common
extensor tendon

Anconeus

Tricipital
aponeurosis
(Fascia covering
Anconeus)

6-62 ELBOW, FROM BEHIND—I

Observe:
1. Triceps is inserted not only into the upper surface of the olecranon but also via the deep fascia covering Anconeus—"tricipital aponeurosis"—into lateral border of olecranon.
2. The subcutaneous and palpable posterior surfaces of the medial epicondyle, lateral epicondyle, and olecranon.
3. The ulnar nerve, also palpable, running subfascially behind the medial epicondyle. Distal to this point it disappears deep to the two heads of Flexor Carpi Ulnaris.
4. The two heads of Flexor Carpi Ulnaris; one arising from the common flexor tendon, the other from the medial border of the olecranon and posterior border of the shaft of the ulna.
5. The continuous linear origin from the humerus of the superficial extensor muscles. These are Brachio-radialis, Extensor Carpi Radialis Longus, common extensor tendon, and Anconeus (Fig. 6-65).

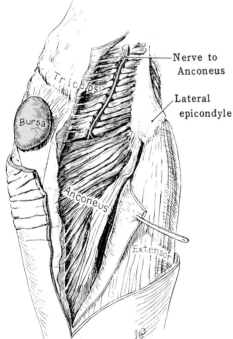

Nerve to
Anconeus

Lateral
epicondyle

Triceps

Bursa

Anconeus

Extensor

6-63 ANCONEUS

As well as Anconeus and its nerve, note the subcutaneous olecranon bursa and the interosseous recurrent artery.

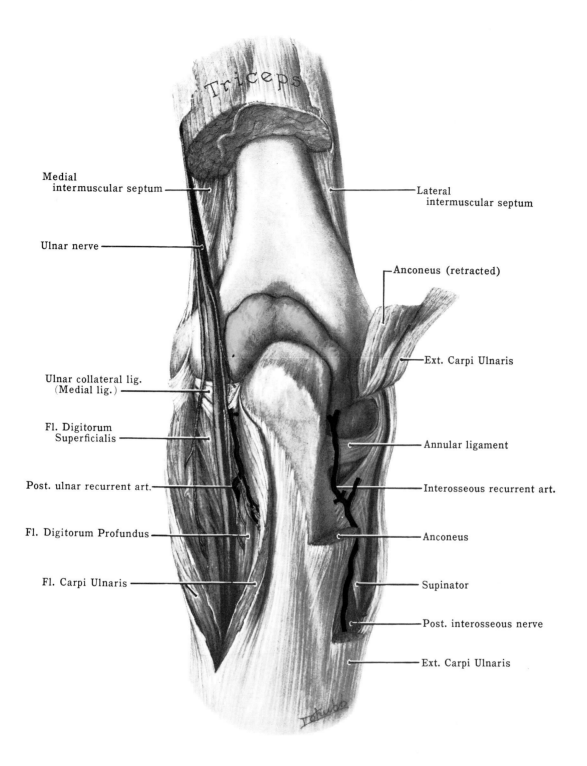

Medial intermuscular septum

Ulnar nerve

Ulnar collateral lig. (Medial lig.)

Fl. Digitorum Superficialis

Post. ulnar recurrent art.

Fl. Digitorum Profundus

Fl. Carpi Ulnaris

Lateral intermuscular septum

Anconeus (retracted)

Ext. Carpi Ulnaris

Annular ligament

Interosseous recurrent art.

Anconeus

Supinator

Post. interosseous nerve

Ext. Carpi Ulnaris

6-64 ELBOW, FROM BEHIND—II

The lower portion of Triceps is removed.

Observe:

1. The ulnar nerve descending: (1st) subfascially, within the posterior compartment of the arm, applied to the medial head of Triceps, and behind the medial epicondyle; (2nd) applied to the medial ligament of the joint; and (3rd) between Flexor Carpi Ulnaris and Flexor Digitorum Profundus.

2. The first branches of the ulnar nerve distributed to Flexor Carpi Ulnaris, half of Profundus, and the joint.

3. Laterally, the synovial membrane protruding below the annular ligament as a sacciform recess, as in Figure 6-60. The joint is here covered with Anconeus and the common extensor tendon, including Extensor Carpi Ulnaris.

4. The posterior interosseous nerve (deep radial nerve), appearing through Supinator $2\frac{1}{2}$ below the head of the radius.

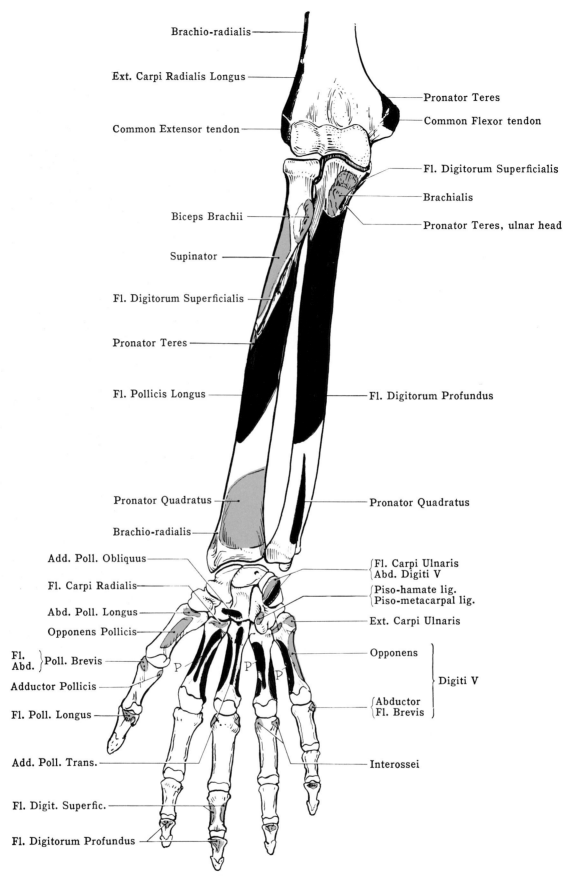

Brachio-radialis

Ext. Carpi Radialis Longus

Pronator Teres

Common Extensor tendon

Common Flexor tendon

Fl. Digitorum Superficialis

Brachialis

Pronator Teres, ulnar head

Biceps Brachii

Supinator

Fl. Digitorum Superficialis

Pronator Teres

Fl. Pollicis Longus

Fl. Digitorum Profundus

Pronator Quadratus

Pronator Quadratus

Brachio-radialis

Add. Poll. Obliquus

Fl. Carpi Ulnaris
Abd. Digiti V

Fl. Carpi Radialis

Piso-hamate lig.
Piso-metacarpal lig.

Abd. Poll. Longus

Ext. Carpi Ulnaris

Opponens Pollicis

Opponens

Fl.
Abd. } Poll. Brevis

Digiti V

Adductor Pollicis

Abductor
Fl. Brevis

Fl. Poll. Longus

Add. Poll. Trans.

Interossei

Fl. Digit. Superfic.

Fl. Digitorum Profundus

6-65A BONES OF THE FOREARM AND HAND SHOWING ATTACHMENTS OF MUSCLES, ANTERIOR VIEW

Note that the origins of the 3 Palmar Interossei are indicated by the letters *P, P, P*, the origins of the 4 Dorsal Interossei by *color* only; the origins of the 3 Thenar and 2 of the Hypothenar muscles are omitted.

For posterior view, see Figure 6-90.

For humerus see Figures 6-35 and 6-36.

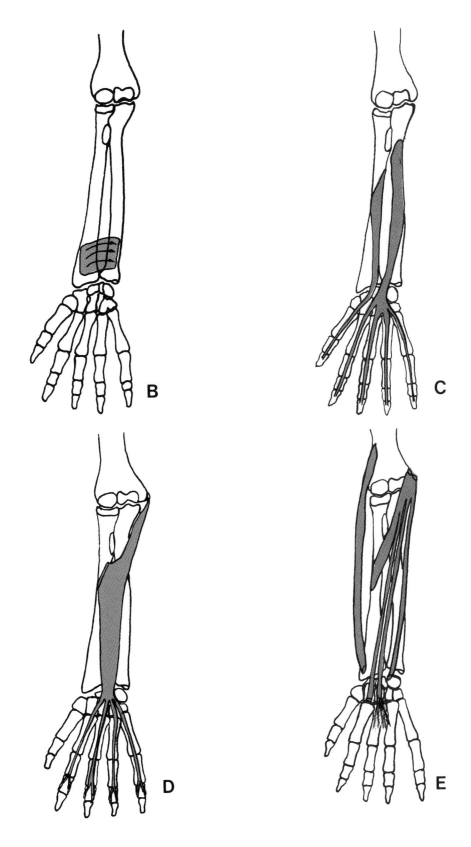

6-65B-E   FOUR LAYERS OF ANTERIOR FOREARM MUSCLES

Examine these four diagrams in relation to the origins and insertions shown on the facing page. From deep to superficial: *B*, Pronator Quadratus; *C*, Flexor pollicis longus, laterally; Flexor digitorum profundus, medially; *D*, Flexor digitorum superficialis; *E*, on the lateral side: Brachioradialis; from the common flexor tendon: Pronator teres, Flexor carpi radialis, Palmaris longus, and Flexor carpi ulnaris.

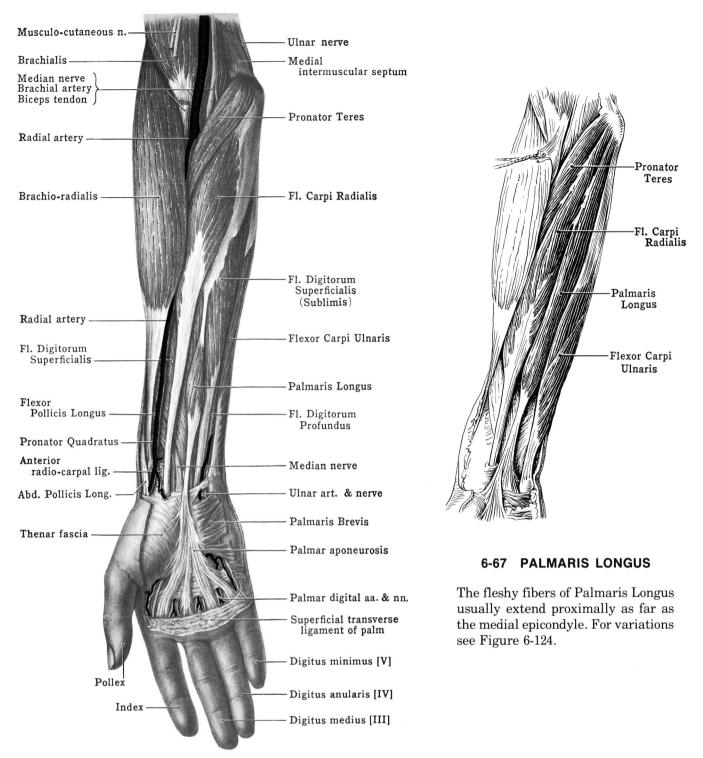

Musculo-cutaneous n.

Brachialis

Median nerve
Brachial artery
Biceps tendon

Radial artery

Brachio-radialis

Radial artery

Fl. Digitorum Superficialis

Flexor Pollicis Longus

Pronator Quadratus

Anterior radio-carpal lig.

Abd. Pollicis Long.

Thenar fascia

Pollex

Index

Ulnar nerve

Medial intermuscular septum

Pronator Teres

Fl. Carpi Radialis

Fl. Digitorum Superficialis (Sublimis)

Flexor Carpi Ulnaris

Palmaris Longus

Fl. Digitorum Profundus

Median nerve

Ulnar art. & nerve

Palmaris Brevis

Palmar aponeurosis

Palmar digital aa. & nn.

Superficial transverse ligament of palm

Digitus minimus [V]

Digitus anularis [IV]

Digitus medius [III]

Pronator Teres

Fl. Carpi Radialis

Palmaris Longus

Flexor Carpi Ulnaris

6-67 PALMARIS LONGUS

The fleshy fibers of Palmaris Longus usually extend proximally as far as the medial epicondyle. For variations see Figure 6-124.

6-66 SUPERFICIAL MUSCLES ON THE FRONT OF THE FOREARM, PALMAR APONEUROSIS

Observe:

1. At the elbow, the brachial artery lying between Biceps tendon and median nerve, and there bifurcating into the radial and ulnar arteries.
2. At the wrist, the radial artery lateral to Flexor Carpi Radialis tendon and the ulnar artery lateral to Flexor Carpi Ulnaris tendon.
3. In the forearm, the radial artery lying between two muscle groups or two motor territories. The muscles lateral to the artery are supplied by the radial nerve; those medial to it, by the median and ulnar nerves. Thus no motor nerve crosses the radial artery.
4. The lateral group of muscles, represented by Brachio-radialis, slightly overlapping the radial artery which otherwise is superficial.

5. The medial or flexor group of muscles is in three layers: superficial (shown here), middle, and deep (Figs. 6-68 and 6-69).
6. The four superficial muscles—Pronator Teres, Flexor Carpi Radialis, Palmaris Longus, and Flexor Carpi Ulnaris—radiating from the medial epicondyle. The muscle of the middle layer Flexor Digitorum Superficialis, is partially in view.
7. Palmaris Longus continued into the palm as the palmar aponeurosis, which receives an accession of fibers from the flexor retinaculum and divides into 4 longitudinal bands, one for each finger. These bands are crossed on their deep surface by transverse fibers.
8. Palmaris Brevis arising from this aponeurosis.

Observe:

1. The oblique origin of Superficialis from (a) the medial epicondyle of the humerus, (b) the ulnar collateral ligament of the elbow, (c) the tubercle of the coronoid process, (d) the anterior oblique line of the radius and perhaps the anterior border below it.

2. Superficialis, which like the 3 muscles in front of it and the 2½ muscles behind it, is supplied by the median nerve.

3. The sequence of muscles crossed by the radial artery. At the wrist the artery crosses the radius and the palmar radiocarpal ligament.

4. The radial artery having muscular and 3 anastomotic branches (recurrent, palmar carpal, and superficial palmar). Neither muscle nor motor nerve crosses the artery, but Brachio-radialis overlaps it.

5. The ulnar artery descending obliquely behind Superficialis to meet and acocompany the ulnar nerve.

6. The ulnar nerve descending vertically near the medial border of Superficialis. It is exposed by splitting the septum between Superficialis and Flexor Carpi Ulnaris.

7. The median nerve descending vertically behind Superficialis, clinging to it, and appearing at its lateral border.

8. The digital tendons at the wrist—III and IV large and side by side, V slender, and II behind (Fig. 6-72).

9. The median artery, here persisting.

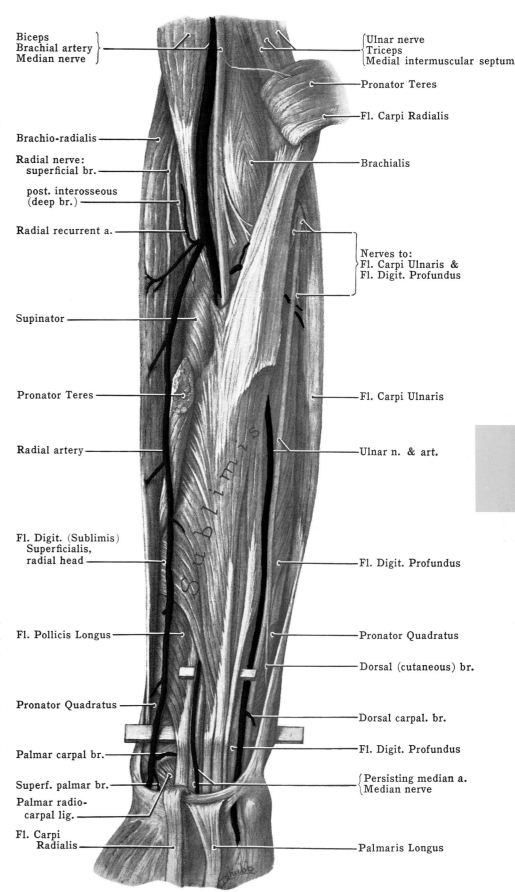

Biceps
Brachial artery
Median nerve

Ulnar nerve
Triceps
Medial intermuscular septum

Pronator Teres

Fl. Carpi Radialis

Brachio-radialis

Radial nerve: superficial br.

post. interosseous (deep br.)

Brachialis

Radial recurrent a.

Nerves to:
Fl. Carpi Ulnaris &
Fl. Digit. Profundus

Supinator

Pronator Teres

Fl. Carpi Ulnaris

Radial artery

Ulnar n. & art.

Fl. Digit. (Sublimis) Superficialis, radial head

Fl. Digit. Profundus

Fl. Pollicis Longus

Pronator Quadratus

Dorsal (cutaneous) br.

Pronator Quadratus

Dorsal carpal. br.

Palmar carpal br.

Fl. Digit. Profundus

Superf. palmar br.

Persisting median a.
Median nerve

Palmar radio-carpal lig.

Fl. Carpi Radialis

Palmaris Longus

6-68   FLEXOR DIGITORUM SUPERFICIALIS (SUBLIMIS) AND RELATED STRUCTURES

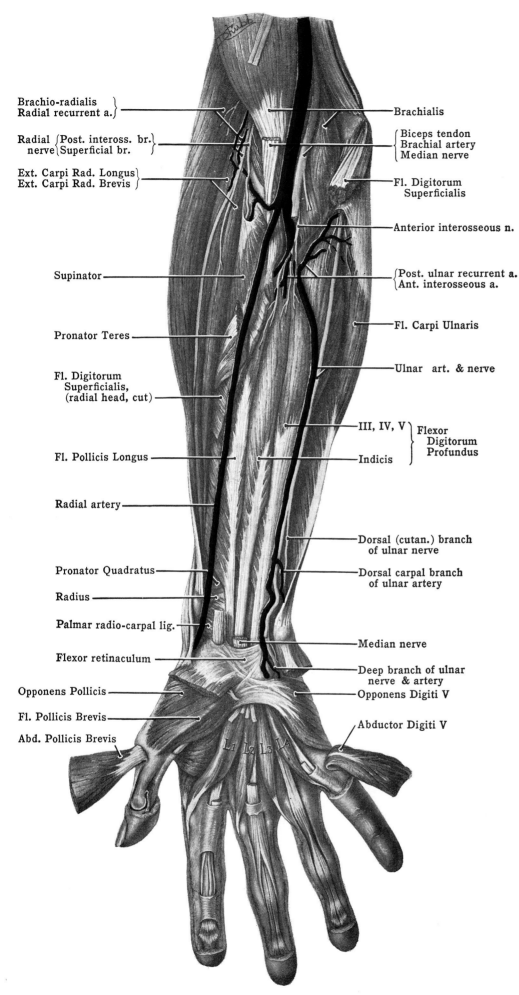

Brachio-radialis
Radial recurrent a.

Radial $\left\{\begin{array}{l}\text{Post. inteross. br.}\\ \text{Superficial br.}\end{array}\right.$
nerve

Ext. Carpi Rad. Longus
Ext. Carpi Rad. Brevis

Supinator

Pronator Teres

Fl. Digitorum
Superficialis,
(radial head, cut)

Fl. Pollicis Longus

Radial artery

Pronator Quadratus

Radius

Palmar radio-carpal lig.

Flexor retinaculum

Opponens Pollicis

Fl. Pollicis Brevis

Abd. Pollicis Brevis

Brachialis

$\left\{\begin{array}{l}\text{Biceps tendon}\\ \text{Brachial artery}\\ \text{Median nerve}\end{array}\right.$

Fl. Digitorum
Superficialis

Anterior interosseous n.

$\left\{\begin{array}{l}\text{Post. ulnar recurrent a.}\\ \text{Ant. interosseous a.}\end{array}\right.$

Fl. Carpi Ulnaris

Ulnar art. & nerve

III, IV, V $\left.\begin{array}{l}\\ \\ \end{array}\right\}$ Flexor
Digitorum
Indicis   Profundus

Dorsal (cutan.) branch
of ulnar nerve

Dorsal carpal branch
of ulnar artery

Median nerve

Deep branch of ulnar
nerve & artery

Opponens Digiti V

Abductor Digiti V

L₁ L₂ L₃ L₄

Observe:

1. The two deep, digital flexor muscles, namely, Flexor Pollicis Longus and Flexor Digitorum Profundus, forming a set of muscles that arises from the flexor aspects of radius, interosseous membrane, and ulna between the origin of Superficialis proximally and Pronator Quadratus distally.

2. The portion of Profundus for the index, here free above the wrist; the portions for digits III, IV, and V, fused.

3. The median nerve crossing in front of the ulnar artery at the elbow and behind the flexor retinaculum at the wrist.

4. The ulnar nerve is sheltered by the medial epicondyle at the elbow and by the pisiform bone at the wrist.

5. The ulnar nerve entering the forearm behind the medial epicondyle, descending on Profundus, joined by the ulnar artery, continuing on Profundus to the wrist, and there passing in front of the flexor retinaculum and lateral to the pisiform it enters the palm. At the elbow it supplies Flexor Carpi Ulnaris and the medial half of Profundus; above the wrist it gives off its dorsal branch.

6. The recurrent, common interosseous, and dorsal carpal branches as well as muscular branches of the ulnar artery.

7. The 4 Lumbricals arising from Profundus tendons, seen better in Figure 6-79.

6-69 DEEP FLEXORS OF THE DIGITS AND THE RELATED STRUCTURES

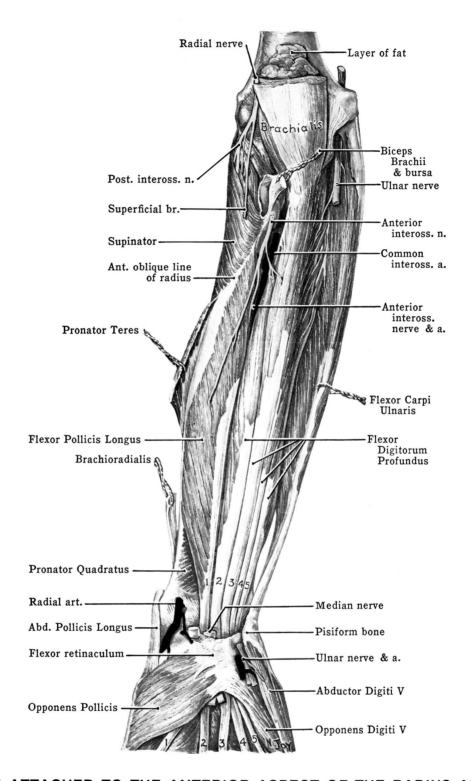

Radial nerve

Layer of fat

Brachialis

Post. inteross. n.

Biceps
Brachii
& bursa

Ulnar nerve

Superficial br.

Anterior
inteross. n.

Supinator

Common
inteross. a.

Ant. oblique line
of radius

Anterior
inteross.
nerve & a.

Pronator Teres

Flexor Carpi
Ulnaris

Flexor Pollicis Longus

Flexor
Digitorum
Profundus

Brachioradialis

Pronator Quadratus

1 2 3 4 5

Radial art.

Median nerve

Abd. Pollicis Longus

Pisiform bone

Flexor retinaculum

Ulnar nerve & a.

Abductor Digiti V

Opponens Pollicis

Opponens Digiti V

1   2   3   4 5 N Joy

6-70   MUSCLES ATTACHED TO THE ANTERIOR ASPECT OF THE RADIUS AND ULNA

Observe:

1. The anterior aspect of the ulna, clothed with (a) Brachialis, which is inserted into the coronoid process (Fig. 6-65), and distal to this with (b) Flexor Digitorum Profundus, which arises as far distally as Pronator Quadratus. Profundus also arising from the upper two-thirds of the medial aspect and with Flexor Carpi Ulnaris arising from the deep fascia.

2. The anterior aspect of the radius, clothed with (a) Supinator above the anterior oblique line and, distal to this, with (b) Flexor Pollicis Longus as far as Pronatur Quadratus. Flexor Digitorum Superficialis, here removed, arises from the the oblique line.

3. The 5 tendons of the deep digital flexors, side by side, converging on the carpal tunnel, and, having traversed it, diverging to the 5 terminal phalanges.

4. Biceps inserted into the medial aspect of radius, hence it can rotate laterally, i.e., supinate, whereas Pronator Teres by invading the lateral surface can rotate medially, i.e., pronate.

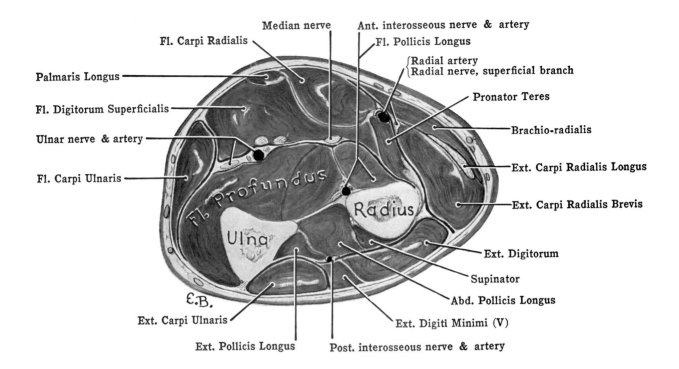

Median nerve
Fl. Carpi Radialis
Ant. interosseous nerve & artery
Fl. Pollicis Longus
Palmaris Longus
Radial artery
Radial nerve, superficial branch
Fl. Digitorum Superficialis
Pronator Teres
Ulnar nerve & artery
Brachio-radialis
Fl. Carpi Ulnaris
Ext. Carpi Radialis Longus
Ext. Carpi Radialis Brevis
Fl. Profundus
Radius
Ulna
Ext. Digitorum
Supinator
Abd. Pollicis Longus
E.B.
Ext. Carpi Ulnaris
Ext. Digiti Minimi (V)
Ext. Pollicis Longus
Post. interosseous nerve & artery

6-71 CROSS-SECTION THROUGH THE MIDDLE OF THE FOREARM (AT LEVEL OF INSERTION OF THE PRONATOR TERES)

Observe:

1. The interosseous membrane stretching from the interosseous (lateral) border of the ulna to a fibro-cartilaginous labrum along the interosseous (medial) border of the radius and spreading far onto the anterior surface (Fig. 6-51) and less far onto the posterior surface.

2. The ulnar nerve and artery and the median nerve lying in the areolar septum between the superficial and deep digital flexors, *i.e.*, between Flexor Digitorum Superficialis and Flexor Digitorum Profundus and Flexor Pollicis Longus. The ulnar nerve, as usual, clinging to Profundus and supplying its medial part; the median nerve clinging to Superficialis and supplying it. The anterior interosseous branch of the median nerve, placed on the skeletal plane.

3. The radial artery overlapped by Brachio-radialis.

4. The superficial branch of the radial nerve following the anterior border of Extensor Carpi Radialis Brevis; the posterior interosseous nerve lying between the superficial and deep layers of digital extensors.

5. Flexor Digitorum Profundus and Flexor Pollicis Longus wrapped around the medial and anterior surfaces of the ulna and the anterior surface of the radius, this being Flexor Territory. Pronator Teres inserted into the lateral surface of the radius, and therefore invading Extensor Territory; hence its ability to pronate.

6. Artificial spaces, readily created: (a) The space deep to Flexor Carpi Ulnaris extending dorsally to where Flexor Carpi Ulnaris and Flexor Digitorum Profundus both arise from the deep fascia and through it from the posterior border of the ulna. (b) The space deep to Extensor Carpi Ulnaris. This muscle arises slightly from the deep fascia and through it from the posterior border of the ulna. (c) The space between the superficial and deep digital flexors, noted in item 2, above. (d) The space between the long extensors of the fingers and the deep muscles of the back of the forearm. (e) The space deep to the three lateral muscles of the forearm (Brachio-radialis, Extensor Carpi Radialis Longus and Brevis).

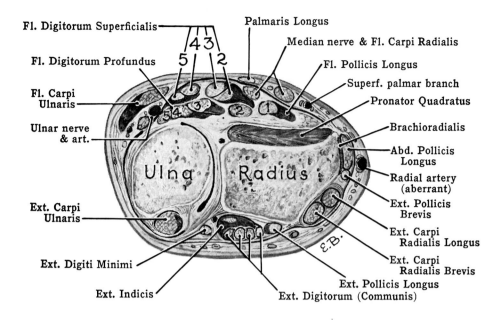

6-72 CROSS-SECTION THROUGH THE FOREARM, ABOVE THE WRIST

Observe:
1. The synovial cavity of the distal radio-ulnar joint.
2. Flexor Carpi Radialis, Palmaris Longus, and Flexor Carpi Ulnaris constituting a surface layer of flexors of the wrist.
3. Deep to these, the long flexors of the digits: (a) the 4 tendons of Flexor Digitorum Superficialis lying two deep, those to the middle and ring fingers being anterior to those to the index and little fingers; (b) the 5 tendons of the deep digital flexors, lying side by side, those to the thumb and index being free.
4. The ulnar nerve and artery under cover of Flexor Carpi Ulnaris

where the pulse of the artery could not be felt. The median nerve at the midpoint on the front of the wrist, deep to Palmaris Longus, and at the lateral border of Flexor Digitorum Superficialis. The radial artery is here aberrant (see Fig. 6-87), so its pulse might be missed.
5. Four tendons on the dorsum of the wrist, large, because, being inserted into metacarpal bones, they work as synergists with the powerful flexors of the digits, whereas the remaining tendons, being extensors of the digits, are slender (Fig. 6-105).

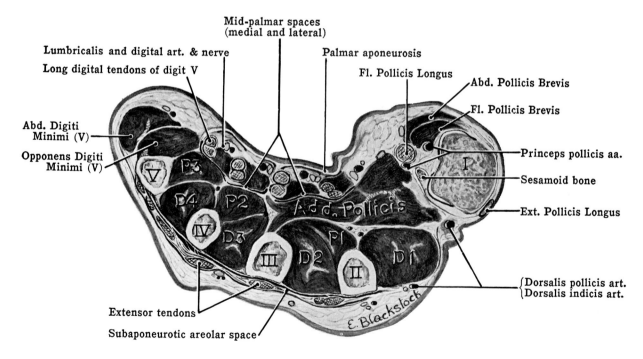

6-73 CROSS-SECTION THROUGH THE MIDDLE OF THE PALM

Observe:
1. The section passes through the head of the first metacarpal and is therefore distal to Opponens Pollicis.
2. The 4 Dorsal Interossei (Abductors) fill the spaces between the 5 metacarpal bones.
3. The 3 Palmar Interossei (Adductors) arise from the palmar aspects of the 2nd, 4th, and 5th metacarpal bones, and the Adductor Pollicis from the palmar aspects of the 3rd.
4. All Interossei and the adductor are supplied by the ulnar nerve.
5. Between the foregoing muscles and the palmar aponeurosis lie the long flexor tendons (superficial and deep) of the 4 fingers, and 4 Lumbricales, and the palmar digital nerves and arteries.
6. Flexor Pollicis Longus tendon, in its synovial sheath, accompanied by its palmar digital arteries and nerves, and passing anterior to the palmar ligament (palmar plate) of the metacarpo-phalangeal joint.
7. The flattened tendons of Extensor Digitorum.

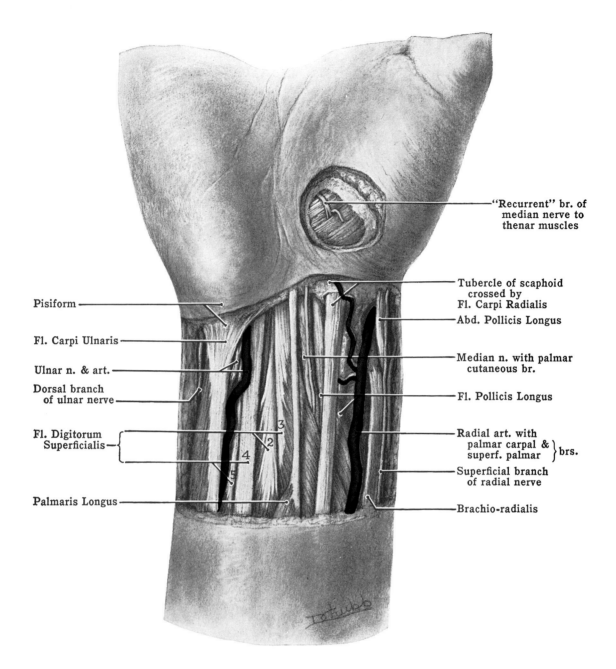

Pisiform

Fl. Carpi Ulnaris

Ulnar n. & art.

Dorsal branch
of ulnar nerve

Fl. Digitorum
Superficialis

Palmaris Longus

"Recurrent" br. of
median nerve to
thenar muscles

Tubercle of scaphoid
crossed by
Fl. Carpi Radialis

Abd. Pollicis Longus

Median n. with palmar
cutaneous br.

Fl. Pollicis Longus

Radial art. with
palmar carpal &
superf. palmar } brs.

Superficial branch
of radial nerve

Brachio-radialis

3
2
4
5

6-74 STRUCTURES AT THE FRONT OF THE WRIST

Observe:

1. The distal skin incision follows the transverse skin crease at the wrist. It crosses the pisiform, to which Flexor Carpi Ulnaris is a guide. At the junction of lateral one-third and medial two-thirds the crease crosses the tubercle of the scaphoid, to which the tendon of Flexor Carpi Radialis is a guide.

2. Palmaris Longus tendon bisects the crease. Deep to its lateral margin is the median nerve.

3. The radial artery disappearing deep to the Abductor.

4. The ulnar nerve and artery sheltered by Flexor Carpi Ulnaris tendon and by the expansion this gives to the flexor retinaculum.

5. Flexor Digitorum Superficialis tendons to digits III and IV somewhat anterior to those to digits II and V.

6. The "recurrent" branch of the median nerve to the thenar muscles lying within a circle whose center is from 2.5 to 4 cm below the tubercle of the scaphoid. Seek it in the proper plane: between deep fascia and muscle fibers.

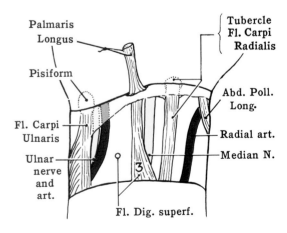

Palmaris
Longus

Pisiform

Fl. Carpi
Ulnaris

Ulnar
nerve
and
art.

Tubercle
Fl. Carpi
Radialis

Abd. Poll.
Long.

Radial art.

Median N.

Fl. Dig. superf.

3

6-75 FRONT OF WRIST

These are the most common deep structures seen by psychiatrists because the wrist is a favorite site for suicide attempts.

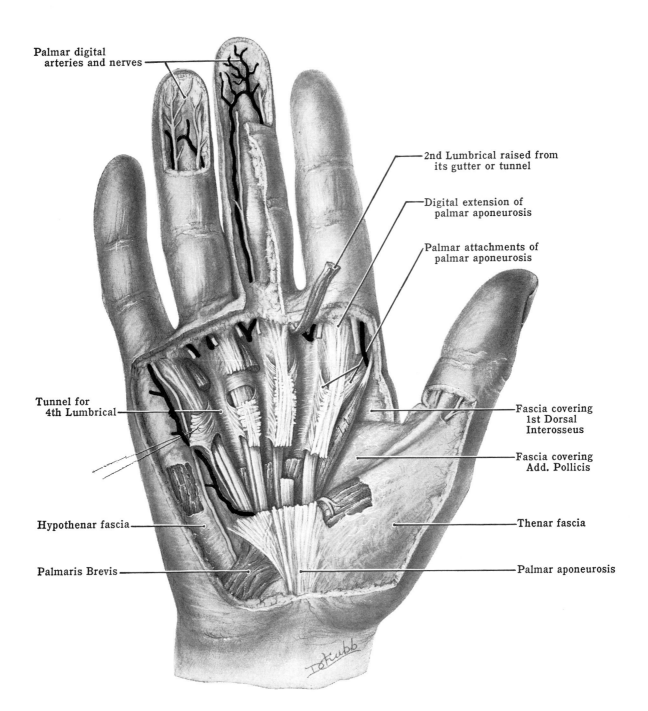

Palmar digital
arteries and nerves

2nd Lumbrical raised from
its gutter or tunnel

Digital extension of
palmar aponeurosis

Palmar attachments of
palmar aponeurosis

Tunnel for
4th Lumbrical

Fascia covering
1st Dorsal
Interosseus

Fascia covering
Add. Pollicis

Hypothenar fascia

Thenar fascia

Palmaris Brevis

Palmar aponeurosis

6-76 ATTACHMENTS OF THE PALMAR APONEUROSIS, DIGITAL VESSELS AND NERVES

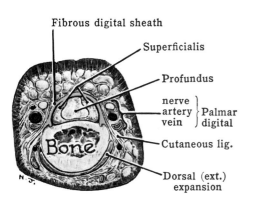

Fibrous digital sheath

Superficialis

Profundus

nerve
artery
vein } Palmar
digital

Cutaneous lig.

Dorsal (ext.)
expansion

Bone

6-77 CROSS-SECTION OF PROXIMAL PHALANX

Note that the palmar digital nerve and vessels
are applied to the fibrous sheath, not to bone.
The skin is thickest on the palmar surface.

Observe:
1. From the palmar aponeurosis a few longitudinal fibers enter the fingers where they
   are lost; the other fibers, forming extensive fibro-areolar septa, pass dorsally, to the
   palmar ligaments (Fig. 6-85), and also, more proximally, to the fascia covering the
   Interossei. Thus two sets of tunnels exist in the distal half of the palm: (a) tunnels for
   long flexor tendons, and (b) tunnels for Lumbricals, digital vessels, and digital nerves.
   The former are continued into the fingers; the latter open on the dorsum of the hand
   behind the web.
2. In a finger, the digital artery and nerve lying on the side of the fibrous digital sheath.
3. The absence of fat deep to the skin creases of the fingers.
4. The 4 palmar spaces: (a) a "thenar," behind the thenar fascia; (b) a "hypothenar,"
   behind the hypothenar fascia; between these (c) a "middle," behind the palmar
   aponeurosis; and (d) a fourth, between Adductor Pollicis and 1st Dorsal Interosseous.
   The middle space contains Superficialis and Profundus tendons, Lumbricals, and
   digital vessels and nerves. It is bounded behind by (a) the fascia covering Interossei
   and Adductor Pollicis, (b) the palmar ligaments, and (c) the deep transverse metacarpal ligaments of the palm (Fig. 6-85).

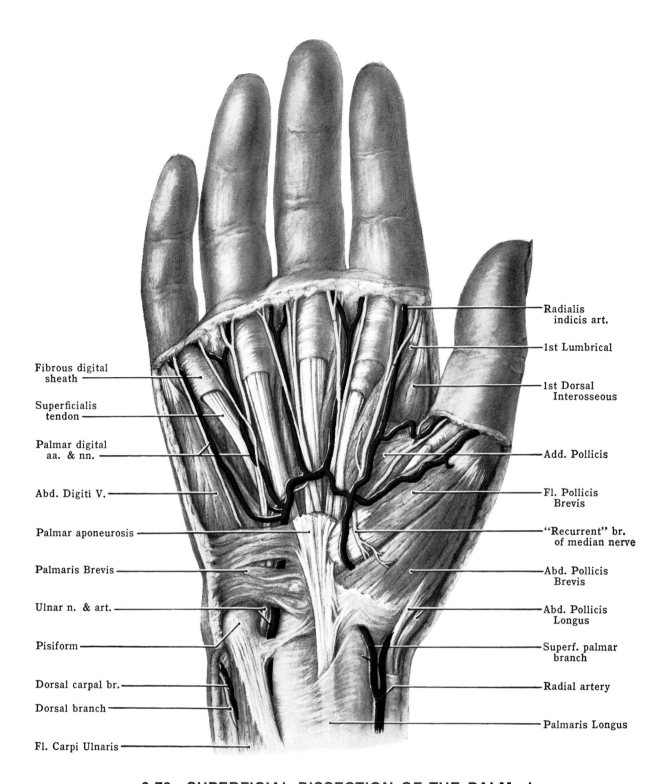

Radialis
indicis art.

1st Lumbrical

1st Dorsal
Interosseous

Add. Pollicis

Fl. Pollicis
Brevis

"Recurrent" br.
of median nerve

Abd. Pollicis
Brevis

Abd. Pollicis
Longus

Superf. palmar
branch

Radial artery

Palmaris Longus

Fibrous digital
sheath

Superficialis
tendon

Palmar digital
aa. & nn.

Abd. Digiti V.

Palmar aponeurosis

Palmaris Brevis

Ulnar n. & art.

Pisiform

Dorsal carpal br.

Dorsal branch

Fl. Carpi Ulnaris

6-78 SUPERFICIAL DISSECTION OF THE PALM—I

Observe:
1. Dissection has removed skin, superficial fascia, the palmar aponeurosis, and the thenar
   and hypothenar fasciae (Fig. 6-76).
2. The superficial palmar arch is formed by the ulnar artery and is completed by the
   superficial palmar branch of the radial artery. Only the foregoing structures and Palmaris
   Brevis cover the arch. It is truly superficial. So likewise are the digital vessels and nerves
   and the "recurrent" branch of the median nerve exposed in Figure 6-74.
3. The four Lumbricals lie behind digital vessels and nerves.
4. The prominent pisiform shelters the ulnar nerve and artery as they pass into the palm.

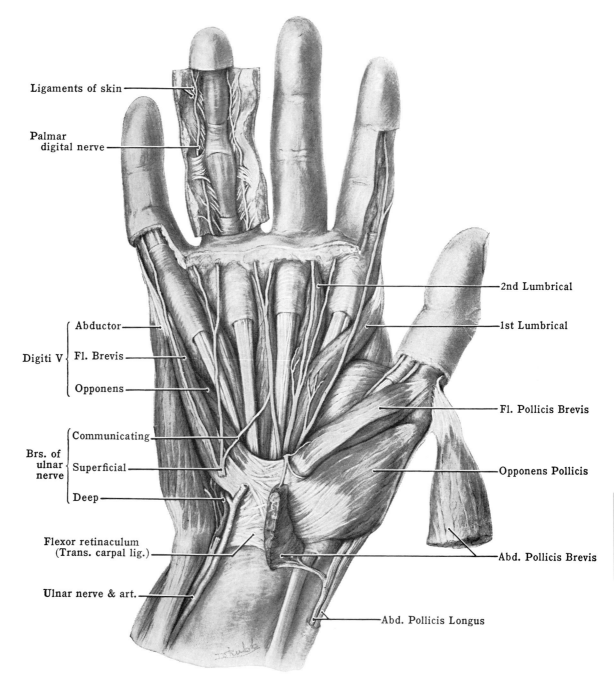

Ligaments of skin

Palmar digital nerve

2nd Lumbrical

1st Lumbrical

Digiti V
- Abductor
- Fl. Brevis
- Opponens

Fl. Pollicis Brevis

Brs. of ulnar nerve
- Communicating
- Superficial
- Deep

Opponens Pollicis

Flexor retinaculum (Trans. carpal lig.)

Abd. Pollicis Brevis

Ulnar nerve & art.

Abd. Pollicis Longus

6-79   SUPERFICIAL DISSECTION OF THE PALM—II

Observe:

1. The 3 thenar and 3 hypothenar muscles arise from the flexor retinaculum and from the 4 marginal carpal bones united by this retinaculum. Abductor Pollicis Brevis arises from Abductor Pollicis Longus tendon also; Abductor Digiti V arises from the pisiform only.

2. The 4 lumbricals arise from the radial sides of the 4 Profundus tendons, and are inserted into the radial sides of the dorsal expansions (Fig. 6-106) of the corresponding digits. The medial 2 Lumbricals, however, also arise from the medial sides of adjacent Profundus tendons; and the 3rd Lumbrical is commonly inserted into two extensor expansions, as here.

3. The median nerve is distributed in the hand to 5 muscles (3 thenar and 2 lumbrical) and provides cutaneous branches to 3½ digits including parts of their dorsal aspects (Fig. 6-99).

4. The ulnar nerve supplies all other short muscles in the hand and provides cutaneous branches to 1½ digits.

5. The skin of the fingers cannot be pulled off like a glove, on account of the oblique fibrous strands, cutaneous ligaments, that moor it to the sides of the 1st and 2nd phalanges and front of the distal phalanges.

See McFarlane, R. M. (1962) Observations on the functional anatomy of the intrinsic muscles of the thumb. *J. Bone Joint Surg.*, 44A: 1073.

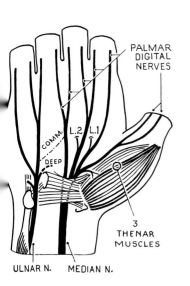

PALMAR DIGITAL NERVES

COMM.
DEEP
L.2 L.1

3 THENAR MUSCLES

ULNAR N.   MEDIAN N.

6-80   NERVE SUPPLY

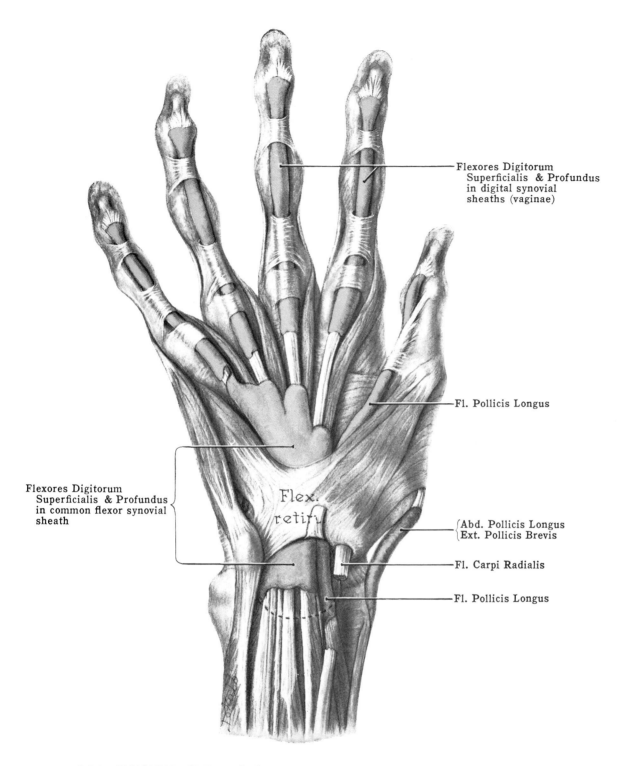

Flexores Digitorum
Superficialis & Profundus
in digital synovial
sheaths (vaginae)

Fl. Pollicis Longus

Flexores Digitorum
Superficialis & Profundus
in common flexor synovial
sheath

Flex.
retin.

{ Abd. Pollicis Longus
{ Ext. Pollicis Brevis

Fl. Carpi Radialis

Fl. Pollicis Longus

6-81 SYNOVIAL SHEATHS OF THE LONG FLEXOR TENDONS OF THE DIGITS

Observe:

1. There are 2 sets: (a) proximal or carpal, behind the flexor retinaculum; (b) distal or digital, behind the fibrous sheaths of the digital flexors.

2. The carpal synovial sheaths, although developmentally separate, unite with one another to form a common flexor sheath, and the carpal sheath of the thumb tendon usually communicates with it. This common flexor sheath extends 1 to 2.5 cm proximal to the flexor retinaculum and extends distally farthest on a thumb and little finger where it is continuous with the distal sheaths.

3. Each digital sheath extends from the proximal end of the palmar ligament or plate (Figs. 6-85 and 6-116) that covers a metacarpal head to the base of a distal phalanx.

4. The flexor tendons play across the very prominent anterior border of the inferior articular surface of the radius; hence the common flexor sheath extends further behind the tendons (broken line) than in front.

5. The sheath of Flexor Carpi Radialis is not injected in this specimen.

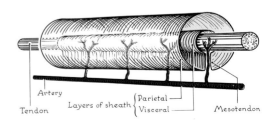

Artery

Tendon   Layers of sheath { Parietal
                              { Visceral   Mesotendon

6-82 SYNOVIAL SHEATHS

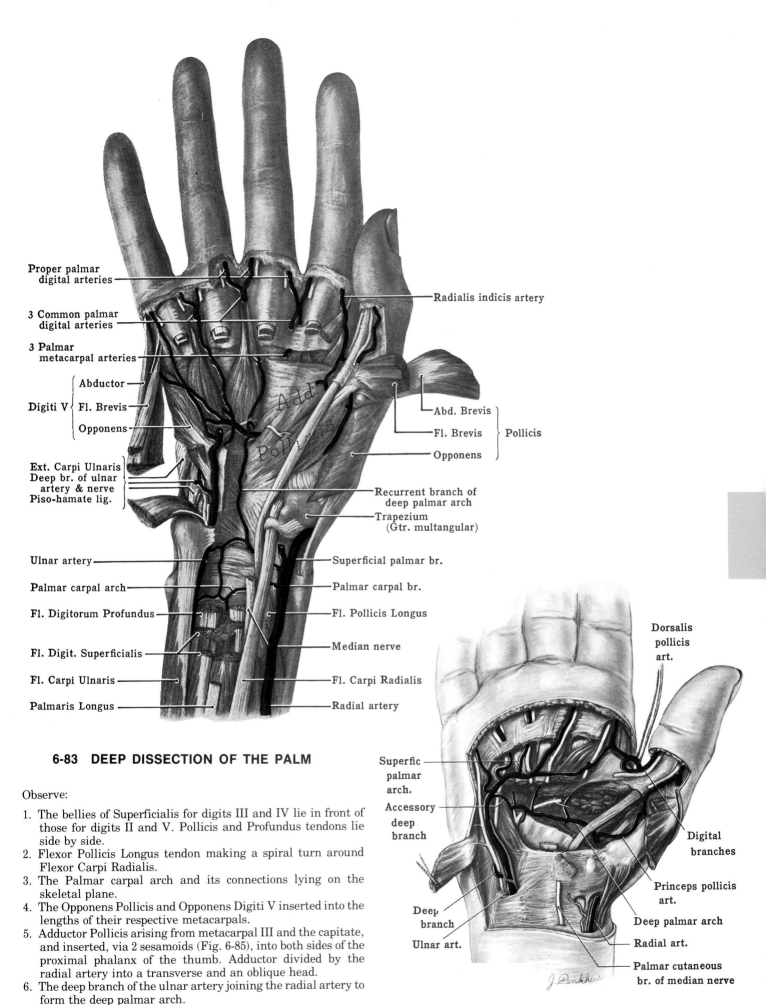

Proper palmar digital arteries

3 Common palmar digital arteries

3 Palmar metacarpal arteries

Digiti V { **Abductor** — **Fl. Brevis** — **Opponens** —

Ext. Carpi Ulnaris
Deep br. of ulnar artery & nerve
Piso-hamate lig.

Ulnar artery

Palmar carpal arch

Fl. Digitorum Profundus

Fl. Digit. Superficialis

Fl. Carpi Ulnaris

Palmaris Longus

Radialis indicis artery

Abd. Brevis
Fl. Brevis } Pollicis
Opponens

Recurrent branch of deep palmar arch
Trapezium (Gtr. multangular)

Superficial palmar br.
Palmar carpal br.
Fl. Pollicis Longus
Median nerve
Fl. Carpi Radialis
Radial artery

6-83 DEEP DISSECTION OF THE PALM

Observe:

1. The bellies of Superficialis for digits III and IV lie in front of those for digits II and V. Pollicis and Profundus tendons lie side by side.
2. Flexor Pollicis Longus tendon making a spiral turn around Flexor Carpi Radialis.
3. The Palmar carpal arch and its connections lying on the skeletal plane.
4. The Opponens Pollicis and Opponens Digiti V inserted into the lengths of their respective metacarpals.
5. Adductor Pollicis arising from metacarpal III and the capitate, and inserted, via 2 sesamoids (Fig. 6-85), into both sides of the proximal phalanx of the thumb. Adductor divided by the radial artery into a transverse and an oblique head.
6. The deep branch of the ulnar artery joining the radial artery to form the deep palmar arch.

Dorsalis pollicis art.

Superfic palmar arch.

Accessory deep branch

Deep branch

Ulnar art.

Digital branches

Princeps pollicis art.

Deep palmar arch

Radial art.

Palmar cutaneous br. of median nerve

6-84 PALMAR ARCHES

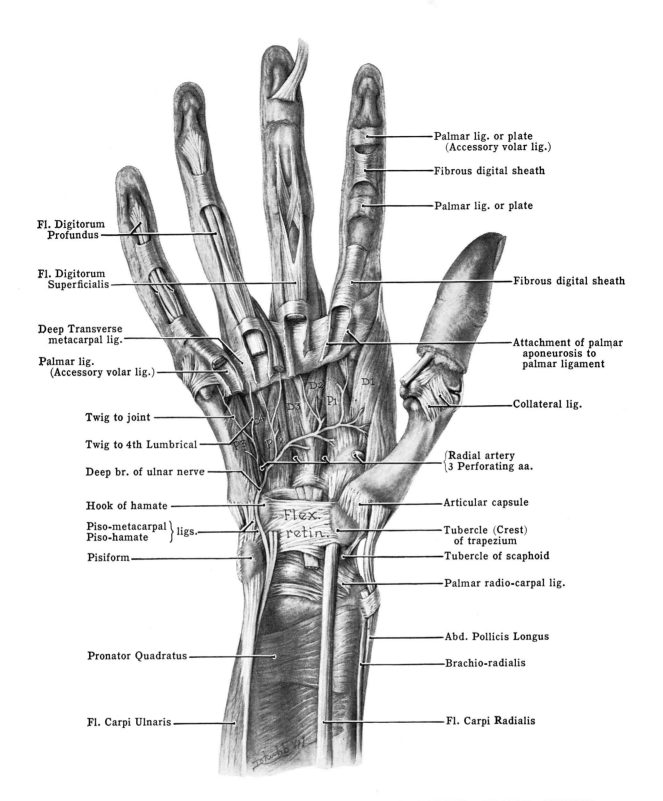

Fl. Digitorum Profundus

Fl. Digitorum Superficialis

Deep Transverse metacarpal lig.

Palmar lig. (Accessory volar lig.)

Twig to joint

Twig to 4th Lumbrical

Deep br. of ulnar nerve

Hook of hamate

Piso-metacarpal
Piso-hamate } ligs.

Pisiform

Pronator Quadratus

Fl. Carpi Ulnaris

Flex. retin.

D2 D1
D3 P1
D4
P3 P2

Palmar lig. or plate (Accessory volar lig.)

Fibrous digital sheath

Palmar lig. or plate

Fibrous digital sheath

Attachment of palmar aponeurosis to palmar ligament

Collateral lig.

Radial artery
3 Perforating aa.

Articular capsule

Tubercle (Crest) of trapezium

Tubercle of scaphoid

Palmar radio-carpal lig.

Abd. Pollicis Longus

Brachio-radialis

Fl. Carpi Radialis

6-85A DEEP DISSECTION OF THE PALM AND DIGITS, ULNAR NERVE

Observe:

1. The flexor retinaculum (transverse carpal ligament) uniting the 4 marginal carpal bones, and having the ulnar nerve in front and the median nerve (not labeled) behind.
2. Flexor Carpi Radialis descending vertically in front of the tubercle of the scaphoid and along the groove on the trapezium to the 2nd metacarpal.
3. Flexor Carpi Ulnaris continuing beyond the pisiform as the piso-hamate and the piso-metacarpal ligament.
4. The loose capsule of the carpo-metacarpal joint of the thumb and the strong collateral ligament of its metacarpo-phalangeal joint.
5. Four Dorsal and 3 Palmar Interossei.
6. The ulnar nerve crossing the hook of the hamate to be distrib-uted by its deep branch to the 3 hypothenar muscles, all 7 Interossei, 2 Lumbricals, Adductor Pollicis, and several joints. The superficial branch supplies Palmaris Brevis and 1½ dig-its.
7. The radial artery appearing in series with the 3 perforating arteries.
8. The palmar ligaments, with the deep transverse metacarpal ligaments uniting them, and the septa from the palmar apo-neurosis attached to them.
9. A lumbrical passing in front of the deep transverse ligament, and Interossei passing behind.
10. On the index, a proximal and distal osseo-fibrous tunnel; on the middle finger, a Superficialis tendon spreading like a V and decussating like an X; on the ring finger, a Profundus tendon; and on the little finger, both tendons.

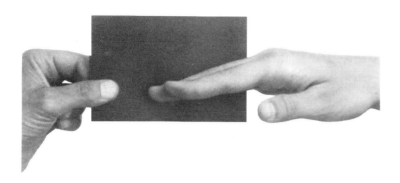

6-85B   FINGER ADDUCTION

Here, active adduction is being tested by forcefully grasping a card between the fingers. The muscles responsible are the palmar interossei, supplied by the deep branch of the ulnar nerve.

6-85C   OPPOSITION

The human thumb is able to touch the tips of each of the other fingers. Movement occurs at the saddle-shaped joint between the proximal end of the first metacarpal and the trapezium. Opponens pollicis is supplied by the median nerve.

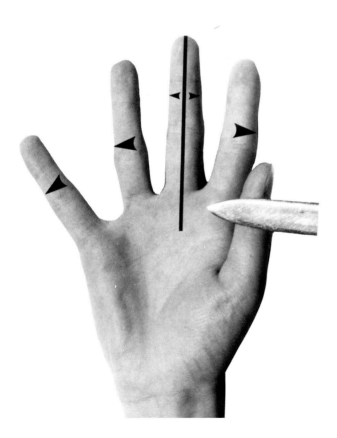

6-85D   ABDUCTION

Abduction of the fingers, by convention, is movement away from the midline of the middle finger as shown by the *arrows*. The middle finger may be abducted by both medial and lateral deviation from the anatomical position. Note that the thumb, at rest, is oriented at right angles to the other fingers. Abduction of the thumb, then, is movement away from the palm by contraction of Abductor pollicis longus (radial nerve) and Abductor pollicis brevis (median nerve.)

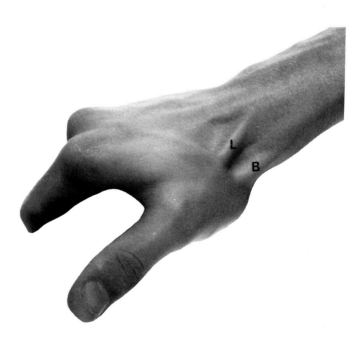

6-85E   ANATOMICAL SNUFFBOX

This triangular depression is bounded by Extensor pollicis longus (*L*) and brevis (*B*) (see Fig. 6-87).

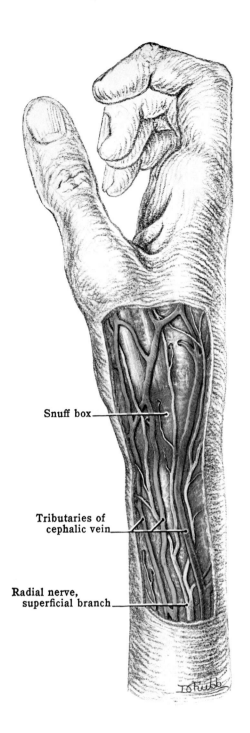

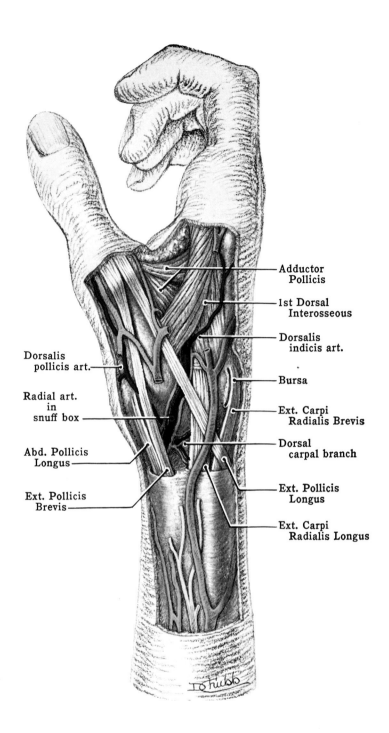

Snuff box

Tributaries of
cephalic vein

Radial nerve,
superficial branch

Dorsalis
pollicis art.

Radial art.
in
snuff box

Abd. Pollicis
Longus

Ext. Pollicis
Brevis

Adductor
Pollicis

1st Dorsal
Interosseous

Dorsalis
indicis art.

Bursa

Ext. Carpi
Radialis Brevis

Dorsal
carpal branch

Ext. Pollicis
Longus

Ext. Carpi
Radialis Longus

6-86 RADIAL ASPECT OF THE WRIST—I

Observe:
1. The depression at the base of the thumb: the "anatomical snuffbox," retaining its name from an archaic habit.
2. Superficial veins and nerves crossing the snuffbox.
3. Perforating veins and articular nerves piercing the deep fascia.

6-87 RADIAL ASPECT OF THE WRIST—II

Observe:
1. Three long tendons of the thumb forming the sides of the snuffbox.
2. The radial artery and its venae comitantes crossing the floor of the snuffbox and disappearing between the two heads of the 1st Dorsal Interosseous.
3. Adductor Pollicis and 1st Dorsal Interosseous, proximal to the web between pollex and index. Both are supplied by the ulnar nerve (Figs. 6-83 and 6-85).

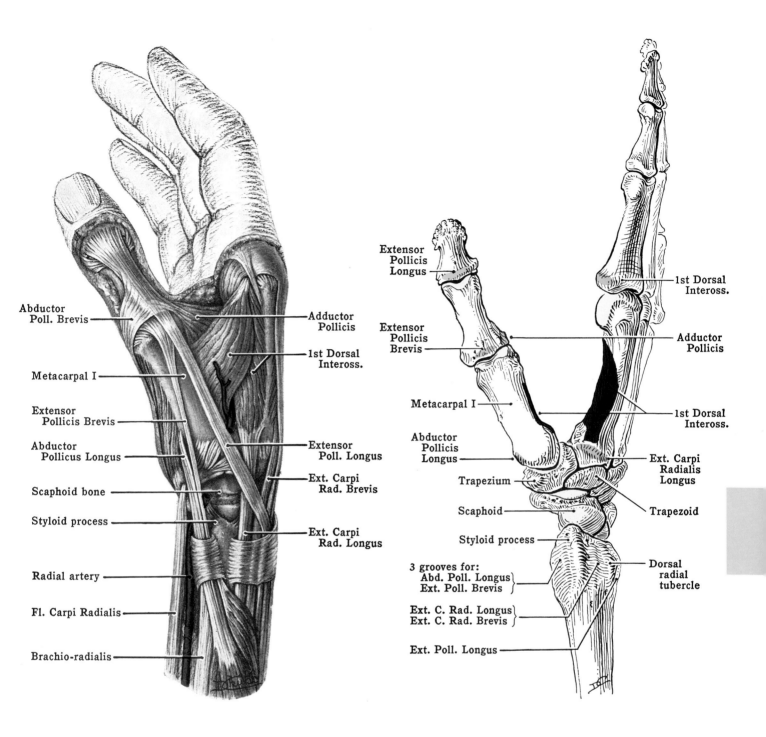

Abductor Poll. Brevis

Metacarpal I

Extensor Pollicis Brevis

Abductor Pollicus Longus

Scaphoid bone

Styloid process

Radial artery

Fl. Carpi Radialis

Brachio-radialis

Adductor Pollicis

1st Dorsal Inteross.

Extensor Poll. Longus

Ext. Carpi Rad. Brevis

Ext. Carpi Rad. Longus

Extensor Pollicis Longus

Extensor Pollicis Brevis

Metacarpal I

Abductor Pollicis Longus

Trapezium

Scaphoid

Styloid process

3 grooves for:
Abd. Poll. Longus
Ext. Poll. Brevis

Ext. C. Rad. Longus
Ext. C. Rad. Brevis

Ext. Poll. Longus

1st Dorsal Inteross.

Adductor Pollicis

1st Dorsal Inteross.

Ext. Carpi Radialis Longus

Trapezoid

Dorsal radial tubercle

6-88   RADIAL ASPECT OF THE WRIST—III

Observe:
1. The scaphoid bone; the wrist joint (and radius) proximal to the scaphoid; and the midcarpal joint (and trapezium and trapezoid) distal to it.
2. The capsule of the 1st carpo-metacarpal joint.
3. The Abductor Pollicis Brevis and Adductor Pollicis partly inserted into the dorsal (extensor) expansion.

6-89   RADIAL ASPECT OF THE WRIST—IV

Observe:
1. The attachments of muscles to bone.
2. Articular surfaces (*yellow*).
3. The anatomical snuffbox, limited proximally by the styloid process of the radius and distally by the base of the metacarpal of the thumb.
4. The 2 lateral marginal bones of the carpus (scaphoid and trapezium) forming the floor of the snuffbox.

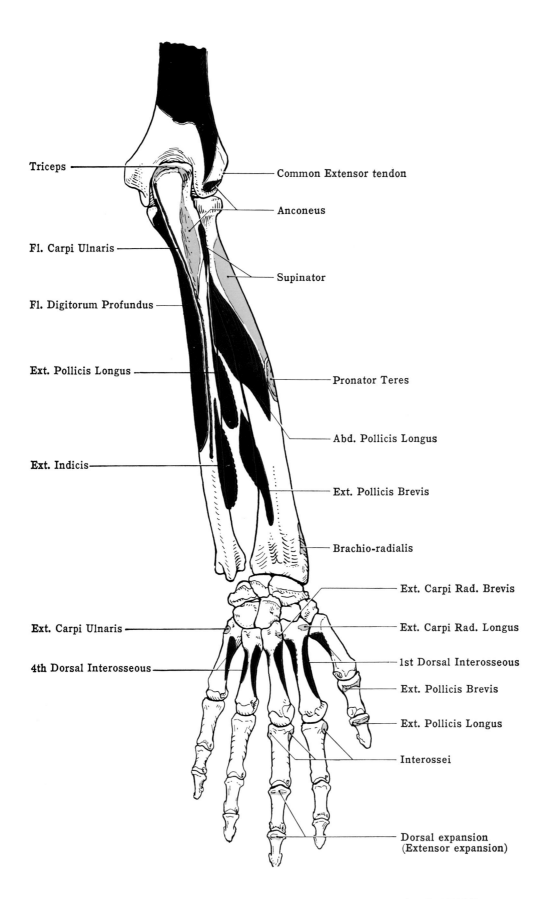

Triceps

Fl. Carpi Ulnaris

Fl. Digitorum Profundus

Ext. Pollicis Longus

Ext. Indicis

Ext. Carpi Ulnaris

4th Dorsal Interosseous

Common Extensor tendon

Anconeus

Supinator

Pronator Teres

Abd. Pollicis Longus

Ext. Pollicis Brevis

Brachio-radialis

Ext. Carpi Rad. Brevis

Ext. Carpi Rad. Longus

1st Dorsal Interosseous

Ext. Pollicis Brevis

Ext. Pollicis Longus

Interossei

Dorsal expansion
(Extensor expansion)

**6-90   BONES OF FOREARM AND HAND SHOWING
ATTACHMENTS OF MUSCLES, POSTERIOR VIEW**

For anterior view, see Figure 6-65.

For attachments to humerus, see Figures 6-35 and 6-36.

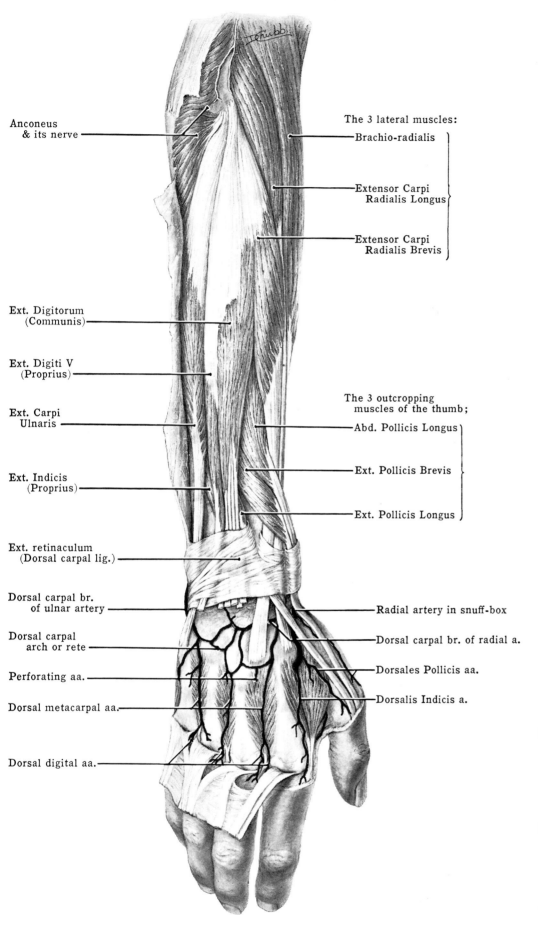

Anconeus & its nerve

The 3 lateral muscles:
Brachio-radialis

Extensor Carpi Radialis Longus

Extensor Carpi Radialis Brevis

Ext. Digitorum (Communis)

Ext. Digiti V (Proprius)

Ext. Carpi Ulnaris

The 3 outcropping muscles of the thumb;
Abd. Pollicis Longus

Ext. Pollicis Brevis

Ext. Indicis (Proprius)

Ext. Pollicis Longus

Ext. retinaculum (Dorsal carpal lig.)

Radial artery in snuff-box

Dorsal carpal br. of ulnar artery

Dorsal carpal br. of radial a.

Dorsal carpal arch or rete

Dorsales Pollicis aa.

Perforating aa.

Dorsalis Indicis a.

Dorsal metacarpal aa.

Dorsal digital aa.

6-91A MUSCLES OF THE EXTENSOR REGION OF THE FOREARM

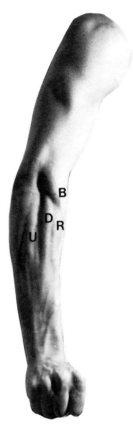

6-91B FOREARM EXTENSORS

Observe in Figure 6-91A:
1. The finger extensors have been reflected without disturbing the arteries since they lie on the skeletal plane.
2. No muscle is attached to the back of a carpal bone. The 3 extensors of the wrist span the carpal bones to reach the bases of metacarpals II, III, and V.
3. The radial artery disappearing between the two heads of the 1st Dorsal Interosseous where it is in series with the 3 perforating arteries.
4. Compare the living muscles in Figure 6-91B with the dissection: B, Brachioradialis: R, Extensor carpi radialis longus and brevis; D, Extensor digitorum; U, Extensor carpi ulnaris.

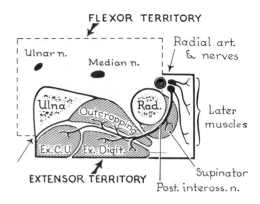

Ulnar n. Median n. Radial art. & nerves

Ulna Outcropping Rad. Later muscles

Ex.C.U. Ex. Digit. Supinator

EXTENSOR TERRITORY Post. inteross. n.

6-92 NERVE SUPPLY

The (*yellow*) flexor territory, supplied by ulnar and median nerves, is separated from extensor territory (radial nerve) by the radial artery laterally and by the posterior, sharp, palpable border of the ulna postero-medially. No motor nerve crosses either line.

Observe:

1. Three muscles of the thumb outcropping between Extensor Carpi Radialis Brevis and Extensor Digitorum: Abductor pollicis longus, Extensor pollicis brevis, and Extensor pollicis longus.

2. The furrow from which the 3 muscles outcrop has been opened widely, up to the lateral epicondyle. It crosses Supinator and is a "line of safety" since the 3 laterally retracted muscles are supplied before the posterior interosseous nerve enters the fleshy tunnel in Supinator, while the others are supplied after it emerges 6 cm below the head of the radius.

3. The tendons of the 3 outcropping muscles of the thumb, or pollex, pass to the epiphyses at the bases of the 3 long bones of the pollex (metacarpal, proximal phalanx, and distal phalanx) (Fig. 10-20). Of these thumb muscles, Extensor Longus is retracted from Extensor Brevis and Abductor Longus by its pulley, the dorsal radial tubercle (Figs. 6-111 and 6-113); hence the "anatomical snuffbox." No tubercle, no snuffbox.

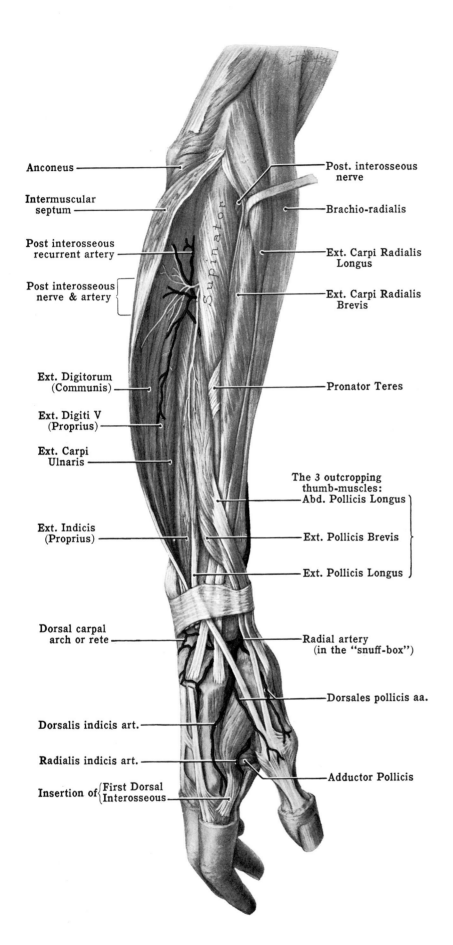

Anconeus

Intermuscular septum

Post interosseous recurrent artery

Post interosseous nerve & artery

Ext. Digitorum (Communis)

Ext. Digiti V (Proprius)

Ext. Carpi Ulnaris

Ext. Indicis (Proprius)

Dorsal carpal arch or rete

Dorsalis indicis art.

Radialis indicis art.

Insertion of {First Dorsal Interosseous}

Post. interosseous nerve

Brachio-radialis

Ext. Carpi Radialis Longus

Ext. Carpi Radialis Brevis

Pronator Teres

The 3 outcropping thumb-muscles:
Abd. Pollicis Longus

Ext. Pollicis Brevis

Ext. Pollicis Longus

Radial artery (in the "snuff-box")

Dorsales pollicis aa.

Adductor Pollicis

6-93 DEEP STRUCTURES AT THE BACK OF THE FOREARM, POSTERO-LATERAL VIEW

I   II   III

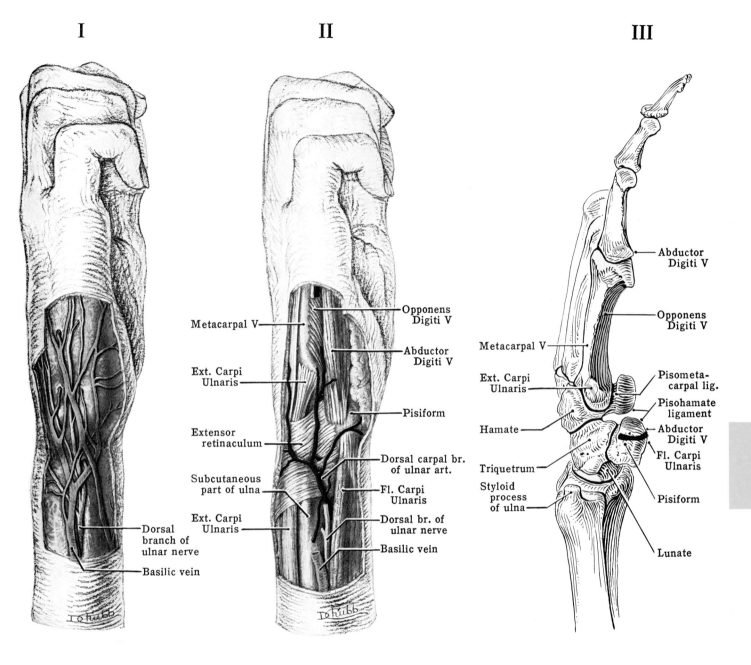

6-94 ULNAR BORDER OF THE WRIST—I

Observe:
1. The superficial veins and their perforating branches.
2. The superficial nerve appearing from under cover of Flexor Carpi Ulnaris.

6-95 ULNAR BORDER OF THE WRIST—II

Note: A vertical incision made along the medial subcutaneous surface of the ulna and along the medial border of the hand passes between two motor territories (Flexor Carpi Ulnaris, Abductor Digiti V, and Opponens Digiti V, supplied by the ulnar nerve; and Extensor Carpi Ulnaris by the posterior interosseous nerve). Superficial veins, nerves, and arteries will be divided, but no motor nerves.

6-96 ULNAR BORDER OF THE WRIST—III

Observe:
1. Attachments of muscles to bone.
2. Extensor Carpi Ulnaris is inserted directly into the base of Metacarpal V.
3. Flexor Carpi Ulnaris is inserted *indirectly* through the medium of the pisiform bone and the piso-metacarpal ligament.

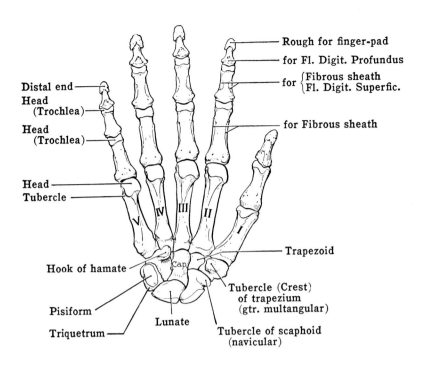

Distal end—
Head
(Trochlea)

Head
(Trochlea)—

Head—
Tubercle—

Rough for finger-pad
for Fl. Digit. Profundus
for {Fibrous sheath
     Fl. Digit. Superfic.

for Fibrous sheath

Hook of hamate—

Pisiform—

Triquetrum—

Cap.

Trapezoid

Tubercle (Crest)
of trapezium
(gtr. multangular)

Tubercle of scaphoid
(navicular)

Lunate

A. PALMAR ASPECT

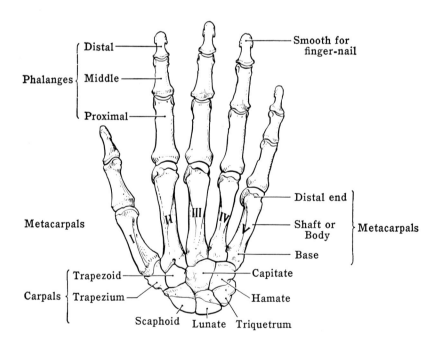

Phalanges {Distal—
           Middle—
           Proximal—

Smooth for
finger-nail

Metacarpals

Carpals {Trapezoid
         Trapezium—

Distal end }
Shaft or   } Metacarpals
Body
Base

Capitate

Hamate

Scaphoid   Lunate   Triquetrum

B. DORSAL ASPECT

6-97 BONES OF THE HAND

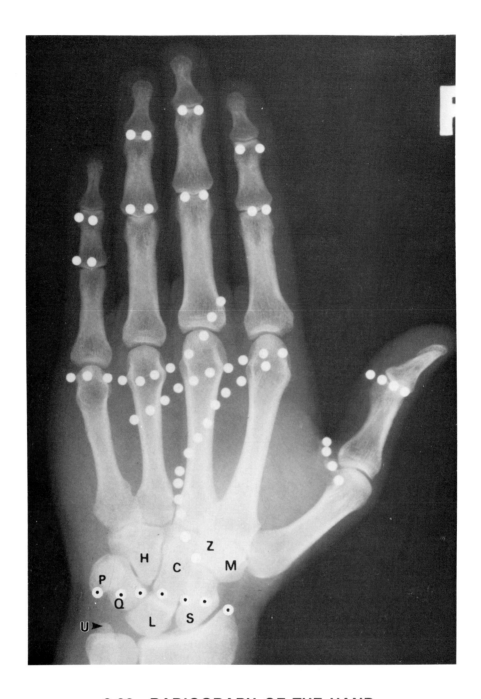

6-98 RADIOGRAPH OF THE HAND

Observe:
1. The 8 carpal bones in two rows. In the distal row: hamate (*H*), capitate (*C*), the largest and first to ossify, trapezoid (*Z*), and trapezium (*M*) which forms a saddle-shaped joint with metacarpal (*I*). In the proximal row: scaphoid (*S*), lunate (*L*), and pisiform (*P*) superimposed on triquetrum (*Q*).
2. The *arrow* pointing to the ulnar styloid process. The apex of the articular disc attaches to a pit at the root of the styloid and its base attaches to the ulnar notch of the radius. Thus the disc separates the distal radio-ulnar joint from the radio-carpal joint.
3. Lead shot has been placed along the palmar and digital skin creases to show their relationship to the joints. Note that the proximal skin crease (identified by *black dots*) crosses the pisiform, the joint between capitate and lunate, the tubercle of the scaphoid, and the radial styloid process.

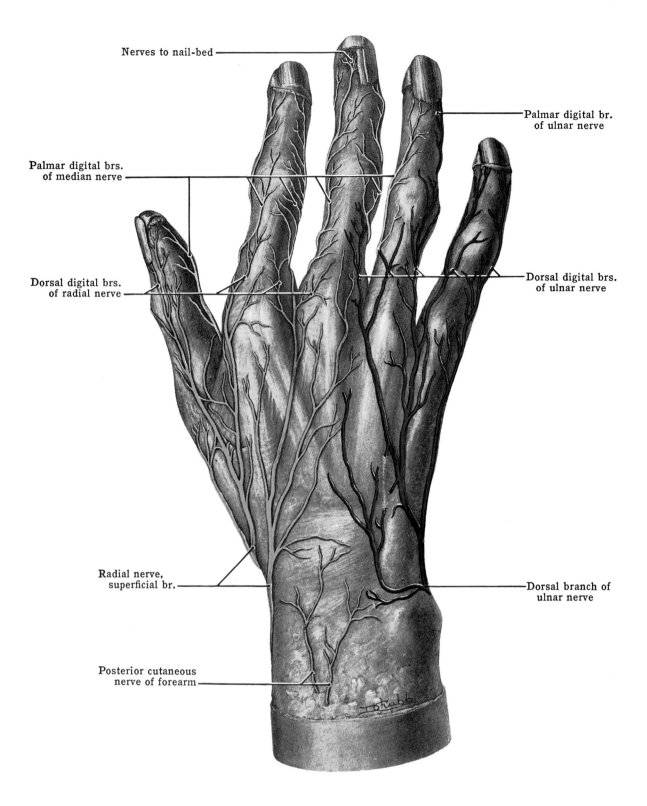

Nerves to nail-bed

Palmar digital brs.
of median nerve

Dorsal digital brs.
of radial nerve

Radial nerve,
superficial br.

Posterior cutaneous
nerve of forearm

Palmar digital br.
of ulnar nerve

Dorsal digital brs.
of ulnar nerve

Dorsal branch of
ulnar nerve

6-99   CUTANEOUS NERVES OF THE DORSUM OF THE HAND

Observe:
1. The radial nerve and the dorsal branch of the ulnar nerve distributed nearly equally and symmetrically on the dorsum of the hand and digits. The radial nerve supplies the radial half of the dorsum and extends on the 2½ digits, in fact, all the way along the first digit. The dorsal branch of the ulnar nerve behaves similarly on the ulnar half. Each distribution is almost a looking-glass distribution of the other.
2. The palmar digital branches of the median and ulnar nerves (Fig. 6-79) alone supply the distal halves of the three middle digits, including the nail beds.
3. Communications between adjacent nerves are numerous.

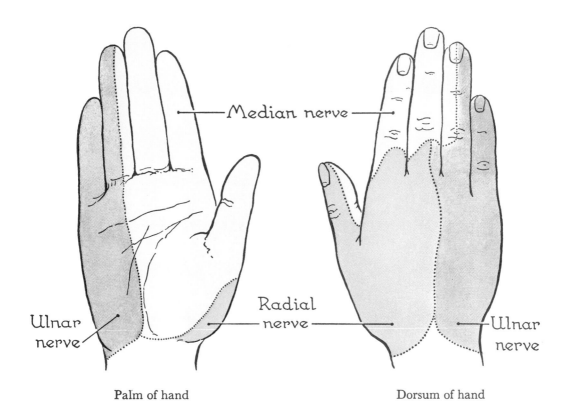

Palm of hand

Dorsum of hand

6-100 DISTRIBUTION OF CUTANEOUS NERVES TO PALM AND DORSUM OF THE HAND

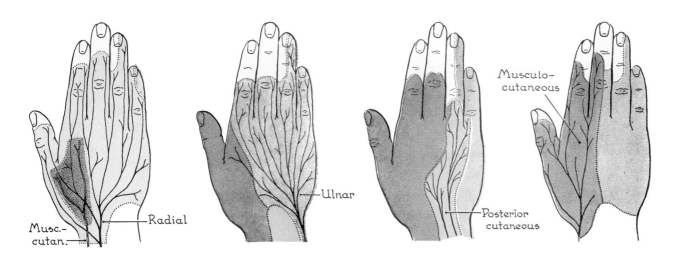

6-101 VARIATIONS IN PATTERN OF CUTANEOUS NERVES IN THE DORSUM OF THE HAND

See Appleton, A. B. (1911) A case of abnormal distribution of the musculocutaneous nerve. *J. Anat. Physiol.*, *46:* 89; Hutton, W. R. (1906) Remarks on the innervation of the dorsus manus with special reference to certain rare abnormalities. *J. Anat. Physiol.*, *40:* 326; Learmonth, J. R. (1919) A variation in the distribution of the "radial nerve." *J. Anat.*, *53:* 371; Stopford, J. S. B. (1918) The variation in distribution of the cutaneous nerves of the hand and digits. *J. Anat.*, *53:* 14.

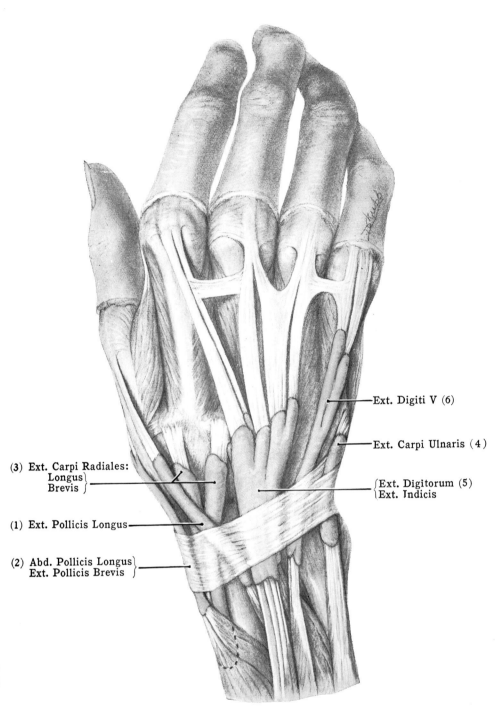

Ext. Digiti V (6)

Ext. Carpi Ulnaris (4)

(3) Ext. Carpi Radiales:
Longus }
Brevis }

{ Ext. Digitorum (5)
{ Ext. Indicis

(1) Ext. Pollicis Longus

(2) Abd. Pollicis Longus }
Ext. Pollicis Brevis }

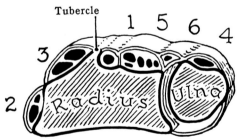

6-103 CROSS-SECTION,
TENDONS ON DORSUM

6-102 SYNOVIAL SHEATHS ON THE DORSUM
OF THE WRIST

Note:
1. These six sheaths occupy the six osseo-fibrous tunnels deep to the extensor retinaculum and contain nine tendons:
   3 for the thumb in sheaths (*1* and *2*);
   3 for extensors of wrist in 2 sheaths (*3* and *4*);
   3 for extensors of fingers in 2 sheaths (*5* and *6*).
2. They may be studied in the foregoing order after the dorsal radial tubercle (Figs. 6-2 and 6-113), which is the pulley for Extensor Pollicis Longus, has been located.
3. The tendons of the three extensors of the wrist are the strongest (Figs. 6-72 and 6-105) because they work synergically with the flexors of the digits.

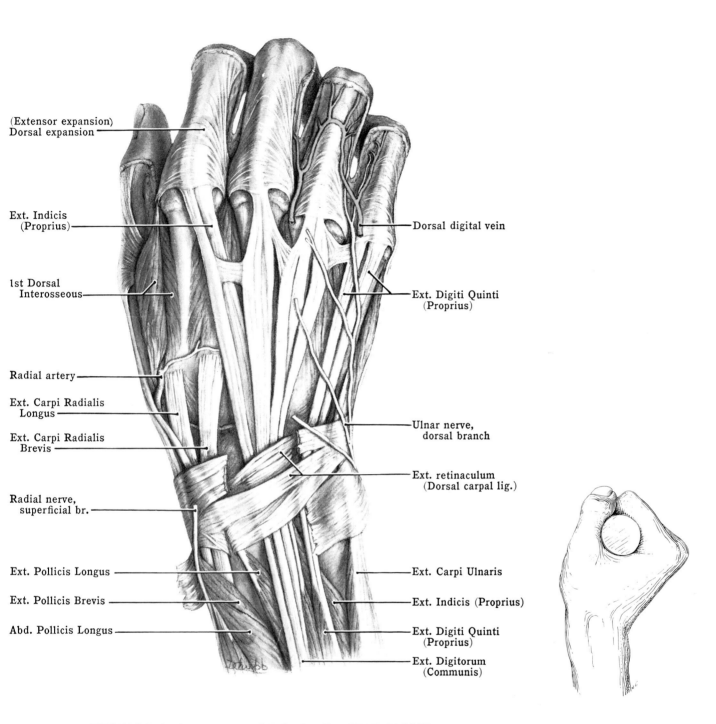

(Extensor expansion)
Dorsal expansion

Ext. Indicis
(Proprius)

1st Dorsal
Interosseous

Radial artery

Ext. Carpi Radialis
Longus

Ext. Carpi Radialis
Brevis

Radial nerve,
superficial br.

Ext. Pollicis Longus

Ext. Pollicis Brevis

Abd. Pollicis Longus

Dorsal digital vein

Ext. Digiti Quinti
(Proprius)

Ulnar nerve,
dorsal branch

Ext. retinaculum
(Dorsal carpal lig.)

Ext. Carpi Ulnaris

Ext. Indicis (Proprius)

Ext. Digiti Quinti
(Proprius)

Ext. Digitorum
(Communis)

6-104 TENDONS ON THE DORSUM OF THE HAND, EXTENSOR RETINACULUM

Observe:

1. The disposition of the tendons of the 9 muscles at the back of the wrist and hand.

2. The deep fascia, here thickened and called the extensor retinaculum, stretching obliquely from one ridge on the radius to another. Medially it passes distal to the ulna to be attached to the pisiform and triquetrum, as depicted in Figure 6-95.

3. The bands, proximal to the knuckles, that connect the tendons of the digital extensors and thereby restrict the independent action of the fingers.

. The digital veins passing to the dorsum of the hand where they are not subjected to pressure.

. The body of the 2nd metacarpal is not covered with an extensor tendon.

6-105 GRASPING

The grasping hand requires an extended wrist. The 3 extensors of the wrist, as synergists, are essential to the digital flexors when grasping, hence their strength (see Fig. 6-72).

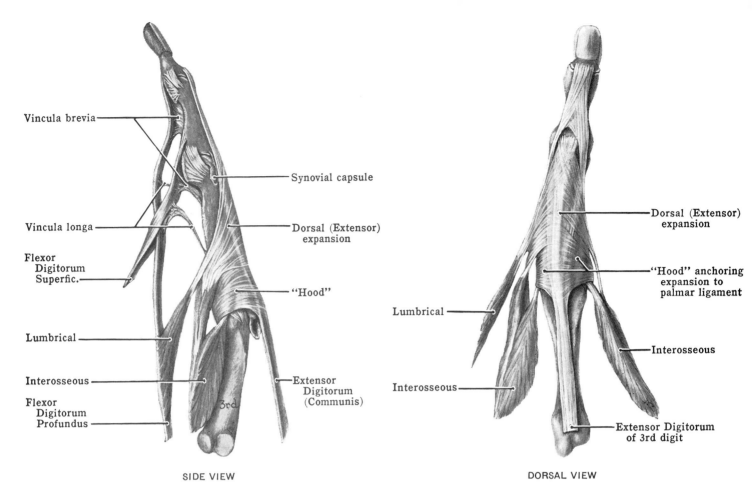

Vincula brevia

Synovial capsule

Vincula longa

Dorsal (Extensor) expansion

Flexor Digitorum Superfic.

"Hood"

Lumbrical

Interosseous

Extensor Digitorum (Communis)

Flexor Digitorum Profundus

3rd.

SIDE VIEW

Dorsal (Extensor) expansion

"Hood" anchoring expansion to palmar ligament

Lumbrical

Interosseous

Interosseous

Extensor Digitorum of 3rd digit

DORSAL VIEW

6-106 EXTENSOR (DORSAL) EXPANSION OF THE MIDDLE DIGIT

Observe:
1. Interossei in part inserted into the bases of the proximal phalanx (Figs. 6-89 to 6-93) and in part into the expansion.
2. Lumbrical inserted wholly into the radial side of the expansion.
3. The hood covering the head of the metacarpal. It is moored to the palmar ligament (Figs. 6-85 and 6-116); hence medial and lateral bowstringing of the extensor tendon and expansion is prevented.
4. The expansion extending to the bases of the middle and distal phalanges, and giving a strong areolar band to the base of the proximal phalanx, here not in view.
5. (On side view) the vincula longa and brevia, which are all that remain of the primitive mesotendons.

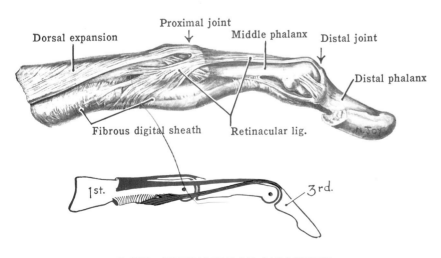

Dorsal expansion

Proximal joint

Middle phalanx

Distal joint

Distal phalanx

Fibrous digital sheath

Retinacular lig.

1st.

3rd.

6-107 RETINACULAR LIGAMENT

Observe:
1. This delicate fibrous band runs from the 1st or proximal phalanx and the fibrous digital sheath obliquely across the 2nd phalanx and the two interphalangeal joints to join the dorsal expansion, and so to the 3rd or distal phalanx.
2. On flexing the distal joint the retinacular ligament becomes taut and pulls the proximal joint into flexion.
3. Similarly, on extending the proximal joint, the distal joint is pulled by the retinacular ligament into nearly complete extension.

See Landsmeer, J. M. F. (1949) The anatomy of the dorsal aponeurosis of the human finger and its functional significance. *Anat. Rec.*, 104: 31.

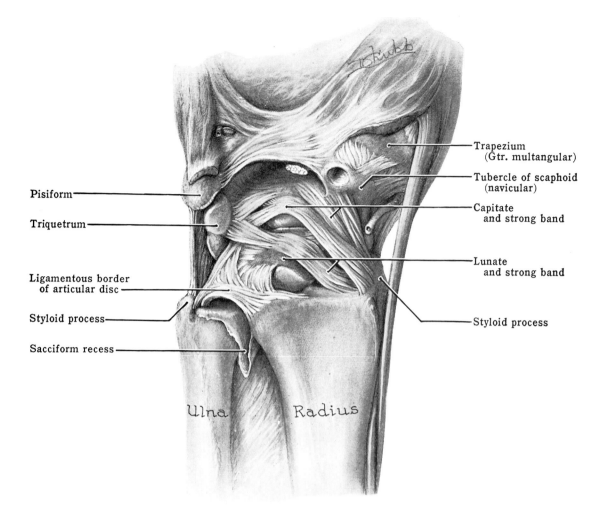

Pisiform

Triquetrum

Ligamentous border
of articular disc

Styloid process

Sacciform recess

Trapezium
(Gtr. multangular)

Tubercle of scaphoid
(navicular)

Capitate
and strong band

Lunate
and strong band

Styloid process

Ulna   Radius

6-108 LIGAMENTS OF THE DISTAL RADIO-ULNAR, RADIO-CARPAL, AND INTERCARPAL JOINTS

The hand is forcibly extended.

Observe:

1. The sacciform recess of the digital radio-ulnar joint (similar to the recess of the proximal joint, Fig. 6-60) and the ligamentous anterior border of the disc.
2. The anterior or palmar ligaments, passing from the radius to the two rows of carpal bones. They are strong, and so directed that the hand shall follow the radius during supination. The dorsal ligaments take the same direction; hence the hand is obedient during pronation also.
3. The proximal articular surface of the triquetrum applied to the medial ligament of the wrist.

6-109 CARPAL TUNNEL

These two sections, proximal and distal, show the thickness of the flexor retinaculum and the proximal articular surfaces.

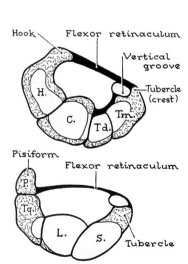

Hook   Flexor retinaculum

Vertical groove

Tubercle (crest)

H.   Tm.

C.   Td.

Pisiform   Flexor retinaculum

P.

Tq.

L.   S.

Tubercle

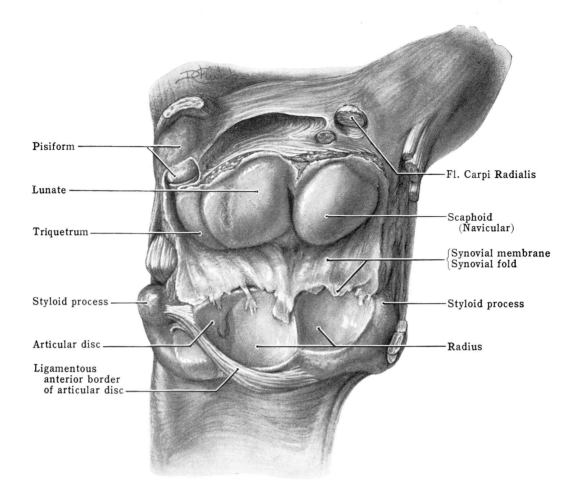

Pisiform

Lunate

Triquetrum

Styloid process

Articular disc

Ligamentous
anterior border
of articular disc

Fl. Carpi Radialis

Scaphoid
(Navicular)

{Synovial membrane
{Synovial fold

Styloid process

Radius

6-110 SURFACES OF THE RADIO-CARPAL OR WRIST JOINT, OPENED FROM FRONT

Observe:
1. The nearly equal proximal articular surfaces of the scaphoid and lunate.
2. The lunate articulating with the radius and the articular disc. Only during adduction of the wrist does the triquetrum come into articulation with the disc.
3. Transparent synovial folds, like cellophane, projecting between the articular surfaces.

4. The perforation in the disc and the associated roughened surface of the lunate. This is a common occurence.
5. The pisiform joint communicating with the radio-carpal joint.
6. The interosseous ligament between the scaphoid and the lunate, partly absorbed. In such a specimen—and they are common—infection could spread widely.

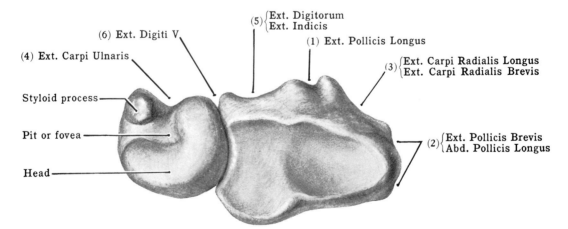

(6) Ext. Digiti V

(4) Ext. Carpi Ulnaris

Styloid process

Pit or fovea

Head

(5){Ext. Digitorum
{Ext. Indicis

(1) Ext. Pollicis Longus

(3){Ext. Carpi Radialis Longus
{Ext. Carpi Radialis Brevis

(2){Ext. Pollicis Brevis
{Abd. Pollicis Longus

6-111 DISTAL ENDS OF RADIUS AND ULNA, FROM BELOW

Observe:
1. The 4 features of the distal end of the ulna, head, fovea, styloid process, and groove for the tendon of Extensor Carpi Ulnaris.
2. The 6 grooves for the 9 tendons at the back of the wrist. Figure 6-102 shows these tendons in their synovial sheaths.

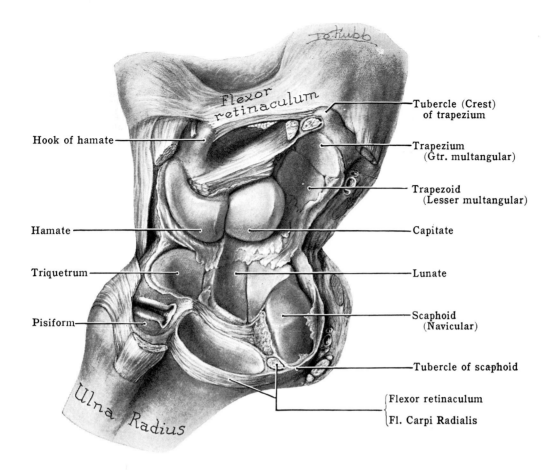

6-112 SURFACES OF THE MIDCARPAL JOINT (TRANSVERSE CARPAL JOINT)

Observe:

1. The flexor retinaculum has been divided.
2. The sinuous surfaces of the opposed bones; the trapezium and trapezoid together presenting a concave, oval surface to the scaphoid; the capitate and hamate together presenting a convex surface to the scaphoid, lunate, and triquetrum, which is slightly broken by the linear facet on the apex of the hamate for its counterpart on the lunate.
3. Synovial folds projecting into the joint.
4. The relative weakness of the proximal part of the flexor retinaculum, which stretches from the movable pisiform to the scaphoid, and the strength of the distal part, which stretches from the hook of the hamate to the tubercle (crest) of the trapezium.

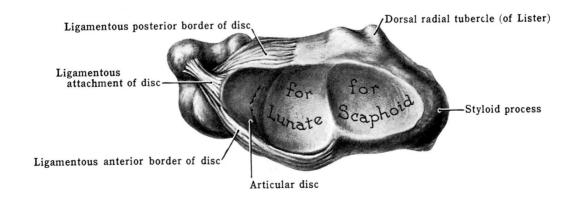

6-113 ARTICULAR DISC OF THE DISTAL RADIO-ULNAR JOINT, FROM BELOW

This disc is the bond of union between the lower ends of the radius and ulna. It is fibro-cartilaginous, smooth, and stiff at the triangular area compressed between the head of the ulna and the lunate bone (Figs. 6-51 and 6-110), but it is ligamentous and pliable elsewhere. The cartilaginous part is commonly fissured, as here, but the ligamentous parts are not.

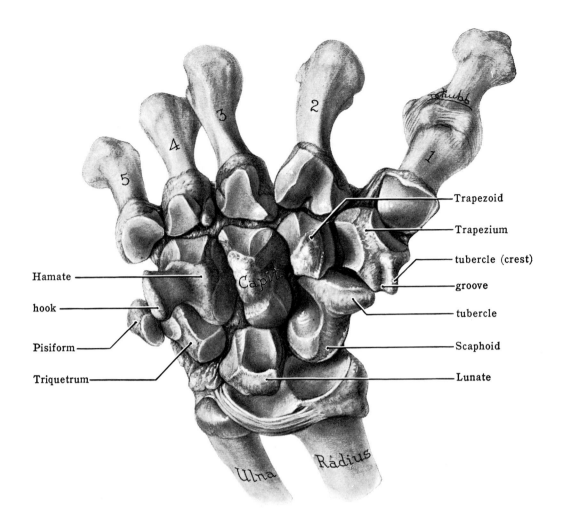

Labels on figure: Trapezoid, Trapezium, tubercle (crest), groove, tubercle, Scaphoid, Lunate, Hamate, hook, Pisiform, Triquetrum, Capit., Ulna, Radius

6-114 CARPAL BONES AND THE BASES OF THE METACARPALS, FRONT VIEW

The dorsal ligaments remain as a binding allowing study of the articular facets.

Observe:
1. One bone (radius) supporting two proximal carpals (scaphoid and lunate); these in turn supporting three distal carpals (trapezium, trapezoid, and capitate) and articulating with the apex of the hamate; the four distal carpals supporting the five metacarpals. The triquetrum is unsupported.
2. The marginal projections (pisiform, hook of hamate, tubercle of scaphoid, and tubercle of trapezium) which afford attachment to the flexor retinaculum.
3. The triquetrum having an isolated facet for the pisiform.
4. Of the eight carpals only the lunate is wider in front than behind.
5. The capitate articulating with three metacarpals (2nd, 3rd, and 4th).
6. The second metacarpal articulating with three carpals (trapezium, trapezoid, and capitate).
7. The basal surfaces of the 2nd and 3rd metacarpals individually about as large as those of the 4th and 5th collectively.
8. The 2nd and 3rd carpo-metacarpal joints are practically immobile; the 1st, saddle-shaped; the 4th and 5th, hinge-shaped.

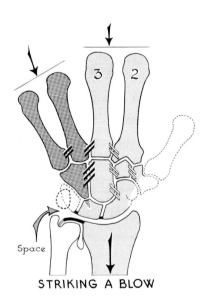

STRIKING A BLOW

6-115 STRIKING A BLOW

On striking a blow, triquetrum being unsupported, the hamate and meta-carpals IV and V rely on ligaments, directed as shown, for resistance and transmission of force.

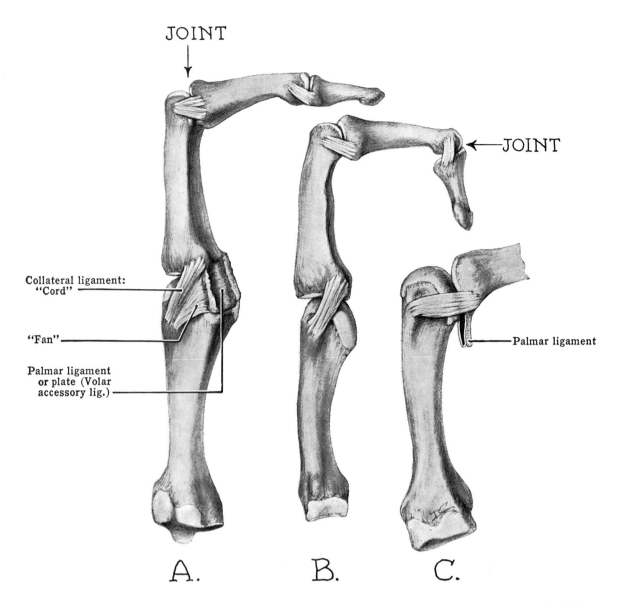

JOINT

JOINT

Collateral ligament:
"Cord"

"Fan"

Palmar ligament
or plate (Volar
accessory lig.)

Palmar ligament

A. B. C.

6-116 METACARPO-PHALANGEAL AND INTERPHALANGEAL JOINTS

Observe:

1. A fibro-cartilaginous plate, the palmar ligament, hanging from the base of the proximal phalanx; fixed to the head of the metacarpal by the weaker, fanlike part, of the collateral ligament; and moving like a visor across the metacarpal head. Figure 6-85 shows the deep transverse metacarpal ligaments that unite the plates, and the insertions of the palmar aponeurosis into them.

2. The extremely strong, cordlike parts of the collateral ligaments of this joint, being eccentrically attached to the metacarpal heads. They are slack during extension and taut during flexion; hence the fingers cannot be spread (abducted) unless the hand is open.

3. The interphalangeal joints have corresponding ligaments but the distal ends of the 1st and 2nd phalanges, being flattened antero-posteriorly and having two small condyles (*cf.* distal end of femur), permit neither adduction nor abduction.

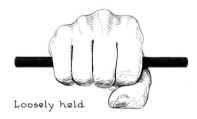

Loosely held

Firmly gripped

6-117 GRIPPING

The flexion allowed at 4th and 5th carpo-metacarpal joints allows the grip to be more secure.

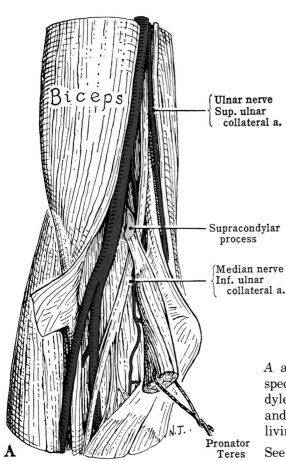

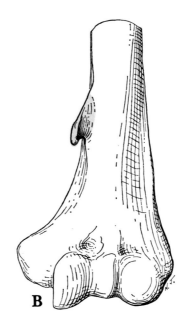
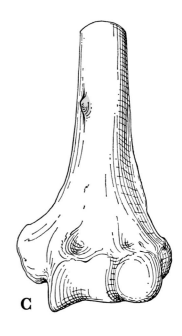

SUPRACONDYLAR PROCESS OF THE HUMERUS

A and *B* are from opposite sides of the same subject. *C* is a left-sided specimen. A fibrous band joins this process or spine to the medial epicondyle. Through the foramen so-formed the median nerve passes, as here, and the brachial artery may go with it. A process was found in 7 of 1000 living subjects.

See Terry, R. J. (1921) A study of the supracondyloid process in the living. *Am. J. Phys. Anhropol.*, 4: 129.

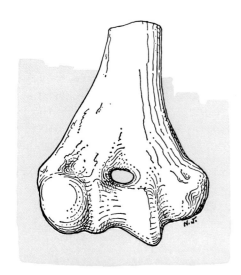

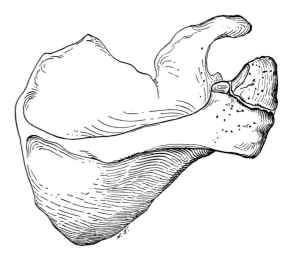

D. SUPRATROCHLEAR FORAMEN

This hole, closed in life by membrane, is more common on the left than on the right, and in females than in males.

See Trotter, M. (1934) Septal apertures in the humerus of American whites and Negroes. *Am. J. Phys. Anthropol.*, 19: 213.

E. UNFUSED ACROMIAL EPIPHYSIS (OS ACROMIALE)

The acromial epiphysis persists in adult life in about 8 per cent of dissecting room subjects. Like other anomalies it is more commonly unilateral than bilateral so comparison x-rays of the opposite side are of little help.

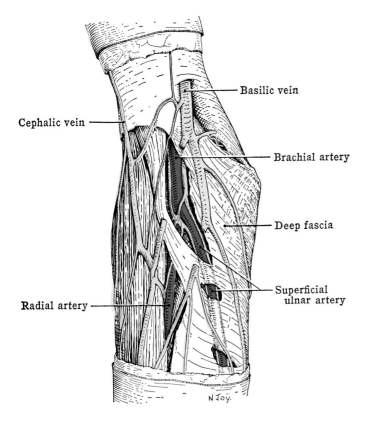

A. SUPERFICIAL ULNAR ARTERY

The ulnar artery descends superficial to the flexor muscles in about 3 per cent of limbs. This must be kept in mind when doing intravenous injections in the cubital area.

See Hazlett, J. W. (1949) The superficial ulnar artery with references to accidental intra-arterial injection. *Can. Med. Assn. J., 61:* 289.

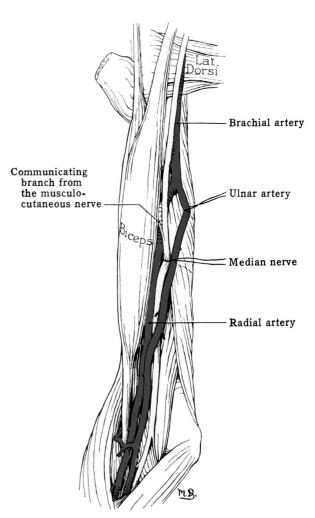

B. HIGH DIVISION OF BRACHIAL ARTERY

In this case, the median nerve passes between the radial and ulnar artery which arise high in the arm. The musculocutaneous and median nerves commonly communicate, as here.

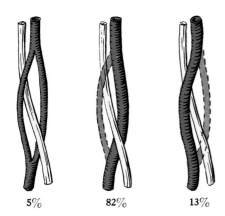

C. MEDIAN NERVE AND BRACHIAL ARTERY

The variable relationship of these two structures may be explained developmentally. In a study of 307 limbs, both primitive brachial arteries persisted in 5 per cent, the posterior in 82 per cent, and the anterior in 13 per cent.

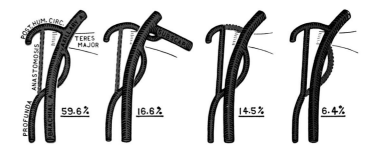

D. CIRCUMFLEX AND PROFUNDA ARTERIES

Four types of variations in origin of the posterior humeral circumflex and profunda brachii arteries; in 2.9 per cent the arteries were otherwise irregular. Percentages are based on 235 specimens.

6-119   ARTERIAL VARIATIONS

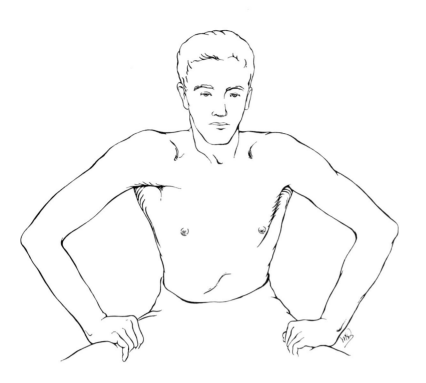

A. ABSENT STERNOCOSTAL HEAD OF RIGHT PECTORALIS MAJOR

In this case, the absence is associated with compensatory hypertrophy of Latissimus Dorsi. It is demonstrated, as here, by pressing downward.

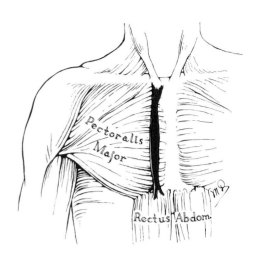

B. STERNALIS

A Sternalis muscle, in line with Rectus Abdominis and Sternomastoid occurs in about 6 per cent of cases.

See Barlow, R. N. (1935) The sternalis muscle in American whites and Negroes. *Anat. Rec.*, *61:* 413.

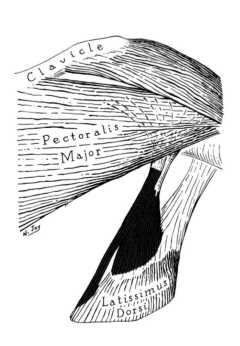

C. AN AXILLARY ARCH

This occasional muscle crosses the base of the axilla superficially and joins Latissimus Dorsi to the tissues deep to Pectoralis Major.

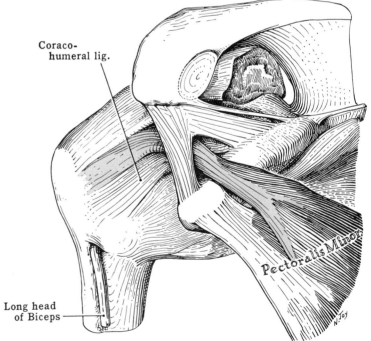

D. PECTORALIS MINOR INSERTION

In about 15 per cent of limbs, part of the tendon passes over the coracoid process and between the two limbs of the coraco-acromial ligament to reinforce the coraco-humeral ligament (Fig. 6-33).

See Seib, G. A. (1938) The musculus pectoralis minor in American whites and American Negroes. *Am. J. Phys. Anthropol.*, *23:* 389.

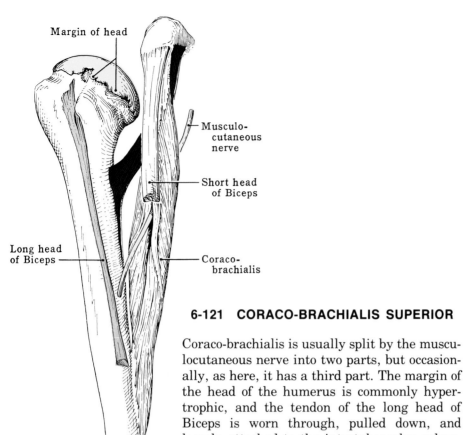

Margin of head

Musculo-cutaneous nerve

Short head of Biceps

Long head of Biceps

Coraco-brachialis

6-121 CORACO-BRACHIALIS SUPERIOR

Coraco-brachialis is usually split by the musculocutaneous nerve into two parts, but occasionally, as here, it has a third part. The margin of the head of the humerus is commonly hypertrophic, and the tendon of the long head of Biceps is worn through, pulled down, and loosely attached to the intertubercular sulcus, as here and in Figure 6-122.

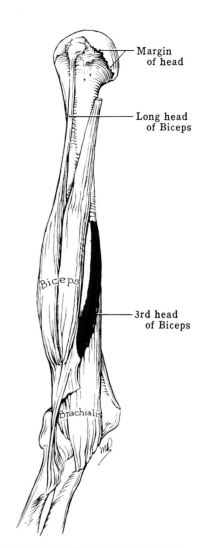

Margin of head

Long head of Biceps

Biceps

3rd head of Biceps

Brachialis

6-122 THIRD HEAD OF BICEPS

A third or humeral head occurs in about 5 per cent of limbs. In this case there is also attrition of the Biceps tendon.

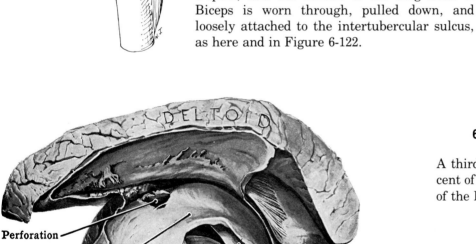

DELTOID

Perforation

Supraspinatus tendon and the capsule of the joint

Perforation

Teres Minor

Coracoid process

Tendon of long head of Biceps seen through perforation

Cut edge of subacromial bursa

Tendon of long head of biceps

6-123 ATTRITION OF SUPRASPINATUS TENDON

As a result of wearing away of the Supraspinatus tendon and underlying capsule, the subacromial bursa and the shoulder joint come into wide open communication. The intracapsular part of the long tendon of the Biceps becomes frayed—even worn away—leaving it adherent to the intertubercular sulcus, as in Figures 6-121 and 6-122.

Of 95 dissecting room subjects of proved age, none of the 18 under 50 years of age had a perforation, but 4 of the 19 between 50 and 60 years and 23 of the 57 over 60 years had perforations. The perforation was bilateral in 11 subjects and unilateral in 14.

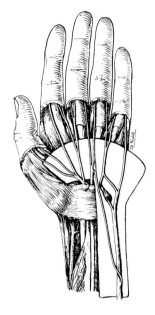

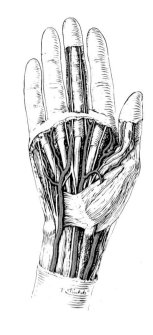

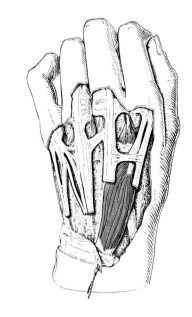

A. LONG COMMUNICATING BRANCH FROM ULNAR TO MEDIAN NERVE

B. PERSISTING MEDIAN ARTERY

C. EXTENSOR DIGITORUM BREVIS

This muscle, constant on the dorsum of the foot, is found occasionally on the dorsum of the hand, usually as a single bundle.

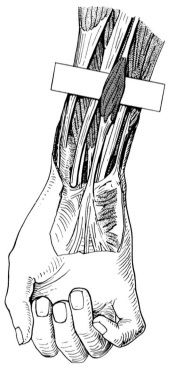

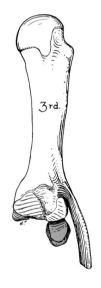

LEFT SIDE   RIGHT SIDE

D. PALMARIS LONGUS

This muscle is absent in about 14 per cent of limbs. The fleshy belly may be much reduced in size as in Figure 6-66.

E. SEPARATE STYLOID PROCESS OF 3RD METACARPAL

F. FUSED LUNATE AND TRIQUETRUM

6-124   VARIATIONS AND ANOMALIES

The Head

CONTENTS

7-1

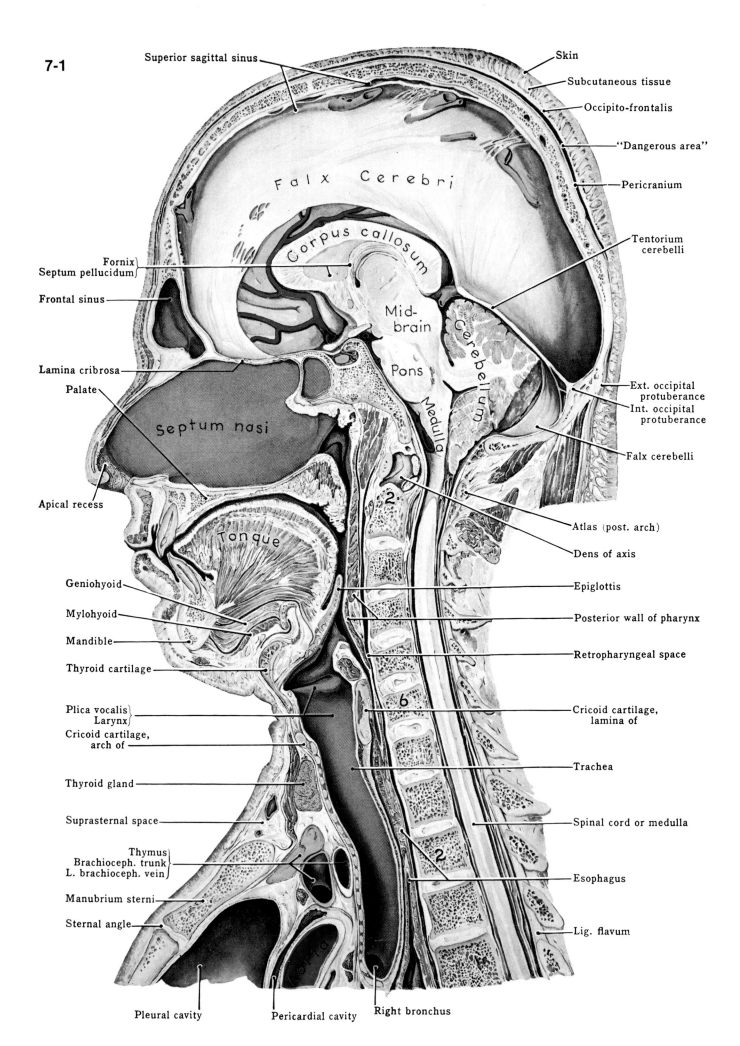

Superior sagittal sinus

Skin

Subcutaneous tissue

Occipito-frontalis

"Dangerous area"

Pericranium

Falx Cerebri

Corpus callosum

Tentorium cerebelli

Fornix
Septum pellucidum

Frontal sinus

Mid-brain

Cerebellum

Pons

Medulla

Lamina cribrosa

Palate

Septum nasi

Ext. occipital protuberance
Int. occipital protuberance

Falx cerebelli

Apical recess

Tongue

2

Atlas (post. arch)

Dens of axis

Geniohyoid

Epiglottis

Mylohyoid

Posterior wall of pharynx

Mandible

Retropharyngeal space

Thyroid cartilage

Plica vocalis
Larynx

6

Cricoid cartilage, lamina of

Cricoid cartilage, arch of

Trachea

Thyroid gland

Suprasternal space

Spinal cord or medulla

Thymus
Brachioceph. trunk
L. brachioceph. vein

2

Esophagus

Manubrium sterni

Sternal angle

Lig. flavum

Pleural cavity

Pericardial cavity

Right bronchus

7-1 HEAD AND NECK, ON MEDIAN SECTION

Observe:

1. The three adherent layers of the scalp—skin, subcutaneous tissue, and Occipito-frontalis muscle with its aponeurosis—separated from the pericranium by a layer of loose areolar tissue through which emissary veins connect the venous sinuses in the skull with the veins of the scalp. Bleeding from these travels freely under Occipito-fontalis, limited only by its attachments: the superior nuchal line behind and the zygomatic arch laterally. In front, it may enter the eyelids because Frontalis attaches to skin, not bone.

2. The external occipital protuberance, nearly level with the internal protuberance, marking the line between scalp and thick bone above and nuchal muscles and thin bone below.

3. Behind the tip or apex of the nose, a shelf above which is the apical recess of the nasal cavity.

4. The nasal septum extending from the apical recess in front to the nasopharynx behind, where it ends in a free posterior border, and from the sievelike lamina cribosa (cribriform plate) above to the palate below.

5. The palate, the anterior two-thirds of which contains bone and is known as the hard palate, and the posterior one-third which contains gland and muscle and is called soft palate. The Levator Palati (in contraction, as it is during the act of swallowing) pulling the soft palate upward and backward (it retracts as well as elevates), thereby closing the oral pharynx (not labeled), which lies below the soft palate, from the nasopharynx which lies above. A small mass, the pharyngeal tonsil, projecting from the roof of the nasopharynx.

6. The Orbicularis Oris in the upper and lower lips, with free margins curved forward.

7. The Geniohyoid passing from the genial tubercle of the mandible to the hyoid bone (not labeled), and above it the Genioglossus (not labeled) radiating into the tongue. The anterior two-thirds of the tongue forming part of the floor of the mouth; the posterior one-third forming the anterior wall of the oral pharynx. Behind the tongue, the epiglottis.

8. The pharynx lying in front of the upper cervical vertebrae whose bodies may be palpated by the tip of a finger in the mouth.

9. The Falx Cerebri, a midline partition made of dura mater.

10. The Corpus Callosum joining right and left halves of the brain.

11. The tentorium cerebelli suspended by the falx cerebri, sloping to the internal occipital protuberance, and forming a floor for the cerebrum and a roof for the cerebellum.

12. The larynx, guarded in front by the thyroid cartilage and extending from the tip of the epiglottis above to the lower border of the cricoid cartilage below, where it becomes the trachea. A horizontal slit that runs posteriorly from the thyroid cartilage separating an upper or false cord from a lower or true vocal cord, the plica vocalis.

13. The 10 cm long trachea, half in the neck and half in the thorax, bifurcating below into a right and a left bronchus, the mouth of the right bronchus being in view.

14. The cut ends of 19 tracheal rings below the arch of the cricoid cartilage which always projects in front of the rings and is therefore palpable and readily identified. It is a valuable landmark. It is also a guide to the level of the 6th cervical vertebra which lies behind it.

15. The isthmus of the thyroid gland crossing several tracheal rings, but leaving the upper one or two uncovered.

16. The brachiocephalic trunk (innominate artery), here as commonly impressing the trachea.

17. The cricoid cartilage lying at the level of the body of the 6th cervical vertebra. At the lower border of this cartilage the larynx becoming the trachea and the pharynx becoming the esophagus. The diameter of the alimentary canal is here at its narrowest and least dilatable part. In the neck the esophagus projects to the left of the trachea; hence the right wall of the upper part of the esophagus is cut longitudinally and no lumen is seen.

18. The retropharyngeal space extending from the level of the atlas downward into the superior mediastinum.

19. The manubrium sterni is 5 cm in length and is its own length of 5 cm from the body of the 2nd thoracic vertebra.

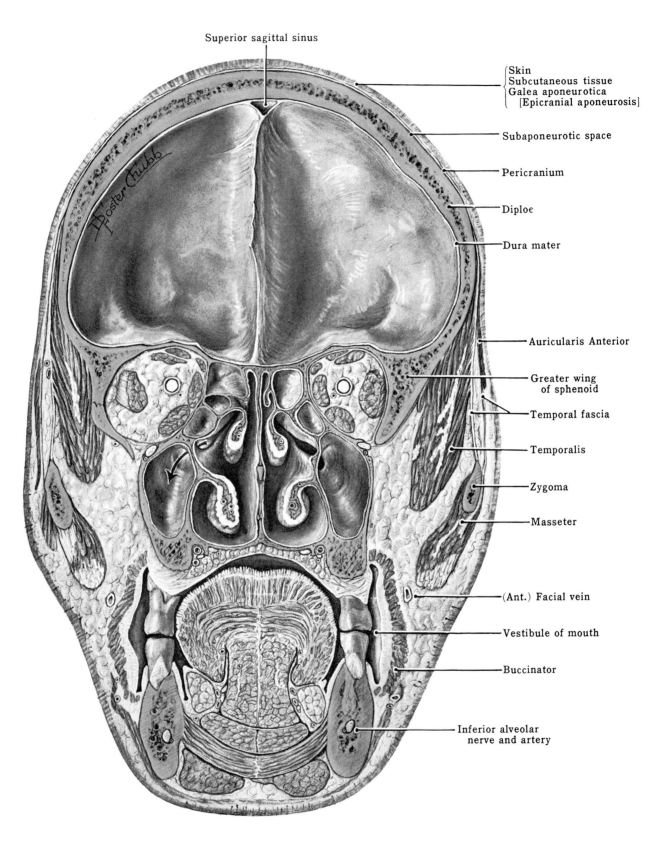

Superior sagittal sinus

Skin
Subcutaneous tissue
Galea aponeurotica
  [Epicranial aponeurosis]

Subaponeurotic space

Pericranium

Diploe

Dura mater

Auricularis Anterior

Greater wing
of sphenoid

Temporal fascia

Temporalis

Zygoma

Masseter

(Ant.) Facial vein

Vestibule of mouth

Buccinator

Inferior alveolar
nerve and artery

7-2 CORONAL SECTION OF THE HEAD

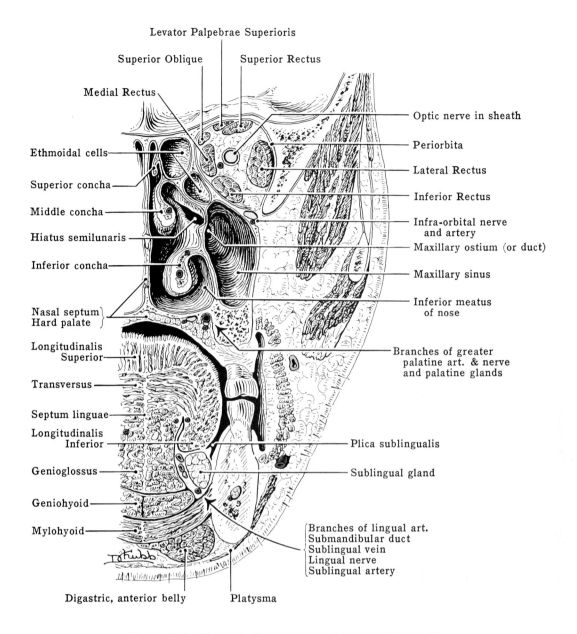

Levator Palpebrae Superioris

Superior Oblique — Superior Rectus

Medial Rectus

Optic nerve in sheath

Ethmoidal cells — Periorbita

Superior concha — Lateral Rectus

Middle concha — Inferior Rectus

Hiatus semilunaris — Infra-orbital nerve and artery

Maxillary ostium (or duct)

Inferior concha — Maxillary sinus

Nasal septum}
Hard palate } — Inferior meatus of nose

Longitudinalis Superior — Branches of greater palatine art. & nerve and palatine glands

Transversus

Septum linguae

Longitudinalis Inferior — Plica sublingualis

Genioglossus — Sublingual gland

Geniohyoid

Mylohyoid — {Branches of lingual art.
Submandibular duct
Sublingual vein
Lingual nerve
Sublingual artery}

Digastric, anterior belly — Platysma

7-3 CORONAL SECTION OF THE HEAD

Use this drawing as a key to Figure 7-2 on the facing page.

Note:

1. The central position of the ethmoid bone whose horizontal component forms the central part of the anterior cranial fossa above and the roof of the nasal cavity below. The suspended ethmoidal air cells give attachment to the superior and middle concha and form part of the medial wall of the orbit. The perpendicular plate of the ethmoid forms part of the nasal septum.

2. The thin orbital plate of the frontal bone forms a roof over the orbit and a floor for the anterior cranial fossa.

3. The palate forms the floor of the nasal cavity and the roof of the oral cavity.

4. The maxillary sinus forms the lower part of the lateral wall of the nose. The middle concha shelters the hiatus semilunaris into which the maxillary ostium opens.

5. In chewing, the tongue pushes food between the molar or millstone teeth into the vestibule and the Buccinator pushes it back again.

6. The Mylohyoid, slung like a hammock between right and left halves of the mandible, supports the structures of the oral cavity.

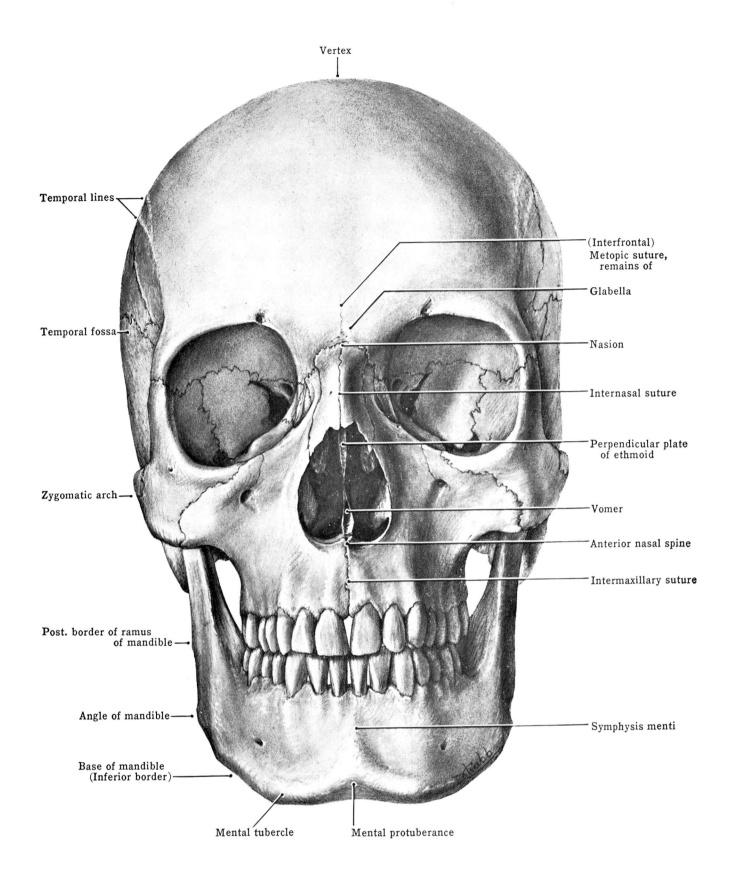

Vertex

Temporal lines

(Interfrontal)
Metopic suture,
remains of

Glabella

Temporal fossa

Nasion

Internasal suture

Perpendicular plate
of ethmoid

Zygomatic arch

Vomer

Anterior nasal spine

Intermaxillary suture

Post. border of ramus
of mandible

Angle of mandible

Symphysis menti

Base of mandible
(Inferior border)

Mental tubercle   Mental protuberance

7-4   SKULL, FRONT VIEW (NORMA FRONTALIS)

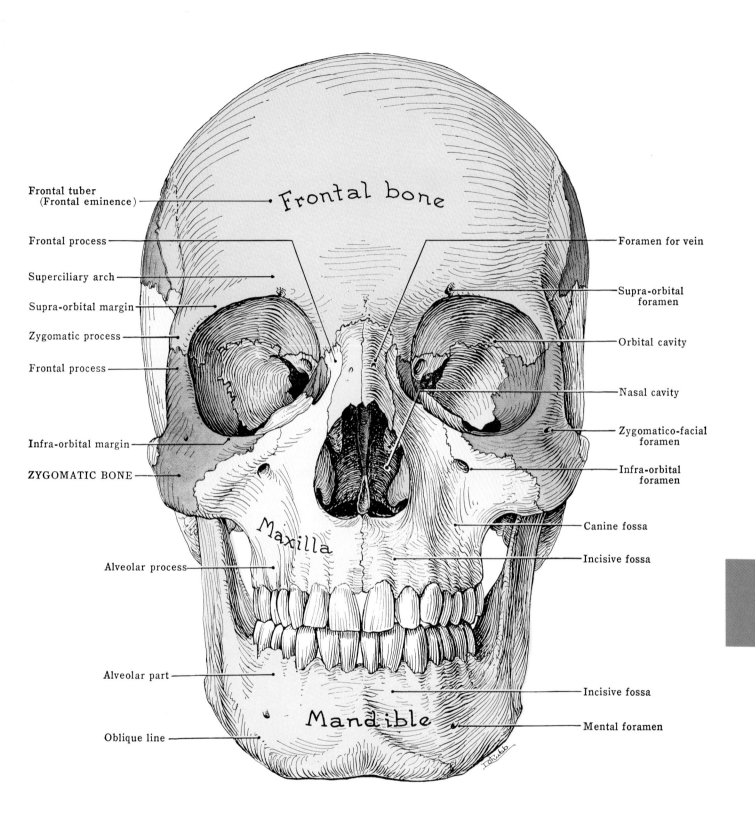

Frontal tuber
(Frontal eminence)

Frontal process

Superciliary arch

Supra-orbital margin

Zygomatic process

Frontal process

Infra-orbital margin

ZYGOMATIC BONE

Alveolar process

Alveolar part

Oblique line

Frontal bone

Foramen for vein

Supra-orbital
foramen

Orbital cavity

Nasal cavity

Zygomatico-facial
foramen

Infra-orbital
foramen

Canine fossa

Incisive fossa

Incisive fossa

Mental foramen

Maxilla

Mandible

7-5 SKULL, FRONT VIEW (NORMA FRONTALIS)

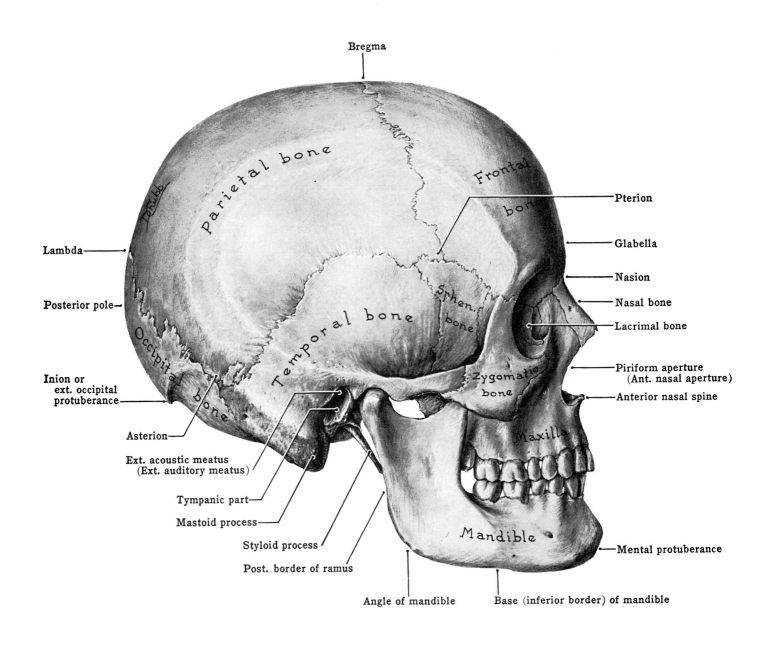

7-6 SKULL, FROM THE SIDE (NORMA LATERALIS)

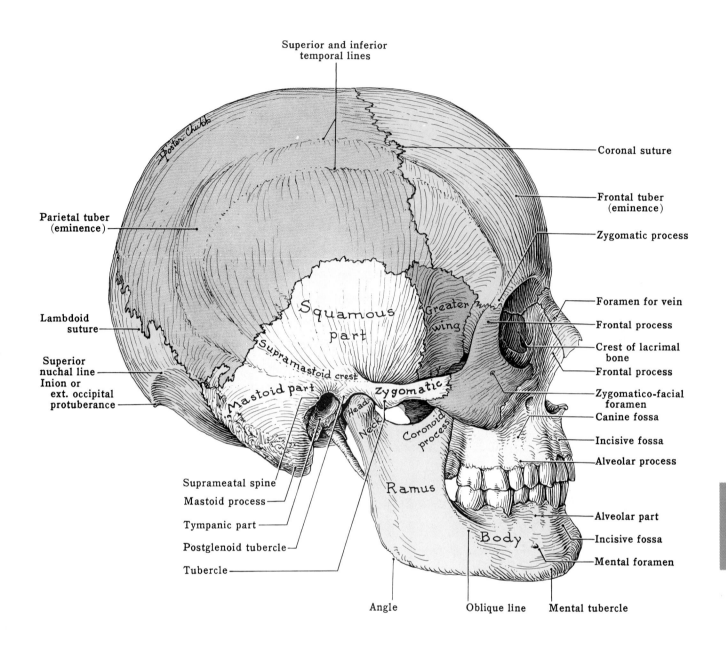

Superior and inferior
temporal lines

Coronal suture

Frontal tuber
(eminence)

Zygomatic process

Parietal tuber
(eminence)

Foramen for vein

Frontal process

Crest of lacrimal
bone

Frontal process

Squamous
part

Greater
wing

Lambdoid
suture

Zygomatico-facial
foramen

Supramastoid crest

Superior
nuchal line

Zygomatic

Canine fossa

Mastoid part

Inion or
ext. occipital
protuberance

Incisive fossa

Head

Coronoid
process

Neck

Alveolar process

Ramus

Supermeatal spine

Mastoid process

Body

Alveolar part

Tympanic part

Incisive fossa

Postglenoid tubercle

Mental foramen

Tubercle

Angle

Oblique line

Mental tubercle

7-7 SKULL, FROM THE SIDE (NORMA LATERALIS)

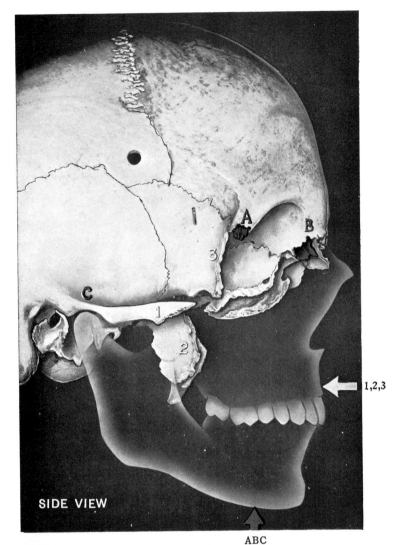

SIDE VIEW

1,2,3 →

↑
ABC

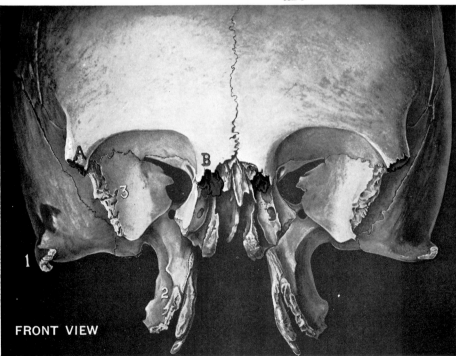

FRONT VIEW

7-8   THE BUTTRESSES OF THE FACE

Observe:

1. The three buttresses (*yellow*) on each side that resist backward displacement of the facial skeleton are:
   1, the zygomatic process of the temporal bone—horizontal.
   2, the pterygoid process of the sphenoid bone—oblique.
   3, the greater wing of the sphenoid bone—vertical.

2. The three buttresses (*red*) on each side that resist upward displacement of the facial skeleton are:
   A, the zygomatic process of the frontal bone.
   B, the nasal part of the frontal bone.
   C, the roof of the mandibular fossa.

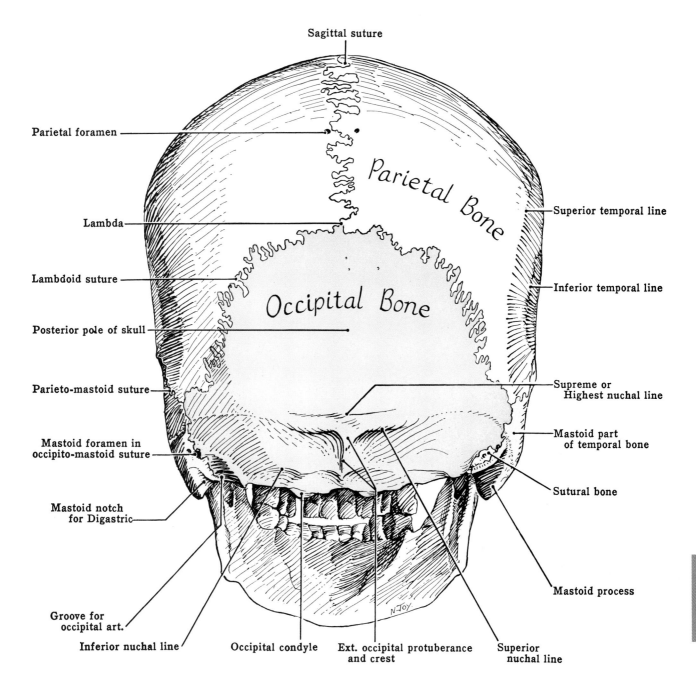

Sagittal suture

Parietal foramen

Parietal Bone

Lambda

Superior temporal line

Lambdoid suture

Occipital Bone

Inferior temporal line

Posterior pole of skull

Parieto-mastoid suture

Supreme or
Highest nuchal line

Mastoid part
of temporal bone

Mastoid foramen in
occipito-mastoid suture

Sutural bone

Mastoid notch
for Digastric

Mastoid process

Groove for
occipital art.

Inferior nuchal line   Occipital condyle   Ext. occipital protuberance   Superior
and crest   nuchal line

7-9   SKULL FROM BEHIND (NORMA OCCIPITALIS)

Observe:
1. The outline is horseshoe-shaped from the tip of one mastoid process over the vertex to the tip of the other.
2. At the base of the skull, the outline is nearly straight from one mastoid process to the other, except where the occipital condyles project downward. On each side, it crosses two grooves (for the origin of the posterior belly of Digastric laterally, and for the occipital artery medially). Between the condyles is the foramen magnum.
3. The surface is convex. Near the center is the lambda. From it a triradiate suture runs: the sagittal (interparietal) upward in the median plane, and the lambdoid (parietooccipital) inferolaterally to the blunt postero-inferior angles of the parietal bones where it bifurcates.
4. On each side are two inconstant foramina for emissary veins and meningeal arteries: parietal and mastoid foramina.
5. Midway between lambda and foramen magnum is the external occipital protuberance or inion. From it the superior nuchal line curves laterally and crosses the lateral aspect of the mastoid, dividing it into a smooth upper and a rough lower part.
6. The surface below the superior nuchal line is the nuchal area for the muscles of the neck or nucha.

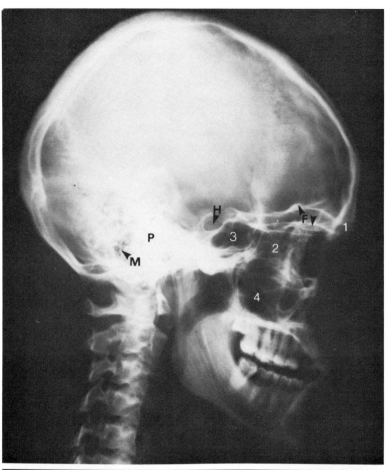

7-10 RADIOGRAPHS OF THE SKULL

A. LATERAL VIEW

Observe:
1. The air sinuses: frontal (*1*), ethmoid (*2*), sphenoid (*3*), and maxillary (*4*).
2. The hypophysial fossa (*H*).
3. The great density of the petrous part of the temporal bone (*P*) and the mastoid air cells (*M*).
4. Right and left orbital plates of the frontal bone are not superimposed and thus the floor of the anterior cranial fossa appears as 2 lines (*F*).

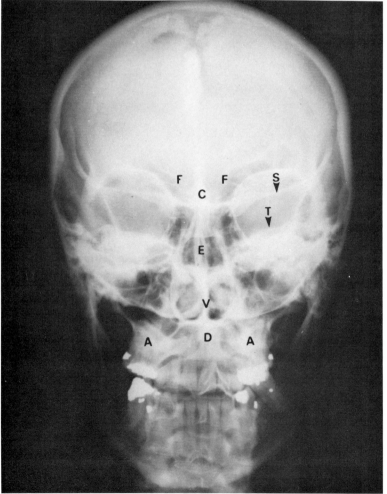

B. POSTERO-ANTERIOR VIEW

Observe:
1. The orbital outline is divided into three horizontal parts by the lesser wing of the sphenoid (*S*) and the upper surface of the petrous part of the temporal bone (*T*).
2. The nasal septum is formed by the perpendicular plate of the ethmoid (*E*) and the vomer (*V*).
3. The crista galli (*C*) and the frontal sinus (*F*).
4. Superimposed on the facial skeleton is the dens (*D*) and the lateral masses of the atlas (*A, A*).

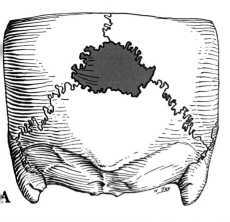 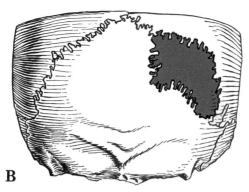 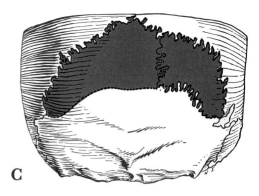

INTERPARIETAL BONE
(OS INCAE)

Occasionally, the several centers of ossification of the interparietal part of the occipital bone fail (*A* and *B*) in part or (*C*) in whole to coalesce with the supraoccipital part. The result is an independent interparietal bone.

Sutural bones (*yellow*) are commonly present, as in this specimen and in *E*, notably at the inferior angles of the parietal bone.

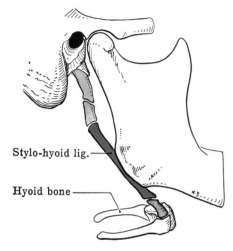

Stylo-hyoid lig.

Hyoid bone

D. OSSIFIED STYLOHYOID LIGAMENT

Here, a chain of four bones attaches the body and greater horn of the hyoid to the skull.

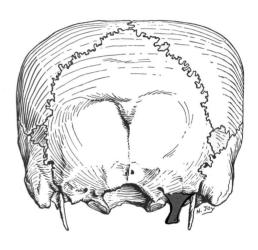

E. PARAMASTOID PROCESS

Rarely, a process descends from the jugular process of the occipital bone toward the transverse process of the atlas, and may articulate with it.

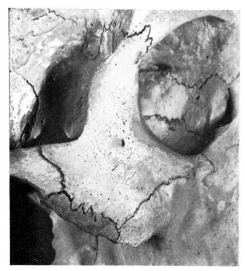

F. OS JAPONICUM

Rarely the zygomatic bone is divided either horizontally or vertically into two parts. In this instance the condition was bilateral.

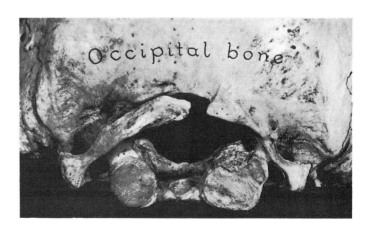

G. OCCIPITALIZATION OF ATLAS

Not uncommonly, varying degrees of fusion of atlas and base of the skull may occur. In this case, the posterior arch is defective and unfused in the midline.

7-11   VARIATIONS AND ANOMALIES

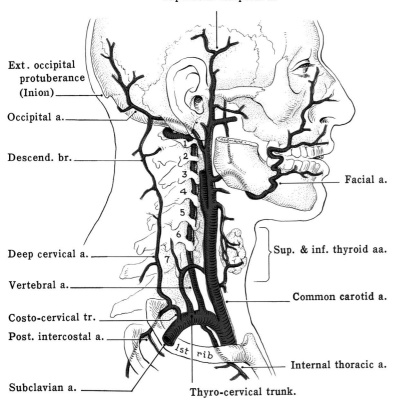

Superficial temporal a.

Ext. occipital protuberance (Inion)

Occipital a.

Descend. br.

Deep cervical a.

Vertebral a.

Costo-cervical tr.

Post. intercostal a.

Subclavian a.

Facial a.

Sup. & inf. thyroid aa.

Common carotid a.

Internal thoracic a.

Thyro-cervical trunk.

1st rib

7-12A THE FOUR CHIEF ARTERIAL CONNECTIONS BETWEEN THE AORTIC ARCH AND THE HEAD

BLOOD SUPPLY OF HEAD

ARTERIES:

Blood supply to the head comes from the common carotid and subclavian arteries. The latter provides the *vertebral artery* which enters the foramen magnum and contributes to the circle of Willis. The common carotid terminates in the internal and external carotids. In general, the *internal carotid* supplies structures inside the skull and the *external carotid* supplies the exterior. However, the ophthalmic artery (from the internal carotid) sends supraorbital and supratrochlear arteries to the forehead. The external carotid provides posterior auricular and occipital arteries, the facial artery, and terminates as maxillary and superficial temporal arteries. Note that the deep cervical artery (from the costo-cervical trunk) anastomoses with the descending branch of the occipital artery and with branches of the vertebral artery.

VEINS:

Within the skull, meningeal, cerebral, cerebellar, and ophthalmic veins drain into venous sinuses which ultimately form right and left internal *jugular veins*. Outside the skull, superficial temporal and maxillary veins form the retromandibular vein whose posterior division unites with the posterior auricular to form the *external jugular vein*. The facial vein receives the anterior division of the retromandibular vein before emptying into the external jugular.

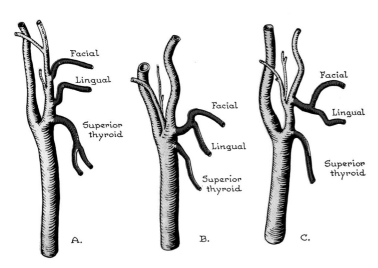

Facial

Lingual

Superior thyroid

A.

Facial

Lingual

Superior thyroid

B.

Facial

Lingual

Superior thyroid

C.

7-12B ORIGIN OF THE LINGUAL ARTERY

Variation in origin was studied in 211 specimens. In 80 per cent, the superior thyroid, lingual, and facial arteries arose separately as in (A) In 20 per cent, the lingual and facial arteries arose from a common stem low down (B) or high on the external carotid artery (C) In one specimen, the superior thyroid and lingual arteries arose from a common stem.

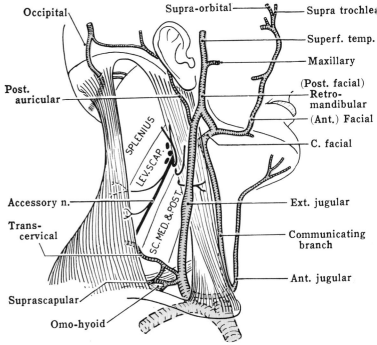

Occipital

Supra-orbital

Supra trochle

Superf. temp.

Maxillary

(Post. facial) Retro-mandibular

(Ant.) Facial

C. facial

Post. auricular

SPLENIUS

LEV. SCAP.

SC. MED. & POST.

Accessory n.

Trans-cervical

Suprascapular

Omo-hyoid

Ext. jugular

Communicating branch

Ant. jugular

B

7-13 SUPERFICIAL VEINS

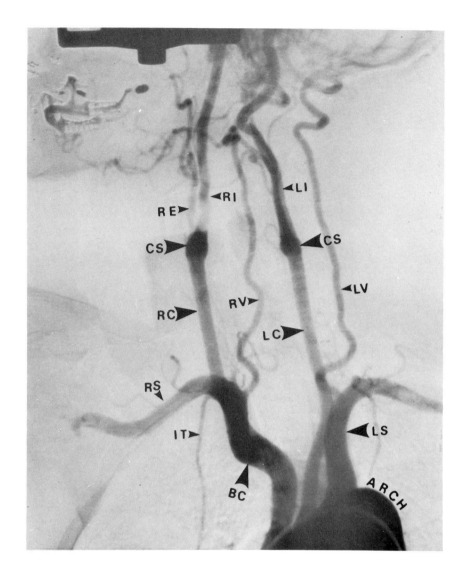

7-14 ARTERIOGRAM OF NECK

This is a positive print of a radiograph with radiopaque material in the arteries. Follow the branches of the aortic arch:

> *BC:* Brachiocephalic
> *LC:* Left common carotid
> *LS:* Left subclavian

The brachiocephalic divides into right subclavian (*RS*) and right common carotid (*RC*) which in turn becomes right external (*RE*) and right internal (*RI*) carotids. The internal thoracic artery (*IT*) is shown arising from the subclavian as is the right vertebral artery (*RV*).

On the *left* is shown the common carotid (*LC*), internal carotid (*LI*), and the tortuous course of the left vertebral artery (*LV*).

> *CS:* Carotid sinus

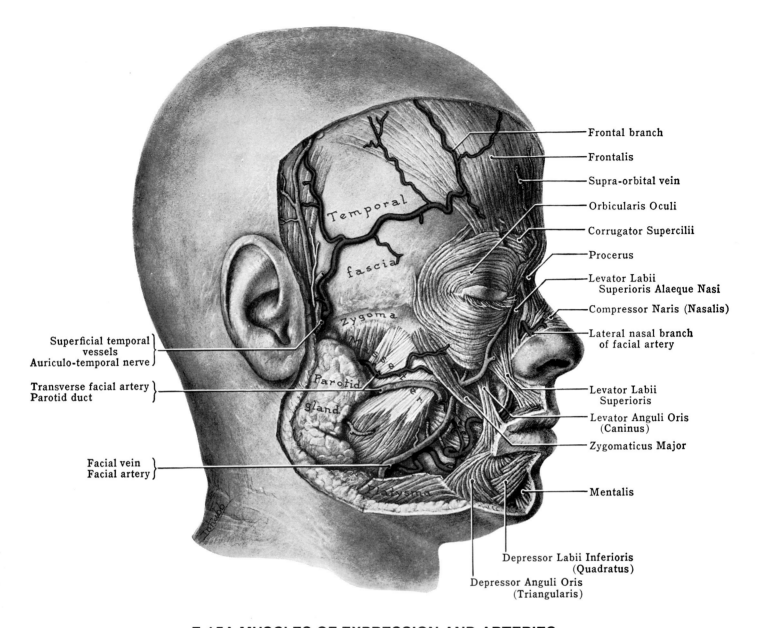

Labels on figure:

- Frontal branch
- Frontalis
- Supra-orbital vein
- Orbicularis Oculi
- Corrugator Supercilii
- Procerus
- Levator Labii Superioris Alaeque Nasi
- Compressor Naris (Nasalis)
- Lateral nasal branch of facial artery
- Levator Labii Superioris
- Levator Anguli Oris (Caninus)
- Zygomaticus Major
- Mentalis
- Depressor Labii Inferioris (Quadratus)
- Depressor Anguli Oris (Triangularis)

- Superficial temporal vessels
- Auriculo-temporal nerve
- Transverse facial artery
- Parotid duct
- Facial vein
- Facial artery

- Temporal
- fascia
- Zygoma
- Parotid gland
- Masseter
- Platysma

7-15A MUSCLES OF EXPRESSION AND ARTERIES OF FACE, SIDE VIEW

The "muscles of expression" are superficial sphincters and dilators of the orifices of the head. All are supplied by the facial nerve (cranial VII).

Observe:

1. Around the eye—Orbicularis Oculi, Corrugator Supercilii, also Frontalis.
2. Around the nose—Compressor Naris, Procerus, and Levator Labii Superioris Alaeque Nasi.
3. Around the mouth—Orbicularis Oris (see Fig. 7-17). In the upper lip—Levator Labii Superioris Alaequae Nasi, Levator Labii Superioris, Zygomaticus Minor, Zygomaticus Major, Levator Anguli Oris. In the lower lip—Platysma (Fig. 9-3), Risorius (not shown), Depressor Anguli Oris, and Depressor Labii Inferioris. In the cheek—Buccinator (Fig. 7-80). In the chin—Mentalis (Fig. 7-17).
4. Around the ear—Auriculares Anterior, Superior, and Posterior (see Fig. 7-16).

   The arteries shown here are:

   a. The facial artery, usually sinuous but here tortuous, crossing the base of the mandible at the anterior border of Masseter, passing within 1 cm of the angle of the mouth, and lying in front of the facial vein which takes a straight and more superficial course.
   b. The transverse facial artery, crossing Masseter between the zygoma and the parotid duct.
   c. Branches or twigs that accompany each branch of the trigeminal (V) nerve, shown in Figure 7-17. Those accompanying V^1 are derived from the ophthalmic branch of the internal carotid artery; those accompanying V^2 and V^3, from the maxillary branch of the external carotid artery.

Palpebral Part

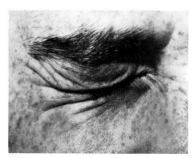

Orbital Part

Orbicularis Oculi

Frontalis

Corrugator Supercilii

Procerus

Nasalis

Risorius

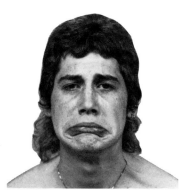

Depressor Anguli Oris

Orbicularis Oris

Zygomaticus Major

Mentalis

7-15B MUSCLES OF EXPRESSION IN ACTION

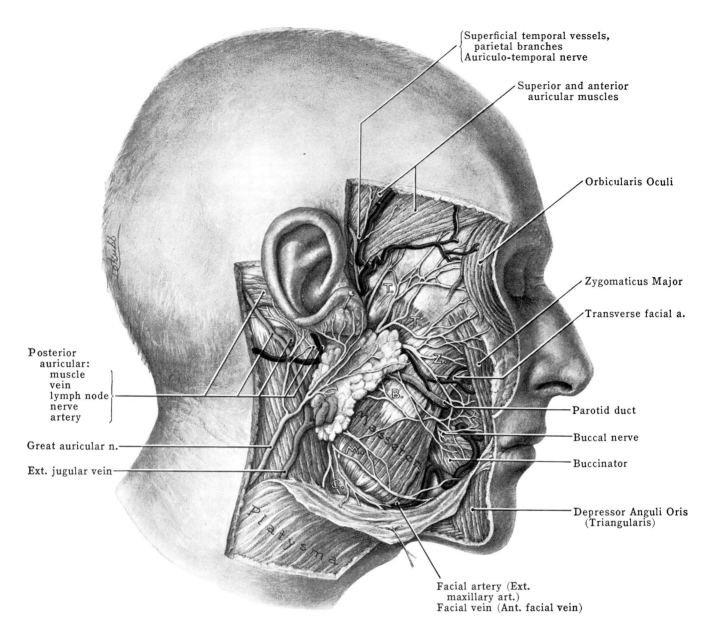

Labels on figure:

Superficial temporal vessels, parietal branches
Auriculo-temporal nerve

Superior and anterior auricular muscles

Orbicularis Oculi

Zygomaticus Major

Transverse facial a.

Parotid duct

Buccal nerve

Buccinator

Depressor Anguli Oris (Triangularis)

Posterior auricular:
muscle
vein
lymph node
nerve
artery

Great auricular n.

Ext. jugular vein

Facial artery (Ext. maxillary art.)
Facial vein (Ant. facial vein)

7-16 FACE: TERMINAL BRANCHES OF THE FACIAL NERVE, SIDE VIEW

Observe:
1. The parotid gland, and the parotid duct crossing Masseter a finger's breadth below the zygomatic arch and turning medially to pierce Buccinator (Fig. 9-55).
2. The facial nerve, motor to the muscles of expression, is in jeopardy in surgery of the parotid gland. Its branches radiate from under cover of the gland like digits from an outstretched hand, anastomosing with each other and with the branches of the trigeminal nerve, and entering the muscles on their deep surfaces except where the muscles are two layers thick, when the deeper layer is entered on its superficial surface, e.g., Buccinator, Levator Anguli Oris. The nerve is divided somewhat arbitrarily into temporal (T), zygomatic (Z), buccal (B), mandibular (M), and cervical (C) branches.
3. The cervical branch, running less than a finger's breadth below the angle of the jaw and, after supplying branches to Platysma, crossing the base of the jaw at the anterior border of Masseter, crossing superficial to the facial vein and facial artery, anastomosing with the mandibular branch, and supplying the muscles of the lower lip and chin.
4. The buccal branch, which is motor to Buccinator, anastomosing with the buccal nerve (a branch of the trigeminal) which is sensory.
5. The greater auricular nerve (C2, C3), here lying 1.3 cm behind the external jugular vein, but commonly in contact with it (Fig. 9-4), and dividing into facial, auricular, and mastoid branches.
6. The posterior auricular nerve (a branch of the facial nerve), which is motor to Auricularis Posterior and Occipitalis, joined by a twig from the great auricular nerve.
7. The retro-auricular (posterior auricular) and superficial parotid lymph nodes.
8. The auriculo-temporal nerve, ascending with the superficial temporal vessels. Their relationships vary (Fig. 7-15); when the nerve is deep, a layer of fascia commonly separates it from the vessels.
9. Masseter is a muscle of mastication and is supplied by the trigeminal nerve.

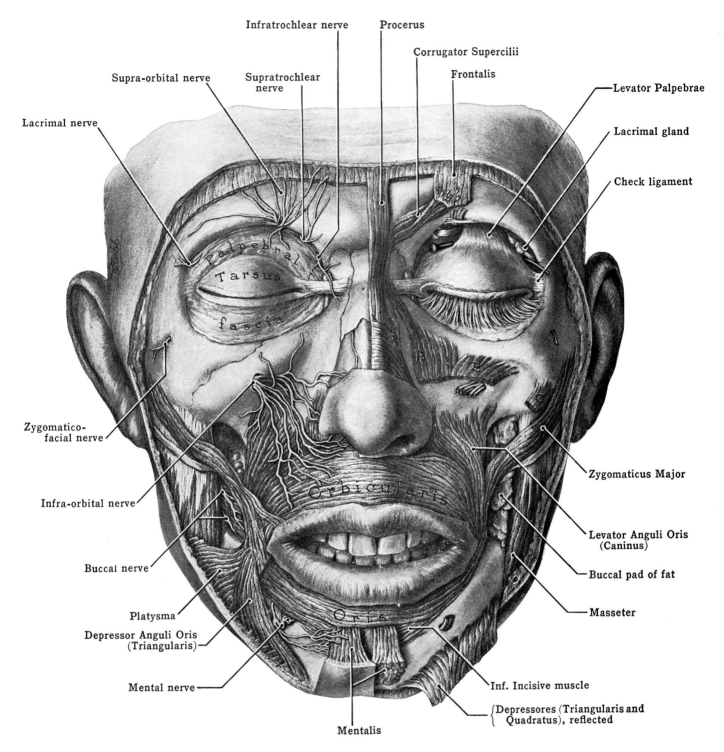

Infratrochlear nerve

Procerus

Corrugator Supercilii

Frontalis

Supra-orbital nerve

Supratrochlear nerve

Levator Palpebrae

Lacrimal nerve

Lacrimal gland

Check ligament

Palpebral

Tarsus

fascia

Zygomatico-facial nerve

Zygomaticus Major

Infra-orbital nerve

Levator Anguli Oris (Caninus)

Buccal nerve

Buccal pad of fat

Platysma

Masseter

Depressor Anguli Oris (Triangularis)

Mental nerve

Inf. Incisive muscle

Depressores (Triangularis and Quadratus), reflected

Mentalis

Orbicularis

Oris

7-17 CUTANEOUS BRANCHES OF TRIGEMINAL NERVE, MUSCLES, EYELID

Observe:

The 5 cutaneous branches of the ophthalmic nerve (V¹). Of these, the external nasal branch is shown but not labeled. Cornea is supplied by the naso-ciliary nerve (Fig. 7-53).

The 3 cutaneous branches of the maxillary nerve (V²): (a) the infra-orbital branch, (b) the zygomatico-facial, and (c) the zygomatico-temporal (Fig. 7-70).

The 3 cutaneous branches of the mandibular nerve (V³): (a) the mental branch, (b) the buccal branch, and (c) the auriculo-temporal branch (Fig. 7-15).

Three sectioned muscles: *A*, Levator Labii Superioris Alaeque Nasi; *B*, Levator Labii Superioris; *C*, Zygomaticus Minor.

The buccal pad of fat, filling the space between Buccina-

tor medially and the ramus of the jaw and Masseter laterally.

6. The palpebral fascia, attached to the orbital margin, and medially passing behind the lacrimal sac to the crest of the lacrimal bone.

7. The medial palpebral ligament, crossing in front of the lacrimal sac, and attaching the elliptical upper and the rodlike lower tarsus to the frontal process of the maxilla.

8. The fan-shaped aponeurosis of Levator Palpebrae Superioris, attached to the front of the superior tarsus. Its medial edge, which is attached behind the lacrimal sac, and its lateral edge, which is attached to a tubercle within the orbital margin, are said to check the overaction of Levator Palpebrae.

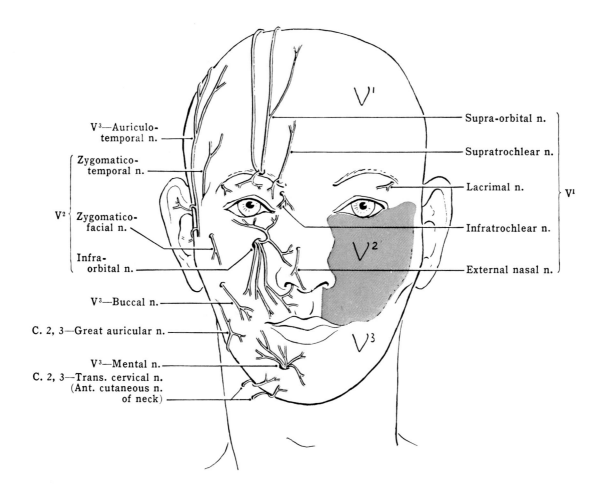

V³—Auriculo-temporal n.

Zygomatico-temporal n.

V²

Zygomatico-facial n.

Infra-orbital n.

V³—Buccal n.

C. 2, 3—Great auricular n.

V³—Mental n.

C. 2, 3—Trans. cervical n. (Ant. cutaneous n. of neck)

V¹

V²

V³

Supra-orbital n.

Supratrochlear n.

Lacrimal n.

Infratrochlear n.

External nasal n.

V¹

7-18   SENSORY NERVES OF FACE, FRONT VIEW

The 3 divisions of the trigeminal nerve (cranial V) correspond in their distribution, nearly but not absolutely, to the 3 embryological regions of the face. Thus, the ophthalmic nerve (V¹) supplies the fronto-nasal process; the maxillary nerve (V²), the maxillary process (colored *pink*); and the mandibular nerve (V³), the mandibular process. They supply the whole thickness of the processes—from skin to mucous surface—indeed, to the median plane (*i.e.,* falx cerebri, nasal septum, and septum of tongue).

Cutaneous branches (supra-orbital and auriculo-temporal) have spread backward in the scalp beyond a line that joins the auricles across the vertex, and there they meet the greater and lesser occipital nerves (Fig. 7-19). The great auricular nerve has spread into the parotid region. The buccal nerve supplies the skin and mucous membrane of the cheek, reaching to the angle of the mouth.

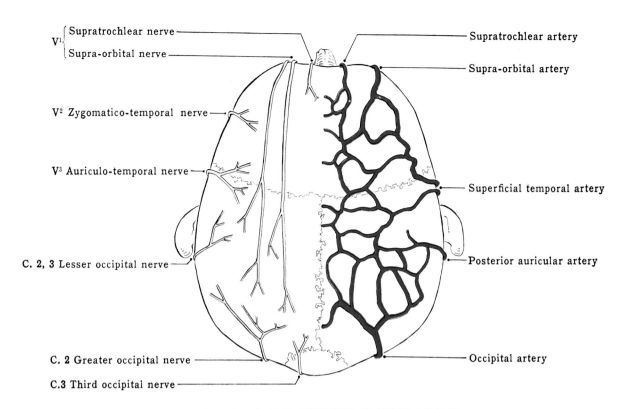

V¹ { Supratrochlear nerve	Supratrochlear artery
Supra-orbital nerve	Supra-orbital artery
V² Zygomatico-temporal nerve	
V³ Auriculo-temporal nerve	Superficial temporal artery
	Posterior auricular artery
C. 2, 3 Lesser occipital nerve	
C. 2 Greater occipital nerve	Occipital artery
C.3 Third occipital nerve	

7-19   ARTERIES AND NERVES OF THE SCALP

Observe:

1. The arteries, anastomosing freely. The supraorbital and supratrochlear are derived from the internal carotid artery via the ophthalmic artery; the three others are branches of the external carotid artery.

2. The nerves, appearing in sequence: V¹, V², V³, ventral rami of C2 and C3, and dorsal rami of C2 and C3 (C1 has no cutaneous branch).

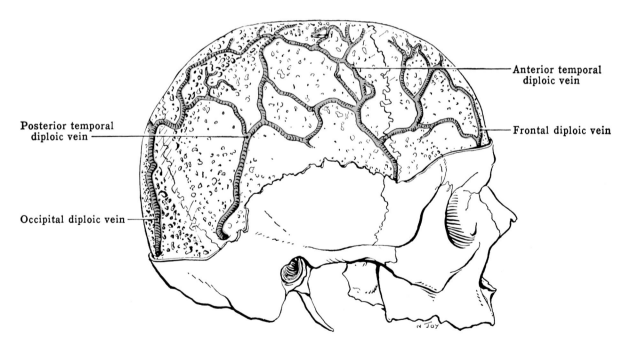

Anterior temporal diploic vein

Posterior temporal diploic vein

Frontal diploic vein

Occipital diploic vein

7-20   DIPLOIC VEINS

Note:

1. The outer table of the skull has been filed away and the channels for the diploic veins thereby opened.
2. Of the four (paired) diploic veins, the frontal opens into the supraorbital vein at the supraorbital notch; the anterior temporal opens into the spheno-parietal sinus; the posterior temporal and the occipital both open into the transverse sinus—but they may open into surface veins.
3. There are no accompanying diploic arteries; the meningeal and pericranial arteries provide the arterial blood.
4. Connections with intracranial and extracranial venous channels allows infection to travel freely between scalp, skull, meninges, and brain.

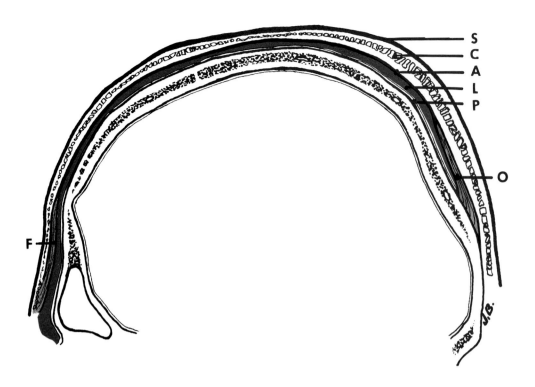

7-21 LAYERS OF THE SCALP

S: Skin
C: Connective tissue
A: Aponeurosis
L: Loose Areolar tissue
P: Periosteum (Pericranium)

The skin is bound tightly to the aponeurosis which is attached to the skull laterally and to the two fleshy muscles for which it serves as intermediate tendon. Occipitalis (*O*) attaches to bone while Frontalis (*F*) attaches superficially to subcutaneous tissue. Thus, blood from a torn vessel may spread widely over the skull deep to the aponeurosis and leak out anteriorly, appearing as bruising in the area of the eyelids.

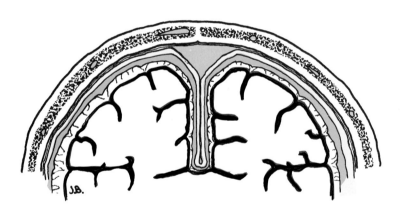

7-22 THE MENINGES

A diagram representing a coronal section through skull and brain. The membranes have been unreasonably thickened to assist in identification.

Observe:
1. The interior of the skull lined by periosteum.
2. The outer tough *dura mater* (*yellow*) encloses venous sinuses by reflecting away from the skull (*e.g.* superior sagittal) or within the free edges of double layers of dura (*e.g.* inferior sagittal sinus in the free edge of the falx cerebri).
3. The *arachnoid mater* (*green*) in contact with the dura and bridging over sulci on the cortical surface.
4. The *pia mater* (*red*), a delicate, intimate investment of the brain.
5. Between dura and arachnoid, a potential *subdural space* into which hemorrhage may occur.
6. Between arachnoid and pia, the subarachnoid space containing cerebrospinal fluid.

7-23A BRAIN IN CORONAL SECTION

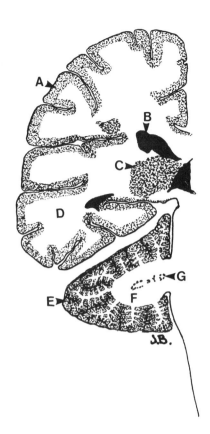

During dissection, the brain is removed and its external features observed. This page provides an introduction to the brain pending its later study in neuroanatomy. The brain consists of three parts: a brain stem continuous below with the spinal cord, and two extravagant outgrowths, the cerebral hemispheres and the cerebellum.

1. The *brain stem* is what remains when the cerebral hemispheres and cerebellum have been removed (see Figures 7-30B and 7-31). It contains ascending and descending fiber tracts and collections of cell bodies. The latter are nuclei of cranial nerves, centers for the control of vital functions, connections with the cerebellum, and association centers for motor and sensory systems.

2. The *cerebral hemispheres* (right and left) are incompletely separated by a deep fissure and joined together by the corpus callosum. The hemispheres are covered by a veneer of gray matter, the cerebral cortex (*A*), containing the cell bodies of about 14 billion neurones. The interior is composed of three features: cavities containing cerebrospinal fluid, the *ventricles* (*B*); collections of gray matter, the *basal ganglia* (*C*); and (mainly) white matter consisting of the processes of neurones forming fiber tracts (*D*). These tracts are of three sorts: *commisural* (linking the two hemispheres with each other); *association* (connecting different parts of the cortex); and *projection* (which establish communication between the cerebral hemispheres and lower centers).

3. The *cerebellum* which occupies the posterior cranial fossa below the tentorium cerebelli. It is connected to the brain stem by three pairs of cerebellar peduncles. Its fissured surface consists of a thick layer of gray matter, the cortex (*E*). The interior contains white matter, fiber tracts traveling to and from the cortex (*F*); and four pair of nuclei (*G*).

7-23B CEREBRAL CORTEX

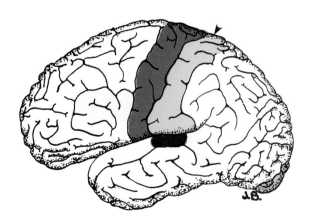

This lateral view of the left cerebral cortex shows that the surface of gray matter consists of folds (gyri) and grooves (sulci). The cerebrum may be roughly divided into lobes in relation to the overlying cranial bones: frontal, parietal, occipital, and temporal. Functional areas may be related to certain regions of the cortex, some of which are shown here. An *arrow* points to a deep groove, the *central sulcus*, which separates the frontal and parietal lobes. Behind this groove is the *general sensory area* (*blue*). The area of cortex devoted to each region of the body is not proportional to the size of that region, but to the density of sensory receptors from that part. Thus, the hand and face have a relatively huge share of the sensory cortex. In front of the central sulcus is the *primary motor area* (*red*). Also seen are parts of the *visual area* (*green*) in the occipital lobe, and of the *auditory area* (*brown*) in the temporal lobe. Shown in *yellow* are two areas important in speech, the *motor speech area* (*of Broca*) in the frontal lobe and the *auditory association cortex* (*of Wernicke*) in the temporal lobe.

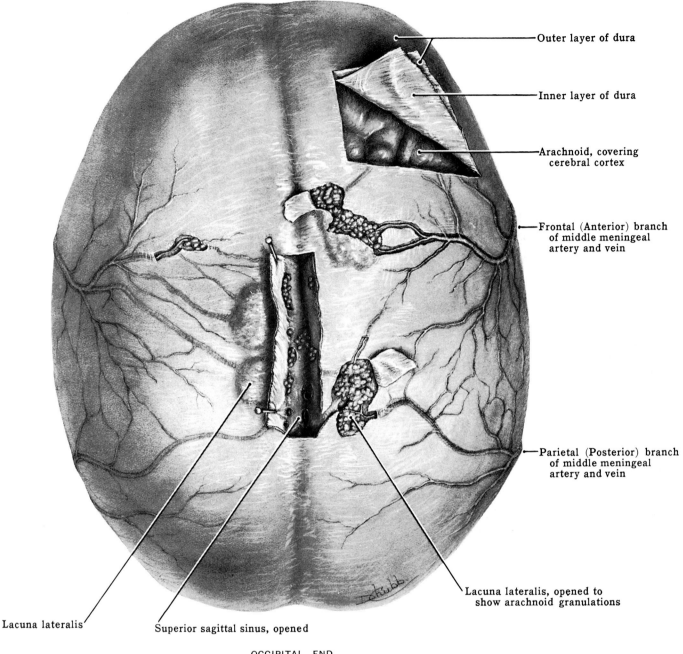

Outer layer of dura

Inner layer of dura

Arachnoid, covering cerebral cortex

Frontal (Anterior) branch of middle meningeal artery and vein

Parietal (Posterior) branch of middle meningeal artery and vein

Lacuna lateralis, opened to show arachnoid granulations

Lacuna lateralis

Superior sagittal sinus, opened

7-24 EXTERNAL SURFACE OF THE DURA MATER: ARACHNOID GRANULATIONS

Observe:

1. The skull cap is removed. In the median plane, the thick roof of the superior sagittal dural sinus is partly pinned aside and, laterally, the thin roofs of two lacunae laterales are reflected.

2. At the right front, an angular flap of both layers of dura is turned forward; the subdural space is thereby opened; and the convolutions of the cerebral cortex are visible through the cobweblike arachnoid mater.

3. The frontal and parietal branches of the middle menin-geal artery, arborizing and anastomosing within the external or periosteal layer of the dura.

4. Each artery lying in a venous channel (middle menin-geal veins) which enlarges above into a lake, lacuna lateralis.

5. Cauliflowerlike masses, arachnoid granulations, dan-gling in the lakes. They are responsible for absorption of cerebrospinal fluid.

6. A channel or channels, draining the lake into the supe-rior sagittal sinus which is triangular on cross-section and has many discrete granulations protruding into it.

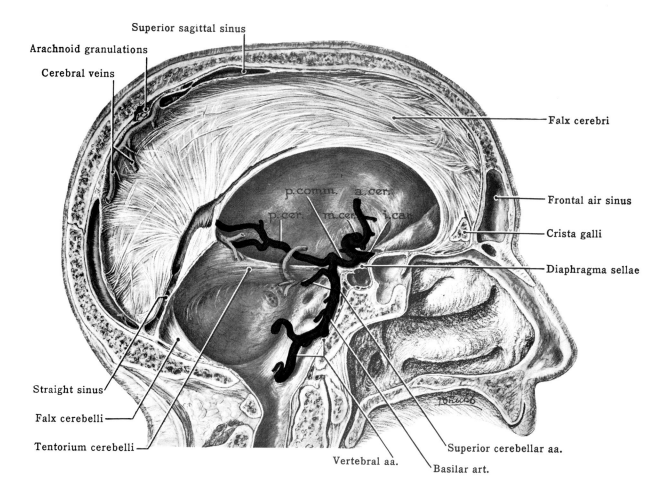

Superior sagittal sinus

Arachnoid granulations

Cerebral veins

Falx cerebri

Frontal air sinus

Crista galli

Diaphragma sellae

p.comm. a.cer.
p.cer. m.cer. i.car.

Straight sinus

Falx cerebelli

Tentorium cerebelli

Superior cerebellar aa.

Vertebral aa.

Basilar art.

7-25 FOLDS OF THE DURA MATER

Observe:
1. The 4 reduplications of the inner layer of the dura mater: 2 sickle-shaped folds, (a) falx cerebri and (b) falx cerebelli, which lie vertically in the median plane; and 2 rooflike folds, (c) tentorium cerebelli and (d) diaphragma sellae, which lie horizontally.
2. The tentorium is perforated by the midbrain; the diaphragma, by the stalk of the hypophysis cerebri.
3. The 2 paired arteries that supply the brain—internal carotid and vertebral.

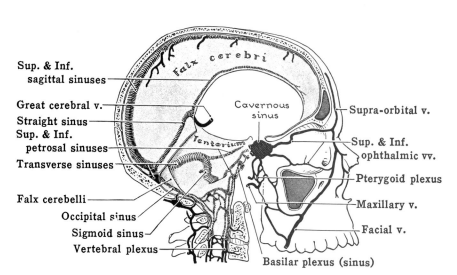

Sup. & Inf. sagittal sinuses

Great cerebral v.

Straight sinus

Sup. & Inf. petrosal sinuses

Transverse sinuses

Falx cerebelli

Occipital sinus

Sigmoid sinus

Vertebral plexus

Falx cerebri

Cavernous sinus

Tentorium

Supra-orbital v.

Sup. & Inf. ophthalmic vv.

Pterygoid plexus

Maxillary v.

Facial v.

Basilar plexus (sinus)

7-26 DIAGRAM OF VENOUS SINUSES OF DURA MATER

Observe:
1. (a) The superior sagittal sinus at the upper border of the falx cerebri; (b) the inferior sagittal sinus in its free border; (c) the great cerebral vein joining the inferior sagittal sinus to form (d) the straight sinus which runs obliquely in the junction between falx cerebri and tentorium; (e) the occipital sinus in the attached border of the falx cerebelli.
2. The superior sagittal sinus usually becomes: right transverse sinus-right sigmoid sinus-right internal jugular vein; the straight sinus behaves similarly on left side.
   For variations, see: Browning, H. (1953) The confluence of dural venous sinuses. *Am. J. Anat., 93: 307*.
3. The cavernous sinus communicating with the veins of the face via the ophthalmic veins and the pterygoid plexus, and emptying via the two petrosal sinuses (Fig. 7-161).
4. The basilar sinus connecting the inferior petrosal sinuses of opposite sides and, like the occipital sinus, communicating below with the vertebral plexus (Fig. 5-36).

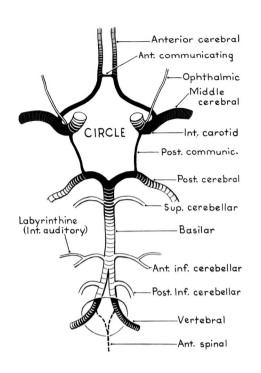

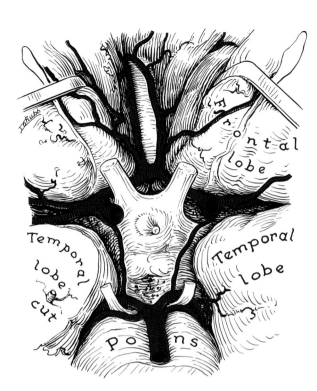

7-27A   CEREBRAL ARTERIAL CIRCLE (OF WILLIS)

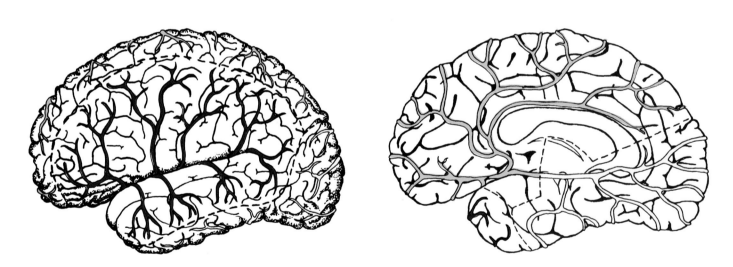

7-27B   BLOOD SUPPLY OF THE BRAIN

Three arterial stems ascend to supply the brain: right and left internal carotids and the basilar artery which results from the union of the two vertebral arteries. Observe that these three stems form an arterial circle (the circle of Willis) at the base of the brain through the linkages provided by the anterior communicating artery and the two posterior communicating arteries. The cerebellum is mainly supplied by branches from the vertebral or basilar arteries. Supply to the cerebrum is shared by the anterior and middle cerebral arteries from the internal carotids, and the posterior cerebral arteries from the basilar. In the schematic diagram above the general areas of supply are shown for the three cerebral arteries: anterior (*green*), middle (*red*), and posterior (*yellow*). Because of functional localization in the cortex, a cerebrovascular accident occurring in one vessel will have predictable results. For example, the visual area in the occipital lobe is supplied by a branch of the posterior cerebral artery; sensory and motor supply to the lower limb is located in the area served by the anterior cerebral artery.

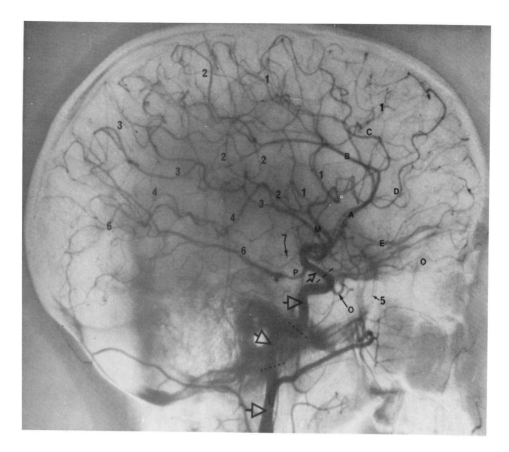

7-28 CAROTID ARTERIOGRAM

A positive print of a radiograph with radiopaque material in the internal carotid artery. (Some external carotid branches are also seen.) *Four arrows* indicate parts of the internal carotid: *cervical*, before entering the skull; *petrous*, within the temporal bone; *cavernous*, within that venous sinus; and *cerebral*, within the cranial subarachnoid space. *A*, anterior cerebral artery; *M*, middle cerebral artery; *P*, posterior communicating artery connecting the internal carotid to the posterior cerebral artery.

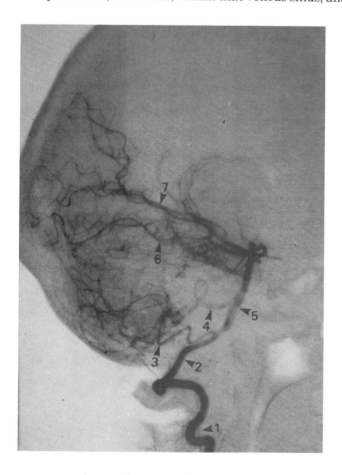

7-29 VERTEBRAL ARTERIOGRAM

In this positive print of a vertebral arteriogram observe:

1. The curve made by the vertebral artery to reach the foramen transversarium of the atlas.
2. The vertebral artery enters the skull through the foramen magnum.
3. Posterior inferior cerebellar artery.
4. Anterior inferior cerebellar artery.
5. The basilar artery formed by the union of right and left vertebral arteries (notice poor filling).
6. Superior cerebellar artery (see how the tiny arterial branches outline the form of the cerebellum).
7. Posterior cerebral artery with branches going to occipital and temporal lobes, including the supply to the visual area of the cortex.

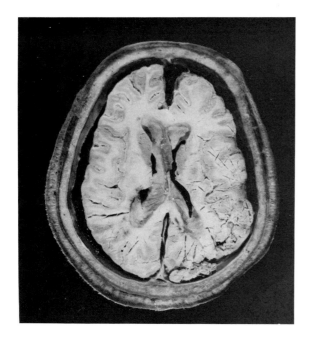
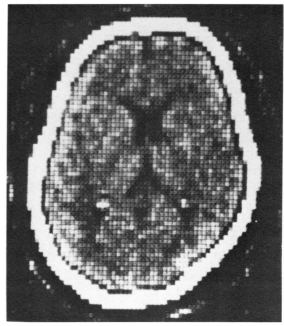

7-30A COMPUTERIZED TOMOGRAPHY OF THE HEAD

On the *left* is a horizontal section through the head of a cadaver for comparison. On the *right* is the printout of a live head scanned in a horizontal plane.

Observe:
1. Reduced density (*dark*) in the area of the cerebral ventricles.
2. Reduced density in the subarachnoid spaces outside the brain.
3. The dense (*white*) skull outline.
4. The intermediate density of brain substance.

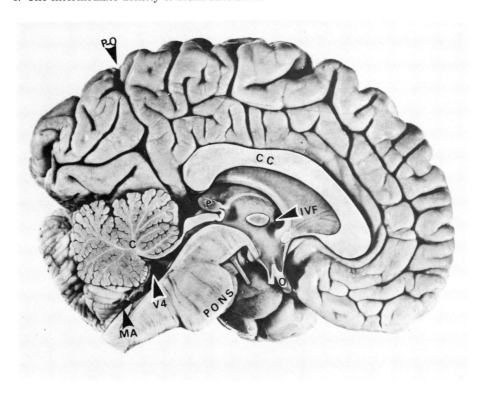

7-30B MEDIAN SAGITTAL SECTION OF BRAIN

Observe:
1. The parieto-occipital sulcus (*P-O*) clearly dividing parietal from occipital lobes.
2. Below the occipital lobe, the intricate branching of the sectioned cerebellum (*C*).
3. The knife has bisected the corpus callosum (*CC*) which is a broad curved band of fibers connecting the two hemispheres; the optic chiasma (*O*); and the brain stem.
4. The interventricular foramen (*IVF*) connecting the lateral with the 3rd ventricle.
5. The median aperture (*MA*) of the 4th ventricle (*V4*) which connects the ventricular system with the subarachnoid space.

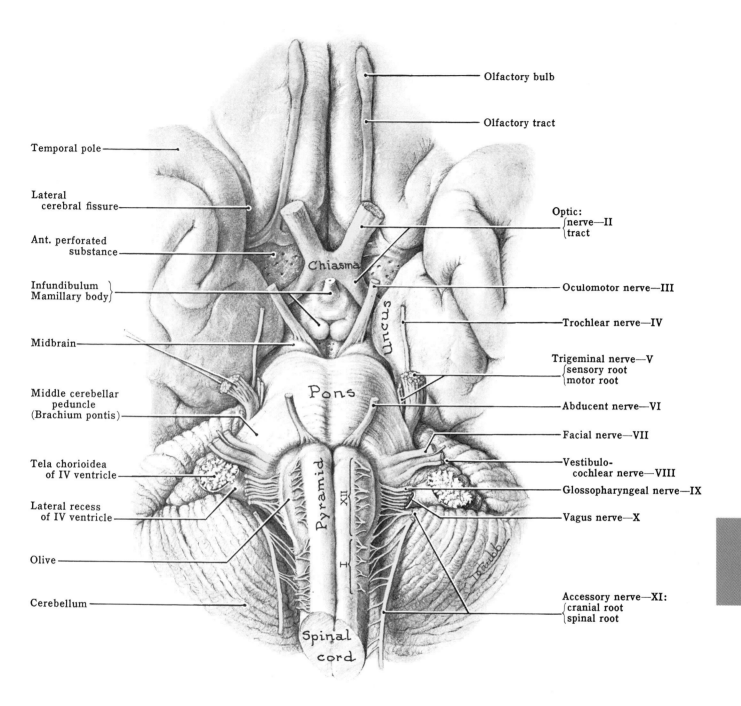

Olfactory bulb

Olfactory tract

Temporal pole

Lateral cerebral fissure

Ant. perforated substance

Infundibulum
Mamillary body

Midbrain

Middle cerebellar peduncle (Brachium pontis)

Tela chorioidea of IV ventricle

Lateral recess of IV ventricle

Olive

Cerebellum

Chiasma

Uncus

Pons

Pyramid

XII

IX

Spinal cord

Optic:
nerve—II
tract

Oculomotor nerve—III

Trochlear nerve—IV

Trigeminal nerve—V
sensory root
motor root

Abducent nerve—VI

Facial nerve—VII

Vestibulo-cochlear nerve—VIII

Glossopharyngeal nerve—IX

Vagus nerve—X

Accessory nerve—XI:
cranial root
spinal root

7-31 BASE OF THE BRAIN: THE SUPERFICIAL ORIGINS OF THE CRANIAL NERVES

Note:
1. The olfactory bulb, in which the olfactory (cranial I) nerves (not shown) end.
2. The superficial origin of the trochlear (cranial IV) nerve is shown in Figure 7-32.
3. The slender nervus intermedius, or so-called sensory root of the facial nerve (not labeled) between the facial (VII) and vestibulo-cochlear (VIII) nerves.
4. The fila of the hypoglossal (XII) nerve, arising between the pyramid and the olive, and in line with the ventral root of the 1st cervical nerve.

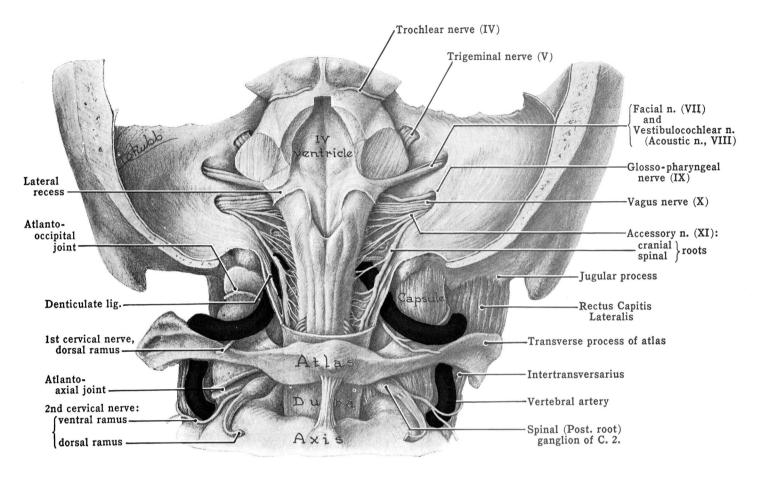

Trochlear nerve (IV)

Trigeminal nerve (V)

Facial n. (VII) and Vestibulocochlear n. (Acoustic n., VIII)

Glosso-pharyngeal nerve (IX)

Vagus nerve (X)

Accessory n. (XI): cranial } roots spinal

Jugular process

Rectus Capitis Lateralis

Transverse process of atlas

Intertransversarius

Vertebral artery

Spinal (Post. root) ganglion of C. 2.

IV ventricle

Capsule

Atlas

Dura

Axis

Lateral recess

Atlanto-occipital joint

Denticulate lig.

1st cervical nerve, dorsal ramus

Atlanto-axial joint

2nd cervical nerve: { ventral ramus

{ dorsal ramus

7-32 CRANIAL NERVES, EXPOSED FROM BEHIND

Observe:
1. The trochlear (IV) nerves, arising from the dorsal aspect of midbrain just below the inferior colliculi; the trigeminal (V) nerves, ascending to enter the mouths of the trigeminal (Meckel's) caves; the facial (VII) and vestibulo-cochlear (VIII) nerves ascending to enter the internal acoustic meatuses; the glossopharyngeal (IX) nerves, piercing the dura mater separately and passing with the vagus (X) and accessory (XI) nerves through the jugular foramina; the fila of the accessory nerves of opposite sides, leaving the medulla and spinal cord asymmetrically.
2. The abducent (VI) nerves are not in view. The hypoglossal (XII) nerves are seen vaguely in front of the spinal roots of nerves XI and just above the vertebral arteries.
3. The transverse process of the atlas, joined to the jugular process of the occipital bone by Rectus Capitis Lateralis, which morphologically is an Intertransverse muscle.
4. The vertebral arteries, raised from their beds on the posterior arch of the atlas.
5. The 1st cervical or suboccipital nerve, here having no sensory component. Its dorsal ramus (suboccipital nerve) passing between the vertebral artery and the posterior arch of the atlas. Its ventral ramus (not labeled), curving around the atlanto-occipital joint.
6. The 2nd cervical nerve, largely sensory, having a large spinal ganglion, a large dorsal ramus (or greater occipital nerve), and a smaller ventral ramus. The fila of its dorsal root are seen just above the cut edge of the dura and behind the spinal root of nerve XI.

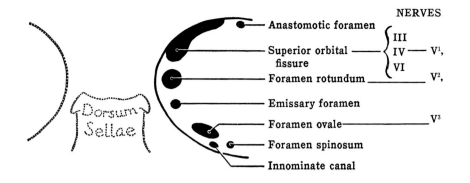

7-33 CRESCENT OF FORAMINA IN THE MIDDLE CRANIAL FOSSA

Note:
1. Seven paired foramina, of which 4 (*black*) are constant and 3 (*red*) are inconstant, open from the sphenoid bone into the middle cranial fossa. Because of the crescentic design made by these foramina, it is simple to relate one foramen to another.
2. Constant foramina: The superior orbital fissure, foramen rotundum, and foramen ovale transmit nerves III, IV, V, and VI. Foramen spinosum transmits the middle meningeal artery.
3. Inconstant foramina: *Anastomotic foramen* transmits a branch from middle meningeal artery to lacrimal artery (Fig. 7-56). *Emissary foramen* transmits a vein from cavernous sinus to pterygoid plexus. *Innominate canal*, in lieu of foramen ovale, transmits lesser petrosal nerve to otic ganglion.

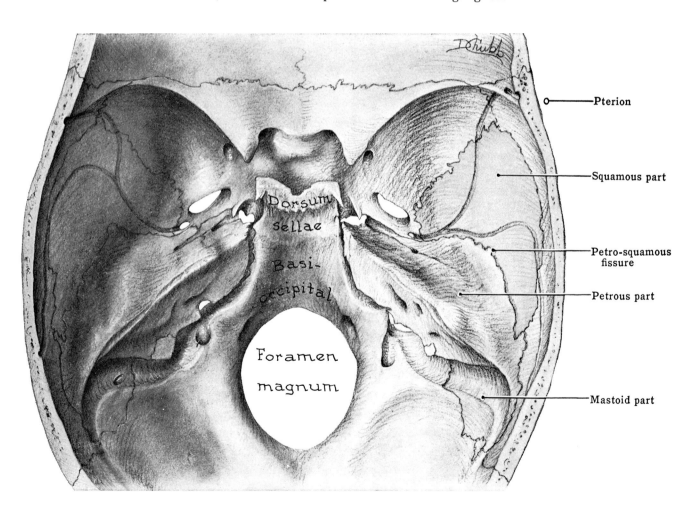

7-34 TEMPORAL BONE, IN THE INTERIOR OF THE BASE OF THE SKULL

See also Figures 7-35 and 7-39.

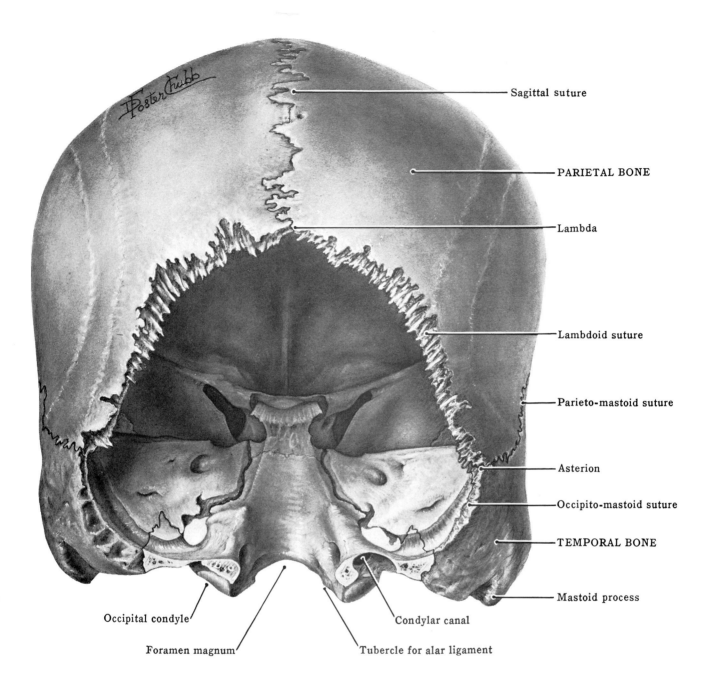

Sagittal suture

PARIETAL BONE

Lambda

Lambdoid suture

Parieto-mastoid suture

Asterion

Occipito-mastoid suture

TEMPORAL BONE

Mastoid process

Occipital condyle

Foramen magnum

Condylar canal

Tubercle for alar ligament

7-35 POSTERIOR CRANIAL FOSSA, FROM BEHIND

Consult Figure 7-36 on facing page. Part of the occipital bone has been removed.

Note:

1. The *dorsum sellae* is the squarish plate of bone rising from the body of the sphenoid. At its superior angles are the posterior clinoid processes.

2. The *clivus* is the sloping surface between the dorsum sellae and the foramen magnum. It is formed by the basilar part of the occipital bone (basi-occipital) with some assistance from the body of the sphenoid, as Figure 10-9 makes clear.

3. The *sulci,* or grooves, for the sigmoid sinus and the inferior petrosal sinus both lead downward to the jugular foramen.

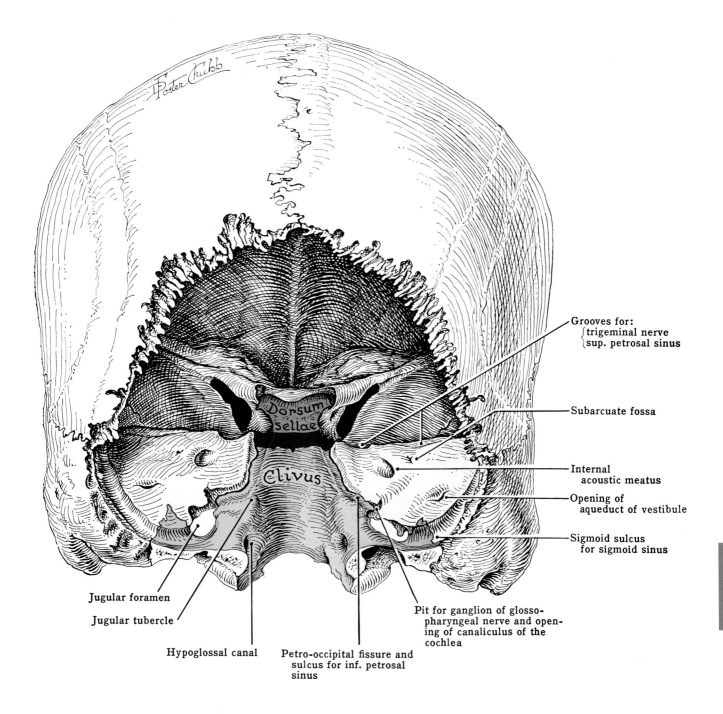

Grooves for:
{ trigeminal nerve
{ sup. petrosal sinus

Subarcuate fossa

Internal
acoustic meatus

Opening of
aqueduct of vestibule

Sigmoid sulcus
for sigmoid sinus

Pit for ganglion of glossopharyngeal nerve and opening of canaliculus of the
cochlea

Jugular foramen

Jugular tubercle

Hypoglossal canal

Petro-occipital fissure and
sulcus for inf. petrosal
sinus

Dorsum
sellae

Clivus

7-36 POSTERIOR CRANIAL FOSSA, FROM BEHIND

For a view of this area from above see Figures 7-38 and 7-39.

Note:

1. That at birth the subarcuate fossa was large and extended laterally, under the arc of the anterior semicircular canal.
2. That the aqueduct of the vestibule opened under the arc of the posterior semicircular canal. This aqueduct transmits the endolymphatic duct (see Figs. 7-162 and 7-167).
3. That the perilymphatic duct (within the canaliculus of the cochlea, Fig. 7-162) opens at the bottom of the pyramidal pit for the glossopharyngeal ganglion. This capillary aqueduct is said to allow the perilymph of the internal ear to mix with the cerebrospinal fluid in the posterior cranial fossa, but there is evidence that it ends as a closed sac.

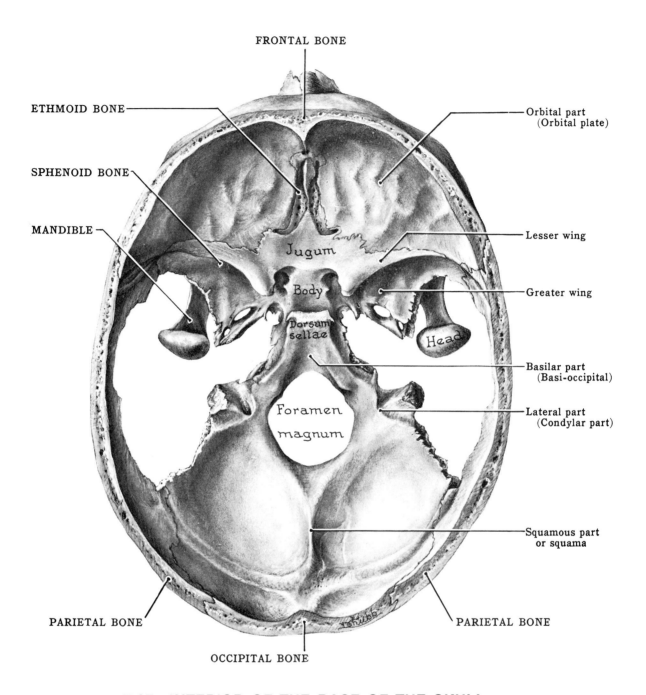

FRONTAL BONE

ETHMOID BONE

SPHENOID BONE

MANDIBLE

Orbital part
(Orbital plate)

Lesser wing

Greater wing

Jugum

Body

Dorsum
sellae

Head

Basilar part
(Basi-occipital)

Lateral part
(Condylar part)

Foramen
magnum

Squamous part
or squama

PARIETAL BONE

PARIETAL BONE

OCCIPITAL BONE

7-37 INTERIOR OF THE BASE OF THE SKULL

Observe:
1. The temporal bones are removed, but the heads of the mandible remain *in situ*.
2. The head of the mandible, lying lateral to the spine of the sphenoid, at which there is a perforation, the foramen spinosum, for the transmission of the middle meningeal vessels.
3. The long axes of the right and left heads, when produced medially, meet near the anterior border of the foramen magnum.
4. The 3 bones contributing to the anterior cranial fossa: orbital plate of frontal, cribriform plate of ethmoid, and lesser wing of sphenoid.
5. The 4 developmental parts of the occipital bone: basilar, right and left lateral, and squamous. See Figure 10-6.

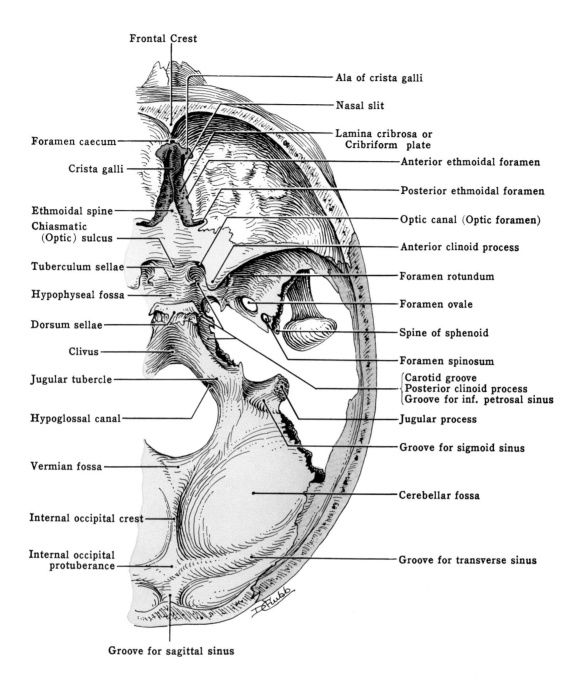

Frontal Crest

Ala of crista galli

Nasal slit

Lamina cribrosa or Cribriform plate

Foramen caecum

Anterior ethmoidal foramen

Crista galli

Posterior ethmoidal foramen

Ethmoidal spine

Optic canal (Optic foramen)

Chiasmatic (Optic) sulcus

Anterior clinoid process

Tuberculum sellae

Foramen rotundum

Hypophyseal fossa

Foramen ovale

Dorsum sellae

Spine of sphenoid

Clivus

Foramen spinosum

Jugular tubercle

{ Carotid groove
Posterior clinoid process
Groove for inf. petrosal sinus

Hypoglossal canal

Jugular process

Groove for sigmoid sinus

Vermian fossa

Cerebellar fossa

Internal occipital crest

Internal occipital protuberance

Groove for transverse sinus

Groove for sagittal sinus

7-38 INTERIOR OF THE BASE OF THE SKULL

Note the following features in the median plane:
1. In the anterior cranial fossa: frontal crest and critsa galli for attachment of the falx cerebri. Between them, the foramen caecum — not usually blind — which transmits a vein connecting the superior sagittal sinus with the veins of the frontal sinus and root of the nose.
2. In the middle cranial fossa: the chiasmatic sulcus leading from one optic canal to the other, but not lodging the optic chiasma (see Fig. 7-54); tuberculum sellae; hypophyseal fossa; and dorsum sellae.
3. In the posterior cranial fossa: clivus, foramen magnum, vermian fossa (for vermis of the cerebellum), internal occipital crest for attachment of the falx cerebelli, and the internal occipital protuberance from which sulci for the transverse sinuses curve laterally.

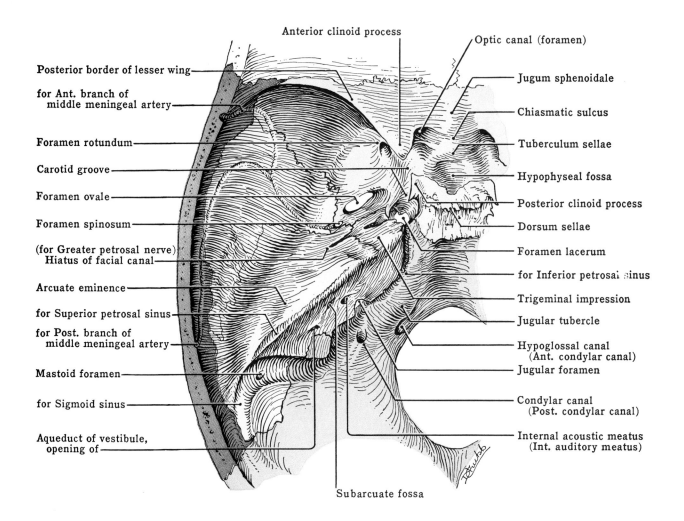

Anterior clinoid process

Optic canal (foramen)

Posterior border of lesser wing

Jugum sphenoidale

for Ant. branch of middle meningeal artery

Chiasmatic sulcus

Foramen rotundum

Tuberculum sellae

Carotid groove

Hypophyseal fossa

Foramen ovale

Posterior clinoid process

Foramen spinosum

Dorsum sellae

(for Greater petrosal nerve) Hiatus of facial canal

Foramen lacerum

for Inferior petrosal sinus

Arcuate eminence

Trigeminal impression

for Superior petrosal sinus

Jugular tubercle

for Post. branch of middle meningeal artery

Hypoglossal canal (Ant. condylar canal)

Mastoid foramen

Jugular foramen

for Sigmoid sinus

Condylar canal (Post. condylar canal)

Aqueduct of vestibule, opening of

Internal acoustic meatus (Int. auditory meatus)

Subarcuate fossa

7-39 MIDDLE AND POSTERIOR CRANIAL FOSSAE, FROM ABOVE

Note:
1. Three features—tuberculum sellae, hypophyseal fossa, and dorsum sellae— constitute the sella turcica or Turkish saddle.
2. Of the two paired clinoid processes for the attachment of the tentorium (Fig. 7-43), the anterior on the lesser wing of the sphenoid is conical; the posterior, on the angle of the dorsum sellae, is beaklike.
3. The foramen lacerum is situated between the hypophyseal fossa and the apex of the petrous bone. There the carotid canal discharges the internal carotid artery into the upper half of the foramen lacerum.

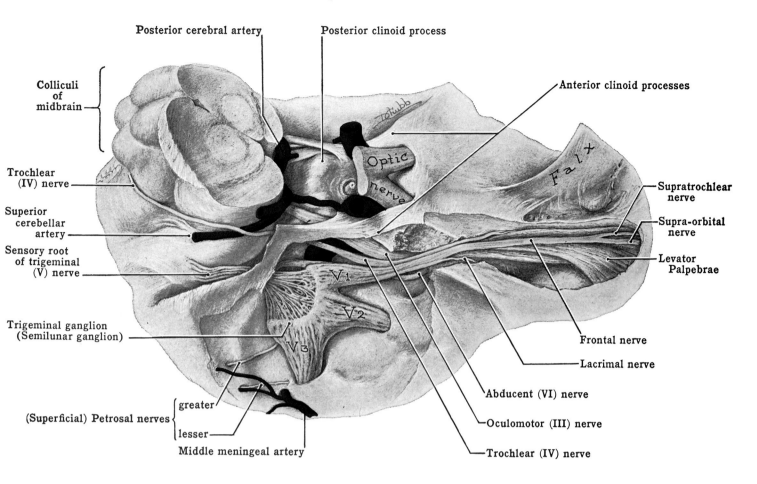

7-40 NERVES IN THE MIDDLE CRANIAL FOSSA—I

The tentorium cerebelli is cut away to reveal the courses of the trochlear and trigeminal nerves in the posterior cranial fossa. The dura is largely removed from the middle fossa. The roof of the orbit is partly removed.

Observe:
1. The trigeminal (semilunar) ganglion and its 3 divisions.
2. The mandibular nerve (V³), dropping down through the foramen ovale into the infratemporal fossa.
3. The maxillary nerve (V²), passing forward through the foramen rotundum into the pterygo-palatine fossa.
4. The ophthalmic nerve (V¹), ascending slightly, closely applied to the trochlear (IV) nerve, and dividing into frontal and lacrimal branches. These 3 (trochlear, frontal, lacrimal) nerves, running forward through the superior orbital fissure and applied to the roof of the orbital cavity (removed).

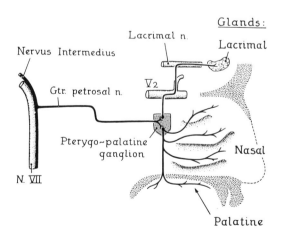

7-41 PTERYGO-PALATINE GANGLION

The greater petrosal nerve brings secretomotor fibers to the pterygo-palatine ganglion for relay and distribution, via branches of the maxillary nerve, to glands (lacrimal, nasal, and palatine) in maxillary territory (the *pink* area of Fig. 7-18).

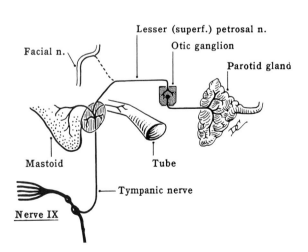

7-42 OTIC GANGLION

The lesser petrosal nerve brings secretomotor fibers to the otic ganglion for relay and distribution to the parotid gland.

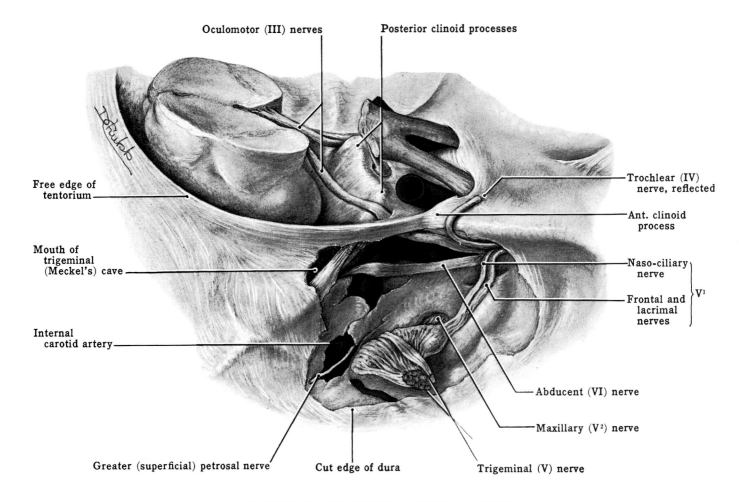

Oculomotor (III) nerves — Posterior clinoid processes

Free edge of tentorium

Mouth of trigeminal (Meckel's) cave

Internal carotid artery

Trochlear (IV) nerve, reflected

Ant. clinoid process

Naso-ciliary nerve } V¹

Frontal and lacrimal nerves }

Abducent (VI) nerve

Maxillary (V²) nerve

Greater (superficial) petrosal nerve   Cut edge of dura   Trigeminal (V) nerve

7-43 NERVES IN THE MIDDLE CRANIAL FOSSA—II

The trigeminal nerve is divided, withdrawn from the mouth of the trigeminal cave, and turned forward. The trochlear nerve also is turned forward.

Observe:

1. The bed of the trigeminal ganglion is partly formed by the greater petrosal nerve and the internal carotid artery, dura intervening.
2. The motor root of nerve V (the nerve to the muscles of mastication), crossing the ganglion diagonally, from medial to lateral side, to join V^3.
3. V^1, giving off the naso-ciliary nerve, and crowding with nerves III, IV, and VI through the superior orbital fissure.
4. The anterior clinoid process, "pulled backward" by the free edge of the tentorium between the optic nerve and internal carotid artery medially, and the oculomotor nerve below.
5. The abducent (VI) nerve, making a right-angled turn at the apex of the petrous bone and then, as it runs horizontally forward, hugging the internal carotid artery which flattens it.
6. The sinuous course of the internal carotid artery.

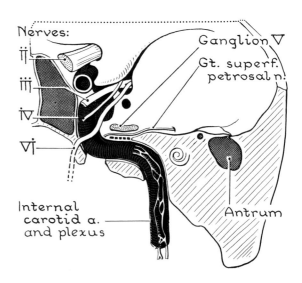

Nerves:

II̅

III̅

IV̅

VI̅

Ganglion ▽

Gt. superf. petrosal n.

Internal carotid a. and plexus

Antrum

7-44 INTERNAL CAROTID ARTERY

The internal carotid artery takes an inverted L-shaped course from the under surface of the petrous bone to its apex. There, at the upper end of foramen lacerum, it enters the cranial cavity and takes an S-shaped course. Note it contacts.

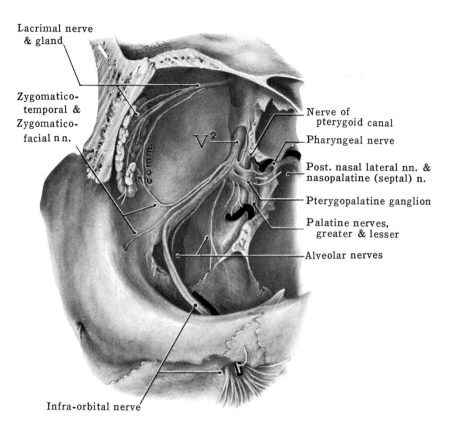

Lacrimal nerve & gland

Zygomatico-temporal & Zygomatico-facial n n.

V²

comm.

Nerve of pterygoid canal

Pharyngeal nerve

Post. nasal lateral nn. & nasopalatine (septal) n.

Pterygopalatine ganglion

Palatine nerves, greater & lesser

Alveolar nerves

Infra-orbital nerve

7-45   MAXILLARY NERVE

Note:
1. This afferent nerve supplies territory extending from skin laterally (*pink* area of Fig. 7-18) to nasal septum medially (see Fig. 7-2). Lateral and medial branches of the nerve are separated by the maxillary sinus. (For lateral branches, see Fig. 8-6; for medial branches, Fig. 7-109.)
2. The greater petrosal nerve, via nerve of pterygoid canal, brings parasympathetic fibers to the pterygo-palatine ganglion, there to be relayed and distributed, with branches of nerve V², as secretomotor fibers (see Fig. 7-41).

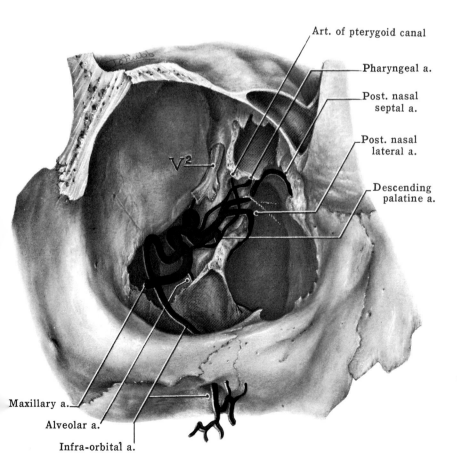

Art. of pterygoid canal

Pharyngeal a.

Post. nasal septal a.

Post. nasal lateral a.

Descending palatine a.

V²

Maxillary a.

Alveolar a.

Infra-orbital a.

7-46   MAXILLARY ARTERY, THIRD PART

Observe:
1. The stem of this artery, arising at the neck of the jaw, is divided into 3 parts by lateral pterygoid (Fig. 7-77).
2. The branches of the 1st part pass through either foramina or canals.
3. The branches of the 2nd part supply muscles of mastication.
4. The branches of 3rd part arise just before and within the pterygo-palatine fossa: (a) infra-orbital, (b) posterior superior alveolar, (c) descending palatine, (d) artery of pterygoid canal, (e) pharyngeal, and (f) spheno-palatine arteries. The descending palatine artery divides into the greater and lesser palatine arteries (Fig. 7-94). The spheno-palatine artery divides into the posterior nasal and posterior lateral nasal arteries (Fig. 7-111).
5. The 3rd part of the artery, often very tortuous, lies ventral to the maxillary nerve and its branches.
6. In this specimen, the artery of the pterygoid canal and the pharyngeal artery make unusually long ascents on the pterygoid process.

For variations see: Pearson, B. W., MacKenzie, R. G., and Goodman, W. S. (1969) The anatomical basis of transantral ligation of the maxillary artery in severe epistaxis. *Laryngoscope, 79:* 969.

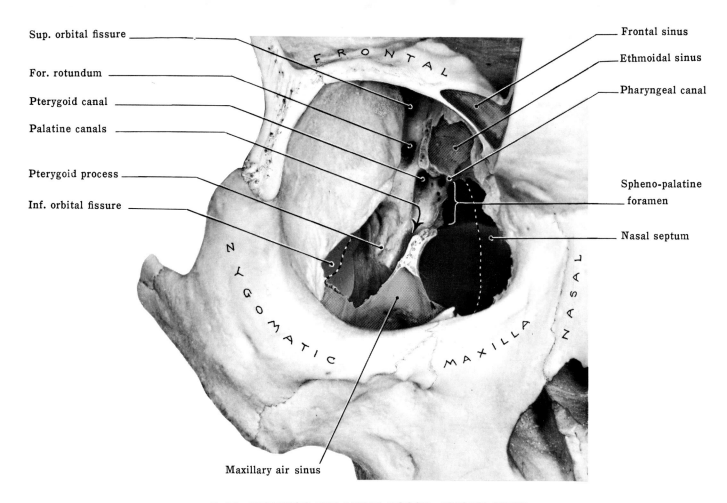

Sup. orbital fissure

For. rotundum

Pterygoid canal

Palatine canals

Pterygoid process

Inf. orbital fissure

FRONTAL

ZYGOMATIC

MAXILLA

NASAL

Frontal sinus

Ethmoidal sinus

Pharyngeal canal

Spheno-palatine foramen

Nasal septum

Maxillary air sinus

7-47 PTERYGO-PALATINE FOSSA, FRONT VIEW

The fossa has been exposed through the floor of the orbit and maxillary sinus. For another front view see Figure 7-133. For lateral view see Figure 7-79. For medial view see Figure 7-113.

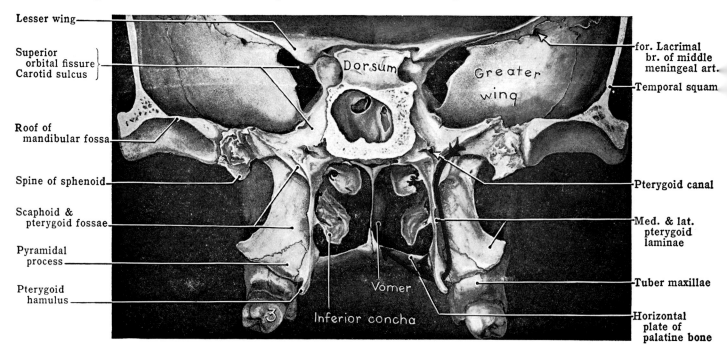

Lesser wing

Superior orbital fissure
Carotid sulcus

Roof of mandibular fossa

Spine of sphenoid

Scaphoid & pterygoid fossae

Pyramidal process

Pterygoid hamulus

Dorsum

Greater wing

Vomer

Inferior concha

3

for. Lacrimal br. of middle meningeal art.

Temporal squam

Pterygoid canal

Med. & lat. pterygoid laminae

Tuber maxillae

Horizontal plate of palatine bone

7-48 A CORONAL SECTION OF THE SKULL

Note:
1. The 2 posterior nasal apertures separated by the vomer, and each bounded below by the horizontal plate of the palatine bone, laterally by the medial pterygoid lamina, and above by the ala of the vomer and the vaginal process of the medial pterygoid lamina (not labeled).

2. The pterygoid fossa, bounded medially by the medial pterygoid lamina, which extends upward to the pterygoid canal and which ends below the level of the palate, as the hamulus (*i.e.*, the pulley of the Tensor Palati). The Tensor (*scarlet arrow*) arises from the canoe-shaped scaphoid fossa.

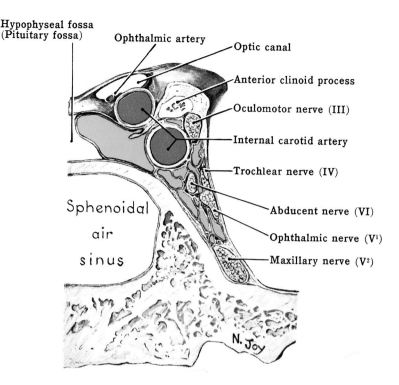

Hypophyseal fossa
(Pituitary fossa)

Ophthalmic artery

Optic canal

Anterior clinoid process

Oculomotor nerve (III)

Internal carotid artery

Trochlear nerve (IV)

Abducent nerve (VI)

Ophthalmic nerve (V¹)

Maxillary nerve (V²)

Sphenoidal air sinus

N. Joy

7-49 CAVERNOUS SINUS, CORONAL SECTION

Observe:

1. This venous sinus is situated at the side of the sphenoidal air sinus and of the hypophyseal fossa.
2. Cranial nerves III, IV, V, and V² are in a sheath in the lateral wall of the sinus.
3. The internal carotid artery surrounded by the internal carotid plexus (not drawn) and the abducent nerve swim through and so are vulnerable in thrombosis of the cavernous sinus.
4. The internal carotid artery, having made an acute bend, is cut twice. This artery and the oculomotor nerve groove the anterior clinoid process.
5. The ophthalmic veins drain into the cavernous sinus.

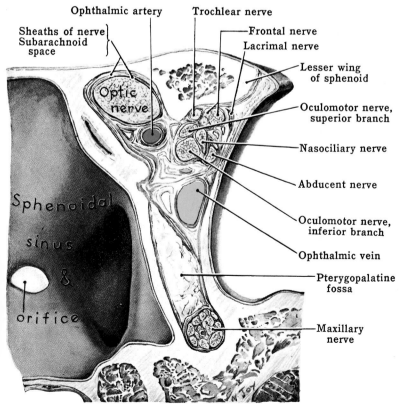

Ophthalmic artery

Trochlear nerve

Sheaths of nerve
Subarachnoid space

Frontal nerve

Lacrimal nerve

Lesser wing of sphenoid

Optic nerve

Oculomotor nerve, superior branch

Nasociliary nerve

Abducent nerve

Oculomotor nerve, inferior branch

Ophthalmic vein

Pterygopalatine fossa

Maxillary nerve

Sphenoidal sinus & orifice

Observe:

1. The optic nerve within its pial, arachnoid, and dural sheaths and the subarachnoid space. The ophthalmic artery emerging from the optic canal.
2. Other nerves crowded together, tightly packed, passing through the medial end of the superior orbital fissure. They are: the upper and the lower branch of the oculomotor nerve (III), the trochlear nerve (IV); the frontal, lacrimal, and naso-ciliary branches of the ophthalmic nerve (V¹); and the abducent nerve (VI).
3. The ophthalmic vein, about to open into the cavernous sinus, in turn to be drained by the superior and inferior petrosal sinuses (Fig. 7-161).

7-50 APEX OF ORBITAL CAVITY, CORONAL SECTION

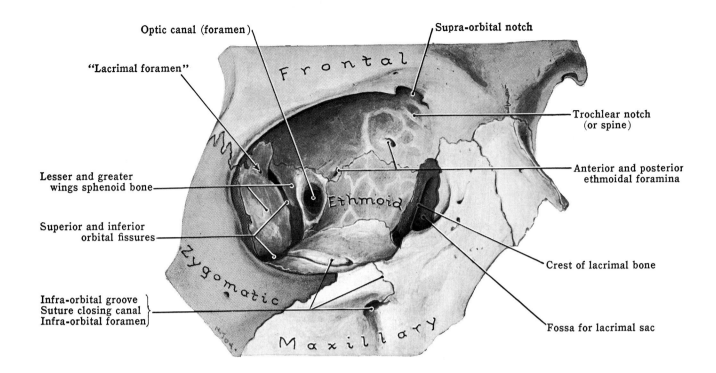

Optic canal (foramen)

Supra-orbital notch

"Lacrimal foramen"

Frontal

Trochlear notch (or spine)

Lesser and greater wings sphenoid bone

Ethmoid

Anterior and posterior ethmoidal foramina

Superior and inferior orbital fissures

Zygomatic

Crest of lacrimal bone

Infra-orbital groove
Suture closing canal
Infra-orbital foramen

Maxillary

Fossa for lacrimal sac

7-51 ORBITAL CAVITY

Observe:

1. The quadrangular orbital margin, at the base of the cavity, to which the frontal, maxillary, and zygomatic bones contribute.

2. The spiral form of the medial part of this margin. It is spiral since the supraorbital margin leads to the crest of the lacrimal bone (posterior lacrimal crest), whereas the infra-orbital margin is continuous with the crest on the frontal process of the maxilla (anterior lacrimal crest).

3. The fossa for the lacrimal sac, between these two crests.

4. The optic canal, situated at the apex of the pear-shaped orbital cavity, and placed between the body of the sphenoid and the two roots of the lesser wing. A straight probe must pass along the lateral wall of the cavity, if it is to traverse the canal.

5. The superior wall or roof, formed by the orbital plate of the frontal bone.

6. The inferior wall or floor, formed by the orbital plate of the maxilla and slightly by the zygomatic bone, and crossed by the infra-orbital groove, the anterior end of which is converted into the infra-orbital canal which ends at the infra-orbital foramen.

7. The stout lateral wall, formed by the frontal process of the zygomatic bone and by the greater wing of the sphenoid. The superior and inferior orbital fissures, together forming a V-shaped fissure which limits the greater wing of the sphenoid.

8. The fragile medial wall, formed by the papery lacrimal bone and the papery orbital plate (lamina papyracea) of the ethmoid bone. The anterior and posterior ethmoidal foramina, which developed in the suture between the frontal and ethmoidal bones, but are now, in this specimen, enveloped by the frontal bone.

9. The "lacrimal foramen," just beyond the superolateral end of the superior orbital fissure, for the anastomosis between the middle meningeal and lacrimal arteries. The zygomatic foramen on the orbital surface of the zygomatic bone is not in view.

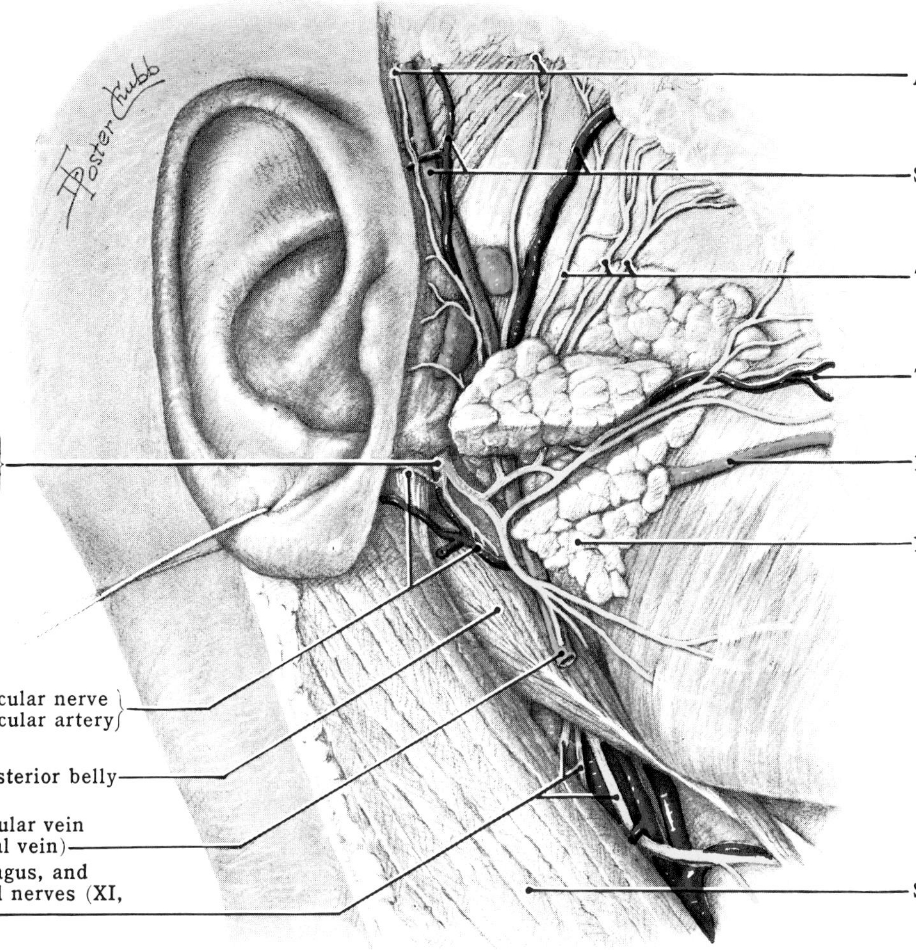
Auriculo-temporal n.
Superficial temporal artery and vein
Temporal branches of facial nerve
Transverse facial a.
Facial nerve
Stylomastoid branch of post. auricular art.
Parotid duct
Parotid gland
Posterior auricular nerve
Posterior auricular artery
Digastric, posterior belly
Retromandibular vein (Post. facial vein)
Accessory, vagus, and hypoglossal nerves (XI, X, XII)
Sternomastoid

7-67 PAROTID REGION

See Figure 7-16 for a more superficial dissection.

Observe:

1. The stem of the facial nerve descending from the stylomastoid foramen for about 1 cm before curving forward to penetrate the deeper part of the parotid gland.
2. The nerve to the posterior belly of Digastric arising from the stem of the facial nerve.
3. The posterior auricular artery giving off a branch, the stylomastoid artery, which accompanies the facial nerve through the stylomastoid foramen into the facial canal.
4. The relatively superficial position of the great landmark in the upper part of the neck, posterior belly of Digastric (Fig. 9-21). Only three structures cross superficial to it: (a) the cervical branch of the facial nerve, (b) branches of the retromandibular vein, and (c) branches of the great auricular nerve shown in Figure 7-16. All other crossing structures cross deep to it.
5. Preauricular lymph nodes.
6. Auricular and temporal branches of the auriculotemporal nerve.
7. Enlargement of the parotid and other salivary glands occurs in certain metabolic and endocrine diseases as well as in nutritional deficiency including anorexia nervosa. The resulting facial swelling is often not reversible. See Walsh, B. T., Croft, C. B., and Katz, J. L. (1981) Anorexia nervosa and salivary gland enlargement. *Int. J. Psychiat. Med. 11(3):* 255–261.

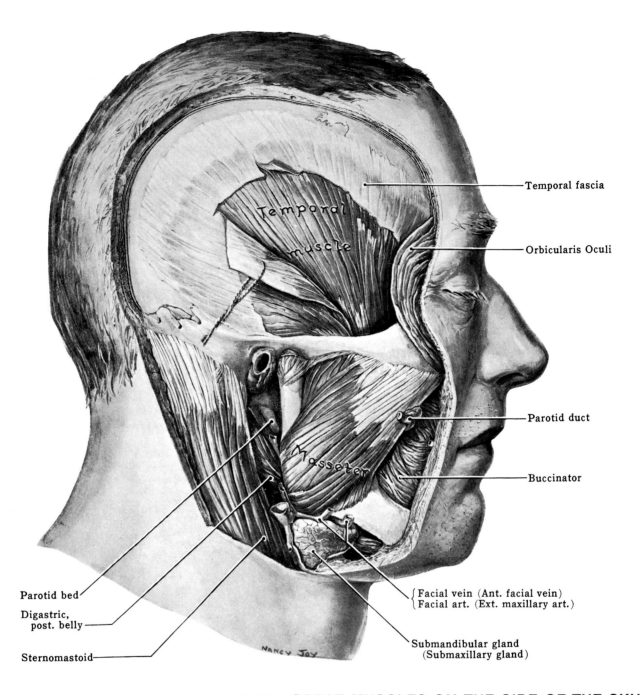

Temporal fascia

Orbicularis Oculi

Parotid duct

Buccinator

Facial vein (Ant. facial vein)
Facial art. (Ext. maxillary art.)

Submandibular gland
(Submaxillary gland)

Parotid bed

Digastric,
post. belly

Sternomastoid

7-68 GREAT MUSCLES ON THE SIDE OF THE SKULL

Observe:

1. The temporal and massetric muscles. Both are supplied by the trigeminal nerve and both close the jaw. Temporalis arises in part from the overlying fascia.
2. Orbicularis Oculi and Buccinator, both supplied by the facial nerve. One closes the eye; the other prevents food from collecting between cheeks and teeth.
3. Sternomastoid, which is the chief flexor of the head and neck, forming the posterior boundary of the parotid region; and Digastric which limits this region below.
4. The submandibular gland, with the facial artery passing deep to it and the facial vein passing superficial.

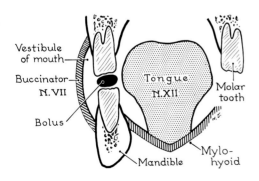

Vestibule
of mouth

Buccinator
N. VII

Bolus

Tongue
N. XII

Molar
tooth

Mandible

Mylo-
hyoid

7-69 CHEWING

The tongue and the Buccinator retain food between the molar teeth during the act of chewing.

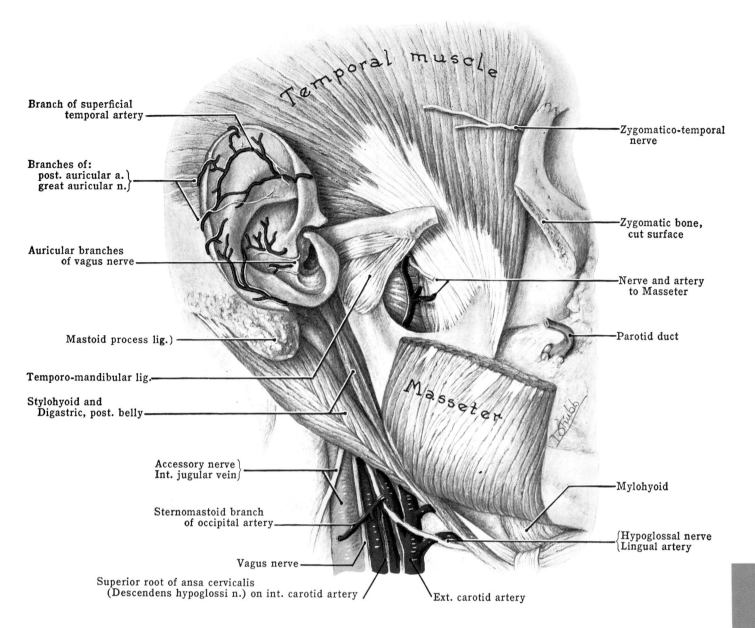

Branch of superficial temporal artery

Branches of:
post. auricular a.
great auricular n.

Auricular branches of vagus nerve

Mastoid process lig.)

Temporo-mandibular lig.

Stylohyoid and Digastric, post. belly

Accessory nerve
Int. jugular vein

Sternomastoid branch of occipital artery

Vagus nerve

Superior root of ansa cervicalis
(Descendens hypoglossi n.) on int. carotid artery

Ext. carotid artery

Zygomatico-temporal nerve

Zygomatic bone, cut surface

Nerve and artery to Masseter

Parotid duct

Mylohyoid

Hypoglossal nerve
Lingual artery

7-70 PAROTID BED, THE TEMPORAL MUSCLE, AURICULAR VESSELS AND NERVES

Observe:
1. The mastoid process, rough where Sternomastoid (and also Splenius Capitis and Longissimus Capitis) has been removed from it.
2. The posterior belly of Digastric, arising deep to the mastoid process and passing deep to the angle of the jaw.
3. The vessels and nerves passing deep to Digastric: the internal jugular vein, internal carotid artery, external carotid artery and its lingual, facial, and occipital branches, and the last three cranial nerves.
4. The Temporal muscle, inserted into the beak-shaped coronoid process.
5. The nerve and artery to Masseter, crossing the mandibular notch behind the Temporal muscle—the nerve appearing from above Lateral Pterygoid and the artery from below it.
6. Vessels and nerves of the auricle, mostly turning round the free borders of the cartilage, but some piercing the cartilage.

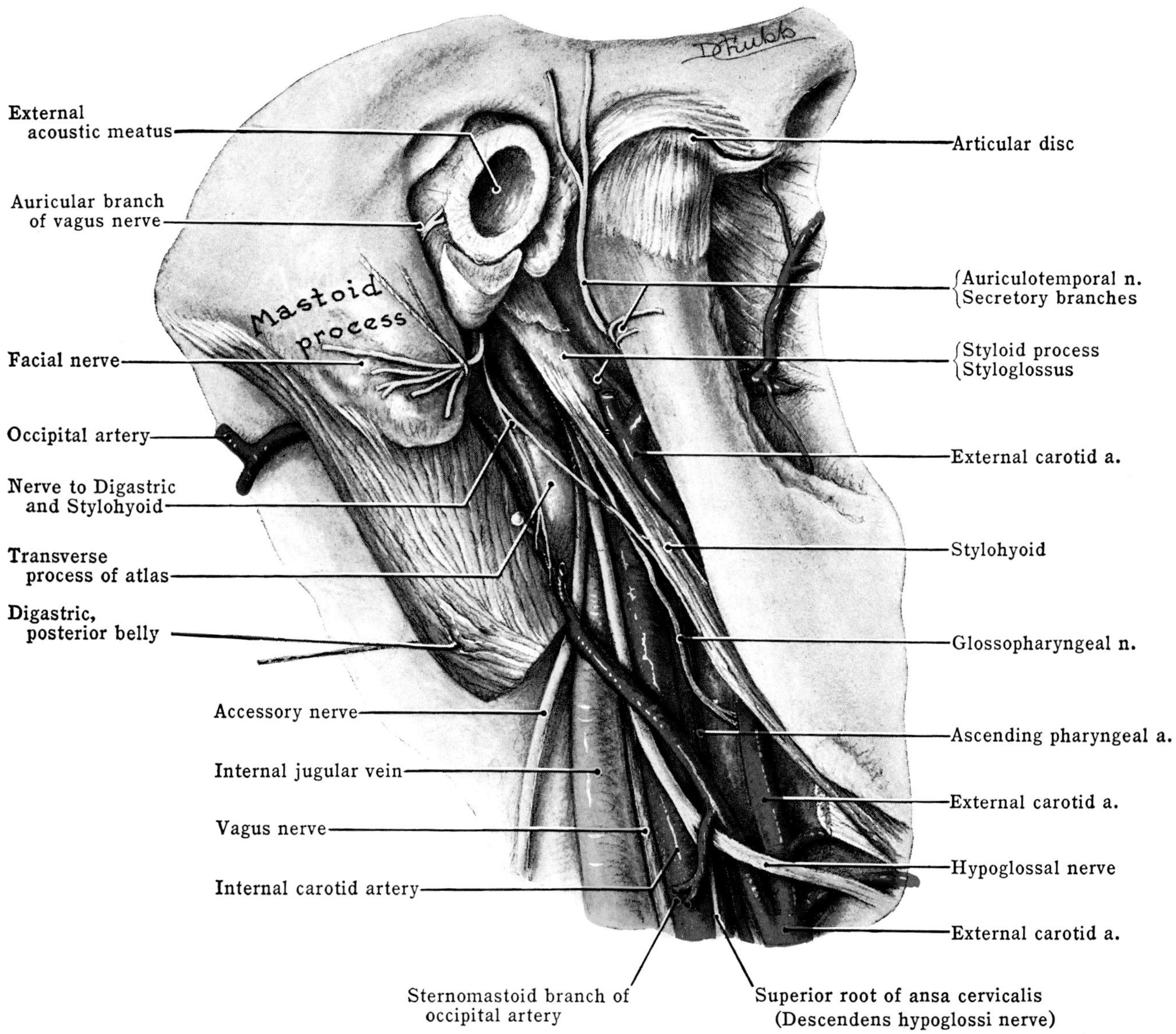
External acoustic meatus
Auricular branch of vagus nerve
Mastoid process
Facial nerve
Occipital artery
Nerve to Digastric and Stylohyoid
Transverse process of atlas
Digastric, posterior belly
Accessory nerve
Internal jugular vein
Vagus nerve
Internal carotid artery
Sternomastoid branch of occipital artery
Superior root of ansa cervicalis (Descendens hypoglossi nerve)
Articular disc
Auriculotemporal n.
Secretory branches
Styloid process
Styloglossus
External carotid a.
Stylohyoid
Glossopharyngeal n.
Ascending pharyngeal a.
External carotid a.
Hypoglossal nerve
External carotid a.

7-71 STRUCTURES DEEP TO THE PAROTID BED

The facial nerve, the posterior belly of Digastric, and the nerve to this belly are retracted, whereas the external carotid artery, Stylohyoid, and the nerve to Stylohyoid remain *in situ*.

Observe:

1. The tip of the transverse process of the atlas, about midway between the tip of the mastoid process and the angle of the jaw.
2. The internal jugular vein, the internal carotid artery, and the last four cranial nerves crossing in front of the transverse process and deep to the styloid process.
3. The strange impression made on the internal jugular vein by the transverse process.
4. The internal and external carotids separated from each other by the styloid process.
5. The last four cranial nerves (XII concealing X) starting to diverge from each other as they cross the transverse process.
6. The two nerves that pass forward to the tongue: (a) IX or glossopharyngeal being above the level of the angle of the jaw and passing between the two carotid arteries, and (b) XII or hypoglossal being below the angle of the jaw and passing superficial to both carotids, and indeed, to all the arteries it meets, except the occipital artery and its sternomastoid branch.
7. The thickness of the skin lining the cartilage of the meatus, and the stems of the two nerves—auricular branch of the vagus and auriculo-temporal—that supply the meatus and the outer surface of the tympanic membrane.

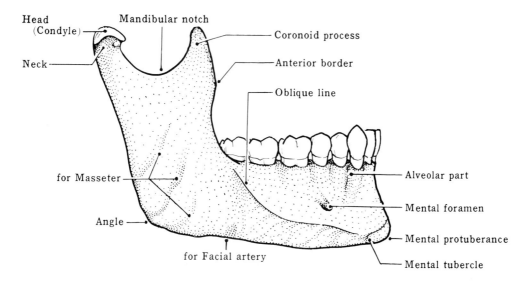

7-72 MANDIBLE, EXTERNAL SURFACE

Observe: The coronoid process and the mental tubercle imperfectly connected by (a) the anterior border of the ramus, and (b) the oblique line that crosses the body diagonally.

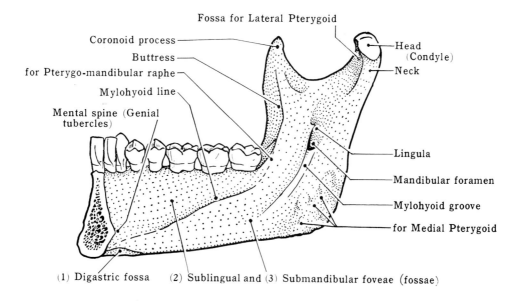

7-73 MANDIBLE, INTERNAL SURFACE

Observe: The coronoid process and the symphysis menti imperfectly connected by (a) the strengthening buttress, (b) the site of attachment of the pterygo-mandibular raphe, and (c) the mylohyoid line which crosses the body diagonally to end between the mental spine and digastric fossa.

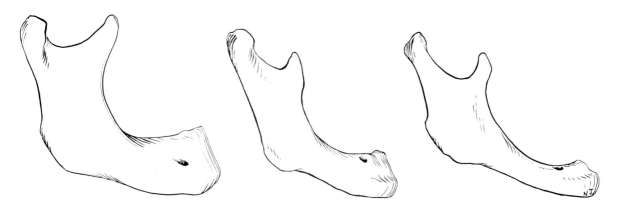

7-74 MENTAL FORAMEN IN EDENTULOUS JAWS

The position of the mental foramen in edentulous jaws varies with the extent of the absorption of the alveolar process. Pressure of a dental prosthesis on a vulnerable mental nerve in an edentulous jaw may produce pain.

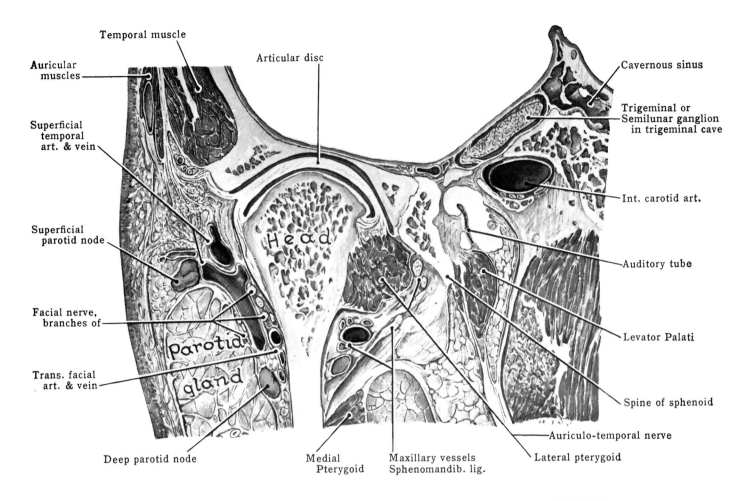

Temporal muscle

Auricular muscles

Articular disc

Cavernous sinus

Trigeminal or Semilunar ganglion in trigeminal cave

Superficial temporal art. & vein

Int. carotid art.

Head

Auditory tube

Superficial parotid node

Facial nerve, branches of

parotid gland

Levator Palati

Trans. facial art. & vein

Spine of sphenoid

Auriculo-temporal nerve

Deep parotid node

Medial Pterygoid

Maxillary vessels Sphenomandib. lig.

Lateral pterygoid

7-75 TEMPOROMANDIBULAR JOINT, CORONAL SECTION

Observe:

1. The articular disc attached to the neck of the jaw medially and laterally, partly in conjunction with the Lateral Pterygoid.
2. The roof of the mandibular fossa, which separates head and disc from the middle cranial fossa, is thin centrally but thick elsewhere.
3. The spine of the sphenoid (Figs. 9-47 and 9-50) at the medial end of the fossa, and the two roots of the auriculo-temporal nerve crossing lateral to it (Fig. 7-81).
4. The auditory tube with closed slitlike lumen, and the Levator Palati lying below it.
5. The trigeminal ganglion in its trigeminal cave, the mouth of which opens below the tentorium (Fig. 5-39); and separated from the internal carotid artery by membrane (bone being deficient) and, therefore, subjected to the pulsations of that artery (Figs. 7-40 and 7-43).
6. The maxillary vessels crossing the neck of the jaw on its medial side.
7. Superficial and deep parotid lymph nodes.
8. The "aponeurotic" tendon of Temporalis, buried in the muscle for it receives fleshy fibers from the temporal fossa on one side and from the temporal fascia on the other (Figs. 7-2 and 7-68).

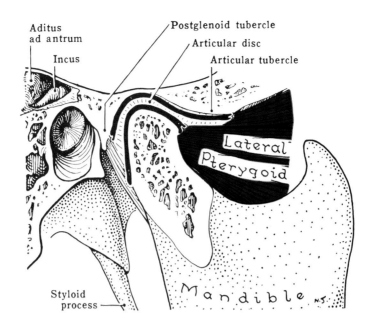

7-76 TEMPOROMANDIBULAR JOINT, SAGITTAL SECTION

Note the articular disc dividing the articular cavity into an upper and a lower compartment, and the Lateral Pterygoid inserted in part into the front of the disc. For articular part of the mandibular fossa and articular tubercle see Figure 7-79. For temporomandibular ligament see Figure 7-70.

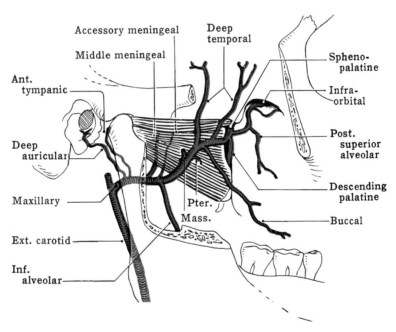

7-77 DIAGRAM OF THE MAXILLARY ARTERY

Observe:

1. This, the larger of the two terminal branches of the external carotid artery, arising at neck of jaw and divided into three parts by Lateral Pterygoid.
2. The branches of the first part pass through foramina or canals: (a) deep auricular to external acoustic meatus, (b) anterior tympanic to tympanum, (c, d) middle and accessory meningeal to cranial cavity, and (e) inferior alveolar to jaw and teeth.
3. The branches of the second part supply muscles, by masseteric, deep temporal, pterygoid and buccal branches.

4. The branches of the third part arise just before and within the pterygo-palatine fossa: (a) posterior superior alveolar, (b) infra-orbital, (c) descending palatine, (d) artery of the pterygoid canal, (e) pharyngeal, and (f) spheno-palatine. These, accompanied by branches of the maxillary nerve, pass through bony canals or foramina. The descending palatine artery divides into a greater and 2 or 3 lesser palatine arteries (Fig. 7-94). The spheno-palatine artery ends as the posterior lateral nasal and posterior nasal septal arteries (Fig. 7-111).

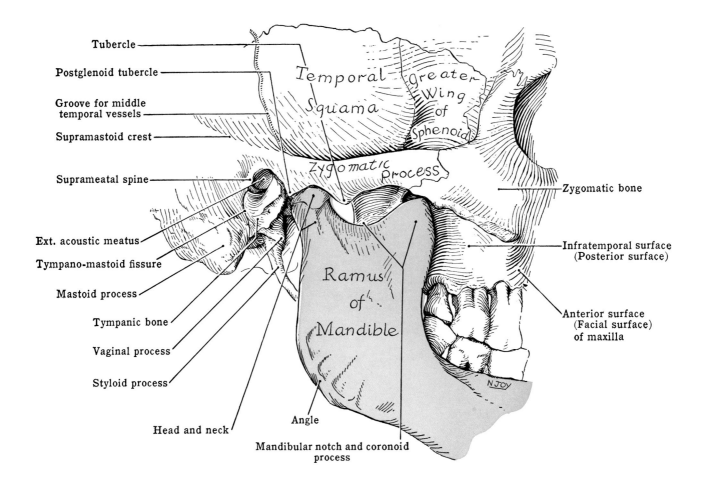

Tubercle
Postglenoid tubercle
Groove for middle temporal vessels
Supramastoid crest
Suprameatal spine
Ext. acoustic meatus
Tympano-mastoid fissure
Mastoid process
Tympanic bone
Vaginal process
Styloid process
Head and neck
Angle
Mandibular notch and coronoid process

Temporal Squama
Greater Wing of Sphenoid
Zygomatic process
Ramus of Mandible
Zygomatic bone
Infratemporal surface (Posterior surface)
Anterior surface (Facial surface) of maxilla

N JOY

7-78 INFRATEMPORAL FOSSA – I: LATERAL WALL

Note:
1. The lateral wall of the infratemporal fossa is the ramus of the mandible.
2. The zygomatic process of the squamous part of the temporal bone plus the zygomatic bone constitute the zygomatic arch. This arch is continued as a buttress downward and forward to the first or second molar tooth. The buttress forms the anterior limit of the infratemporal fossa and separates it from the facial aspect of the skull.
3. The zygomatic process lies at the boundary line between the temporal fossa above and the infratemporal fossa below.
4. Below the tubercle of the zygomatic process and in front of the neck of the jaw there is a clear passage across the base of the skull through which a pencil can be passed. See Figure 9-49.

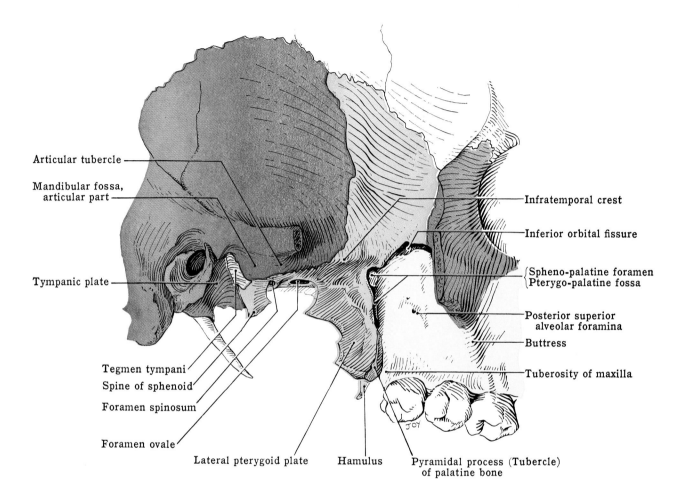

Articular tubercle

Mandibular fossa, articular part

Tympanic plate

Tegmen tympani
Spine of sphenoid
Foramen spinosum

Foramen ovale

Lateral pterygoid plate

Hamulus

Pyramidal process (Tubercle) of palatine bone

Infratemporal crest

Inferior orbital fissure

Spheno-palatine foramen
Pterygo-palatine fossa

Posterior superior alveolar foramina

Buttress

Tuberosity of maxilla

7-79 INFRATEMPORAL FOSSA—II: ROOF AND THE MEDIAL AND LATERAL WALLS

Note:
1. The medial wall of the fossa is formed by the lateral pterygoid plate.
2. The posterior free border of this plate, when followed upward, leads to the foramen ovale in the roof of the fossa. Behind the foramen ovale, at the root of the spine of the sphenoid, is the foramen spinosum (Figs. 7-38 and 9-47). The roof is separated from the temporal fossa by the infratemporal crest.
3. Below, the anterior border of the lateral plate is separated from the maxilla by the pyramidal process of the palatine bone which is insinuated as a buffer between the two (Figs. 9-47 and 9-48). Above, the border is free and forms the posterior limit of the pterygo-maxillary fissure, which is the entrance to the pterygo-palatine fossa on the medial wall of which can be seen the spheno-palatine foramen which leads to the nasal cavity.
4. The rounded anterior wall of the fossa is the infratemporal surface of the maxilla, which is of eggshell thickness, is limited above by the inferior orbital fissure, and is pierced by two (or more) posterior superior alveolar foramina for the vessels and nerves of the same name.

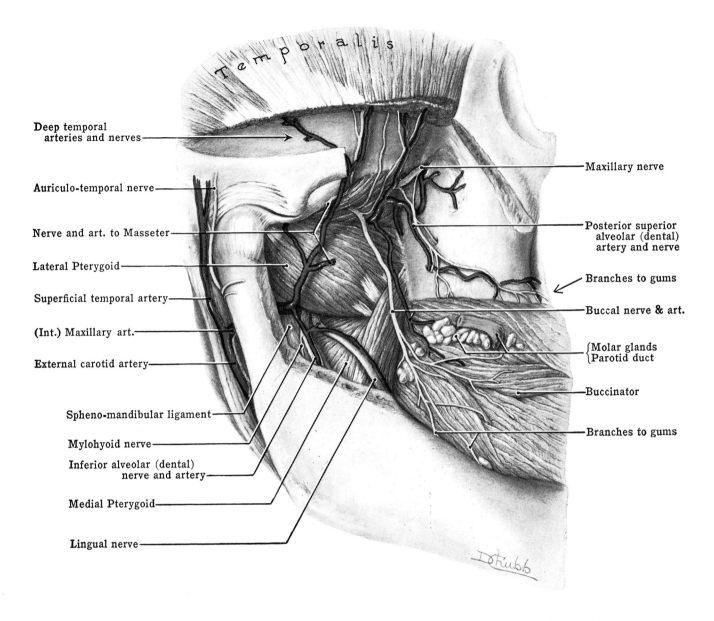

Deep temporal arteries and nerves

Auriculo-temporal nerve

Nerve and art. to Masseter

Lateral Pterygoid

Superficial temporal artery

(Int.) Maxillary art.

External carotid artery

Spheno-mandibular ligament

Mylohyoid nerve

Inferior alveolar (dental) nerve and artery

Medial Pterygoid

Lingual nerve

Temporalis

Maxillary nerve

Posterior superior alveolar (dental) artery and nerve

Branches to gums

Buccal nerve & art.

Molar glands
Parotid duct

Buccinator

Branches to gums

7-80 INFRATEMPORAL REGION—III: SUPERFICIAL DISSECTION

Observe:

1. Three muscles: lateral pterygoid, medial pterygoid, and buccinator.
2. The maxillary nerve (V^2) becoming the infraorbital nerve and passing through the inferior orbital fissure after giving off the posterior superior alveolar nerve.
3. Branches of the mandibular nerve (V^3), both sensory and motor.
4. The maxillary artery, the larger of two end branches of the external carotid, divided into three parts by its relationship to the lateral pterygoid muscle. The first part sends branches accompanying branches of V^3, the second part supplies blood to the muscles

of the region, and the third part sends branches accompanying branches of V^2.
5. Buccinator is pierced by the parotid duct, the ducts of molar glands, and branches (sensory) of the buccal nerve.
6. The lateral pterygoid muscle arising by two heads, from the roof and medial wall of the infratemporal fossa. It inserts into the articular disc of the jaw joint and into the front of the neck of the mandible. The upper head is active during jaw closing while the lower head is active during jaw opening and protrusion. See Grant, P.G. (1973) "Lateral pterygoid: two muscles?" *Am. J. Anat. 138:* 1-10.

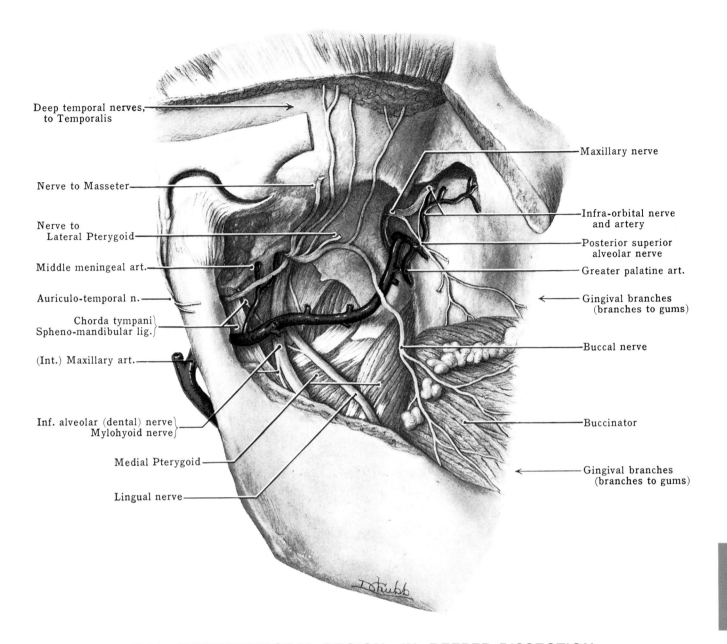

Deep temporal nerves, to Temporalis

Nerve to Masseter

Nerve to Lateral Pterygoid

Middle meningeal art.

Auriculo-temporal n.

Chorda tympani
Spheno-mandibular lig.

(Int.) Maxillary art.

Inf. alveolar (dental) nerve
Mylohyoid nerve

Medial Pterygoid

Lingual nerve

Maxillary nerve

Infra-orbital nerve and artery

Posterior superior alveolar nerve

Greater palatine art.

Gingival branches (branches to gums)

Buccal nerve

Buccinator

Gingival branches (branches to gums)

7-81   INFRATEMPORAL REGION—IV: DEEPER DISSECTION

The Lateral Pterygoid and most branches of the maxillary artery have been removed.

Observe:

1. Medial Pterygoid arising from the medial surface of the lateral pterygoid plate and having a small superficial head which arises from the pyramidal process of the palatine bone (Fig. 7-79).
2. The spheno-mandibular ligament, which, as a fascial band, descends from near the spine of the sphenoid to the lingula of the mandible (Fig. 7-72).
3. The maxillary artery and the auriculo-temporal nerve passing between the ligament and the neck of the jaw.
4. The mandibular nerve (V^3) entering the infratemporal fossa through the roof, via the foramen ovale which also transmits the accessory meningeal artery (not labeled).

5. The middle meningeal artery and vein passing through the roof via the foramen spinosum.
6. The inferior alveolar and lingual nerves descending on Medial Pterygoid. The former giving off the mylohyoid nerve (to Mylohyoid and anterior belly of Digastric); the latter receiving the chorda tympani (which carries secretory fibers and fibers of taste).
7. The nerves to 4 muscles of mastication: Masseter, Temporal, and Lateral Pterygoid, which are labeled, and the nerve to Medial Pterygoid which is not labeled. Note that the buccal branch of the mandibular nerve is sensory. The buccal branch of the facial nerve is the motor supply to Buccinator.
8. The maxillary nerve (V^2) becoming the infra-orbital nerve which enters the infra-orbital groove at the inferior orbital fissure.

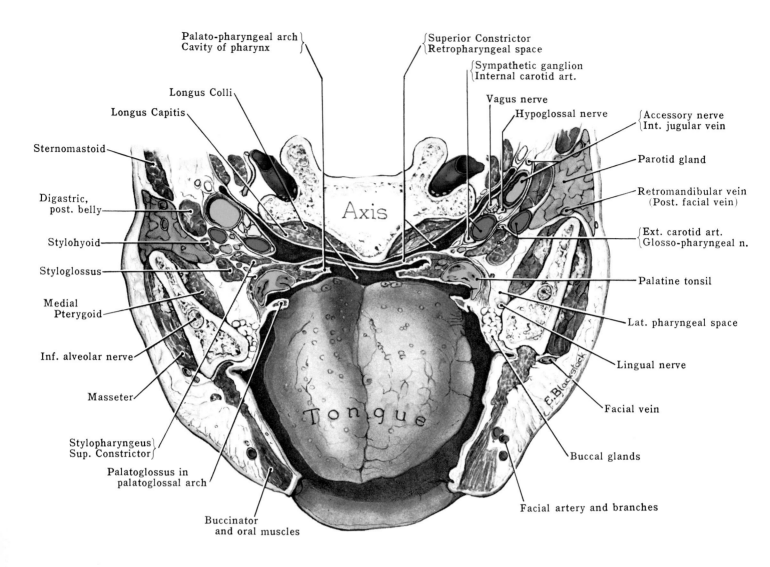

Palato-pharyngeal arch} Cavity of pharynx }
{Superior Constrictor Retropharyngeal space
{Sympathetic ganglion Internal carotid art.
Longus Colli
Vagus nerve
Hypoglossal nerve
{Accessory nerve Int. jugular vein
Longus Capitis
Sternomastoid
Parotid gland
Digastric, post. belly
Retromandibular vein (Post. facial vein)
Stylohyoid
{Ext. carotid art. Glosso-pharyngeal n.
Styloglossus
Palatine tonsil
Medial Pterygoid
Lat. pharyngeal space
Inf. alveolar nerve
Lingual nerve
Masseter
Facial vein
Stylopharyngeus} Sup. Constrictor}
Buccal glands
Palatoglossus in palatoglossal arch
Facial artery and branches
Buccinator and oral muscles

Axis

Tongue

E. Blackstock

7-82 CROSS-SECTION PASSING THROUGH THE MOUTH

Observe:

1. The parotid gland filling its wedge-shaped bed or mold. The Digastric and Stylohyoid intervening between the parotid gland and the great vessels and nerves of the neck.
2. The Masseter inserted into the outer surface of the ramus of the mandible and the Medial Pterygoid inserted into the inner.
3. The lingual nerve in contact with the ramus of the mandible.
4. Anterior to the ribbonlike Palatoglossus and its arch is the mouth; behind it is the pharynx.
5. The pharynx is flattened antero-posteriorly, and the tonsil is in its wall.
6. The tonsil bed formed by Superior Constrictor and Palato-pharyngeus, an areolar space intervening, and limited in front and behind by the palatine arches. The carotid arteries are well behind the bed.
7. The retropharyngeal space, here opened up, which allows the pharnyx to contract and relax during swallowing. It is closed laterally at the carotid sheath and is limited posteriorly by the prevertebral fascia.
8. The three styloid muscles: Stylohyoid, Styloglossus, and Stylopharyngeus, Stylopharyngeus being deepest and about to blend ith Palato-pharyngeus (Fig. 9-60). All three muscles arise from the styloid process. Their names reveal their insertions. Each is supplied by a different cranial nerve: Stylohyoid, VII; Stylopharyngeus, IX; and Styloglossus, XII.

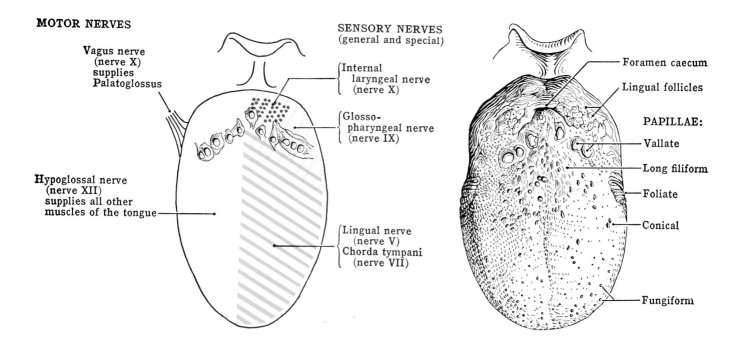

Vagus nerve
(nerve X)
supplies
Palatoglossus

SENSORY NERVES
(general and special)

Internal
laryngeal nerve
(nerve X)

Glosso-
pharyngeal nerve
(nerve IX)

Hypoglossal nerve
(nerve XII)
supplies all other
muscles of the tongue

Lingual nerve
(nerve V)
Chorda tympani
(nerve VII)

Foramen caecum

Lingual follicles

PAPILLAE:

Vallate

Long filiform

Foliate

Conical

Fungiform

7-83   NERVE SUPPLY TO THE TONGUE   **7-84   DORSUM OF THE TONGUE**

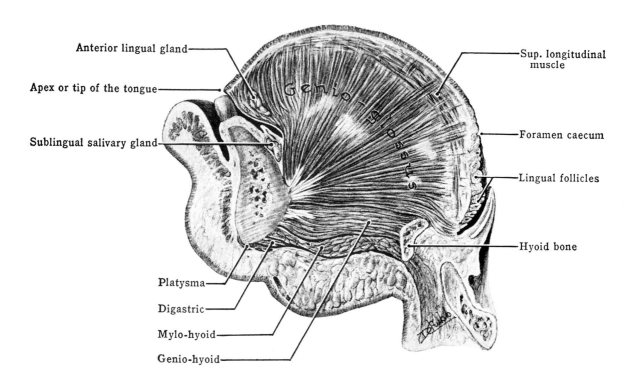

Anterior lingual gland

Apex or tip of the tongue

Sublingual salivary gland

Sup. longitudinal
muscle

Foramen caecum

Lingual follicles

Hyoid bone

Platysma

Digastric

Mylo-hyoid

Genio-hyoid

7-85   TONGUE AND FLOOR OF THE MOUTH, MEDIAN SECTION

Observe:
1. The tongue is composed mainly of muscles: *extrinsic* (which alter the *position* of the tongue) and *intrinsic* (which alter its *shape*). In this illustration, extrinsic muscles are represented by Genioglossus and intrinsic by the Superior Longitudinal muscle.
2. The foramen caecum, which is the patent upper end of the primitive thyro-glossal duct, and the limbs of the V-shaped sulcus terminalis, which diverge from the foramen, lie slightly behind the vallate papillae, and demarcate the developmentally different, posterior one-third of the tongue from the anterior two-thirds.
3. The anterior lingual gland, covered with a layer of muscle. The several ducts of this mixed muco-serous gland open below the tongue, but are not in view.
4. Lingual follicles of lymphoid tissues constituting the lingual tonsil.
5. In Figures 7-83 and 7-84: the vallate papillae, variable in number, and four other shapes of papillae.

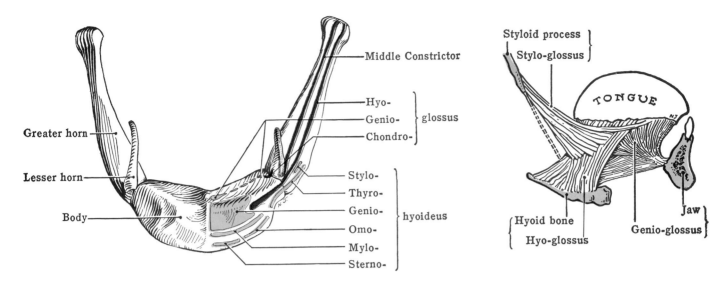

7-86 HYOID BONE, ATTACHMENTS OF MUSCLES

7-87 EXTRINSIC MUSCLES OF THE TONGUE

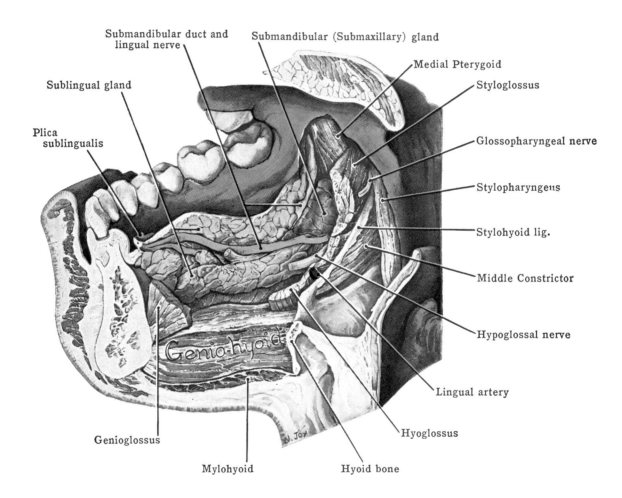

7-88 FLOOR AND SIDE OF MOUTH, FROM WHICH TONGUE IS EXCISED

Observe:

1. Undisturbed: Geniohyoid inferiorly, Middle Constrictor posteriorly, and the cut edge of the mucous membrane superiorly.
2. Three divided muscles: Genioglossus anteriorly, Hyoglossus inferiorly, and Styloglossus posteriorly.
3. Other 3 divided structures: lingual and hypoglossal nerves, and lingual artery. The lingual nerve appearing between Medial Pterygoid and the ramus of the mandible and making three-quarters of a spiral around the submandibular duct, being first superolateral, then in turn lateral, inferior, medial, and superomedial. The hypoglossal nerve separated from the lingual artery by Hyoglossus.
4. The deep or oral part of the submandibular gland in the angle between the lingual nerve and the submandibular duct, which separate it from the sublingual gland. The orifice of the duct is seen at the anterior end of the plica sublingualis.
5. The submandibular duct adhering to the medial side of the sublingual gland, and here receiving, as it sometimes does, a large accessory duct from the lower part of the sublingual gland. For the lesser ducts see Figure 9-18.

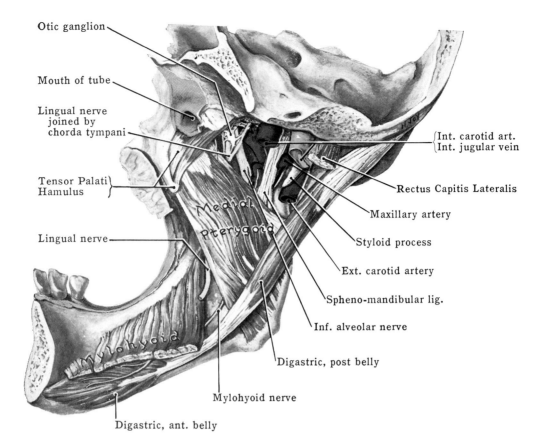

Otic ganglion
Mouth of tube
Lingual nerve
joined by
chorda tympani
Tensor Palati
Hamulus
Lingual nerve
Medial Pterygoid
Mylohyoid
Digastric, ant. belly
Mylohyoid nerve
Int. carotid art.
Int. jugular vein
Rectus Capitis Lateralis
Maxillary artery
Styloid process
Ext. carotid artery
Spheno-mandibular lig.
Inf. alveolar nerve
Digastric, post belly

7-89 OTIC GANGLION, MEDIAL VIEW

Observe:
1. Mylohyoid with a thick, free, posterior border thinning anteriorly where below the origins of the genial muscles it must be almost functionless and may be deficient, as here, with a resulting thin, free, anterior border.
2. Medial Pterygoid taking much the same direction on the medial side of the ramus as Masseter takes on the lateral.
3. Tensor Palati, here sending some fibers to the hamulus.
4. The lingual nerve joined above Medial Pterygoid by the chorda tympani and appearing in the mouth at the anterior border of that muscle.

5. The otic ganglion lying medial to the mandibular nerve, and between foramen ovale above and Medial Pterygoid below. Tensor Palati usually covers the ganglion (Fig. 8-9). The otic ganglion receives sensory fibers from the auriculo-temporal branch of V^3, sympathetic fibers from the plexus on the middle meningeal artery, and contains the synapse of parasympathetic fibers from the lesser superficial petrosal branch of IX. It does two queer things: connects with VII and allows the motor fibers to the Tensors to pass through it. It is distributed with the auriculo-temporal nerve.

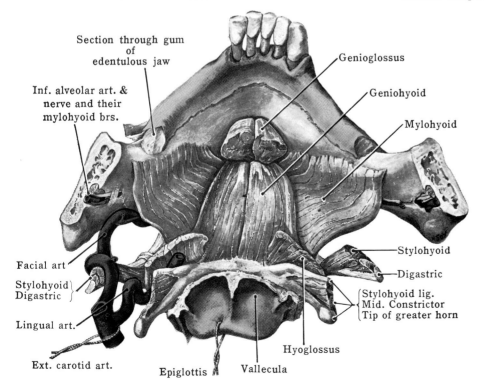

Section through gum
of
edentulous jaw
Inf. alveolar art. &
nerve and their
mylohyoid brs.
Facial art
Stylohyoid
Digastric
Lingual art.
Ext. carotid art.
Epiglottis
Vallecula
Hyoglossus
Stylohyoid lig.
Mid. Constrictor
Tip of greater horn
Digastric
Stylohyoid
Mylohyoid
Geniohyoid
Genioglossus

7-90 MUSCLES OF THE FLOOR OF THE MOUTH

Observe:
1. Geniohyoid, paired, triangular, and occupying a horizontal plane, with apex at the mental spine, base at the body of the hyoid bone, medial border in contact with its fellow, and lateral border in contact with Mylohyoid.
2. Mylohyoid arising from the mylohyoid line of the jaw (Fig. 7-73); having a thick, free, posterior border; thinning as it is traced forward; and ending in a delicate, free, anterior border as it nears the origin of the genial muscles.

For a view of Mylohyoid from below see Figure 9-20.

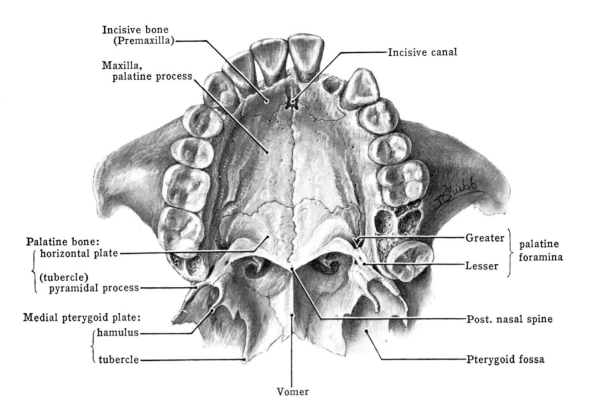

Incisive bone (Premaxilla)

Incisive canal

Maxilla, palatine process

Palatine bone:
  horizontal plate
  (tubercle) pyramidal process

Greater } palatine foramina
Lesser }

Medial pterygoid plate:
  hamulus
  tubercle

Post. nasal spine

Pterygoid fossa

Vomer

7-91 PALATE—I

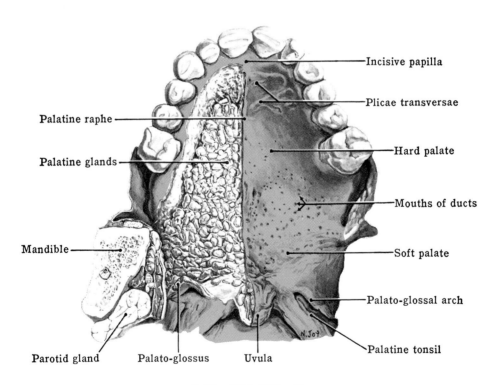

Palatine raphe

Incisive papilla

Plicae transversae

Palatine glands

Hard palate

Mouths of ducts

Mandible

Soft palate

Palato-glossal arch

Palatine tonsil

Parotid gland   Palato-glossus   Uvula

7-92 PALATE—II

Observe:

1. On the *right side* of the page—the transverse palatine folds, which are so conspicuous in the dog; and the orifices of the ducts of the palatine glands which give the mucous membrane an orange skin appearance.

2. On the *left side*—the palatine glands forming a very thick layer in the soft palate, a thin one in the hard palate, and absent in the region of the incisive bone and anterior part of the median raphe.

3. Posteriorly, the palate ending medially in the uvula, and on each side in the palato-pharyngeal arch. The Palato-glossus and the palato-glossal arch extending to the undersurface of the soft palate.

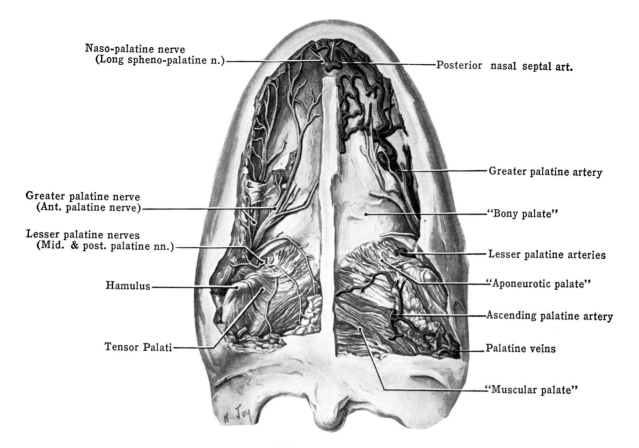

Naso-palatine nerve
(Long spheno-palatine n.)

Posterior nasal septal art.

Greater palatine artery

Greater palatine nerve
(Ant. palatine nerve)

"Bony palate"

Lesser palatine nerves
(Mid. & post. palatine nn.)

Lesser palatine arteries

"Aponeurotic palate"

Hamulus

Ascending palatine artery

Palatine veins

Tensor Palati

"Muscular palate"

N. Joy

7-93 PALATE—III

Observe:
1. The palate having bony, aponeurotic, and muscular parts.
2. Tensor Palati hooking around the hamulus to join the palatine aponeurosis.
3. Septa from the aponeurosis enclosing palatine glands.
4. A crest on the bony palate, having a branch of the greater palatine nerve on each side and the artery on the lateral side.
5. The lateral branch of the nerve expended mainly on the gums, the medial branch on the hard palate, the naso-palatine nerve in the incisive region, and the lesser palatine nerves in the soft palate.
6. Four palatine arteries, 2 on the hard palate and 2 on the soft: (a) greater palatine, (b) posterior septal (Fig. 7-111), and (c) lesser palatine and (d) ascending palatine (Fig. 7-158).

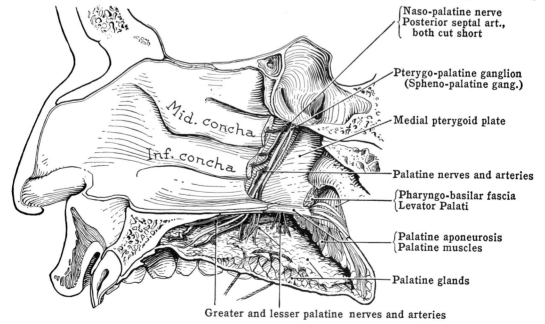

Naso-palatine nerve
Posterior septal art., both cut short

Pterygo-palatine ganglion
(Spheno-palatine gang.)

Mid. concha

Inf. concha

Medial pterygoid plate

Palatine nerves and arteries

Pharyngo-basilar fascia
Levator Palati

Palatine aponeurosis
Palatine muscles

Palatine glands

Greater and lesser palatine nerves and arteries

7-94 PALATE—IV

Observe:
1. The mucous membrane, containing a layer of mucous glands, has been separated by blunt dissection. The layer of glands is thin on the bony palate where it is part of a mucoperiosteum; it is thickest on the aponeurotic part; and it is less thick on the muscular part (Fig. 9-57).
2. The posterior ends (1 cm) of the middle and inferior conchae are cut through. These and the muco-periosteum are peeled off the side wall of the nose as far as the posterior border of the medial pterygoid plate. The papery perpendicular plate of the palatine bone is broken through and the palatine nerves and arteries are thereby exposed.

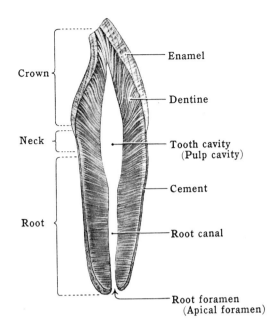

Crown

Neck

Root

Enamel

Dentine

Tooth cavity
(Pulp cavity)

Cement

Root canal

Root foramen
(Apical foramen)

Crown

Neck

Root

Enamel

Dentine

Tooth cavity
(Pulp cavity)

Cement

Root canal

Root foramen
(Apical foramen)

**7-95 INCISOR TOOTH,
LONGITUDINAL SECTION**

**7-96 MOLAR TOOTH,
LONGITUDINAL SECTION**

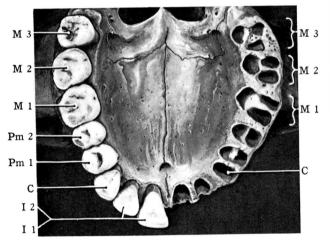

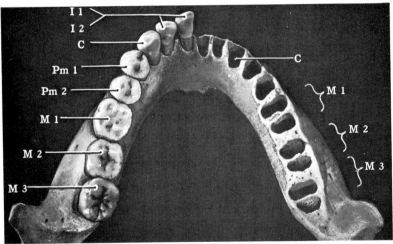

7-97 PERMANENT TEETH AND THEIR SOCKETS

Observe:

1. There are 32 permanent teeth, of which 8 are on each side of each dental arch — 2 incisors, 1 canine, 2 premolars, and 3 molars. Hence the formula reads:

$$\frac{3.2.1.2.}{3.2.1.2.} \quad \frac{2.1.2.3.}{2.1.2.3.}$$

2. Upper or maxillary incisor teeth are larger than lower or mandibular incisor teeth. The upper central incisors are the largest of the incisors and the lower central are the smallest. In each dental arch the 1st molar tooth is usually the largest molar and the 3rd molar is the smallest, although the 3rd lower molar may be very large, as here.

3. Crowns: An incisor tooth has a cutting edge; a canine tooth (cuspid) has one cusp on its crown; a premolar tooth (bicuspid) has 2 (or 3) cusps; and a molar tooth has from 3 to 5 cusps. The crowns of the upper molars are either square or rhomboidal. The 1st usually has 4 cusps; the 2nd either 4 or 3, and the 3rd 3. The crowns of the lower molars are oblong. The 1st usually has 5 cusps, the 2nd 4, and the 3rd from 3 to 5. The crowns are, here, somewhat worn.

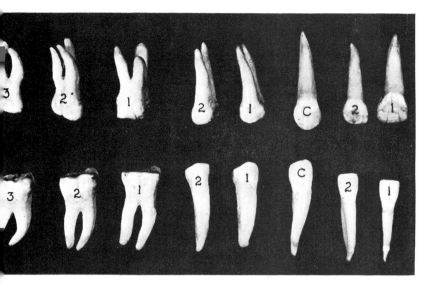

Observe:
1. Lower or mandibular teeth: The incisor, canine, and premolar (bicuspid) teeth have each one root, whereas the molars have each 2 roots, a mesial (anterior) and a distal (posterior). The mesial roots generally have 2 root canals. The roots are flattened mesiodistally (*i.e.,* from central incisor backward to the 3rd molar). The sockets for the lower incisor teeth are near the labial surface of the mandible, whereas those for the lower molars are near the lingual surface.
2. Upper or maxillary teeth: The incisor and canine teeth have each one root. The premolars have each either one or 2 roots, the first premolar usually having 2, a lingual and a labial, and the second usually having 1 — although sometimes, as here, both premolars have 2 roots. Each of the 3 molars has 3 roots, one being lingual and 2 being labial (buccal). The roots are flattened mesio-distally, except the root of the central incisor and the lingual root of each of the 3 molars which are circular on cross-section.

7-98   ROOTS OF PERMANENT TEETH, BUCCAL VIEW

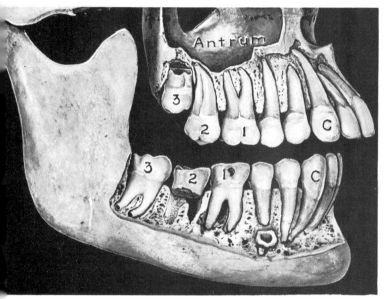

Observe:
1. The upper canine ("eye tooth") has the longest root; indeed, this is the longest tooth.
2. The roots of the upper premolars, in this specimen, are at some distance from the maxillary sinus or antrum, but the roots of the 3 molars almost penetrate into the sinus (see Figs. 7-138 and 7-141).
3. The root of the 2nd lower premolar, very long in this specimen, does not usually extend below the level of the mental foramen.
4. The roots of the 2nd lower molar have been removed, thereby revealing the cribriform nature of the wall of a socket.
5. The upper and lower 3rd molars are not yet fully developed. The lower is more advanced than the upper — the root foramina of the lower are still large, whereas the roots of the upper have not yet formed.

7-99   PERMANENT TEETH, ROOTS EXPOSED

Observe:
1. The upper and lower teeth begin flush, or nearly so, on one side and end flush on the other.
2. The lower central incisor is the smallest of the incisors, and the 3rd upper molar is the smallest of the molars. Except for these 2 teeth — the first in the lower row and the last in the upper — all teeth, when in occlusion, bite on two opposing teeth.
3. The upper dental arch overlaps the lower dental arch.
4. The lower incisors bite against the lingual surface of the upper incisors (Fig. 7-1). As a variant there may be a considerable overbite (Fig. 7-99).

100   PERMANENT TEETH, IN OCCLUSION

Observe:
1. There are 20 primary or deciduous teeth, 5 being in each half of the mandible and 5 in each maxilla. They are named: central incisor, lateral incisor, canine, 1st molar, and 2nd molar. The formula reads:

$$\frac{2.1.2.}{2.1.2.} \quad \frac{2.1.2.}{2.1.2.}$$

2. Of these 20 primary teeth the first to erupt through the gums are the lower central incisors, about the 6th month, and the last to erupt are the 2nd upper molars, about the end of the 2nd year. The 3 roots of the upper or maxillary molars and the 2 roots of the lower or mandibular molars are spread to grasp the developing permanent premolars (Fig. 7-104).
3. Primary teeth differ from permanent teeth in being smaller and whiter ("milk teeth"). The molars have more bulbous crowns and more divergent roots.

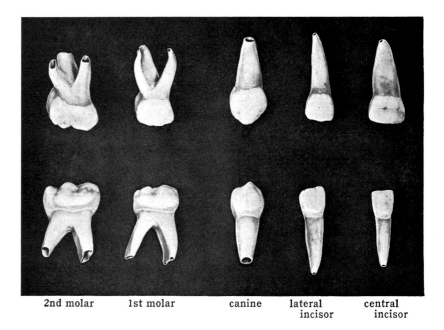

| 2nd molar | 1st molar | canine | lateral incisor | central incisor |

7-101 PRIMARY TEETH

Observe:
1. The canines have not fully erupted, the 2nd molars have just started to erupt—the sequence of eruption being incisors, 1st molars, canines, 2nd molars.
2. The 2nd molars have much larger crowns than the 1st molars.
3. The socket for the 3-pronged root of the 1st upper molar is seen.
4. The foramina, seen on the lingual side of the primary incisors, lead to the alveoli for the permanent incisors.
5. Permanent teeth are colored *yellow*. The crowns of the unerupted 1st and 2nd permanent molars are partly visible.

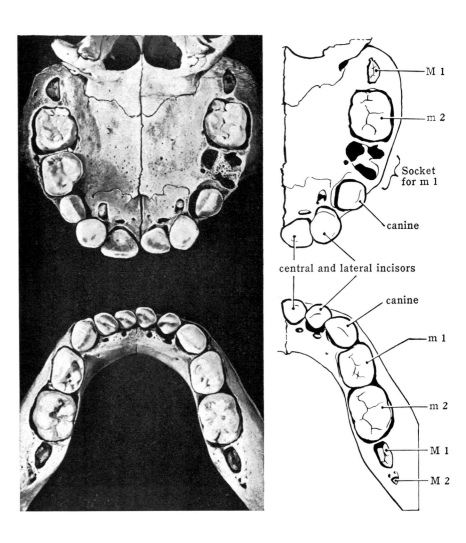

7-102 PRIMARY DENTITION, AGED UNDER 2 YEARS

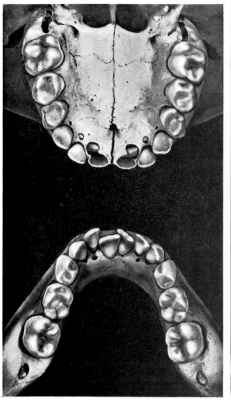

A. AGED 6 TO 7 YEARS

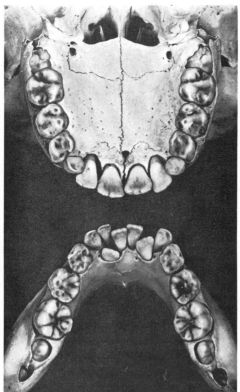

B. AGED 8 YEARS

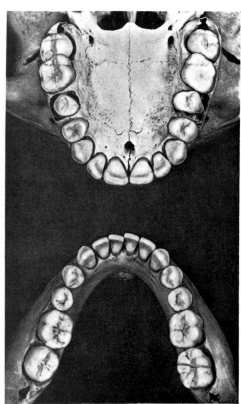

C. AGED 12 YEARS

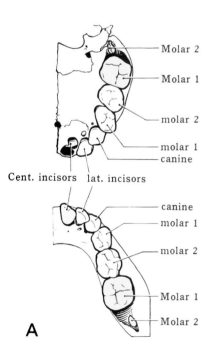

A

Molar 2
Molar 1
molar 2
molar 1
canine
Cent. incisors   lat. incisors
canine
molar 1
molar 2
Molar 1
Molar 2

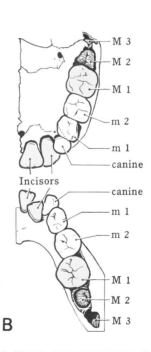

B

M 3
M 2
M 1
m 2
m 1
canine
Incisors
canine
m 1
m 2
M 1
M 2
M 3

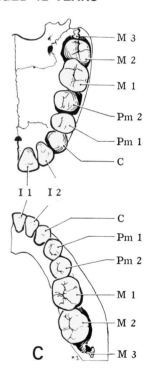

C

M 3
M 2
M 1
Pm 2
Pm 1
C
I 1   I 2
C
Pm 1
Pm 2
M 1
M 2
M 3

7-103   PROGRESS IN THE ERUPTION OF PERMANENT TEETH

Between the 6th and 12th years the primary teeth are shed and are succeeded by permanent teeth.

A. The 1st molars (6th-year molars) have fully erupted. The primary central incisors have been shed. The lower central incisors have nearly fully erupted and the upper central incisors are moving downward into the empty sockets.

B. All the permanent incisors have erupted, the upper and lower central and the upper lateral fully and the lower lateral partially. Note that the alveolus has not yet closed around the upper lateral incisors. Here, the root of the left lower lateral primary incisor has not been resorbed, so the tooth has not been shed.

C. The 20 primary teeth have been replaced by 20 permanent teeth, and the 1st molars and the 2nd molars (12th-year molars) have erupted. But, the canines, 2nd premolars, and 2nd molars—especially those in the upper jaw—have not erupted fully nor have their bony sockets closed around them. By the age of 12, 28 permanent teeth are in evidence. The last 4 teeth, the 3rd molars, may erupt any time after this, or never.

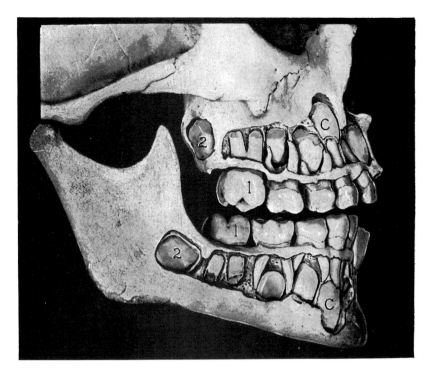

A

Alveolar bone has been ground away from the specimen shown in Figure 7-103A, aged 6 to 7 years. Permanent teeth are shown in *yellow*. The crowns of the permanent teeth are fully formed at the time of eruption (*i.e.*, they grow no larger). The permanent incisors and canines develop and erupt on the lingual side of the primary incisors and canines, the lateral incisors being the most posterior. Indeed, the upper lateral incisors extend into the bony palate. The premolar teeth develop between the spread roots of the primary molars. The permanent molars have no predecessors; they erupt distal to the deciduous teeth in the dental arch.

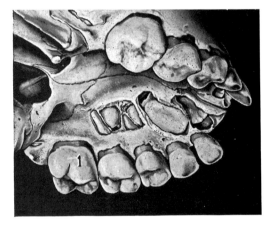

B

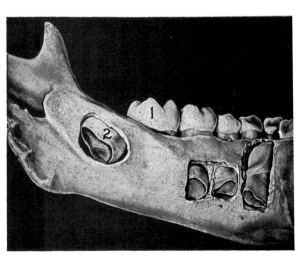

C

7-104   MIXED DENTITION

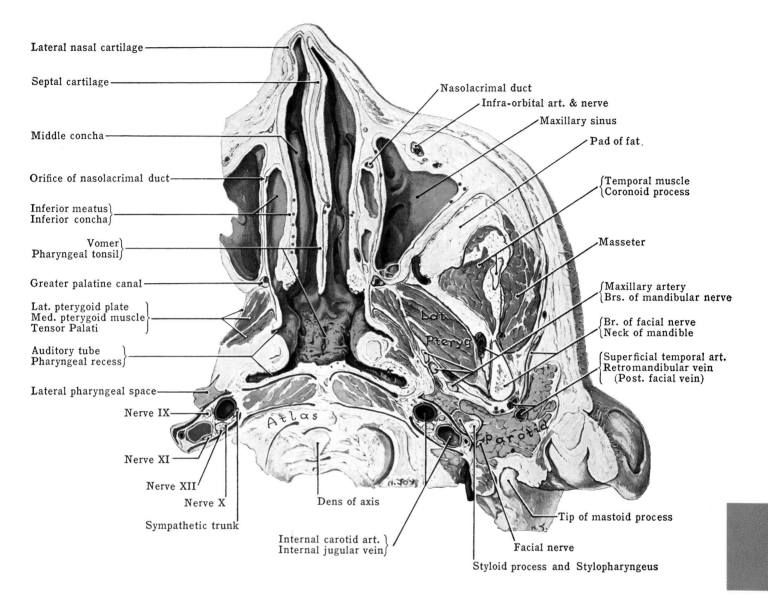

Lateral nasal cartilage

Septal cartilage

Middle concha

Orifice of nasolacrimal duct

Inferior meatus
Inferior concha

Vomer
Pharyngeal tonsil

Greater palatine canal

Lat. pterygoid plate
Med. pterygoid muscle
Tensor Palati

Auditory tube
Pharyngeal recess

Lateral pharyngeal space

Nerve IX

Nerve XI

Nerve XII

Nerve X

Sympathetic trunk

Nasolacrimal duct
Infra-orbital art. & nerve
Maxillary sinus
Pad of fat

Temporal muscle
Coronoid process

Masseter

Maxillary artery
Brs. of mandibular nerve

Br. of facial nerve
Neck of mandible

Superficial temporal art.
Retromandibular vein
(Post. facial vein)

Tip of mastoid process

Dens of axis

Internal carotid art.
Internal jugular vein

Facial nerve

Styloid process and Stylopharyngeus

7-105 CROSS-SECTION PASSING THROUGH NASAL CAVITIES, FROM BELOW

Observe:

1. The septal cartilage continuous with the lateral nasal cartilages.
2. An unusual spur from the nasal septum pushing into the middle concha.
3. The pharyngeal tonsil somewhat overgrown (adenoids).
4. The nasolacrimal duct within the mucous membrane of the inferior meatus on one side, and its orifice on the other side.
5. The posterior ends of the inferior conchae, overgrown.
6. The pharyngeal recess spreading widely behind the mouth of the tube.
7. The mucous membrane of one maxillary sinus much thickened. The infra-orbital nerve and artery in front, the greater palatine nerve and artery behind.
8. The carotid sheath at the base of the skull, postero-lateral to the pharynx, containing the internal carotid, the internal jugular, and 5 nerves (Fig. 9-53).

9. Around the temporal muscle, fatty tissue which is continuous with the buccal pad of fat.
10. Lateral Pterygoid passing obliquely from the medial wall (lateral pterygoid plate) of the infratemporal fossa to the lateral wall (neck of mandible).
11. The parotid gland, its deep relations, and the facial nerve largely deep to it.
12. The lateral pharyngeal fatty-areolar "space" (*pale green*) having the pharynx medially, the carotid sheath posteriorly, the styloid process and its 3 muscles (one shown at this high level) laterally, branches of the mandibular nerve in a fascial septum anteriorly, and separated from the pharyngeal process of the parotid gland by the stylomandibular ligament (not labeled).

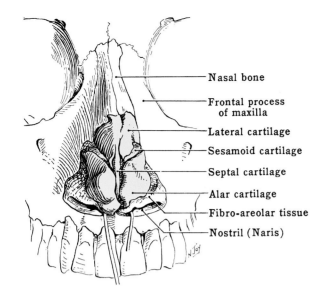

Nasal bone
Frontal process of maxilla
Lateral cartilage
Sesamoid cartilage
Septal cartilage
Alar cartilage
Fibro-areolar tissue
Nostril (Naris)

A

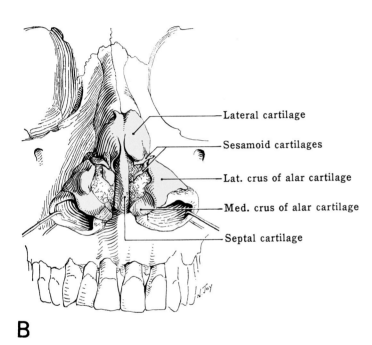

Lateral cartilage
Sesamoid cartilages
Lat. crus of alar cartilage
Med. crus of alar cartilage
Septal cartilage

B

7-106 CARTILAGES OF THE NOSE

Observe:
1. In *A*, the alar cartilages have been pulled down to expose the sesamoid cartilages. In *B*, the alar cartilages are separated by dissection and retracted laterally.
2. The lateral nasal cartilages, fixed by suture to the nasal bones and continuous with the septal cartilage.
3. The alar nasal cartilages, free, movable, and U-shaped.
4. The medial crus of the right and of the left U, when in apposition, forming part of the septum of the nose.
5. The lower part of the ala of the nose, formed of fibro-areolar tissue; *cf.* the lobule of the auricle.

Note: The nasal cartilages are hyaline cartilage; the cartilage of the auricle is elastic cartilage.

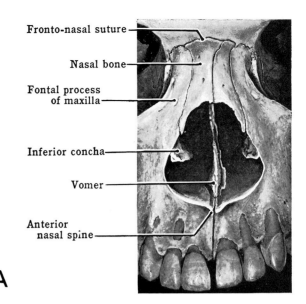

Fronto-nasal suture
Nasal bone
Fontal process of maxilla
Inferior concha
Vomer
Anterior nasal spine

A

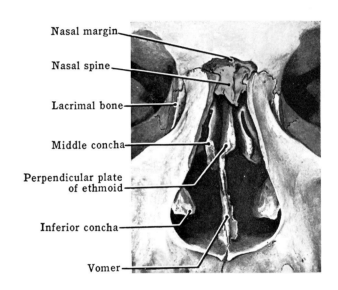

Nasal margin
Nasal spine
Lacrimal bone
Middle concha
Perpendicular plate of ethmoid
Inferior concha
Vomer

B

7-107 BONES OF THE NOSE

Observe:
A. The margin of the piriform aperture, or bony anterior nasal aperture, is sharp and is formed by the maxillae and the nasal bones.
B. On removing the nasal bones the areas on the frontal processes of the maxillae *(yellow)* and on the frontal bone *(blue)* that articulate with and buttress the nasal bones are seen; so is the nasal septum (Fig. 7-108).

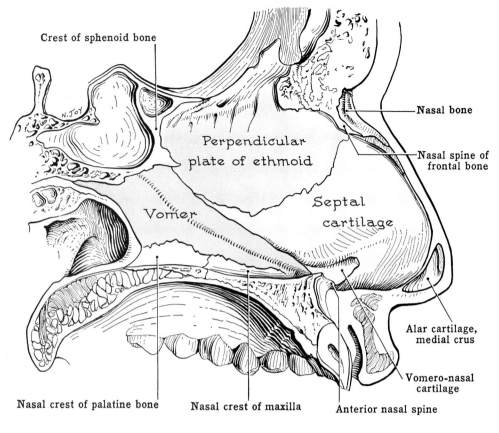

Crest of sphenoid bone

Perpendicular plate of ethmoid

Nasal bone

Nasal spine of frontal bone

Vomer

Septal cartilage

Alar cartilage, medial crus

Vomero-nasal cartilage

Nasal crest of palatine bone

Nasal crest of maxilla

Anterior nasal spine

7-108 SEPTUM OF NOSE

Observe:

1. Like the palate, the septum of the nose has a hard part and a soft or mobile part. The skeleton or basis of the hard septum consists of 3 parts—perpendicular plate of ethmoid, septal cartilage, vomer— and, around the circumference of these, the adjacent bones (frontal, nasal, maxillary, palatine, and sphenoid) make minor contributions.

2. The mobile septum comprises (a) the medial limbs (crura) of the U-shaped alar cartilages (Figs. 7-106 and 7-118), and (b) the skin and soft tissues between the tip of the nose and the anterior nasal spine. Behind the vomer, an extension of the muco-periosteum of the septum forms a second, although unimportant, mobile septum.

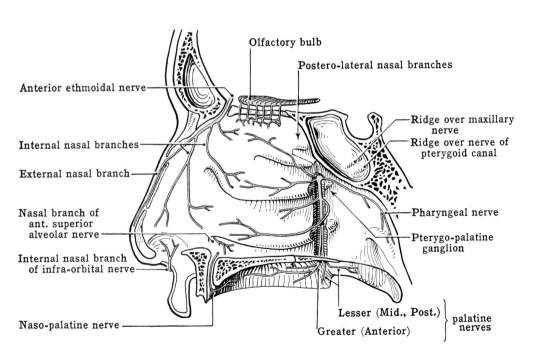

Olfactory bulb

Postero-lateral nasal branches

Anterior ethmoidal nerve

Internal nasal branches

External nasal branch

Nasal branch of ant. superior alveolar nerve

Internal nasal branch of infra-orbital nerve

Naso-palatine nerve

Ridge over maxillary nerve

Ridge over nerve of pterygoid canal

Pharyngeal nerve

Pterygo-palatine ganglion

Lesser (Mid., Post.)
Greater (Anterior) } palatine nerves

7-109 NERVE SUPPLY TO LATERAL WALL OF NASAL CAVITY

See also Figure 8-6.

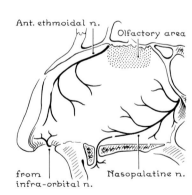

Ant. ethmoidal n.

Olfactory area

from infra-orbital n.

Nasopalatine n.

7-110 NERVES OF NASAL SEPTUM

The pterygopalatine ganglion sends the naso-palatine nerve through the spheno-palatine foramen, the greater and lesser palatine nerves through canals of the same name, and the pharyngeal nerve through the pharyngeal (palato-vaginal) canal.

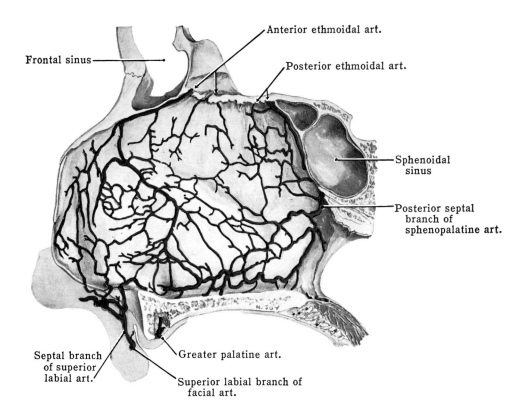

A. ARTERIES TO SEPTUM

Frontal sinus

Anterior ethmoidal art.

Posterior ethmoidal art.

Sphenoidal sinus

Posterior septal branch of sphenopalatine art.

Greater palatine art.

Septal branch of superior labial art.

Superior labial branch of facial art.

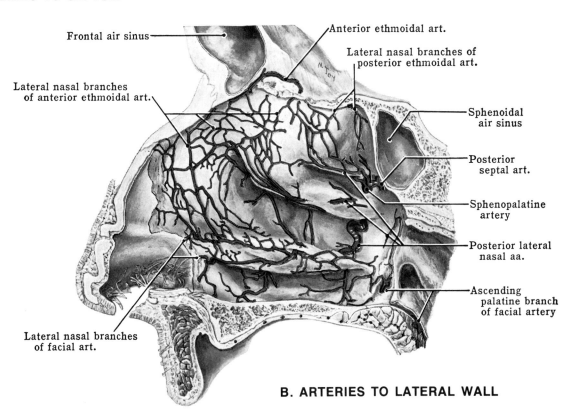

Frontal air sinus

Anterior ethmoidal art.

Lateral nasal branches of posterior ethmoidal art.

Lateral nasal branches of anterior ethmoidal art.

Sphenoidal air sinus

Posterior septal art.

Sphenopalatine artery

Posterior lateral nasal aa.

Ascending palatine branch of facial artery

Lateral nasal branches of facial art.

B. ARTERIES TO LATERAL WALL

7-111 ARTERIES TO WALLS OF NASAL CAVITY

Observe:
1. Anterior ethmoidal and posterior ethmoidal branches of the ophthalmic artery entering through the cribriform plate. Only these are derived from the internal carotid, all others being from the external carotid.
2. The spheno-palatine artery is the main supply. Entering through the spheno-palatine foramen (Fig. 7-113), it sends (a) lateral nasal branches forward on both surfaces of the conchae, partly in bony canals, and (b) the posterior septal artery which crosses the roof of the nose below the anterior part of the floor of the sphenoidal sinus and anastomoses through the incisive foramen.
3. The facial artery via 3 branches: (a) the septal branch of the superior labial, (b) twigs from the lateral nasal that pierce the ala to supply the vestibular region; and (c) anastomoses from the ascending palatine.

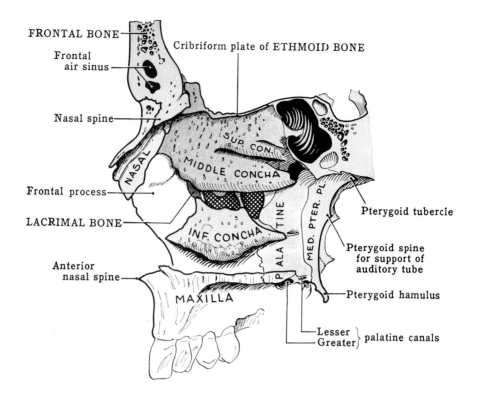

7-112 BONES OF LATERAL WALL OF NASAL CAVITY—I

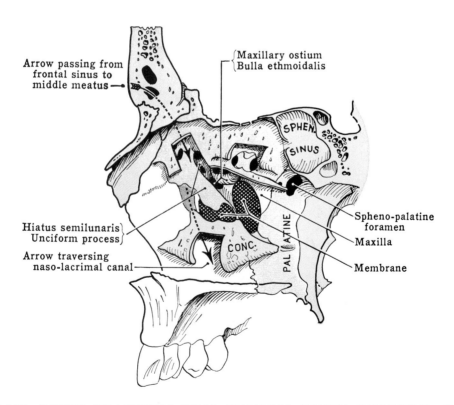

7-113 BONES OF LATERAL WALL OF NASAL CAVITY, DISSECTED—II

Note:
1. The superior and middle conchae are parts of the ethmoid bone, whereas the inferior concha is a bone of itself.
2. The pterygoid process is part of the sphenoid bone (Fig. 7-133). Its medial plate forms the posterior part of the lateral wall of the bony nasal cavity.
3. The fragile, perpendicular plate of the palatine bone has a notch at its upper border, which, when in articulation with the body of the sphenoid bone, forms the spheno-palatine foramen (Fig. 7-137).
4. The groove on the nasal bone is for the external nasal branch of the anterior ethmoidal nerve (Fig. 7-109).
5. The hiatus semilunaris is bounded below by the unciform process of the ethmoid bone. The maxillary ostium opens on to the hiatus. Accessory maxillary ostia may occur in the membrane that closes the maxillary hiatus.

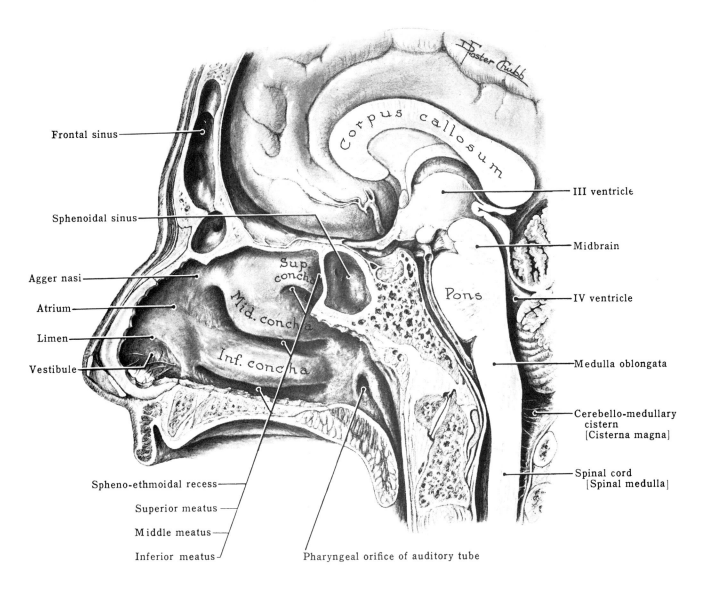

Labels on figure:

Frontal sinus

Sphenoidal sinus

Agger nasi

Atrium

Limen

Vestibule

Corpus callosum

III ventricle

Midbrain

IV ventricle

Medulla oblongata

Cerebello-medullary cistern [Cisterna magna]

Spinal cord [Spinal medulla]

Pons

Sup. concha

Mid. concha

Inf. concha

Spheno-ethmoidal recess

Superior meatus

Middle meatus

Inferior meatus

Pharyngeal orifice of auditory tube

7-114 LATERAL WALL OF NASAL CAVITY—I

Observe:

1. The vestibule, above the nostril or naris and in front of the inferior meatus. The hairs growing from its skin-lined surface, spreading in all directions.

2. The atrium, above the vestibule and in front of the middle meatus.

3. The inferior and middle conchae, curving downward and medially from the lateral wall, dividing it into 3 nearly equal parts, and covering the inferior and middle meatuses respectively.

4. The superior concha, small and in front of the sphenoidal sinus. The middle concha, with an angled lower border, ending below the sphenoidal sinus. The inferior concha, with a slightly curved lower border, ending below the middle concha, about 1 cm in front of the orifice of the tube—*i.e.*, about the width of the medial pterygoid plate (Fig. 7-112).

5. The floor of the nose, inclined lightly downward and backward, at the level of the atlas.

6. The roof comprising: (a) an anterior sloping part, corresponding to the bridge of the nose; (b) an intermediate horizontal part, formed by the delicate cribriform plate; (c) a perpendicular part in front of the sphenoidal sinus, and (d) a curved part, below the sinus, which is confluent with the roof of the naso-pharynx.

7. The pons and the fourth ventricle of the brain at the level of the sphenoidal sinus.

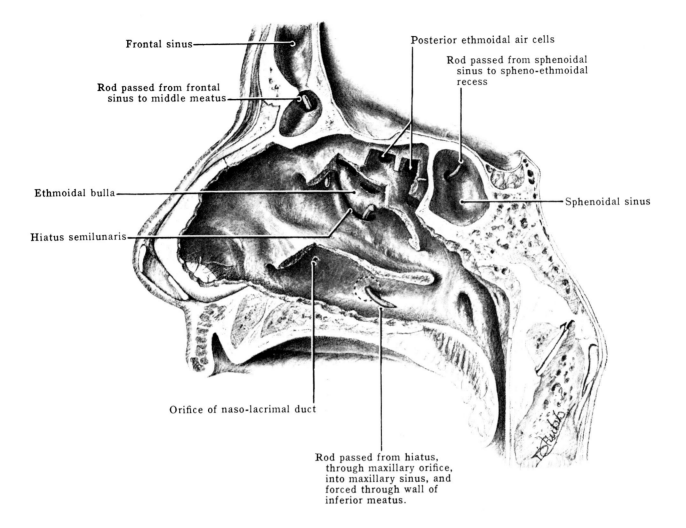

Frontal sinus

Rod passed from frontal
sinus to middle meatus

Ethmoidal bulla

Hiatus semilunaris

Orifice of naso-lacrimal duct

Posterior ethmoidal air cells

Rod passed from sphenoidal
sinus to spheno-ethmoidal
recess

Sphenoidal sinus

Rod passed from hiatus,
through maxillary orifice,
into maxillary sinus, and
forced through wall of
inferior meatus.

7-115 LATERAL WALL OF NASAL CAVITY, DISSECTED – II

Parts of the superior, middle, and inferior conchae are cut away.

Observe:
1. The sphenoidal sinus in the body of the sphenoid bone. Its orifice, above the middle of its anterior wall, opens into the spheno-ethmoidal recess.
2. The orifices of posterior ethmoidal cells open into the superior meatus.
3. A cell, in this specimen, opening onto the upper surface of the ethmoidal bulla.
4. The attachment of the inferior concha, steep in its anterior one-third, but gently sloping in its posterior two-thirds. The orifice of the naso-lacrimal duct, a short (variable) distance below the angle of union of the anterior one-third and posterior two-thirds.
5. The sharp probe forced through the thinnest portion of the medial wall of the maxillary sinus, well above the level of the floor of the nasal cavity.

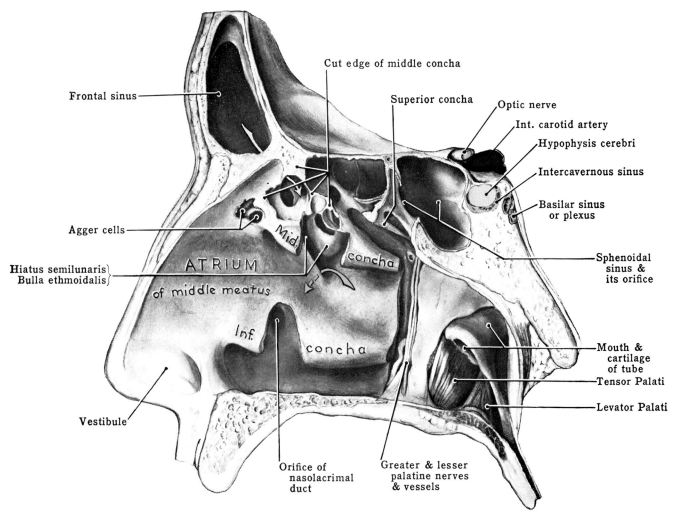

7-116   PARANASAL AIR SINUSES AND HYPOPHYSIS CEREBRI

Note:
1. The frontal sinus with its outlet at its lowest point, leading into the middle meatus medial to the hiatus semilunaris. The hiatus ends blindly in front as an anterior ethmoidal cell, and posteriorly as the maxillary orifice, indicated by an *arrow*.
2. The sphenoidal sinus of average size, and with a very large orifice.
3. The relationship of the Hypophysis cerebri. For anatomical variations see Bergland, R. M., Bronson, S. R., and Torack, R. M. (1968) *J. Neurosurg. 28:* 93–99.

Note:
1. In addition to the primary or normal ostium (not in view), there are here present 4 secondary or acquired ostia resulting from the breaking down of the membrane shown in *cross-hatching* in Figure 7-113.
2. The septum between the right and left sphenoidal sinus, here occupies the median plane—it is usually deflected to one side or other.

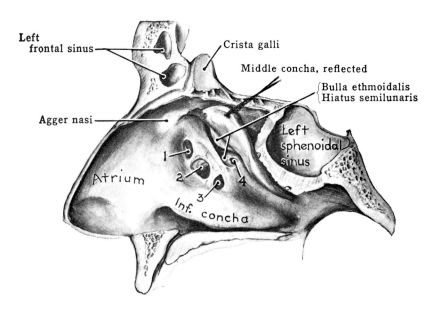

7-117   ACCESSORY MAXILLARY ORIFICES

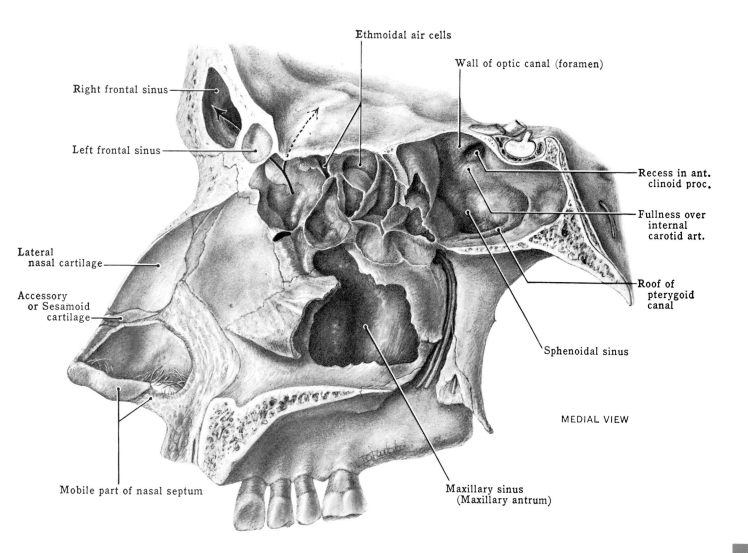

Ethmoidal air cells

Wall of optic canal (foramen)

Right frontal sinus

Left frontal sinus

Recess in ant. clinoid proc.

Fullness over internal carotid art.

Lateral nasal cartilage

Accessory or Sesamoid cartilage

Roof of pterygoid canal

Sphenoidal sinus

MEDIAL VIEW

Mobile part of nasal septum

Maxillary sinus (Maxillary antrum)

7-118 PARANASAL AIR SINUSES, OPENED

Observe:

1. The ethmoidal cells *(pink)*, collectively called a sinus, like a honeycomb, has the thin orbital plate of the frontal bone for a roof (Figs. 7-53 and 7-120).
2. An anterior ethmoidal cell *(blue)* invading the diploe of the squama of the frontal bone to become a frontal sinus. It is ethmoidal in origin, but frontal in location. An offshoot *(broken arrow)* invades the orbital plate of the frontal bone.
3. The sphenoidal sinus *(blue)* in this specimen is very extensive—compare with Figure 7-115—extending (a) backward below the hypophysis cerebri to the dorsum sellae, (b) laterally, below the optic nerve, into the anterior clinoid process, and (c) downward to the pterygoid process, but leaving the pterygoid canal rising as a ridge on the floor of the sinus.
4. The maxillary sinus *(yellow)* is pyramidal in shape. Its base (largely nibbled away) contributes to the lateral wall of the nasal cavity, its apex is in the zygomatic process, and its orifice is at its highest point.

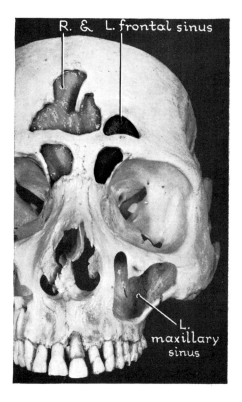

R. & L. frontal sinus

L. maxillary sinus

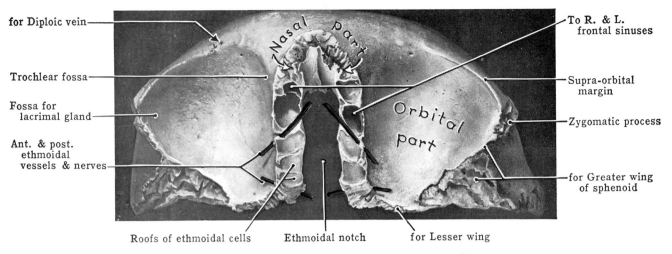

for Diploic vein

Trochlear fossa

Fossa for
lacrimal gland

Ant. & post.
ethmoidal
vessels & nerves

Nasal part

Orbital part

To R. & L.
frontal sinuses

Supra-orbital
margin

Zygomatic process

for Greater wing
of sphenoid

Roofs of ethmoidal cells   Ethmoidal notch   for Lesser wing

7-120 FRONTAL BONE, INFERIOR ASPECT

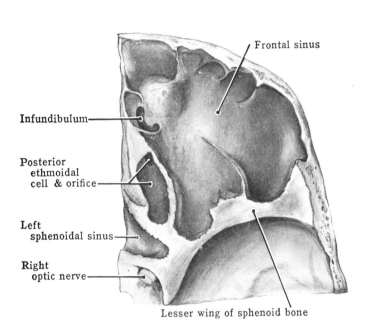

Frontal sinus

Infundibulum

Posterior
ethmoidal
cell & orifice

Left
sphenoidal sinus

Right
optic nerve

Lesser wing of sphenoid bone

7-121   EXTENSIVE FRONTAL SINUS

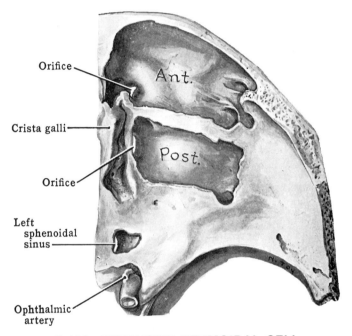

Orifice

Crista galli

Orifice

Left
sphenoidal
sinus

Ophthalmic
artery

Ant.

Post.

7-122   EXTENSIVE ETHMOIDAL CELL

Note:
1. This sinus occupies the entire orbital part of the frontal bone.
2. An anterior ethmoidal cell bulges domelike into the floor of the frontal sinus.
3. The left sphenoidal sinus crosses the median plane and helps to form the roof of the right optic canal.

This cell occupies the posterior half of the orbital part of the frontal bone, behind an average-sized frontal sinus.

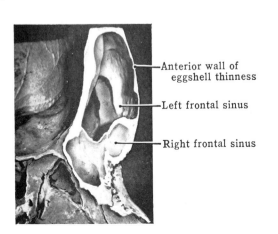

Anterior wall of
eggshell thinness

Left frontal sinus

Right frontal sinus

7-123 FRONTAL SINUS, THIN ANTERIOR WALL

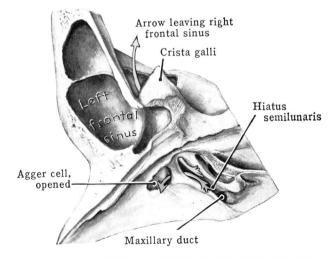

Arrow leaving right
frontal sinus

Crista galli

Left
frontal
sinus

Hiatus
semilunaris

Agger cell,
opened

Maxillary duct

7-124   FRONTAL SINUS IN CRISTA GALLI

Note:
1. The left sinus here extends far across the median plane.
2. The right sinus opens at the summit of the hiatus.
3. The agger cell is here a diverticulum from the hiatus.

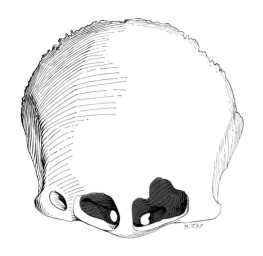

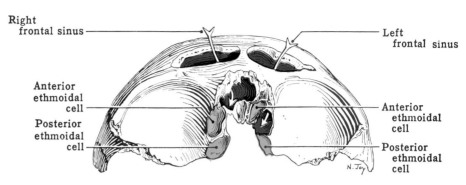

Right frontal sinus — Left frontal sinus

Anterior ethmoidal cell — Anterior ethmoidal cell

Posterior ethmoidal cell — Posterior ethmoidal cell

7-125   FRONTAL AIR SINUSES, FROM THE FRONT

The orifices of the sinuses are at the lowest points of the sinuses.

7-126   FRONTAL AIR SINUSES, FROM BELOW

The right frontal air sinus is here, as usual, an extension of an anterior ethmoidal cell. The corresponding left cell (*blue*) is small, but the next cell behind it has invaded the diploe of the frontal bone and so become a frontal sinus.

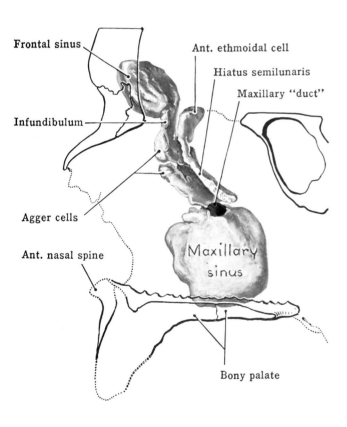

Frontal sinus

Ant. ethmoidal cell

Hiatus semilunaris

Maxillary "duct"

Infundibulum

Agger cells

Ant. nasal spine

Maxillary sinus

Bony palate

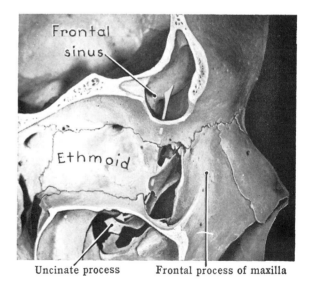

Frontal sinus

Ethmoid

Uncinate process   Frontal process of maxilla

7-127   CAST, FRONTAL AND MAXILLARY SINUSES, MEDIAL VIEW

Note:
1. The frontal sinus lies above the orbital cavity and has its opening at its lowest point, whereas the maxillary sinus lies below the orbital cavity and has its opening on a level with its highest point.
2. Had this frontal sinus failed to develop, the anterior ethmoidal cell (shown) would have taken its place in default.

7-128   FRONTONASAL DUCT

This is the specimen shown in Figure 7-135 after removal of the lacrimal bone. The *yellow arrow* suggests that, if there were fluid in the frontal sinus, it would drain through the infundibulum ethmoidale (situated at the lowest point of the sinus) into the hiatus semilunaris and thence into the maxillary sinus via its orifice or duct (Fig. 7-127), situated on a level with its highest point.

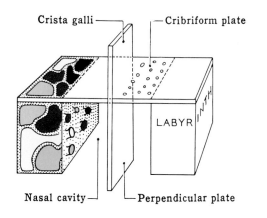

Crista galli — — Cribriform plate

LABYRINTH

Nasal cavity — — Perpendicular plate

The ethmoidal air cells may be likened to a number of rubber balloons projecting into an oblong box and variously inflated to the full capacity of the box. Indeed, one (occasionally more) of the anterior balloons bursts through the lid of the box (*i.e.*, the roof of the ethmoidal labyrinth) and invades the neighboring territory (*i.e.*, the frontal bone) to a variable extent and acquires the name "frontal air sinus."

7-129 SCHEME OF THE ETHMOIDAL AIR CELLS

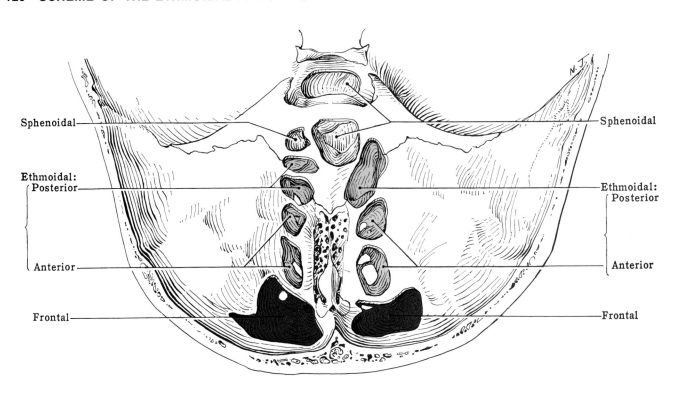

Sphenoidal — — Sphenoidal

Ethmoidal:
Posterior — — Ethmoidal: Posterior

Anterior — — Anterior

Frontal — — Frontal

7-130 AIR SINUSES SURROUNDING THE CRIBRIFORM PLATE, FROM ABOVE

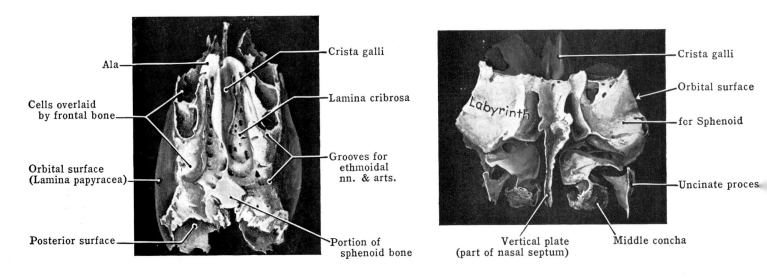

Ala — — Crista galli

Cells overlaid by frontal bone — — Lamina cribrosa

Orbital surface (Lamina papyracea) — — Grooves for ethmoidal nn. & arts.

Posterior surface — — Portion of sphenoid bone

7-131 ETHMOID BONE, SUPERIOR ASPECT

Labyrinth

Crista galli — — Orbital surface

for Sphenoid

Uncinate proces

Vertical plate (part of nasal septum) — — Middle concha

7-132 ETHMOID BONE, POSTERIOR ASPECT

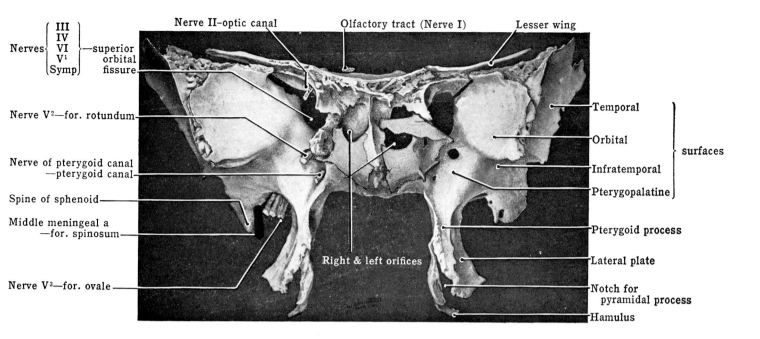

7-133 SPHENOID BONE OF AN ADULT, FRONT VIEW

Note: On each side 6 of the 12 cranial nerves are closely related to the sphenoid, nerve V piercing it in 3 divisions. The nerve of the pterygoid canal and the middle meningeal artery pierce the bone. The parts colored *pink* are the sphenoidal conchae.

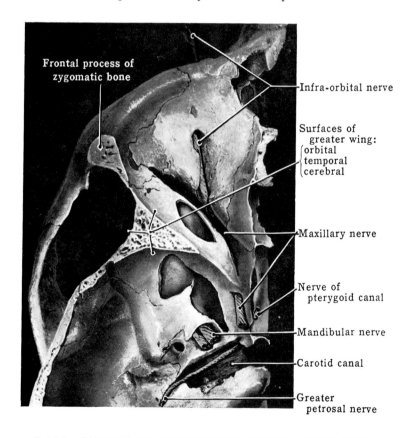

7-134 EXTENSIVE SPHENOIDAL SINUS, LEFT SIDE

Note:

1. This very extensive sinus extends laterally between the maxillary nerve and the nerve of the pterygoid canal, inflating the greater wing of the sphenoid bone and the pterygoid process. It extends forward into the lateral wall of the orbital cavity, backward to the mandibular nerve which passes through the foramen ovale, laterally to the temporal fossa, and downward to the roof and medial walls of the infratemporal fossa.

2. The greater (superficial) petrosal nerve is seen leaving the hiatus facialis, running above the carotid canal, descending through the foramen lacerum, and, when joined by sympathetic fibers (not shown), it becomes the nerve of the pterygoid canal (Fig. 8-7).

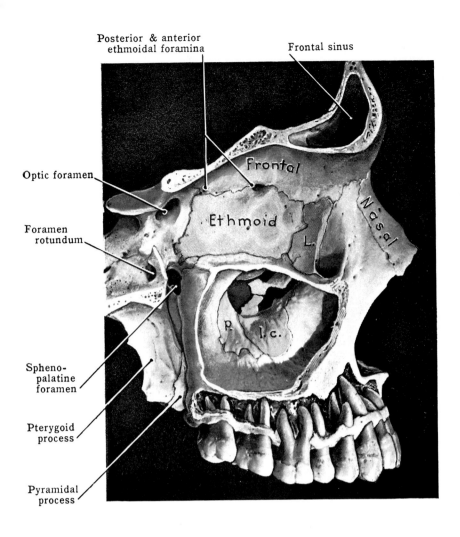

I.c., inferior concha (*green*); *L.*, lacrimal bone (*blue*); *P.*, palatine bone (*yellow*).

Note:
1. The site of the hiatus semilunaris between the bulla of the ethmoid bone above and the uncinate (=hooklike) process below.
2. The pterygo-palatine fossa between pterygoid process, maxilla, palatine bone, and sphenoid bone. The foramen rotundum opens into the fossa from the middle cranial fossa and the spheno-palatine foramen opens into the nasal cavity.

7-135 MEDIAL WALL OF ORBITAL CAVITY AND MAXILLARY SINUS

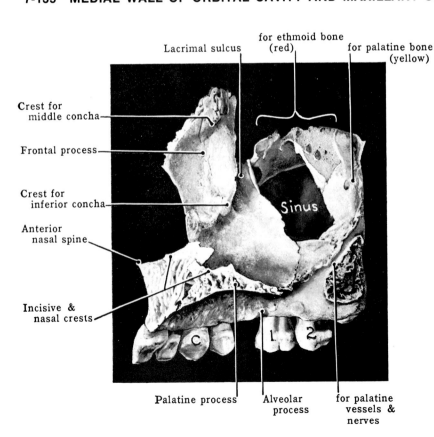

7-136 MEDIAL ASPECT OF MAXILLA

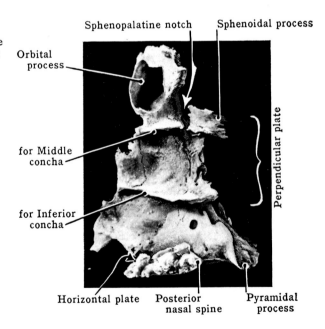

7-137 PALATINE BONE, MEDIAL ASPECT

Note:
1. The pyramidal process has 2 grooves for the pterygoid process.
2. The cell in the orbital process is an extension of either the maxillary sinus, an ethmoidal cell, or the sphenoidal sinus.

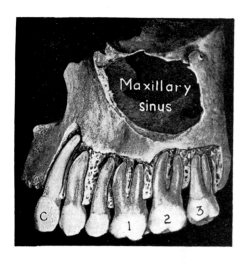

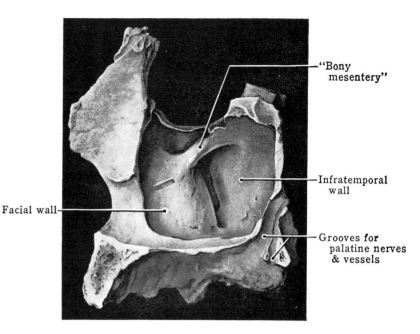

7-138 MAXILLARY SINUS, LATERAL VIEW

The roots of the teeth are, here, at some distance from the sinus.

Contrast this figure with Figures 7-99 and 7-135, where the roots of the teeth come closer to the floor of the bony sinus, and with Figure 7-141 where they penetrate it.

7-139 MAXILLARY SINUS, MEDIAL VIEW

Note:
1. The "bony mesentery" for the infra-orbital nerve and vessels, here projecting not from the roof, but from the lateral wall.
2. The anterior or facial wall pushing far into the sinus.
3. The grooves for the greater and lesser palatine nerves.

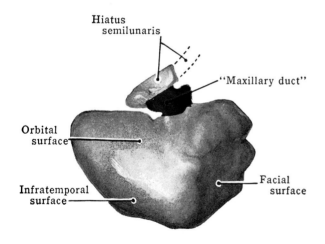

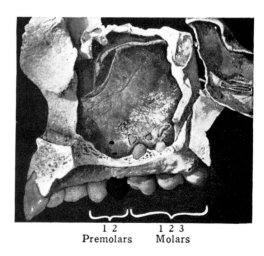

7-140 CAST OF RIGHT MAXILLARY SINUS, LATERAL VIEW

The so-called orifice or ostium, which opens into the hiatus semilunaris, is usually a duct, in shape flattened oval or slitlike, and of the following average dimensions: length of the duct along its anterior wall, 6.0 mm, and along its posterior wall, 3.5 mm; similarly, the long diameter of the oval, 6.0 mm, and the short diameter, 3.5 mm.

7-141 MAXILLARY SINUS, MEDIAL VIEW

Observe that in this specimen the roots of three teeth penetrate the bony floor of the sinus.

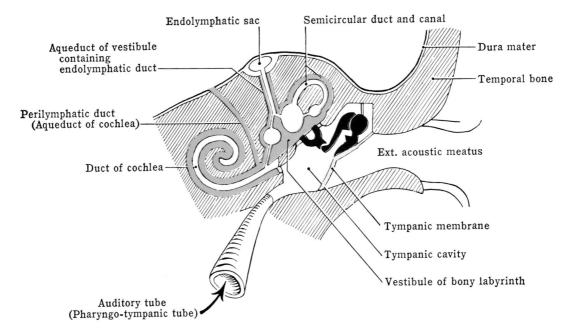

7-142 GENERAL SCHEME OF THE EAR

The ear is divisible into 3 parts—external, middle, internal. The *external ear* comprises (a) the auricle and (b) the external acoustic meatus, the medial end of which is closed by the tympanic membrane.

The *middle ear* or tympanum lies between the tympanic membrane and the internal ear. Three ossicles (*red*)—malleus, incus, stapes—stretch from the lateral to the medial wall of the tympanum. Of these, the malleus is attached to the tympanic membrane; the stapes is attached by an annular ligament to the oval opening, called the fenestra vestibuli; and the incus connects these two ossicles.

The auditory tube opens into the anterior wall of the tympanic cavity; the aditus ad antrum opens from the epitympanic recess backward to the mastoid antrum (Fig. 7-157).

The *internal ear* comprises a closed system of membranous tubes and bulbs, called the membranous labyrinth, which are filled with fluid, called endolymph, and are bathed in surrounding fluid called perilymph (*blue*). The perilymph is contained within the bony labyrinth, but this system is not a closed one, for, as represented here, perilymph and cerebrospinal fluid are confluent in the posterior cranial fossa. There is, however, good evidence that this system also is closed, the perilymphatic duct ending as a perilymphatic sac.

When the tympanic membrane vibrates, the malleus vibrates with it and transmits the vibrations via the incus to the stapes. The stapes, being attached to the margins of the fenestra vestibuli (oval window) by an annular ligament, transmits the vibrations to the perilymph within the vestibule.

A secondary tympanic membrane which closes the fenestra cochleae (round window) receiving the vibrations transmitted to the incompressible perilymph is itself made to vibrate in turn.

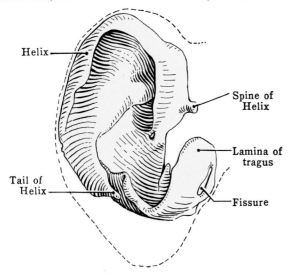

A. CARTILAGE OF RIGHT AURICLE

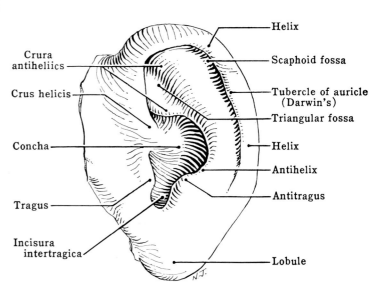

B. LEFT AURICLE

7-143 THE AURICLE

Note:
1. The cartilage of the auricle is *yellow* or elastic cartilage. It is continuous with the cartilage of the external acoustic meatus. It does not extend into the lobule, the basis of which is fibro-areolar tissue.
2. There being very little subcutaneous tissue on the lateral surface of the auricle, the skin adheres to the cartilage and follows its irregularities. On the cranial surface, there being fine muscles and more subcutaneous tissue, the skin is movable.
3. The foramen at the root of the crus of the cartilage of the helix transmits an artery (Fig. 7-70); the fissure in the cartilage of the external meatus is an unchondrified area, closed with fibrous tissue (Fig. 7-158). For arteries and nerves, see Figure 7-70.

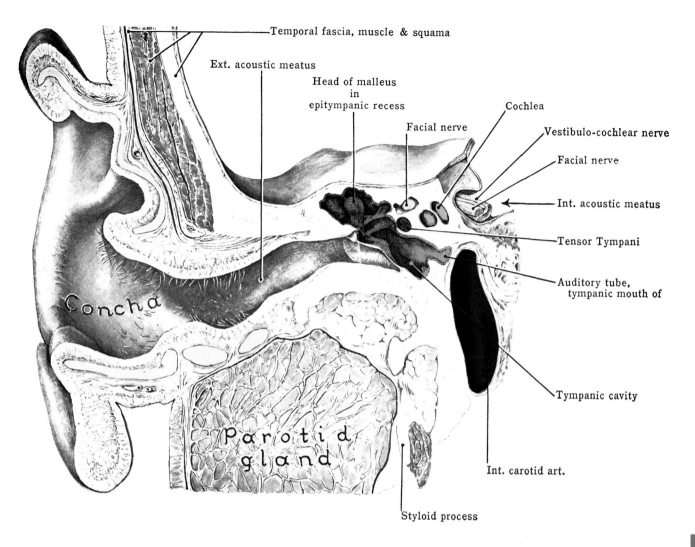

Temporal fascia, muscle & squama

Ext. acoustic meatus

Head of malleus
in
epitympanic recess

Facial nerve

Cochlea

Vestibulo-cochlear nerve

Facial nerve

Int. acoustic meatus

Tensor Tympani

Auditory tube,
tympanic mouth of

Tympanic cavity

Int. carotid art.

Styloid process

Concha

Parotid gland

7-144 EAR ON CORONAL SECTION, ANTERIOR VIEW

The inner ear is tinted *blue*; the mucous membrane of the middle ear is *pink*.

Observe:

1. The external acoustic (auditory) meatus which from tragus to eardrum is 3 cm long, half the length being cartilaginous and half bony. It is narrowest near the drum due to the rise on the floor, hence the "well" where fluid might collect at the medial end of the meatus.
2. The cartilaginous or mobile part of the external meatus, lined with thick skin and having hairs and the mouths of many glands. The bony part is lined with a thin epithelium which adheres to the periosteum and also forms the outermost layer of the tympanic membrane.
3. The obliquity of the tympanic membrane which meets the roof of the meatus at an obtuse angle and the floor at an acute one.
4. The middle ear or tympanic cavity, extending above the level of the drum as the epitympanic recess, and the recess extending laterally above the bony meatus.
5. The tympanic cavity widest above, narrow below, and narrowest at the level of the umbo where the membrane is indrawn and faces the promontory of the cochlea.
6. The thin shell of bone covering the facial nerve. The grooved anterior crus of the stapes and the anterior half of its base closing the fenestra vestibuli. The long axis of the stapes inclined upward and medially—not lying horizontally.
7. The lateral canal, above the facial nerve (Fig. 7-147).

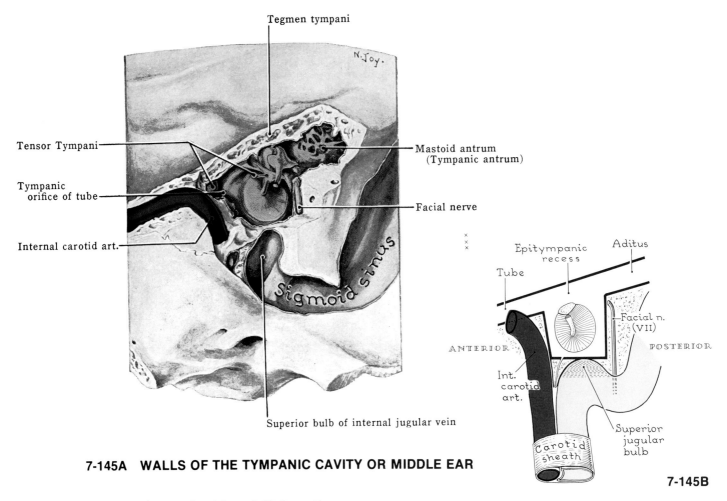

Tegmen tympani

N. Joy.

Tensor Tympani

Tympanic orifice of tube

Internal carotid art.

Mastoid antrum (Tympanic antrum)

Facial nerve

Sigmoid sinus

Superior bulb of internal jugular vein

7-145A   WALLS OF THE TYMPANIC CAVITY OR MIDDLE EAR

Epitympanic recess

Aditus

Tube

Facial n. (VII)

ANTERIOR

POSTERIOR

Int. carotid art.

Superior jugular bulb

Carotid sheath

7-145B

The specimen was dissected with a drill from the medial aspect. The *inset* simplifies the dissection to emphasize the relations of the tympanic cavity.

Observe:

1. The tegmen tympani forming the roof of the tube, tympanic cavity, and antrum, here fairly thick but commonly papery in thinness.

2. The internal carotid artery as the main feature of the anterior wall; the internal jugular vein the main feature of the floor; and the facial nerve the main feature of the posterior wall.

3. The superolateral part of the anterior wall leading to the auditory tube and Tensor Tympani; the superolateral part of the posterior wall leading to the mastoid antrum.

4. The tympanic membrane forming much of the lateral wall. Above it is the epitympanic recess in which are housed the greater parts of the malleus and incus.

See: Donaldson, J. A. and Anson, B. J. (1974) Surgical anatomy of the facial nerve. *Otolaryng. Clin. N. Am.*, 7: 289.

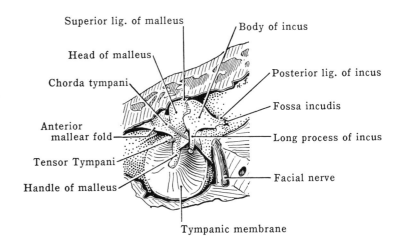

Superior lig. of malleus

Body of incus

Head of malleus

Chorda tympani

Anterior mallear fold

Tensor Tympani

Handle of malleus

Posterior lig. of incus

Fossa incudis

Long process of incus

Facial nerve

Tympanic membrane

7-146   MIDDLE EAR

Observe:

1. A ligament suspending the malleus from the tegmen tympani. The short process of the incus moored by ligaments to the fossa incudis in the floor of the aditus to the antrum.

2. The chorda tympani passing between incus and malleus, carrying taste fibers from the anterior two-thirds of the tongue and parasympathetic fibers to the submandibular ganglion.

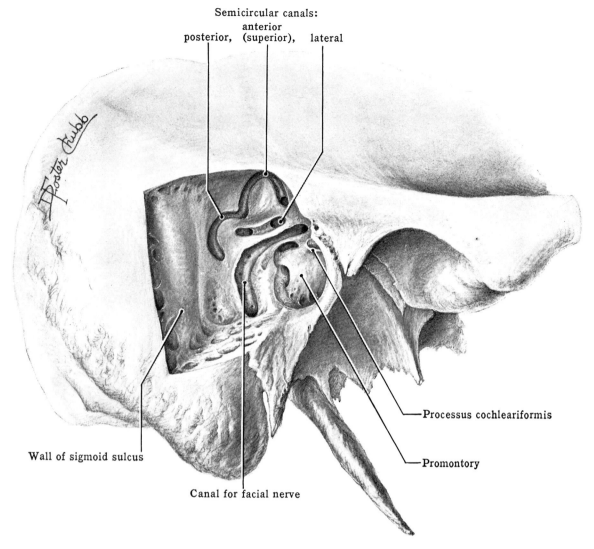

Semicircular canals:
anterior
posterior,   (superior),   lateral

Wall of sigmoid sulcus

Canal for facial nerve

Processus cochleariformis

Promontory

7-147 SEMICIRCULAR CANALS AND MEDIAL WALL OF TYMPANIC CAVITY, LATERAL VIEW

The mastoid cells, seen in coronal section in Figure 7-161, are largely re-moved and the inner table of the skull, here forming the semitubular wall of the sigmoid sulcus, is exposed. The posterior wall of the external acoustic meatus and the mastoid antrum are removed.

Observe:
1. The 3 semicircular canals, lying in the 3 planes of space, are opened.
2. The anterior and posterior canals, placed vertically and at right angles to each other. The lateral canal, placed horizontally and at right angles to the anterior and posterior canals.
3. The medial end of the anterior canal and the upper end of the posterior canal, uniting to form the crus commune. The lateral canal lying in the medial wall of the aditus ad antrum.
4. The features of the medial or labyrinthine wall of the tympanic cavity:
   a. The promontory, lying 2 mm deep to the umbo, and overlying the basal turn of the cochlea.
   b. The processus cochleariformis, at the end of the canal for Tensor Tympani. It acts as a pulley for the Tensor.
   c. The fenestra vestibuli, close behind the pulley and medial to it. It is closed by the footplate of the stapes, which is bound to its margin by an annular ligament.
   d. The fossula leading to the fenestra cochlea, below the vestibular win-dow and separated from it by a rounded bar which projects backward from the promontory.
   e. The facial canal (opened) running horizontally backward, between the vestibular window and the lateral semicircular canal, to the junction of the medial and posterior walls, then descending in the posterior wall to its orifice, the stylomastoid foramen.

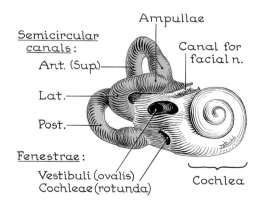

Ampullae

Semicircular canals:
Ant. (Sup.)
Lat.
Post.

Canal for facial n.

Fenestrae:
Vestibuli (ovalis)
Cochleae (rotunda)

Cochlea

7-148 BONY INNER EAR, LATERAL VIEW

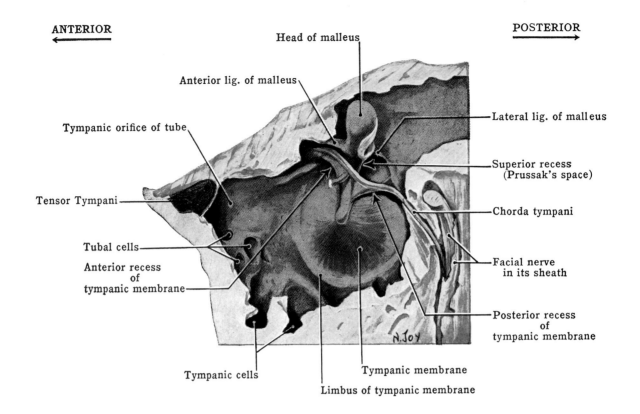

Head of malleus

Anterior lig. of malleus

Tympanic orifice of tube

Tensor Tympani

Tubal cells

Anterior recess of tympanic membrane

Tympanic cells

Limbus of tympanic membrane

Tympanic membrane

Lateral lig. of malleus

Superior recess (Prussak's space)

Chorda tympani

Facial nerve in its sheath

Posterior recess of tympanic membrane

N. Joy

7-149 LATERAL WALL OF THE TYMPANIC CAVITY, MEDIAL VIEW

Observe:
1. The oval tympanic membrane, with a greater vertical than horizontal diameter (9 mm × 8 mm).
2. The handle of the malleus incorporated in the membrane, its end being at the umbo.
3. The anterior process of the malleus anchored forward by the anterior ligament.
4. The facial nerve within its tough periosteal tube. The chorda tympani leaving the facial nerve, and (a) lying within 2 crescentic folds of mucous membrane, (b) crossing the neck of the malleus above the tendon of Tensor, and (c) following the anterior process and anterior ligament.
5. The 3 recesses of the membrane: anterior, posterior, and superior.
6. The fibro-cartilaginous margin or limbus of the membrane which fastens it to the sulcus in the tympanic bone.

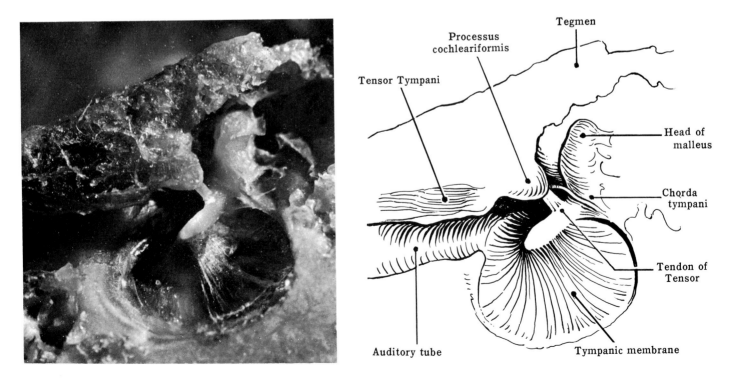

Processus cochleariformis

Tegmen

Tensor Tympani

Head of malleus

Chorda tympani

Tendon of Tensor

Auditory tube

Tympanic membrane

7-150 TENDON OF TENSOR TYMPANI PASSING FROM MEDIAL TO LATERAL WALL

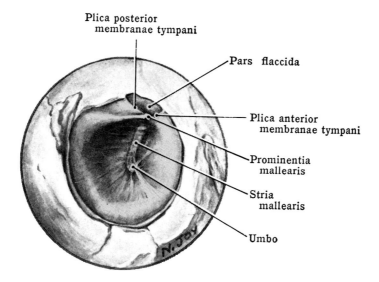

Plica posterior
membranae tympani

Pars flaccida

Plica anterior
membranae tympani

Prominentia
mallearis

Stria
mallearis

Umbo

7-151 TYMPANIC MEMBRANE, LATERAL VIEW

Observe:

1. The tympanic membrane, oval rather than round, and shaped like a funnel with rolled rim and a depressed part, called the umbo, at the tip of the handle of the malleus which is situated antero-inferior to the center of the membrane.
2. The stria mallearis, which overlies the handle of the malleus, extending upward to the prominentia, which overlies the lateral process of the malleus.
3. Above the prominentia the membrane is thin and is called the pars flaccida. The pars flaccida lacks the radial and circular fibers present in the remainder of the membrane (pars tensa). The junction between the two parts, flaccid and tense, is marked by an anterior and a posterior line which run from the prominentia to the free ends of the horseshoe-shaped tympanic ring (Fig. 10-10).
4. The pars flaccida forms the lateral wall of the superior recess (Fig. 7-149) of the tympanic cavity.

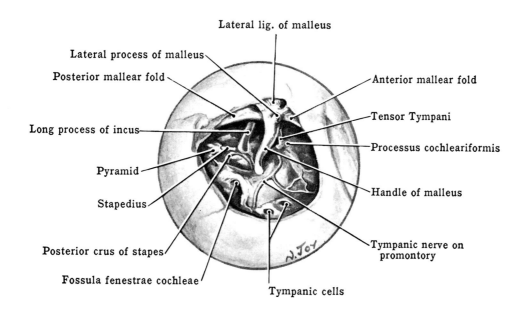

Lateral lig. of malleus

Lateral process of malleus

Posterior mallear fold

Long process of incus

Pyramid

Stapedius

Posterior crus of stapes

Fossula fenestrae cochleae

Anterior mallear fold

Tensor Tympani

Processus cochleariformis

Handle of malleus

Tympanic nerve on promontory

Tympanic cells

7-152 TYMPANIC CAVITY AFTER REMOVAL OF THE TYMPANIC MEMBRANE, INFERO-LATERAL VIEW

Observe:

1. The direction of the handle of the malleus and of the long process of the incus which lies behind it. The posterior and anterior mallear folds of mucous membrane in which the chorda tympani passes between the two bones (Fig. 7-149).
2. The fullness of the promontory with grooves for the tympanic nerve (a branch of the glossopharyngeal nerve) and its connections.
3. The end of a fossula at the deep end of which is the fenestra cochleae or round window, closed by the secondary tympanic membrane (not in view).
4. The tendon of Stapedius passing forward to the neck of the stapes; the tendon of Tensor Tympani passing laterally to the neck of the malleus.
5. The lateral ligament of the malleus and the neck of the malleus forming the medial wall of the superior recess of the tympanic membrane.

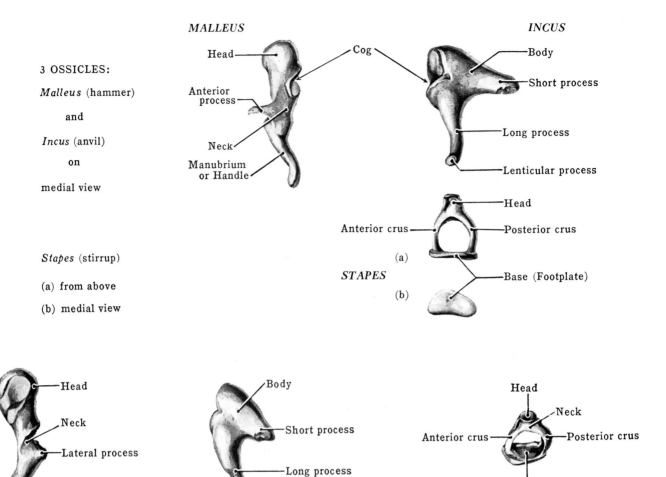

3 OSSICLES:

Malleus (hammer)

and

Incus (anvil)

on

medial view

Stapes (stirrup)

(a) from above

(b) medial view

MALLEUS

Head — Cog

Anterior process

Neck

Manubrium or Handle

INCUS

Body

Short process

Long process

Lenticular process

Head

Anterior crus — Posterior crus

(a)

STAPES

(b)

Base (Footplate)

Head

Neck

Lateral process

Handle (Manubrium)

MALLEUS
on postero-medial view

Body

Short process

Long process

Lenticular process

INCUS
on postero-medial view

Head

Neck

Anterior crus — Posterior crus

Base (Footplate)

STAPES
on supero-lateral view

7-153 OSSICLES OF THE MIDDLE EAR

Observe:
1. The head of the malleus and the body and short process of the incus lie in the epitympanic recess.
2. The saddle-shaped articular surface of the head of the malleus and the reciprocally saddle-shaped articular surface of the body of the incus form the incudo-mallear synovial joint.
3. The anterior process of the malleus and the short process of the incus (it might better have been called the posterior process) are in line and are moored fore and aft by ligaments.
4. The handle of the malleus, from lateral process to tip, is embedded in the tympanic membrane.
5. The end of the long (vertical) process of the incus has a convex articular facet for articulation with the head of the stapes, at the incudo-stapedial synovial joint.
6. The hole in the stapes in the embryo transmits an artery, the stapedial artery. It is now closed by an obturator. The upper border of the footplate is convex and is deeper anteriorly than posteriorly. The two crura are grooved. The anterior crus is the more slender and straighter and it is fixed to a small area on the plate. The posterior crus is attached to the whole depth of the plate.

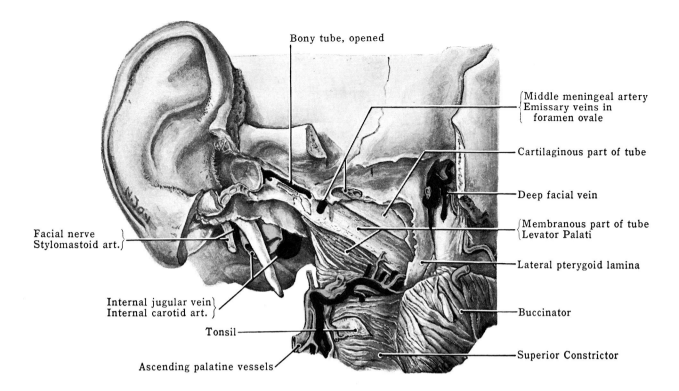

Bony tube, opened

Middle meningeal artery
Emissary veins in
foramen ovale

Cartilaginous part of tube

Deep facial vein

Membranous part of tube
Levator Palati

Lateral pterygoid lamina

Buccinator

Superior Constrictor

Facial nerve
Stylomastoid art.

Internal jugular vein
Internal carotid art.

Tonsil

Ascending palatine vessels

7-154   AUDITORY TUBE (PHARYNGO-TYMPANIC TUBE), LATERAL VIEW

Note:
1. Tensor Palati has been removed.
2. The tonsil, in this specimen, bulging through the Superior Constrictor.
3. The cartilaginous part of the tube resting on a spine on the medial pterygoid lamina; the membranous part "resting on" Levator Palati.
4. Tube, Levator, and vessels crossing the upper border of Superior Constrictor.
5. Emissary veins from the cavernous sinus in the foramen ovale and the deep facial vein connecting the maxillary and facial veins.

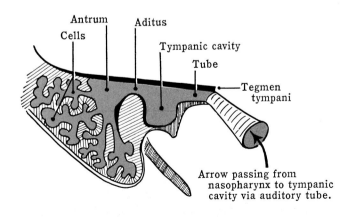

Antrum
Cells
Aditus
Tympanic cavity
Tube
Tegmen tympani

Arrow passing from nasopharynx to tympanic cavity via auditory tube.

7-155   DIAGRAM OF TEGMEN TYMPANI

The mastoid air cells are in communication with the outside air via the mastoid antrum, aditus ad antrum, tympanic cavity, and bony and cartilaginous parts of the auditory tube.

A thin plate of bone, called the tegmen tympani, forms a roof for these.

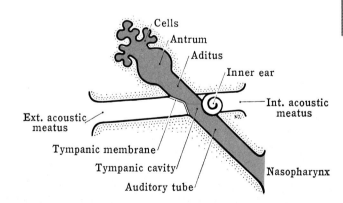

Cells
Antrum
Aditus
Inner ear
Int. acoustic meatus
Ext. acoustic meatus
Tympanic membrane
Tympanic cavity
Auditory tube
Nasopharynx

7-156   SCHEME OF MEATUSES AND AIRWAY

This illustrates that the line of the external and internal meatus intersects at the tympanic cavity with the line of the airway from mastoid cells to nasopharynx.

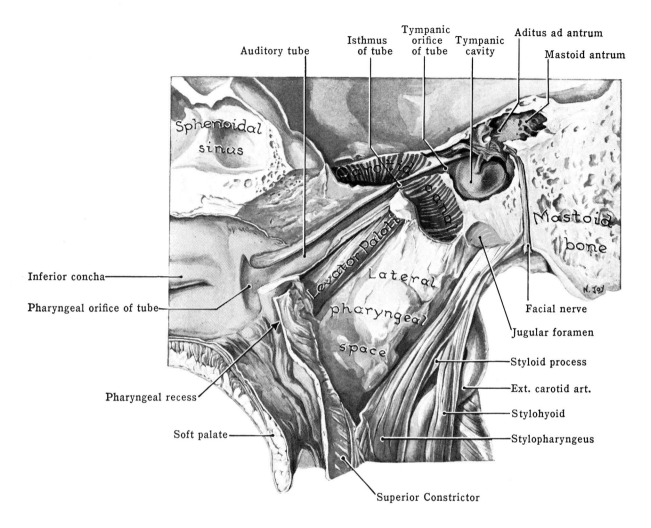

Labels on figure:
Auditory tube · Isthmus of tube · Tympanic orifice of tube · Tympanic cavity · Aditus ad antrum · Mastoid antrum

Sphenoidal sinus

Inferior concha

Pharyngeal orifice of tube

Levator Palati

Lateral pharyngeal space

Mastoid bone

Facial nerve

Jugular foramen

Pharyngeal recess

Soft palate

Styloid process

Ext. carotid art.

Stylohyoid

Stylopharyngeus

Superior Constrictor

N. 19*

7-157 AUDITORY TUBE (PHARYNGO-TYMPANIC TUBE), EXPOSED FROM THE MEDIAL OR PHARYNGEAL ASPECT

Observe:
1. The general direction of the tube—upward, backward, and laterally from nasopharynx to tympanic cavity.
2. The funnel-shaped pharyngeal orifice of the tube, situated 1 cm behind the inferior concha of the nose.
3. The cartilaginous part of the tube, 2.5 cm long, resting throughout its length on Levator Palati, but affording it almost no origin.
4. The bony part of the tube passing lateral to the carotid canal, about 1 cm long, narrow at the isthmus where it joins the cartilaginous part, wider at its tympanic orifice, and less steep than the cartilaginous part.
5. Tensor Tympani, lying above a bony ledge, called the processus cochleariformis, and inserted into the neck of the malleus.
6. The chorda tympani lying in a "mesentery," the anterior and posterior mallear folds, and the anterior and posterior recesses of the tympanic membrane lateral to the respective folds.
7. The anterior mallear fold acting as a mesentery for Tensor Tympani also, and continuous with a fold that passes forward from the head of the malleus.
8. The upper half of the lateral pharyngeal space, seen on cross-section in Figures 7-82 and 7-105.

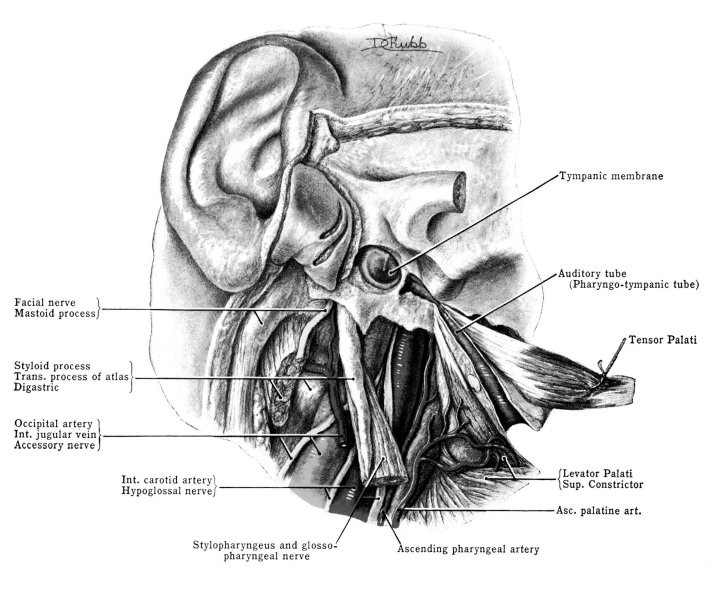

Tympanic membrane

Auditory tube
(Pharyngo-tympanic tube)

Tensor Palati

Facial nerve
Mastoid process

Styloid process
Trans. process of atlas
Digastric

Occipital artery
Int. jugular vein
Accessory nerve

Levator Palati
Sup. Constrictor

Asc. palatine art.

Int. carotid artery
Hypoglossal nerve

Stylopharyngeus and glossopharyngeal nerve

Ascending pharyngeal artery

7-158   AUDITORY TUBE (PHARYNGO-TYMPANIC TUBE)

Note:
1. The anterior wall of the cartilaginous part of the external acoustic meatus has two fissures—two unchondrified areas—which are closed with fibrous tissue.
2. The anterior wall of the bony part of the meatus has been ground away.
3. The tympanic membrane lies at the bottom of the meatus. It faces laterally, downward, and forward as though to catch sounds reflected from the ground, as one advances. The central part of the membrane is indrawn (toward the tympanic cavity), the umbo or bottom of the concavity being at the end of the handle of the malleus. The peripheral part of the membrane is thickened at its attachment to the tympanic bone.
4. The auditory tube has been opened from pharyngeal end to tympanic end by excising its lateral membranous wall, but Tensor Palati, which arises in part from the tube, was first reflected. The tympanic segment of the tube is short and bony; the pharyngeal segment is long and cartilaginous. The two segments meet at an angle, and there the tube is narrowest. The lateral wall of the pharyngeal or cartilaginous segment of the tube is, in reality, fibrous; this wall has been excised.
5. The ascending palatine branch of the facial artery ascends on the outer surface of Superior Constrictor to its free upper border and then descends on its medial surface with Levator Palati.
6. The last 4 cranial nerves are close together between the internal jugular vein and the internal carotid artery. The vagus, however, is concealed by the hypoglossal nerve.
7. In this specimen, the stylo-mastoid artery (not labeled), which accompanies the facial nerve, springs from the occipital artery, not the posterior auricular artery as in Figure 7-67.

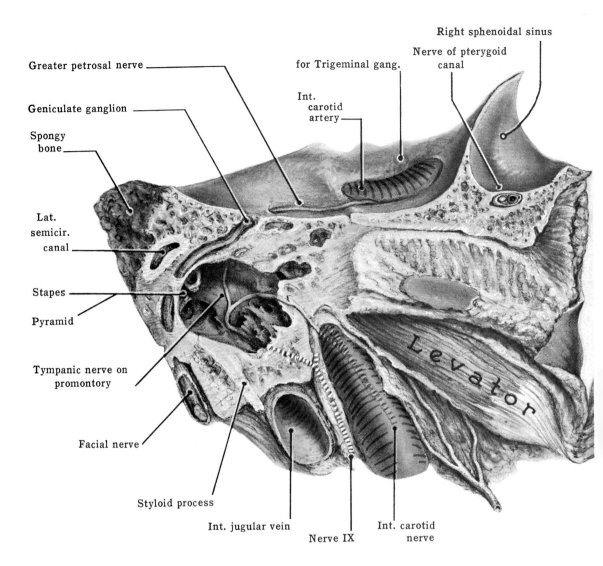

Greater petrosal nerve

Geniculate ganglion

Spongy bone

Lat. semicir. canal

Stapes

Pyramid

Tympanic nerve on promontory

Facial nerve

Styloid process

Int. jugular vein

Nerve IX

Int. carotid nerve

for Trigeminal gang.

Int. carotid artery

Right sphenoidal sinus

Nerve of pterygoid canal

Levator

7-159 AUDITORY TUBE AND TYMPANIC CAVITY, RIGHT SIDE

The cut surfaces of this longitudinally split specimen is shown on these two facing pages. The procedure used was modified after Laurenson, R. D. (1965) A rapid method of dissecting the middle ear. *Anat. Rec., 151:* 503.

The squamous and mastoid parts of the temporal bone are sawn across coronally from suprameatal spine (Fig. 7-7), through the mastoid antrum, into the posterior cranial fossa. The posterior part of the bone is then discarded.

The thin roof (tegmen) of the antrum ad aditus (Fig. 7-155) is nibbled away until the incus comes into view (Fig. 7-163). The *incus* is now picked from its articulation with *malleus* laterally and *stapes* medially.

A probe, passed from the pharynx up the auditory tube, until arrested at the isthmus, will serve as a directional guide.

Identify the *internal carotid artery* medially, beneath the trigeminal ganglion (Figs. 7-40 and 7-43) at the foramen lacerum, and the *middle meningeal artery* laterally, at the foramen spinosum (Fig. 7-39).

A saw cut from the gap left by the incus to the space between the two arteries (carotid and meningeal) will pass between the *greater and lesser petrosal nerves,* being parallel, and continue into the tube.

The incus having been removed, the only structure that crosses the path of the saw is

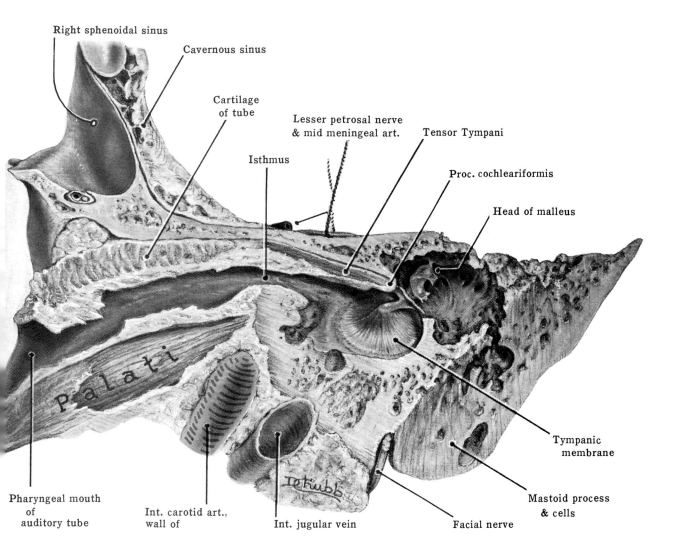

Right sphenoidal sinus

Cavernous sinus

Cartilage
of tube

Lesser petrosal nerve
& mid meningeal art.

Tensor Tympani

Isthmus

Proc. cochleariformis

Head of malleus

Palati

Tympanic
membrane

Pharyngeal mouth
of
auditory tube

Int. carotid art.,
wall of

Int. jugular vein

Facial nerve

Mastoid process
& cells

SPLIT LONGITUDINALLY INTO LATERAL AND MEDIAL PARTS

Tensor Tympani tendon, which passes from medial to lateral wall. In this specimen a shaving of the medial wall (containing the fleshy Tensor in its semicanal and the processus cochleariformis) was included with the lateral part, leaving the tendon intact.

The lateral wall of the cavity is dominated by the tympanic membrane, handle of malleus and chorda, tympani nerve (Fig. 7-149).

The medial wall has a broad bulging, the promontory, which overlies the 1st turn of the cochlea (Figs. 7-147 and 7-165). On it the tympanic nerve (Fig. 8-9) and carotico-tympanic branches of the internal carotid nerve (Fig. 7-44) form the tympanic plexus, which supplies the neighborhood and gives off the lesser petrosal nerve (Fig. 8-9).

Structures, divided and seen on both medial and lateral parts:
a. Levator Palati, supporting the tube.
b. Auditory tube, cartilaginous above and medially; membranous below and laterally (Fig. 7-157).
c. Right sphenoidal sinus, with the pterygoid canal below it.
d. Internal carotid artery.
e. Internal jugular vein.
f. Facial nerve.
g. A petrosal nerve, either greater or lesser.

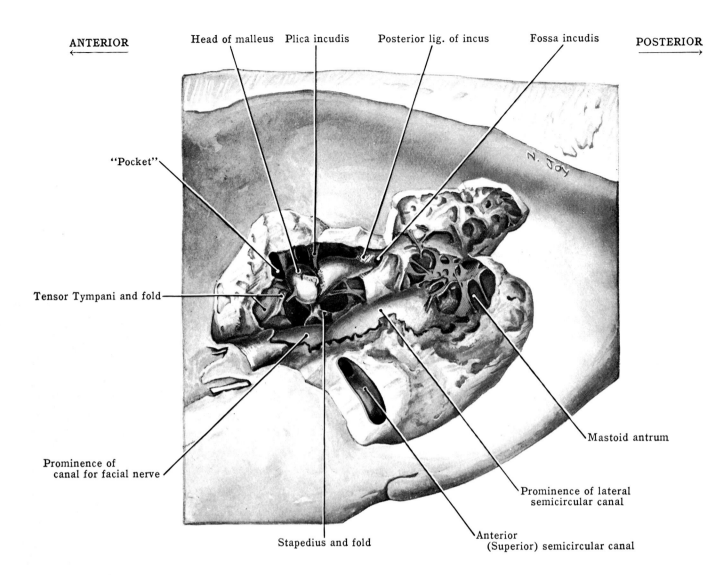

ANTERIOR · Head of malleus · Plica incudis · Posterior lig. of incus · Fossa incudis · POSTERIOR

"Pocket"

Tensor Tympani and fold

Prominence of
canal for facial nerve

Stapedius and fold

Anterior
(Superior) semicircular canal

Prominence of lateral
semicircular canal

Mastoid antrum

7-160 TYMPANIC CAVITY AND MASTOID ANTRUM, FROM ABOVE

The bony roof, or tegmen tympani, has been removed with the aid of an electric drill.

Observe:

1. Extensive folds, strands, "mesenteries," and pockets of mucous membrane.
2. The mesentery for Tensor Tympani (commonly perforated) and the mesentery for Stapedius and stapes.
3. The head of the malleus and the body and short crus of the incus in the epitympanic recess. The short process of the incus moored by two ligamentous bands to the sides of the fossa incudis on the floor of the aditus ad antrum.
4. The strand from the body of the incus to the lateral wall is commonly an extensive fold as in Figure 7-163 and the result is a pocket. The superior ligament of the malleus (not labeled) is cut short.

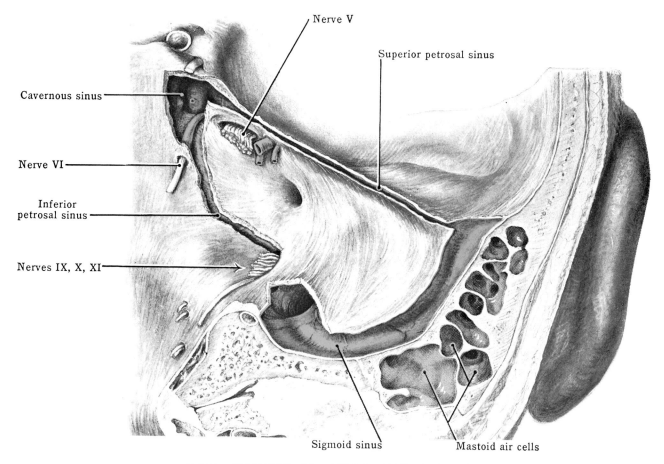

Nerve V

Superior petrosal sinus

Cavernous sinus

Nerve VI

Inferior petrosal sinus

Nerves IX, X, XI

Sigmoid sinus

Mastoid air cells

7-161 MASTOID AIR CELLS—DURAL SINUSES

Observe:

1. Mastoid cells (*pink*), lined with mucous membrane, and occupying the diploe (Fig. 7-155). The branching ducts through which these cells communicate with the antrum. The two lowest cells almost bursting through the outer table of the skull.

2. The posterior surface of the petrous bone encircled with 3 sinuses—sigmoid, superior petrosal, and inferior pe-

trosal. The two petrosal sinuses draining the cavernous sinus.

3. The superior petrosal sinus, at the attached margin of the tentorium, bridging nerve V. The sigmoid sinus, nerves IX, X, and XI, and the inferior petrosal sinus disappearing into the jugular foramen.

4. Nerve VI, passing through the inferior petrosal sinus and bending forward to enter the cavernous sinus (Fig. 7-43).

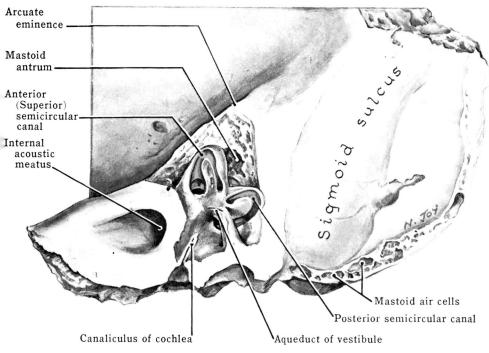

Arcuate eminence

Mastoid antrum

Anterior (Superior) semicircular canal

Internal acoustic meatus

Sigmoid sulcus

N. Joy

Mastoid air cells

Posterior semicircular canal

Canaliculus of cochlea

Aqueduct of vestibule

Observe:

1. The anterior semicircular canal, set vertically below the arcuate eminence, and making a right angle with the posterior surface of the petrous bone.

2. The posterior semicircular canal, nearly parallel to the posterior surface of the bone, close to that surface, and only 5 mm from the sigmoid sulcus.

3. The aqueduct of the vestibule, which contains the ductus endolymphaticus and opens medial to the posterior canal.

4. The canaliculus of the cochlea, which contains the perilymphatic duct (aqueduct of cochlea) and opens at the apex of the depression for the ganglion of nerve IX.

See: Young, M. W. (1952) The termination of the perilymphatic duct. *Anat. Rec.,* 112: 404.

7-162 SEMICIRCULAR CANALS AND THE AQUEDUCTS, POSTERO-SUPERIOR VIEW

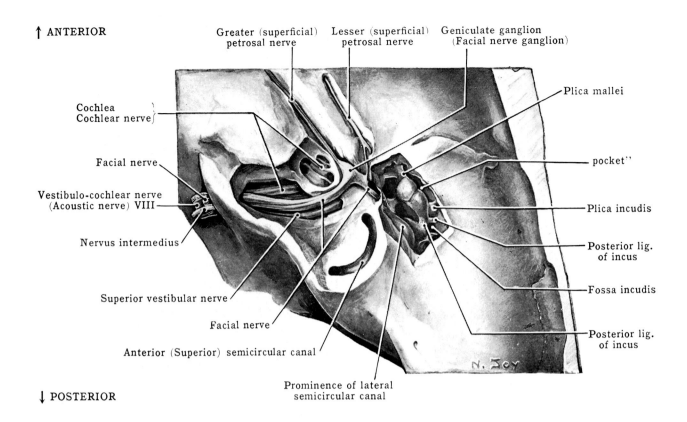

↑ ANTERIOR

Greater (superficial) petrosal nerve

Lesser (superficial) petrosal nerve

Geniculate ganglion (Facial nerve ganglion)

Cochlea
Cochlear nerve

Facial nerve

Vestibulo-cochlear nerve (Acoustic nerve) VIII

Nervus intermedius

Superior vestibular nerve

Facial nerve

Anterior (Superior) semicircular canal

Plica mallei

pocket"

Plica incudis

Posterior lig. of incus

Fossa incudis

Posterior lig. of incus

Prominence of lateral semicircular canal

↓ POSTERIOR

7-163 GENICULATE GANGLION, FROM ABOVE

Observe:
1. The facial nerve, the nervus intermedius, and the vestibulocochlear nerve, entering and traversing the internal acoustic meatus. The facial nerve, joined by the nervus intermedius, running close behind the cochlea and, therefore, across the roof of the vestibule (Fig. 7-165) to the geniculate ganglion and at the ganglion making a right angle bend, called the genu, and then curving downward and backward within the bony facial canal, whose papery lateral wall separates it from the tympanic cavity.
2. The petrosal branch of the middle meningeal artery,

which enters the canal at the hiatus (Fig. 7-41), running with the nerve.
3. The geniculate ganglion, which is the cell station of fibers of general sensation and of taste (Fig. 8-7), situated at the genu and in line with the internal acoustic meatus. Through the ganglion run forward fibers of the greater (superficial) petrosal nerve on their way to the pterygo-palatine ganglion. From the facial nerve, beyond the ganglion, goes a communicating branch to the lesser (superficial) petrosal nerve on its way to the otic ganglion. Further on, but not in view, the chorda tympani leaves the facial nerve and joins the lingual which conducts it to the submandibular ganglion.

MEDIAL

LATERAL

for Facial nerve

Crista transversa

for Cochlear nerve

To ⎰ Utricle
⎱ Anterior ampulla
⎱ Lateral ampulla ⎰ via Superior vestibular nerve

To Saccule

To Posterior ampulla ⎱ via Inferior vestibular nerve

7-164 FUNDUS OF THE INTERNAL ACOUSTIC MEATUS

In this specimen the walls of the meatus have been ground away.

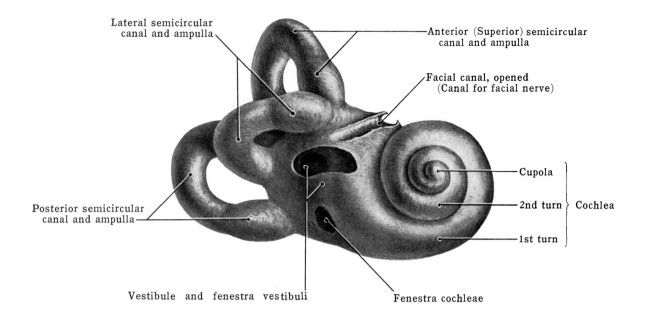

Lateral semicircular
canal and ampulla

Anterior (Superior) semicircular
canal and ampulla

Facial canal, opened
(Canal for facial nerve)

Cupola

2nd turn } Cochlea

1st turn

Posterior semicircular
canal and ampulla

Vestibule and fenestra vestibuli

Fenestra cochleae

7-165 BONY LABYRINTH, LATERAL VIEW, RIGHT SIDE

Observe:

1. The 3 parts of the bony internal ear or bony labyrinth: cochlea, in front; vestibule in the middle; semicircular canals, behind.
2. The 2½ turns or coils of the cochlea. The first or basal coil, which lies deep to the medial wall of the tympanic cavity, communicating with the tympanic cavity through the fenestra cochleae (round window). In life this fenestra is closed by the secondary tympanic membrane.
3. The vestibule, crossed above by the facial canal and communicating with the tympanic cavity through the fenestra vestibuli

(oval window). In life this window is closed by the base or footpiece of the stapes.

4. The 3 semicircular canals—anterior, posterior, and lateral. The anterior and posterior canals set vertically at a right angle to each other; the lateral canal set horizontally and at a right angle to the two others. Each canal, forming about two-thirds of a circle, and each having an ampulla at one end. The lateral canal is the shortest; the posterior canal is the longest. (See Figs. 7-147 and 7-162.)

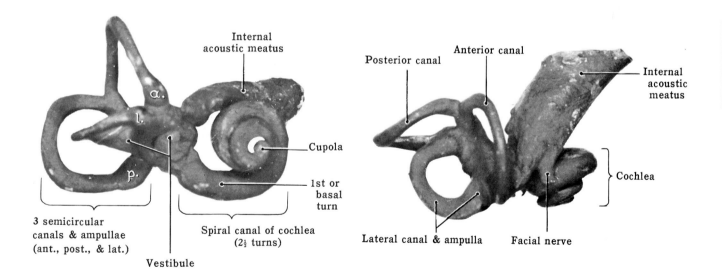

Internal
acoustic meatus

a.

l.

p.

Cupola

1st or
basal
turn

3 semicircular
canals & ampullae
(ant., post., & lat.)

Spiral canal of cochlea
(2½ turns)

Vestibule

Posterior canal

Anterior canal

Internal
acoustic
meatus

Cochlea

Lateral canal & ampulla

Facial nerve

7-166 PLASTIC CAST OF INTERIOR OF BONY LABYRINTH, LATERAL VIEW AND FROM ABOVE

Note:

1. The length of this cast, from the anterior end of the cochlea to the posterior end of the posterior semicircular canal is 18 mm.
2. The casts of the semicircular canals are flattened, or compressed, from side to side.
3. Each of the three canals has two ends—a simple and an ampullary (or dilated). These open into the vestibule by 5 openings, the simple ends of the two vertical canals having a common crus.

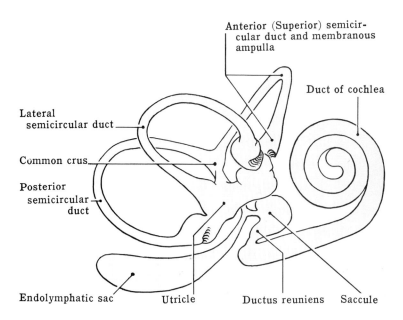

Anterior (Superior) semicircular duct and membranous ampulla

Duct of cochlea

Lateral semicircular duct

Common crus

Posterior semicircular duct

Endolymphatic sac

Utricle

Ductus reuniens

Saccule

Note:
1. The membranous labyrinth or membranous internal ear is contained within the bony labyrinth or bony internal ear. It is a closed system of ducts and chambers, filled with endolymph and surrounded with, or bathed in, perilymph.
2. It has 3 parts—the duct of the cochlea, within the cochlea; the saccule and the utricle, within the vestibule; and the 3 semicircular ducts, within the 3 semicircular canals.
3. One end of the duct of the cochlea is closed; the other end communicates with the saccule through the ductus reuniens.
4. The saccule in turn communicates with the utricle through the utriculo-saccular duct (not labeled). From this duct springs the endolymphatic duct, which occupies the aqueduct of the vestibule and ends in the endolymphatic sac (Fig. 7-142). The 3 semicircular ducts have 5 openings into the utricle.

7-167 MEMBRANOUS LABYRINTH, LATERAL VIEW, RIGHT SIDE

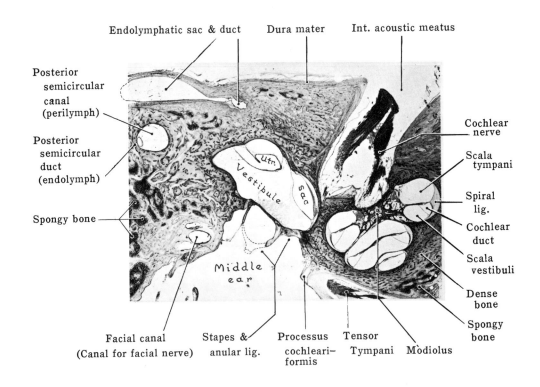

Endolymphatic sac & duct

Dura mater

Int. acoustic meatus

Posterior semicircular canal (perilymph)

Posterior semicircular duct (endolymph)

Spongy bone

Utr.

Vestibule

Sac

Middle ear

Cochlear nerve

Scala tympani

Spiral lig.

Cochlear duct

Scala vestibuli

Dense bone

Spongy bone

Facial canal (Canal for facial nerve)

Stapes & anular lig.

Processus cochleariformis

Tensor Tympani

Modiolus

7-168 LABYRINTH, OR INNER EAR, ON CROSS-SECTION

Observe:
1. At the *top*, the endolymphatic sac lying extradurally.
2. The stapes is broken. Its anular ligament, extending from its basis (footplate) to the fenestra vestibuli, is short behind and long in front, so it moves more like a door than a piston.
3. The utricle, pulled away from its recess in the vestibule. So is the saccule, but to a lesser extent.
4. The modiolus, which is the central bony core of the cochlea, resembles a screw nail; its spiral bony lamina (or thread, Fig. 7-163) is attached by the spiral membrane (basilar membrane) to the peripherally placed spiral ligament.
5. The cochlear duct, triangular on cross-section, has for its sides

(a) spiral membrane, (b) spiral ligament, and (c) vestibular ligament which is delicate.
6. Above the cochlear duct (scala media) is the scala vestibuli, which leads off the vestibule; below it is the scala tympani, which leads to the fenestra cochleae which is closed by the secondary tympanic membrane.
7. The cochlear duct and 3 semicircular ducts are attached to the convex sides of their respective bony canals; that is, they are not completely surrounded with perilymph, as is commonly represented in diagrams.
8. The bony labyrinth is composed of dense bone largely embedded in spongy bone. This makes it possible to define it.

The Cranial Nerves

SECTION 8

CONTENTS

8-1 OUTLINE OF THE CRANIAL NERVES

No.	Name	Special Sense	Sensory	Motor	Parasym-pathetic
I	Olfactory	*			
II	Optic	*			
III	Oculomotor			*	*
IV	Trochlear			*	
V	Trigeminal		*	*	
VI	Abducent			*	
VII	Facial	*	(*)	*	*
VIII	Stato-acoustic	*			
IX	Glosso-pharyngeal	*	*	*	*
X	Vagus	(*)	*	*	*
XI	Accessory			*	
XII	Hypoglossal			*	

Note that there are four modalities which may be carried by cranial nerves. Three nerves carry special sense only (I, II, VIII) and have no motor component. Four nerves (III, VII, IX, and X) carry parasympathetic fibers to smooth muscles and glands.

There are four autonomic ganglia in the head: ciliary, pterygopalatine, otic, and submandibular.

Each receives three types of fibers:
a. Sensory: from a branch of the trigeminal nerve.
b. Parasympathetic: from cranial nerves III, VII, or IX. These nerves synapse in the ganglion.
c. Sympathetic: from the sympathetic trunk, hitchhiking on the wall of the closest artery.

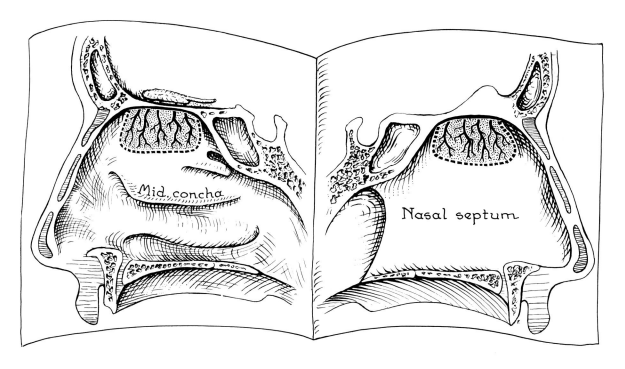

8-2 DISTRIBUTION OF THE OLFACTORY NERVE

CRANIAL NERVE I

In the roof of the nasal cavity, an area of yellowish brown mucous membrane contains the olfactory receptors. From here, 15 to 20 fine bundles of nerve fibers pierce the cribriform plate to enter the anterior cranial fossa and synapse in the olfactory bulb. The olfactory tract passes backward to the brain.

The discovery of unilateral loss of the sense of smell in a patient may indicate a lesion in the anterior cranial fossa. "Olfactory hallucinations" may occur when there is a lesion of the brain in the general area of the uncus.

The olfactory area is usually much smaller than that shown here, and it is irregular in outline as a result of streamerlike invasion by nonolfactory, ciliated, columnar epithelium. The decrease in size is believed to result mainly from the destruction of the sensory olfactory neurons in the course of recurring infections of the nasal mucosa.

A study of the olfactory nerves in 143 adults (over 21 years of age) revealed that only 12 per cent had a full complement of olfactory nerve fibers, that 8 per cent had lost all fibers on one side, and that 5 per cent had lost all fibers on both sides.

There is considerable variation in the number of olfactory nerve fibers in individuals of a given age, but on the average there is a loss of 1 per cent of fibers per year during postnatal life; *i.e.,* at the age of 50 years the average person has lost 50 per cent of fibers and, at the age of 75 years, 75 per cent of fibers.

See Smith, C. G. (1941) Incidence of atrophy of the olfactory nerves in man. *Arch. Otolaryng., 34:* 533.

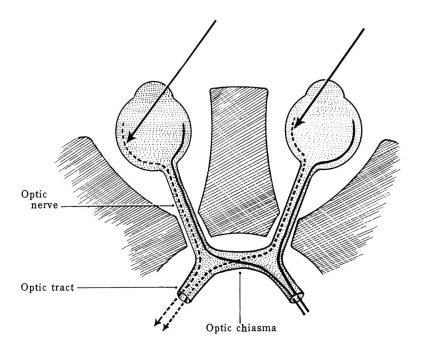

Optic nerve

Optic tract

Optic chiasma

8-3 DISTRIBUTION OF THE OPTIC NERVE

CRANIAL NERVE II

This diagram of a horizontal section through the visual apparatus shows that neurons from the retina of the eyeball travel through the *optic nerve* to the *optic chiasma* where some fibers cross the midline and join the *optic tract* of the opposite side on their way to the visual area of the brain. Note that it is the fibers from the inner or nasal half of the retina which cross over in the chiasma. The *large arrows* represent rays of light from the *right* half of this person's *field of vision* stimulating receptors in the *left* half of the retina of both eyes and so reaching the brain through the *left* optic tract.

Thus a section through the right optic nerve would result in blindness of the right eye; a section through the right optic tract would eliminate vision from left visual fields of both eyes; and a section through the optic chiasma would reduce peripheral vision. Remember that the hypophysis cerebri (pituitary gland) lies just behind the optic chiasma and expansion of this gland by a tumor would put pressure on these crossing over fibers.

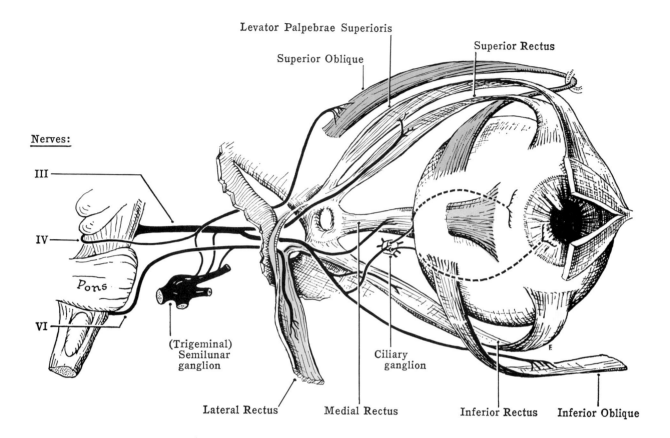

Nerves:

III

IV

Pons

VI

Levator Palpebrae Superioris

Superior Oblique

Superior Rectus

(Trigeminal)
Semilunar
ganglion

Ciliary
ganglion

Lateral Rectus

Medial Rectus

Inferior Rectus

Inferior Oblique

8-4 DISTRIBUTION OF THE OCULOMOTOR, TROCHLEAR, AND ABDUCENT NERVES

CRANIAL NERVES III, IV, VI

These 3 motor nerves, after receiving proprioceptive fibers from the trigeminal nerve, supply the orbital muscles. Nerves IV and VI each supply one muscle and nerve III supplies the remaining five muscles.

The trochlear nerve supplies Superior Oblique—the muscle that passes through a trochlea or pulley; the abducent nerve supplies Lateral Rectus—the muscle that abducts; and the oculomotor nerve supplies Levator Palpebrae Superioris, Superior Rectus, Medial Rectus, Inferior Rectus, and Inferior Oblique. So all three nerves carry fibers which are motor to the striated extraocular muscles.

In addition, the oculomotor nerve carries fibers which are preganglionic, parasympathetic, and motor to smooth muscle. These fibers pass to the ciliary ganglion where they synapse and are distributed via short ciliary nerves to the Sphincter Pupillae (causing constriction of the pupil) and to the ciliary muscle (resulting in a more convex lens.)

Not shown here is the sympathetic contribution to the ciliary ganglion and to the Dilator Pupillae.

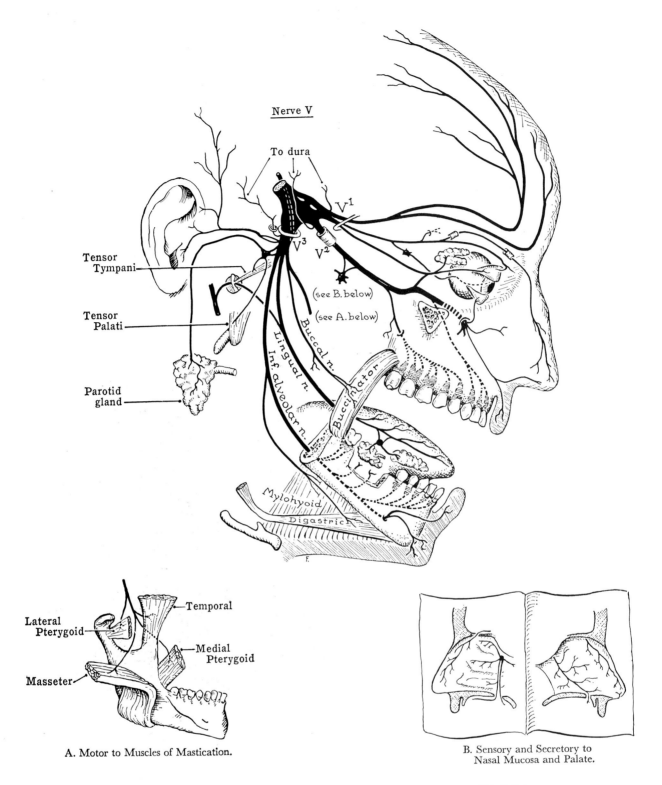

Nerve V

To dura

V¹

V³

V²

(see B. below)

(see A. below)

Tensor Tympani

Tensor Palati

Parotid gland

Buccal n.

Lingual n.

Inf. alveolar n.

Buccinator

Mylohyoid

Digastric

Lateral Pterygoid

Temporal

Medial Pterygoid

Masseter

A. Motor to Muscles of Mastication.

B. Sensory and Secretory to Nasal Mucosa and Palate.

8-5 DISTRIBUTION OF THE TRIGEMINAL NERVE

CRANIAL NERVE V

The trigeminal nerve has three divisions: V¹, the ophthalmic nerve; V², the maxillary nerve; V³, the mandibular nerve.

All three divisions are sensory. Their cutaneous distribution is shown in Figure 7-18. However, each division supplies not the skin surface only, but the whole thickness of tissue from skin to mucous membrane. Each of the three divisions sends a twig to the dura mater: V¹ to the tentorium cerebelli, V² and V³ to the floor and side wall of the middle cranial fossa. Each of the three divisions provides the sensory component to an autonomic ganglion: V¹ to the ciliary, V² to the pterygopalatine, and V³ to the submandibular and otic.

In addition to its sensory component, V³ is motor to four pairs of muscles:
a. the two large elevators of the mandible, Temporalis and Masseter.
b. the two Pterygoid muscles, Medial and Lateral.
c. the two Tensors, Palati and Tympani.
d. the two muscles of the floor of the mouth, Mylohyoid and anterior belly of Digastric.

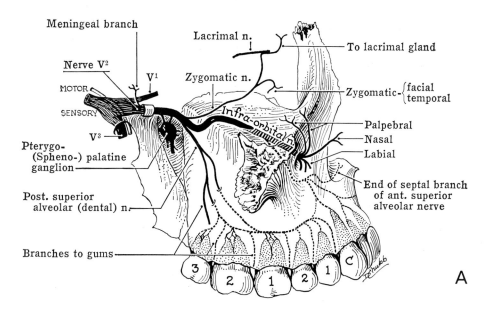

Figure labels (A):
- Meningeal branch
- Nerve V²
- MOTOR
- SENSORY
- V¹
- V³
- Lacrimal n.
- Zygomatic n.
- Infra-orbital n.
- To lacrimal gland
- Zygomatic-{facial temporal
- Palpebral
- Nasal
- Labial
- End of septal branch of ant. superior alveolar nerve
- Pterygo- (Spheno-) palatine ganglion
- Post. superior alveolar (dental) n.
- Branches to gums
- 3 2 1 2 1 C

A

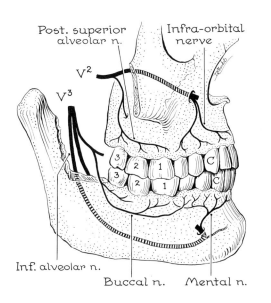

Figure labels (B):
- Post. superior alveolar n.
- Infra-orbital nerve
- V²
- V³
- 3 2 1 C / 3 2 1 C
- Inf. alveolar n.
- Buccal n.
- Mental n.

B

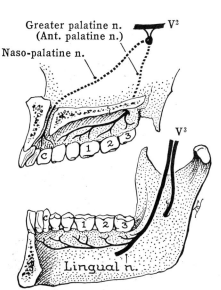

Figure labels (C):
- Greater palatine n. (Ant. palatine n.)
- V²
- Naso-palatine n.
- V³
- C 1 2 3
- c 1 2 3
- Lingual n.

C

8-6 ADDITIONAL DIAGRAMS OF V² AND V³

The ophthalmic nerve (V¹) is sensory: (a) to the eyeball and cornea via the ciliary nerves (Fig. 7-53), hence, if paralyzed, the ocular conjunctiva is insensitive to touch; (b) to the frontal, ethmoidal, and sphenoidal air sinuses via the supra-orbital and ethmoidal nerves; and (c) to the skin and conjunctival surfaces of the upper eyelid and to the skin and mucous surfaces of the external nose (Fig. 7-109).

The maxillary nerve (V²) is sensory: (a) to the upper teeth and gums (see below); (b) to the face, both surfaces of the lower lid, the skin of the side and vestibule of the nose, and both surfaces of the upper lip (Fig. 7-17); (c) via the pterygopalatine ganglion (Fig. 7-109) to the mucoperiosteum of the nasal cavity, palate, and roof of the pharynx; and (d) to the maxillary, ethmoidal, and sphenoidal air sinuses; (e) secretory fibers from this ganglion pass with the zygomatic and then with the lacrimal nerve to the lacrimal gland (Fig. 8-6).

The mandibular nerve (V³) is motor: (a) to the 4 muscles of mastication—but not to Buccinator; (b) to the 2 Tensors (Tympani and Palati) via the otic ganglion; and (c) to Mylohyoid and anterior belly of Digastric. It is sensory: (a) to the lower teeth and gums (see below); (b) to both surfaces of the lower lip by the mental nerve; (c) to the auricle and temporal region by the auriculotemporal nerve which also sends twigs to the external meatus and outer surface of the eardrum, and conveys secretory fibers from the otic ganglion to the parotid gland; (d) to the mucous membrane of the cheek by the buccal nerve (Fig. 7-80); and (e) to the anterior two-thirds of the tongue, floor of the mouth, and gums by the lingual nerve which also distributes the chorda tympani.

The nerve supply to gums and teeth is shown above. Nerve V² subserves sensation to the upper gums by 4 branches and to the upper teeth by 2 branches, namely, the posterior and anterior superior alveolar (dental) nerves. Nerve V³ subserves the lower gums by 3 branches and the lower teeth by one branch, namely, inferior alveolar. The gums are also supplied by twigs that perforate the alveoli. The territory of any of these gingival and dental nerves may either be extended or contracted; e.g., twigs of the mental and lingual nerves may cross the median plane to supply the gums of the opposite side, and the inferior alveolar nerves may decussate in the mandibular canal to supply the incisors of the opposite side.

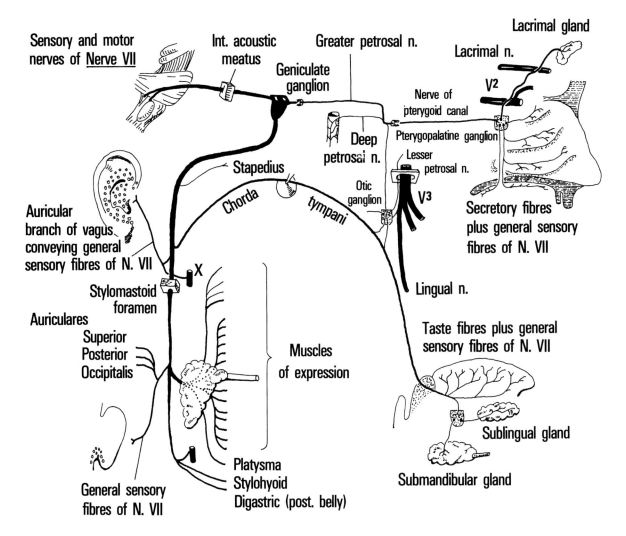

Sensory and motor nerves of <u>Nerve VII</u>

Int. acoustic meatus

Greater petrosal n.

Geniculate ganglion

Lacrimal gland

Lacrimal n.

V2

Nerve of pterygoid canal

Deep petrosal n.

Pterygopalatine ganglion

Stapedius

Chorda

tympani

Otic ganglion

Lesser petrosal n.

V3

Auricular branch of vagus conveying general sensory fibres of N. VII

Secretory fibres plus general sensory fibres of N. VII

X

Stylomastoid foramen

Lingual n.

Auriculares

Superior
Posterior
Occipitalis

Muscles of expression

Taste fibres plus general sensory fibres of N. VII

Sublingual gland

Platysma
Stylohyoid
Digastric (post. belly)

Submandibular gland

General sensory fibres of N. VII

8-7   DISTRIBUTION OF THE FACIAL NERVE

CRANIAL NERVE VII

All four modalities are carried by the facial nerve:

Motor: To the "muscles of expression," the superficial muscles around the eye, nose, mouth, and ear; of the scalp above and the platysma below. It also supplies Stylohyoid and posterior belly of Digastric, as well as Stapedius. It does not supply Levator Palpebrae; it does supply Buccinator.

Special Sense: Taste fibers, with cell stations in the geniculate ganglion, pass (a) from the palate nonstop through the pterygopalatine ganglion, nerve of the pterygoid canal, and greater petrosal nerve to the geniculate ganglion; and (b) from the anterior two-thirds of the tongue two routes are followed: (1) via the chorda tympani to the facial nerve and so to the geniculate ganglion, and (2) by a branch of the chorda that traverses the otic ganglion to join the greater petrosal nerve and so to the geniculate ganglion. As evidence of this double route is the fact that the chorda tympani may be cut without any loss of taste, whereas cutting the greater petrosal nerve may result in loss of taste.

See Schwartz, H. G., and Wedell, G. (1938) Observations on the pathways transmitting the sensation of taste. *Brain, 61:* 99.

Parasympathetic: Secretory (1) via the greater superficial petrosal nerve and the nerve of the ptergoid canal to the pterygopalatine ganglion, thence by relay to the glands of the nose and palate and to the lacrimal gland; (2) via the chorda tympani (a) to the submandibular (submaxillary) ganglion whence fibers are relayed to the submandibular and sublingual salivary glands; and, (b) via its connection with the otic ganglion, it activates the parotid gland.

Sensory: supplies general sensation to a small area of the external meatus and the auricle. Its role in deep sensation from the face has not been clearly demonstrated.

Of all the nerves of the body, the facial nerve is the most frequently paralyzed. An upper motor neuron lesion results in paralysis of the superficial muscles on the lower two-thirds of the face on the opposite side. A lower motor neuron lesion produces paralysis of the whole face on the same side.

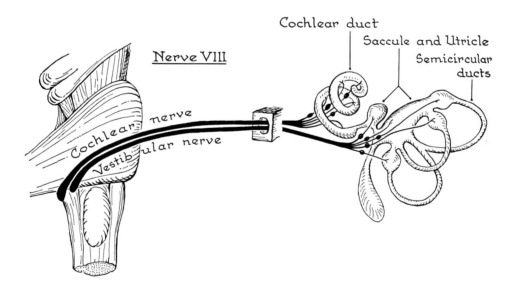

8-8 DISTRIBUTION OF THE VESTIBULO-COCHLEAR NERVE

CRANIAL NERVE VIII

This nerve has two parts: (a) the cochlear nerve, or nerve of hearing, whose fibers transmit impulses from the spiral organ of Corti in the cochlear duct; and (b) the vestibular nerve, or nerve of balancing, whose fibers transmit impulses from the maculae of the saccule and utricle and in the ampullae of the three semicircular ducts.

CRANIAL NERVE IX

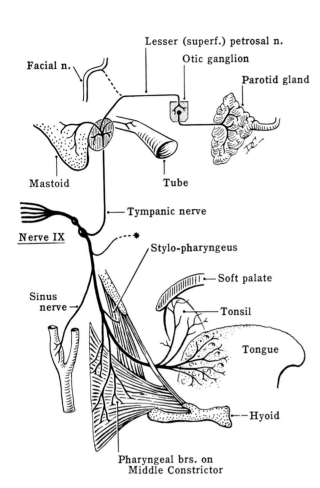

8-9A

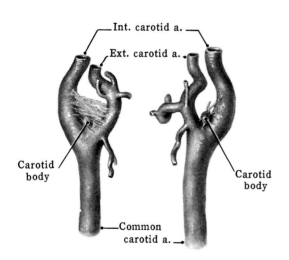

8-9B CAROTID BODY

Carotid body, viewed from behind in two stages. This particular body appeared black from engorged surface veins and, so, was easily recognized.

This nerve does all four things, but sparingly.

a. It is *motor* to one muscle, Stylopharyngeus.

b. Its *parasympathetic* component supplies secretory fibers through the otic ganglion to the parotid gland.

c. It provides the special sense of *taste* to the posterior third of the tongue including the vallate papillae (see Fig. 7-83).

d. *General sensory fibers* supply almost the entire one-half of the pharyngeal wall, including the oro-pharyngeal isthmus (*i.e.*, undersurface of the soft palate, tonsil, pharyngeal arches, and posterior third of the tongue). They also supply the dorsum of the soft palate, the auditory tube, tympanum, medial surface of the eardrum, mastoid antrum, and mastoid air cells. The sinus nerve is afferent from the carotid sinus (which responds to pressure changes within the artery) and the carotid body (which responds to falling PO_2 or rising PCO_2 in the blood).

Some additional details:

(1) The glossopharyngeal nerve, like the facial nerve, activates each of the three large salivary glands. (2) Clinical evidence is undecided as to the share taken by nerves VII, IX, and X in conveying sensation from the auricle and external meatus and in supplying the muscles of the palate. (3) It has been observed that cutting the chorda tympani reduces permanently the secretion not only of the submandibular gland but also of the parotid. Cutting the glossopharyngeal nerve above the connecting branch it sends to the nerve to Digastric (posterior belly) also reduces secretion in the three large salivary glands. Hence, it is surmised that secretory fibers travel down nerve IX, through the connecting branch to the nerve to Digastric (indicated by a *star* in Fig. 8-9*A*, and shown in Fig. 8-7), thence up the stem of nerve VII and along the chorda (a) to the submandibular ganglion where the impulses are relayed to the submandibular and sublingual glands, and (b) to the otic ganglion where the impulses are relayed to the parotid gland.

See Reichert, F. L., and Poth, E. J. (1933) Recent knowledge regarding the physiology of the glossopharyngeal nerve in man with analysis of its sensory, motor, gustatory and secretory functions. *Bull. Johns Hopkins Hosp.*, 53: 131.

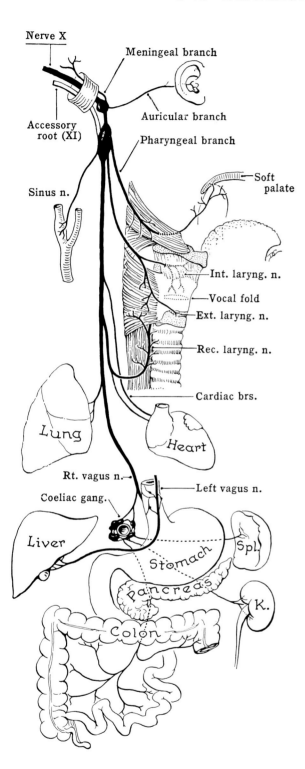

CRANIAL NERVE X

The vagus nerve, the wanderer, is:

(1) *Motor to all smooth muscle*, (2) *secretory* to all glands, and (3) *afferent* from all mucous surfaces in the following parts — pharynx (lowest part), larynx, trachea, bronchi, and lungs; esophagus (entire), stomach, and gut down to the left colic flexure; liver, gallbladder, and bile passages; pancreas and pancreatic ducts; and perhaps spleen and kidney, (4) *motor* to all muscles of the larynx, all muscles of the pharynx (except Stylopharyngeus), and all the muscles of the palate (except Tensor Palati), (5) the conveyor of *taste* from the few taste buds about the epiglottis, (6) *inhibitory* to cardiac muscle, (7) *sensory* to the outer surface of the eardrum, the external acoustic meatus, and the back of the auricle.

Branches arise from the vagus thus:

In the jugular fossa — (a) a meningeal branch to the dura of the posterior cranial fossa; and (b) an auricular branch (Figs. 7-70 and 7-71).

In the neck — (a) the pharyngeal branch is motor to Superior and Middle Constrictors and muscles of the soft palate; (b) the superior laryngeal nerve, via the internal laryngeal nerve, is sensory to the larynx above the vocal cords and to the lowest part of the pharynx (Fig. 9-64) and, via the external laryngeal nerve, motor to Inferior Constrictor and Cricothyroid (Figs. 9-42 and 9-63), (c) a twig (sinus nerve) to the carotid sinus, and (d) two cardiac branches.

In the thorax — (a) the recurrent nerve sends a motor branch to Inferior Constrictor, is motor to all the laryngeal muscles (excepting Cricothyroid), and is both afferent and efferent to the larynx below the level of the cords, as well as to the upper part of the esophagus; (b) cardiac branches; (c) pulmonary branches; and (d) the esophageal plexus.

In the abdomen — see Figure 2-115.

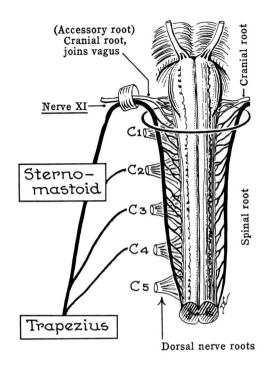

8-11 DISTRIBUTION OF THE ACCESSORY NERVE

CRANIAL NERVE XI

The cranial root of this nerve is accessory to the vagus by providing part of its motor component.

The spinal root of the accessory nerve, joined by fibers from the ventral ramus of C2, supplies Sternomastoid and, joined by fibers from the ventral rami of C3 and C4, supplies Trapezius. There is clinical evidence (both surgical and medical) that these contributions from C2, C3, and C4 convey motor as well as sensory fibers.

See Haymaker, W., and Woodhall, B. (1953) *Peripheral Nerve Injuries*, 2nd ed. W. B. Saunders Company, Philadelphia.

The spinal root of the accessory nerve usually passes through the dorsal root ganglion of C1 and may receive sensory fibers from it.

See Pearson, A. A. (1938) The spinal accessory nerve in human embryos. *J. Comp. Neurol.*, 68: 243.

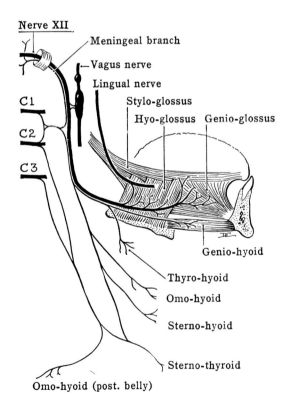

8-12 DISTRIBUTION OF THE HYPOGLOSSAL NERVE

CRANIAL NERVE XII

This efferent nerve supplies all the intrinsic (longitudinal, transverse, and vertical) and extrinsic (Styloglossus, Hyoglossus, and Genioglossus) muscles of the tongue, Palatoglossus excepted.

It receives a mixed (motor and sensory) branch from the loop between the ventral rami of C1 and C2. The sensory or afferent fibers in part take a recurrent course and end in the dura mater of the posterior cranial fossa. The motor or efferent branch supplies Geniohyoid and Thyrohyoid, and it provides a descending branch which unites with a descending branch of C2 and C3 to form a loop, the ansa cervicalis. This and the ansa supply the remaining depressor muscles of the hyoid bone.

9

The
Neck

CONTENTS

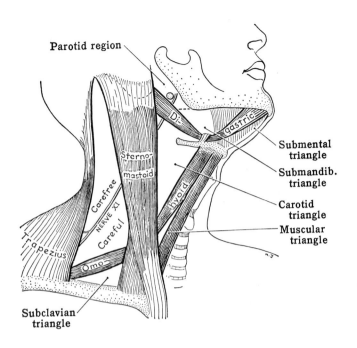

9-1A TRIANGLES OF THE NECK

For descriptive purposes, the neck has been traditionally divided into triangular areas.

The obliquely set Sternomastoid divides the side of the neck into an anterior and a posterior triangle.

The *anterior triangle* is bounded by Sternomastoid, the median line of the neck, and the lower border of the mandible. It is subdivided into 3 small triangles: *submandibular, carotid,* and *muscular.* The posterior belly of Digastric (and Stylohyoid) separates the carotid triangle from the submandibular triangle; the superior belly of Omohyoid separates the carotid triangle from the muscular triangle. The region between the anterior bellies of the Digastrics and the body of the hyoid bone is the (unpaired) *submental* triangle.

The *posterior triangle* is bounded by Trapezius, Sternomastoid, and the middle third of the clavicle (Figs. 9-4 to 9-8). It is divisible into a *subclavian* (supraclavicular) and an *occipital* triangle by the inferior belly of Omohyoid, but of much greater significance is the fact that it is divided by the accessory nerve (nerve XI) into nearly equal upper and lower parts. Of these, the upper contains little of importance, but the lower contains numerous structures of great importance. Hence, above the nerve your dissection may be care-free, whereas below it you must proceed very carefully.

9-1B POSTERIOR TRIANGLE

The posterior triangle is bounded by the Sternocleidomastoid (*S*), the middle third of the clavicle, and the Trapezius. As demonstrated here, the Sternocleidomastoid turns the head in the *opposite* direction.

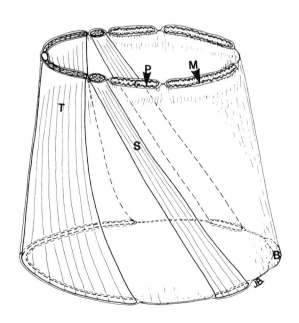

9-1C INVESTING FASCIA

This diagram illustrates that the investing fascia forms a complete collar around the neck, attaching to bone above and below. On each side it splits to surround muscles: Trapezius (*T*) and Sternocleidomastoid (*S*), and glands: Parotid (*P*) and Submandibular (*M*). In the absence of muscle to surround, the two layers fuse to form windows looking into anterior and posterior triangles. Separation of the two layers attaching to the manubrium produces the *suprasternal space* (*B*).

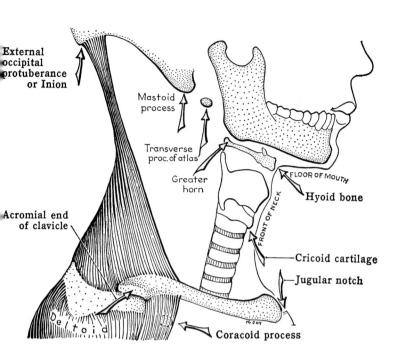

9-2A BONY LANDMARKS OF THE NECK

Note:

1. The inion and the mastoid process (and the superior nuchal line uniting them) are created by the downward pull of Trapezius and Sternomastoid.
2. The transverse process of the atlas, being the most prominent of the cervical transverse processes, is felt with the fingertip on pressing upward between the angle of the jaw and the mastoid process.
3. The body of the hyoid bone lies at the angle between the floor of the mouth and the front of the neck.

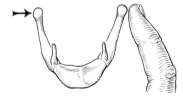

The greater horn of one side of the hyoid bone is palpable only when the greater horn of the opposite side is steadied.
. The arch of the cricoid cartilage projects beyond the rings of the trachea (Figs. 9-77 and 9-65), and is thereby readily identified in life, on running the fingertip upward. It is the guide to the level of C6, where so many things happen.
. The jugular (suprasternal) notch is visible and palpable between the medial ends of the clavicles.
. The lateral end of the clavicle, being thicker than the acromion, is palpable on pressing medially.
. The coracoid process, located 2.5 cm below the clavicle, under the edge of the Deltoid, is palpable on pressing laterally with the finger in the deltopectoral triangle.

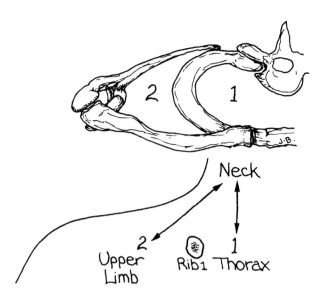

9-2B NECK DOORWAYS

Two diagrams which show two doorways:

1. The superior thoracic inlet: an oval space bounded by the first thoracic vertebra, manubrium, and the first two ribs. Traffic between neck and thorax passes through here.
2. The triangular doorway to the axilla formed by first rib, scapula, and clavicle. It is in communication with the neck. Structures passing between axilla and thorax hook over the first rib.

9-2C BASE OF THE NECK

Forced inspiration against a closed glottis exposes part of the outline of the superior thoracic inlet.

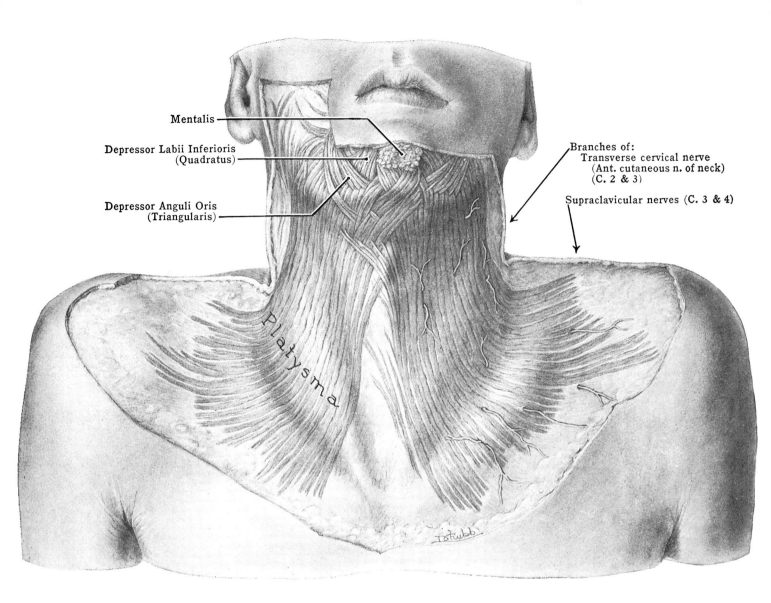

Mentalis

Depressor Labii Inferioris
(Quadratus)

Depressor Anguli Oris
(Triangularis)

Branches of:
Transverse cervical nerve
(Ant. cutaneous n. of neck)
(C. 2 & 3)

Supraclavicular nerves (C. 3 & 4)

Platysma

9-3A PLATYSMA

Observe the dissection above and the inset on the left:

1. The platysma, spreading subcutaneously like a sheet, pierced by cutaneous nerves, crossing the whole length of the lower border of the mandible above, crossing the whole length of the clavicle below, and extending downward to the level of the 1st or 2nd rib and to (or toward) the acromion.

2. The anterior borders of the two Platysmas, decussating behind the chin in the submental region and below that free and diverging, and so leaving the median part of the neck uncovered.

3. Its posterior border, free, covering the antero-inferior part of the posterior triangle, and continuing upward across the lower border of the jaw to the angle of the mouth.

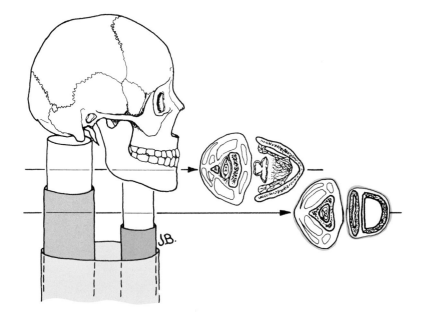

9-3B UNITS OF THE NECK

Note that the neck is "packaged" in two major units: an *anterior* visceral unit including food and air passageways coated with *pretracheal* fascia (*green*) and a *posterior* vertebral unit consisting of spinal cord, vertebrae, and muscles coated with *prevertebral* fascia (*blue*). The outer wrapping is the *investing* fascia (*yellow*).

9-3C FASCIA OF THE NECK

Observe in this diagram of a horizontal section through the neck:

Pretracheal (*PT*): a thin sheath covering the thyroid gland.

Carotid sheath (*C*): surrounding the carotid artery, vagus nerve, and (loosely) the internal jugular vein.

Prevertebral (*PV*): sheaths the muscles associated with the vertebrae. As components of the brachial plexus emerge in their gutter between Scalenus anterior and medius they carry an investment of this fascia forming the axillary sheath.

Investing (*IF*): surrounds the neck (Fig. 9-1C). The *arrow* points to its two fused layers, a window for viewing the posterior triangle.

Other structures shown:

1. Esophagus
2. Trachea
3. Thyroid gland
4. Trapezius
5. Sternomastoid
6. Sternohyoid
7. Omohyoid
8. Sternothyroid
9. Splenius capitis
10. Levator scapulae

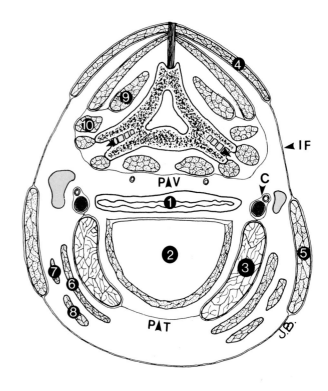

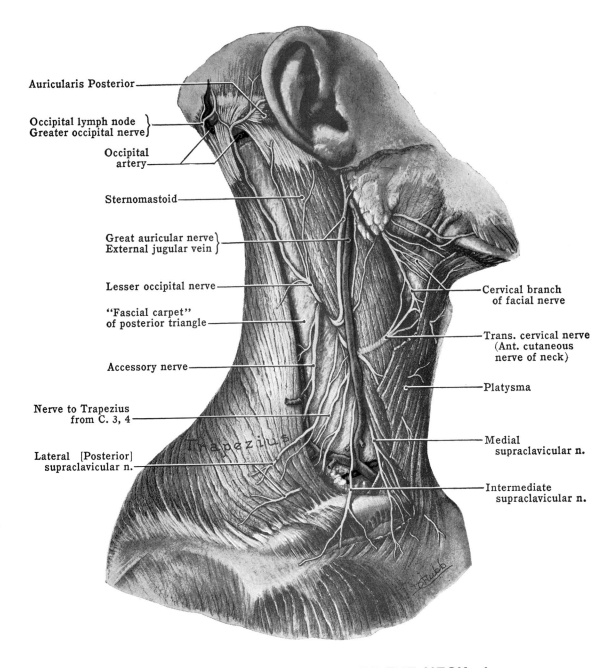

Auricularis Posterior

Occipital lymph node
Greater occipital nerve

Occipital
artery

Sternomastoid

Great auricular nerve
External jugular vein

Lesser occipital nerve

"Fascial carpet"
of posterior triangle

Accessory nerve

Nerve to Trapezius
from C. 3, 4

Lateral [Posterior]
supraclavicular n.

Cervical branch
of facial nerve

Trans. cervical nerve
(Ant. cutaneous
nerve of neck)

Platysma

Medial
supraclavicular n.

Intermediate
supraclavicular n.

Trapezius

9-4 POSTERIOR TRIANGLE OF THE NECK—I

SUPERFICIAL STRUCTURES

Observe:
1. The 3 sides of the triangle: Trapezius, Sternomastoid, middle third of the clavicle. The apex is where the aponeuroses of the two muscles blend a little below the superior nuchal line.
2. Platysma (partly cut away), covering the lower part of the triangle. For entire Platysma, see Figure 9-3.
3. The external jugular vein, descending vertically from behind the angle of the jaw, across Sternomastoid to its posterior border where, an inch above the clavicle, it pierces the investing deep fascia.
4. The "fascial carpet" that covers the muscular floor.
5. The accessory nerve—the only motor nerve superficial to the "fascial carpet"—descending within the deep fascia, and disappearing 2 fingers' breadth or more above the clavicle.
6. The cutaneous nerves (C2, C3, C4), radiating from the posterior border of Sternomastoid, below the accessory nerve.

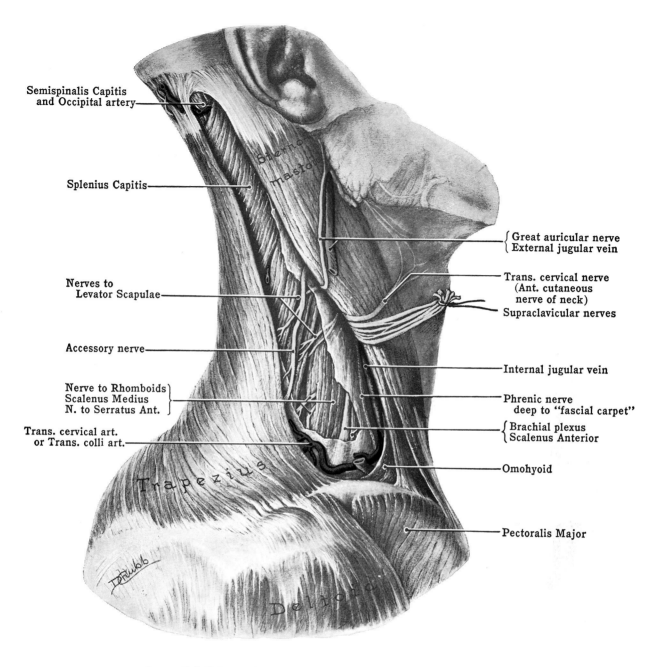

Semispinalis Capitis
and Occipital artery

Splenius Capitis

Nerves to
Levator Scapulae

Accessory nerve

Nerve to Rhomboids
Scalenus Medius
N. to Serratus Ant.

Trans. cervical art.
or Trans. colli art.

Great auricular nerve
External jugular vein

Trans. cervical nerve
(Ant. cutaneous
nerve of neck)
Supraclavicular nerves

Internal jugular vein

Phrenic nerve
deep to "fascial carpet"

Brachial plexus
Scalenus Anterior

Omohyoid

Pectoralis Major

9-5   POSTERIOR TRIANGLE OF THE NECK—II

MOTOR NERVES DEEP TO FASCIAL CARPET

Observe

1. The muscles forming the floor of the upper part of the triangle (Semispinalis Capitis, Splenius, Levator Scapulae). Those forming the lower part are shown in Figure 9-7.

2. The accessory nerve (*i.e.*, nerve to Sternomastoid and Trapezius) (Fig. 8-11), lying along Levator Scapulae but separated by the "fascial carpet."

3. Three motor nerves to upper limb muscles, lying (a) in the plane between the fascial carpet and the muscular floor, and (b) between the accessory nerve above and the brachial plexus below. They are: nerves to Levator Scapulae (C3, C4), to Rhomboids (C5), and to Serratus Anterior (C5, C6; the branch from C7 lies protected behind the plexus, Fig. 6-28).

4. Two structures of surgical importance, situated just beyond the geometrical confines of the triangle: (a) a fourth motor nerve, the phrenic nerve to the Diaphragm (C3, C4, C5), placed between carpet and floor; (b) the internal jugular vein, superficial to the carpet.

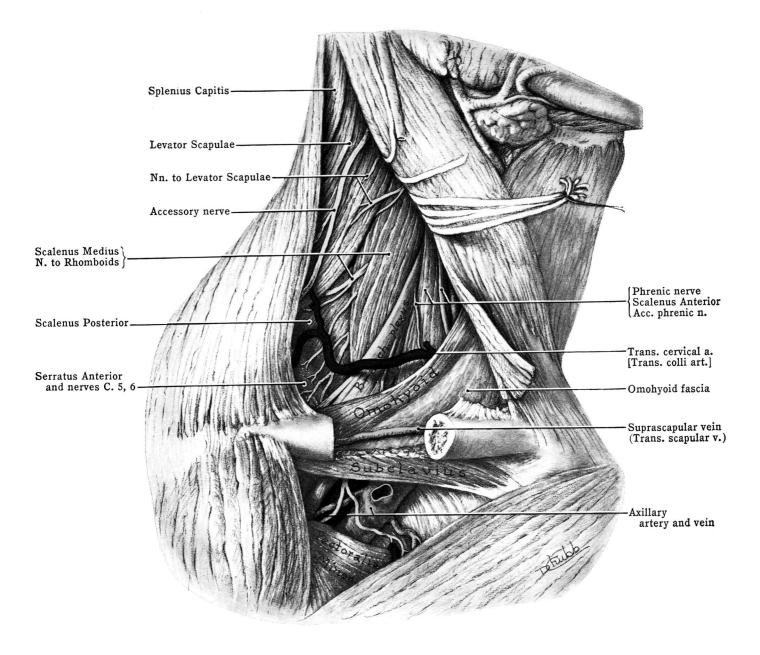

Splenius Capitis

Levator Scapulae

Nn. to Levator Scapulae

Accessory nerve

Scalenus Medius
N. to Rhomboids

Scalenus Posterior

Serratus Anterior
and nerves C. 5, 6

Phrenic nerve
Scalenus Anterior
Acc. phrenic n.

Trans. cervical a.
[Trans. colli art.]

Omohyoid fascia

Suprascapular vein
(Trans. scapular v.)

Axillary
artery and vein

9-6 POSTERIOR TRIANGLE OF THE NECK—III

OMOHYOID AND ITS FASCIA

The clavicular head of Pectoralis Major and part of the clavicle have been excised.

Observe:

1. The posterior belly of Omohyoid, held down by a sheet of "Omohyoid" fascia to the fascia ensheathing Subclavius; and the resulting pocket between this fascia posteriorly and the investing deep fascia and Sternomastoid anteriorly (Fig. 9-15).

2. The brachial plexus, appearing between Scalenus Anterior and Scalenus Medius and here, as commonly, giving off an accessory phrenic nerve from C5.

3. The posterior border of Scalenus Anterior, nearly parallel to the posterior border of Sternomastoid, and slightly behind it; hence one is a guide to the other.

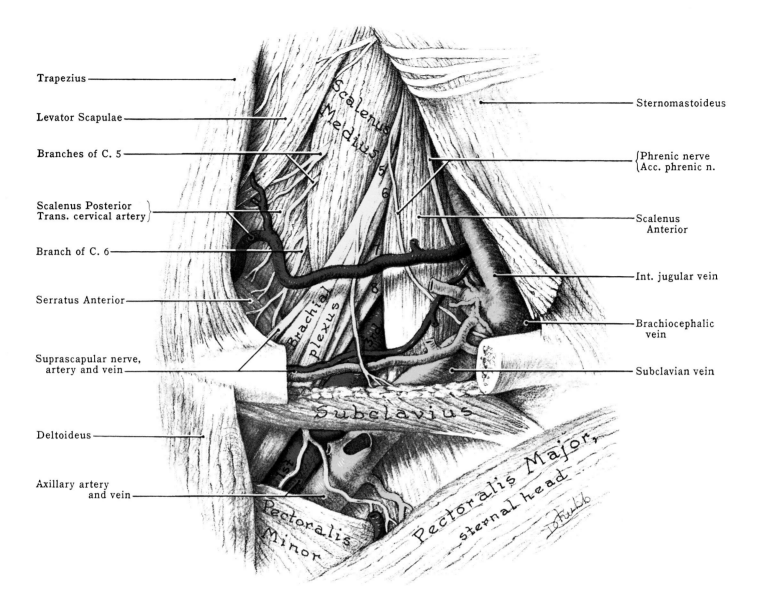

Trapezius

Levator Scapulae

Branches of C. 5

Scalenus Posterior
Trans. cervical artery

Branch of C. 6

Serratus Anterior

Suprascapular nerve,
artery and vein

Deltoideus

Axillary artery
and vein

Sternomastoideus

Phrenic nerve
Acc. phrenic n.

Scalenus
Anterior

Int. jugular vein

Brachiocephalic
vein

Subclavian vein

9-7 POSTERIOR TRIANGLE OF THE NECK – IV

BRACHIAL PLEXUS AND SUBCLAVIAN VESSELS

Observe:
1. The 3rd part of the subclavian artery and the 1st part of the axillary artery
   are labeled.
2. The muscles forming the floor of the lower part of the triangle (Scaleni
   Posterior, Medius, and Anterior and Serratus Anterior).
3. The brachial plexus and subclavian artery, appearing between Scalenus
   Medius and Scalenus Anterior. The lowest root of the plexus (T1) is concealed
   by the 3rd part of the artery.
4. The suprascapular nerve, found by following the lateral border of the plexus
   caudally.
5. The subclavian vein, hardly rising above the level of the clavicle, and
   separated from the 2nd part of the subclavian artery by Scalenus Anterior.
6. Subclavius, unimportant as a muscle, but valuable as a buffer between a
   fractured clavicle and the subclavian vessels. By voluntarily forcing the
   upper limb backward and downward, the clavicle and Subclavius compress
   the subclavian vessels against the first rib and so arrest the pulse.

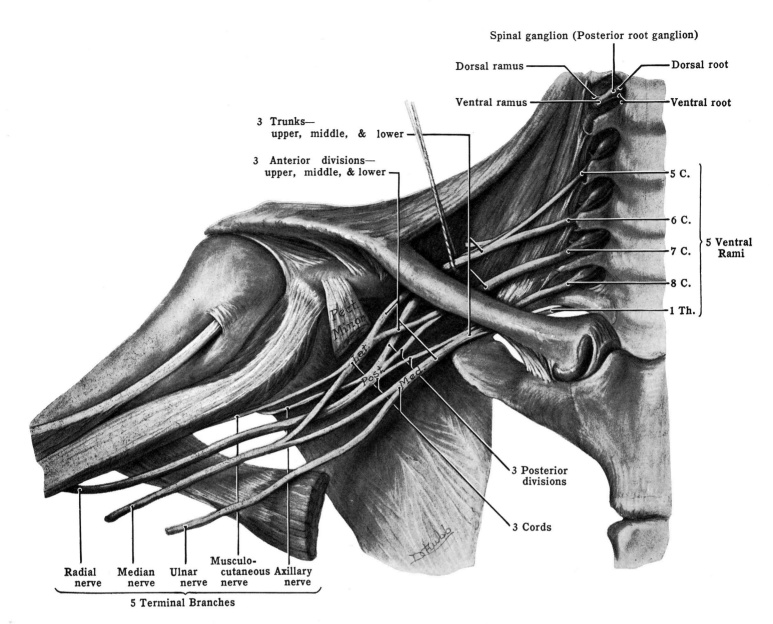

Spinal ganglion (Posterior root ganglion)

Dorsal ramus

Ventral ramus

Dorsal root

Ventral root

3 Trunks—
upper, middle, & lower

3 Anterior divisions—
upper, middle, & lower

5 C.

6 C.

7 C.

8 C.

1 Th.

5 Ventral
Rami

Pect.
Minor

Lat.

Post.

Med.

3 Posterior
divisions

3 Cords

Radial
nerve

Median
nerve

Ulnar
nerve

Musculo-
cutaneous
nerve

Axillary
nerve

5 Terminal Branches

9-8 BRACHIAL PLEXUS

Observe:
1. A dorsal (sensory) root of a spinal nerve is larger than a ventral (motor) root.
2. The 2 roots uniting beyond the ganglion to form a very short mixed spinal nerve.
3. The mixed nerve at once dividing into a small dorsal ramus and a large ventral ramus.
4. The 5 ventral *rami* forming the brachial plexus (of these the middle ramus, C7, is the largest, and C5 and Th1 the smallest).
5. The 5 rami uniting to form the 3 *trunks* of the plexus.
6. Each trunk dividing into 2 *divisions*, an anterior and a posterior.
7. From the divisions 3 *cords* resulting.
8. The 3 cords lying behind Pectoralis Minor.
9. The posterior cord and the radial nerve are derived from segments C5, C6, C7, and C8, and slightly from Th1.
10. The median nerve is derived from segments C6, C7, C8, and Th1, and slightly from C5.

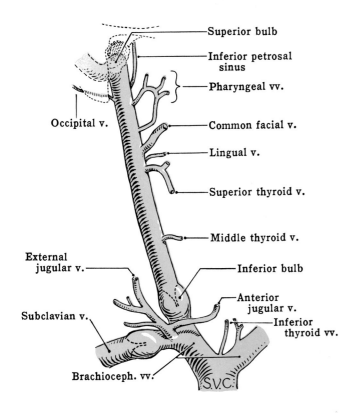

- Superior bulb
- Inferior petrosal sinus
- Pharyngeal vv.
- Occipital v.
- Common facial v.
- Lingual v.
- Superior thyroid v.
- Middle thyroid v.
- External jugular v.
- Inferior bulb
- Subclavian v.
- Anterior jugular v.
- Inferior thyroid vv.
- Brachioceph. vv.
- S.V.C.

9-9 INTERNAL JUGULAR VEIN AND ITS TRIBUTARIES

Note the dilation or bulb at each end of the internal jugular vein. The superior jugular bulb is separated from the floor of the middle ear by a delicate bony plate. The inferior jugular bulb, like the corresponding bulb at the end of the subclavian vein, contains a bicuspid valve which permits the flow of blood toward the heart. There are no valves in the brachiocephalic veins or in the superior vena cava.

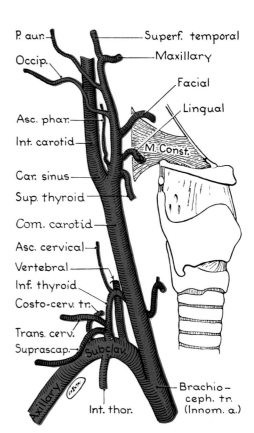

- P. aur.
- Superf. temporal
- Occip.
- Maxillary
- Facial
- Lingual
- Asc. phar.
- Int. carotid
- M. Const.
- Car. sinus
- Sup. thyroid
- Com. carotid
- Asc. cervical
- Vertebral
- Inf. thyroid
- Costo-cerv. tr.
- Trans. cerv.
- Suprascap.
- Subclav.
- Axillary
- Int. thor.
- Brachio-ceph. tr. (Innom. a.)

9-10 SUBCLAVIAN AND CAROTID ARTERIES AND THEIR BRANCHES

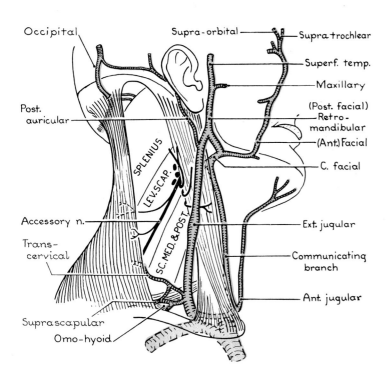

- Occipital
- Supra-orbital
- Supra-trochlear
- Superf. temp.
- Maxillary
- (Post. facial) Retro-mandibular
- Post. auricular
- (Ant.) Facial
- C. facial
- SPLENIUS
- LEV. SCAP.
- Accessory n.
- SC. MED. & POST.
- Ext. jugular
- Trans-cervical
- Communicating branch
- Suprascapular
- Ant. jugular
- Omo-hyoid

9-11 SUPERFICIAL VEINS

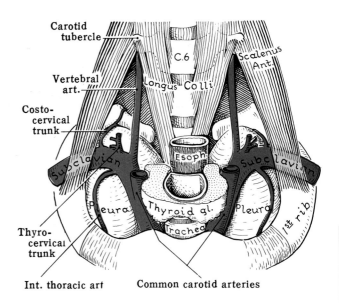

- Carotid tubercle
- C.6
- Scalenus Ant.
- Vertebral art.
- Longus Colli
- Costo-cervical trunk
- Subclavian
- Esoph.
- Subclavian
- Pleura
- Pleura
- 1st rib
- Thyroid gl.
- Trachea
- Thyro-cervical trunk
- Int. thoracic art
- Common carotid arteries

9-12 TRIANGLE OF VERTEBRAL ARTERY

The base is the first part of the subclavian artery; the other two sides are Longus Colli and Scalenus Anterior. The vertebral artery exits at the apex to enter the foramen transversarium of C6.

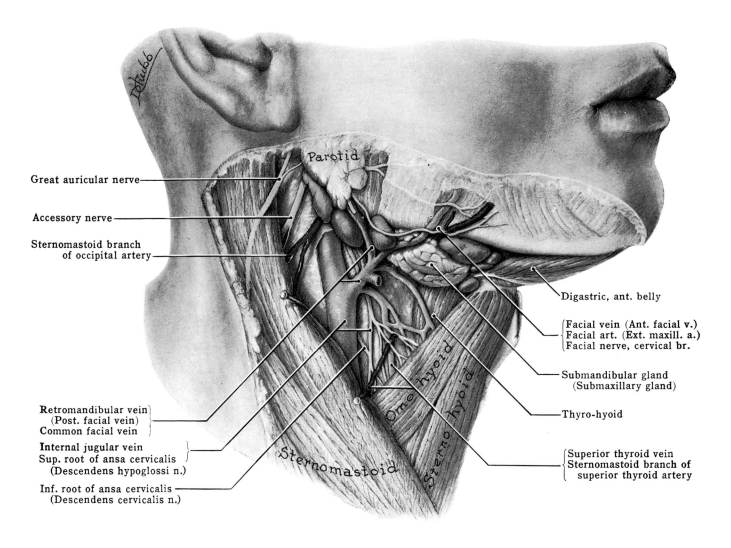

Great auricular nerve

Accessory nerve

Sternomastoid branch
of occipital artery

Parotid

Digastric, ant. belly

Facial vein (Ant. facial v.)
Facial art. (Ext. maxill. a.)
Facial nerve, cervical br.

Submandibular gland
(Submaxillary gland)

Thyro-hyoid

Retromandibular vein
(Post. facial vein)
Common facial vein

Internal jugular vein
Sup. root of ansa cervicalis
(Descendens hypoglossi n.)

Inf. root of ansa cervicalis
(Descendens cervicalis n.)

Superior thyroid vein
Sternomastoid branch of
superior thyroid artery

9-13   ANTERIOR TRIANGLE OF THE NECK – I

SUPERFICIAL DISSECTION

Observe:

1. The accessory nerve entering the deep surface of Sternomastoid about 5 cm below the tip of the mastoid process, and joined along its lower border by the sternomastoid branch of the occipital artery.
2. The internal jugular vein joined in front by several veins, notably the common facial vein, about the level of the hyoid bone.
3. Branches of the ansa cervicalis (ansa hypoglossi) passing deep to Omohyoid.
4. The sternomastoid branch of the superior thyroid artery descending near the upper border of Omohyoid.
5. The submandibular gland and lymph nodes overflowing the submandibular triangle (digastric triangle). The retromandibular and facial veins running superficial to the gland.

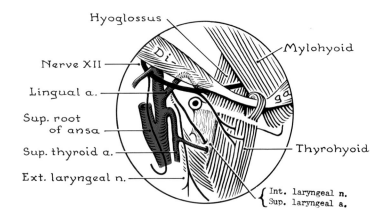

Hyoglossus

Mylohyoid

Nerve XII

Lingual a.

Sup. root
of ansa

Sup. thyroid a.

Ext. laryngeal n.

Thyrohyoid

Int. laryngeal n.
Sup. laryngeal a.

9-14   HYOID AS REFERENCE

The tip of the greater horn is the reference point for many structures—muscles, nerves, and arteries.

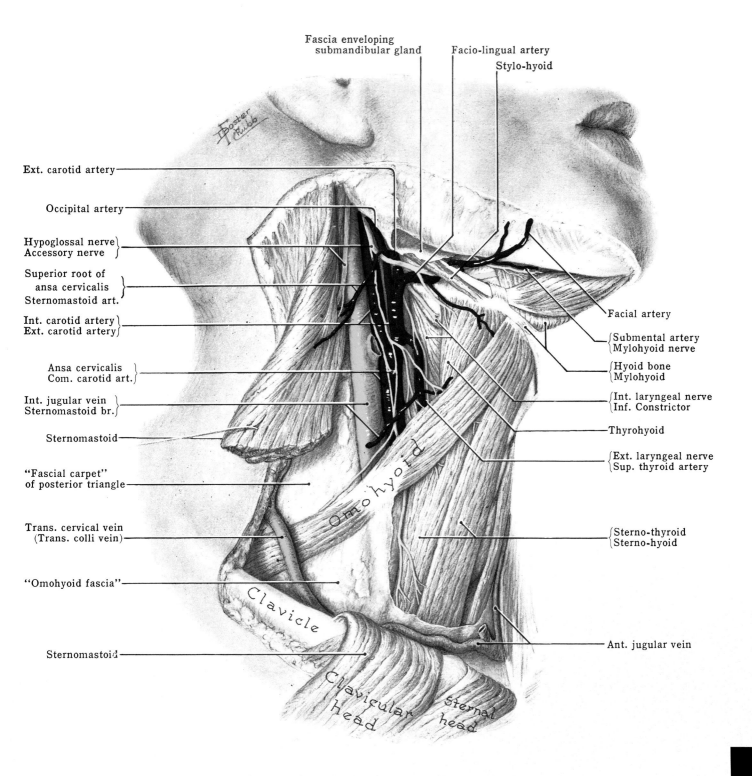

Fascia enveloping submandibular gland

Facio-lingual artery

Stylo-hyoid

Ext. carotid artery

Occipital artery

Hypoglossal nerve
Accessory nerve

Superior root of ansa cervicalis
Sternomastoid art.

Int. carotid artery
Ext. carotid artery

Ansa cervicalis
Com. carotid art.

Int. jugular vein
Sternomastoid br.

Sternomastoid

"Fascial carpet" of posterior triangle

Trans. cervical vein (Trans. colli vein)

"Omohyoid fascia"

Sternomastoid

Facial artery

Submental artery
Mylohyoid nerve

Hyoid bone
Mylohyoid

Int. laryngeal nerve
Inf. Constrictor

Thyrohyoid

Ext. laryngeal nerve
Sup. thyroid artery

Sterno-thyroid
Sterno-hyoid

Ant. jugular vein

Omohyoid

Clavicle

Clavicular head

Sternal head

9-15 ANTERIOR TRIANGLE OF THE NECK—II

DEEPER DISSECTION

Observe:

1. The intermediate tendon of Digastric held down to the hyoid bone by a fascial sling; the intermediate tendon of Omohyoid similarly held down to the clavicle.

2. The remains of the fascial sheath of the submandibular gland which posteriorly, as the stylomandibular ligament, separates this gland from the parotid gland.

3. The facial and lingual arteries, here arising by a common stem, passing deep to Stylohyoid and Digastric to enter the submandibular triangle. The facial artery here giving off its submental branch, which accompanies the mylohyoid nerve.

4. Thyrohyoid and Inferior Constrictor forming the floor or medial wall of the carotid triangle.

5. The hypoglossal nerve curving into the carotid triangle and curving out again, passing deep to Digastric twice, crossing the internal and external carotid arteries, and giving off two branches from its convex side: the superior root of the ansa cervicalis and the nerve to Thyrohyoid.

6. The internal and external laryngeal nerves appearing deep to the external carotid artery.

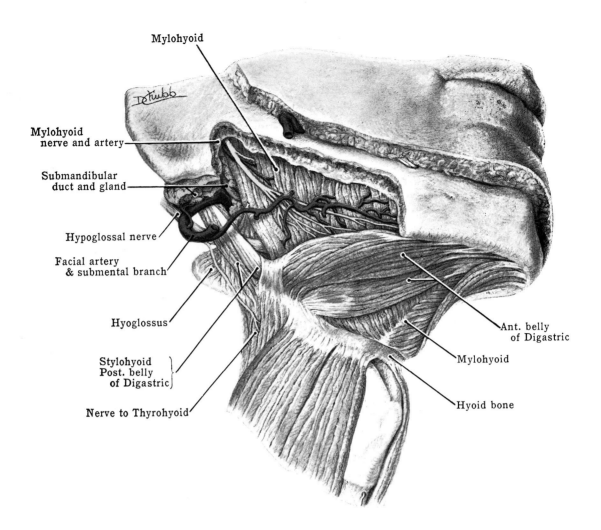

Mylohyoid

Mylohyoid
nerve and artery

Submandibular
duct and gland

Hypoglossal nerve

Facial artery
& submental branch

Hyoglossus

Stylohyoid
Post. belly
of Digastric

Nerve to Thyrohyoid

Ant. belly
of Digastric

Mylohyoid

Hyoid bone

9-16   SUPRAHYOID REGION—I

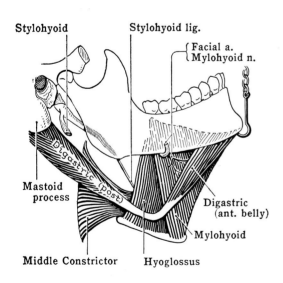

Stylohyoid

Stylohyoid lig.

Facial a.
Mylohyoid n.

Mastoid
process

Digastric
(ant. belly)

Mylohyoid

Middle Constrictor

Hyoglossus

9-17   SUPRAHYOID MUSCLES

Note that the muscles are in 4 layers: Digastric,
Mylohyoid, Hyoglossus, and Middle Constrictor.

Observe:
1. The medial wall of the submandibular (digastric) triangle.
2. Stylohyoid and the posterior belly of Digastric forming the posterior side of the triangle; the facial artery arching over these. The anterior belly of the Digastric forming the anterior side. Here this belly has an extra origin from the hyoid bone.
3. Mylohyoid forming the medial wall of the triangle and having a free, thick posterior border.
4. The mylohyoid nerve, which supplies Mylohyoid and anterior belly of Digastric, accompanied by the mylohyoid branch of the inferior alveolar artery posteriorly and by the submental branch of the facial artery anteriorly.
5. The hypoglossal nerve, the submandibular gland, and the submandibular duct passing forward deep to the posterior border of Mylohyoid.

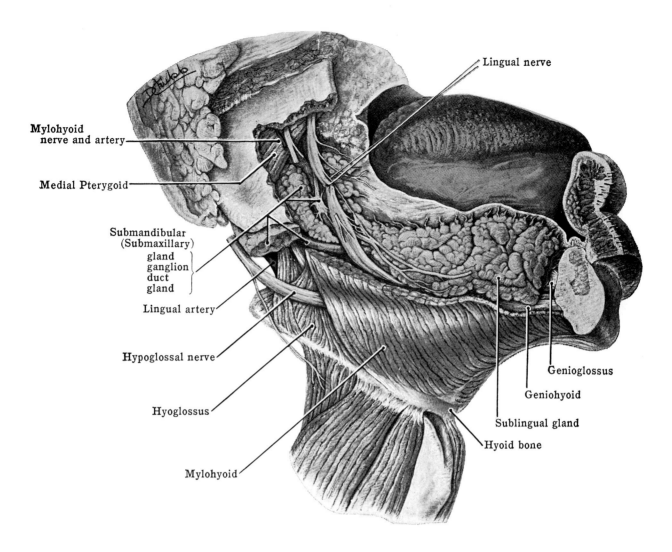

Lingual nerve

Mylohyoid nerve and artery

Medial Pterygoid

Submandibular (Submaxillary)
  gland
  ganglion
  duct
  gland

Lingual artery

Hypoglossal nerve

Hyoglossus

Mylohyoid

Genioglossus

Geniohyoid

Sublingual gland

Hyoid bone

9-18 SUPRAHYOID REGION—II

Observe:
1. The cut surface of Mylohyoid becoming progressively thinner as traced forward.
2. The sublingual salivary gland, almond-shaped, almost touching its fellow of the opposite side behind the symphysis menti and in contact with the deep part of the submandibular gland posteriorly. (For medial view, Fig. 7-88.)
3. The dozen or more fine ducts passing from the upper border of the sublingual gland to open on the plica sublingualis.
4. Several individual or detached lobules of the sublingual gland, each having a fine duct, behind the main mass of the gland, and labial glands in the lip (unlabeled).
5. The mylohyoid nerve and artery (cut short) and the lingual nerve clamped between Medial Pterygoid and the ramus of the mandible.
6. The lingual nerve lying between the sublingual gland and the deep or oral part of the submandibular gland. The submandibular ganglion is suspended from this nerve, and various branches leave it.

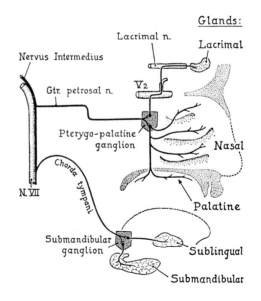

Glands:

Lacrimal n.

Lacrimal

Nervus Intermedius

Gtr. petrosal n.

V_2

Pterygo-palatine ganglion

Nasal

N. VII

Chorda tympani

Palatine

Submandibular ganglion

Sublingual

Submandibular

9-19 SECRETOMOTOR FIBERS

Nerve VII, via the greater petrosal nerve, brings secretomotor fibers to pterygo-palatine ganglion and via chorda tympani to submandibular ganglion; one to be relayed and distributed to glands above the oral cavity, the other to glands below it. (See Fig. 7-41.)

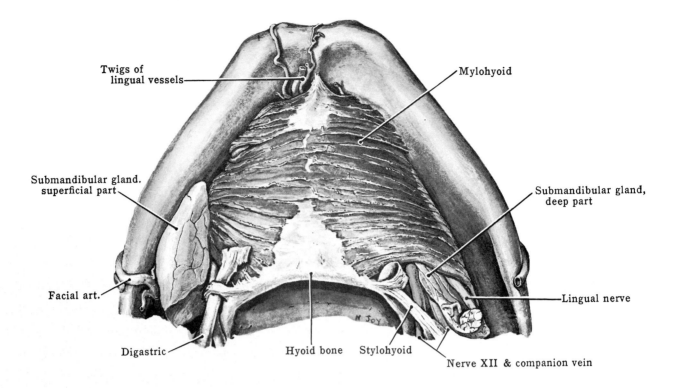

Twigs of lingual vessels

Mylohyoid

Submandibular gland, superficial part

Submandibular gland, deep part

Facial art.

Lingual nerve

Digastric

Hyoid bone

Stylohyoid

Nerve XII & companion vein

9-20 FLOOR OF THE MOUTH, FROM BELOW—I

MYLOHYOIDS

The anterior bellies of the Digastrics have been removed.

Observe:

1. The right and left Mylohyoids, which together form the "oral diaphragm," arising from the mylohyoid line of the jaw (Fig. 7-73), and inserted into an indefinite median raphe and into the hyoid bone (Fig. 7-86).
2. The submandibular gland turning round the posterior border of Mylohyoid.
3. The hypoglossal nerve and its companion vein passing deep to the same posterior border; and high up the lingual nerve applied to the jaw.

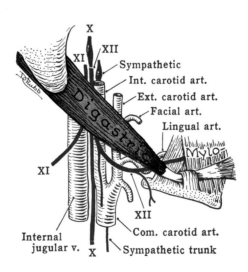

X

XI

XII

Sympathetic

Int. carotid art.

Ext. carotid art.

Facial art.

Lingual art.

Digastric

Mylo

XI

XII

Internal jugular v.

X

Com. carotid art.

Sympathetic trunk

9-21 POSTERIOR BELLY OF DIGASTRIC

Note:

1. The superficial and key position of this muscle which runs from the mastoid process to the hyoid bone and crosses deep to the angle of the jaw.
2. All vessels and nerves cross deep to this belly except for: (a) cervical branches of the facial nerve, (b) facial branches of the great auricular nerve, and (c) the external jugular vein and its connections (Figs. 7-16 and 9-11).

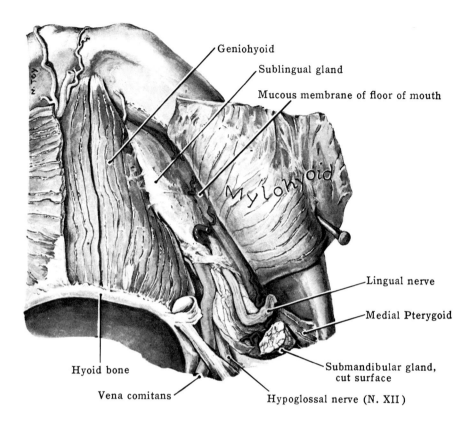

Geniohyoid

Sublingual gland

Mucous membrane of floor of mouth

Mylohyoid

Lingual nerve

Medial Pterygoid

Submandibular gland, cut surface

Hyoid bone

Vena comitans

Hypoglossal nerve (N. XII)

9-22 FLOOR OF THE MOUTH, FROM BELOW—II

GENIOHYOIDS

(The left Mylohyoid and part of the right are reflected.)

Observe:

1. The Geniohyoid, triangular, in contact with its fellow, and extending from the mental spine of the jaw to the front of the body of the hyoid bone.
2. The structures seen in Figure 9-20 followed forward: a companion vein (distended), hypoglossal nerve, deep part of gland, and lingual nerve (appearing at anterior border of Medial Pterygoid).
3. The areolar covered sublingual gland, and lateral to it the mucous membrane of the mouth with twigs of the sublingual artery.

For mouth from medial side and from above, see Figures 7-88 to 7-90.

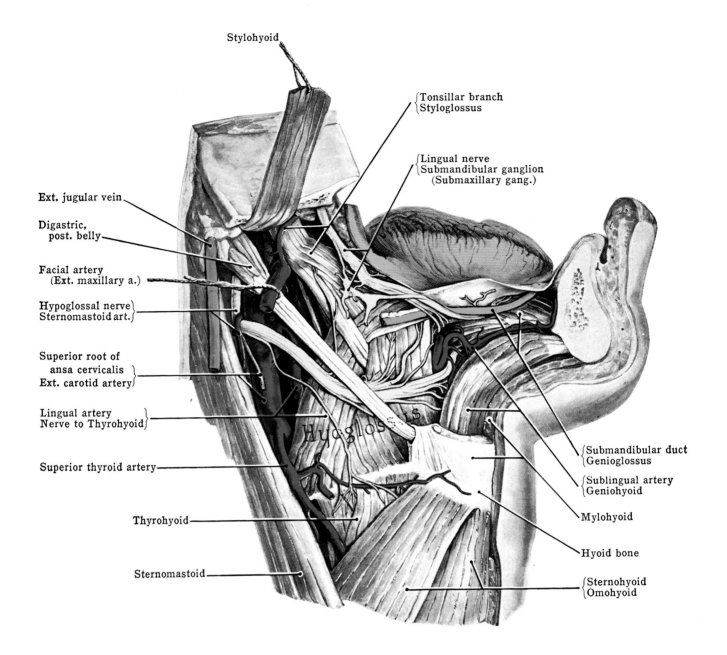

Stylohyoid

Tonsillar branch
Styloglossus

Lingual nerve
Submandibular ganglion
(Submaxillary gang.)

Ext. jugular vein

Digastric,
post. belly

Facial artery
(Ext. maxillary a.)

Hypoglossal nerve
Sternomastoid art.

Superior root of
ansa cervicalis
Ext. carotid artery

Lingual artery
Nerve to Thyrohyoid

Superior thyroid artery

Thyrohyoid

Sternomastoid

Submandibular duct
Genioglossus

Sublingual artery
Geniohyoid

Mylohyoid

Hyoid bone

Sternohyoid
Omohyoid

9-23 SUPRAHYOID REGION—III

Observe:
1. Stylohyoid pulled up. Posterior belly of Digastric left *in situ*, as a landmark.
2. Hyoglossus ascending from the greater horn and body of the hyoid bone to the side of the tongue; Styloglossus, postero-superiorly, crossed by the tonsillar branch of the facial artery and interdigitating with bundles of Hyoglossus; and Genioglossus, anteriorly, fanning out into the tongue. These are 3 extrinsic muscles of the tongue, all supplied by the hypoglossal nerve.
3. The hypoglossal nerve, crossed twice by Digastric, crossing twice the lingual artery, and supplying all the muscles of the tongue, both extrinsic and intrinsic, Palatoglossus excepted. The branches of the hypoglossal nerve before the 2nd crossing of Digastric, leaving its lower border; hence, the wise dissector works along the upper border.
4. The submandibular duct running forward, across Hyoglossus and Genioglossus, to its orifice.
5. The lingual nerve in contact with the jaw posteriorly, making a partial spiral around the submandibular duct, and ending in the tongue. The submandibular ganglion suspended from the nerve, and twigs leaving the nerve to supply mucuous membrane.
6. The 1st part of the lingual artery behind Hyoglossus, and the 3rd part in front.

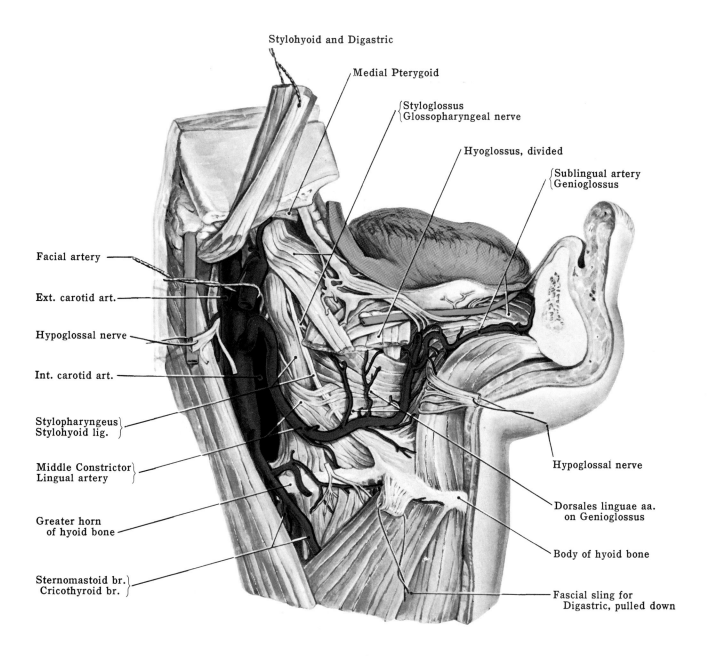

Stylohyoid and Digastric

Medial Pterygoid

Styloglossus
Glossopharyngeal nerve

Hyoglossus, divided

Sublingual artery
Genioglossus

Facial artery

Ext. carotid art.

Hypoglossal nerve

Int. carotid art.

Stylopharyngeus
Stylohyoid lig.

Middle Constrictor
Lingual artery

Greater horn
of hyoid bone

Sternomastoid br.
Cricothyroid br.

Hypoglossal nerve

Dorsales linguae aa.
on Genioglossus

Body of hyoid bone

Fascial sling for
Digastric, pulled down

9-24 SUPRAHYOID REGION—IV

Stylohyoid and posterior belly of Digastric are pulled up, the hypoglossal nerve is divided and thrown backward and forward, and Hypoglossus is mostly removed.

Observe:

1. The 1st part of the lingual artery, as in Figure 9-23.
2. The 2nd part, deep to Hyoglossus, parallel to greater horn of hyoid bone, and lying on Middle Constrictor, stylohyoid ligament, and Genioglossus.
3. The 3rd part, ascending at the anterior border of Hyoglossus, which partly overlaps it, and turning into the tongue as the profunda linguae artery (see Fig. 7-3).
4. The branches of the lingual artery: (a) muscular, (b) dorsales linguae from the 2nd part which reach the tonsil bed, and (c) the sublingual artery which supplies the sublingual gland and the front of the floor of the mouth.
5. Stylopharyngeus, appearing deep to Styloglossus and disappearing deep to Middle Constrictor.
6. The glossopharyngeal nerve, not making the usual spiral descent lateral to Stylopharyngeus as in Figure 9-55 but descending medial to it.

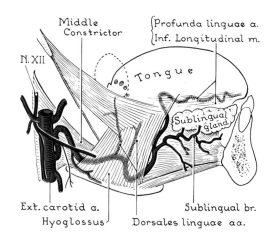

Middle
Constrictor

Profunda linguae a.
Inf. Longitudinal m.

N. XII

Tongue

Sublingual
gland

Ext. carotid a.
Hyoglossus

Sublingual br.

Dorsales linguae aa.

9-25 LINGUAL ARTERY

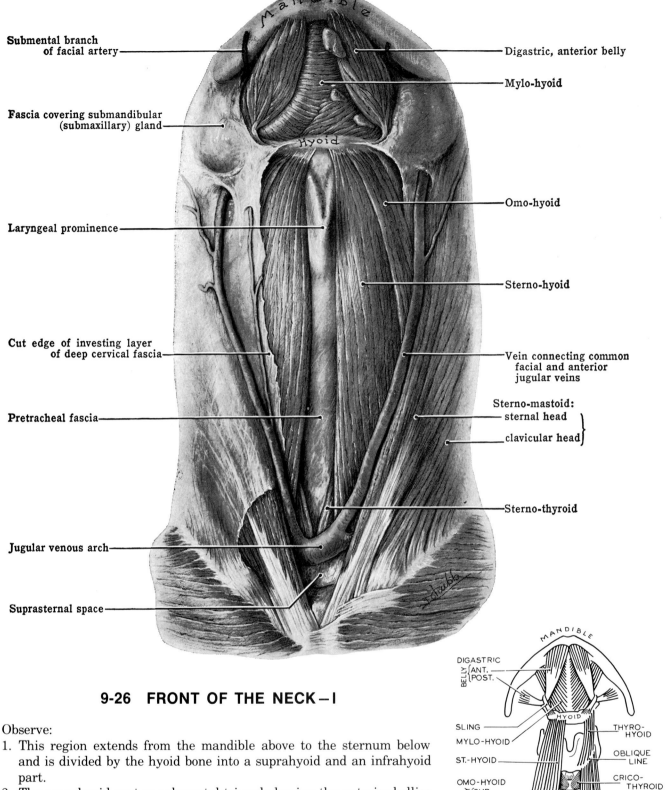

Submental branch of facial artery

Fascia covering submandibular (submaxillary) gland

Laryngeal prominence

Cut edge of investing layer of deep cervical fascia

Pretracheal fascia

Jugular venous arch

Suprasternal space

Mandible

Hyoid

Digastric, anterior belly

Mylo-hyoid

Omo-hyoid

Sterno-hyoid

Vein connecting common facial and anterior jugular veins

Sterno-mastoid:
sternal head
clavicular head

Sterno-thyroid

9-26 FRONT OF THE NECK—I

Observe:

1. This region extends from the mandible above to the sternum below and is divided by the hyoid bone into a suprahyoid and an infrahyoid part.

2. The suprahyoid part or submental triangle having the anterior bellies of Digastrics for its sides, the hyoid bone for its base, the Mylohyoid for its floor, and some submental lymph nodes for contents. Actually, the submental triangle is part of the floor of the mouth (Fig. 7-85).

3. The infrahyoid part, shaped like an elongated diamond, and bounded on each side by Sternohyoid above and Sternothyroid below.

4. The suprasternal (fascial) space, containing a cross-connecting vein called the jugular venous arch. In this specimen the anterior jugular veins were absent, as such, in the median part of the neck, but were present above the clavicles.

5. The connecting vein along the anterior border of Sternomastoid.

MANDIBLE

DIGASTRIC BELLY { ANT. POST.

SLING

MYLO-HYOID

ST.-HYOID

OMO-HYOID BELLY { SUP. INF.

SLING

THYRO-HYOID

OBLIQUE LINE

CRICO-THYROID

STERNO-THYROID

CLAVICLE

1st C.C.

STERNUM

9-27 INFRAHYOID MUSCLES

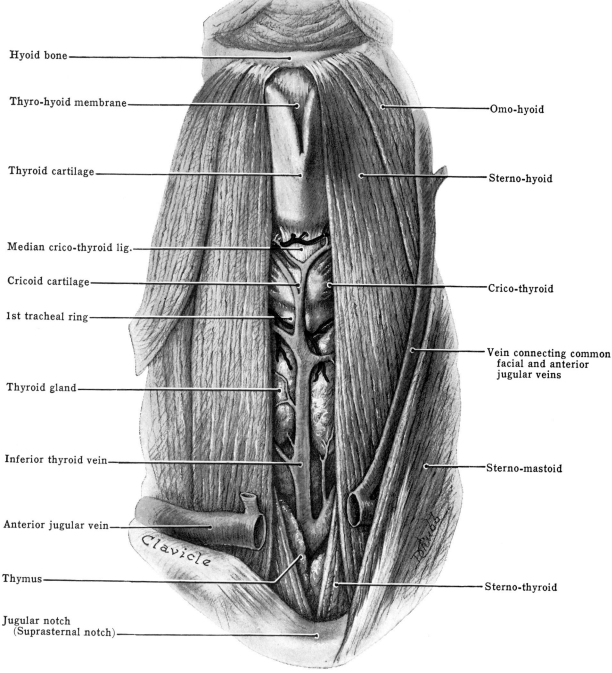

Hyoid bone

Thyro-hyoid membrane

Thyroid cartilage

Median crico-thyroid lig.

Cricoid cartilage

1st tracheal ring

Thyroid gland

Inferior thyroid vein

Anterior jugular vein

Thymus

Jugular notch
(Suprasternal notch)

Omo-hyoid

Sterno-hyoid

Crico-thyroid

Vein connecting common
facial and anterior
jugular veins

Sterno-mastoid

Sterno-thyroid

Clavicle

9-28 FRONT OF THE NECK—II

Observe:

1. The structures in the median plane labeled on the *left side* of the picture.
2. The unlabeled anastomoses between the cricothyroid arteries, between the superior thryoid arteries, and between the anterior jugular veins (excised).
3. The enlarged thymus projecting upward from the thorax.
4. The two superficial depressors of the larynx ("strap muscles"): Omohyoid (superior belly) and Sternohyoid.

9-29 ANSA CERVICALIS

The upper branch of this nerve to Sternohyoid is not usually present.

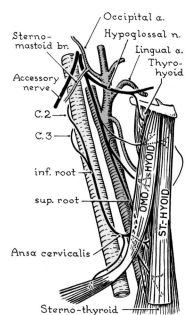

Occipital a.

Sterno-mastoid br.

Hypoglossal n.

Lingual a.

Thyro-hyoid

Accessory nerve

C.2

C.3

inf. root

sup. root

Ansa cervicalis

OMO-HYOID

ST-HYOID

Sterno-thyroid

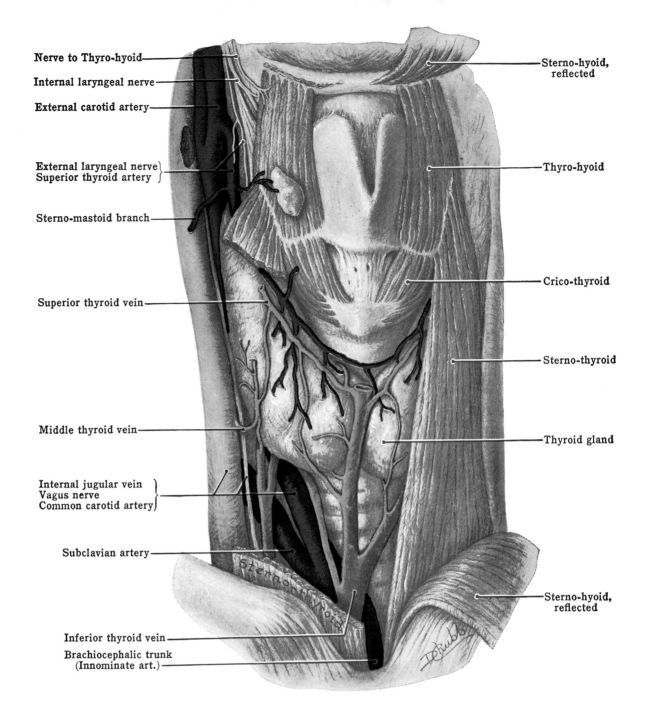

Nerve to Thyro-hyoid

Internal laryngeal nerve

External carotid artery

External laryngeal nerve
Superior thyroid artery

Sterno-mastoid branch

Superior thyroid vein

Middle thyroid vein

Internal jugular vein
Vagus nerve
Common carotid artery

Subclavian artery

Inferior thyroid vein

Brachiocephalic trunk
(Innominate art.)

Sterno-hyoid,
reflected

Thyro-hyoid

Crico-thyroid

Sterno-thyroid

Thyroid gland

Sterno-hyoid,
reflected

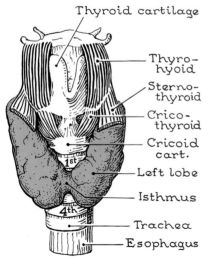

9-31 THYROID GLAND

Thyroid cartilage

Thyro-
hyoid

Sterno-
thyroid

Crico-
thyroid

Cricoid
cart.

1st

Left lobe

Isthmus

4th

Trachea

Esophagus

9-30   FRONT OF THE NECK—III

THYROID GLAND

On the *left side* the two superficial Depressors (Sternohyoid and Omohyoid) are reflected and the two deep Depressors (Sternothyroid and Thyrohyoid) are thereby uncovered. On the *right side*, Sternothyroid is largely excised.

Observe:

1. The two lobes of the thyroid gland, united across the median plane by an isthmus.

2. The surface network of veins on the gland, drained by the superior, middle, and inferior thyroid veins.

3. The right lobe of the gland, overlying the common carotid artery.

4. The upper pole of the right lobe, pushing the superior thyroid artery into the angle between the attachments of Sternothyroid and Inferior Constrictor to the thyroid cartilage, and, therefore, against the external laryngeal nerve (see also Figs. 9-32 and 9-37).

5. An accessory thyroid gland, or detached lobule, occasionally present.

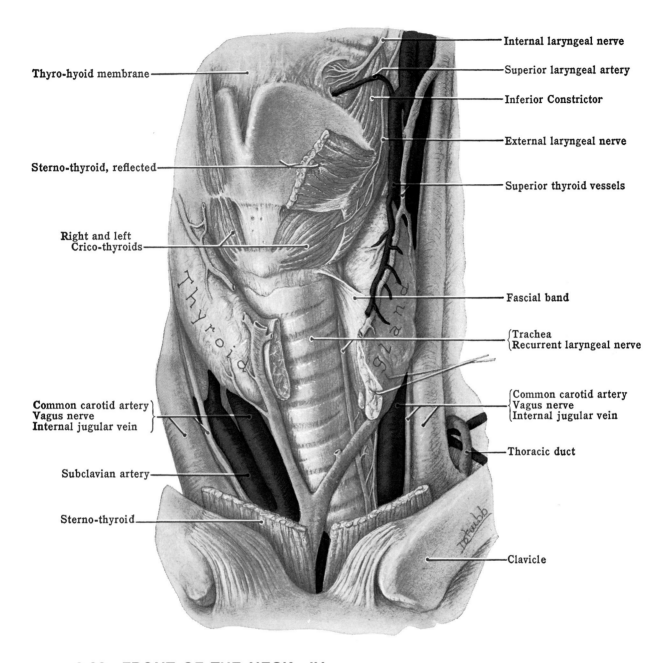

Thyro-hyoid membrane

Sterno-thyroid, reflected

Right and left
Crico-thyroids

Common carotid artery
Vagus nerve
Internal jugular vein

Subclavian artery

Sterno-thyroid

Internal laryngeal nerve

Superior laryngeal artery

Inferior Constrictor

External laryngeal nerve

Superior thyroid vessels

Fascial band

Trachea
Recurrent laryngeal nerve

Common carotid artery
Vagus nerve
Internal jugular vein

Thoracic duct

Clavicle

9-32   FRONT OF THE NECK—IV

The isthmus of the thyroid gland is divided and the left lobe is retracted.

Observe:

1. The retaining fascial band, attaching the capsule of the thyroid gland to the cricotracheal ligament and cricoid cartilage.

2. The left recurrent laryngeal nerve, on the side of the trachea, just in front of the angle between the trachea and esophagus, and behind the retaining band.

3. The internal laryngeal nerve, running along the upper border of Inferior Constrictor, and piercing the thyrohyoid membrane as several branches.

4. The external laryngeal nerve, applied to Inferior Constrictor, running along the anterior border of the superior thyroid artery, passing deep to the insertion of Sternothyroid, and giving twigs to Inferior Constrictor and piercing it before ending in Cricothyroid.

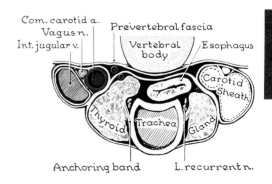

Com. carotid a.
Vagus n.
Int. jugular v.

Prevertebral fascia

Vertebral body

Esophagus

Carotid Sheath

Thyroid

Trachea

Gland

Anchoring band

L. recurrent n.

9-33   RELATIONS OF THYROID GLAND

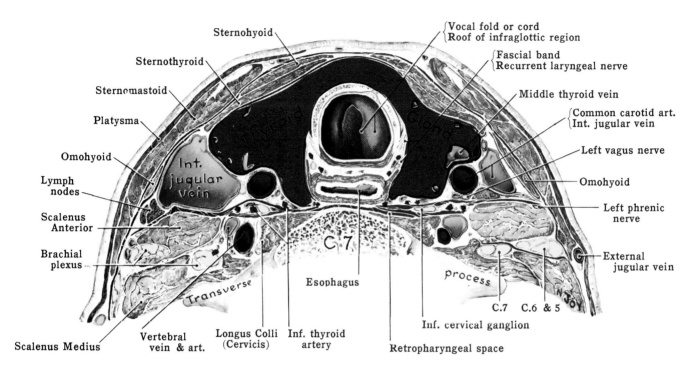

Labels on figure:

Sternohyoid
Sternothyroid
Sternomastoid
Platysma
Omohyoid
Lymph nodes
Scalenus Anterior
Brachial plexus
Scalenus Medius
Vertebral vein & art.
Longus Colli (Cervicis)
Inf. thyroid artery
Esophagus
Transverse
C 7
Retropharyngeal space
Inf. cervical ganglion
C.7   C.6 & 5
process
Vocal fold or cord
Roof of infraglottic region
Fascial band
Recurrent laryngeal nerve
Middle thyroid vein
Common carotid art.
Int. jugular vein
Left vagus nerve
Omohyoid
Left phrenic nerve
External jugular vein
Int. jugular vein
Thyroid Gland

9-34   CROSS-SECTION OF NECK THROUGH THYROID GLAND, FROM BELOW

Observe:

1. The thyroid gland, within its sheath, asymmetrically enlarged and overflowing the carotid sheath and its contents (common carotid artery, internal jugular vein, and vagus nerve) on one side and thrusting it laterally on the other.

2. The internal jugular veins, of unequal size as sometimes happens, and usually unequal vertebral arteries.

3. The retropharyngeal space of loose areolar tissue, extending far laterally behind the carotid sheath. The approach to the space is from the posterior border to Sternomastoid.

4. Scalenus Anterior deep to the posterior border of Sternomastoid.

5. The vertebral artery and vein near the apex of the "triangle of the vertebral artery" (Fig. 9-12) between Longus Colli and Scalenus Anterior.

6. The brachial plexus passing infero-laterally between Scalenus Anterior and Scalenus Medius (see Fig. 9-7).

7. The inferior thyroid artery (divided twice) and the middle cervical ganglion on a plane between the carotid sheath and the vertebral artery.

8. The fascial band that retains the thyroid gland and, behind it, the recurrent laryngeal nerve and the inferior laryngeal artery (see Fig. 9-32).

9. The vocal folds and the conus elasticus (crico-vocal membranes), covered with mucous membrane and having the same shape as the tentorium cerebelli (Fig. 9-52); hence, air expelled forcibly from the lung would blow the vocal folds apart.

10. Note that the rich blood supply of the thyroid gland is from the Superior thyroid artery, a branch of the External Carotid, which enters it superficially, and the Inferior thyroid artery, a branch of the thyrocervical trunk of the subclavian artery, which enters the deep surface of the gland.

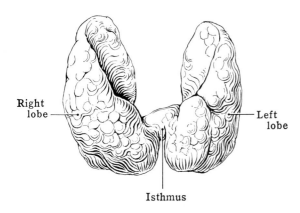

A. NORMAL THYROID GLAND

Consisting of a right and a left lobe and a connecting isthmus.

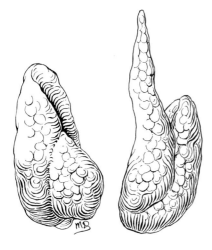

C. PYRAMIDAL LOBE. ABSENCE OF ISTHMUS

About 50 per cent of glands have a pyramidal lobe which extends from near the isthmus to, or toward, the hyoid bone. The isthmus is occasionally absent, the gland being then in two parts.

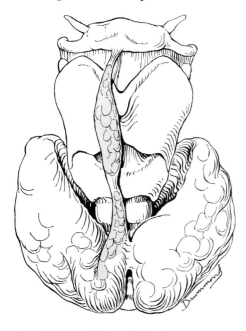

D. ACCESSORY THYROID TISSUE

This may occur along the course of the thyroglossal duct.

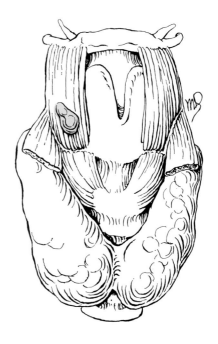

B. AN ACCESSORY THYROID GLAND

An accessory gland may occur between the levels of the suprahyoid region and the aortic arch. See Figure 9-30.

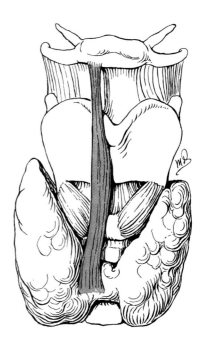

E. LEVATOR GLANDULAE THYROIDEAE

This errant slip of infrahyoid musculature is sometimes present.

9-35 THYROID GLAND, VARIATIONS

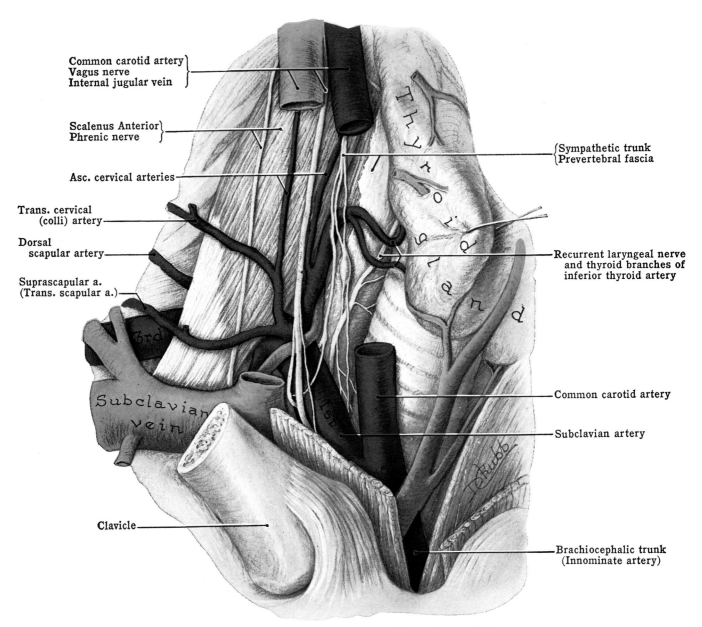

Common carotid artery ⎫
Vagus nerve
Internal jugular vein ⎭

Scalenus Anterior ⎫
Phrenic nerve ⎭

Asc. cervical arteries

Trans. cervical
(colli) artery

Dorsal
scapular artery

Suprascapular a.
(Trans. scapular a.)

Subclavian
vein

Clavicle

⎧Sympathetic trunk
⎩Prevertebral fascia

Recurrent laryngeal **nerve**
and thyroid branches of
inferior thyroid artery

Common carotid artery

Subclavian artery

Brachiocephalic trunk
(Innominate artery)

9-36 ROOT OF THE NECK, RIGHT SIDE

The clavicle is removed, sections are taken from the common carotid artery
and internal jugular vein, the right lobe of the thyroid gland is retracted.

Observe:

1. The brachiocephalic trunk dividing behind the sternoclavicular joint into the right common carotid and right subclavian arteries.

2. Scalenus Anterior, dividing the subclavian artery into three parts—1st, 2nd, and 3rd—and separating the 2nd part from the subclavian vein. This vein, lying antero-inferior to the artery joins the internal jugular vein at the medial border of Scalenus Anterior to form the brachiocephalic vein.

3. Running vertically on the obliquely placed Scalenus Anterior: (a) the common carotid artery, internal jugular vein, and vagus nerve (inside the carotid sheath); (b) the sympathetic trunk behind the common carotid artery (outside the sheath); (c) the ascending cervical artery (here represented by two vessels); and (d) most lateral of all, the phrenic nerve which is clamped down by two arteries and, therefore, cannot be mistaken for the vagus nerve.

4. The vagus, crossing the 1st part of the subclavian artery, and giving off an (inferior) cardiac branch and the recurrent laryngeal nerve. The latter (a) recurring below the artery, (b) crossing behind the common carotid artery on its way to the side of the trachea, and (c) giving twigs to the trachea and esophagus, and receiving twings from the sympathetic.

5. The sympathetic trunk, throwing a fine loop, the ansa subclavia, around the 1st part of the artery. The middle cervical ganglion is not labeled.

6. The thyro-cervical trunk (not labeled), dividing into the inferior thyroid artery which takes an S-shaped course, and the transverse cervical and suprascapular arteries which cross Scalenus Anterior.

7. The deep or dorsal scapular branch of the transverse cervical artery, here and commonly, springing from the 2nd or 3rd part of the subclavian artery. The vertebral vein (not labeled) crossing the 1st part of the artery.

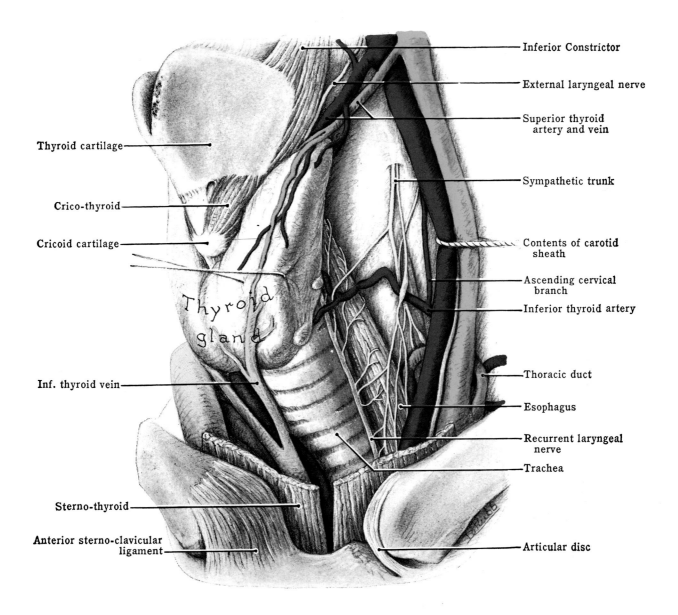

Thyroid cartilage

Crico-thyroid

Cricoid cartilage

Thyroid gland

Inf. thyroid vein

Sterno-thyroid

Anterior sterno-clavicular ligament

Inferior Constrictor

External laryngeal nerve

Superior thyroid artery and vein

Sympathetic trunk

Contents of carotid sheath

Ascending cervical branch

Inferior thyroid artery

Thoracic duct

Esophagus

Recurrent laryngeal nerve

Trachea

Articular disc

9-37 ROOT OF THE NECK, LEFT SIDE

Observe:
1. The three structures contained in the carotid sheath (internal jugular vein, common carotid artery, and vagus nerve), retracted.
2. The esophagus, bulging to the left of the trachea. It does not bulge to the right.
3. The left recurrent nerve, ascending on the side of the trachea just in front of the angle between the trachea and esophagus, giving twigs to the esophagus and trachea (not in view), and receiving twigs from the sympathetic.
4. The thoracic duct, passing from the side of the esophagus to its termination (Figs. 9-40 and 1-83) and, in so doing, arching immediately behind the 3 structures contained in the carotid sheath.
5. The middle cervical (sympathetic) ganglion, here in 2 parts: one in front of the inferior thyroid artery; the other, just above the thoracic duct, is called the vertebral ganglion.

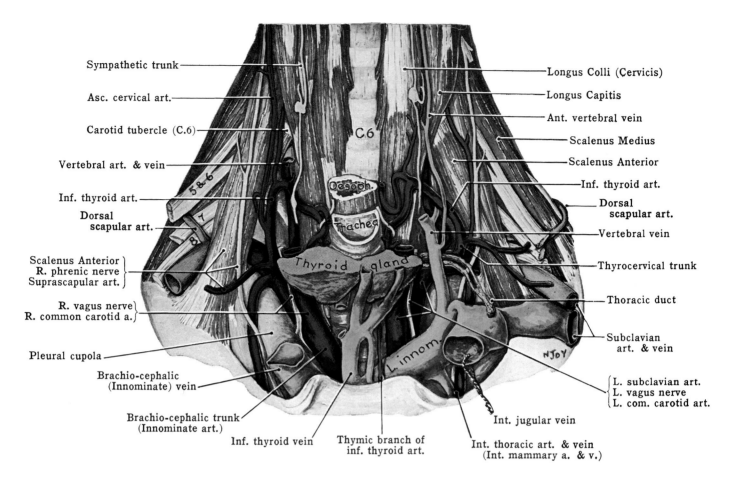

Sympathetic trunk

Asc. cervical art.

Carotid tubercle (C.6)

Vertebral art. & vein

Inf. thyroid art.

Dorsal scapular art.

Scalenus Anterior ⎱
R. phrenic nerve ⎰
Suprascapular art. ⎰

R. vagus nerve ⎱
R. common carotid a. ⎰

Pleural cupola

Brachio-cephalic (Innominate) vein

Brachio-cephalic trunk (Innominate art.)

Inf. thyroid vein

Thymic branch of inf. thyroid art.

Longus Colli (Cervicis)

Longus Capitis

Ant. vertebral vein

Scalenus Medius

Scalenus Anterior

Inf. thyroid art.

Dorsal scapular art.

Vertebral vein

Thyrocervical trunk

Thoracic duct

Subclavian art. & vein

⎰ L. subclavian art.
⎱ L. vagus nerve
⎱ L. com. carotid art.

Int. jugular vein

Int. thoracic art. & vein (Int. mammary a. & v.)

9-38 ROOT OF THE NECK, VIEWED OBLIQUELY FROM ABOVE

Observe:

1. *Laterally:* The pleural cupola, rising 1¹/₂ inches above the sternal end of the 1st rib. The subclavian artery, which arches over the pleura, divided into 3 unequal parts by Scalenus Anterior. The 3rd part of the artery and the brachial plexus appearing between Scalenus Anterior and Scalenus Medius, the lowest trunk of the plexus (C8 and T1) being behind the artery.

2. The phrenic nerve, descending almost vertically and crossing the obliquely running Scalenus Anterior to which it is clamped by the suprascapular artery. On leaving Scalenus Anterior it lies on the pleura and crosses the internal thoracic artery before meeting the right brachiocephalic vein.

3. *In the Median Plane:* The esophagus applied to the vertebral column, the trachea applied to the esophagus, the thyroid gland applied to the trachea and overlapping the common carotid arteries. The inferior thyroid veins descending to the left brachiocephalic (innominate) vein, and an occasional branch of the inferior thyroid artery descending to the thymus).

4. *Lateral to These Median Structures:* "The triangle of the vertebral artery" bounded laterally by Scalenus Anterior, and medially by Longus Colli. The apex where these 2 muscles meet at the carotid tubercle (anterior tubercle of transverse process of C6); the base being the 1st part of the subclavian artery. The vertebral artery, which ascends from base to apex, dividing the triangle into 2 nearly equal parts.

Anterior to the triangle, the carotid sheath and its 3 contents (artery, vein, and nerve). On the *left side,* between sheath and vertebral artery, and therefore below the level of the carotid tubercle, 2 vessels arching in opposite directions—the inferior thyroid artery and the thoracic duct. The artery of both sides arching medially, and on the *left side* the duct arching laterally. The artery reaches the thyroid gland as 2 branches, an upper and a lower. The recurrent laryngeal nerve bears a varying relationship to these 2 terminal arteries (being either anterior to, or posterior to, or between them), here, between them.

The duct pulled down by the reflected internal jugular vein in which it ends. The arching duct lying immediately behind the carotid sheath and its 3 contents and sometimes, as here, a 4th structure, namely, the vertebral vein.

5. The right and left vagus nerves descending on the lateral side of the corresponding common carotid artery. The right vagus is conducted to the subclavian artery which it crosses, giving off its recurrent nerve (not labeled) as it does so; the left vagus is conducted to the aortic arch (not in view) where it behaves similarly. The vagus nerves, free, and not clamped down as are the phrenic nerves.

6. Scalenus Anterior intervening between the subclavian artery and vein.

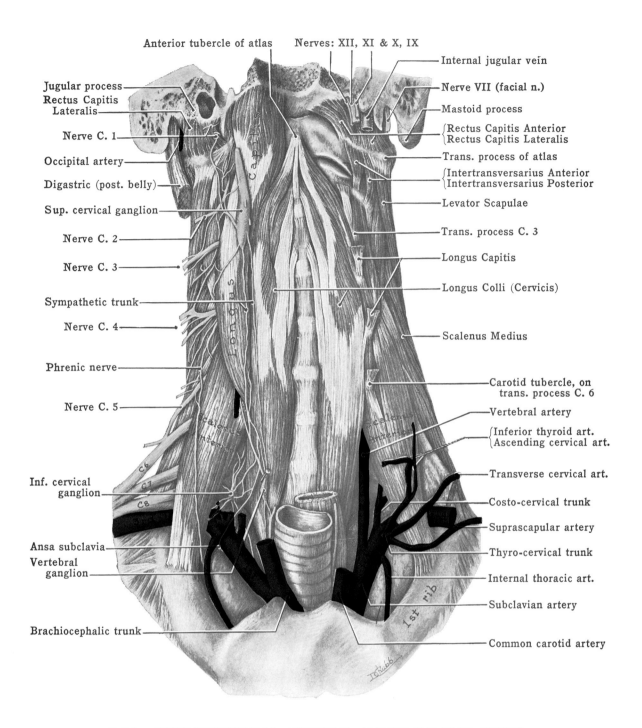

9-39 PREVERTEBRAL REGION: ROOT OF THE NECK

On the *right side*, Longus Capitis is removed.

Observe:

1. The prevertebral and deep lateral muscles of the neck. Of these muscles, 3—Scalenus Anterior, Longus Capitis, and Longus Colli—are attached to the anterior tubercles of the transverse processes of vertebrae C3, C4, C5, and C6; hence, the prominence of these tubercles, as seen in Figures 5-6 and 5-7.

2. The transverse process of the atlas joined to the transverse process of the axis by Intertransverse muscles and joined similarly to the "transverse process" of the occipital bone (*i.e.*, jugular process) by Rectus Capitis Lateralis which morphologically is an Intertransverse muscle. Note: the internal jugular vein crosses these structures.

3. The cervical plexus arising from ventral rami, C1, C2, C3, and C4; the brachial plexus from C5, C6, C7, C8, and T1.

4. The sympathetic trunk and ganglia and its gray rami communicantes.

5. The subclavian artery and its branches.

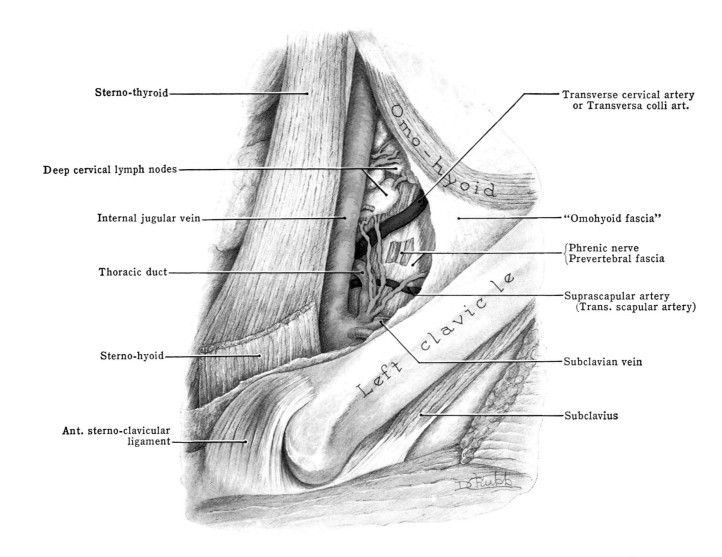

Sterno-thyroid

Deep cervical lymph nodes

Internal jugular vein

Thoracic duct

Sterno-hyoid

Ant. sterno-clavicular ligament

Omo-hyoid

Transverse cervical artery or Transversa colli art.

"Omohyoid fascia"

Phrenic nerve
Prevertebral fascia

Suprascapular artery
(Trans. scapular artery)

Left clavicle

Subclavian vein

Subclavius

9-40　TERMINATION OF THE THORACIC DUCT

Sternomastoid and the enveloping fascia of the neck are removed. The deeper layer of fascia, "the Omohyoid fascia," is partly snipped away.

Observe:

1. "The Omohyoid fascia" continuous in front of Sternohyoid and across the suprasternal space with the fascia of the opposite side.
2. The thoracic duct, receiving a tributary from the nodes of the neck (jugular trunk), and another from the nodes of upper limb (subclavian trunk), and ending in the angle between the internal jugular and subclavian veins. (For thoracic course, Figs. 1-47 and 1-83.)
3. The phrenic nerve, descending in naked contact with Scalenus Anterior, *i.e.*, deep to Scalenus Anterior fascia, which is a lateral prolongation of the prevertebral fascia, and clamped down by the 2 arteries labeled.

Phrenic n.
Inf. Thyroid a.
Thoracic duct
C.6
Recurrent laryngeal n.
Contents of carotid sheath

9-41　THORACIC DUCT IN NECK

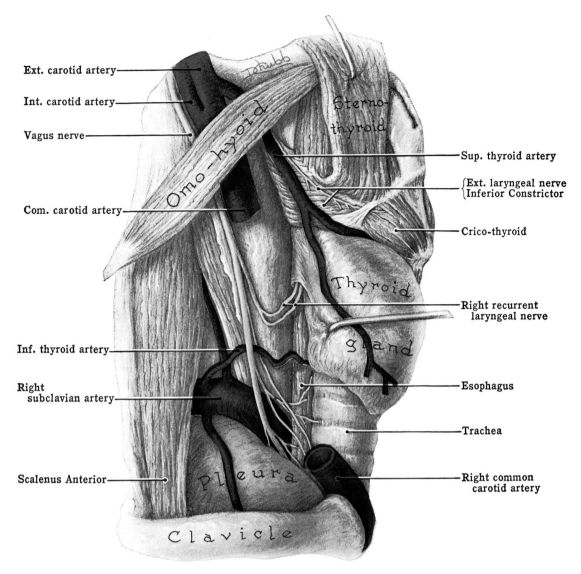

Ext. carotid artery

Int. carotid artery

Vagus nerve

Com. carotid artery

Sup. thyroid artery

Ext. laryngeal nerve
Inferior Constrictor

Crico-thyroid

Right recurrent
laryngeal nerve

Inf. thyroid artery

Right
subclavian artery

Esophagus

Trachea

Scalenus Anterior

Right common
carotid artery

9-42 ANOMALOUS RIGHT RECURRENT LARYNGEAL NERVE

This illustration is from the same subject as Figure 1-72. Occasionally the right subclavian artery springs directly from the aortic arch, as its fourth branch, and passes behind the trachea and esophagus. For embryological reasons, shown in Figure 1-71, the right recurrent nerve, having no artery around which to recur, takes an almost direct course to the larynx. As would be expected, many of its esophageal and tracheal branches then spring directly from the parent vagus nerve.

Note: The inferior thyroid artery here springs directly from the subclavian artery. The vertebral and internal thoracic arteries are not labeled.

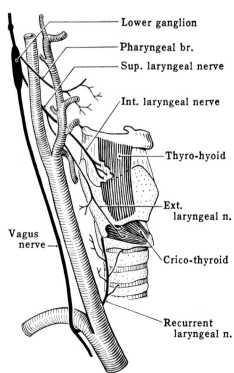

Lower ganglion

Pharyngeal br.

Sup. laryngeal nerve

Int. laryngeal nerve

Thyro-hyoid

Ext.
laryngeal n.

Vagus
nerve

Crico-thyroid

Recurrent
laryngeal n.

9-43 LARYNGEAL BRANCHES OF RIGHT VAGUS NERVE

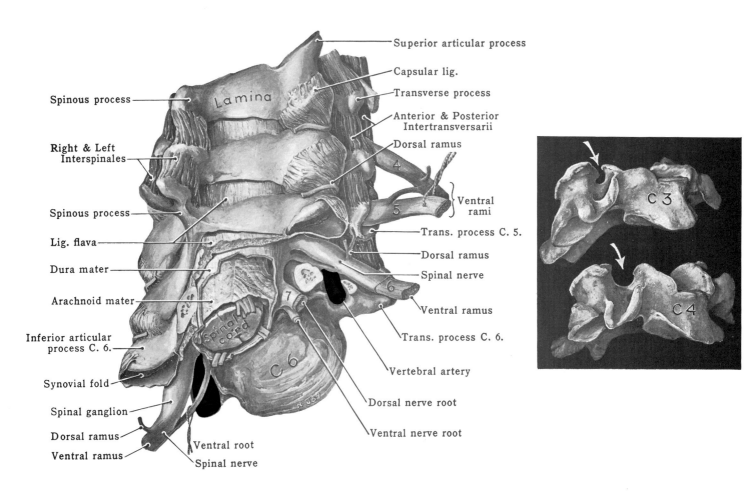

Superior articular process
Capsular lig.
Transverse process
Anterior & Posterior Intertransversarii
Dorsal ramus
Ventral rami
Trans. process C. 5.
Dorsal ramus
Spinal nerve
Ventral ramus
Trans. process C. 6.
Vertebral artery
Dorsal nerve root
Ventral nerve root

Spinous process
Right & Left Interspinales
Spinous process
Lig. flava
Dura mater
Arachnoid mater
Inferior articular process C. 6.
Synovial fold
Spinal ganglion
Dorsal ramus
Ventral ramus
Ventral root
Spinal nerve

9-44 A CERVICAL NERVE, *IN SITU*

Observe:

1. Two paired Interspinales uniting the bifid ends of 3 spinous processes.
2. Capsular ligaments uniting articular processes.
3. Anterior and Posterior Intertransversarii uniting anterior and posterior tubercles of transverse processes.
4. The ligamentum flavum attached to the upper border of one lamina and to the ventral surface of the lamina next above, and extending transversely from one articular capsule to the other.
5. Dura mater applied to ligamentum flavum; arachnoid mater applied to dura and separated from pia mater, which clothes the medulla or cord, by cerebrospinal fluid, which has escaped.
6. The fila of the dorsal nerve roots (7th and 8th) leaving the spinal medulla in single file. One filum of C8 joining C7.
7. Dura (and arachnoid) carried distally as a covering for the roots, nerve, and rami and gradually fading away.
8. A dorsal nerve root is larger than an ventral root; its swelling is the spinal ganglion. The 2 roots, each in a separate dural sheath, uniting beyond the ganglion to form a mixed spinal nerve. The nerve, about 1 cm long, dividing into a small dorsal and a large ventral ramus.
9. The roots and the nerve crossing behind the vertebral artery. The dorsal ramus of the spinal nerve curving dorsally, applied to the root of a superior articular process, and passing medial to a Posterior Intertransverse muscle. The ventral ramus resting on the transverse process, which is grooved to support it, and emerging between an anterior and a posterior intertransverse muscle. C5 raised from its grooved bed.
10. *Inset:* On vertebra C3, in contrast with C4, bony overgrowths at intervertebral foramen (on superior articular process and body) may commonly constrict the roots of the nerve.
11. Synovial folds projecting between articular surfaces, as they do elsewhere.

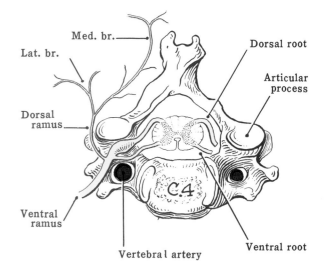

Med. br.
Lat. br.
Dorsal ramus
Ventral ramus
Vertebral artery
Dorsal root
Articular process
Ventral root
C.4

9-45 SPINAL END OF A CERVICAL NERVE

This diagram reminds us of the vulnerability of vertebral artery, spinal cord, and nerve roots to arthritic expansion from articular processes and the vertebral body, especially the lateral edge of the upper surface of the body: the unco-vertebral joint.

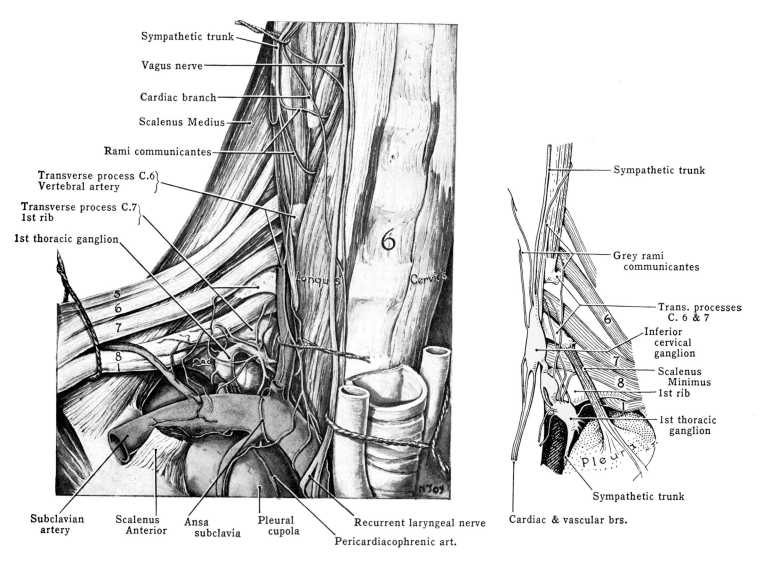

Sympathetic trunk

Vagus nerve

Cardiac branch

Scalenus Medius

Rami communicantes

Transverse process C.6
Vertebral artery

Transverse process C.7
1st rib

1st thoracic ganglion

Sympathetic trunk

Grey rami
communicantes

Trans. processes
C. 6 & 7

Inferior
cervical
ganglion

Scalenus
Minimus

1st rib

1st thoracic
ganglion

Subclavian
artery

Scalenus
Anterior

Ansa
subclavia

Pleural
cupola

Recurrent laryngeal nerve

Pericardiacophrenic art.

Cardiac & vascular brs.

Sympathetic trunk

9-46 STELLATE GANGLION (CERVICOTHORACIC GANGLION)

The pleura has been depressed and the vertebral artery retracted.

Observe:

1. The lowest trunk of the branchial plexus (C8 and T1) raised from the groove it occupies on the 1st rib behind the subclavian artery. The dorsal scapular artery (not labeled) here arises from the 3rd part of the artery and passes through the plexus (Fig. 9-85).

2. The sympathetic trunk (retracted laterally) sending a communicating branch to the vagus, and gray rami communicates (postganglionic fibers) to the roots of the cervical nerves, and cardiac branches.

3. The vertebral artery, retracted medially in order to uncover the stellate ganglion (the combined inferior cervical and 1st thoracic ganglia) which rests behind it on the 1st and 2nd ribs.

4. Fine branches passing from the ganglion to nerves C7, C8, and T1, and to adjacent arteries.

5. The vertebral ganglion, quite small, just below the thread that retracts the vertebral artery. From this ganglion the sympathetic trunk passing behind the vertebral artery to join the stellate ganglion, but before so doing sending a slender branch around the subclavian artery as the ansa subclavia.

6. The illustration on the *right* is a tracing of a photograph of the left side of the same specimen, revealing a very different pattern. Thus, the inferior cervical ganglion occupies its more usual position between the transverse process of C7 and the 1st rib, ganglion T1 being on and below the 1st rib. Scalenus Minimus (Scalenus Pleuralis) here present (also in Fig. 9-83).

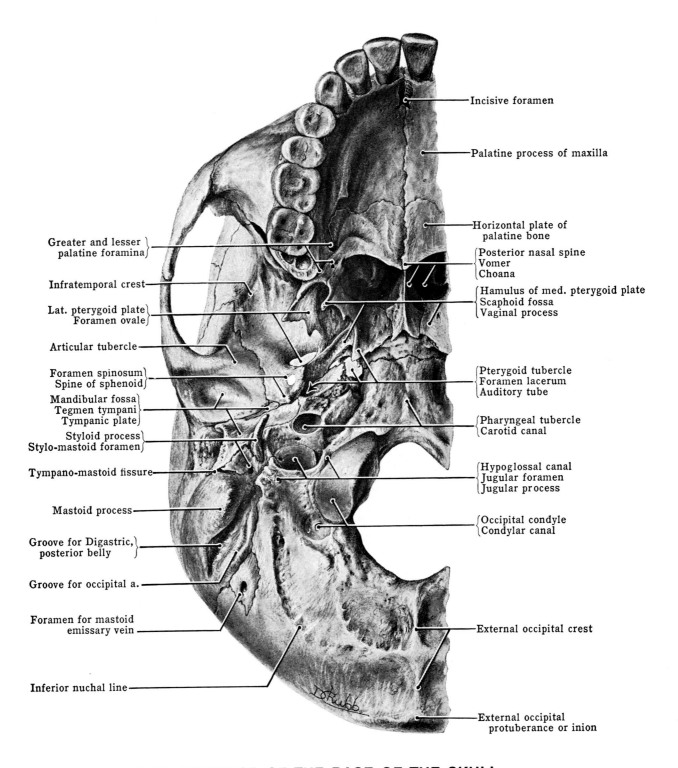

Incisive foramen

Palatine process of maxilla

Horizontal plate of palatine bone

Posterior nasal spine
Vomer
Choana

Hamulus of med. pterygoid plate
Scaphoid fossa
Vaginal process

Pterygoid tubercle
Foramen lacerum
Auditory tube

Pharyngeal tubercle
Carotid canal

Hypoglossal canal
Jugular foramen
Jugular process

Occipital condyle
Condylar canal

External occipital crest

External occipital protuberance or inion

Greater and lesser palatine foramina

Infratemporal crest

Lat. pterygoid plate
Foramen ovale

Articular tubercle

Foramen spinosum
Spine of sphenoid

Mandibular fossa
Tegmen tympani
Tympanic plate

Styloid process
Stylo-mastoid foramen

Tympano-mastoid fissure

Mastoid process

Groove for Digastric, posterior belly

Groove for occipital a.

Foramen for mastoid emissary vein

Inferior nuchal line

9-47 EXTERIOR OF THE BASE OF THE SKULL

Note:
1. The unlabeled foramen between the carotid canal and the jugular foramen which transmits the tympanic branch of the glossopharyngeal nerve (Figs. 7-146 and 8-9).
2. The unlabeled foramen on the lateral wall of the jugular foramen which transmits the auricular branch of the vagus nerve, shown in Figure 7-71. The auricular branch enters this foramen, crosses the stylomastoid foramen, and leaves through the tympanomastoid fissure.
3. The foramen of the mastoid emissary vein, in this specimen, lies in a sutural bone.

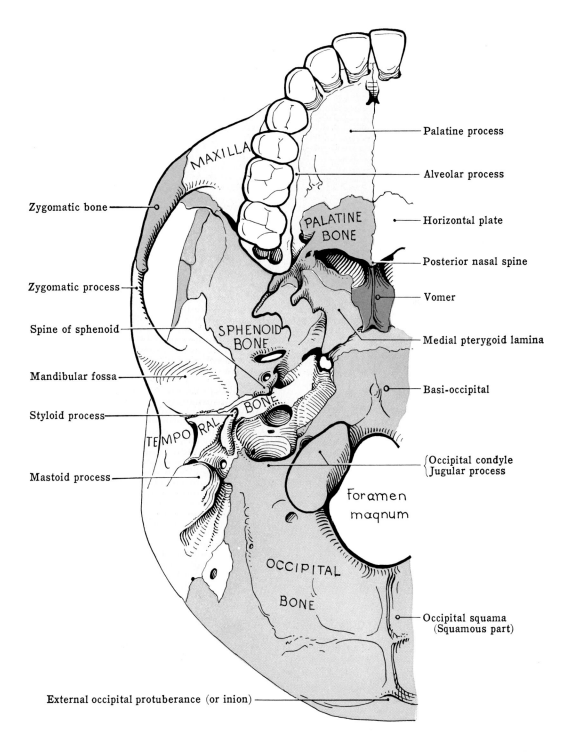

Palatine process

Alveolar process

Horizontal plate

Posterior nasal spine

Vomer

Medial pterygoid lamina

Basi-occipital

Occipital condyle
Jugular process

Foramen magnum

Occipital squama
(Squamous part)

Zygomatic bone

Zygomatic process

Spine of sphenoid

Mandibular fossa

Styloid process

Mastoid process

External occipital protuberance (or inion)

MAXILLA

PALATINE BONE

SPHENOID BONE

TEMPORAL BONE

OCCIPITAL BONE

9-48 BONES OF THE EXTERIOR OF THE BASE OF THE SKULL

Certain ligaments are associated with the base of the skull:

Ligament	From:	To:
Sphenomandibular	Spine of the sphenoid	Lingula of the mandible
Pterygomandibular	Hamulus on medial pterygoid plate	Posterior end of the mylohyoid line
Stylomandibular	Styloid process	Angle of the mandible
Stylohyoid	Tip of the styloid	Lesser horn of hyoid bone

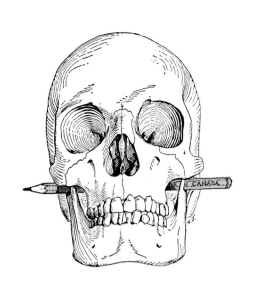

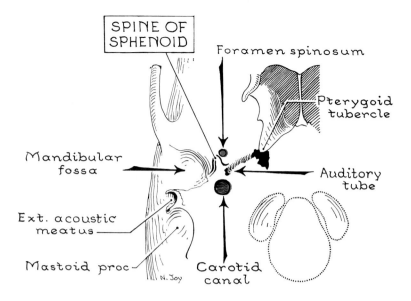

9-49 ANTERIOR TRANSVERSE LINE

A pencil can be passed through both mandibular notches and across the base of the skull. The pencil lies behind the pterygoid plates and occupies the anterior transverse line shown in Figure 9-51.

9-50 SPINE OF THE SPHENOID

A blow delivered to the side of the jaw does not drive the jaw under the cranium, because the head of the jaw would require to descend the steep medial wall of the manidbular fossa and the spine of the sphenoid in which that wall ends. Like a sentinel, the spine "keeps guard over" 4 strategic points: (a) anteriorly the foramen spinosum for the middle meningeal artery; (b) posteriorly, the entrance to the carotid canal; (c) medially, the entrance to the bony auditory tube; and (d) laterally, the mandibular fossa which lodges the head of the mandible, which therefore is the surface landmark to the spine.

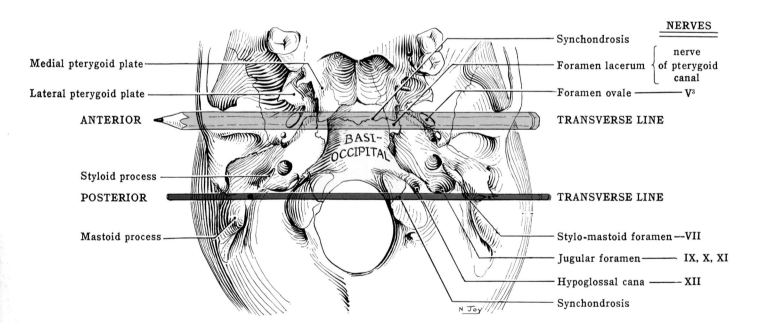

9-51 ANTERIOR AND POSTERIOR TRANSVERSE LINES

Note:
1. The anterior transverse line described in Figure 9-49.
2. The posterior transverse line stretches across the base of the skull between the mastoid and the styloid processes of the two sides. The features it crosses are labeled.
3. The foramen ovale at the root of the lateral pterygoid plate.
4. The foramen lacerum at the root of the medial pterygoid plate.
5. The synchondrosis between the basioccipital and the body of the sphenoid.

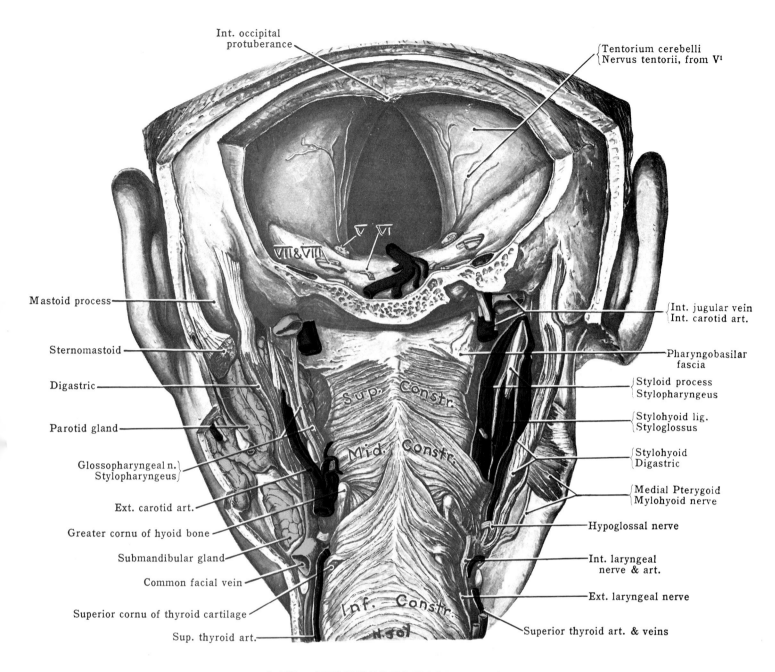

Int. occipital protuberance

Tentorium cerebelli
Nervus tentorii, from V¹

Mastoid process

Int. jugular vein
Int. carotid art.

Sternomastoid

Pharyngobasilar fascia

Digastric

Styloid process
Stylopharyngeus

Parotid gland

Stylohyoid lig.
Styloglossus

Glossopharyngeal n.
Stylopharyngeus

Stylohyoid
Digastric

Ext. carotid art.

Medial Pterygoid
Mylohyoid nerve

Greater cornu of hyoid bone

Hypoglossal nerve

Submandibular gland

Int. laryngeal nerve & art.

Common facial vein

Ext. laryngeal nerve

Superior cornu of thyroid cartilage

Superior thyroid art. & veins

Sup. thyroid art.

Sup. Constr.
Mid. Constr.
Inf. Constr.

VII & VIII   V   VI

9-52   TENTORIUM FROM BELOW:
PHARYNX AND PAROTID GLAND, FROM BEHIND

Observe:

1. The tentorium cerebelli, forming the roof of the posterior cranial fossa, and its nerve. The basioccipital, sawn across.
2. The basilar artery, formed by the union of the two unequal vertebral arteries, ascending on the dorsum sellae.
3. The three pharyngeal constrictor muscles nestle within each other. Thus, the Inferior overlaps the Middle and the Middle overlaps the Superior. This posterior aspect is flat or even slightly concave.
4. *On the right side:* The styloid process with its three muscles: *Stylopharyngeus* passing from the medial side of the styloid process forward and medially to the interval between Superior and Middle Constrictors. *Stylohyoid* passing from the lateral side forward and laterally to be split on its way to the hyoid bone by

the Digastric. *Styloglossus* passing from the anterior aspect of the process and the ligament medially and forward to its insertion into the side of the tongue.

5. *On the left side:* The glossopharyngeal nerve making a spiral around Stylopharyngeus and both entering the pharyngeal wall.
6. The posterior surface of the parotid gland grooved: by the mastoid process, below this by Sternomastoid, and more medially by Digastric, the styloid process, and the 3 muscles that arise from it (see cross-sections in Figs. 2-82 and 7-105). The parotid gland extending from the skin almost to the pharyngeal wall and separated from the submandibular gland merely by a strong layer of fascia. From the deep fascia processes extend between the lobules of the parotid gland but the submaxillary gland is free within its fascial sheath.

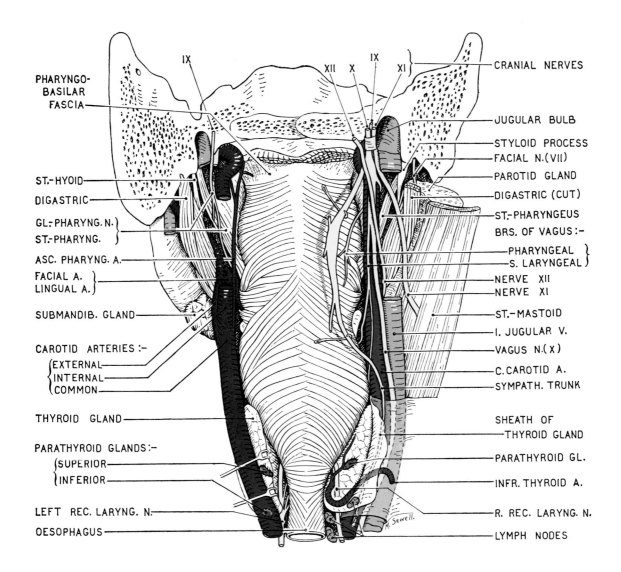

PHARYNGO-
BASILAR
FASCIA

IX XII X IX XI } — CRANIAL NERVES

JUGULAR BULB
STYLOID PROCESS
FACIAL N.(VII)
PAROTID GLAND
DIGASTRIC (CUT)
ST.-PHARYNGEUS
BRS. OF VAGUS :-
  PHARYNGEAL }
  S. LARYNGEAL }
NERVE XII
NERVE XI
ST.-MASTOID
I. JUGULAR V.
VAGUS N.(X)
C.CAROTID A.
SYMPATH. TRUNK
SHEATH OF
THYROID GLAND
PARATHYROID GL.
INFR. THYROID A.
R. REC. LARYNG. N.
LYMPH NODES

ST.-HYOID
DIGASTRIC
GL.-PHARYNG. N. }
ST.-PHARYNG. }
ASC. PHARYNG. A.
FACIAL A. }
LINGUAL A. }
SUBMANDIB. GLAND
CAROTID ARTERIES :-
 (EXTERNAL
 (INTERNAL
 (COMMON
THYROID GLAND
PARATHYROID GLANDS :-
 (SUPERIOR
 (INFERIOR
LEFT REC. LARYNG. N.
OESOPHAGUS

M. Sewell.

9-53 PHARYNX AND THE LAST FOUR CRANIAL NERVES, FROM BEHIND

Observe:

1. The narrowest and least distensible part of the alimentary canal, where the pharynx becomes the esophagus.
2. Inferior Constrictor of the pharynx overlapping Middle Constrictor, and Middle overlapping Superior.
3. Between the Superior Constrictor and the base of the skull, the semilunar area on each side where the pharyngobasilar fascia can be seen attaching the pharynx to the basioccipital bone.
4. The nerves and vein that emerge from the foramina on "the posterior transverse line of the skull" (Fig. 9-51) and which give this line its importance. They are: (a) facial nerve, (b) internal jugular vein, and (c) last four cranial nerves. Of the 4 nerves: IX lies anterior to X and XI; and XII, which is the most medial, makes a half-spiral dorsal to X and they both descend.
5. The internal carotid artery lies just behind the midpoint of the oblique line (Fig. 9-50) and therefore anterior to the structures on the posterior line. Lying posterior to the artery are the sympathetic trunk and the (elongated) superior cervical ganglion from which fibers, called the internal carotid nerve, accompany the artery into the skull.

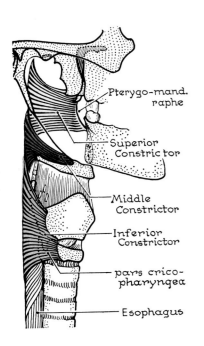

Pterygo-mand.
raphe
Superior
Constrictor
Middle
Constrictor
Inferior
Constrictor
pars crico-
pharyngea
Esophagus

9-54 CONSTRICTORS OF THE PHARYNX

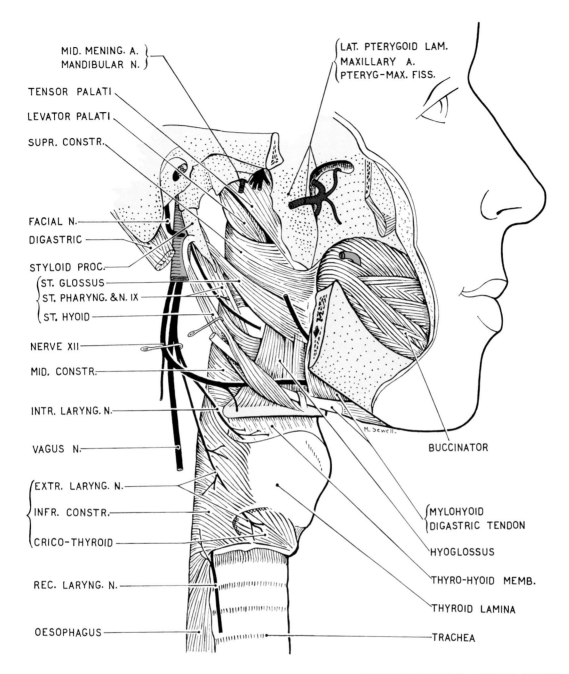

MID. MENING. A.
MANDIBULAR N.

LAT. PTERYGOID LAM.
MAXILLARY A.
PTERYG-MAX. FISS.

TENSOR PALATI

LEVATOR PALATI

SUPR. CONSTR.

FACIAL N.
DIGASTRIC

STYLOID PROC.
ST. GLOSSUS
ST. PHARYNG. & N. IX
ST. HYOID

NERVE XII

MID. CONSTR.

INTR. LARYNG. N.

VAGUS N.

EXTR. LARYNG. N.

INFR. CONSTR.

CRICO-THYROID

REC. LARYNG. N.

OESOPHAGUS

M. Sewell

BUCCINATOR

MYLOHYOID
DIGASTRIC TENDON

HYOGLOSSUS

THYRO-HYOID MEMB.

THYROID LAMINA

TRACHEA

9-55 PHARYNGEAL MUSCLES AND THE BUCCINATOR, SIDE VIEW

Note that there are four gaps in the pharyngeal musculature allowing the entry of structures:

1. Above the Superior Constrictor: Levator Palati and auditory tube.

2. Between Superior and Middle Constrictors: Stylopharyngeus, IX and Stylohyoid ligament.

3. Between Middle and Inferior Constrictors: Internal laryngeal nerve and superior laryngeal artery and nerve (not shown).

4. Below Inferior Constrictor: Recurrent laryngeal nerve and Inferior Laryngeal artery and nerve (not shown).

Observe also:
1. Superior Constrictor and Buccinator arise from opposite sides of the pterygo-mandibular raphe.
2. Middle Constrictor overlapped by Hyoglossus, and Hyoglossus in turn overlapped by Mylohyoid.
3. Tensor and Levator Palati, behind the lateral pterygoid lamina or plate, helping to form the medial wall of the infratemporal region.
4. The styloid process and the 3 muscles that arise from it: Styloglossus, Stylopharyngeus, Stylohyoid.
5. Styloglossus interdigitating with Hyoglossus on the side of the tongue.
6. Stylohyoid split by the tendon of Diagastric and attached to the hyoid bone.

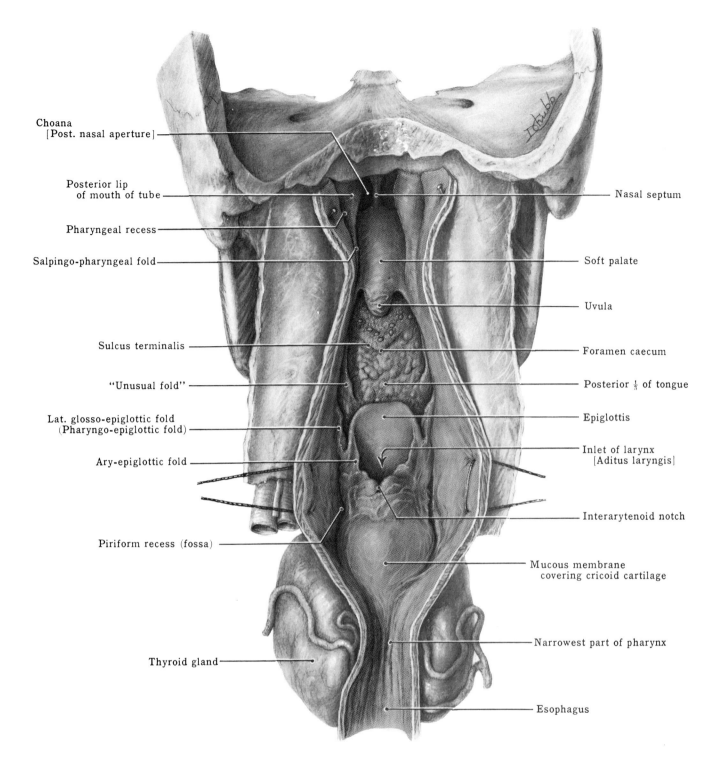

Choana
[Post. nasal aperture]

Posterior lip
of mouth of tube

Pharyngeal recess

Salpingo-pharyngeal fold

Sulcus terminalis

"Unusual fold"

Lat. glosso-epiglottic fold
(Pharyngo-epiglottic fold)

Ary-epiglottic fold

Piriform recess (fossa)

Thyroid gland

Nasal septum

Soft palate

Uvula

Foramen caecum

Posterior ⅓ of tongue

Epiglottis

Inlet of larynx
[Aditus laryngis]

Interarytenoid notch

Mucous membrane
covering cricoid cartilage

Narrowest part of pharynx

Esophagus

9-56 INTERIOR OF THE PHARYNX, FROM BEHIND

Observe:

1. The pharynx extending from the base of the skull to the lower border of the cricoid cartilage where it narrows to become the esophagus.

2. The soft palate ending postero-inferiorly in the uvula, and the larynx ending above at the tip of the epiglottis.

3. The 3 parts of the pharynx: nasal, oral, laryngeal.

4. The nasal part (naso-pharynx) lying above the level of the soft palate and continuous in front, through the choanae, with the nasal cavities.

5. The oral part, lying between the levels of the soft palate and larynx, communicating in front with the oral cavity, and having the posterior one-third of the tongue as its anterior wall. This part of the tongue is studded with lymph follicles (collectively called the lingual tonsil), and is demarcated from the anterior two-thirds by the foramen caecum and the V-shaped sulcus terminalis.

6. The laryngeal part lying behind the larynx, and communicating with the cavity of the larynx through the oblique inlet or aditus. On each side of the inlet and separated from it by the ary-epiglottic fold is a piriform recess.

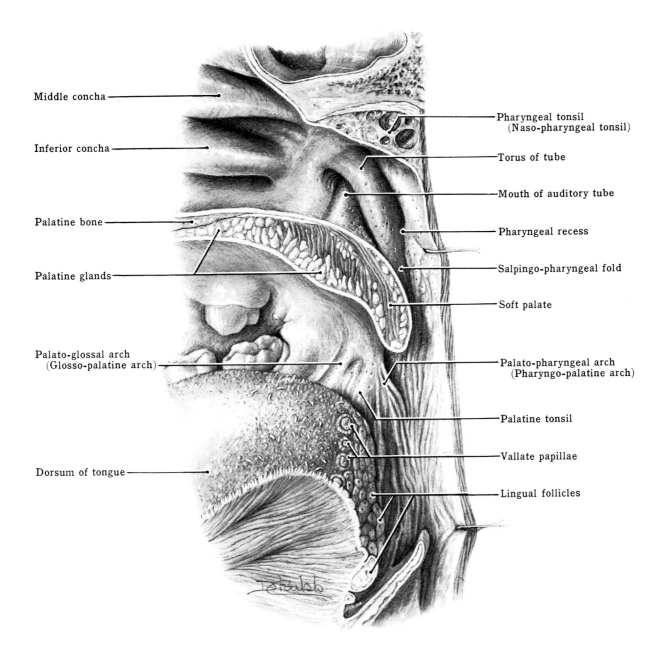

Middle concha

Inferior concha

Palatine bone

Palatine glands

Palato-glossal arch
(Glosso-palatine arch)

Dorsum of tongue

Pharyngeal tonsil
(Naso-pharyngeal tonsil)

Torus of tube

Mouth of auditory tube

Pharyngeal recess

Salpingo-pharyngeal fold

Soft palate

Palato-pharyngeal arch
(Pharyngo-palatine arch)

Palatine tonsil

Vallate papillae

Lingual follicles

9-57 INTERIOR OF THE PHARYNX, SIDE VIEW

Observe:

1. The prominent torus (superior and posterior lips) of the auditory tube and the salpingo-pharyngeal fold, which descends from the torus.

2. **The location of the orifice of the tube—about 1 cm** behind the inferior concha.

3. The ridge (torus levatorius) produced by Levator Palati which has the appearance of being poured out of the tube.

4. The deep pharyngeal recess behind the torus of the tube.

5. The numerous pinpoint orifices of the ducts of mucous glands about the torus and elsewhere.

6. The pharyngeal tonsil, better seen in Figure 7-105 where deep clefts extend into the lymphoid tissue.

7. The considerable proportion of glandular tissue in the soft palate and its disposition.

8. The palatoglossal and the palatopharyngeal arches which, in this specimen, are not the sharp folds commonly seen, and the somewhat inconspicuous palatine tonsil between them.

9. The lingual follicles, each with the duct of a mucous gland opening on to its surface. Collectively, the follicles are known as the lingual tonsil.

10. Although not shown here, it is instructive to summarize the sensory supply of the pharynx. The pharyngeal branch of the maxillary nerve (Fig. 7-109) attends to the roof, the lesser palatine nerves (Fig. 7-93) to the anterior parts of the soft palate, and the internal laryngeal nerve to the lowest part of the pharynx (Fig. 9-64) and somewhat to the surroundings of the laryngeal orifice (Fig. 7-83).

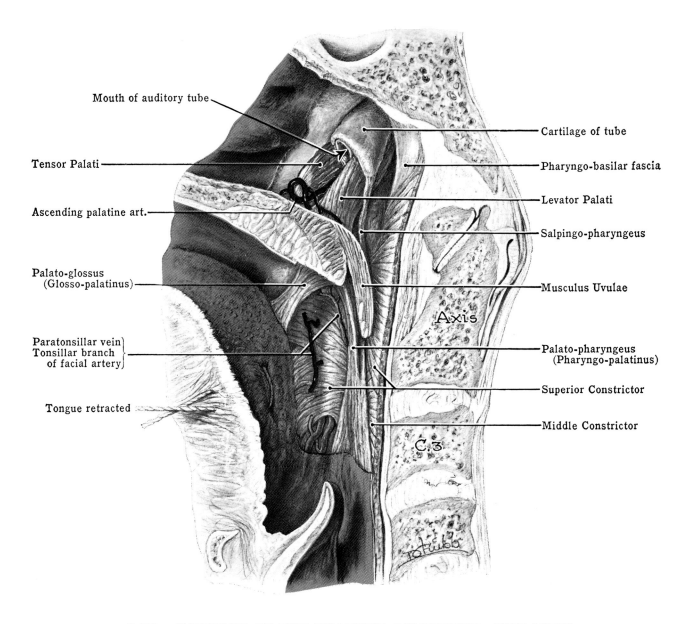

Mouth of auditory tube

Tensor Palati

Ascending palatine art.

Palato-glossus
(Glosso-palatinus)

Paratonsillar vein
Tonsillar branch
of facial artery

Tongue retracted

Cartilage of tube

Pharyngo-basilar fascia

Levator Palati

Salpingo-pharyngeus

Musculus Uvulae

Axis

Palato-pharyngeus
(Pharyngo-palatinus)

Superior Constrictor

Middle Constrictor

C. 3

9-58 INTERIOR OF THE PHARYNX DISSECTED, SIDE VIEW

The palatine and pharyngeal tonsils and the mucous membrane are removed. The submucous pharyngobasilar fascia, which attaches the pharynx to the basilar part of the occipital bone, is thick above and thin below. It too has been removed, except at the upper arched border of Superior Constrictor.

Observe:

1. The curved cartilage of the auditory tube; its free upper and posterior lips and the pharyngeal orifice of the tube; and Salpingopharyngeus descending from the posterior lip to join Palatopharyngeus.

2. The ascending palatine branch of the facial artery descending with Levator Palati to the soft palate (Fig. 7-158).

3. The 5 paired muscles of the palate: Tensor Palati (see Fig. 9-59); Levator Palati providing most of the muscle fibers seen in the cross-section of the soft palate; Musculus Uvulae, a fingerlike bundle, arising largely from the palatine aponeurosis at the posterior nasal spine (Fig. 9-91); Palatoglossus, here a substantial band, but commonly a wisp of muscle with free anterior and posterior borders (Fig. 7-82); and Palatopharyngeus, described with Figure 9-60.

4. The tonsil bed from which a thin sheet of pharyngobasilar fascia has been removed, thereby exposing Palatopharyngeus and Superior Constrictor.

Note: The bed of the palatine tonsil extends far into the soft palate. The tonsillar branch of the facial artery is here long and large. The paratonsillar vein, descending from the soft palate to join the pharyngeal plexus of veins, is a close lateral relation of the tonsil.

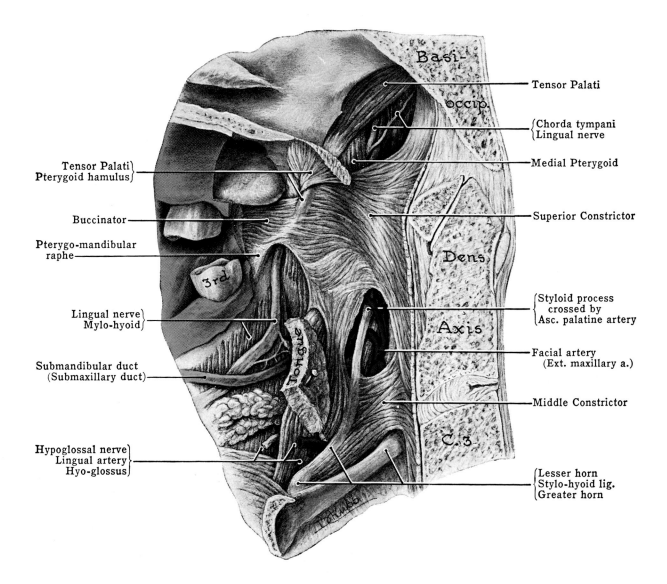

Labels on figure:
Basi-occip.
Tensor Palati
Chorda tympani / Lingual nerve
Medial Pterygoid
Superior Constrictor
Dens
Styloid process crossed by Asc. palatine artery
Axis
Facial artery (Ext. maxillary a.)
Middle Constrictor
C.3
Lesser horn / Stylo-hyoid lig. / Greater horn

Tensor Palati / Pterygoid hamulus
Buccinator
Pterygo-mandibular raphe
3rd
Lingual nerve / Mylo-hyoid
Tongue
Submandibular duct (Submaxillary duct)
Hypoglossal nerve / Lingual artery / Hyo-glossus
Tonsil

9-59 SUPERIOR AND MIDDLE CONSTRICTORS OF THE PHARYNX, FROM WITHIN

Observe:

1. Superior Constrictor arising from the pterygomandibular raphe which unites it to Buccinator, and from the bone at each end of the raphe (the hamulus of the medial pterygoid plate superiorly and the mandible inferiorly) and also from the root of the tongue.
2. The arched upper and lower borders of Superior Constrictor extending to the median plane where the muscle meets its fellow of the opposite side.
3. Middle Constrictor arising from the angle formed by the greater and lesser horns or cornua of the hyoid bone and from the stylohyoid ligament. In this specimen, the styloid process is long and is therefore a lateral relation of the tonsil.
4. The facial artery arching over the posterior belly of Digastric, and the loop of the lingual artery just below it.
5. The tendon of Tensor Palati hooking around the hamulus and then ascending to blend with the palatine aponeurosis.
6. The lingual nerve joined by the chorda tympani, disappearing at the posterior border of Medial Pterygoid, reappearing at the anterior border, and applied to the mandible.

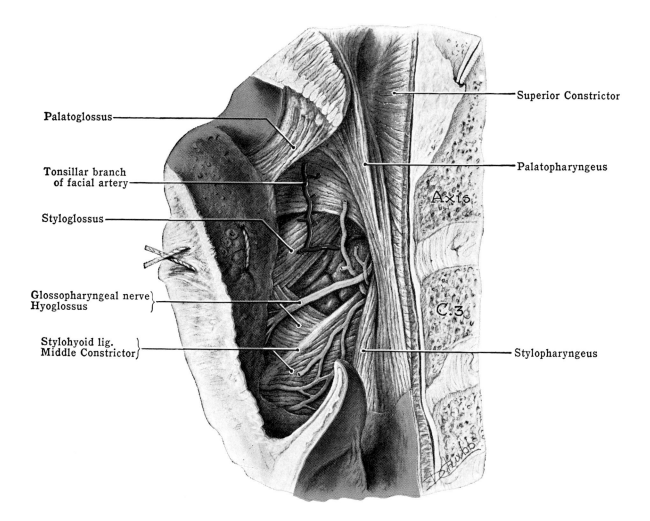

Palatoglossus

Tonsillar branch
of facial artery

Styloglossus

Glossopharyngeal nerve
Hyoglossus

Stylohyoid lig.
Middle Constrictor

Superior Constrictor

Palatopharyngeus

Axis

C-3

Stylopharyngeus

9-60   DEEP DISSECTION OF THE TONSIL BED

The tongue is pulled forward and the lower or lingual origin of Superior
Constrictor is cut away.

Observe:

1. Styloglossus passing to the anterior two-thirds of the tongue, where its bundles interdigitate with those of Hyoglossus, and therefore anterior to the
   glossopharyngeal nerve which is passing to the posterior one-third of the
   tongue. This nerve, in turn, lying anterior to Stylopharyngeus which descends along the anterior border of Palatopharyngeus.
2. The tonsillar branch of the facial artery, here sending a large branch (cut
   short) to accompany the glossopharyngeal nerve to the tongue. Lateral to
   the artery and the paratonsillar vein the submandibular salivary gland is
   seen.
3. Palatopharyngeus and Stylopharyngeus form the longitudinal coat of the
   pharynx; the Constrictors form the circular coat.

Note: In the region of the tonsil bed, Palatopharyngeus is commonly not well
differentiated from Superior Constrictor, which makes it difficult or even
arbitrary to decide where the lower border of Palatopharyngeus is. Elsewhere
its borders are easily defined.

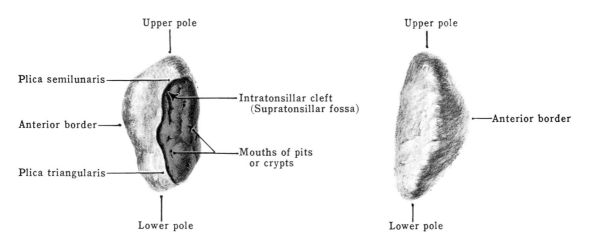

Upper pole

Plica semilunaris

Intratonsillar cleft
(Supratonsillar fossa)

Anterior border

Mouths of pits
or crypts

Plica triangularis

Lower pole

Upper pole

Anterior border

Lower pole

9-61 PALATINE TONSIL ("THE TONSIL"), MEDIAL AND LATERAL VIEWS

Observe:
1. The long axis running vertically.
2. The fibrous capsule forming the lateral or attached surface of the tonsil. In removing the tonsil, the loose areolar tissue lying between the capsule and the thin pharyngo-basilar fascia, which forms the immediate bed of the tonsil, was easily torn through.
3. The capsule extending round the anterior border and slightly over the medial surface as a thin, free fold, covered with mucous membrane on both surfaces. The upper part of this fold is called the plica semilunaris; the lower, the plica triangularis.
4. On the medial or free surface, the dozen stellate orifices of the test tube-like crypts, which extend right through the organ to the capsule. The intratonsillar cleft, which extends toward the upper pole.

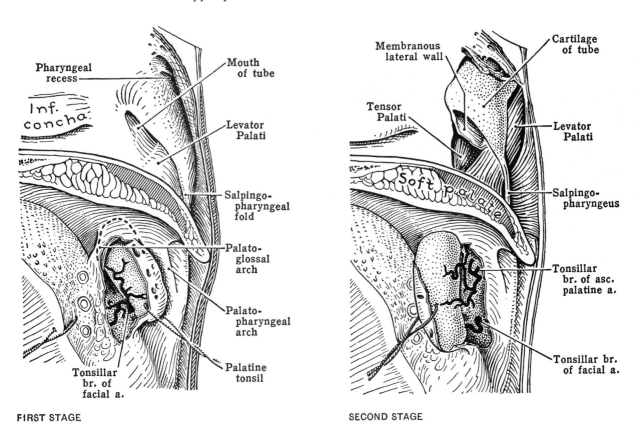

FIRST STAGE

Pharyngeal recess

Inf. concha

Mouth of tube

Levator Palati

Salpingo-pharyngeal fold

Palato-glossal arch

Palato-pharyngeal arch

Palatine tonsil

Tonsillar br. of facial a.

SECOND STAGE

Membranous lateral wall

Cartilage of tube

Tensor Palati

Levator Palati

Salpingo-pharyngeus

Tonsillar br. of asc. palatine a.

Tonsillar br. of facial a.

Soft palate

9-62 REMOVAL OF THE TONSIL. THE ARTERIAL SUPPLY

The mucous membrane has been incised along the palatoglossal arch, and the areolar space lateral to the fibrous capsule of the tonsil has been entered.

With the point and the rounded handle of the knife, the anterior border of the tonsil has been freed and the upper part, which extends far into the soft palate, has been shelled out. The mucous membrane along the palatopharyngeal arch is now cut through.

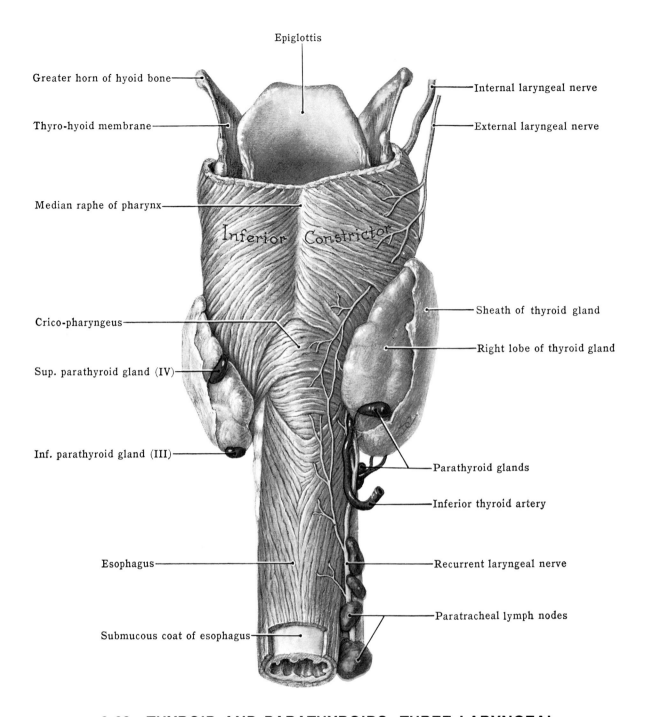

Epiglottis

Greater horn of hyoid bone

Internal laryngeal nerve

Thyro-hyoid membrane

External laryngeal nerve

Median raphe of pharynx

Inferior Constrictor

Sheath of thyroid gland

Crico-pharyngeus

Right lobe of thyroid gland

Sup. parathyroid gland (IV)

Inf. parathyroid gland (III)

Parathyroid glands

Inferior thyroid artery

Esophagus

Recurrent laryngeal nerve

Paratracheal lymph nodes

Submucous coat of esophagus

9-63 THYROID AND PARATHYROIDS, THREE LARYNGEAL NERVES, FROM BEHIND

Observe:

1. The right and left lobes of the thyroid gland, unequal in size, applied to the Inferior Constrictor of the pharynx, the trachea, and the esophagus.
2. The superior parathyroid gland, here, as usual, fusiform in shape and lying in a crevice on the posterior border of the lateral lobe of the thyroid gland. The inferior gland, more circular and applied to the lower pole of the thyroid gland. On the *right side*, both parathyroid glands are rather low, the inferior gland being altogether below the thyroid gland.
3. The internal laryngeal nerve, which is sensory (see Fig. 9-64).
4. The external laryngeal nerve, supplying Inferior Constrictor and Cricothyroid (see Figs. 9-32 and 9-42).
5. The recurrent laryngeal nerve, which is mixed, supplying esophagus, trachea (Fig. 9-37), and Inferior Constrictor, dividing into 2 branches which ascend variously related to the branches of the inferior thyroid artery. The end branch of the recurrent laryngeal nerve is officially named "inferior laryngeal nerve" and it accompanies the inferior laryngeal artery (Fig. 9-80) into the larynx.

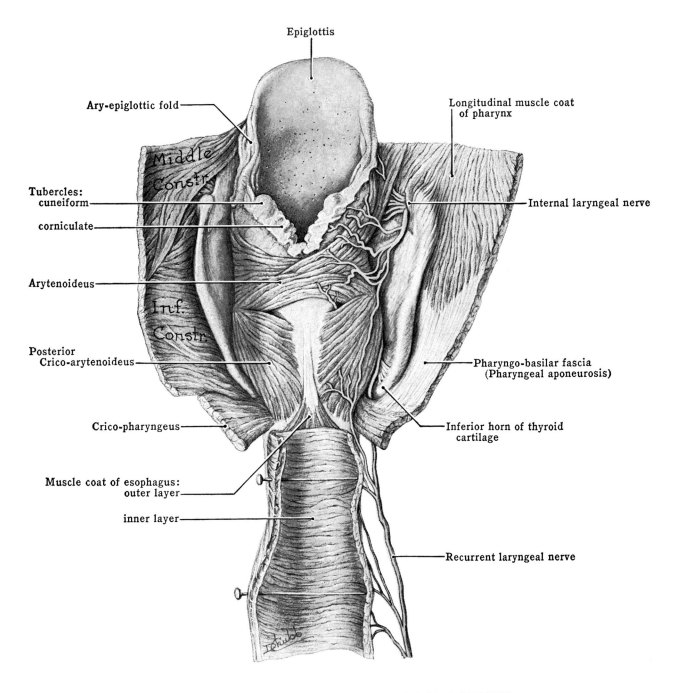

Epiglottis

Ary-epiglottic fold

Longitudinal muscle coat
of pharynx

Middle Constr.

Tubercles:
cuneiform

corniculate

Internal laryngeal nerve

Arytenoideus

Inf. Constr.

Posterior
Crico-arytenoideus

Pharyngo-basilar fascia
(Pharyngeal aponeurosis)

Crico-pharyngeus

Inferior horn of thyroid
cartilage

Muscle coat of esophagus:
outer layer

inner layer

Recurrent laryngeal nerve

9-64   MUSCLES OF THE PHARYNX, LARYNX,
AND ESOPHAGUS, POSTERIOR VIEW

The mucous membrane of the pharynx and esophagus is removed; the left Palatopharyngeus also is removed and the Constrictors are thereby uncovered.

Observe:

1. On the epiglottis, the pinpoint orifices of the glands that occupy the pits on the epiglottic cartilage (Fig. 9-72).

2. Palatopharyngeus and Stylopharyngeus together constituting the inner or longitudinal muscle coat of the pharynx (Figs. 9-58 and 9-60), inserted into the pharyngobasilar fascia and thyroid cartilage.

3. The esophagus, having inner or circularly arranged muscle fibers and outer or longitudinally arranged fibers, the latter suspending the esophagus from the cricoid cartilage.

4. Inferior Constrictor, attached not to the posterior border of the thyroid cartilage but to the oblique line and the tubercles (Fig. 9-65). Its lowest fibers, called Cricopharyngeus, which act as a sphincter, attached to the cricoid cartilage.

5. The fan shape of Cricoarytenoideus Posterior. Its upper fibers rotate the arytenoid cartilage laterally; its lower fibers pull the cartilage downward.

6. Artytenoideus (Interarytenoideus), having transverse fibers and also oblique fibers which are continued into the ary-epiglottic fold as Ary-epiglotticus.

7. The recurrent laryngeal nerve (mixed—motor and sensory), entering the larynx as two branches, of which the anterior runs immediately behind the cricothyroid joint.

8. The internal laryngeal nerve (sensory) piercing the thyro-hyoid membrane as several diverging branches.

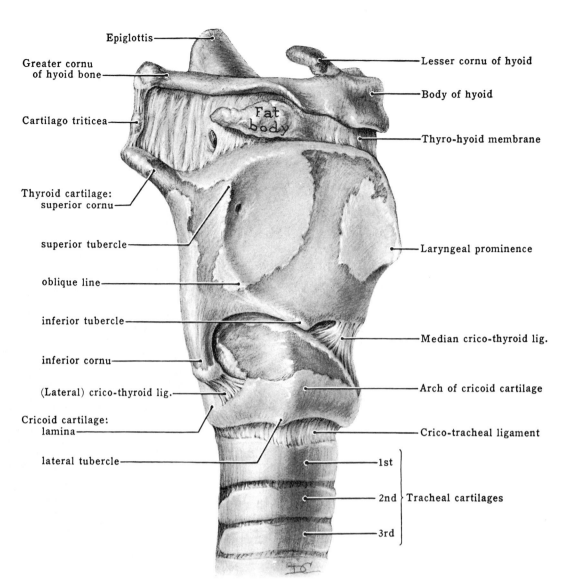

Labels (clockwise from top):
Epiglottis — Lesser cornu of hyoid — Body of hyoid — Thyro-hyoid membrane — Laryngeal prominence — Median crico-thyroid lig. — Arch of cricoid cartilage — Crico-tracheal ligament — 1st / 2nd / 3rd Tracheal cartilages — lateral tubercle — Cricoid cartilage: lamina — (Lateral) crico-thyroid lig. — inferior cornu — inferior tubercle — oblique line — superior tubercle — Thyroid cartilage: superior cornu — Cartilago triticea — Greater cornu of hyoid bone — Fat body

9-65 SKELETON OF THE LARYNX, SIDE VIEW

The larynx extends vertically from the tip of the epiglottis to the lower border of the cricoid cartilage. The hyoid bone is not regarded as part of the larynx.

Observe:

1. The lesser cornu of the hyoid bone, still partly cartilaginous, and not yet fused with the body.

The thyroid and cricoid cartilages, on the other hand, partly ossified.

2. The right lamina of the thyroid cartilage, projecting anteriorly above the point of union with its fellow to form the laryngeal prominence. Its posterior border, prolonged into a superior and an inferior cornu; of these, the inferior articulates with the cricoid cartilage. The oblique line (for the attachment of 3 muscles—Inferior Constrictor, Sternothyroid, Thyrohyoid) curving from the superior tubercle to the inferior tubercle.

3. The cricoid cartilage, having 2 parts—an arch anteriorly, and a lamina posteriorly.

4. The thyrohyoid membrane: (a) attaching the whole length of the upper border of the thyroid lamina to the upper border (not lower; See Fig. 9-74) of the body and greater cornu of the hyoid bone; (b) thickened posteriorly to form the thyrohyoid ligament which contains a nodule of cartilage; (c) pierced by the internal laryngeal nerve and companion vessels; and (d) evaginated by a fat body.

5. The median cricothyroid ligament, uniting the median parts of the adjacent borders of the cricoid and thyroid cartilages. The remainder of the lower border of the thyroid cartilage gives attachment to the Cricothyroid (Fig. 9-68), whereas the remainder of the upper border of the arch of the cricoid cartilage gives attachment to the Lateral Cricoarytenoid (Fig. 9-73) and the cricothyroid ligament (Fig. 9-74).

6. The upper border of the arch of the cricoid, inclined; the lower border, resembling that of the thyroid cartilage and projecting anteriorly beyond the trachea. By this projecting feature the cricoid cartilage can be identified in life.

Female

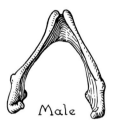

Male

9-66 THYROID CARTILAGES

Compare the angles formed by the thyroid cartilages—male and female—with those of the pubic arches in Figures 3-1 and 3-3.

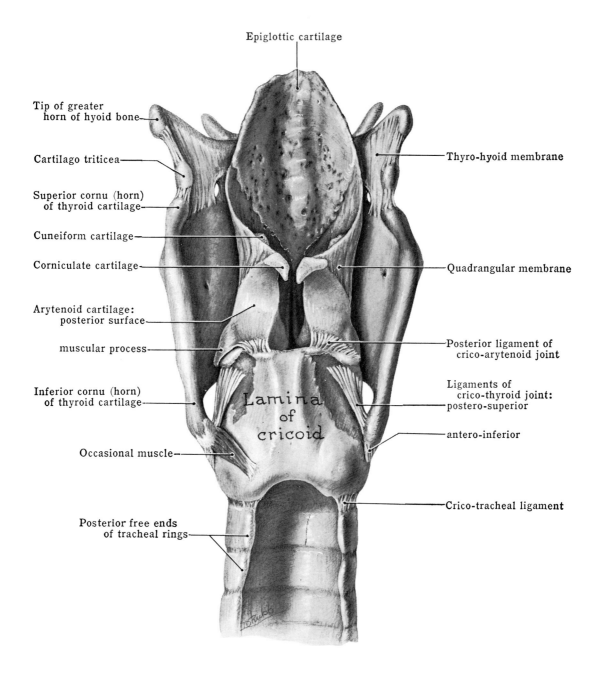

Epiglottic cartilage

Tip of greater
horn of hyoid bone

Cartilago triticea

Superior cornu (horn)
of thyroid cartilage

Cuneiform cartilage

Corniculate cartilage

Arytenoid cartilage:
posterior surface

muscular process

Inferior cornu (horn)
of thyroid cartilage

Occasional muscle

Posterior free ends
of tracheal rings

Thyro-hyoid membrane

Quadrangular membrane

Posterior ligament of
crico-arytenoid joint

Ligaments of
crico-thyroid joint:
postero-superior

antero-inferior

Crico-tracheal ligament

Lamina
of
cricoid

9-67   SKELETON OF THE LARYNX, FROM BEHIND

Observe:
1. The thyroid cartilage, shielding the smaller cartilages of the larynx (epiglottic, arytenoid, corniculate, and cuneiform). The hyoid bone—although not a part of the larynx—likewise shields the upper part of the epiglottic cartilage.
2. The rounded posterior border of the thyroid cartilage, prolonged into an upper and a lower cornu: the lower cornu articulating with the cricoid cartilage at a synovial joint (Figs. 9-73 and 9-74), the capsule of which is reinforced by 2 distinct mooring bands (postero-superior and antero-inferior) and not uncommonly by a muscle, the Keratocricoid, as shown.
3. The quadrangular membrane, connecting the border of the epiglottic cartilage to the arytenoid and corniculate cartilages, having a free upper border, and ending below as the vestibular ligament (Fig. 9-78).
4. The posterior, concave surface of the three-sided, pyramidal arytenoid cartilage.

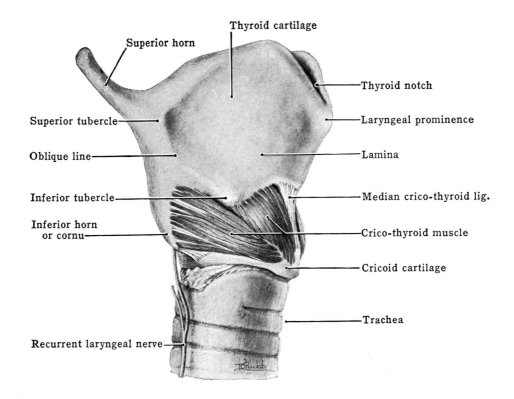

9-68 THYROID CARTILAGE, CRICOTHYROIDEUS, SIDE VIEW

The Cricothyroid arises from the outer surface of the arch of the cricoid cartilage and has 2 parts: (a) a straight, which is inserted into the lower border of the lamina of the thyroid cartilage, and (b) an oblique, which is inserted into the anterior border of the inferior horn.

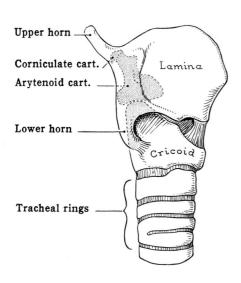

9-69 LARYNGEAL SKELETON

The thyroid cartilage shields the arytenoid cartilage and the upper part of the cricoid cartilage on which the arytenoid rests.

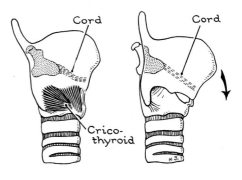

9-70 ACTION OF CRICOTHYROIDS

The Cricothyroids, acting on the Cricothyroid joints, render the vocal cords taut.

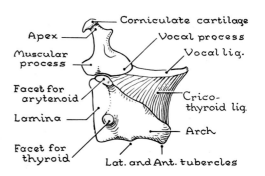

9-71 CRICOTHYROID AND VOCAL LIGAMENTS, SIDE VIEW

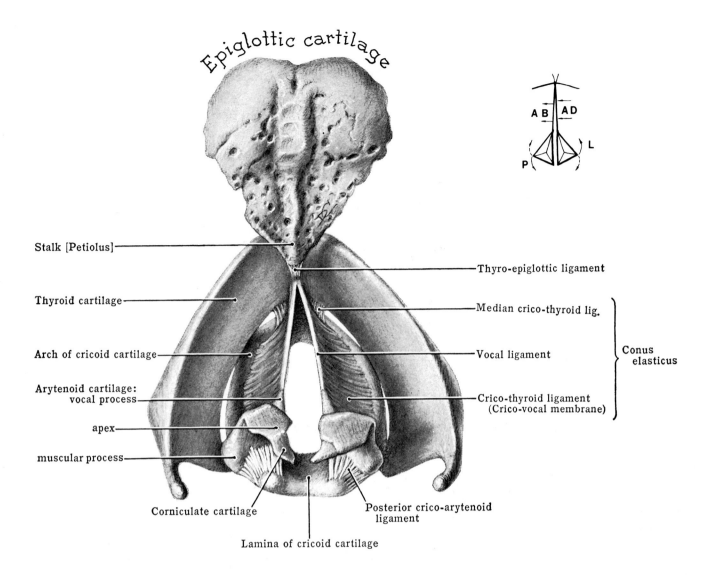

Epiglottic cartilage

Stalk [Petiolus]
Thyroid cartilage
Arch of cricoid cartilage
Arytenoid cartilage: vocal process
apex
muscular process

Thyro-epiglottic ligament
Median crico-thyroid lig.
Vocal ligament
Crico-thyroid ligament (Crico-vocal membrane)

} Conus elasticus

Corniculate cartilage
Posterior crico-arytenoid ligament
Lamina of cricoid cartilage

9-72 SKELETON OF THE LARYNX, FROM ABOVE

Observe:

1. The right and the left lamina of the thyroid cartilage, united anteriorly at an angle of about 60° in the male and 90° in the female (*cf.* the subpubic angle, Figs. 3-1 and 3-3).

2. The epiglottic cartilage, shaped like a bicycle saddle, pitted for mucous glands, and attached at its apex by ligamentous fibers to the angle of the thyroid cartilage above the vocal ligaments.

3. The paired arytenoid cartilages, having a blunt apex prolonged as the corniculate cartilage; a rounded, lateral, basal angle called the muscular process; and a sharp, anterior basal angle called the vocal process, for the attachment of the vocal ligament.

4. The strong posterior cricoarytenoid ligament, which prevents the arytenoid cartilage from falling into the larynx.

5. The vocal ligament, which forms the skeleton of the vocal fold, extending from the vocal process to the "angle" of the thyroid cartilage, and there joining its fellow below the thyroepiglottic ligament.

6. The cricothyroid ligament blending in front with the median cricothyroid ligament (Fig. 9-65), and sweeping upward from the upper border of the arch of the cricoid cartilage to the vocal ligament. Hence, when the vocal ligaments are in apposition, the membranes of opposite sides form a roof for the infraglottic section of the larynx below them. The 3 ligaments—median cricothyroid, cricothyroid, and vocal—are sometimes referred to bilaterally as the *conus elasticus*, but the term is indefinite.

7. In the *inset:* two muscles attach to the muscular process of the arytenoids: the Lateral (*L*) and Posterior (*P*) Cricoa. ⁺enoids. Contractions of the Lateral muscle *adducts* the cords; contraction of the Posterior muscle *abducts* the cords.

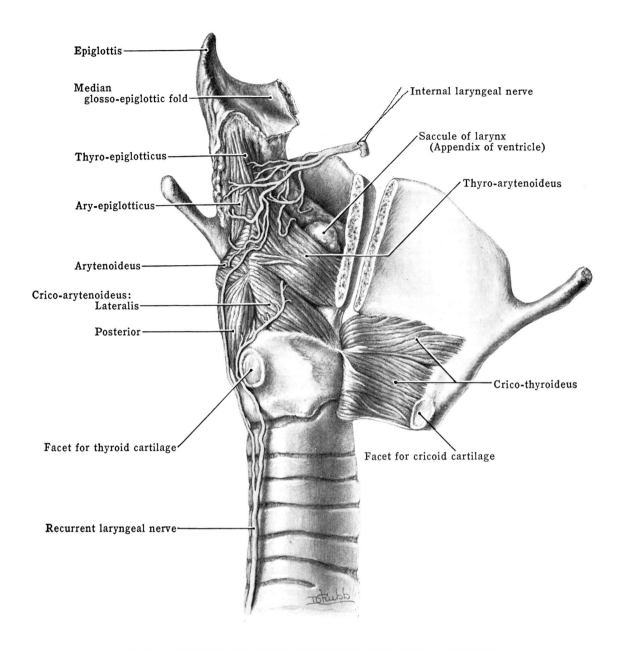

Epiglottis

Median glosso-epiglottic fold

Thyro-epiglotticus

Ary-epiglotticus

Arytenoideus

Crico-arytenoideus:
Lateralis

Posterior

Facet for thyroid cartilage

Recurrent laryngeal nerve

Internal laryngeal nerve

Saccule of larynx
(Appendix of ventricle)

Thyro-arytenoideus

Crico-thyroideus

Facet for cricoid cartilage

9-73 MUSCLES AND NERVES OF THE LARYNX, CRICOTHYROID JOINT, SIDE VIEW

The thyroid cartilage is sawn through on the right of the median plane; the cricothyroid joint is laid open; the right lamina of the thyroid cartilage is turned forward, stripping Cricothyroideus off the arch of the cricoid cartilage.

Observe:
1. Cricoarytenoideus Lateralis, arising from the upper border of the arch of the cricoid cartilage, and inserted with Cricoarytenoideus Posterior into the muscular process of the arytenoid cartilage.
2. Thyroarytenoideus, inserted with Arytenoideus into the lateral border of the arytenoid cartilage. Its upper most fibers continue to (or toward) the epiglottis as Thyroepiglotticus.
3. The blind upper end of the laryngeal saccule, see Figure 9-77.
4. The internal and recurrent laryngeal nerves, described with Figure 9-64.

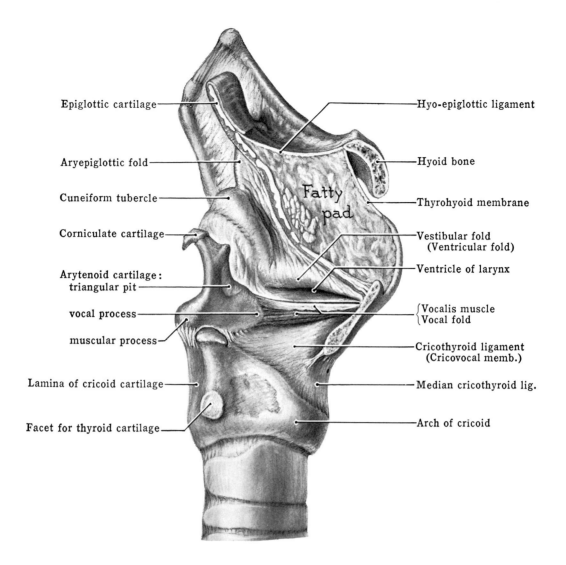

Epiglottic cartilage

Aryepiglottic fold

Cuneiform tubercle

Corniculate cartilage

Arytenoid cartilage:
triangular pit

vocal process

muscular process

Lamina of cricoid cartilage

Facet for thyroid cartilage

Fatty pad

Hyo-epiglottic ligament

Hyoid bone

Thyrohyoid membrane

Vestibular fold
(Ventricular fold)

Ventricle of larynx

Vocalis muscle
Vocal fold

Cricothyroid ligament
(Cricovocal memb.)

Median cricothyroid lig.

Arch of cricoid

9-74 LARYNX, SIDE VIEW

Above the vocal folds (vocal cords), the larynx is sectioned near the median plane and the interior of its left side is seen. Below this level, the right side of the larynx is dissected.

Observe:

1. The hyoepiglottic ligament and the thyrohyoid membrane, both attached to the upper part of the body of the hyoid bone. The space behind the body of the hyoid for the subhyoid bursa.

2. The fatty pad and the collection of glands (not labeled) filling the triangular space between ligament, membrane, and epiglottic cartilage.

3. The antero-lateral surface of the arytenoid cartilage and most of the features of this cartilage, including the pit for the attachment of the vestibular ligament and of the cuneiform cartilage. Figure 9-67 shows the concave posterior surface (covered with Arytenoideus). Figure 9-78 shows that flat medial surface (covered with mucous membrane).

4. The lateral aspect of the cricoid cartilage—the raised circular facet for the inferior cornu of the thyroid cartilage, separating lamina from arch; above this, the sloping facet for the arytenoid cartilage; the nearly horizontal lower border; and the oblique upper border of the arch.

5. The triangular membrane, called the cricothyroid ligament, having the vocal ligament (Fig. 9-78) for its upper border, blending with the median cricothyroid ligament antero-inferiorly, attached in front to the angle of the thyroid cartilage between these two structures, and below to the cricoid cartilage.

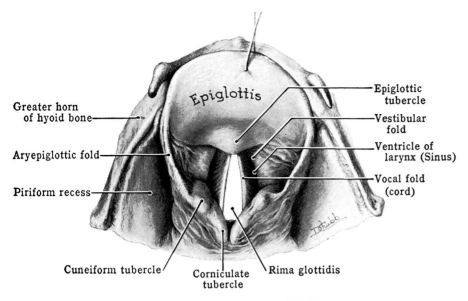

Greater horn of hyoid bone

Aryepiglottic fold

Piriform recess

Epiglottis

Cuneiform tubercle

Corniculate tubercle

Rima glottidis

Epiglottic tubercle

Vestibular fold

Ventricle of larynx (Sinus)

Vocal fold (cord)

9-75 LARYNX, FROM ABOVE

Observe:

1. The inlet or aditus to the larynx, bounded in front by the free, curved edge of the epiglottis; behind, by the arytenoid cartilages, the corniculate cartilages which cap them and the interarytenoid fold which unites them; and, on each side, by the aryepiglottic fold, which contains the upper end of the cuneiform cartilage.

2. The vocal folds, closer together than the vestibular folds (false cords) and, therefore, visible below them. The sharpness of the vocal folds; the fullness of the vestibular folds.

3. The mucous membrane, smooth and adherent in the epiglottic region and over the vocal folds, but loose and even wrinkled posteriorly, where movement is free.

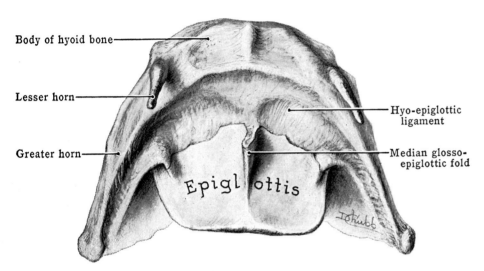

Body of hyoid bone

Lesser horn

Greater horn

Epiglottis

Hyo-epiglottic ligament

Median glosso-epiglottic fold

9-76 HYOEPIGLOTTIC LIGAMENT, FROM ABOVE

Observe:

1. The hyoepiglottic ligament, uniting the epiglottic cartilage to the hyoid bone.

2. The three parts of the hyoid bone, and the asymmetry of the greater and lesser horns or cornua of opposite sides (Fig. 7-86).

The laryngeal saccule is a diverticulum of the ventricle of the larynx which may perforate the thyrohyoid membrane. It is variable in size. Here, as in Figure 9-73, it is large. In certain apes it is enormous.

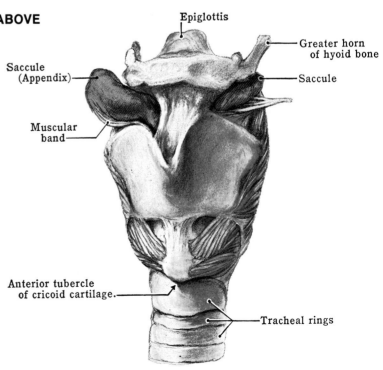

Epiglottis

Greater horn of hyoid bone

Saccule (Appendix)

Saccule

Muscular band

Anterior tubercle of cricoid cartilage

Tracheal rings

FRONT VIEW

9-77 LARGE LARYNGEAL SACCULES

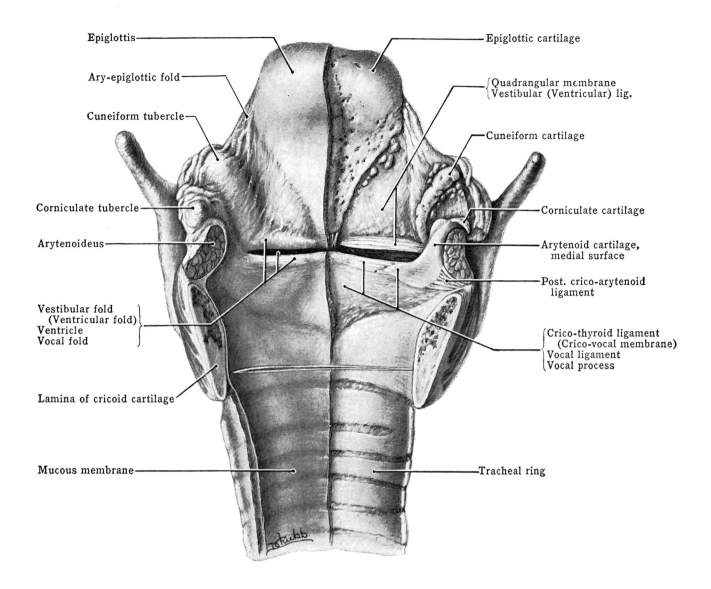

Epiglottis

Ary-epiglottic fold

Cuneiform tubercle

Corniculate tubercle

Arytenoideus

Vestibular fold
(Ventricular fold)
Ventricle
Vocal fold

Lamina of cricoid cartilage

Mucous membrane

Epiglottic cartilage

{Quadrangular membrane
{Vestibular (Ventricular) lig.

Cuneiform cartilage

Corniculate cartilage

Arytenoid cartilage,
medial surface

Post. crico-arytenoid
ligament

{Crico-thyroid ligament
{ (Crico-vocal membrane)
{Vocal ligament
{Vocal process

Tracheal ring

9-78 INTERIOR OF THE LARYNX, POSTERIOR VIEW

The posterior wall of the larynx is split medianly, and the two sides are held apart. On the *left side,* the mucous membrane, which is the innermost coat of the larynx, is intact; on the *right side,* the mucous and submucous coats are peeled off and the next coat, consisting of cartilages, ligaments, and fibroelastic membrane, is thereby laid bare.

Observe:
1. Arytenoideus and the lamina of the cricoid cartilage, divided posteriorly.
2. The entrance to the larynx is oblique; the lower limit, at the lower border of the cricoid cartilage where the trachea begins, is horizontal.
3. The 3 compartments of the larynx: (a) the uppermost compartment of vestibule, above the level of the vestibular folds; (b) the middle, between the levels of the vestibular and vocal folds, and having a right and a left canoe-shaped depression, the ventricles; and (c) the lowest or infraglottic cavity, below the level of the vocal folds.
4. The mucous membrane, particularly smooth and adherent over the epiglottic cartilage and vocal ligaments; and particularly loose and wrinkled about the arytenoid cartilages, where movement is free.
5. The two parts of the fibroelastic membrane: (a) an upper quadrangular, and (b) a lower triangular. The upper part, the quadrangular membrane, is thickened below to form the vestibular ligament. The lower part, the cricothyroid ligament (conus elasticus), begins below as the strong median cricothyroid ligament and ends above as the vocal ligament. Between the vocal and the vestibular ligament, the membrane, lined with mucous membrane, is evaginated to form the wall of the ventricle.
6. The cuneiform cartilage: (a) more club-shaped than wedge-shaped, (b) composed of elastic cartilage and pitted for, and surrounded with, glands, like the epiglottic cartilage of which it is a detached part, and (c) attached to the arytenoid cartilage beside the posterior end of the vestibular ligament.
7. The posterior ligament of the cricoarytenoid joint, anchoring the arytenoid cartilage. The flat, medial, submucous surface of the arytenoid cartilage.

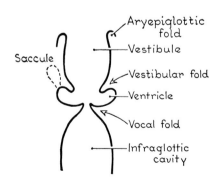

Aryepiglottic fold

Saccule

Vestibule

Vestibular fold

Ventricle

Vocal fold

Infraglottic cavity

9-79 COMPARTMENTS OF LARYNX, CORONAL SECTION

These are: a vestibule, a middle compartment having a right and a left ventricle, and an infraglottic cavity.

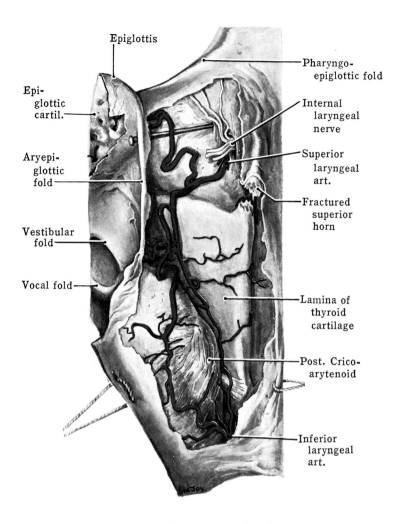

Epiglottis

Epi-
glottic
cartil.

Aryepi-
glottic
fold

Vestibular
fold

Vocal fold

Pharyngo-
epiglottic fold

Internal
laryngeal
nerve

Superior
laryngeal
art.

Fractured
superior
horn

Lamina of
thyroid
cartilage

Post. Crico-
arytenoid

Inferior
laryngeal
art.

Note the anastomoses between superior and inferior laryngeal arteries (which are branches of superior and inferior thyroid arteries). Arterial twigs pierce the epiglottic cartilage at the sites of the pits for glands.

9-80   LARYNGEAL ARTERIES

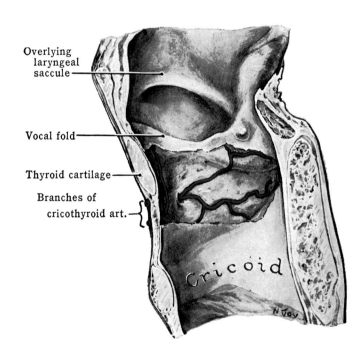

Overlying
laryngeal
saccule

Vocal fold

Thyroid cartilage

Branches of
cricothyroid art.

cricoid

9-81   DISTRIBUTION OF CRICOTHYROID ARTERY

This branch of the superior thyroid artery anastomoses with its fellow in front of the median cricothyroid ligament as shown in Figure 9-28. Lymph vessels accompany it.

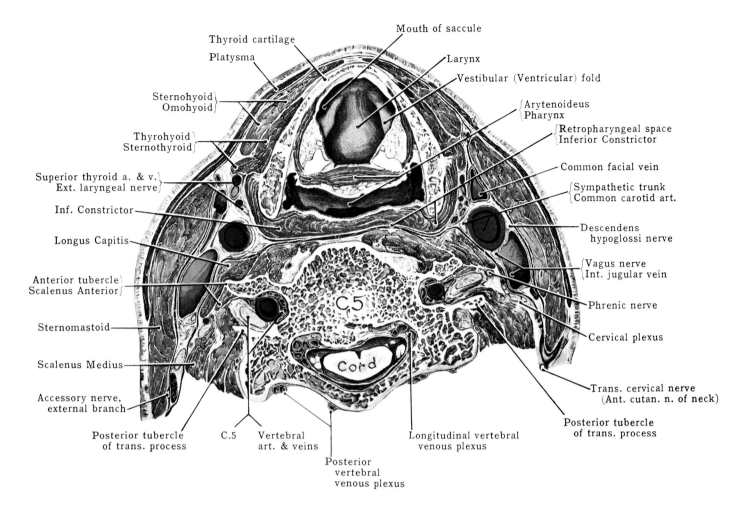

Labels (clockwise from top):
Mouth of saccule
Thyroid cartilage
Platysma
Sternohyoid } Omohyoid
Thyrohyoid } Sternothyroid
Superior thyroid a. & v. } Ext. laryngeal nerve
Inf. Constrictor
Longus Capitis
Anterior tubercle } Scalenus Anterior
Sternomastoid
Scalenus Medius
Accessory nerve, external branch
Posterior tubercle of trans. process
C.5
Vertebral art. & veins
Posterior vertebral venous plexus
Longitudinal vertebral venous plexus
Posterior tubercle of trans. process
Trans. cervical nerve (Ant. cutan. n. of neck)
Cervical plexus
Phrenic nerve
Vagus nerve } Int. jugular vein
Descendens hypoglossi nerve
Sympathetic trunk } Common carotid art.
Common facial vein
Retropharyngeal space } Inferior Constrictor
Arytenoideus } Pharynx
Vestibular (Ventricular) fold
Larynx
C.5
Cord

9-82 CROSS-SECTION OF THE NECK, THROUGH MIDDLE OF LARYNX, FROM BELOW

Observe:

1. The thyroid cartilage shielding the larynx and the pharynx.
2. The vestibular folds, seen from below, and lateral to them the mouths of the saccules of the larynx.
3. Arytenoideus (cut obliquely, hence appearing wide) attached to the posterior surface of the arytenoid cartilage and in continuity with Thyroarytenoideus (not labeled).
4. Inferior Constrictor, curving round the posterior borders of the laminae of the thyroid cartilage to be attached to the oblique line (Fig. 9-65). Sternothyroid and Thyrohyoid sharing the oblique line (Figs. 9-30 and 9-32).
5. The superior thyroid vessels and the external laryngeal nerve applied to Inferior Constrictor.
6. The 3 contents of the carotid sheath: the common carotid artery, internal jugular vein, and, in the posterior angle between them, the vagus nerve.
7. The sympathetic trunk, postero-medial to the carotid artery and medial to the vagus. The superior root of ansa cervicalis (descendens hypoglossi nerve) in front of the carotid artery.
8. The retropharyngeal space, between the pharyngeal fascia, which covers Inferior Constrictor, and the prevertebral fascia, which covers Longi Colli and Capitis. The areolar space extending laterally to the carotid sheath and readily opened up beyond it. The phrenic nerve deep to the prevertebral fascia.
9. The vetebral artery, surrounded with a plexus of veins which (inferiorly becomes the vertebral vein) and the ventral ramus of a cervical nerve (C5) crossing behind it.
10. Internal and external parts of the vertebral venous plexus.

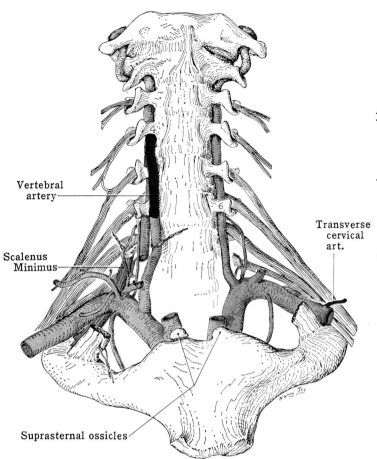

Vertebral artery

Scalenus Minimus

Suprasternal ossicles

Transverse cervical art.

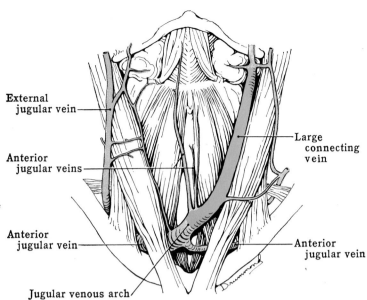

External jugular vein

Anterior jugular veins

Anterior jugular vein

Jugular venous arch

Large connecting vein

Anterior jugular vein

9-84 LARGE CONNECTING VEIN

A small vein that lies along the anterior border of Sternomastoid and connects the common facial vein to the anterior jugular vein may attain great size. It may, indeed, be greater than even the internal jugular vein and it is a times mistaken for it.

9-83 ABNORMAL VERTEBRAL ARTERY, SUPRASTERNAL OSSICLES, SCALENUS MINIMUS

1. The vertebral artery, in 6.4 per cent of 1000 half-heads, enters not the foramen of the 6th cervical transverse process but another—the 5th in 4.5 per cent, the 7th in 1.2 per cent, and the 4th as here in 0.7 per cent.

See Adachi, B. (1928) Anteria vertebralis. In *Das Arteriensystem der Japaner,* vol. 1, p. 138. Kyoto.

2. Suprasternal ossicles, which range in size between that of a small shot and an average female lunate bone, were found either paired or singly, and either separate from the manubrium or fused to it, in 6.8 per cent of 544 white adult sterna examined by x-rays.

See Cobb, W. M. (1937) The ossa suprasternalia in whites and American Negroes and the form of the superior border of the manubrium. *J. Anat. (Part 2),* 71: 245.

3. Scalenus Minimus, passing between subclavian artery and branchial plexus, is commonly present. (See also Fig. 9-46.)
4. Aberrant transverse cervical artery; see Figure 9-85.

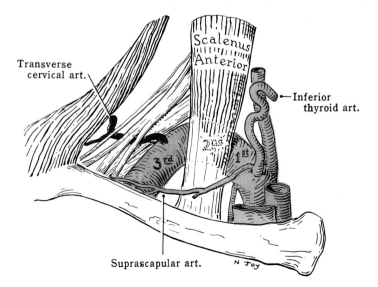

Transverse cervical art.

Scalenus Anterior

Inferior thyroid art.

3rd

2nd

1st

Suprascapular art.

9-85 BRANCHES OF THE THIRD PART OF THE SUBCLAVIAN ARTERY

Almost as a rule some branch springs from the 3rd part of this artery, or else from its short 2nd part, and passes laterally through the branchial plexus, usually behind nerve segments 5 and 6 or 5, 6, and 7 (Figs. 9-36, 9-38, and 9-46).

Thus:
a. "The dorsal scapular artery" (artery to Rhomboids, or descending branch of transverse cervical artery) so behaves in about 50 per cent of cases;
b. The transverse cervical artery in about 20 per cent of cases, when it may be said to spring from the dorsal scapular artery rather than vice versa; and
c. The suprascapular artery in about 10 per cent of cases.

See Huelke, D. F. (1958) A study of the transverse cervical and dorsal scapular arteries. *Anat. Rec.,* 132: 233.

10

The Child

CONTENTS

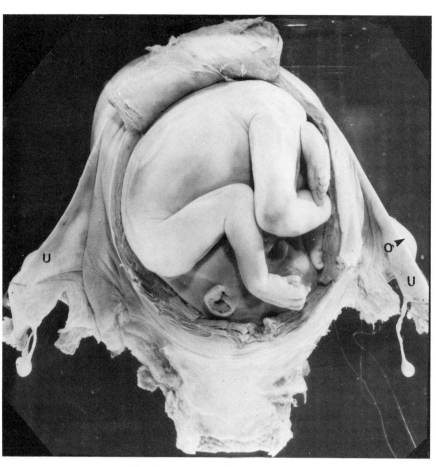

10-1A A FETUS *IN UTERO*

The anterior wall of the uterus has been removed to expose the fetus. For orientation, the uterine tubes have been labeled (*U*). The left ovary (*O*) is peering over the edge of its tube.

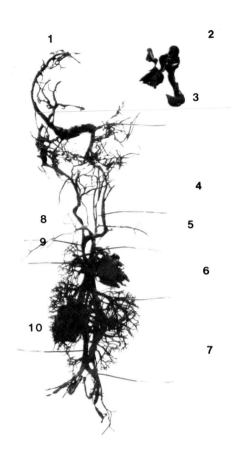

10-1B FETAL CAST

Key: *1.* Superior sagittal sinus
2. Lateral ventricles
3. Transverse sinus
4. L. Internal jugular vein
5. Arch of aorta
6. Hepatic veins
7. L. Common iliac vessels
8. Superior vena cava
9. R. Pulmonary artery
10. Inferior vena cava

10-1C FETAL MEMBRANES

Freshly delivered placenta, fetal surface with amnion/chorion attached and tortuous umbilical cord emerging from the sac.

10-1D PLACENTAL VESSELS

The fetal surface of a cast of the placental vessels. Arteries (*green*) radiate from the attachment of the cord. Veins (*red*) are deeper and larger.

10-1E PLACENTA

The uterine surface of a fresh placenta showing its granular surface divided by deep furrows into 15 to 30 lobes inaccurately called cotyledons.

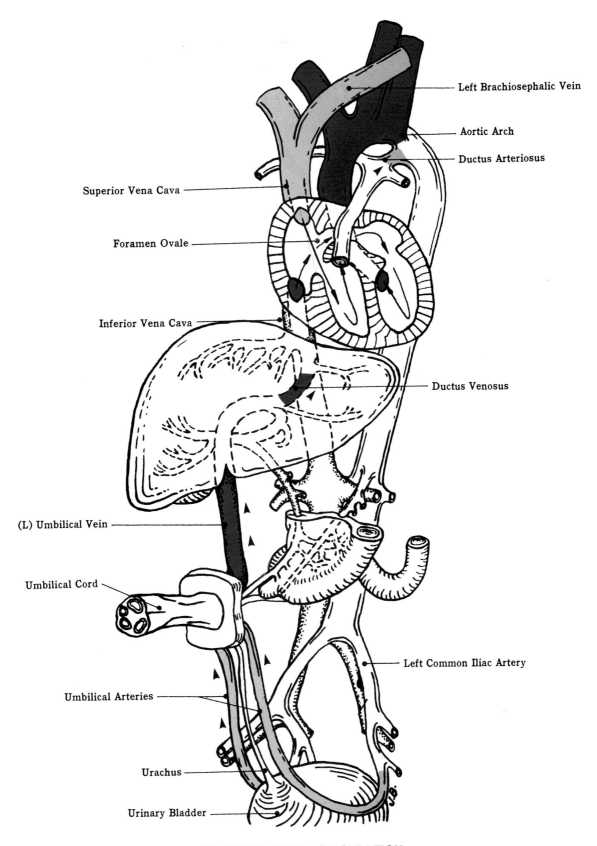

10-1F THE FETAL CIRCULATION

Follow blood passing to the placenta through the umbilical cord via the umbilical arteries. It returns in the umbilical vein, short circuits to the liver through the ductus venosus which takes it to the IVC which empties into the right atrium where it passes through the foramen ovale into the left atrium, left ventricle, and thence to the aorta for systemic circulation.

Blood entering the right atrium from the SVC passes to the right ventricle into the pulmonary trunk. The ductus arteriosus shunts it into the aorta where it descends with the oxygenated blood, eventually reaching the umbilical arteries (branches of the internal iliac) to complete the cycle.

This fetus has been prepared by staining with alizarin which binds to calcified tissue, clearing, and removing viscera. To determine its age we need the following information:

1. The shafts (diaphyses) of long bones ossify during intrauterine life beginning with the clavicle about the 5th week.
2. Only two epiphyses appear in the fetus (late): distal femur and proximal tibia.
3. Vertebral bodies begin to ossify in the lower thoracic region by the 10th week; cervical and sacral extremes not until the 4th month.
4. Ribs begin to ossify by the 8th week; the sternum not until the 5th month.
5. The long bones of the hand appear in sequence during the 9th, 10th, and 11th weeks: metacarpals and distal phalanges, proximal phalanges, and middle phalanges. No carpal bones have ossified at birth.
6. Long bones of the foot also appear in a similar sequence but a few weeks later. Two tarsal bones ossify prior to birth: calcaneus (3rd month) and talus (6th month).

10-2A FETAL OSSIFICATION

10-2B KNEE OF NEWBORN

This longitudinal section through the knee of a newborn shows ossified femoral (*F*) and tibial (*T*) diaphyses. Secondary centers of ossification in their adjacent epiphyses are marked by *arrows*. The patella (*P*), alas, does not ossify until well after the 3rd year.

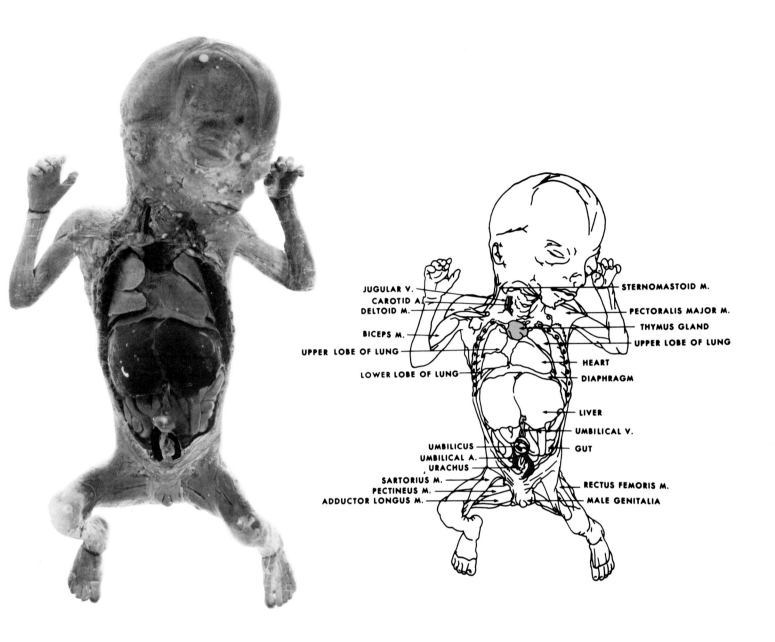

10-2C   DISSECTION OF FETUS

Observe:
1. The relatively large head and relatively small limbs.
2. The huge thymus gland occupying the anterior mediastinum.
3. The small lungs which have never been inflated.
4. The large, transversely oriented heart.
5. The enormous liver with its left lobe almost as large as its right.
6. Paired umbilical arteries passing to the umbilical cord. The umbilical vein returning from the placenta and traveling toward the liver.

See: Gasser, R. F. (1975) *Atlas of Human Embryos*. Harper & Row, Hagerstown; Parke, W. W. (1975) *Photographic Atlas of Fetal Anatomy*. University Park Press, Baltimore.

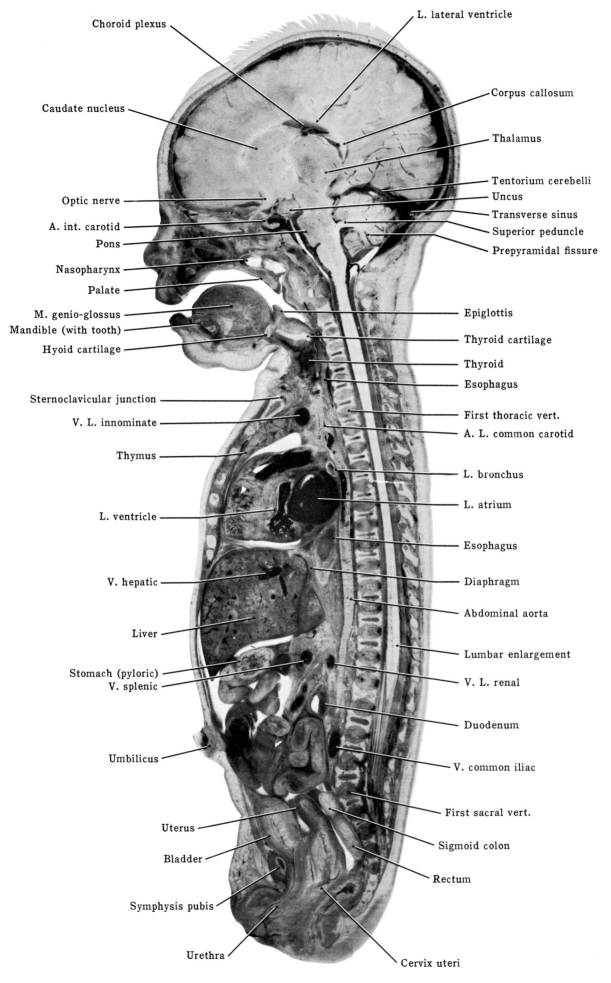

Choroid plexus

L. lateral ventricle

Caudate nucleus

Corpus callosum

Thalamus

Tentorium cerebelli

Uncus

Optic nerve

Transverse sinus

A. int. carotid

Superior peduncle

Pons

Prepyramidal fissure

Nasopharynx

Palate

M. genio-glossus

Epiglottis

Mandible (with tooth)

Thyroid cartilage

Hyoid cartilage

Thyroid

Esophagus

Sternoclavicular junction

First thoracic vert.

V. L. innominate

A. L. common carotid

Thymus

L. bronchus

L. atrium

L. ventricle

Esophagus

V. hepatic

Diaphragm

Abdominal aorta

Liver

Lumbar enlargement

Stomach (pyloric)

V. L. renal

V. splenic

Duodenum

Umbilicus

V. common iliac

First sacral vert.

Uterus

Sigmoid colon

Bladder

Rectum

Symphysis pubis

Urethra

Cervix uteri

10-3 MEDIAN SECTION OF A NEWBORN

MEDIAN SECTION OF A NEWBORN

SEE FIGURE 10-3 ON FACING PAGE (Courtesy of Prof. C. H. Sawyer)

The size and position of the various structures at birth may be compared with those of the adult.

Observe:
1. The cranium is relatively large, the face is small, the neck is short.
2. The thymus is huge and occupies the anterior mediastinum.
3. The liver is large and extends far below the costal margin.
4. The pelvis is small; thus bladder and uterus are abdominal structures.
5. The spinal cord extends further caudally than in the adult.

See: Crelin, E. S. (1973) *Functional Anatomy of the Newborn*. Yale University Press, New Haven; Falkner, F., Ed. (1966) *Human Development*. W. B. Saunders, Philadelphia.

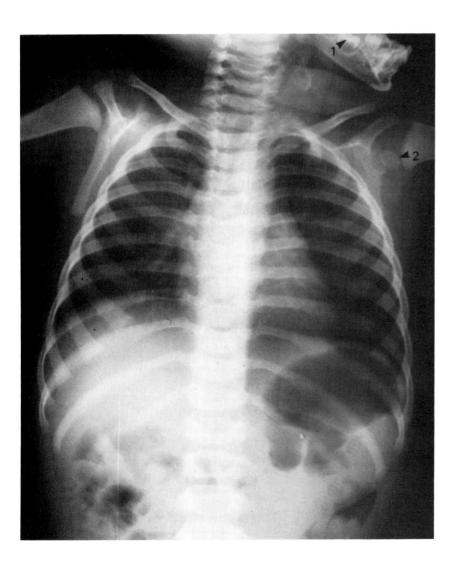

Observe:
1. The calcifying crowns of deciduous teeth.
2. The epiphysis for the head of the humerus.
3. The disposition of the ribs, more horizontal than in the adult.
4. The large, globular heart outline.
5. The large radiopaque area of the liver.

10-4 RADIOGRAPH OF AN INFANT

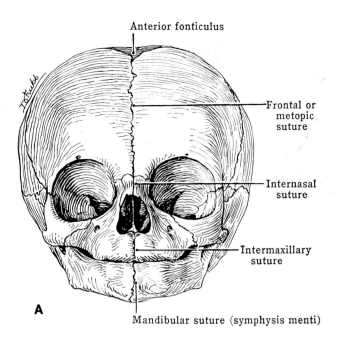

Anterior fonticulus

Frontal or metopic suture

Internasal suture

Intermaxillary suture

Mandibular suture (symphysis menti)

A

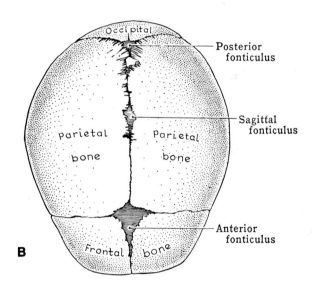

Occipital

Posterior fonticulus

Sagittal fonticulus

parietal bone

Parietal bone

Anterior fonticulus

Frontal bone

B

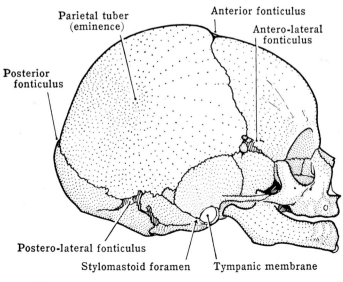

Parietal tuber (eminence)

Anterior fonticulus

Antero-lateral fonticulus

Posterior fonticulus

Postero-lateral fonticulus

Stylomastoid foramen

Tympanic membrane

C

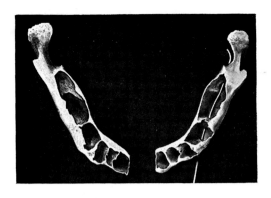

D MANDIBLE AT BIRTH FROM ABOVE

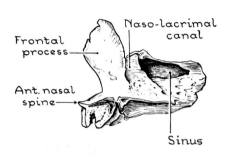

Naso-lacrimal canal

Frontal process

Ant. nasal spine

Sinus

E MAXILLA AT BIRTH, MEDIAL VEIW

10-5 SKULL AT BIRTH

Note:

1. The face is very small because the teeth have not erupted and the air sinuses are rudimentary.

2. Ossification of the frontal and parietal bones begins near their center and spreads in centrifugal waves. It has not yet reached the angles of the bones and so membranous fonticuli (fontanelles) are present. The large four-sided anterior fonticulus is ossified by about 18 months, the three-sided posterior fonticulus by 6 months, and the tiny lateral fonticuli soon after birth. The anterior fonticulus is useful in assessing intracranial pressure in the infant and provides easy access to the superior sagittal venous sinus for blood samples.

3. The suture between right and left frontal bones becomes obliterated by the 6th year. It persists as the metopic suture in some adults.

4. The temporal bone is in four parts. There is no mastoid process overshadowing the stylomastoid foramen; thus the facial nerve is close to the skin surface. The external acoustic meatus is formed by the narrow tympanic ring; thus the tympanic membrane is close to the examiner's otoscope.

5. The two halves of the mandible fuse during the 2nd year. Five alveoli are present on each side for the unerupted deciduous teeth. The course of the inferior alveolar nerve is shown in *D*.

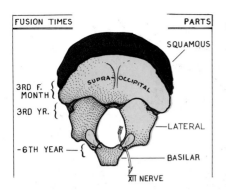

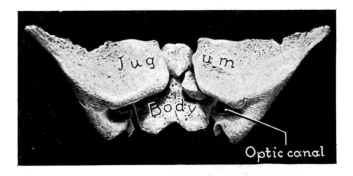

10-6 OCCIPITAL BONE AT BIRTH

At birth the bone is in 4 parts. *Blue areas* develop in cartilage; the *red area* develops in membrane. The squamous part has a double origin. The occipital condyles are on both lateral and basilar parts.

10-7 LESSER WINGS OF SPHENOID, BEFORE BIRTH, POSTERIOR VIEW

Note:
1. These wings, like sliding doors, close above the body of the sphenoid to form the jugum or yoke.
2. The posterior edge of the jugum becomes the anterior edge of the chiasmatic sulcus, which leads to the optic canals.
3. The left canal is not yet pinched off from the superior orbital fissure, but the right is.

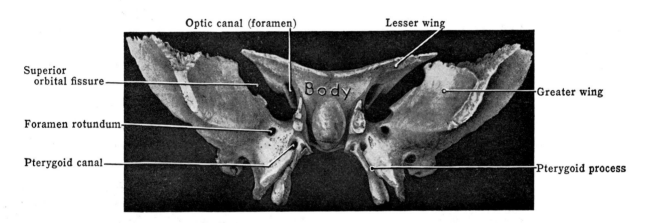

10-8 SPHENOID BONE OF A YOUNG CHILD, FRONT VIEW

The right and left sphenoidal sinuses do not invade the body until about the 4th year.

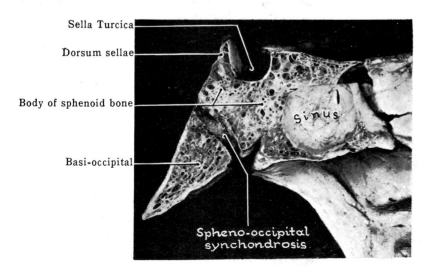

This is a paramedian section of a child aged 10 years. The base of the skull grows in length at this synchondrosis. In general, fusion begins in the male between 13 and 16 years and is completed by 20; in the female fusion begins between 11 and 14 years and is completed by 17.

See: Powell, T. V., and Brodie, A. G. (1963) Closure of spheno-occipital synchondrosis. *Anat. Rec., 147:* 15.

10-9 SPHENO-OCCIPITAL SYNCHONDROSIS (BASILAR SUTURE)

Note:
1. The 3 parts of the temporal bone—squamous, petrous, tympanic—separable at birth.
2. The mandibular fossa is shallow. The postauditory process, descending from the squamous part, closes the mastoid antrum laterally.
3. No mastoid process has yet appeared on the petrous part, there being as yet no mastoid cells.
4. The tympanic part is a ring that is incomplete above. No tympanic plate has yet grown to give length to the external acoustic meatus; hence the tympanic membrane is close to the surface of the skull.

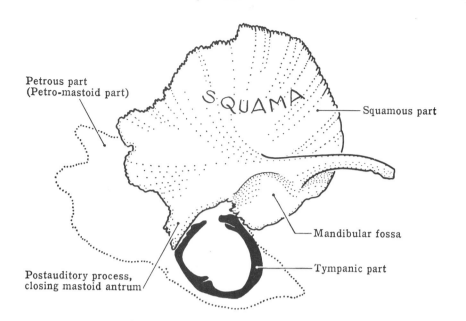

10-10A   TEMPORAL BONE AT BIRTH, LATERAL VIEW

Note:
1. A mastoid antrum is present and has been opened.
2. There are no mastoid air cells, thus no mastoid process.
3. The tympanic ring is still incomplete above. It has no length, but sprouting from its lateral border are two spines.
4. The stylo-mastoid foramen, from which the facial nerve emerges, is near the skin surface.
5. The tympanic cavity is large.

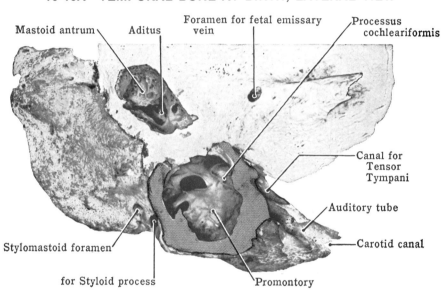

10-10B   TEMPORAL BONE DURING THE FIRST YEAR OF LIFE

Note:
1. The mastoid antrum and the epitympanic recess have been opened.
2. The mastoid process is appearing. The walls of the antrum are like a honeycomb.
3. The enlarging spines on the tympanic ring are nearer to meeting and so to enclosing a temporary foramen (closed by membrane) in what is becoming the tympanic plate. This plate will form the anterior, the inferior, and part of the posterior wall of the external acoustic meatus, as shown in Figure 7-79.

See: Anderson, J. E. (1960) *The Development of the Tympanic Plate*. National Museum of Canada, Bulletin No. *180:* 143.

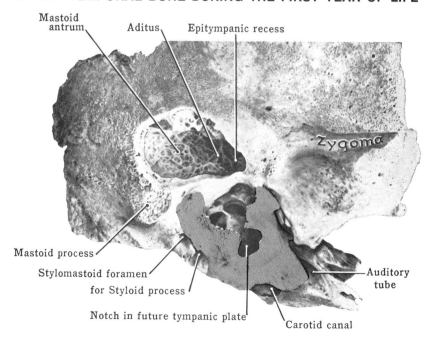

10-10C   TEMPORAL BONE DURING THE THIRD OR FOURTH YEAR

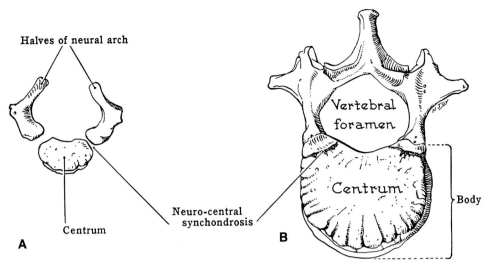

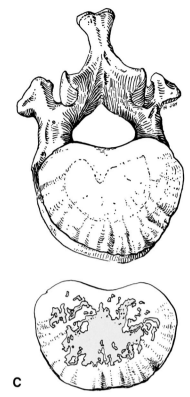

10-11 DEVELOPMENT OF VERTEBRAE

A. At birth, a vertebra consists of three bony parts, united by hyaline cartilage.

B. From age 2, the halves of each neural arch begin to fuse to each other, from lumbar to cervical region. From about age 7, the arches fuse to the centrum in sequence from cervical to lumbar regions.

C. During puberty, secondary centers of ossification appear for the tips of spinous and transverse processes. Epiphyseal plates for the body, as shown here, consist of a plate of hyaline cartilage and circumferential bony ring.

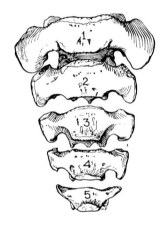

10-12 SACRUM IN YOUTH

The costal elements (Fig. 5-3) begin to fuse with each other about puberty. The bodies begin to fuse with each other from below upward about the 17th to 18th year, fusion being complete by the 23rd year. A gap, however, may persist between the 3rd and 2nd bodies until the 24th year, and between the 2nd and 1st until the 33rd year.

See: McKern, T. W., and Stewart, T. D. (1957) *Skeletal Age Changes in Young American Males.* Quartermaster Research & Development Center, Natick, Ma.

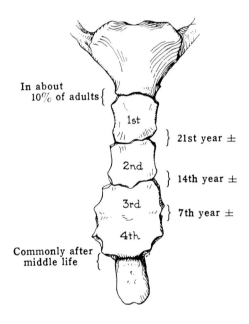

10-13 YOUNG STERNUM

Times of synostosis are shown. The four sternebrae fuse together from below upward. Numbers 4, 3, and 2 are usually fused together not later than the 17th or 18th year, and 2 to 1 not later than the 23rd year. They may, however, be delayed. Synostosis of manubrium and body was found in about 10 per cent of adults, aged 30 to 80 years. It is apparently unrelated to age.

See: Trotter, M. (1934) Synostosis between manubrium and body of the sternum in whites and Negroes. *Am. J. Phys. Anthropol., 18:* 439.

10-14   BONES OF UPPER LIMB AT BIRTH

The diaphyses of long bones and the scapula are ossified, but the epiphyses and carpal bones have not. The coracoid process, the medial border of the scapula, and the acromion are cartilaginous. Times given in Figures 10-15 to 10-19 unless otherwise noted are for males and derived from: McKern, T. W. and Stewart, T. D. (1957) *Skeletal Age Changes in Young American Males*. Quartermaster Research & Development Center, Natick Ma; Francis, C. C., and Werle, P. P. (1939) The appearance of centers of ossification from birth to 5 years. *Anat. Rec.*, *24:* 273; Francis, C. C. (1940) The appearance of centers of ossification from 6 to 15 years. *Anat. Rec.*, *27:* 127.

10-15   EPIPHYSIS OF THE CLAVICLE

The clavicle has a thin epiphysis covering its sternal end. This may start to fuse at any time between the 18th and 25th year. Between the 23rd and 31st years (mostly between 26th and 29th) all undergo terminal fusion. It is the last of the long bone epiphyses to fuse.

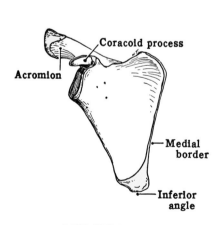

AGED ABOUT 3 YEARS.

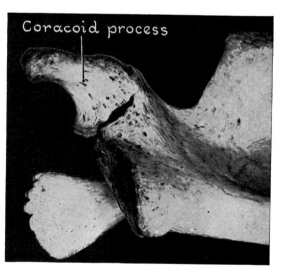

THE CORACOID PROCESS, FUSING,
AGED ABOUT 15 YEARS.

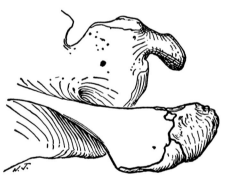

AGED ABOUT 17 YEARS.

10-16   OSSIFICATION OF THE SCAPULA

The *Coracoid process.* A single ossific center appears in the coracoid during the 1st year; another center for the upper end of the glenoid cavity and adjacent part of the coracoid appears about the 10th year. These fuse with the scapula about the 15th year.

The *acromion* has two ossific centers; the *medial border* and the *inferior angle* have separate centers. These appear about puberty and usually start to fuse before the 17th year. Fusion of all three is usually complete between the 18th and 20th years and always by the 23rd year.

The acromial epiphysis may persist into adult life as shown in Figure 6-118.

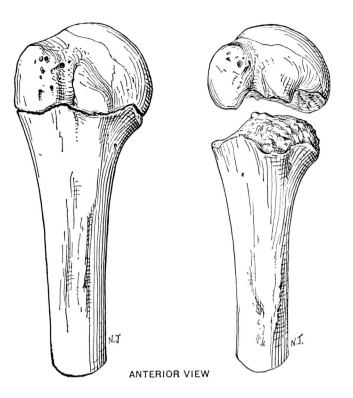

ANTERIOR VIEW

10-17 HUMERUS, PROXIMAL EPIPHYSIS

The proximal epiphysis fits onto the billowy conical end of the diaphysis. It develops from three centers, (a) for the head during the 1st year, (b) for the greater tubercle before the end of the 3rd year, and (c) for the lesser tubercle before the 5th year. These fuse to form a single mass before the 7th year. This proximal epiphysis is in some cases completely fused to the diaphysis by the 17th to 18th year and in all cases by the 24th year. The epiphyseal plate lies well above the level of the surgical neck but it cuts through the head medially. The articular cartilage of the head extends for 2 or 3 mm onto the diaphysis; hence the epiphyseal plate cuts into the shoulder joint.

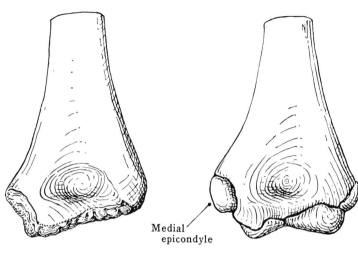

Medial epicondyle

10-18 HUMERUS, DISTAL EPIPHYSES

The lower border of the diaphysis is oblique and lies well below the levels of the radial, coronoid, and olecranon fossae. Its surface is ridged and pitted. There are 4 centers of ossification: for (a) lateral epicondyle, (b) capitulum and lateral part of trochlea, (c) medial part of trochlea, and (d) medial epicondyle. The first three appear about the 12th, 1st, and 10th years respectively and fuse into a single mass which is completely fused to the diaphysis before the end of the 17th year. The epiphysis for the medial epicondyle appears about the 6th year and is completely fused to the side of the diaphysis by the end of the 19th year.

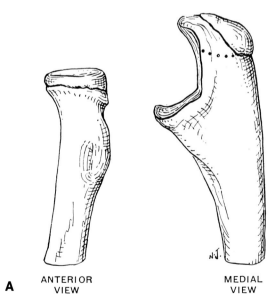

A   ANTERIOR VIEW   MEDIAL VIEW

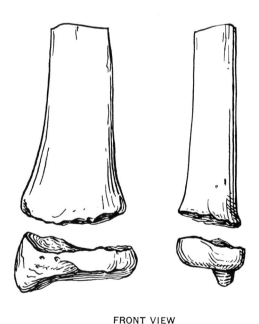

B   FRONT VIEW

10-19 EPIPHYSES OF RADIUS AND ULNA

A. Proximal ends: Each bone has one center, for olecranon and head of radius. The olecranon (the traction epiphysis of the triceps) appears about the 11th year. It may be quite large and cut into the trochlear notch (see *dotted line*). The center for the head of the radius appears about the 7th year and, like the olecranon epiphysis, is usually completely fused by the 17th and 18th year and always by the 19th year.

B. Distal ends: The epiphyseal plates of the ulna and radius lie at the same level and cut into the distal radio-ulnar joint. The center for the radius appears at the end of the 1st year; that for the ulna appears about the 7th year. Early stages of fusion occur from the 17th to 20th year; thereafter all cases are either fused or in late stages of fusion, fusion being always complete by the 23rd year.

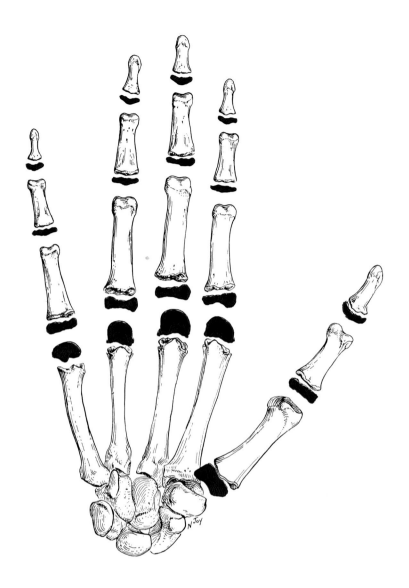

10-20 EPIPHYSES OF HAND

Note:

1. Phalenges have a single *proximal* epiphysis.
2. Metacarpals 2, 3, 4, and 5 have single *distal* epiphyses. The 1st metacarpal behaves as a phalanx by having a *proximal* epiphysis. Short-lived epiphyses may appear at the other ends of metacarpals 1 and/or 2 as in Figure 10-21A.
3. The capitate starts to ossify soon after birth. Figure 10-20B shows the spiral sequence of ossification of the carpals with approximate age in years. There are individual and sex differences in sequence and timing.

See: Greulich, W. W., and Pyle, S. I. (1959) *Radiographic Atlas of Skeletal Development of the Hand and Wrist*, 2nd ed. Stanford University Press, Stanford, CA; Pyle, S. I., Waterhouse, A. M., and Greulich, W. W. (1971) *A Radiographic Standard of Reference for the Growing Hand and Wrist*. Case Western Reserve University Press, Chicago.

10-20B CARPAL OSSIFICATION

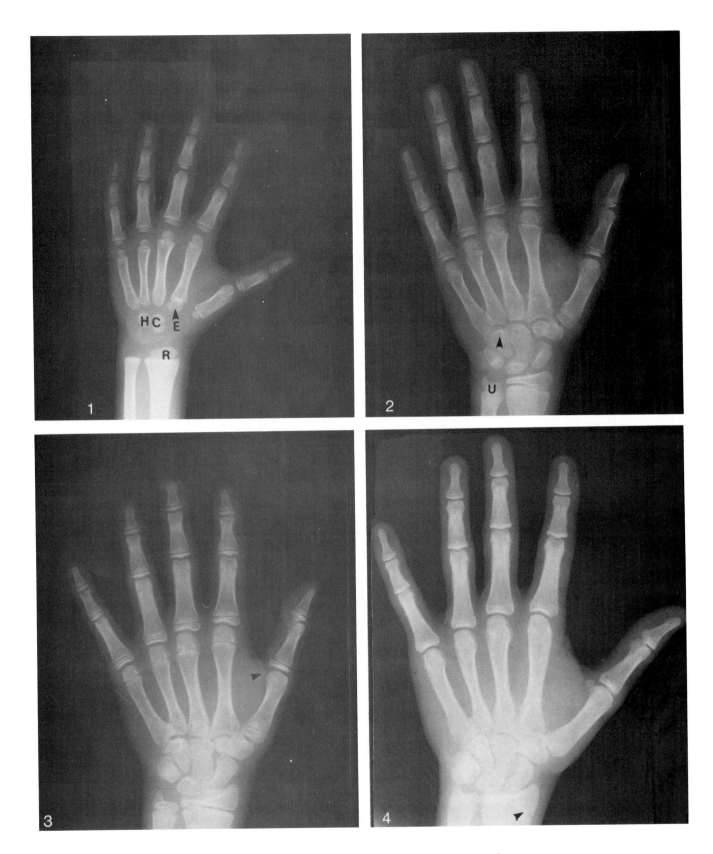

10-21   HAND-WRIST RADIOGRAPHS

A radiograph of the hand and wrist is commonly used to assess skeletal age. For routine clinical work, the radiograph is compared with a series of standards in an atlas. (See references on facing page.)

In these four radiographs observe:

1. A 2-year-old child. Capitate (*C*) and hamate (*H*) are ossified and large. The distal epiphysis (*R*) is present. The *arrow* points to an uncommon proximal epiphysis on the 2nd metacarpal.
2. A 12-year-old. All carpal bones are ossified and epiphyses are growing in size. The *arrow* points to the hook of the hamate which may now be visualized. The styloid process of the ulna (*U*) has ossified.
3. A 13-year-old. Rapid growth occurs with the onset of puberty. The *arrow* points to the sesamoid bone in adductor pollicis.
4. A 17-year-old. Growth almost complete. Finger and hand epiphyses have fused. The *arrow* points to the closing radial epiphysis.

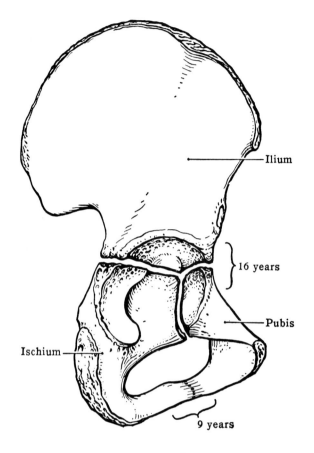

10-22 BONES OF LOWER LIMB AT BIRTH

The hip bone is in 3 primary parts—ilium, ischium, and pubis. The diaphyses of the long bones are well ossified. Certain epiphyses and certain tarsal bones have started to ossify:·the distal epiphysis of the femur and the proximal epiphysis of the tibia; the calcaneus, talus, and cuboid. In Figures 10-23 to 10-28, times given are for males and references are those cited in Figure 10-14.

10-23 HIP BONE IN YOUTH, EXTERNAL ASPECT

The three elements of the hip bone meet in the acetabulum at a triradiate synchondrosis. Of these, the pubis contributes least to the acetabulum, the ilium next least, and the ischium most, including the nonarticular part.

One or more centers of ossification appear in the triradiate cartilage about the 12th year. One of these, the os acetabuli, extends widely over the acetabular surface of the pubic bone.

Secondary centers of ossification appear: (a) along the whole length of the iliac crest (attachment of the 3 flat abdominal muscles); (b) at the anterior inferior spine (origin of Rectus Femoris); (c) at the ischial tuberosity (origin of the ham muscles); and also (d) at the symphysis pubis, hence the wavy surfaces. These appear about puberty. Fusion may start as early as the 17th and 18th year and is complete before the 23rd year.

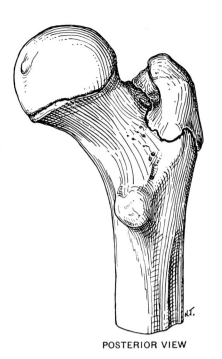

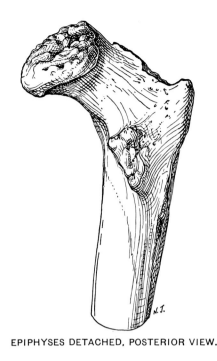

POSTERIOR VIEW   EPIPHYSES DETACHED, POSTERIOR VIEW.

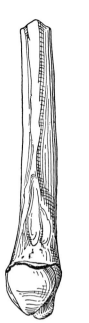

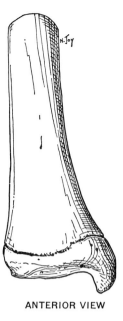

ANTERIOR VIEW

MEDIAL VIEW

10-24   EPIPHYSES AT THE PROXIMAL END OF THE FEMUR

The epiphysis of the head fits like a cap onto the slightly conical end of the diaphysis. The apposed surfaces are billowy—pitted and ridged— as at other epiphyses. The articular cartilage of the head is carried for a few millimeters onto the inferior aspect of the neck. (Compare the humerus in Fig. 10-17.) The epiphysis of the greater trochanter lies above the site of confluence of the shaft and neck. That of the lesser trochanter is a thick scale.

The epiphysis of the head begins to ossify during the 1st year; that of the greater trochanter before the 5th year; and that of the lesser trochanter before the 14th year. These have in most cases fused completely with the shaft before the end of the 18th year, and in all cases by the 20th.

10-26   DISTAL EPIPHYSES OF THE TIBIA AND FIBULA

The epiphyseal plate of the fibula lies at the level of the ankle joint, and therefore necessarily below the level of the epiphyseal plate of the tibia. The tibial epiphysis starts to ossify during the 1st year; the fibula during the 1st or 2nd year. Fusion is usually complete by the 17th or 18th year, and in all cases by the 20th.

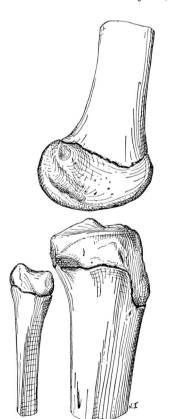

10-25   EPIPHYSES AT THE KNEE JOINT

The epiphyseal plate splits the adductor tubercle, skims above the patellar trochlea, shaves the cartilage of the lateral condyle, and follows along the intercondylar line.

The proximal epiphyseal plate of the tibia shaves the fibular facet and passes below the groove for the semimembranous tendon. Anteriorly, it dips down like a tongue to include the tuberosity. The epiphyseal plate of the fibula lies well above the neck.

The distal epiphysis of the femur begins to ossify before birth; that of the tibia soon follows it; that of the fibula begins to ossify before the 5th year. These 3 epiphyses at the knee have usually completely fused with their respective diaphyses before the end of the 19th year, and always by the 22nd—that of the tibia may be the 23rd. A separate center for the tibial tuberosity may appear before puberty.

See: Pyle, S. I., and Hoerr, N. L. (1971) *A Radiographic Standard of Reference for the Growing Knee.* Charles C Thomas, Springfield, IL.

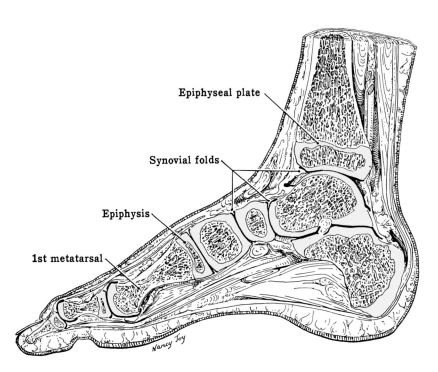

Epiphyseal plate

Synovial folds

Epiphysis

1st metatarsal

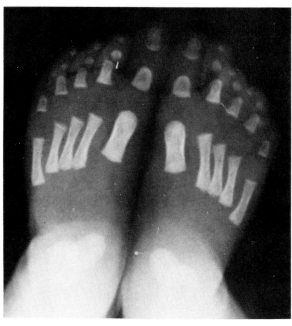

10-28A   RADIOGRAPH OF FOOT AT BIRTH

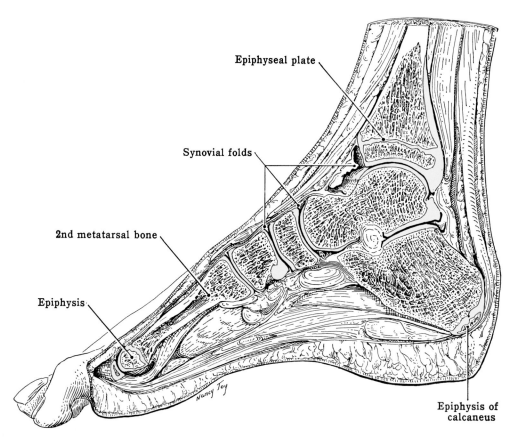

Epiphyseal plate

Synovial folds

2nd metatarsal bone

Epiphysis

Epiphysis of
calcaneus

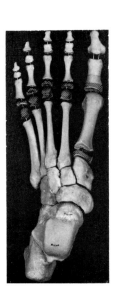

10-28B   OSSIFICATION

OF THE FOOT

10-27   LONGITUDINAL SECTIONS THROUGH THE FEET OF CHILDREN AGED 4 AND 10 YEARS

Observe:

1. In the younger foot, epiphyses of long bones (tibia, metatarsals, and phalanges) ossify like short bones (carpals and tarsals), the ossific centers being enveloped in cartilage. Ossification has already extended to the surface of the larger tarsal bones.

2. In the older foot, ossification has spread to the dorsal and plantar surfaces of all the tarsal bones in view, and cartilage persists on the articular surfaces only.

3. The traction epiphysis of the calcaneus for the tendo Achillis and plantar aponeurosis starts to ossify between the ages of 6 and 10 years.

4. The 1st metatarsal bone behaves as a phalanx in that its epiphysis is at its base and not at its head like the 2nd and other metatarsal bones (*cf.* the hand, Fig. 10-20*A*).

5. Synovial folds project between the articular cartilages where they are liable to be pinched.

Guide to Anatomical Terminology

In times when scholars had a good knowledge of Latin (and Greek), anatomical terminology assisted in visualizing and remembering structures. "Clavicle" being derived from "clavicula," a little key, identified the collar bone. "Teres" = round, thus "ligamentum teres" reminded immediately of the shape of the ligament. Today, terminology is a hazard rather than a help. The following few suggestions may assist your entry into a new vocabulary. You will find the following paperback both helpful and amusing:

Squires, B.P. (1981) *Basic Terms of Anatomy and Physiology*, W.B. Saunders, Philadelphia.

PLURAL FORMS
Nouns in Latin as in English usually show whether they are singular or plural by their endings. Although there are exceptions, the following rules will indicate the correct form of Latin plural in most instances.

1. If the singular ends in 'a,' the plural ends in 'ae.'

   Examples:

ala-alae	axilla-axillae	bursa-bursae
cauda-caudae	fossa-fossae	fovea-foveae
lamina-laminae	uvula-uvulae	vagina-vaginae
vallecula-valleculae	vertebra-vertebrae	

2. If the singular ends in '-um' or '-on,' the plural ends in '-a.'

   Examples:

cingulum-cingula	ganglion-ganglia
infundibulum-infundibula	labium-labia
retinaculum-retinacula	tegmentum-tegmenta
tuberculum-tubercula	vestibulum-vestibula

3. If the singular ends in 'us,' most plurals end in '-i.'

   Examples:

ductus-ducti	fundus-fundi	funiculus-funiculi
gyrus-gyri	hilus-hili	isthmus-isthmi
ramus-rami	sulcus-sulci	

   Some are irregular: corpus-corpora, viscus-viscera, crus-crura.

4. Less common endings:

Singular Ending	Plural Ending	
'-is'	'-es'	pelvis-pelves, diaphysis-diaphyses
'-en'	'-ina'	foramen-foramina, lumen-lumina
'-u'	'-ua'	genu-genua, cornu-cornua
'-ex'	'-ices'	apex-apices
'-ix'	'-ices'	appendix-appendices
'-anx'	'-anges'	phalanx-phalanges

5. Sometimes we anglicize:
   ulna: ulnas (rather than ulnae)
   rectum: rectums (rather than recta)
   aorta: aortas (rather than aortae)
   penis: penises (rather than penes)

6. Abbreviation for plurals:

a. = artery	aa. = arteries
m. = muscle	mm. = muscles
n. = nerve	nn. = nerves
v. = vein	vv. = veins

COMPARISON OF ADJECTIVES
Just as in English, Latin adjectives show differences of degree by differences of ending.

Examples:

English:	wide	wider	widest
Latin:	latus	latior	latissimus

Thus, "fascia lata" means "wide fascia" while "latissimus dorsi" muscle means "widest muscle of the back."
The following irregular forms are also much used in anatomical nomenclature.

magnus (large)	major (larger)	maximus (largest)
perva (small)	minor (smaller)	minimus (smallest)

Examples:
teres major muscle = larger round muscle
teres minor muscle = smaller round muscle
gluteus major muscle = largest gluteal muscle
gluteus minor muscle = smallest gluteal muscle

PREFICES AND SUFFICES
Many medical terms consist of a central root word to which another component has been fused either before the root word (a prefix), or after the root word (a suffix).

Here is a list of some common prefices:

Prefix	Meaning	Examples
a(n)-	no, not	anencephaly = without a head
ab-	away from	abductor = leading away
ad-	toward	adductor = leading to
anti-	against	antimalarial = against malaria
apo-	away from	apocrine = pouring out
de-	down	depressor = pressing down
e-, ek-, ex-	out	evagination = outpouching
en-, (endo-)	in, within	endogenous = created within
epi-	upon	epinephron = upon the kidney
hyper-	above, over	hypertrophy = overgrowth
hypo-	under	hypodermic = under the skin
infra-	below	infra-orbital = below the orbit
par-(a)	beside	parotid = beside the ear
peri-(a)	around	pericardium = around the heart
pre-/pro-	before	prevertebral = in front of the vertebrae
retro-	behind	retro-peritoneal = behind the peritoneum
supra-	above	supra-hepatic = above the liver
sub-	below	sub-cortical = below the cortex
trans-	across, through	transaortic = through the aorta

A common anatomical suffix is -form = like:
vermiform = like a worm.
However, the Greek suffix for like (-oid) is common:
scaphoid = like a ship.

CASES
Latin uses endings to denote six "cases," singular and plural. This places the word in its proper position in a sentence. The nominative case denotes the subject of a sentence. The genitive case denotes "of the. . ."

Examples:
scapula (nominative), scapulae (genitive)
   levator scapulae muscle = elevator *of the* scapula.
uterus (nominative), uteric (genitive)
   cervic uteri = neck *of the* uterus.
femur (nominative), femoris (genitive)
   arteria profunda femoris = deep artery *of the* femur.

INDEX

References are to Figure Numbers

The chief references to any item are printed in bold type